MW00843370

WILEY SERIES IN MICROWAVE AND OPTICAL ENGINEERING

KAI CHANG, Series Editor
Texas A&M University

A complete list of the titles in this series appears at the end of this volume.

ENABLING TECHNOLOGIES FOR HIGH SPECTRAL-EFFICIENCY COHERENT OPTICAL COMMUNICATION NETWORKS

Edited by

XIANG ZHOU
CHONGJIN XIE

WILEY

Copyright © 2016 by John Wiley & Sons, Inc. All rights reserved

Published by John Wiley & Sons, Inc., Hoboken, New Jersey
Published simultaneously in Canada

No part of this publication may be reproduced, stored in a retrieval system, or transmitted in any form or by any means, electronic, mechanical, photocopying, recording, scanning, or otherwise, except as permitted under Section 107 or 108 of the 1976 United States Copyright Act, without either the prior written permission of the Publisher, or authorization through payment of the appropriate per-copy fee to the Copyright Clearance Center, Inc., 222 Rosewood Drive, Danvers, MA 01923, (978) 750-8400, fax (978) 750-4470, or on the web at www.copyright.com. Requests to the Publisher for permission should be addressed to the Permissions Department, John Wiley & Sons, Inc., 111 River Street, Hoboken, NJ 07030, (201) 748-6011, fax (201) 748-6008, or online at http://www.wiley.com/go/permission.

Limit of Liability/Disclaimer of Warranty: While the publisher and author have used their best efforts in preparing this book, they make no representations or warranties with respect to the accuracy or completeness of the contents of this book and specifically disclaim any implied warranties of merchantability or fitness for a particular purpose. No warranty may be created or extended by sales representatives or written sales materials. The advice and strategies contained herein may not be suitable for your situation. You should consult with a professional where appropriate. Neither the publisher nor author shall be liable for any loss of profit or any other commercial damages, including but not limited to special, incidental, consequential, or other damages.

For general information on our other products and services or for technical support, please contact our Customer Care Department within the United States at (800) 762-2974, outside the United States at (317) 572-3993 or fax (317) 572-4002.

Wiley also publishes its books in a variety of electronic formats. Some content that appears in print may not be available in electronic formats. For more information about Wiley products, visit our web site at www.wiley.com.

Library of Congress Cataloging-in-Publication Data:

Names: Zhou, Xiang and Xie, Chongjin (Writers on optical communications) editors.
Title: Enabling technologies for high spectral-efficiency coherent optical communication networks / edited by Xiang Zhou and Chongjin Xie.
Description: Hoboken, New Jersey : John Wiley & Sons, Inc., [2016] | Includes bibliographical references and index.
Identifiers: LCCN 2015034521 | ISBN 9781118714768 (cloth)
Subjects: LCSH: Optical fiber communication. | Laser communication systems.
Classification: LCC TK5103.592.F52 E53 2016 | DDC 621.382/7–dc23 LC record available at http://lccn.loc.gov/2015034521

ISBN: 9781118714768 (Hardback)

Printed in the United States of America

10 9 8 7 6 5 4 3 2 1

CONTENTS

LIST OF CONTRIBUTORS

Erik Agrell, The Communication Systems Group, Department of Signals and Systems, Chalmers University of Technology, Gothenburg, Sweden

Jean-Christophe Antona, Bell Laboratories, Nokia, Nozay, France

Gabriella Bosco, Dipartimento di Elettronica e Telecomunicazioni (DET), Politecnico di Torino, Torino, Italy

Stephan ten Brink, Institute of Telecommunications, University of Stuttgart, Stuttgart, Germany

Andrea Carena, Dipartimento di Elettronica, Politecnico di Torino, Torino, Italy

Sethumadhavan Chandrasekhar, Bell Laboratories, Nokia, Holmdel, NJ, USA

Di Che, Department of Electrical and Electronic Engineering, The University of Melbourne, Melbourne, VIC, Australia

Xi Chen, Department of Electrical and Electronic Engineering, The University of Melbourne, Melbourne, VIC, Australia

Po Dong, Bell Laboratories, Nokia, Holmdel, NJ, USA

Zhenhua Dong, Photonics Research Center, Department of Electrical Engineering, The Hong Kong Polytechnic University, Hung Hom, Kowloon, Hong Kong

Nicolas K. Fontaine, Bell Laboratories, Nokia, Holmdel, NJ, USA

Fabrizio Forghieri, Cisco Photonics, Vimercate, Italy

Qian Hu, Department of Electrical and Electronic Engineering, The University of Melbourne, Melbourne, VIC, Australia

Yanchao Jiang, College of Information Engineering, Dalian University, Dalian, China

Magnus Karlsson, Photonics Laboratory, Department of Microtechnology and Nanoscience, Chalmers University of Technology, Gothenburg, Sweden

Faisal N. Khan, School of Electrical and Electronic Engineering, Engineering Campus, Universiti Sains Malaysia, Penang, Malaysia

Alan Pak Tao Lau, Photonics Research Center, Department of Electrical Engineering, The Hong Kong Polytechnic University, Hung Hom, Kowloon, Hong Kong

Patricia Layec, Bell Laboratories, Nokia, Nozay, France

Andreas Leven, Bell Laboratories, Nokia, Stuttgart, Germany

An Li, Department of Electrical and Electronic Engineering, The University of Melbourne, Melbourne, VIC, Australia

Guifang Li, CREOL, The College of Optics and Photonics, University of Central Florida, Orlando, FL, USA

Chao Lu, Photonics Research Center, Department of Electronic and Information Engineering, The Hong Kong Polytechnic University, Hung Hom, Kowloon, Hong Kong

Annalisa Morea, Bell Laboratories, Nokia, Nozay, France

Timo Pfau, DSP and Optics, Acacia Communications Inc., Maynard, MA, USA

Pierluigi Poggiolini, Dipartimento di Elettronica, Politecnico di Torino, Torino, Italy

Yvan Pointurier, Bell Laboratories, Nokia, Nozay, France

Roland Ryf, Bell Laboratories, Nokia, Holmdel, NJ, USA

Seb Savory, Department of Engineering, University of Cambridge, Cambridge, United Kingdom

Laurent Schmalen, Bell Laboratories, Nokia, Stuttgart, Germany

William Shieh, Department of Electrical and Electronic Engineering, The University of Melbourne, Melbourne, VIC, Australia

Han Sun, Infinera, Ottawa, Canada

Kuang-Tsan Wu, Infinera, Ottawa, Canada

Chongjin Xie, R&D Lab, Ali Infrastructure Service, Alibaba Group, Santa Clara, CA, USA

Xiang Zhou, Platform advanced technology, Google Inc., Mountain View, CA, USA

PREFACE

Coherent detection attracted lots of attention for optical communications in the 1980s and was considered an effective technology to achieve high sensitivity and thus longer distances between repeaters. Partly due to the difficulties in real-time implementation and partly due to the advent of optical amplifiers in the 1990s, the advantage of coherent detection in receiver sensitivity diminished, and direct detection became the main detection technology for optical communication systems until 2010.

With the advances in high-speed electronics and digital signal processing (DSP) technology, coherent detection revived in later 2000s and has fundamentally changed the optical communication industry. One distinct feature of today's digital coherent optical communication systems is the massive use of DSP. Not only carrier frequency and phase recovery and polarization tracking, the main obstacles for analog implementations of coherent detection in earlier years, can be realized using DSP, most impairments in optical transmission systems can be compensated in the electrical domain with DSP as well, which significantly simplifies the management and increases the flexibility of optical communication networks.

Coherent detection not just increases receiver sensitivity but, more importantly, enables the modulation of information on phase and polarization, and thus has the ability to greatly increase the spectral efficiency. With the rapid advances of key technologies such as advanced modulation, coding, and DSP, significant progress has been made on digital coherent optical communications in the past few years. Digital coherent detection has become the main technology for high-speed optical transport networks. The purpose of this book is to present a comprehensive coverage of key technology advances in high spectral efficiency fiber-optic communication systems and networks, enabled by the use of coherent detection and DSP. Each chapter includes basic theories and up-to-date technology advancements and the authors of all the chapters are leading and active researchers and experts in their subject topics. We hope this book will be found valuable not only for researchers and engineers, but for graduate students as well.

CHONGJIN XIE
JANUARY 2015
SAN MATEO, CA 94402

XIANG ZHOU
JANUARY 2015
SUNNY VALE, CA 94087

1

INTRODUCTION

Xiang Zhou[1] and Chongjin Xie[2]

[1]*Platform advanced technology, Google Inc, Mountain View, CA, USA*
[2]*R&D Lab, Ali Infrastructure Service, Alibaba Group, Santa Clara, CA, USA*

1.1 HIGH-CAPACITY FIBER TRANSMISSION TECHNOLOGY EVOLUTION

Since the first demonstration of an optical fiber transmission system in 1977 [1], the demands for higher capacity and longer reach have always been the dominant driver behind the evolution of this new communication technology. In less than four decades, single-fiber transmission capacity has increased by more than five orders of magnitude, from the early 45 Mb/s, using direct modulation and direct detection [2], to more than 8.8 Tb/s by using the digital coherent optical transmission technology [3]. In the meantime, optical transmission reach has increased from only a few kilometers to more than 10,000 km [4]. Such dramatic growth in capacity and reach has been enabled by a series of major breakthroughs in device, subsystem, and system techniques, including lasers, modulators, fibers, optical amplifiers, and photodetectors, as well as various modulation, coding, and channel impairment management methods.

The first generation of optical fiber communications was developed during the late 1970s, operating near 0.8 μm using GaAs semiconductor lasers [2] and multimode fibers (MMF). Although the total capacity of the first commercial system was only running at 45 Mb/s, with an optical reach or repeater spacing of 10 km, this capacity is now much greater than that of comparable coax systems (assuming identical reach or repeater spacing).

With breakthroughs in InGaAsP semiconductor lasers/photodetectors and single-mode fiber manufacturing technologies, the *second generation* shifted the

Enabling Technologies for High Spectral-efficiency Coherent Optical Communication Networks,
First Edition. Edited by Xiang Zhou and Chongjin Xie.
© 2016 John Wiley & Sons, Inc. Published 2016 by John Wiley & Sons, Inc.

wavelength to 1.3 μm by taking advantage of the low attenuation (<1 dB/km) and low dispersion of single-mode fibers. A laboratory experiment in 1981 demonstrated transmission at 2 Gb/s over 44 km of single-mode fiber [5]. By 1987, second-generation optical fiber communication systems, operating at bit rates of up to 1.7 Gb/s with a repeater spacing of about 50 km, were commercially available.

The optical transmission reach of second-generation fiber communication systems was limited by fiber losses at the operating wavelength of 1.3 μm (typically 0.5 dB/km). Losses of silica fibers approached minimum near 1.55 μm. Indeed, a 0.2-dB/km loss was realized in 1979 in this spectral region [6]. However, the introduction of *third-generation* systems operating at 1.55 μm was delayed by large fiber dispersion near 1.55 μm. Conventional InGaAsP semiconductor lasers (with Fabry–Perot type resonators) could not be used because of pulse spreading occurring as a result of simultaneous oscillation in several longitudinal modes. Two methods were developed to overcome the dispersion problem: (i) a dispersion-shifted fiber was designed to minimize the dispersion near 1.55 μm and (ii) a single longitudinal mode laser, that is the widely used distributed feedback (DFB) laser, was developed to limit the spectral width. By using these two methods together, bit rates up to 4 Gb/s over distances in excess of 100 km were successfully demonstrated in 1985 [7]. Third-generation fiber communication systems operating at 2.5 Gb/s became available commercially in 1990 with a typical optical reach of 60–70 km. Such systems are capable of operating at a bit rate of up to 10 Gb/s [8].

To further increase optical transmission reach and reduce the number of costly optical–electrical–optical (O–E–O) repeaters for long distance transmission, efforts were focused on coherent optical transmission technology during the late 1980s. The purpose was to improve optical receiver sensitivity by using a local oscillator (LO) to amplify the received optical signal. The potential benefits of coherent transmission technology were demonstrated in many system experiments [9]. However, commercial introduction of such systems was postponed with the advent of erbium-doped fiber amplifiers (EDFAs) in 1989. The *fourth generation* of fiber communication systems makes use of optical amplification for increasing O–E–O repeater spacing and of wavelength-division multiplexing (WDM) for increasing total capacity. The advent of the WDM technique in combination with EDFAs started a revolution that resulted in doubling of the system capacity every 6 months or so and led to optical communication systems operating at >1 Tb/s by 2001. In most WDM systems, fiber losses are compensated for by spacing EDFAs 60–80 km apart. EDFAs were developed after 1985 and became available commercially by 1990. A 1991 experiment showed the possibility of data transmission over 21,000 km at 2.5 Gb/s, and over 14,300 km at 5 Gb/s, using a recirculating-loop configuration [10]. This performance proved that an amplifier-based, all-optical, submarine transmission system was feasible for intercontinental communications. By 1996, not only had transmission over 11,300 km at a bit rate of 5 Gb/s been demonstrated by using actual submarine cables [11], but commercial trans-Atlantic and trans-Pacific cable systems also became available. Since then, a large number of submarine fiber communication systems have been deployed worldwide.

In the late 1990s and early 2000s, several efforts were made to further increase single-fiber capacity. The first effort focused on increasing system capacity by transmitting more and more channels through WDM. This was mainly achieved by reducing channel bandwidth through (i) better control of the laser wavelength stability and (ii) development of dense wavelength multiplexing and demultiplexing devices. At the same time, new kinds of amplification schemes had also been explored, as the conventional EDFA wavelength window, known as the C band, only covers the wavelength range of 1.53–1.57 μm. The amplifier bandwidth was extended on both the long- and short-wavelength sides, resulting in the L and S bands, respectively. The Raman amplification technique, which can be used to amplify signals in all S, C, and L wavelength bands, had also been intensely investigated. The second effort attempted to increase the bit rate of each channel within the WDM signal. Starting in 2000, many experiments used channels operating at 40 Gb/s. Such systems require high-performance optical modulator as well as extremely careful management of fiber chromatic dispersion (CD), polarization-mode dispersion (PMD) and fiber nonlinearity [12]. To better manage fiber CD, dispersion compensating fiber (DCF) has been developed and various dispersion management methods have also been explored to better manage fiber nonlinearity. These efforts led in 2000 to a 3.28-Tb/s experiment in which 82 channels, each operating at 40 Gb/s, were transmitted over 3000 km. Within a year, the system capacity was increased to nearly 11 Tb/s (273 WDM channels, each operating at 40 Gb/s) but the transmission distance was limited to 117 km [13]. In another record experiment, 300 channels, each operating at 11.6 Gb/s, were transmitted over 7380 km [14]. Commercial terrestrial systems with the capacity of 1.6 Tb/s were available by the end of 2000.

Until early 2000s, all the commercial optical transmission systems used the same direct modulation and direct detection on/off keying non-return-to-zero (NRZ) modulation format. The impressive fiber capacity growth was mainly achieved by advancement in photonics technologies, although forward error correction (FEC) coding also played a significant role in extending the reach for 10 Gb/s per channel WDM systems. Starting from 40 Gb/s per channel WDM systems, it became evident that more spectrally efficient modulation formats were needed to further increase the fiber capacity to meet the ever-growing bandwidth demands.

High spectral-efficiency (SE) modulation formats can effectively increase the aggregate capacity without resorting to expanding the optical bandwidth, which is largely limited by optical amplifier bandwidth. Using high-SE modulation formats also help reduce transceiver speed requirements. Furthermore, high-SE systems are generally more tolerant of fiber CD and PMD, since they use smaller bandwidths for the same bit rate. CD and PMD tolerance are particularly attractive for high-bit-rate transmission, since dispersion tolerance is reduced by a factor of 4 for a factor-of-2 increase in bit-per-symbol [15].

Early efforts in achieving high SE used direct detection. The first widely investigated modulation format with SE $>$ 1 bit/symbol was the optical differential quaternary phase-shift keying (DQPSK) with differential detection. This is a constant intensity modulation format, which can transmit 2 bits/symbol, corresponding to a theoretical SE of 2 bits/s/Hz [16, 17]. This modulation format also exhibits excellent fiber nonlinearity tolerance due to the nature of constant intensity. To go beyond 2 bit/s/Hz, polarization-division multiplexing (PDM) has been suggested to further increase SE in combination with DQPSK [18]. However, as the state of polarization of the light wave is not preserved during transmission, dynamic polarization control is required at the receiver to recover the transmitted signals.

The need for higher SE and the advancement in digital signal processing (DSP) eventually revives coherent optical communication. The concept of digital coherent communication was proposed by several research groups around 2004–2005 [19–22]. Quickly, this technology was recognized as the best technology for 40 Gb/s, 100 Gb/s and beyond WDM transmission systems, mostly due to the following reasons: (i) coherent technology preserves both amplitude and phase information, allowing all four dimensions of an optical field (in-phase and quadrature components in each of the two orthogonal polarizations) to be retained for information coding and thus offering much greater spectral efficiency than intensity-modulated direct detection (IMDD) systems; (ii) coherent technologies include powerful DSP that helps to solve the problems of chromatic and polarization-mode dispersion suffered by IMDD systems above 10 Gb/s, and thereby deliver vastly increased capacity over the same, or even better distances; and (iii) coherent detection offers better sensitivity than IMDD systems.

The advent of digital coherent detection has resulted in remarkable SE and fiber capacity improvement in the past few years. In lab experiments, the SE of optical communication systems has been increased from 0.8 b/s/Hz to more than 14.0 b/s/Hz [23] in single-mode fiber, and >100 Tb/s single-fiber capacity has been demonstrated [24]. The use of digital coherent detection technology also enables us to explore a few new avenues to further increase the optical network capacity or performance. For example, fiber capacity can be further increased by using few-mode fibers through mode-division multiplexing (MDM), which is enabled by coherent detection and DSP. Coherent technology and DSP also enable rate-adaptable optical transmission, which is critical for future elastic optical networks. Since coherent detection offers higher receiver sensitivity than direct detection, this technology may also facilitate the development of silicon-photonics-based photonic integration technologies, which suffer a significantly higher optical loss than the conventional discrete optical systems.

1.2 FUNDAMENTALS OF COHERENT TRANSMISSION TECHNOLOGY

1.2.1 Concept of Coherent Detection

In coherent optical communication, information is encoded onto the electrical field of a lightwave; decoding entails the direct measurement of the complex electrical field. To measure the complex electrical field of lightwave, the incoming data signal (after fiber transmission) interferes with a local oscillator (LO) in an optical 90° hybrid as schematically shown in Figure 1.1. If the balanced detectors in the upper branches measure the real part of the input data signal, the lower branches, with the

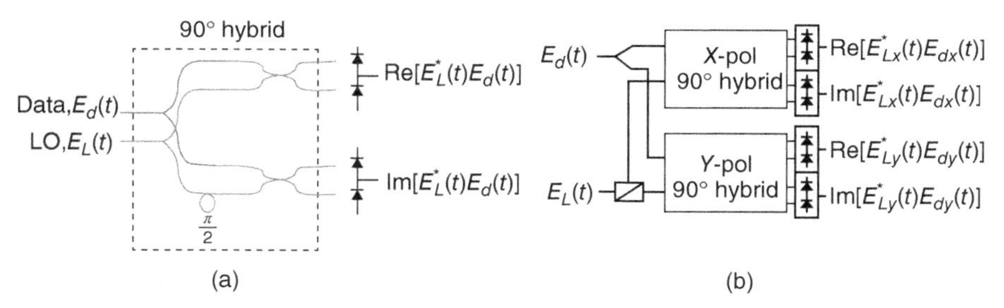

FIGURE 1.1 Coherent detection principle illustration. (a) Phase-diverse coherent detection. (b) Polarization- and phase-diverse coherent detection.

LO phase delayed by 90°, will measure the imaginary part of the input data signal. For reliable measurement of the complex field of the data signal, the LO must be locked in both phase and polarization with the incoming data. In order to realize phase and polarization synchronization in the electrical domain through DSP, a polarization- and phase-diverse receiver is required as is shown in Figure 1.1(b). Such a receiver will project the baseband complex electrical field of the incoming signal into a four-dimensional space vector using the LO as the reference frame.

A coherent receiver requires careful phase and polarization management, which turned out to be the main obstacle for the practical implementation of a coherent receiver using optical-based management methods. The state of polarization of the lightwave is random in the fiber. Dynamic control of the state of polarization of the incoming data signal is required so that it matches that of the LO. Each dynamic polarization controller is bulky and expensive [25], and for WDM systems, each channel needs a dedicated dynamic polarization controller. The difficulty in polarization-management alone severely limits the practicality of coherent receivers, and phase locking is challenging as well. All coherent modulation formats with phase encoding are usually carrier suppressed; therefore, conventional techniques such as injection locking and optical phase-locked loops cannot be directly used to lock the phase of the LO. Instead, decision-directed phase-locked loops must be employed [26, 27]. At high symbol rates, the delays allowed in the phase-locked loop are so small that it becomes impractical [27].

But the advancement of high-speed DSP changed the whole picture. By digitizing the coherently detected optical signals, both phase and polarization can be managed in the electrical domain through advanced DSP. Coherent detection in conjunction with DSP also enables compensation of several major fiber-optic transmission impairments, opening up new possibilities that are shaping the future of optical transmission and networking technology.

1.2.2 Digital Signal Processing

Figure 1.2 shows the functional block diagrams for a typical DSP-enabled coherent transmitter (a) and receiver (b). In principle, the coherent transceiver shown in

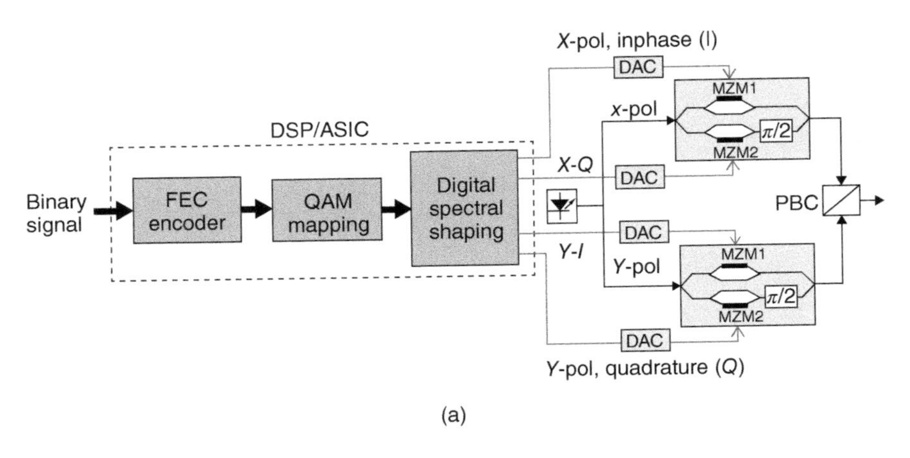

(a)

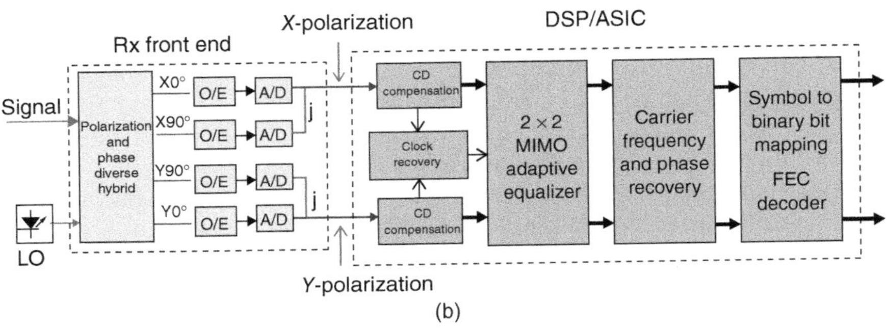

(b)

FIGURE 1.2 DSP-enabled coherent transmitter (a) and receiver (b). CD: chromatic dispersion. ASIC: application-specific integrated circuit.

Figure 1.2 can be used to generate and detect any four-dimensional coded-modulation formats. For the DSP-enabled transmitter shown in Figure 1.2(a), the binary client signal first goes through an FEC encoder, and then the FEC-coded binary signals are mapped into the desired multilevel modulated symbols such as the common quadrature amplitude modulation (QAM) symbols. After that, various digital spectral shaping techniques may be applied to the QAM-mapped signals to improve the transmission performance or to reduce transmission impairments. For example, the Nyquist pulse-shaping technique, which is presented in detail in Chapter 4, can be an effective method to improve the WDM spectral efficiency without resorting to the use of higher-order modulation formats. After digital spectral shaping, the in-phase and quadrature components of the digital QAM signal are converted into two analog signals, which are used to drive an *I*/*Q* modulator to up-convert the baseband electrical signal into an optical signal for transmission. For such a DSP-enabled transmitter, a single *I*/*Q* modulator can be used to generate various QAM formats.

Figure 1.2(b) shows a typical digital coherent receiver. The incoming optical field is coherently mixed with a local oscillator through a polarization- and phase-diverse

90° hybrid. This hybrid separates the in-phase and quadrature components of the received optical field in both X- and Y-polarizations, which are then detected by four balanced photodetectors. The detected analog electrical signals are digitized by four analog-to-digital converters (ADCs) and the digitized signals are then sent to a DSP unit. For such a digital coherent receiver, the front end can be used to receive any quadrature amplitude-modulated signal, because modulation-specific demodulation and decoding are carried out in the DSP unit.

The post-transmission DSP consists of five major functional blocks: (i) fiber CD compensation, (ii) clock recovery, (iii) 2×2 multiple-input-multiple-output (MIMO) adaptive equalization, (iv) carrier frequency and phase recovery, and (v) QAM and FEC decoding. Fiber CD is typically compensated for by using a frequency domain-based phase-only digital spectral shaping technique, and this function can be moved to the transmitter side or split between the transmitter and receiver for ultra-long-haul transmission, where the required computational load may be too heavy for either a single transmitter or receiver DSP chip. The 2×2 adaptive equalization performs automatic polarization tracking, polarization-mode dispersion, and residual CD compensation. This adaptive equalization also helps mitigate impairments from narrow-band filtering effects from the reconfigurable optical add-drop multiplexers (ROADMs) that are widely deployed in today's wavelength-routing optical networks.

Both the transmitter and the receiver DSP units are usually implemented in application-specific integrated circuits (ASICs) for best overall performance (i.e., footprint, power consumption, latency, etc.). Because the computational requirement is substantial for a high-coding-gain soft-decision FEC, an independent ASIC dedicated to FEC has been used in the first generation 100-Gb/s per wavelength coherent transmission systems.

1.2.3 Key Devices

Digital coherent optical transmission technology opens new opportunities to increase the fiber capacity by employing spectrally efficient higher-order modulation formats such as PDM-16QAM, PDM-32QAM, and PDM-64QAM. But these higher-order modulation formats not only require a higher signal-to-noise ratio (SNR), but also become much less tolerant to various impairments from optical devices along the optical links.

One critical device to enable high-SE transmission is narrow linewidth lasers. The widely used DFB laser typically exhibits a Lorentz-type linewidth of about 1 MHz, which is too large to be used for these higher-order modulation formats. External cavity-based tunable lasers exhibit a much more narrow linewidth (~100 kHz) and have been extensively used in recent high-SE transmission system demonstrations. Recently, some progress has been made toward reducing the linewidth of DFB lasers and DBR (distributed Bragg reflector) lasers by employing improved cavity design and a laser linewidth of <500 kHz has been reported [28, 29].

Another critical device for high-SE systems is the so-called I/Q modulator as shown in Figure 1.2(a). It basically consists of two Mach–Zehnder modulators

(MZMs) built in a parallel configuration, with one MZM for in-phase signal modulation and another MZM for quadrature-phase signal modulation. Such an *I*/*Q* modulator in combination with DSP and digital-to-analog converters (DACs) can be used to generate arbitrary QAM. Since a regular MZM exhibits a nonlinear cosine or sine transfer function, digital precompensation of the nonlinear MZM transfer function may be needed in a practical system. Alternatively, some efforts have been made toward developing linear *I*/*Q* modulators [30].

As shown in Figures 1.1 and 1.2, a polarization- and phase-diverse coherent mixer or a hybrid in combination with four pair of balanced photodetectors is needed in a coherent receiver. It should be noted that the four balanced photodetectors may be replaced by four single-ended photodetectors for lower-order modulation formats such as PDM-QPSK, as long as the optical power of the LO is significantly higher than the received optical signal. Several technologies have been used to develop low-loss, small footprint coherent hybrid, including free-space optics, InP or silicon photonics-based photonic integration technology.

To digitize the received analog electrical signal, high-speed ADCs are critical for modern high-speed coherent systems. By using 28-nm CMOS and a successive-approximation-register (SAR)-based architecture, ADCs with >92 Gs/s sampling rate, 8-bit digital resolution, and >25 GHz analog bandwidth have been commercially available since 2014.

1.3 OUTLINE OF THIS BOOK

This book contains 16 chapters. Chapter 2 reviews the modulation formats, starting from basic definitions and performance metrics for modulation formats that are common in the literature to more complicated high-dimensional coded modulation and spectrally efficient modulation. Chapter 3 focuses on detection and error correction technologies for coherent optical communication systems. The chapter shows that the use of differential coding does not decrease capacity and describes how capacity-approaching coding schemes based on LDPC and spatially coupled LDPC codes can be constructed by combining iterative demodulation and decoding. Chapters 4 and 5 are devoted to two spectral-efficient multiplexing techniques, Nyquist WDM and orthogonal-frequency-division multiplexing (OFDM), which use the orthogonality feature either in the time domain or the frequency domain to achieve close to symbol rate channel spacings. In Chapter 6, polarization and nonlinear impairments in coherent optical communication systems are discussed, including PMD and polarization-dependent loss (PDL) impairments and interchannel nonlinear effects in dispersion-managed systems with different configurations. The fiber nonlinear effects in a non-dispersion-managed system is covered in Chapter 7, which shows that fiber nonlinearities in such systems can be accurately described by some analytical models such as GN-EGN model. The next two chapters present impairment equalization and mitigation techniques. Chapter 8 describes linear impairment equalization and Chapter 9 discusses various nonlinear mitigation techniques. Signal synchronization is covered in Chapters 10 and 11, with Chapter 10 focusing on the methods and techniques used to recover timing synchronization and Chapter 11 on carrier phase and frequency recovery in modern high-speed coherent systems. Chapter 12 describes the main constraints put on the DSP algorithms by the hardware structure, and gives a brief overview on technologies and challenges for prototype and commercial real-time implementations of coherent receivers. Chapter 13 addresses the fundamental concepts and recent progress of photonic integration, with a special emphasis on InP- and silicon-based photonic integrated technologies. To increase network efficiency and flexibility, elastic optical network technology and optical performance monitoring have attracted further attention. These are the subjects of Chapters 14 and 15. Chapter 16 discusses spatial-division multiplexing and MIMO processing technology, a potential solution to solve the capacity limit of single-mode fibers.

REFERENCES

1. Alwayn V. Optical Network Design and Implementation. Cisco Press; 2004.
2. Agraval GP. Fiber-Optic Communication Systems. 3rd ed. Wiley Interscience; 2002.
3. Cai J-X, Cai Y, Davidson C, Foursa D, Lucero A, Sinkin O, Patterson W, Pilipetskii A, Mohs G, Bergano N. Transmission of 96x100G pre-filtered PDM-RZ-QPSK channels with 300% spectral efficiency over 10,608 km and 400% spectral efficiency over 4,368 km. Proceedings of OFC 2010, paper PDPB10; Mar 2010.
4. Zhou X, Nelson LE, Magill P, Isaac R, Zhu B, Peckham DW, Borel P, Carlson K. 12,000 km transmission of 100 GHz spaced, 8×495-Gb/s PDM time-domain hybrid QPSK-8QAM signals. Proceedings of OFC 2013, paper OTu2B.4; Mar 2013.
5. Yamada JI, Machida S, Kimura T. 2 Gbit/s optical transmission experiments at 1.3 with 44 km single-mode fibre. *Electron Lett* 1981;17:479–480.
6. Miya T, Terunuma Y, Hosaka T, Miyoshita T. Ultimate low-loss single-mode fiber at 1.55. *Electron Lett* 1979;15:106.
7. Gnauck AH, Kasper BL, Linke RA, Dawson RW, Koch TL, Bridges TJ, Burkhardt EG, Yen RT, Wilt DP, Campbell JC, Nelson KC, Cohen LG. 4-Gbit/s transmission over 103 km of optical fiber using a novel electronic multiplexer/demultiplexer. *J Lightwave Technol* 1985;3:1032–1035.
8. Mohrdiek S, Burkhard H, Steinhagen F, Hillmer H, Losch R, Schlapp W, Gobel R. 10-Gb/s standard fiber transmission using directly modulated 1.55-μm quantum-well DFB lasers. *IEEE Photon. Technol. Lett.* 1995;7(1):1357–1359.
9. Linke RA, Gnauck AH. High-capacity coherent lightwave systems. *J Lightwave Technol* 1988;6:1750–1769.
10. Bergano NS, Aspell J, Davidson CR, Trischitta PR, Nyman BM, Kerfoot FW. Bit error rate measurements of 14000 km 5 Gbit/s fibre-amplifier transmission system using circulating loop. *Electron Lett* 1991;27:1889–1890.
11. Otani T, Goto K, Abe H, Tanaka M, Yamamoto H, Wakabayashi H. 5.3 Gbit/s 11300 km data transmission using actual submarine cables and repeaters. *Electron Lett* 1995;31:380–381.
12. Gnauck AH, Tkach RW, Chraplyvy AR, Li T. High-capacity optical transmission systems. *J Lightwave Technol* 2008;26(9):1032–1045.

13. Fukuchi K, Kasamatsu T, Morie M, Ohhira R, Ito T, Sekiya K, Ogasahara D, Ono T. 0.92 Tbit/s (273 X 40 Gbit/s) triple-band/ultra-dense WDM optical repeated transmission experiment. Proceedings of OFC 2001, paper PD24; Mar 2001.
14. Vareille G, Pitel F, Marcerou JF. 3-Tbit/s (300×11.6Gbit/s) transmission over 7380 km using C+L band with 25 GHz channel spacing and NRZ format. Proceedings of OFC 2001, paper PD22; Mar 2001.
15. Li G. Recent advances in coherent optical communication. *Adv Opt Photon* 2009;1(2):279–307.
16. Griffin R, Carter AC. Optical differential quadrature phase-shift key (oDQPSK) for high capacity optical transmission. Proceedings of OFC 2002, paper WX6; Mar 2002.
17. Tokle T, Davidson CR, Nissov M, Cai JX, Foursa D, Pilipetskii A. 6500 km transmission of RZ-DQPSK WDM signals. *Electron Lett* 2004;40:444–445.
18. Cho PS, Harston G, Kerr CJ, Greenblatt AS, Kaplan A, Achiam Y, Levy-Yurista G, Margalit M, Gross Y, Khurgin JB. Investigation of 2-b/ s/Hz 40-gb/ s DWDM transmission over 4100 km SMF-28 fiber using RZ-DQPSK and polarization multiplexing. *IEEE Photon Technol Lett* 2004;16:656–658.
19. Taylor MG. Coherent detection method using DSP for demodulation of signal and subsequent equalization of propagation impairments. *IEEE Photon Technol Lett* 2004;16(2): 674–676.
20. Noe R. PLL-free synchronous QPSK polarization multiplex/diversity receiver concept with digital I&Q baseband processing. *IEEE Photon Technol Lett* 2005;17:887–889.
21. Kikuchi K. Phase-diversity homodyne detection of multilevel optical modulation with digital carrier phase estimation. *IEEE J Sel Top Quantum Electron* 2006;12:563–570.
22. Han Y, Li G. Coherent optical communication using polarization multiple-input-multiple-output. *Opt Express* 2005;13:7527–7534.
23. Omiya T, Yoshida M, Nakazawa M. 400 Gbit/s 256 QAM-OFDM transmission over 720 km with a 14 bit/s/Hz spectral efficiency by using high-resolution FDE. *Opt Express* 2013;21(3):2632–2641.
24. Qian D, Huang M, Ip E, Huang Y, Shao Y, Hu J, Wang T. 101.7-Tb/s (370×294-Gb/s) PDM-128QAM-OFDM transmission over 3×55-km SSMF using pilot-based phase noise mitigation. Proceedings of OFC 2011, paper PDPB5; Mar 2011.
25. Noe R, Sandel D, Yoshida-Dierolf M, Hinz S, Mirvoda V, Schopflin A, Glingener C, Gottwald E, Scheerer C, Fischer G, Weyrauch T, Haase W. Polarization mode dispersion compensation at 10, 20, and 40 Gb/s with various optical equalizers. *J Lightwave Technol* 1999;17:1602–1616.
26. Barry JR, Kahn JM. Carrier synchronization for homodyne and heterodyne-detection of optical quadriphase-shift keying. *J Lightwave Technol* 1992;10:1939–1951.
27. Kazovsky L. Balanced phase-locked loops for optical homodyne receivers: performance analysis, design considerations, and laser linewidth requirements. *J Lightwave Technol* 1986;4:182–195.
28. Kobayashi G, Kiyota K, Kimoto T, Mukaihara T. Narrow linewidth tunable light source integrated with distributed reflector laser array. Proceedings of OFC 2014, paper Tu2H.2; 9–13 Mar 2014.

29. Larson MC, Feng Y, Koh PC, Huang X, Moewe M, Semakov A, Patwardhan A, Chiu E, Bhardwaj A, Chan K, Lu J, Bajwa S, Duncan K. Narrow linewidth high power thermally tuned sampled-grating distributed Bragg reflector laser. Proceedings of OFC 2013, paper OTh3I.4; Mar 2013.
30. Kaneko A, Yamazaki H, Miyamoto Y. Linear optical modulator. Proceedings of OFC 2014, paper W3K.5; Mar 2014.

2

MULTIDIMENSIONAL OPTIMIZED OPTICAL MODULATION FORMATS

MAGNUS KARLSSON[1] AND ERIK AGRELL[2]

[1]*Photonics Laboratory, Department of Microtechnology and Nanoscience, Chalmers University of Technology, Gothenburg, Sweden*
[2]*The Communication Systems Group, Department of Signals and Systems, Chalmers University of Technology, Gothenburg, Sweden*

2.1 INTRODUCTION

The development of advanced digital signal processing (DSP) to enable intradyne coherent optical receivers [1–3] caused a paradigm shift within optical communications, and there is little doubt that the future of optical transport will be coherent. Coherent receivers ideally map the optical signal to the electrical domain, which enables a lot of novel advanced communication algorithms to be implemented in optical links, for example, digital equalization and advanced modulation. One of the most profound developments was that intradyne receivers enabled all four quadratures of the optical signal (or in optical terms amplitude, phase, and polarization states) to be modulated and detected. This was realized already in the early 1990s when Betti et al. investigated the modulation of all four quadratures in optical links [4–7]. Even if coherent detection was demonstrated already in 1990 by Derr [8], it was too complicated to be commercially interesting and the research faded.

As optical transmission systems had traditionally used rudimentary modulation (typically on-off-keying (OOK) or differential phase-shift-keying (DPSK) [9]), the coherent receivers meant great opportunity to study novel modulation formats, tai-

Enabling Technologies for High Spectral-efficiency Coherent Optical Communication Networks, First Edition. Edited by Xiang Zhou and Chongjin Xie.
© 2016 John Wiley & Sons, Inc. Published 2016 by John Wiley & Sons, Inc.

lored for the emerging coherent optical links. The first such format used was the polarization-multiplexed quadrature shift keying (PM-QPSK) [2, 3], which in its simplest form is binary phase-shift keying (BPSK) in all four quadratures in parallel. As coding and modulation are key building blocks in the design of any communication link, it is a natural first approach to separate them and study the performance of each block separately. Most of the research reviewed and presented in this chapter deals only with the modulation format, but we emphasize that it is only part of the problem in designing a good optical transmission link. The second part is to add forward error correcting (FEC) codes, preferably tailored and co-optimized with the modulation formats, an area often referred to as *coded modulation*. However, a discussion on that topic is beyond the scope of this chapter, and we refer the interested reader to [10, 11] for a recent overview and introduction.

The choice of modulation format in a link is crucial in that it sets an upper limit on the achievable *spectral efficiency*, which loosely speaking measures how well the channel real estate (bandwidth and signaling dimensions) are utilized. The addition of FEC will always reduce the spectral efficiency (but with the crucial benefit of increasing the noise tolerance). Nevertheless, there is a deep relation between coding and modulation. Specifically, all FEC codes can be interpreted as a multidimensional modulation format by considering a sequence of time slots as dimensions. The converse does not necessarily hold. Although many multidimensional modulation formats, in particular those with a regular structure, can be interpreted as a low-dimensional modulation format in combination with an FEC code, this is not always the case. The relation between modulation and coding is discussed further in Sections 2.2.1 and 2.4.

In the choice of modulation format, there is an inherent threefold trade-off between the spectral efficiency, the noise tolerance, and complexity of the format. In this chapter, we aim to shed some light on these trade-offs, by investigating relatively simple, low-dimensional formats in four dimensions. Such research is not new; 4D formats were investigated already in the 1970s by Welti and Lee [12] and by Zetterberg and Brändström [13]. Also, the work by Biglieri [14] contains some of the 4D formats that we discuss in this chapter, as well as discussions on lattices and lattice cuts, which we also cover. The novelty is the application to the optical channel with its specifics and trade-offs when it comes to signal generation, transmission, and detection. Therefore, we devote quite some effort to review and describe implementations and experiments.

This chapter is organized as follows. In the next section, we give basic definitions and performance metrics for modulation formats that are common in the literature. In Section 2.3, the most interesting formats and their performances are theoretically described and characterized. Next, in Section 2.4, we study how low-dimensional codes can be used to extend the known formats to higher dimensions and spectral efficiencies. In Section 2.5, the relatively large body of experimental work done on multidimensional modulation in coherent links that has been done in the last few years is reviewed, and finally Section 2.6 concludes.

2.2 FUNDAMENTALS OF DIGITAL MODULATION

An optical communication channel, like any other physical propagation or storage medium, is what in communication theory is called a *waveform channel*, which communicates a time-varying voltage (or electric field) from one point to another. If the channel is used to transmit digital data, then there are only a finite number of possible waveforms of a given length, and every such waveform corresponds to a certain sequence of bits. The process of mapping bits into waveforms and vice versa is called *digital modulation*. This can be done in a multitude of ways, depending on the type of channel. Some common optical system models, and their preferred modulation techniques, are reviewed in Section 2.2.1. In Section 2.2.2, we discuss several optical channel models that have been proposed to account for the fiber propagation effects.

In order to compare modulation formats and select a suitable one for implementation in a particular communication system, a *performance metric* is needed. There is a multitude of such metrics, for a variety of purposes. A modulation format that is superior in one sense may very well be inferior in another. This is the topic of Section 2.2.3.

2.2.1 System Models

A *multidimensional channel* is one that offers the possibility of transmitting multiple waveforms simultaneously. These waveforms could consist of the two quadratures of an amplitude- and phase-modulated light wave, the two polarizations, multiple wavelengths in a wavelength-division multiplexed (WDM) system, multiple modes, or multiple cores. Each of the parallel waveforms can be thought of residing in one dimension. The traditional paradigm, and the least complex solution, is to transmit independent data on all of these dimensions. However, improved performance can be obtained by encoding data jointly on several dimensions, that is, by *multidimensional modulation*. This improvement is most prominent if the waveforms interfere with each other during transmission, but significant gains can be achieved even if the waveforms are transmitted independently. The topic of multidimensional modulation is revisited in Section 2.5.5.

The mapping of bits into waveforms can be thought of as a three-step process. First, redundant bits are added to the payload. This overhead serves several purposes: to indicate a frame structure, which allows the interpretation of the received bit stream as a sequence of data packets to provide address information for proper routing and to provide error resilience via FEC. These functions, albeit crucial for the operation of an optical communication network, are all outside the scope of the present chapter.

Second, m bits at a time are mapped into a *symbol*, which is a vector in an N-dimensional space. The set $\mathcal{X}$ of all $M = 2^m$ symbols are called a *constellation*. This is the single most important entity in the definition of a modulation format; indeed, it is so important that the term "modulation format" is sometimes used as a synonym for constellation.

Third, the sequence of symbols is mapped into a set of waveforms. The standard way to do this is via a *linear modulator*. Denoting the sequence of N-dimensional symbols with $\boldsymbol{x}[k]$, for $k = \dots, 0, 1, 2, \dots$, the vector of N waveforms is computed as

$$\boldsymbol{x}(t) = \sum_k \boldsymbol{x}[k]\phi(t - kT) \tag{2.1}$$

where T is the symbol time and $\phi(t)$ is a given pulse shape.

At this point, it should be emphasized that the discrete-time sequence $\boldsymbol{x}[k]$ is fundamentally different from its continuous-time counterpart $\boldsymbol{x}(t)$ and they should not be confused with each other. The waveforms $\boldsymbol{x}(t)$ needs to be considered in order to analyze signal spectra as well as propagation effects such as distortions, filtering, added noise, and other hardware limitations. On the contrary, the sequence $\boldsymbol{x}[k]$ is the quantity of interest to analyze bit error rate (BER) and symbol error rate (SER), mutual information, channel capacity, etc.

The vector $\boldsymbol{x}(t)$ represents N baseband waveforms. Each of these waveforms are now multiplied with a carrier, for transmission over an N-dimensional channel, which, as explained in the beginning of this subsection, consists of multiple quadratures, polarizations, wavelengths, modes, and/or cores.

At the receiver side, the reverse operations are performed using a coherent receiver. First, the symbol clock, carrier phase, and polarization are recovered using either blind or pilot-aided estimation algorithms [15–17]. A balanced detector now outputs the N received baseband waveforms, represented by the vector $\boldsymbol{y}(t)$, which should hopefully resemble $\boldsymbol{x}(t)$.

Second, the waveforms are filtered and sampled. The obtained sequence of N-dimensional vectors is

$$\boldsymbol{y}[k] = \int_{-\infty}^{\infty} \boldsymbol{y}(t)h(kT - t)dt \tag{2.2}$$

for $k = \dots, 0, 1, 2, \dots$, where $h(t)$ is the impulse response of the receiver filter. The received symbol sequence $\hat{\boldsymbol{x}}[k]$ is now determined by identifying, independently for each k, the point in $\mathcal{X}$ closest to $\boldsymbol{y}[k]$, in some well-defined sense that depends on the channel model. Ideally, the receiver filter is chosen as a *matched filter* $h(t) \sim \phi(T_d - t)$, where T_d is the processing delay. Furthermore, the pulse $\phi(t)$ is chosen to satisfy the *T-orthogonality criterion*

$$\int_{-\infty}^{\infty} \phi(t)\phi(t - kT)dt = 0, \text{ for all integers } \quad k \neq 0 \tag{2.3}$$

which avoids intersymbol interference for linear channels, that is, $\boldsymbol{y}[k]$ depends on $\boldsymbol{x}[k]$ but not on $\boldsymbol{x}[k \pm 1], \boldsymbol{x}[k \pm 2], \dots$

Third and last, the received bit sequence is obtained by concatenating the bits corresponding to each symbol. Then, the digital overhead is removed, which includes the operations of FEC decoding and frame synchronization.

It is also possible to consider blocks of K symbols $\boldsymbol{x}[k], \boldsymbol{x}[k + 1], \dots, \boldsymbol{x}[k + K - 1]$ as a supersymbol, taken from a constellation of NK dimensions. In general, this technique improves the performance at the cost of a higher transmitter and receiver complexity. A similar effect can be achieved at a more manageable complexity by applying an FEC code before modulation. Specifically, if a block code with codeword length $n = mK$ is applied to the bit stream before modulation, the resulting symbol sequence can be regarded either as a sequence of dependent N-dimensional symbols or as a sequence of independent NK-dimensional supersymbols. We see examples of such NK-dimensional constellations designed from standard FEC codes in Section 2.4.

2.2.2 Channel Models

A complication for optical links is that the fiber propagation of the signal waveform is conventionally modeled with a nonlinear partial differential equation, the nonlinear Schrödinger equation (NLSE), where fiber dispersion, nonlinearities, and amplifier noise distort the signal. This is not the desired discrete-time model that a communication engineer would like to have when designing the coding and modulation algorithms. There are generally three problems associated with taking the fiber propagation to a usable discrete-time model. (i) To correctly model the transitions between symbols and waveforms (discrete and continuous time). Usually, the transmitter is modeled as a continuous pulse source multiplied with discrete data in each symbol time, ignoring the sum in (2.1). This works fairly well, but one may have unwanted intersymbol interference in the symbol borders that is often neglected. The receiver side, going from the continuous waveform to a discrete data sample, is often modeled as an integrate-and-dump filter, that is, restricting the integral in (2.2) to an interval of length T. This is not penalty-free, and it is theoretically complicated when the signal spectrum is distorted or broadened so one cannot guarantee matched filtering or sampling without aliasing. (ii) The NLSE and fiber transmission is *nonlinear* in the general case, and often operated in a regime where the nonlinearity cannot be neglected. In this case, the received signal is generally affected by intersymbol interference even if (2.3) is satisfied and linear ISI in the channel is removed. (iii) The coherent receiver should have negligible distortions, that is, operate in a regime (strong local oscillator with low phase noise) where it linearly maps the optical field to the electrical domain for sampling and detection. In addition, perfect timing synchronization and compensation for channel impairments are assumed.

Often these problems are neglected, which leads to the standard additive white Gaussian noise (AWGN) model for coherent links, where the signal is only distorted by additive amplifier spontaneous emission (ASE) noise [18, 19]. Good agreement between simulations and experiments is evidence that this approach works reasonably well for many systems.

Of the above-mentioned problems, the nonlinearity is the most serious one, but thanks to the recent developments of the *Gaussian noise (GN) model* [20–22], it can be dealt with by a simple extension of the AWGN model. The GN model applies to links with strong dispersive broadening during propagation and electronic dispersion compensation in the receiver. Then, the impact of the nonlinearity can be accurately modeled as AWGN with a variance proportional to the average signal power cubed, which was first observed by Splett et al. in 1993 [23]. In such links, the presented format optimizations (which rely on the noise being uniform in all dimensions) will still work well. The GN model is known to agree well with experiments and to be a useful system design tool, but the usefulness for, for example, capacity estimates in nonlinear links can be questioned [24].

A second model accounting for fiber nonlinearities is the *nonlinear phase-noise model* [25, pp. 157, 225]. This applies to links where the dispersion is negligible, for example, with optical in-line compensation and/or low baudrates. Then, the nonlinear self-phase modulation will, together with the ASE noise, lead to constellations with a spiraling shape. The model has also been extended to dual polarizations by Beygi et al. [26].

2.2.3 Constellations and Their Performance Metrics

The starting point for digital modulation theory is, since long before the invention of fiber-optic communications, the scenario consisting of an AWGN channel, no coding, optimal detection (maximum likelihood, ML), and asymptotically low error probability. In this scenario, the BER and SER are both proportional to $Q(d/\sqrt{2N_0})$ [27, 28], where $Q = (2\pi)^{-1/2} \int_x^\infty \exp(-z^2/2)dz$ is the Gaussian Q function, d is the minimum Euclidean distance between points in the constellation, and N_0 is the noise power spectral density. Modulation formats are, therefore, traditionally designed in order to maximize (a normalized version of) the minimum distance d. Nevertheless, such modulation formats are often applied even in scenarios where the minimum distance does not govern the performance, such as for nonGaussian or nonlinear channels, in coded systems, with suboptimal receivers, or at nonasymptotic error probabilities.

The following performance metrics are often used to quantify the performance of modulation formats [28].

Spectral Efficiency The spectral efficiency or normalized bit rate is defined as [29, 30]

$$\beta = \frac{\log_2 M}{N/2}$$

where N and M, as defined in Section 2.2.1, give the number of dimensions and constellation points, respectively. The spectral efficiency gives the number of bits per channel use, where every (complex) channel use involves two dimensions. It also gives the bit rate per bandwidth, in bit/s/Hz, if Nyquist signaling is applied (sinc pulse shaping). A related quantity is $\beta/2$, which gives the number of bits per dimension, and can be interpreted as the data rate per bandwidth in bit/s/Hz, if rectangular pulse shaping is applied and bandwidth is defined as the width of the spectral main lobe.

Average and Peak Symbol Energy The average symbol energy, also called the second moment or the mean squared Euclidean norm, is

$$E = \frac{1}{M} \sum_{\boldsymbol{x} \in \mathcal{X}} |\boldsymbol{x}|^2$$

and the peak symbol energy is

$$E_{\max} = \max_{x \in \mathcal{X}} |\boldsymbol{x}|^2$$

If the pulse $\phi(t)$ in (2.1) satisfies (2.3), then

$$\lim_{n\to\infty} \frac{1}{2nT} \int_{-nT}^{nT} |\boldsymbol{x}(t)|^2 dt \sim \frac{E}{T} \int_{-\infty}^{\infty} \phi^2(t) dt$$

that is, the continuous-time average energy is proportional to the discrete-time average energy E. Unfortunately, there exists no analogous relation between the continuous-time and discrete-time peak energies. Constellation designs based on $E_{\max}$ tend nevertheless to be relatively good also in terms of the continuous-time peak energy, but not necessarily optimal [31].

Average Bit Energy $E_{\mathrm{b}} = E/\log_2 M$ gives the average energy needed to transit one bit of information.

Constellation Figure of Merit The constellation figure of merit (CFM) is defined as [29, 30]

$$CFM = \frac{d^2 N}{2E}$$

This is, assuming AWGN, no coding, optimal detection (maximum likelihood), and asymptotically high signal-to-noise ratio (SNR; low error probability), the relevant power metric if modulation formats are compared at the same *bandwidth*.

Power Efficiency The (asymptotic) power efficiency is [32, eq. (5.8)], [27]

$$\gamma = \frac{d^2}{4E_{\mathrm{b}}} = \frac{\beta CFM}{4} \tag{2.4}$$

This is, under the same conditions as for the CFM, the relevant power metric if modulation formats are compared at the same *bit rate*.

Gain The gain is quantified with respect to a baseline modulation format at the same spectral efficiency β, commonly chosen as *pulse-amplitude modulation* (PAM) [29, 30]. A PAM constellation has

$$CFM_{\mathrm{PAM}} = \frac{6}{2^{\beta} - 1}$$

and

$$\gamma_{\mathrm{PAM}} = \frac{3\beta}{2(2^{\beta} - 1)} \tag{2.5}$$

Multidimensional extensions of PAM such as quadrature-amplitude modulation (QAM) and polarization-multiplexed (PM) QAM have the same CFM and γ.

Geometrically, the baseline constellations represent cubic subsets of the cubic lattice. The gain is defined as

$$G = \frac{CFM}{CFM_{\mathrm{PAM}}} = \frac{\gamma}{\gamma_{\mathrm{PAM}}}$$

also for spectral efficiencies β for which no PAM constellation exists.

Mutual Information, MI The mutual information is defined as

$$I(\boldsymbol{X};\boldsymbol{Y}) = \iint f(\boldsymbol{x},\boldsymbol{y}) \log_2 \frac{f(\boldsymbol{x},\boldsymbol{y})}{f(\boldsymbol{x})f(\boldsymbol{y})} d\boldsymbol{x}\, d\boldsymbol{y} \tag{2.6}$$

where $\boldsymbol{X}$ and $\boldsymbol{Y}$ are the channel inputs and outputs, respectively, and f denotes the distribution of the stochastic variables indicated by its arguments.

Complexity Finally, some words should be said about complexity. It is one of the most important figures of merit, and it should be considered in any implementation, in order to keep the latency, energy consumption, and cost within reasonable levels. Nevertheless, it is one of the hardest parameters to quantify numerically, depending not only on the modulation format but also on the transmitter and receiver algorithms as well as the hardware platform. As a crude rule of thumb, the complexity increases with the dimension, number of points, and irregularity of the constellation.

2.3 MODULATION FORMATS AND THEIR IDEAL PERFORMANCE

In this section, we briefly review the various modulation formats and format optimizations that have been presented in the literature. Without doubt, the most commonly used formats are the PAM formats, based on the cubic lattice, possibly in N dimensions. Their performance is well known and stated in Section 2.2.3. Their popularity is mostly due to their simplicity of generation and detection, but if some of that simplicity is sacrificed, much better performance (in terms of noise tolerance or spectral efficiency) can be achieved. The formats presented in this section are examples of that.

We extensively discuss *format optimization* later in the text. It is important to emphasize that the outcome of such an optimization is heavily dependent on *what* is optimized and *which constraints* are assumed under the optimization. The simplest and most common scenario is to assume AWGN, no coding, optimal detection (ML), and asymptotically high SNR (low error probability). This ideal scenario is studied in this section. Modulation optimization for some specific nonlinear and nonGaussian channel models is summarized in Section 2.3.2.

In the limit of high SNR, the formats with the lowest SER can be found from optimized packings of solid spheres [27, 31, 33]. For a constant dimensionality and number of spheres, such packing optimization can be done by either minimizing the average distance of the spheres from the origin (the average second moment E) or by minimizing the maximum distance (the maximum symbol energy $E_{\max}$). To

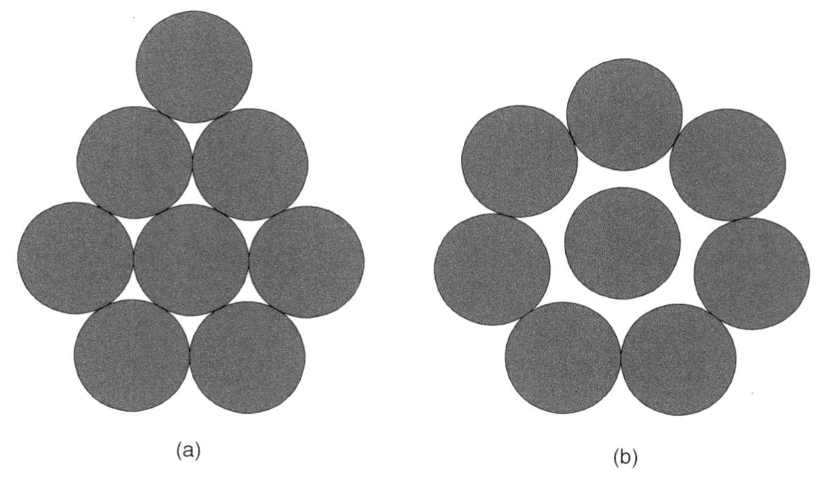

FIGURE 2.1 The cluster $C_{2,8}$ (a) and the ball $\mathcal{B}_{2,8}$ (b).

emphasize this difference, the constellation of M spheres with minimum E in dimension N is called the *cluster* $C_{N,M}$, and the constellation with lowest $E_{\max}$ is called the *ball* $\mathcal{B}_{N,M}$. Sometimes the clusters and balls coincide, but in general they do not. A simple example of the latter arises for 8 points in 2D, as shown in Figure 2.1. This example also shows that the balls may be nonunique, as the center point is loose, and can be freely moved without affecting $E_{\max}$.

In addition to the balls and clusters, one can also compare different formats at the same bit rate (where γ is the relevant metric), or at the same bandwidth (where *CFM* is used). Both cases are discussed in Section 2.3.1. The balls and their relevance were discussed in Ref. [31] and are briefly touched upon in Section 2.3.2.

2.3.1 Format Optimizations and Comparisons

This and the next few sections focus mainly on the *clusters*, that is, the N-dimensional, M-point constellations that minimize the average symbol energy (second moment) E. Tables with coordinates of those constellations are given in, for example, [34] for 2D clusters and [35] for 3D and 4D clusters. These and other constellations are available online [28]. All these are numerically optimized results, presented as tables of coordinates. Some of the most interesting constellations are presented in exact analytic form in Refs [27, 31]. Quite often, the clusters possess some symmetry that facilitates a nice coordinate description.

In the limit of many points, the clusters will be spherical cuts from the regular *lattices* that are known to be the best packings in the given dimension. The best packing lattices are only known exactly in dimensions 2, 3, 4, 8, and 24, and they are listed in Table 2.1, together with their *densities*, Δ, which denotes the fraction of N-dimensional space that is filled by packing nonoverlapping spheres at the lattice

TABLE 2.1 Known densest lattices, their number of nearest neighbors, and densities

Dimensions	Densest lattice	Neighbors (kissing number)	Density (Δ)
2	A_2	6	$\frac{\pi}{2\sqrt{3}} = 0.91$
3	A_3	12	$\frac{\pi}{3\sqrt{2}} = 0.74$
4	D_4	24	$\frac{\pi^2}{16} = 0.62$
8	E_8	240	$\frac{\pi^4}{384} = 0.25$
24	Λ_{24}	196,560	$\frac{\pi^{12}}{12!} = 0.0019$

points. The power efficiency for a spherical cut of M lattice points in N-dimensional space can, if M is sufficiently large, be well approximated as [36, eq. (32)]

$$\gamma_{\text{lat}} = \log_2(M)\left(1 + \frac{2}{N}\right)\left(\frac{\Delta}{M}\right)^{\frac{2}{N}} \tag{2.7}$$

This expression is derived by assuming a uniform point density in the spherical cut. This approach can be expected to be better with increasing M, significantly exceeding the nearest neighbor number, so that many lattice cells are enclosed in the cut. If an N-dimensional hypercubic cut is carried out rather than a spherical cut, a penalty of $\pi e/6 = 1.53$ dB (the so-called *shaping gain*) is sacrificed for large N. In a similar manner, we have the *CFM* and gain for the lattices as

$$CFM_{\text{lat}} = 2(N+2)\left(\frac{\Delta}{M}\right)^{\frac{2}{N}} \tag{2.8}$$

$$G_{\text{lat}} = \frac{N+2}{3}\left(M^{\frac{2}{N}} - 1\right)\left(\frac{\Delta}{M}\right)^{\frac{2}{N}} \tag{2.9}$$

2.3.1.1 ***General Properties of the Metrics*** Properties of the best-known clusters, for $N = 2$, 4, and 8 and selected values of M, are shown in Figure 2.2. The coordinates of the clusters are available online [28]. We conjecture that these clusters are all optimal for their values of N and M.[1]

The spectral efficiency β is shown versus the three power measures: *CFM*, power efficiency γ, and gain G. This also shows the qualitatively different behavior of the three metrics (*CFM*, γ, and G). We now discuss the general behavior of these metrics with spectral efficiency β (or M, since $\beta \sim \log_2(M)$).

[1] The conjecture does not extend to $N = 8$ and $M > 128$. These clusters, which are also included in Figure 2.2, are unpublished and not as well optimized as those on [28].

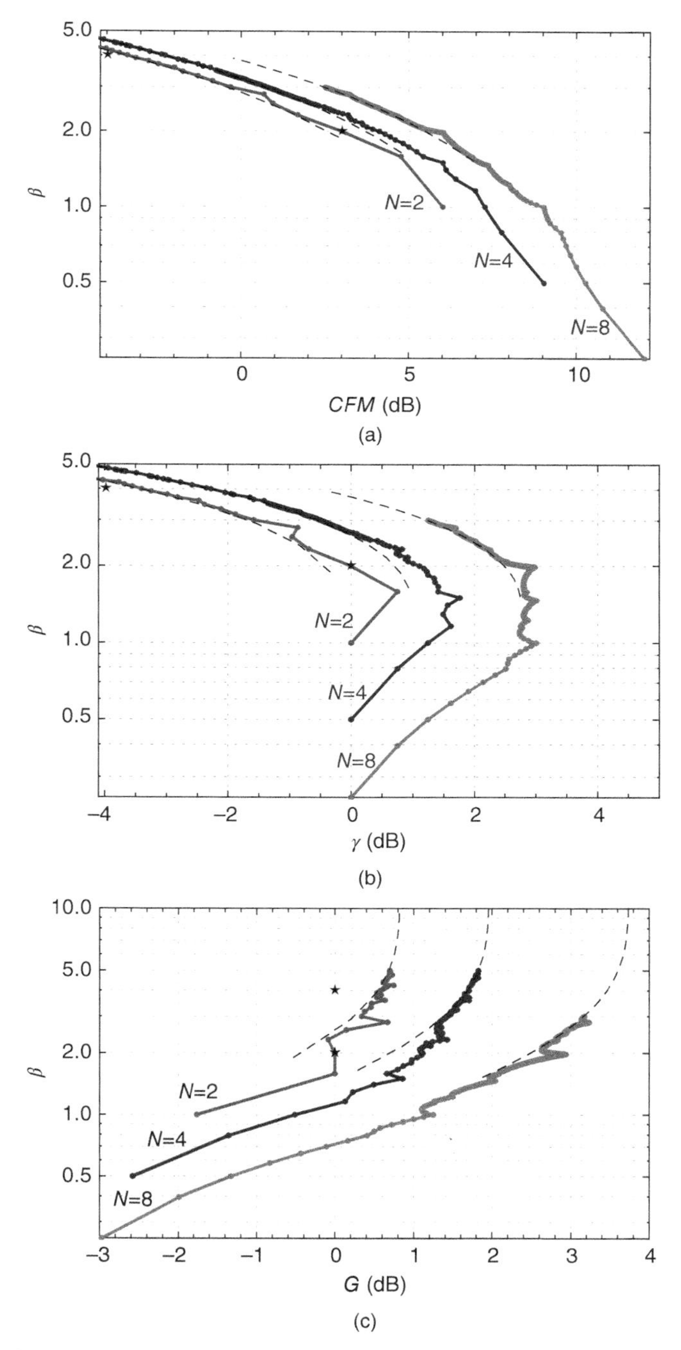

FIGURE 2.2 Spectral efficiency of the best-known clusters, plotted versus *CFM*, γ, and G, respectively. Dimension $N = 2$, 4, and 8 as marked. The performance of the A_2, D_4, and E_8 lattices using (2.7)–(2.9) is shown with dashed lines. The stars show BPSK (corresponding to QPSK in 2D and PM-QPSK in 4D) and 4-PAM (corresponding to 16QAM in 2D and PM-16QAM in 4D).

TABLE 2.2 Known maxima for γ and their optimum number of points M_{opt}

Dimensions (N)	γ_{max} (dB)	M_{opt}	$\gamma_{max,lat}$ (dB)	$M_{opt,lat}$
2	0.75	3	–0.15	2.8
3	1.25	4	0.36	4.5
4	1.76	8	0.98	7.3
8	3.01	16	2.73	54

For comparison are shown the optima based on the asymptotic lattice expression (2.7).

The $CFM \sim 1/E(M)$ decreases monotonically with spectral efficiency β, as it compares formats at the same bandwidth (same baudrate), thus showing essentially how the second moment $E(M)$ increases with M. For large M, one can expect the clusters to behave as lattice packings, and the CFM to decrease as $\sim M^{-2/N}$ according to (2.7).

The γ, on the contrary, weighs in the data rate by multiplying CFM with $\log_2(M)$, giving it a dependence $\gamma \sim \log_2(M)/E(M)$. It can be shown that γ always increases up to at least the simplex ($M = N + 1$). However, for large M, the dependence is the lattice's $\sim \log_2(M)M^{-2/N}$, which will eventually decrease with M, and we conclude that for every dimension $N > 1$, γ has a maximum γ_{max} at some value M_{opt}. The values of γ_{max} and M_{opt} are only known, or conjectured, for $N = 2, 3, 4, 8$ and listed in Table 2.2. Not much is known about the general dependence of γ_{max} and M_{opt} on the dimension N. However, a crude approximation can be obtained from the lattice expression, and maximizing $\log_2(M)M^{-2/N}$ for real M. This optimum is

$$M_{opt,lat} = \exp(N/2) \tag{2.10}$$

$$\gamma_{max,lat} = \frac{N+2}{2e\log_2(2)}\Delta^{\frac{2}{N}} \tag{2.11}$$

These values are compared with the exact known values [2] in Table 2.2, and the agreement is surprisingly accurate, given the rough approximation involved by approximating the discrete points with the homogeneous lattice distributions. It is also interesting to note that $M_{opt,lat}$ corresponds to $\beta_{opt,lat} = 1/\log_2(2) = 1.44$ bits per symbol per dimension pair, independently of N.

The gain G is defined as the performance relative to the cubic-lattice PAM constellations (QPSK, 16QAM, PM-QPSK, PM-16QAM, etc.), which all have $G = 0$. The clusters show a rapid improvement over the cubic lattice as β increases, as is clear from Figure 2.2(c). At high spectral-efficiencies, the gain will approach the asymptote given by

$$G_{max} = \frac{N+2}{3}\Delta^{\frac{2}{N}} \tag{2.12}$$

which is 0.84, 1.97, and 3.72 dB in the respective 2D, 4D, and 8D cases.

[2] With "exact known values" we mean, similar to that for most other sphere packing problems, that they are found from extensive numerical optimization. No formal proofs of optimality are known.

*2.3.1.2 **Two-Dimensional Formats*** The 2D clusters are in almost all cases part of the hexagonal lattice A_2, which is the densest packing of many spheres in 2D space. The only exception is $M = 4$, for which every rhombic constellation with vertex angle between 60° and 120°, including the square constellation (QPSK), have the same average symbol energy E as a four-point subset of A_2. Second moments of clusters for M up to 500 are listed in Ref. [34].

Foschini et al. [33] found the optimum 2D clusters in the cases of practical interest ($M = 8, M = 16$) by numerical optimization already in 1974, but clearly these results have not taken on in the community, and there are at least three reasons for this: (i) the noninteger coordinates make a practical implementation more difficult, (ii) the gains G over square QAM constellations are never more than 0.84 dB according to (2.10), and (iii) (less important) the hexagonal constellations do not lend themselves to a straightforward bit-to-symbol mapping. QAM constellations are, therefore, dominating in practical 2D systems.

The full set of 2D clusters up to $M = 32$ are shown as the $N = 2$ line for the three metrics (γ, CFM and G) in Figure 2.2. The most common formats QPSK and 16QAM are shown as stars in Figure 2.2, and in Figure 2.2(c) they are references at $G = 0$. In the limit of many points, the 2D clusters have performance close to the A_2 lattice (shown with a dashed line), which is not surprising since they *are* cuts from this lattice, as shown by Graham and Sloane [34].

The highest γ is seen to arise for $M = 3$ (3-PSK), at $\beta = 3/2$. However, as for all the other 2D clusters (except for QPSK), it has seen limited use, although being discussed in the literature [37, 38].

*2.3.1.3 **Four-Dimensional Formats*** The 4D clusters $C_{4,M}$ are shown in Figure 2.2 as the $N = 4$ line. Similar optimizations are found in, for example, [39] as well as in online listings of coordinates by, for example, Sloane et al. [35]. A more detailed description, including exact coordinate descriptions of some interesting clusters, was provided in Refs [27, 31]. For communication purposes, the powers of two, $M = 8, 16, 32 \cdots$, are of particular interest, and they are discussed separately later.

In general, the optimum, or nearly optimum, 4D constellations that are subsets of the D_4 lattice are easier to implement than the corresponding 2D clusters, since the D_4 lattice is a subset of the regular cubic (integer) lattice $\mathbb{Z}^4$. They will thus have a better opportunity to find wide use than the 2D clusters. Also, higher gains G are attainable in 4D than 2D.

A few specific cases have caused interest in the research community, and are discussed separately later, namely $M = 4, 8, 16$, and 24, as well as the higher powers of 2.

3D Simplex The best packing of 4 points in 4D, the cluster $C_{4,4}$, is to put them in a regular tetrahedron, also known as the 3D *simplex* [40, p. 178]. Obviously, this is not a 4D object at all, since at least 5 points are required to span a 4D object, but it is the best packing of 4 points in all dimensions $N \geq 3$. Moreover, numerical evidence indicates that all clusters $C_{N,M}$, where $M \leq N + 1$, are the M-ary simplices. In optical communications, this format was proposed and evaluated by Dochhan et al. [41] as an alternative to PM-BPSK, over which it has a 1.25 dB asymptotic sensitivity gain.

PS-QPSK The maximum γ in 4D occurs for $M = 8$, as originally pointed out in Ref. [42]. Geometrically, the format is the 4D *cross-polytope*, and also known in the communications community as 8-ary *biorthogonal modulation*. The biorthogonal (or cross-polytope) formats consist of all permutations and signs of signal vectors with zeroes at all coordinates except one [40, p. 178]. Exact SER expressions for all biorthogonal formats are given in [43, eq. (4.102), 44]. Gray mapping is not possible for biorthogonal formats, but assuming the "obvious" bit-to-symbol mapping that flips all bits between opposing symbol pairs $\pm 1, 0, 0, 0 \ldots$, an exact expression for the BER was given in [43, p. 203].

The 8-ary biorthogonal format was originally proposed for optical coherent systems by Betti et al. [5], although it had been considered for communications much earlier [12, 13, 45]. It can even be considered as a special case of permutation modulation, introduced already by Slepian [46].

In 4D, the cross-polytope can take on many representations [42]; in addition to the permutations of $\pm(1, 0, 0, 0)$, it can be regarded as the odd (or even) parity subset of the 4D cube (PM-QPSK). It can thus also be seen as resulting from a parity-check code applied to the standard PM-QPSK [47, 48], as is discussed in Section 2.4.1. The strength lies in that it loses less in spectral efficiency than it gains in sensitivity over PM-QPSK, so compared at the same bit rate, it gains $\gamma = 3/2$ or 1.76 dB in power efficiency. At a finite BER of 10^{-3}, its gain is around 1 dB.

Transmission simulations of PS-QPSK in nonlinearly limited fiber links were presented in, for example, [49–51]. The general result is that the power efficiency improvement over conventional PM-QPSK can be translated into a reach extension or increased amplifier span losses, which has also been seen in experiments as discussed in Section 2.5.

6PolSK-QPSK/24-cell The 24-cell is a four-dimensional polytope that is, according to Coxeter [52], "... a peculiarity of four-dimensional space ... having no analogue [in dimensions] above or below." The constellation consists of 24 vertices equally spaced from the origin and each other, and plays an important geometric role of being the Voronoi cell of the D_4-lattice, as well as the 4D kissing constellation, the latter being proved relatively recently [53]. The kissing constellation, consisting of the 24-cell and a point at the origin, is also the cluster $C_{4,25}$, notably a local maximum in the γ versus β plot, Figure 2.2(b). The cluster $C_{4,24}$ is *not* the 24-cell, but $C_{4,25}$ with an outer point removed and centered at the center of gravity.

Nevertheless, the 24-cell $C_{24\text{-cell}}$ performs quite well as a format in its own right. The points can be given as the hypercube in union with the cross-polytope, that is,

$$C_{24\text{-cell}} \in \{(\pm 1, \pm 1, \pm 1, \pm 1), (\pm 2, 0, 0, 0)\}$$

taken with all permutations and sign selections. An alternative, rotated and rescaled, representation is all permutations and sign selections of $(\pm 1, \pm 1, 0, 0)$. The 24-cell has a spectral efficiency $\beta = \log_2(24)/2 = 2.29$ and power efficiency $\gamma = \log_2(24)/4 = 1.15$ (or 0.59 dB). An exact expression of its SER was given in Ref. [27] and derived in Ref. [44]. It was suggested for communications in Refs [12, 13, 54]. In coherent optical communications, it was first proposed by Bülow et al. [55], and later identified as the 24-cell in Ref. [27]. In optical links, it is realized by transmitting QPSK in one of six different polarization states (x, y, $\pm 45°$, right/left-hand circular) [55], and hence referred to as 6-polarization shift-keying (6PolSK)-QPSK.

The 24-ary nature of 6PolSK-QPSK makes the bit-to-symbol mapping nontrivial, although a scheme was proposed in Ref. [27] based on mapping 9 bits to two subsequent 6PolSK-QPSK symbols. The resulting format has a slightly reduced γ of 0.51 dB and spectral efficiency of 2.25 bit/symbol/polarization [27].

M-SP-QAM and the D_4-Lattice As was mentioned earlier, the D_4 lattice is the densest packing of many points in 4D space [36]. It is, therefore, of importance when finding useful modulation formats for 4D transmission lines. There are systematic and low-complexity ways of doing this, rather than resorting to sphere packing optimizations and the above-mentioned clusters. The idea is to cut finite portions from the D_4 lattice by using a cubic or spherical cut. The former is easiest in implementations, but the latter is better from a theoretical perspective, which is ultimately, for many points, 1.5 dB better, as was discussed earlier.

In general, the D_4 lattice can be defined, for example, as all points with integer coordinates that sum to an even number. It can be obtained from the cubic (integer) lattice $\mathbb{Z}^4$ in two ways, either by *reduction*, or *extension* [56, 57]. The reduction scheme is similar to Ungerboeck's *set partitioning*, (SP), introduced for trellis-coded modulation [58]. The idea is to remove half of the points in $\mathbb{Z}^4$ points, for example, those with odd parity. Thus one has

$$D_4 = \left\{ (k_1, k_2, k_3, k_4) \in \mathbb{Z}^4 \mid \sum k_i = \text{even} \right\} \tag{2.13}$$

The extension scheme is instead to start from $\mathbb{Z}^4$ add a shifted variant of $\mathbb{Z}^4$ half an integer in every dimension, that is,

$$\mathbb{Z}^4 \cup \mathbb{Z}^4 + \left(\frac{1}{2}, \frac{1}{2}, \frac{1}{2}, \frac{1}{2} \right)$$

However, if the center-of-mass should remain at the origin (which is most efficient from a power efficiency point of view), it is better to shift the two cubic lattices an equal amount in opposite directions, that is

$$D_4^* = \mathbb{Z}^4 - \left(\frac{1}{4}, \frac{1}{4}, \frac{1}{4}, \frac{1}{4} \right) \cup \mathbb{Z}^4 + \left(\frac{1}{4}, \frac{1}{4}, \frac{1}{4}, \frac{1}{4} \right) \tag{2.14}$$

Both methods give the D_4 lattice (apart from a rescaling), and both are useful when obtaining power efficient modulation formats, especially with high spectral efficiency.

The reduction scheme was originally suggested for optical communications by Coelho and Hanik [59] who called the resulting formats M-ary set-partitioned QAM or M-SP-QAM. Later, Karlsson and Agrell [57] extended the concept to the whole hierarchy of formats obtainable from extension or reduction of the standard rectangular QAM formats.

In the reduction process, the minimum distance squared is increased by a factor of two, at the expense of losing 1 bit per symbol. Applying this to PM-QPSK leads to the PS-QPSK format with a gain of $2 \times 3/4 = 3/2 = 1.76$ dB over PM-QPSK. Applying the same technique to PM-16QAM is more attractive, leading to a gain of $2 \times 7/8 = 7/4 = 2.43$ dB over PM-16QAM. This format is called 128-SP-QAM, and after being introduced by Coelho and Hanik [59] it was studied in simulations of nonlinear transmission by Renaudier et al. [60] and Sjödin et al. [61]. The latter paper also discussed the problem of bit-to-symbol mapping and maximum-likelihood decoding for the format.

By using the extension scheme on PM-QPSK, one obtains the 32-SP-QAM format [57, 60], which has $\gamma = 0\,\text{dB}$, that is, the same power efficiency as PM-QPSK, but transmitting 5 bits per symbol rather than 4. By using extension and reduction for known QAM formats, an M-SP-QAM hierarchy with $M = 8, 32, 128, 512, 2048 \ldots$ can be realized, and in the recent review article by Fischer et al. [62], more properties of these formats are given, including, for example, mutual information. The following relations between power efficiency and spectral efficiency, corresponding to (2.5), can be derived for the SP-QAM hierarchy [57]

$$\gamma_{\text{SP-QAM}} = \frac{3\beta}{2^{\beta+\frac{1}{2}} - 1} \tag{2.15}$$

$$\gamma_{\text{SP-QAM}} = \frac{3\beta}{2^{\beta+\frac{1}{2}} - \frac{1}{2}} \tag{2.16}$$

where (2.15) holds for SP-QAM formats obtained by reduction of a rectangular PAM format and (2.16) for extension.

Other 4D Formats of Interest The 16-ary cluster $C_{4,16}$ is 1.11 dB better (in the γ sense) than the hypercube (PM-QPSK), but the coordinates are not very nice. Layered along one coordinate axis, it consists of a 3D octahedron and a 3D cube, sandwiched between two single points. Explicit forms of the coordinates were given in Refs [31, 63]. The mutual information reveals only a marginal improvement over PM-QPSK (see Section 2.4.3 and [63]), although it has received some experimental interest as discussed in Section 2.5.

Another improvement over the PM-QPSK format was proposed by Sjödin et al. [64]. Referred to as *subset-optimized PM-QPSK* (SO-PM-QPSK), the idea was to improve PM-QPSK by rescaling one (e.g., the even-parity) subset and leaving the other unchanged. By optimizing the rescaling to 1.618 (the golden ratio), a 0.44 dB γ improvement over PM-QPSK can be obtained. ML decoding schemes, SER, and BER for this format are given in Ref. [64].

It is possible to obtain a nice symmetric 256-point format by cutting the D_4 lattice with a spherical cut around a deep hole, as pointed out, for example in Refs [12, 65] and recently described and experimentally implemented by Eriksson et al. [66]. The levels comprise all 4D vectors that lie within a radius of 6, whose coordinates are odd integers and where the coordinate sum is a multiple of 4. Remarkably, this is exactly 256 vectors, and the $\gamma = 16/27 = -2.27$ dB, and it is quite likely the most power-efficient 256-ary constellation in 4D.

2.3.1.4 Eight- and Higher-Dimensional Formats

The 8D clusters $C_{4,M}$ are shown in Figure 2.2 as $N = 8$ line. The maximum power efficiency, $\gamma_{\max}$, is 3.01 dB and occurs for $C_{8,16}$, which is the 8D cross-polytope, or biorthogonal 16-ary modulation. Interestingly, almost the same γ is obtained for $M = 58$ and 241 [28]. Expressions for the SER and BER of this format can be found in Ref. [43], and it was recently experimentally realized, as discussed in Section 2.5.

For higher spectral-efficiency constellations, the best 8D lattice packing is given by the E_8 lattice [36], which can be obtained from the D_8-lattice in union with a shifted D_8, that is,

$$E_8 = D_8 \cup D_8 + \left(\frac{1}{2}, \frac{1}{2}, \frac{1}{2}, \frac{1}{2}, \frac{1}{2}, \frac{1}{2}, \frac{1}{2}, \frac{1}{2}\right)$$

where D_8 is defined in analogy with (2.12).

Koike-Akino et al. [67] and Millar et al. [68] discussed two ways of obtaining 8D constellations, namely by cutting (spherical) parts from the E_8 lattice and using known block codes. The latter is discussed in detail Section 2.4. In Reference [67], they classified a few promising 8D modulation formats ($M = 128, 256$) in terms of γ as well as in terms of nonlinear transmission reach. In Reference [68], they went deeper and generalized the study to even higher dimensions as well, for example, 6D, 16D, and 24D. In 16D, the Barnes–Wall lattice is known to be the densest, and a promising 16D constellation with $M = 2^{11}$ points was found from cuts of this lattice, which might be the $\gamma_{\max}$ of 16D, even if further studies are required before this can be settled. It would, interestingly, be in agreement with the approximate expression (2.9). In 24D, the Leech lattice (and the associated Golay code) was used. More details on the latter, including experiments, were given in Refs [69, 70].

2.3.1.5 PPM-Based Formats

Pulse-position modulation (PPM) is a well-known technique to increase power efficiency at the expense of spectral efficiency. The idea is to frame 2^p symbols in time to a K-ary "supersymbol," of which one slot is selected for the transmission of a single pulse. One can in this way transmit $\log_2(K)$ bits per supersymbol. Is has been suggested to combine PPM with higher-order modulation formats to a hybrid K-PPM-M-QAM format by transmitting modulated data in the selected PPM slot [71, 72]. In Reference [47], the efficiency γ of various such formats were theoretically calculated and compared. It was pointed out that PS-QPSK is equivalent to 2-PPM-QPSK—it is just another set of four dimensions. To use PPM (i.e., subsequent symbol slots) instead of polarization is often an easier way of realizing more dimensions, and in fact such formats come closer to being an FEC code. Recently, the PPM idea was further generalized to allow supersymbols with an arbitrary number of nonzero slots (instead of just one) [73]. For example, with inverse PPM, iPPM, the idea is to transmit in all symbols but one in the PPM frame. In this way, formats could be realized that have both higher spectral efficiency and higher sensitivity than PM-QPSK. An example is 8iPPM-QPSK, which has $\beta = 2.13$ and $\gamma = 0.84$ dB.

2.3.2 Optimized Formats in Nonlinear Channels

Most formats discussed earlier have been optimized for the linear AWGN channel. However, as pointed out, the fiber is nonlinear, and often systems are operated in a weakly nonlinear regime, where the signal power is optimized as a trade-off between SNR and nonlinear distortions. What can then be done for the nonlinear channel in term of format optimization?

Within the GN model, the format optimization can be essentially the same as for the linear AWGN model assumed earlier, since the noise will be uniform and approximately Gaussian in all dimensions. Obviously, the formats based on optimizing minimum distance is reasonable for very high SNRs, and in a model with limited SNR (as the GN model) one would need to optimize at a constant SNR. This can be done, but is more computationally demanding. Not much has been published in that area, apart from the work by Foschini et al. [33].

One could also argue (see the discussion in Ref. [31]) that the balls would be better than the clusters, since they suffer less penalties for average power-limitations than clusters do for maximum power-limited channels [31]. Then again, in the GN model the average signal power is relevant, which speaks in favor of the clusters. Moreover, as discussed in section 2.2.3, there is no simple mapping between peak-power limits in discrete and continuous time. The former is easier to analyze, the latter makes more physical sense. Therefore, a more rigorous comparison between balls and clusters in nonlinear links remains to be done.

Format optimizations have also been done for nonGaussian channels, such as the phase-noise channel model used in, for example, [21, 74–76]. Lau and Kahn [74] compared various 4-point constellations, and managed to improve the nonlinear tolerance significantly by going from QPSK to a constellation with 3-PAM plus a fourth point further out. In References [21, 76], constellations in the nonlinear regime were considered. A comparison was made between constellations with points on 2 and up to 5 different radii. By optimizing 16-point constellations, a few decibels of increased nonlinear tolerance was seen.

In this context, the recent work by Kayhan and Montorsi [77] on constellation optimization should be mentioned, although they considered a linear phase-noise channel model. As Foschini et al. [33], they considered finite SNR, but used a different target function for the optimization process (approximations and variants of mutual information).

In Reference [75], *satellite constellations* were introduced, in order to show that the channel capacity of any channel (linear or nonlinear) may not decrease with signal power. They are formed by taking a standard format, for example, 8-PSK, and moving one point far out from the rest. This yields a constellation whose minimum SER, as well as maximum MI, occurs at a high average power, which can be made arbitrarily high by moving the lone point (satellite) further from the rest. A similar trick was used by Steiner [78] when optimizing formats in the low-SNR regime.

2.4 COMBINATIONS OF CODING AND MODULATION

So far, the comparisons of modulation formats in this chapter have concerned uncoded transmission. Modern optical communication systems, however, often include some kind of FEC coding [79–81]. For best system performance, the code should influence the choice of modulation format. For example, a modulation format with high spectral efficiency may require a lower-rate code (better error protection capability) than a modulation format with lower spectral efficiency.

This section discusses optimization of modulation formats in coded systems. We distinguish between three cases, depending on the type of decoder employed, which pose quite different requirements on the choice of modulation format. The three cases are soft-decision decoding, hard-decision decoding, and iterative decoding, which loosely correspond to weak, medium, and strong coding, respectively. Most of this section is devoted to the first case, which is more intimately connected to the problem of constellation design.

2.4.1 Soft-Decision Decoding

We here consider the application of a relatively short-length, well-structured block code, such as a single-parity check code, Hamming code, or Reed–Muller code. The employment of such simple codes implies low latency, simple encoding and decoding hardware, and hence low energy consumption. Nevertheless, significant improvements over uncoded transmission can be obtained.

If the decoder is an optimal soft-decision decoder, in the sense that it finds the codeword that is closest to the received word in Euclidean distance, then the combination of code and modulation can be regarded as a single higher-dimensional modulation operation. This approach has been developed extensively in the communications literature [29, 82] [43, Sec. 4.1] and more recently in an optical context [68].

To be precise, let C be a binary linear block code with parameters (n, k, d_{H}), where n is the total number of bits per codeword, k is the number of information bits per codewords, and d_{H} is the minimum Hamming distance. The code rate is k/n. If this code is used in combination with a (low-dimensional) constellation $\tilde{\mathcal{X}}$ with dimension $\tilde{N}$ and size $\tilde{M}$, then $\log_2 \tilde{M}$ bits are needed to index each point in the constellation, and the k information bits in a codeword suffice to index a block of $k/\log_2 \tilde{M}$ constellation points (assuming that this is an integer). This block can be regarded as a point in a larger constellation $\mathcal{X}$ with parameters

$$N = \frac{n\tilde{N}}{\log_2 \tilde{M}}$$

$$M = \tilde{M}^{k/\log_2 \tilde{M}} = 2^k$$

There exists no general expression for calculating the minimum Euclidean distance d from the parameters of C and $\tilde{\mathcal{X}}$. It depends heavily on the mapping from bits to symbols, which needs to be done with some care.

The simplest, and most common, special case is to let $\tilde{\mathcal{X}}$ be a BPSK constellation $\tilde{\mathcal{X}} = \{\pm\sqrt{\tilde{E}}\}$, where $\tilde{E}$ is the symbol energy, with parameters $\tilde{N} = 1$ and $\tilde{M} = 2$. This yields, for every code C, a constellation $\mathcal{X}$ with parameters

$$
\begin{aligned}
N &= n \\
M &= 2^k \\
d^2 &= 4\tilde{E}d_{\mathrm{H}} \\
E &= n\tilde{E}
\end{aligned}
\tag{2.17}
$$

Geometrically, the constellation $\mathcal{X}$ resides on the vertices of an n-dimensional hypercube. The binary code is used to select a subset of these vertices.

Consider, for example, the 4-bit single-parity check code with parameters $(n, k, d_{\mathrm{H}}) = (4, 3, 2)$. This yields, by (2.17) and (2.4), a four-dimensional constellation with 8 points and power efficiency $\gamma = d^2\log_2 M/(4E) = d_{\mathrm{H}}k/n = 3/2 = 1.76\,\mathrm{dB}$. This constellation is identical to the PS-QPSK constellation, described in Section 2.3.1.3 [27].

Using this approach, a large number of high-dimensional constellations can be designed from standard binary block codes. An attractive family of codes for this purpose is the Reed–Muller (RM) codes. These codes have rather good performance (albeit not optimal) at short block lengths n, and furthermore, there exist fast encoding and decoding algorithms, alleviating the need for table look-up [83, Sec. 4.3]. An RM code is specified by two integer parameters, u and r, chosen such that $u \geq 1$ and $0 \leq r \leq u$. The parameters of the RM(r, u) code are $(n, k, d_{\mathrm{H}}) = (2^u, \sum_{i=0}^{r}\binom{u}{i}, 2^{u-r})$. Special cases are repetition codes ($r = 0$), single-parity check codes ($r = u - 1$), and the universe code (i.e., uncoded transmission, $r = u$).

The parameters of some constellations obtained from Reed–Muller (RM) codes are illustrated in Figure 2.3. The obtained constellations are apparently quite competitive, compared with the best-known constellations at the same dimensions and spectral efficiencies, and in several instances the RM codes actually yield the best-known constellations. For any $\beta < 2$, arbitrarily high *CFM* and γ can be obtained by choosing suitable RM code parameters. This makes RM codes attractive instruments for constellation design, especially since low-complexity coding and decoding algorithms are known.

Analogous curves are presented in Figure 2.4 for Hamming codes, the Golay code, and their extended versions. It turns out that the Hamming codes yield a constant *CFM* of 7.78 dB, regardless of their size, and similarly, the extended Hamming codes yield *CFM* = 9.03 dB. The power efficiency increases with the codeword length n and reaches asymptotically 4.77 dB for Hamming codes and 6.02 dB for extended Hamming codes. The extended Golay code also reaches $\gamma = 6.02$ dB, at a lower spectral efficiency [69].

Theoretically, nothing prevents us from designing extremely high-dimensional constellations by applying the methods of the previous section to codes with long codewords. Consider, for example, the ubiquitous (255,239) Reed–Solomon (RS)

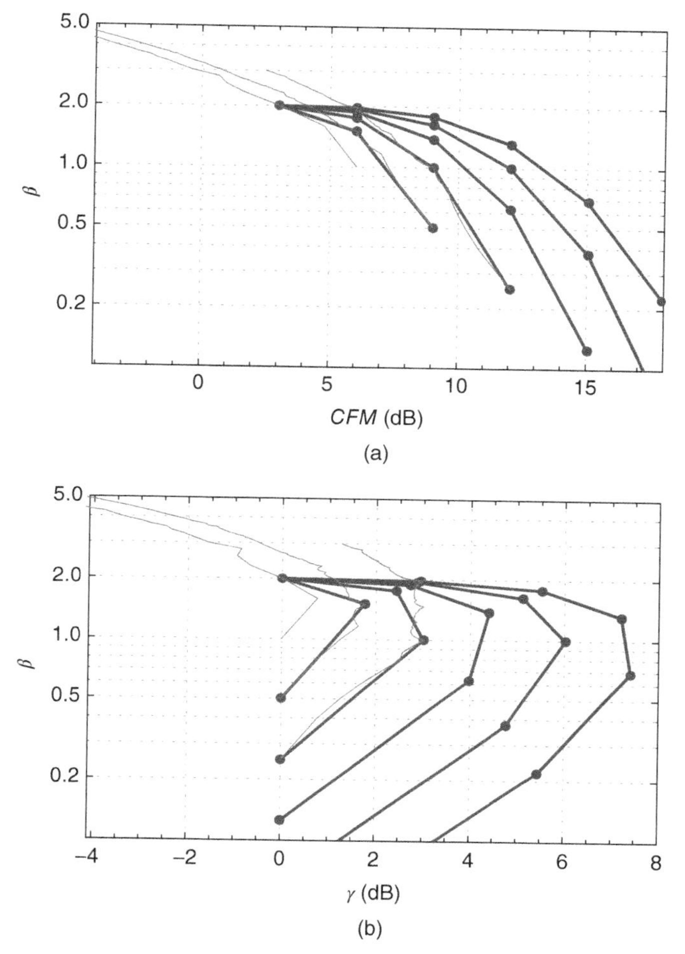

FIGURE 2.3 Reed–Muller codes with BPSK modulation. Codeword lengths $n = 4, 8, 16, 32, 64$ (thick curves, from left to right). Minimum distance $d_{\mathrm{H}} = 1, 2, 4, \ldots, n$ (dots on each curve). Dimension $N = n$. The best-known clusters from Figure 2.2 are shown for reference (thin).

code, which was standardized by the ITU-T in 2000 [84]. This code encodes 239 information bytes into blocks of 255 transmitted bytes, and it has an error-correction capability of 8 bytes. This error-correcting capability corresponds to 8 bit errors in the worst case, but if the bit errors come in burst, many more than 8 bit errors can be corrected, as long as the errors do not affect more than 8 bytes in total.

The parameters of the (255,239) RS code, converted from bytes to bits, are $(n, k, d_{\mathrm{H}}) = (2040, 1912, 17)$. This code can, if combined with BPSK modulation as in (2.17), be regarded as a 2040-dimensional constellation with parameters $\beta = 1.87$, $CFM = 15.31$ dB, $\gamma = 12.02$ dB, and $G = 11.79$ dB. Compared with the curves

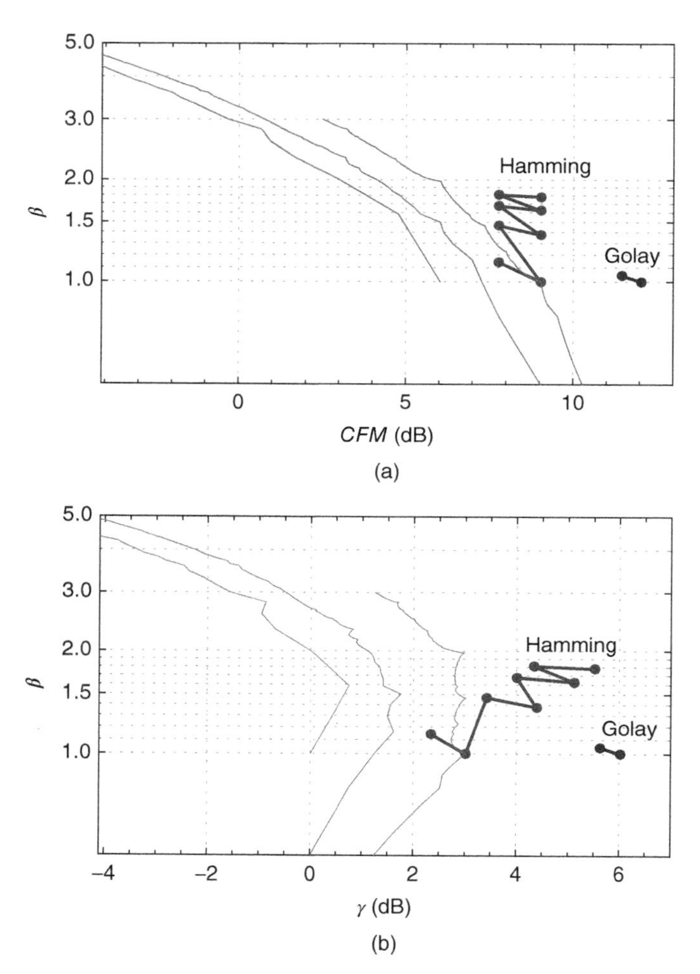

FIGURE 2.4 Hamming codes and their extended versions with BPSK modulation, with codeword lengths $n = 7, 8, 15, 16, 31, 32, 63, 64$ (connected with thick lines). Golay ($n = 23$) and extended Golay code ($n = 24$), also with BPSK modulation (connected with thick line). Dimension $N = n$.

in Figures 2.2–2.4, this constellation has an impressive performance, falling far to the right of any of the curves. This exemplifies the essence of coding, which is to improve the power efficiency by increasing the dimensionality.

One could ask whether it makes sense to consider a 2040-dimensional constellation with $M = 3.7 \cdot 10^{575}$ points. Clearly, it is not possible to enumerate or store all the points. However, if soft-decision decoding is used in the receiver, the performance predicted by the constellation analysis mentioned earlier is indeed achievable, for well-structured codes such as RM and RS. Generalizing, we conclude that any low-dimensional modulation scheme with soft-decision FEC is equivalent to a high-dimensional modulation scheme without FEC.

In all these cases, the spectral efficiency β never goes above 2, which is the spectral efficiency for uncoded BPSK. This is a serious limitation in practical optical system implementations, where higher and higher spectral-efficiencies are being targeted nowadays. To circumvent this limitation, one must employ a multilevel constellation instead of BPSK. This leads to a simple type of coded modulation [11]. In the following, we give some simple examples, based on single-parity check codes and RM codes.

A natural extension to BPSK is to let $\tilde{\mathcal{X}}$ be a regular PAM constellations with $\tilde{M} = 2^{\tilde{m}}$ points. If the distance between two neighboring PAM points is $\tilde{d}$, then the average symbol energy of $\tilde{\mathcal{X}}$ is $E = (\tilde{M}^2 - 1)\tilde{d}^2/12$. As explained earlier, the k information bits in a codeword are divided into groups of $\tilde{m}$ bits, thus indexing a block of $k/\tilde{m}$ PAM symbols. This block of PAM symbols constitutes an N-dimensional constellation $\mathcal{X}$ with parameters

$$\begin{aligned} N &= \frac{n}{\tilde{m}} \\ M &= 2^k \\ d^2 &\geq \tilde{d}^2 d_{\mathrm{H}} \\ E &\leq \frac{n\tilde{E}}{\tilde{m}} \end{aligned} \tag{2.18}$$

for a suitably chosen (Gray-coded) mapping from bits to PAM symbols.

An important special case arises by applying a single-parity check code with parameters $(n, k, d_{\mathrm{H}}) = (\tilde{m}N, \tilde{m}N - 1, 2)$. The codeword length n is chosen so that the obtained constellation $\mathcal{X}$ is N-dimensional. The obtained constellations are plotted in Figure 2.5 for various values of $\tilde{m}$ and N. The 4D case has been studied in optical communications under the name SP-QAM (see Section 2.3.1.3). If, for example, the code with parameters (8,7,2) is mapped to a Gray-coded 4-PAM constellations, then the parameters of the resulting 128-point constellations are, by (2.18), $d^2 \geq 2\tilde{d}^2$ and $E \leq 4\tilde{E} = 5\tilde{d}^2$, which yields $\gamma \geq 7/10 = -1.55$ dB, at a spectral efficiency of 7/2. This modulation format is represented by one of the dots in Figure 2.5(b). As n increases, the gain G of SP-QAM converges to 1.51 and 2.26 for $N = 4$ and 8, respectively [57], which should be compared with the maximum possible gains 1.97 and 3.72 dB, respectively, in (2.10). The asymptotic gain as N and $\tilde{M}$ both approach infinity is 3 dB. There is no gain to be harvested in 2D by this method.

The same types of codes as in Figures 2.3 and 2.4 were applied to 4-PAM, which yielded the results in Figures 2.6 and 2.7. The obtained constellations are relatively weak, compared with the best-known constellations at the same dimensions and spectral efficiencies. Nevertheless, the results show that it is in principle to achieve arbitrarily high *CFM* and γ at any $\beta < 4$, if the codeword length is increased sufficiently.

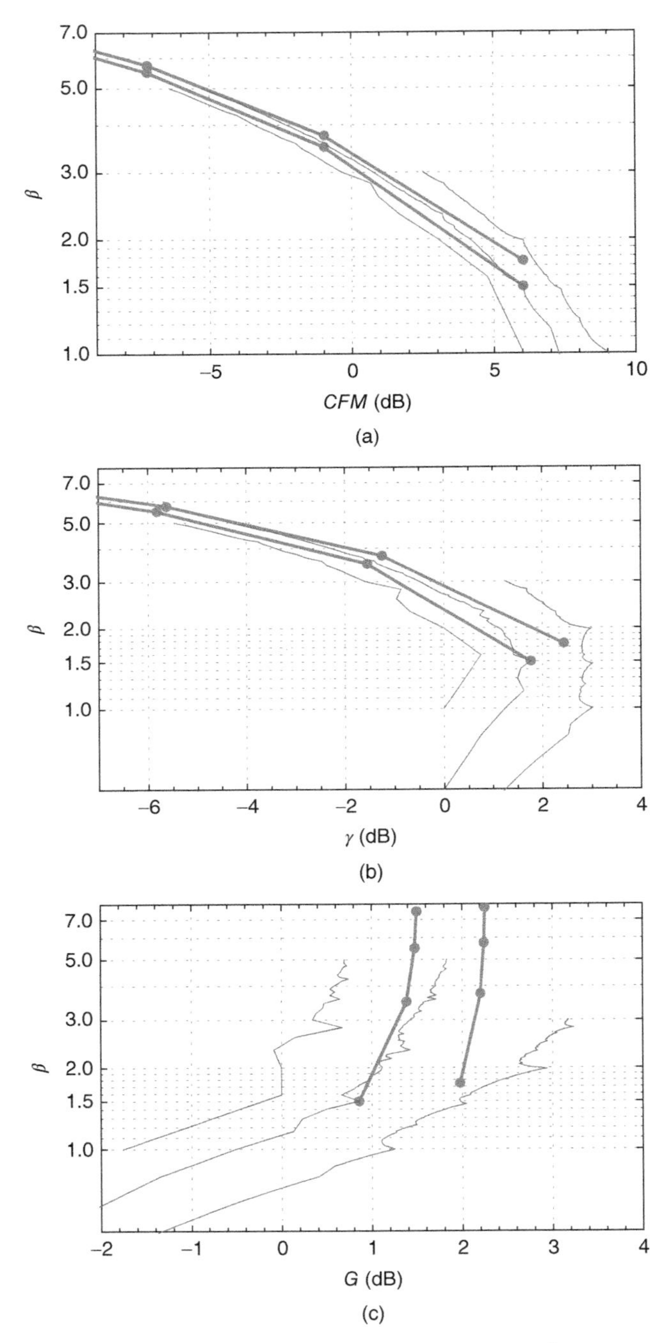

FIGURE 2.5 Single-parity check codes mapped to Gray-coded $\tilde{M}$-PAM, where the code lengths n is chosen so that the dimension $N = n/\log_2(\tilde{M})$ is an integer. Dimensions $N = 4, 8$ (from left to right). PAM alphabet size $\tilde{M} = 2, 4, 8, 16$ (points on each curve). The left curve ($N = 4$) represents SP-QAM, see Section 2.3.1.3.

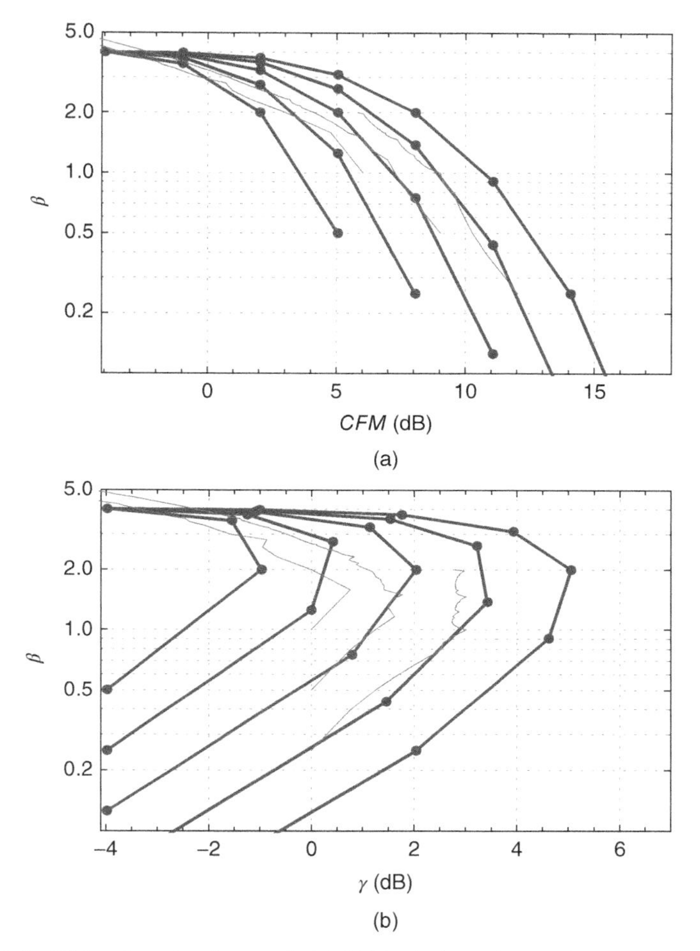

FIGURE 2.6 RM codes with Gray-coded 4-PAM. Codewords lengths $n = 8, 16, 32, 64, 128$ (from left to right). Minimum distance $d_{\mathrm{H}} = 1, 2, 4, \ldots, n$ (dots on each curve). Dimension $N = n/2$.

2.4.2 Hard-Decision Decoding

Most commercially deployed long-haul fiber-optical communication systems use hard-decision decoding, which can be realized at a significantly lower hardware complexity than soft-decision decoding. The decoder can be implemented using binary logic, with no need for analog-to-digital conversion. This in turn admits the use of stronger codes (longer codeword lengths). Reed–Solomon codes are the most popular codes in this context, but BCH (Bose–Chaudhuri–Hocquenghem) codes, Hamming codes, and convolutional codes have also been considered. The interested reader is referred to [79] for a survey of the field, which is outside the scope of this chapter.

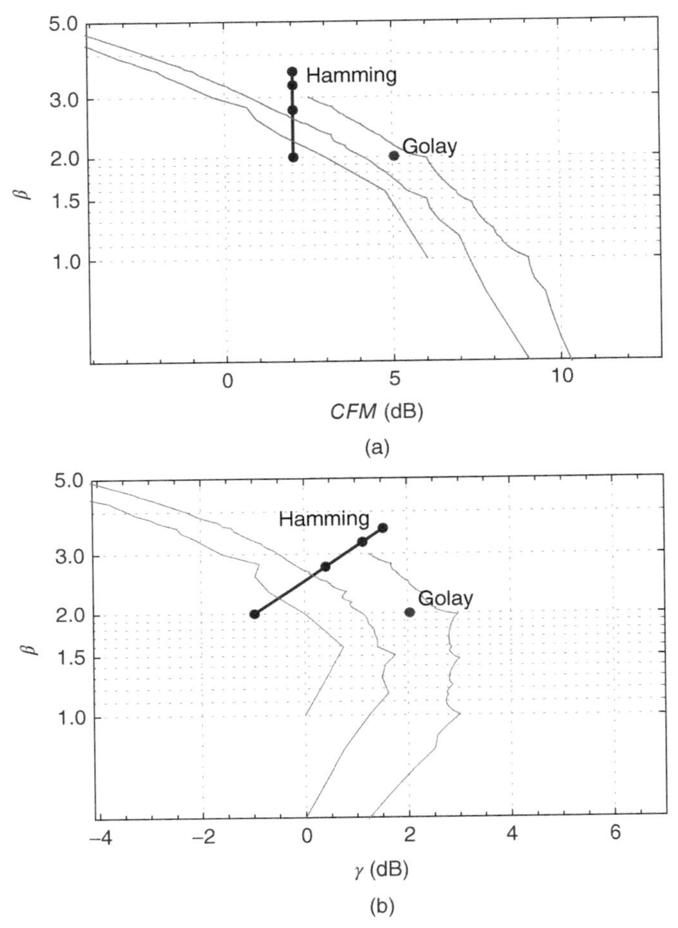

FIGURE 2.7 Extended Hamming codes ($n = 8, 16, 32, 64$) and extended Golay code ($n = 24$) with Gray-coded 4-PAM. Dimension $N = n/2$.

In a system with hard-decision decoding, the geometric framework in the previous section makes less sense. In this case, performance metrics based on the minimum Euclidean distance are misleading, and modulation and coding should be kept separate in the analysis. The standard system design method is to choose a modulation format that guarantees a certain BER, the so-called FEC limit, which is typically in the range of 10^{-3} to 10^{-4} [84], and trust the FEC to bring down the BER to a negligible level. This design principle is very popular in practice, since it decouples the FEC from the rest of the system, which facilitates experimental work. Two main weaknesses with this standard approach is that it offers no simple mechanism to optimize the code rate (varying the FEC limit) and that it does not account for the bursty nature of errors.

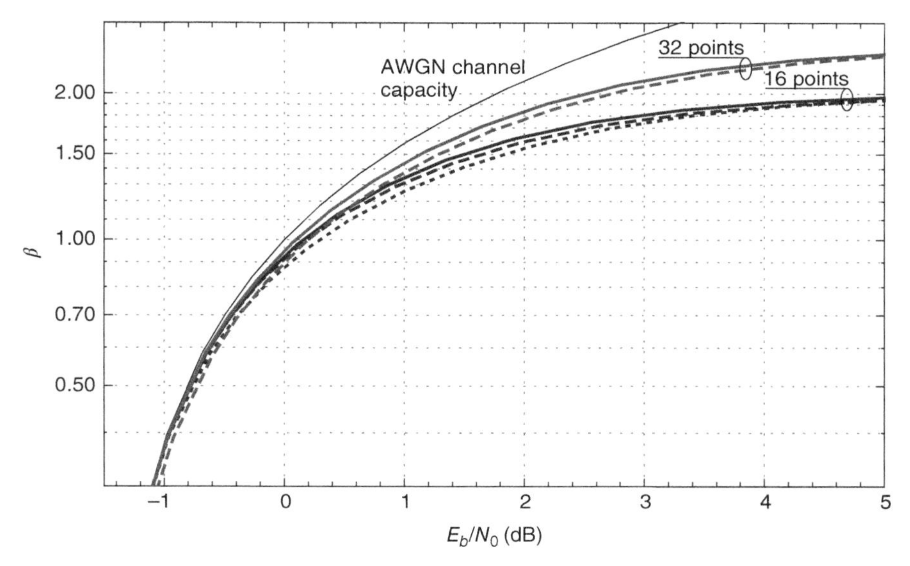

FIGURE 2.8 Spectral efficiency β, calculated as mutual information per dimension pair, as a function of the SNR per bit for selected 4D formats. Solid thick: best known constellations $C_{4,M}$ for M = 16 and 32. Dashed: PM-QPSK and 32-SP-QAM. Dotted: SO-PM-QPSK.

2.4.3 Iterative Decoding

Modern codes such as low-density-parity-check (LDPC) codes and turbo codes have revolutionized wireless communications, and an equally promising potential is envisioned in optical communications [80, 81]. These codes are typically very long (in the order of 10,000 bits) and have a pseudorandom structure. Algebraic decoding would be far too complex for such codes, but there have been devised efficient iterative decoding algorithms, which gradually improves an estimate of the transmitted codeword, using either soft or hard decisions. Such decoders are not guaranteed to find the optimal codeword, but nevertheless these codes have excellent performance. In some cases, they even approach the maximum spectral efficiencies predicted by Shannon in 1948 [85].

Shannon proved that the achievable data rate of a given modulation format is upper-bounded by the MI, defined in (2.6). Furthermore, recent research has shown that a performance very close to the MI is achievable using long LDPC codes and soft-decision iterative decoding. The spectral efficiency in bits per dimension pair is plotted in Figure 2.8 for some common 4D constellations, as a function of the SNR per bit. The input $\boldsymbol{X}$ is drawn uniformly from a constellation $\mathcal{X}$, given by the modulation format, and a memoryless AWGN channel is assumed. The spectral efficiency is here calculated as $\beta = I(\boldsymbol{X};\boldsymbol{Y})/(N/2)$ and the SNR per bit as E_b/N_0, where N_0 is the power spectral density of the Gaussian noise.

Although the plot only includes a small number of constellations, several interesting conclusions can be drawn. At a given target β, it is practically always beneficial in terms of power efficiency to increase the number of points in the constellation. The fact that the minimum distance decreases is fully compensated for by using a lower code rate (higher overhead). The gain obtained by increasing the number of points is, however, negligible at low β, where practically any constellation performs close to capacity.

If the number of points is kept constant, then the constellations designed for optimum uncoded performance tend to be good also in terms of MI. The gains in decibels are, however, less than the corresponding gains in terms of γ or CFM. Interestingly, PM-QPSK is better than SO-PM-QPSK for all β values, in contrast to the uncoded performance where SO-PM-QPSK is 0.44 dB better (see section 2.3.1.3). This effect is even more pronounced in systems with bit-wise receivers, where PM-QPSK is the best-known constellation, despite its simple structure, gaining approximately 1 dB over the cluster $C_{4,16}$ [86].

2.5 EXPERIMENTAL WORK

To experimentally demonstrate four- and higher-dimensional modulation formats, one needs to be able to simultaneously access all dimensions in the transmitter and receiver. Depending on how the dimensions are physically realized in the channel (e.g., time, frequency, or spatial dimensions), this can be more or less complicated, as the used dimensions must be synchronized and not drift between symbols. This often requires tailored DSP algorithms for the considered modulation formats.

In this section, we review the experimental work done on mainly 4D formats, where the four dimensions are the conventional four quadratures (I/Q in each of the x and y polarizations). We divide the discussion into (i) realizations of the transmitter and transmission link properties (Section 2.5.1) and (ii) the receiver algorithms (Section 2.5.2), including DSP and decoding, with a summary table in Section 2.5.3. Then, in Section 2.5.4, we discuss format detection, that is, how to simply determine the transmitted symbol from the received 4D vector, without resorting to a full search of the Euclidean distances to all points in the whole constellation. We finally discuss alternative ways of extending dimensions in signal space in Section 2.5.5 from a complexity and implementation perspective.

2.5.1 Transmitter Realizations and Transmission Experiments

We later describe the experimental work in similar order as in the above-mentioned theory (Section 2.3.1.3).

3D Simplex Dochhan et al. [41] proposed and demonstrated the 3D simplex (tetrahedron) transmission, by using a four-channel digital-to-analog converter (DAC) driving a conventional PM-QPSK modulator. The format transmits QPSK in one polarization and BPSK in the other, thus leaving one quadrature unmodulated.

Of the resulting eight levels (forming a cube), the four with odd parity were selected, giving the desired tetrahedron. The symbol rate was 16 Gbaud, corresponding to a data rate of 32 Gbit/s, which was transmitted over 300 km of single-mode fiber. The back-to-back sensitivity was approximately 1 dB better than PM-BPSK, in agreement with theory, and the nonlinear robustness was similar to PM-BPSK. In Reference [87], Yamazaki et al. developed a simple integrated modulator structure for this format.

PS-QPSK The first experimental realization of a 4D format in fiber-optic transmission, was probably the demonstration by Sjödin et al. of PS-QPSK at 30 Gbit/s in 2011 [88]. In this experiment, a conventional I/Q-modulator for QPSK was used, and then the data were split into two arms, driven by a pair of Mach–Zehnder amplitude modulators in a push–pull constellation, meaning that either one or the other arm was blocked. Then, the two arms were multiplexed together by a polarization combiner. In this way, two bits were encoded in the QPSK symbol and the third in the choice of polarization. Similar transmitter structures were used also by other groups around this time, for example, Millar et al. [89] and Nelson et al. [90].

An alternative transmitter setup was used by Fischer et al. [91] who used a PM-QPSK transmitter with a programmable bit-pattern generator, driving the 4 bits with a preprogrammed pattern. Three of the bit streams were driven by uncorrelated (delayed) pseudo-random sequences, and the fourth was formed as a parity (exclusive OR, XOR) bit from these three sequences. This transmitter was also used by Renaudier et al. [92], who also introduced a timing offset between the two polarizations to facilitate the receiver DSP, as is discussed in Section 2.5.2.

In Reference [93], Yamazaki et al. presented an integrated modulator optimized for PS-QPSK, which could directly generate PS-QPSK in a single device driven by three binary drive signals. It has the additional benefit of avoiding the inherent 3 dB coupling loss in the I/Q modulators when the I and Q quadratures are mixed.

Following the initial demonstration of single-channel transmission [88] came a stream of experimental demonstrations of PS-QPSK; first single-channel demonstrations at higher data rates, for example, 42 Gbit/s [89, 94] and 112 Gbit/s [91, 95], and then WDM experiments over ultralong distances (thousands of kilometers) [89, 90, 95–97]. The general conclusion was that the improvement in transmission distance predicted in simulations [49, 50] was experimentally verified. Typically, PS-QPSK achieved 10–25% longer transmission reach (at a BER=10^{-3}) than PM-QPSK at the same data rate [89, 97, 98], both in single-channel systems and with WDM. Masalkina et al. [99] used PS-QPSK in a 20 Gbit/s orthogonal frequency-division multiplexed (OFDM) transmission experiment. Lavery et al. [95] demonstrated that digital back-propagation could extend the reach for 112 Gbit/s PS-QPSK from 4600 km to 5600 km.

6PolSK-QPSK Experimental generation of 6PolSK-QPSK was first demonstrated in 2012 independently by Buchali and Bülow [100] and by Fischer et al. [101]. The transmitter structure used in these experiments was based on a 4-channel 28-Gbaud DAC driving a dual-polarization I/Q modulator at the three levels $\{-1, 0, 1\}$. By forming all 24 permutations and sign selections of the vector $(0, 0, \pm 1, \pm 1)$, the

24-cell is obtained. The symbol-to-bit mapping followed the suggestion in Ref. [27], by mapping 9 bits to 2 subsequent symbols. In back-to-back measurement, Fischer et al. [101] demonstrated 2.1 and 3.4 dB implementation penalties for PM-QPSK and 6PolSK-QPSK, respectively, for 28 Gbaud and a BER of around 10^{-3}. This corresponds to an extra implementation penalty for 6PolSK-QPSK of 1.3 dB. In Reference [102], Ding et al. generated 6PolSK-QPSK by using a single dual-drive (i.e., not I/Q) Mach–Zehnder modulator in each polarization.

In transmission, Fischer et al. [101] demonstrated transmission of 19 WDM channels over 3400 km, which is less than the 4800 km of PM-QPSK. However, by applying a rate 455/511 Reed–Solomon code to 6PolSK-QPSK, making the bit rates the same for both formats, the transmission distance gap was partly bridged. A later extension of this work by Tanimura et al. [103] used a more advanced setup, including pre-emphasis to compensate for DAC imperfections, investigated the nonlinear tolerance in some detail. It was found that an inner FEC could be beneficial in removing bursty errors due to nonlinearities.

One conclusion from these works was that the lack of Gray mapping for the 6PolSK-QPSK induces an extra penalty relative to PM-QPSK, which is troublesome in the nonasymptotic regime.

M-SP-QAM According to predictions and simulations in, for example, [57, 59, 61, 92], the SP-QAM formats emerged as an interesting 4D format generalization with increased spectral efficiency over PM-QPSK. In 2013, 128-SP-QAM was demonstrated by three independent groups [104–106]. The transmitter structures in these experiments were similar; based on pre-programmable DAC:s using 8-bit streams of which one is a parity check bit.

Eriksson et al. [104, 107] used 12 Gbaud for 128-SP-QAM and compared with 10.5 Gbaud PM-16QAM to obtain the same data rate of 84 Gbit/s. Both single channel and 9 WDM channel (25 GHz separation) transmission were compared. PM-16QAM had a slightly higher implementation penalty (2.1 dB) relative to 128-SP-QAM (1.5 dB), attributed to the improved Euclidean distance of 128-SP-QAM. The back-to-back sensitivity improvement of 128-SP-QAM was 1.9 dB at the same bit rate, and 2.9 dB at the same symbol rate, in close agreement with theoretical expectations. In transmission, the reach was 1300 km (for PM-16QAM in a WDM system) and 2000 km (for 128 SP-QAM, also in WDM), that is, a 54 % improvement. It was also concluded that PM-16QAM has a larger penalty when going from single channel to WDM than 128-SP-QAM.

Zhang et al. [106, 108] used 128 SP-QAM (denoted "half-4D-16QAM") over 294 channels (16.64 Gbaud at 17 GHz separation), covering the full C-band, at 104 Gbit/s each to achieve 30.58 Tb/s. The transmitter used bit-interleaved coded modulation (BICM) together with a 20% overhead LDPC code. They also used DACs for Nyquist channel shaping to reduce interchannel crosstalk. These data were then transmitted over 7230 km, making it (to that date) the experiment with the highest bit rate-times-distance product of 221 Pb/s × km. In a later experiment [108], digital back-propagation was used to increase the transmission distance to 10,300 km, but at a reduced data rate of 21 Tb/s.

Renaudier et al. [105] demonstrated both 32-SP-QAM and 128-SP-QAM at 28 Gbaud. In back-to-back experiments, they achieved 0.8 and 1.5 dB implementation penalties, respectively, relative to the AWGN theory. Then, they propagated 16 channels in a circulating loop constellation, at optimized power (which was the same for all formats), and measured transmission distance. At a BER of $4 \cdot 10^{-3}$, they could propagate PM-QPSK, 32-SP-QAM, 128-SP-QAM, and PM-16QAM, respectively, over 18,000, 14,000, 7000, and 4000 km.

At ECOC 2013, two independent groups [109, 110] compared 32-SP-QAM with another spectral efficiency $\beta = 2.5$-bits/symbol/pol scheme, namely hybrid PM-QPSK/PM-8QAM. The hybrid scheme means that half of the transmitted symbols are PM-QPSK ($\beta = 2$) and half are PM-8QAM ($\beta = 3$). In Reference [109], 65 symbols of PM-QPSK were followed by 65 symbols of PM-8QAM. In Reference [110], the interleaving was every second symbol, but done in both polarizations in a staggered manner, so that each symbol slot contained one polarization of QPSK and one polarization of 8QAM, and then the next symbol swapped formats in the polarizations. The 8QAM format used in both papers was the star-shaped constellation consisting of two QPSK constellations at different radii, rotated 45° relative to each other.[3] Both studies concluded (in line with theoretical predictions) that 32-SP-QAM had better performance in terms of transmission distance.

The 32- and 128-SP-QAM were also demonstrated in few-mode fiber transmission over 42 km by van Uden et al. [111], but then using pairwise time slots, rather than polarizations, to set up the 4D space, and subsequently propagating 4D symbols independently in two polarizations and three spatial modes. Also, the PS-QPSK counterpart, denoted time-switched (TS)-QPSK, was implemented. The transmitter setup consisted of a four-channel programmable DAC at 28 Gbaud, followed by a polarization- and mode-multiplexing stage. One conclusion from this experiment was that the 4D formats have less implementation penalty than their cubic-lattice counterparts (e.g., when comparing 128-SP-QAM with PM-16QAM).

The record SP-QAM experiments in terms of constellation size are impressive 512- and 2048-SP-QAM, which are obtained from the related PAM constellations are PM-32QAM (cross constellation) and PM-64QAM, as recently demonstrated by Fischer et al. [112]. The formats were generated by carefully co-optimizing multi-level analog-to-digital converters with I/Q modulators, but the 2048 case suffered, not surprisingly, from 3 to 6 dB implementation penalty. The 512-SP-QAM had a more moderate 2 dB of implementation penalty.

More Complex 4D Formats The $C_{4,16}$ cluster was experimentally demonstrated independently at OFC 2013 by Karout et al. [113] and by Bülow et al. [114]. The experiment [113] was based on an optical OFDM link, with 81 subcarriers generated by 2 synchronized DACs, comparing $C_{4,16}$ with PM-QPSK. The signal bandwidth was 6.5 GHz, and the resulting data rate of 25.6 Gbit/s was obtained. A small overhead was

[3]This format is erroneously referred to as the optimum (in the γ sense) 8-ary 2D format in [40, p. 277], that is, as the cluster $C_{2,8}$. However, as shown by Foschini et al. in 1974 [33], $C_{2,8}$ is a hexagonal structure comprising the 7-point kissing constellation plus an extra point.

allocated for training sequences and guard-band. A small performance gain for $C_{4,16}$ could be seen, in good agreement with theoretical BER curves. Only back-to-back measurements, that is, no transmission, were carried out in this demonstration. The experiment [114] used a baseband signal realized by a 28-Gbaud DAC in the transmitter. The data were transmitted 480 km, but the theoretical performance gain of $C_{4,16}$ was shadowed by a larger implementation penalty than PM-QPSK. The experiments indicated an increased nonlinear tolerance of $C_{4,16}$, but this was not conclusive.

An extension of 6PolSK-QPSK to 8PolSK-QPSK was proposed and demonstrated by Chagnon et al. [115]. This format uses eight different polarizations, put on the cube corners in Stokes space, and each with QPSK modulation. The format was generated with a four-channel DAC driving a dual-polarization I/Q modulator, yielding a data rate of 129 Gbit/s. The reach was 3800 km (including WDM channel loading), well above the 2800 km of the PM-8QAM it was benchmarked against.

Bülow et al. [116] also demonstrated experiments of formats obtained from spherical cuts from the D_4-lattice, with $M = 64$ and $M = 256$. Power efficiencies over PM-8QAM and PM 16QAM of 1.5–1.7 dB were reported. The experiments also included BICM using a 17 % overhead LDPC code with these formats, and PM-64QAM showed a 35 % reach improvement over PM-8QAM.

Eriksson et al. demonstrated the 256-ary D_4 format, in Ref. [66], and compared it with PM-16QAM at 56 Gbaud. As could be expected, the D_4 format was better back-to-back, for low ($<10^{-3}$) bit error rates, but for higher BERs (corresponding to transmission distances above 1500 km) PM-16QAM had the edge.

Higher-Dimensional Formats As the only way of improving performance (increasing both power efficiency and spectral efficiency) beyond the limits of 4D formats is to increase the dimensionality of the constellations, there has been work carried out in that direction as well. A natural extension is to move to eight dimensions (8D), and biorthogonal 8D modulation was implemented and evaluated by Eriksson et al. [117, 118]. In Reference [117], the eight dimensions were formed by using two phase-locked neighboring frequency (or wavelength) channels, and then performing coherent detection of both 4D channels in parallel. The transmitter in Ref. [117] was a generalization of the corresponding transmitter for PS-QPSK, that is, a PS-QPSK transmitter in cascade with a pair of push–pull modulators to select frequency. The format was referred to as 4-ary frequency- and polarization-switched QPSK, 4FPS-QPSK. The transmission properties (at 10 Gbaud) showed a reach of 14,000 km for 4FPS-QPSK, compared with 7500 km for PM-QPSK.

In Reference [118], PPM (see Section 2.3.1.5 and later) was used instead of frequency to realize the eight dimensions. A pair of two PS-QPSK symbols formed a PPM frame, giving a 2PPM-PS-QPSK format, equivalent to the 8D biorthogonal format. The transmitter becomes notably simpler, requiring a single modulator to select time slot, followed by a PS-QPSK transmitter. The implementation penalty at a BER of 10^{-3} and a symbol rate of 21.4 Gbaud was 0.6 dB, slightly more than the 0.3 dB for PM-QPSK. The data rate was 85.6 Gbit/s for both formats, meaning that the 2PPM-PS-QPSK format used 42 GHz of bandwidth, twice that of the PM-QPSK transmission. The transmission reach was almost doubled; 2PPM-PS-QPSK reached

12,300 km and PM-QPSK 6700 km, which is in reasonable agreement with a simple GN-model-based theory, predicting a doubling of the reach with a 3 dB improved sensitivity.

Shiner et al. [119] also demonstrated the 8D biorthogonal format by using two subsequent temporal symbols to for the eight dimensions. Their experiment demonstrated transmission over 5000 km at 35 Gbaud, including WDM channels. They also rotated the constellation aiming to reduce the nonlinear effects, noticing a 1 dB improvement in the nonlinear system margin.

In a spatial-division multiplexed context, and in particular multi-core fibers, Puttnam et al. [120] investigated modulation over several cores in parallel. Especially, formats based on a single-parity-check scheme (as outlined earlier) showed good results. The idea was to coherently transmit 4 bits per symbol in K parallel cores, and then use a single bit as a parity check bit, giving a total rat$4K - 1$ parallel BPSK streams. It may be shown that this scheme has both the spectral efficiency and the power efficiency equal to $2 - 1/(2K)$. The special case $K = 1$ is equivalent to PS-QPSK. An experiment at 10 Gbaud over 28 km of multi-core fiber demonstrating the concept for $K = 7$ cores were performed, and the results were in good agreement with theoretical expectations.

PPM Implementations Liu et al. [71, 121, 122] suggested combining pulse position modulation (PPM) with PM-QPSK to show a record sensitivity for data transmission of 16-PPM at 2.5 Gbit/s [71] and 4-PPM at 6.23 Gbit/s [121]. The main benefit of these formats are the increased power efficiency, requiring only 2 photons per bit at 2.5 Gbit/s, making them suitable in, for example, single-span long-distance links that demands high sensitivity. The transmitter was a standard PM-QPSK transmitter, but driven with a more complex data signal, including PPM framing and also some synchronization overhead signals. The transmission was done over a single span of 370 km ultra-large-effective area fiber, with a total loss of 69 dB.

Slightly outside the main topic of this chapter, we could mention the free-space link experiment by Ludwig et al. [123], which used 64-PPM overlaid with PS-QPSK to require only 2.2 photons per bit for data transmission at 0.56 Gbit/s.

In this context, we shall also reiterate the experiments by Sjödin [124], Eriksson et al. [118], and van Uden et al. [111], who implemented PS-QPSK as 2-PPM-QPSK, that is, by using two adjacent symbol time frames instead of two polarizations.

2.5.2 Receiver Realizations and Digital Signal Processing

The conventional coherent receiver DSP for PM-QPSK operates according to the following flow. (See [15, 125] for more details and examples.)

1. Compensation for static errors in the receiver front end (timing skew, power imbalance, I/Q phase error, usually using Gram–Schmidt orthogonalization).
2. Static channel equalization (mainly chromatic dispersion compensation, usually with an finite impulse response (FIR) filter).

3. Dynamic channel equalization (mainly polarization tracking and polarization mode dispersion compensation, usually using the constant modulus algorithm (CMA)).
4. Interpolation and timing (clock) recovery.
5. Frequency estimation.
6. Carrier phase estimation (usually using the Viterbi and Viterbi (VV) algorithm)
7. Symbol estimation and decoding.

Traditionally, *blind algorithms* have been mostly used to estimate and compensate for channel effects, meaning that the modulation format is known, but the exact symbol transmitted at each time is unknown. Much is based on the knowledge of the used format, so when changing modulation format, a number of these steps have to be altered or modified. Most "sensitive" in this respect are the final stages: symbol estimation and decoding, and the carrier phase estimation. For example, when moving from PM-QPSK to PM-16QAM, these stages obviously need to be modified. However, also the dynamic channel estimation stage (the CMA) may need to be modified when changing format. For PS-QPSK, this is the case; the CMA needs to be modified, whereas the other stages can be kept the same as for PM-QPSK. More recently, however, there is a research trend toward the use of *nonblind* schemes, where the channel estimation is based on known signals, *training sequences*, which are transmitted regularly. Especially in optical OFDM systems, this is popular, but also in conventional baseband transmission it is becoming used [126, 127]. In the following, we describe how these stages need to be modified for some of the 4D formats.

3D Simplex The experiment of Dochhan et al. [41] used more or less standard coherent receiver algorithms, although slightly modified CMA and phase-tracking algorithms were used. A blind algorithm similar to the one presented by Yan et al. [128] was used together with the standard CMA for polarization tracking.

PS-QPSK Naively, one might think that since PS-QPSK is a subset of PM-QPSK, the conventional DSP should work also for PS-QPSK. This is true for all DSP stages except the polarization equalization, the CMA, which needs to be modified. This dynamic equalizer has a cost function J that is minimized in an iterative process, aiming to optimize FIR-filter coefficients in a Jones-matrix-like filter. Figuratively speaking, the cost function is minimized when the detected samples lie on a circle of unit radius in both polarizations. This scheme works surprisingly well also to compensate for transmission impairments such as polarization mode dispersion (PMD) and polarization dependent losses (PDL), and in fact it also works (albeit with reduced performance) "out-of-the-box" for formats without a constant, but with a nonzero, modulus, such as PM-16QAM.

However, for PS-QPSK it fails, due to an ambiguity making the cost function minimum nonunique. In addition, it requires the polarizations to be independently phase-tracked by two separate VV algorithms [92]. A number of ways to resolve this issue have been reported. Johannisson et al. [129] suggested a modified cost function J according to

$$J = E[(|E_x|^2 + |E_y|^2 - P)^2 + 2q|E_x|^2|E_y|^2)] \tag{2.19}$$

where $E[]$ denotes the expectation operator (i.e., averaging over a number of symbols), $E_{x,y}$ are the complex amplitudes of the symbols in the x and y polarizations, and P is the total signal power. The parameter q is set to -1 for PM-QPSK and $+1$ for PS-QPSK. This enables the same CMA to be used with both formats, by just changing the parameter q in the cost function, which should facilitate implementations.

Alternative CMA approaches were independently suggested. Millar and Savory [130] proposed a modified CMA where the magnitude of the polarization components is compared, and the weaker allowed to reach zero. This scheme was also used in Refs [90, 91, 94, 97].

An experiment using OFDM with PS-QPSK [99] and an outer LDPC code showed that a 4D demapper in the iterative decoder loop gave increased decoder performance.

Renaudier et al. [92] proposed instead to use a combined transmitter/receiver-based solution for the CMA ambiguity. By introducing a time offset equal to an integer number of symbol times between the two polarization components in an XOR-based PM-QPSK transmitter, the polarization ambiguity outlined earlier is suppressed and the received signal "looks" like a PM-QPSK signal to the receiver CMA. Note that his offset must be larger than maximum differential delay of the multi-tap CMA filter. Then, the standard PM-QPSK CMA can be used in the receiver, but a modified VV algorithm must be used instead [92]. Also, Alreesh et al. [131] presented an alternative VV algorithm for PS-QPSK, enabling joint phase tracking in both polarizations.

6PolSK-QPSK The coherent receivers for 6PolSK-QPSK followed the standard coherent receiver, with modifications for the polarization and phase-tracking algorithms. The implementation of Buchali and Bülow [100] to detect 6PolSK-QPSK used a modified CMA with a new cost function for the polarization tracking and a more complicated VV scheme, raising the signals to the eighth power for the phase estimation.

Fischer et al. [101] used a pilot sequence with 1.1 % overhead, which also helped with the local oscillator frequency offset estimation, to do the polarization tracking, thus avoiding a CMA. For phase estimation, they used a standard VV algorithm, but on a subset of the detected symbols. Tanimura et al. [103] also considered nonlinear compensation via digital back-propagation as well as with an inner Reed–Solomon FEC.

In addition to the experiments [100, 101, 103] discussed earlier, a couple of later extensions of the 6PolSK-QPSK work should be mentioned. In Reference [132], Bülow calculated mutual information and estimated its performance with an outer FEC. In Reference [133], Bülow and Masalkina investigated coded modulation based on, among other formats, 6PolSK-QPSK. Chen et al. [134] extended 6PolSK-QPSK to a 32-point constellation, enabling a 5-bits-to-symbol mapping by extending the constellation with 8 additional points outside the 24-cell. Even if this format was shown to have a sensitivity improvement over, for example, star-shaped 8QAM, it has gives a penalty relative to better 4D constellations with 32 points such as 32-SP-QAM.

M-SP-QAM The DSP required for 32- and 128-SP-QAM is very similar to the requirements for PM-16QAM, which is not obvious, but it works according to simulations [61] and experiments [105, 117]. In implementations with the same receiver DSP (intended for PM-16QAM), one generally finds that 32- and 128-SP-QAM have less implementation penalty, which is likely due to the better separation of levels [105, 117].

Sun et al. [109] compared the linewidth tolerance of 32-SP-QAM and the hybrid PM-QPSK/8QAM format and found it to be comparable, provided that the VV algorithm was slightly modified for 32-SP-QAM.

The work by Zhang et al. [106, 108] also added coded modulation (of the BICM flavor) to the use of 128-SP-QAM with an LDPC(18360,15300) code and an interleaver over 30 independent bit streams demultiplexed from 2 separate wavelengths. The decoding process then used 10 inner and 5 outer iterations to achieve the required performance. In Reference [108], the performance was further improved by adding nonlinear back-propagation to the DSP, thus enabling longer transmission distances.

In van Uden's work on few-mode transmission [111], the six spatial channels (two polarization modes in three fiber modes) were first optically polarization- and mode-demultiplexed, and then in a 6×6 MIMO (multiple input multiple output) structure optimized by an iterative least mean squares scheme, before entering the coherent phase estimator, demapper, and BER counter. This MIMO structure can be seen as a generalization of the CMA scheme in a conventional coherent receiver. The receiver was also simplified by the fact that the 4D modulation was performed in the time domain, rather than the polarization domain (which will not affect the theoretical spectral efficiency or sensitivity for linear transmission).

More Complex Formats The $C_{4,16}$ cluster was transmitted using OFDM in Ref. [113], and the synchronization issues, such as polarization and phase tracking, are then addressed in the OFDM receiver, using the channel estimators in the OFDM DSP. For example, one of the OFDM subcarriers was left unmodulated to act as a pilot channel, aiding the synchronization DSP. In addition, three training symbols were used in every 512 OFDM-symbol block. The overall spectral efficiency was thus 3.82 bits/symbol/pol, rather than 4 for the raw format. On the contrary, PM-QPSK was transmitted over the same OFDM channel, so the two formats were compared in a fair fashion. The implementation of Bülow et al. [114] to detect $C_{4,16}$ (referred to as "OPT16") was a more conventional coherent baseband receiver. The CMA was similar to the one used in their 6PolSK-QPSK experiments. The phase tracking was carried out by using a decision-directed least mean squares scheme, aided by a training sequence. After the training sequence set up the initial starting point, the decision-directed scheme could take over.

Higher-Dimensional Formats In the 8D implementation of frequency and polarization-switched QPSK [117], the receiver used only one local oscillator (centered between the two channels) and one optical front-end to detect both channels. The two channels were filtered in DSP, down-converted to baseband and then processed in parallel using the standard DSP flow. The CMA had to be modified with a power threshold to estimate in which frequency a given symbol was sent. The PPM implementation of the 8D format [118] was simpler in that it used a standard coherent DSP throughout, with the exception of the CMA that used the power threshold to judge which PPM frame was used.

PPM Implementations The experiments by Liu et al. [71, 121, 122] on PPM made extensive use of pilot sequences both for frame synchronization and channel estimation. The frame structure consisted of 3 frames (48 symbol slots) of training sequences followed by 16,100 PPM symbol slots. The pilot sequences helped with both polarization tracking and phase estimation. The rate overhead due to this was small, less than 1 %.

2.5.3 Formats Overview

In Table 2.3, we summarize the first proposals and implementations of some relevant optical 4D modulation formats.

TABLE 2.3 First experimental demonstrations and theoretical proposals of 4D modulation formats in coherent optical transmission links

Format	Dimensions (N)	Size (M)	Theoretical proposal	Experimental demo
3D simplex	3	4	Dochhan et al. [41]	Dochhan et al. [41]
PS-QPSK	4	8	Welti and Lee [12], Betti et al. [5], Karlsson and Agrell [42]	Sjödin et al. [88], Millar et al. [89], Nelson et al. [90], Fischer et al. [91]
$C_{4,16}$	4	16	Agrell and Karlsson [27]	Karout et al. [113], Bülow et al. [114]
6PolSK-QPSK	4	24	Bülow [55], Agrell and Karlsson [27]	Buchali and Bülow [100], Fischer et al. [101]
32-SP-QAM	4	32	Coelho and Hanik [59]	Renaudier et al. [105], Sun et al. [109], Rios-Muller et al. [110]
128-SP-QAM	4	128	Coelho and Hanik [59]	Eriksson et al. [104, 107], Zhang et al. [106], Renaudier et al. [105]
4FPS-QPSK 2PPM-PS-QPSK	8	16	Eriksson et al. [73, 117]	Eriksson et al. [73, 117]

2.5.4 Symbol Detection

For nonregular and high-order constellations, the *detection* process, that is, determining which of the constellation points that was transmitted, can be quite cumbersome and computationally intensive. The best detector, in the sense of minimizing the SER, is the ML detector. In the special case of an AWGN channel, the ML detector computes the Euclidean distance between a received vector $\boldsymbol{r}$ and all constellation points in $\mathcal{X}$ and picks the symbol with the smallest distance, that is,

$$\hat{x} = \arg \min_{\boldsymbol{x} \in \mathcal{X}} (|\boldsymbol{r} - \boldsymbol{x}|) \tag{2.20}$$

This will always work, but requires M distance calculations and a number of comparisons, which is computationally costly, and one, therefore, seeks easier detectors in practice. In this section, we describe how this detection is done efficiently for some of the common 4D constellations. We start by describing lattices, and then give examples for the specific formats.

For the n-dimensional integer (or cubic) lattice $\mathbb{Z}^n$, the problem is particularly simple; the ML detection is equivalent to just rounding each component of the received vector to the nearest integer. As shown by Conway and Sloane [56], this procedure can be modified to work for other lattices of interest, such as D_4 and E_8.

For the D_4 lattice (2.14), one can use the following algorithm [56]: (i) decode $\boldsymbol{r}$ to the nearest point in $\mathbb{Z}^4$, and check the parity of the lattice point (i.e., the modulo-2 sum of its coordinates). (ii) If the parity is odd, round the component of $\boldsymbol{r}$ that is farthest away from its closest integer to its second closest integer. The algorithm is described for 128-SP-QAM in Ref. [61] and generalizes straightforwardly to all other SP-QAM formats obtained by reduction. An equally simple algorithm applies to SP-QAM formats obtained by expansion [56]. These algorithms can be generalized to the E_8 lattice as well [56, 135].

For the simpler formats such as PS-QPSK, the symbol detection is straightforward, but depends on the form after synchronization and phase tracking. For the set-partitioned-form, that is, $\pm\{(1, 1, 1, 1), (1, 1, -1, -1), (1, -1, 1, -1), (1, -1, -1, 1)\}$, one can use the above-mentioned D_4 scheme. For the polarization-switched form, where the constellation points are $\{(\pm 1, \pm 1, 0, 0), (0, 0, \pm 1, \pm 1)\}$ one can use the following steps: (i) determine the polarization by comparing $|\Re(E_x)| + |\Im(E_x)|$ with $|\Re(E_y)| + |\Im(E_y)|$ and (ii) detect QPSK as usual in the chosen polarization. It is noteworthy and somewhat surprising that comparing the magnitude of the complex numbers $|E_x|$ and $|E_y|$ for determining the polarization is *suboptimal* in contrast with this scheme. As a simple example of this, consider the received vector $\boldsymbol{r} = (0.9, 0.1, 0.6, 0.5)$, or $E_x = 0.9 + i0.1$, $E_y = 0.5 + i0.6$. The closest PS-QPSK point in the Euclidean distance metric is (0,0,1,1), that is, an ML symbol detector selects the y polarization, despite the received power in x being $|E_x|^2 = 0.82$, which is higher than the power in y, being $|E_y|^2 = 0.61$.

For the 24-cell (6PolSK-QPSK) in the form $\{(\pm 1, \pm 1, \pm 1, \pm 1), (\pm 2, 0, 0, 0)\}$, the following detector is optimal: (i) Find the maximum of $\{(|r_1| + |r_2| + |r_3| + |r_4|)/2, |r_1|, |r_2|, |r_3|, |r_4|\}$, where $(r_1, r_2, r_3, r_4) = \boldsymbol{r}$. (ii) If the first is maximum, proceed with standard PM-QPSK detection to return $(\text{sgn}(r_1), \text{sgn}(r_2), \text{sgn}(r_3), \text{sgn}(r_4))$. If one of the last four are largest, take the sign of that component and multiply with 2.

2.5.5 Realizing Dimensions

The dimensionality can be taken as one metric of complexity, and often one chooses to compare modulation formats of the same dimensionality, that is, keeping the complexity similar. Shannon proved that an arbitrarily small SER is achievable for any channel, assuming that the spectral efficiency is below a certain threshold, the *channel capacity*, and that the dimensionality goes to infinity [85, 136]. The work was later extended to quantify how the SER decreases with dimension [137].

The introduction of a channel (FEC) code is the obvious, and prevailing, way of increasing the dimensionality in communications. If used with BPSK mapping (see Section 2.4.1), the dimensionality is simply the number n of bits in the FEC frame. There may, however, be practical reasons for keeping the value n down; the latency, complexity, and power consumption incurred by the FEC will increase with n. Moreover, even if a good FEC code is used that can tolerate a low SNR, the pre-decoder signal may be so distorted by noise that DSP and synchronization algorithms will limit the performance rather than the ideal FEC ability.

Thus, in practical systems, it will make sense also to use the degrees of freedom (DOF) of the transmission channel to increase the dimensionality and the constellation distances. However, this often leads to issues with crosstalk and synchronization that need to be resolved in the receiver. Later, we briefly discuss the practical implementation challenges by increasing the dimensionality via the physical DOFs.

We should emphasize that the DOFs used as signaling dimensions usually have independent noise sources, which simplifies their usage for the AWGN channel model, as the noise can be modeled as a hyperspherical cloud around the transmitted symbol.

Quadratures Every carrier wave has two DOFs that can be modulated independently, that is, the two quadratures usually described by the real and imaginary parts of a complex phasor, or as the "sine" and "cosine"-components of the wave. An alternative decomposition is the amplitude and phase of the wave. Up to around 2000, the amplitude was the only DOF used in commercial optical links, and it still is in short-haul links, due to the cost and complexity associated with modulating the optical phase. To reliably detect the optical phase, a coherent or differential-phase receiver is required. The differential receiver is simpler in optical hardware, but has extra an penalty relative to the coherent counterpart and is limited to mainly PSK modulation.

Since the intradyne coherent receiver was demonstrated [3], the differential optical receivers have faded away. Coherent receivers are challenging due to the required rapid phase tracking on a microsecond time scale, but with the development of fast electronics and DSPs, they are becoming increasingly common and cost effective, and will likely prevail in future optical links. Ways of reducing the coherent receiver complexity by, for example, co-propagating a local oscillator carrier (so-called self-homodyning [138]) have been proposed.

The quadrature dimensions are, just as polarizations, and contrary to the time- and space-related dimensions discussed later, not scalable to more than 2.

Polarization Electromagnetic waves have a vector property not seen in longitudinal waves (e.g., acoustic waves) or transverse matter waves (e.g., water waves, oscillating strings). The easiest description is that of two independent carrier waves, with orthogonally directed vector field components. They are often referred to as the x and y polarization states, but there are other orthogonal decompositions possible as well. Usually, a "polarization state" refers to the relative amplitude and phase between these two waves. Thus, one has two ways of describing the 4 DOFs of the classical electromagnetic wave, either with "amplitude," "phase," and "polarization state" or with the I and Q quadratures in the x and y polarizations. The former is the traditional description used in optics, and the latter is the most attractive one used in communications, as those DOFs form a Cartesian system.

The polarization state in a fiber link is slowly (second to millisecond time scale) drifting due to imperfections, micro- and macrobendings, thermal changes, fiber movements etc., and the use of polarization dimensions will thus require polarization tracking in the receiver. A commercially attractive and low-cost polarization tracker was not available before the intradyne coherent receiver, and as a result, polarization was not actively used for modulation in commercial systems (although studied, e.g., in Ref. [139]). Also, the intersymbol interference problems related to polarization mode dispersion were long regarded as a significant obstacle (also in conventional links that were not polarization modulated), but elegantly resolved by the adaptive CMA filter in the coherent receiver.

The combined use of quadratures and polarization leads to a real 4D constellation space that is the basis for coherent signaling. It is noteworthy that some key problems and issues with 4D modulation are unresolved or only recently being explored, for example, channel modeling via 4D rotations [140] and modulation format optimization [27, 42].

Time With an AWGN channel, a T-orthogonal pulse (2.3), and a matched-filter receiver, adjacent symbols in time will have independent noise, and can be thus be framed to a "supersymbol" with dimensionality equal to the number of symbols used (possibly times the dimensionality of each symbol). Such a supersymbol can now be modulated with a higher-dimensional format. As detailed in Section 2.4.1, when using a simple format such as 1D OOK or BPSK, this is equivalent to applying an FEC frame with a binary code. This clearly highlights the close relationship between modulation and coding; indeed, there is no clear-cut distinction between coding and modulation in communication theory.

Nonetheless, from a practical and implementation perspective, it makes sense to distinguish between modulation and coding, as they are implemented with very different hardware, meeting different challenges. The time-multiplexing described earlier (of which PPM discussed in section 2.3.1.5 is one special case) is particularly simple and attractive, since the phase and symbol time synchronization essentially comes for free. The frame synchronization needs to be resolved, however, and usually requires a test sequence and/or use of a specific transmission protocol.

The temporal dimension and its simplicity thus forms a simple playground for testing new formats without challenging synchronization issues. For example, this was used in the experiment by van Uden et al. [111], where two time slots were used to form a 4D symbol rather than the two polarization states, simplifying receiver DSP significantly.

Frequency/Wavelength Different channels transmitting at adjacent wavelengths (frequencies) can be used to form multidimensional supersymbols. The concept of "superchannels" were introduced to denote such multi-wavelength channels, which are routed and detected as one entity [141]. To make use of correlated modulation to increase signal space dimensionality with such superchannels has not yet been realized, but should be possible.

In a similar vein, one could perform joint detection of multi-wavelength channels in a WDM link, thus enabling multidimensional modulation and coding. This meets practical problems with temporal (walk-off related) synchronization and phase synchronization, as well as signal ambiguities of the independent wavelengths, so it has not yet been widely used.

A few limited cases have been reported though, for example, the 8D format by Eriksson et al. [117] detecting two wavelengths with the same local oscillator.

Space With the recent interest in spatial-division multiplexing, that is, the use of waveguide modes and/or parallel waveguides to increase the data rates of optical links, it seems natural to try to further increase the capacity by moving to *joint* transmission over parallel modes and/or fibers, that is, increasing signaling dimensionality by making use of also the spatial DOFs. These can be (i) the different modes in a multimode fiber, (ii) different cores in a multi-core fiber, (iii) entirely separate fibers, or a combination of these.

However, spatial multiplexing is associated with severe practical challenges. For example, the different modes in a multimode fiber have different group velocities and hence a significant differential mode delay will arise that needs to be compensated for. Even worse, bends and fiber imperfections gives rise to modal crosstalk that mixes the delayed modes, further complicating the reception of individual modes. As a result, MIMO signal processing is needed to compensate for the modal crosstalk (see, e.g., [111]), which is very DSP-heavy.

The single-mode (linear-polarized, LP_{01}-mode) fiber has, as was outlined earlier, four dimensions due the polarization and quadratures. The next mode in the weakly guiding, step-index, circular fiber mode hierarchy, the (linearly polarized) LP_{11}-mode, is doubly degenerate. This means that it can be of two orientations that are mathematically orthogonal, usually referred to as LP_{11a} and LP_{11b}. A linear combination of these modes is popularly referred to as the lowest orbital angular momentum

(OAM) mode, with azimuthal index ± 1. Thus, OAM modes fall well within the conventional modal description, and offer limited novelty and no principal extension of the existing DOFs, as recently shown for radio frequency (RF) transmission [142].

This means that inclusion of the next higher-order mode in a circular fiber totals to three orthogonal modes, of 4D each, leading to a 12D space. Format optimization in this 12D space has been discussed by Bülow et al. [143].

A more straightforward approach seems to be to use different (single-mode) cores in a multicore fiber to increase the dimensionality. This relaxes the crosstalk and walk-off penalties. Novel formats exploiting these DOFs were recently discussed by Eriksson et al. [73] and some initial experiments were reported by Puttnam et al. [120, 144].

Spatial Frequencies Due to the space–time duality [145], which states that free-space propagation of optical beams is similar to dispersive propagation of optical pulses, one could just as well use spatial frequencies as temporal frequencies. Spatial frequencies translate into physically different propagation directions, so their use would be of most interest in free-space rather than in guided-wave propagation. A deeper discussion of this topic is, therefore, not included in this chapter. It suffices to say that the challenges and solutions connected with spatial frequency (physical beam direction in the paraxial propagation limit) are similar to the MIMO technologies used in wireless communications, cf. [142].

2.6 SUMMARY AND CONCLUSIONS

In this chapter, we have overviewed the relatively large body of work (experimental and theoretical) on modulation formats for optical coherent links that have emerged over the last 5 years. We have also shown the performance limits of formats in 2D, 4D, and 8D by reviewing sphere packing simulations, lattice-cuts, and code-based format design. To reach higher dimensions, formats based on codes are probably the most straightforward approach rather than numerically optimized sphere packings, and we showed a few examples of this as well.

The results summarized in this chapter are somewhat idealized, in that they to a large extent (i) neglect the impact of FEC, (ii) focus on the asymptotic high-SNR behavior (low SER and BER), and (iii) emphasize the AWGN channel. This is the classical starting point for modulation theory research and should be viewed as a first step toward a fuller understanding of optical link design. Broadening the scope in all three directions are presently active areas of research.

First, to separate modulation format and coding is (while unnecessary from the information theorist's point of view) necessary from the system engineer's perspective. The modulation format dictates the transmitter and receiver hardware optoelectronics, the complexity, and thus to some extent the cost. It dictates the DSP and the complexity of the receiver electronics, especially if blind equalization is used. It also dictates the attainable spectral efficiency. Thus, when selecting formats, one makes critical system choices and it is important for all the trade-offs made in system design to know how well the formats behave—even if it is an ideal or asymptotic behavior.

Moreover, there are communication applications that are latency-critical, where the use of FEC is prohibited or at least limited, and there good formats are very important. Examples may be control systems, video conferencing or telephony, and transfer of stock market trading data.

Another issue, seldom studied or emphasized, is the performance of synchronization algorithms, channel estimation algorithms, and adaptive equalizers for various formats. Those tend to operate worse at low SNR (which might arise if strong FEC codes are used), but better if constellations with well-separated points are used. In such situations, the clusters or lattice-based constellations described in this chapter might be a better choice than standard QAM.

An important application for the use of many different formats is the emergence of *elastic networking*. In future optical networks, where an increased *flexibility* is desired (often called elastic optical networks), one strives to adapt the data rate provided to customers after the available bandwidth, SNR, and demand. In such systems, it is of great value to be able to switch between different modulation formats to provide the sought flexibility, and a good overview of the performance and trade-offs of formats is needed. A recent overview of 4D modulation formats from this perspective was recently provided by Fischer et al. [62].

Second, the study of asymptotic performance metrics should be complemented by studies of the performance of modulation formats at more practically relevant SNR values. Unfortunately, no analytic instruments are available for this purpose. To numerically optimize multidimensional modulation formats for specific SNRs is computationally complex, and it may be difficult to get the full picture. Asymptotic metrics such as γ, CFM, and G are attractive in that they give one number to compare formats by, which quickly can be used to compare and select formats. They also provide an intuitive interpretation in terms of sphere packing, and they give an upper limit of the performance gains, which is valuable. However, once a set of formats are selected, they should be compared via SER or BER simulations as well as complexity estimates. Even if the asymptotic gains are never achieved, the current design paradigm in optics is to compare formats with the same FEC, which has a waterfall region around 10^{-3} for the uncoded SER, and then, quite often, the optimized multidimensional formats will outperform (e.g., in term of system reach) the standard QAM formats.

Third, more advanced channel models to go beyond the AWGN model are currently being studied (as discussed in Section 2.3.2) and depending on their nature, the results from AWGN modeling may or may not be useful. For example, the GN model, in links offering high SNR, can benefit from AWGN-optimized constellations, whereas nonlinear phase-noise channels benefit from radically different constellations.

Finally, we note that, as we touched upon previously in this chapter, the price we pay when going to higher-dimensional formats (and codes) is *complexity*, which is a notoriously difficult quantity to quantify. Even if it can be quantified in terms of component cost, number of floating point operations, chip area for DSP implementation, or dimensionality of codes and formats, it is hard to provide generic results, since the most reliable complexity metrics are implementation-specific. Nevertheless, the complexity factor must enter the system design and hopefully the contents of this chapter may somewhat aid system designers when dealing with the trilemma of limited spectral efficiency, low noise tolerance, and complexity.

REFERENCES

1. Pfau T, Hoffmann S, Peveling R, Ibrahim S, Adamczyk O, Porrmann M, Bhandare S, Noé R, Achiam Y. Synchronous QPSK transmission at 1.6 Gbit/s with standard DFB lasers and real-time digital receiver. *Electron Lett* 2006;42(20):1175–1176.
2. Charlet G, Salsi M, Renaudier J, Pardo O, Mardoyan H, Bigo S. Performance comparison of singly-polarised and polarisation-multiplexed coherent transmission at 10 Gbauds under linear impairments. *Electron Lett* 2007;43(20):1109–1111.
3. Sun H, Wu K, Roberts K. Real-time measurements of a 40 Gb/s coherent system. *Opt Express* 2008;16(2):873–879.
4. Betti S, Curti F, De Marchis G, Iannone E. Exploiting fibre optics transmission capacity: 4-quadrature multilevel signalling. *Electron Lett* 1990;26(14):992–993.
5. Betti S, Curti F, De Marchis G, Iannone E. A novel multilevel coherent optical system: 4-quadrature signaling. *J Lightwave Technol* 1991;9(4):514–523.
6. Betti S, De Marchis G, Iannone E, Lazzaro P. Homodyne optical coherent systems based on polarization modulation. *J Lightwave Technol* 1991;9(10):1314–1320.
7. Cusani R, Iannone E, Salonico A, Todaro M. An efficient multilevel coherent optical system: M-4Q-QAM. *J Lightwave Technol* 1992;10(6):777–786.
8. Derr F. Optical QPSK homodyne transmission of 280 Mbit/s. *Electron Lett* 1990;26:401.
9. Gnauck AH, Winzer PJ. Optical phase-shift-keyed transmission. *J Lightwave Technol* 2005;23(1):115.
10. Beygi L, Agrell E, Kahn JM, Karlsson M. Rate-adaptive coded modulation for fiber-optic communications. *J Lightwave Technol* 2013;32(2):333–343.
11. Beygi L, Agrell E, Kahn JM, Karlsson M. Coded modulation for fiber-optic networks: toward better tradeoff between signal processing complexity and optical transparent reach. *IEEE Signal Process Mag* 2014;31(2):93–103.
12. Welti G, Lee J. Digital transmission with coherent four-dimensional modulation. *IEEE Trans Inf Theory* 1974;20(4):497–502.
13. Zetterberg L, Brändström H. Codes for combined phase and amplitude modulated signals in a four-dimensional space. *IEEE Trans Commun* 1977;25(9):943–950.
14. Biglieri E. Advanced modulation formats for satellite communications. In: Hagenauer J, editor. Advanced Methods for Satellite and Deep Space Communications. Berlin: Springer-Verlag; 1992. p 61–80.
15. Savory SJ. Digital coherent optical receivers: algorithms and subsystems. *IEEE J Sel Top Quantum Electron* 2010;16(5):1164–1179.
16. Zhou X. Efficient clock and carrier recovery algorithms for single-carrier coherent optical systems. *IEEE Signal Process Mag* 2014;31(2):35–45.

17. Lau APT, Gao Y, Sui Q, Wang D, Zhuge Q, Morsy-Osman MH, Chagnon M, Xu X, Lu C, Plant DV. Advanced DSP techniques enabling high spectral efficiency and flexible transmissions. *IEEE Signal Process Mag* 2014;31(2):82–92.
18. Ip E, Lau APT, Barros DJF, Kahn JM. Coherent detection in optical fiber systems. *Opt Express* 2008;16(2):753–791. (erratum, 2008;16(26):21943).
19. Kikuchi K, Tsukamoto S. Evaluation of sensitivity of the digital coherent receiver. *J Lightwave Technol* 2008;26(13):1817–1822.
20. Poggiolini P. The gn model of non-linear propagation in uncompensated coherent optical systems. *J Lightwave Technol* 2012;30(24):3857–3879.
21. Beygi L, Agrell E, Karlsson M. Optimization of 16-point ring constellations in the presence of nonlinear phase noise. Optical Fiber Communication Conference; 2011 p OThO.4.
22. Johannisson P, Karlsson M. Perturbation analysis of nonlinear propagation in a strongly dispersive optical communication system. *J Lightwave Technol* 2013;31(8):1273–1282.
23. Splett A, Kurtzke C, Petermann K. Ultimate transmission capacity of amplified optical fiber communication systems taking into account fiber nonlinearities. European Conference and Exhibition on Optical Communication; Montreux, Switzerland; 1993.
24. Agrell E, Alvarado A, Durisi G, Karlsson M. Capacity of a nonlinear optical channel with finite memory. *J Lightwave Technol* 2014;32(16):2862–2876.
25. Ho K-P. Phase-Modulated Optical Communication Systems. New York: Springer-Verlag; 2005.
26. Beygi L, Agrell E, Karlsson M, Johannisson P. Signal statistics in fiber-optical channels with polarization multiplexing and self-phase modulation. *J Lightwave Technol* 2011;29(16):2379–2386.
27. Agrell E, Karlsson M. Power-efficient modulation formats in coherent transmission systems. *J Lightwave Technol* 2009;27(22):5115–5126.
28. Agrell E. Database of sphere packings; 2014. Online: http://codes.se/packings. Accessed 2015 Sept 24.
29. Forney GD Jr. Multidimensional constellations—Part I: introduction, figures of merit, and generalized cross constellations. *IEEE J Sel Areas Commun* 1989;7(6):877–892.
30. Kschischang FR. Optimal nonuniform signaling for Gaussian channels. *IEEE Trans Inf Theory* 1993;39(3):913–929.
31. Karlsson M, Agrell E. Power-efficient modulation schemes. In: Kumar S, editor. Impact of Nonlinearities on Fiber Optic Communications. Chapter 5. New York: Springer-Verlag; 2011. p 219–252.
32. Benedetto S, Biglieri E. Principles of Digital Transmission: With Wireless Applications. New York: Kluwer Academic Publishers; 1999.
33. Foschini G, Gitlin R, Weinstein S. Optimization of two-dimensional signal constellations in the presence of gaussian noise. *IEEE Trans Commun* 1974;22(1):28–38.
34. Graham RL, Sloane NJA. Penny-packing and two-dimensional codes. *Discrete Comput Geom* 1990;5(1):1–11.
35. Sloane NJA, Hardin RH, Duff TS, Conway JH. Minimal-energy clusters. Online: http://neilsloane.com/cluster. Accessed 2015 Sept 24.
36. Conway JH, Sloane NJA. Sphere Packings, Lattices and Groups. 3rd ed. New York: Springer-Verlag; 1998.

37. Pierce JR. Comparison of three-phase modulation with two-phase and four-phase modulation. *IEEE Trans Commun* 1980;COM-28(7):1098–1099.
38. Ekanayake N, Tjhung T. On ternary phase-shift keyed signaling. *IEEE Trans Inf Theory* 1982;28(4):658–660.
39. Porath J-E, Aulin T. Design of multidimensional signal constellations. *IEE Proc Commun* 2003;150(5):317–323.
40. Proakis JG. Digital Communication. Boston (MA): McGraw-Hill; 2001.
41. Dochhan A, Griesser H, Eiselt M. First experimental demonstration of a 3-dimensional simplex modulation format showing improved OSNR performance compared to DP-BPSK. Optical Fiber Communication Conference; 2013. p JTh2A.40.
42. Karlsson M, Agrell E. Which is the most power-efficient modulation format in optical links? *Opt Express* 2009;17(13):10814–10819.
43. Simon MK, Hinedi SM, Lindsey WC. Digital Communication Techniques: Signal Design and Detection. Englewood Cliffs (NJ): PTR Prentice Hall; 1995.
44. Agrell E, Karlsson M. On the symbol error probability of regular polytopes. *IEEE Trans Inf Theory* 2011;57(6):3411–3415.
45. Saha D, Birdsall TG. Quadrature-quadrature phase-shift keying. *IEEE Trans Commun* 1989;37(5):437–448.
46. Slepian D. Permutation modulation. *Proc IEEE* 1965;53(3):228–236.
47. Karlsson M, Agrell E. Multilevel pulse-position modulation for optical power-efficient communication. *Opt Express* 2011;19(26, SI):B799–B804.
48. Krongold B, Pfau T, Kaneda N, Lee SCJ. Comparison between PS-QPSK and PDM-QPSK with equal rate and bandwidth. *IEEE Photonics Technol Lett* 2012;24(3):203–205.
49. Poggiolini P, Bosco G, Carena A, Curri V, Forghieri F. Performance evaluation of coherent WDM PS-QPSK (HEXA) accounting for non-linear fiber propagation effects. *Opt Express* 2010;18(11):11360–11371.
50. Serena P, Vannucci A, Bononi A. The performance of polarization switched-QPSK (PS-QPSK) in dispersion managed WDM transmissions. European Conference on Optical Communication; 2010. p Th.10.E.2.
51. Serena P, Rossi N, Bononi A. PDM-iRZ-QPSK vs. PS-QPSK at 100 Gbit/s over dispersion-managed links. *Opt Express* 2012;20(7):7895–7900.
52. Coxeter HSM Regular Polytopes. New York: Dover Publication; 1973.
53. Musin O. The kissing number in four dimensions. *Ann Math* 2008;168:1–32.
54. Taricco G, Biglieri E, Castellani V. Applicability of four-dimensional modulations to digital satellites: a simulation study. Proceedings of IEEE Global Telecommunications Conference, Volume 4; 1993. p 28–34.
55. Bülow H. Polarization QAM modulation (POL-QAM) for coherent detection schemes. Optical Fiber Communication Conference; 2009. p OWG.2.
56. Conway JH, Sloane NJA. Fast quantizing and decoding and algorithms for lattice quantizers and codes. *IEEE Trans Inf Theory* 1982;28(2):227–232.
57. Karlsson M, Agrell E. Spectrally efficient four-dimensional modulation. Optical Fiber Communication Conference; Optical Society of America; 2012. p OTu2C.1.
58. Ungerboeck G. Channel coding with multilevel/phase signals. *IEEE Trans Inf Theory* 1982;28(1):55–67.

59. Coelho L, Hanik N. Global optimization of fiber-optic communication systems using four-dimensional modulation formats. European Conference on Optical Communication; 2011. p Mo.2.B.4.
60. Renaudier J, Voicila A, Bertran-Pardo O, Rival O, Karlsson M, Charlet G, Bigo S. Comparison of set-partitioned two-polarization 16QAM formats with PDM-QPSK and PDM-8QAM for optical transmission systems with error-correction coding. European Conference on Optical Communication; Optical Society of America; 2012. p We.1.C.5.
61. Sjödin M, Johannisson P, Li J, Agrell E, Andrekson PA, Karlsson M. Comparison of 128-SP-QAM with PM-16-QAM. *Opt Express* 2012;20(8):8356–8366.
62. Fischer JK, Alreesh S, Elschner R, Frey F, Nölle M, Schmidt-Langhorst C, Schubert C. Bandwidth-variable transceivers based on four-dimensional modulation formats. *J Lightwave Technol* 2014;32(16):2886–2895.
63. Karlsson M, Agrell E. Four-dimensional optimized constellations for coherent optical transmission systems. European Conference on Optical Communication; 2010. p We.8.C.3.
64. Sjödin M, Agrell E, Karlsson M. Subset-optimized polarization-multiplexed PSK for fiber-optic communications. *IEEE Commun Lett* 2013;17(5):838–840.
65. Conway JH, Sloane NJA. A fast encoding method for lattice codes and quantizers. *IEEE Trans Inf Theory* 1983;29(6):820–824.
66. Eriksson TA, Alreesh S, Schmidt-Langhorst C, Frey F, Berenguer PW, Schubert C, Fischer JK, Andrekson PA, Karlsson M, Agrell E. Experimental investigation of a four-dimensional 256-ary lattice-based modulation format. Optical Fiber Communication Conference; 2015. p W4K.3.
67. Koike-Akino T, Millar DS, Kojima K, Parsons K. Eight-dimensional modulation for coherent optical communications. European Conference and Exhibition on Optical Communication; 2013. p Tu.3.C.3.
68. Millar DS, Koike-Akino T, Arık SO, Kojima K, Parsons K, Yoshida T, Sugihara T. High-dimensional modulation for coherent optical communications systems. *Opt Express* 2014;22(7):8798–8812.
69. Millar DS, Koike-Akino T, Kojima K, Parsons K. A 24-dimensional modulation format achieving 6 dB asymptotic power efficiency. Signal Processing in Photonic Communications; Optical Society of America; 2013. p SPM3D.6.
70. Millar DS, Koike-Akino T, Maher R, Lavery D, Paskov M, Kojima K, Parsons K, Thomsen BC, Savory SJ, Bayvel P. Experimental demonstration of 24-dimensional extended Golay coded modulation with LDPC. Optical Fiber Communication Conference; Optical Society of America; 2014. p M3A.5.
71. Liu X, Wood TH, Tkach R, Chandrasekhar S. Demonstration of record sensitivity in an optically pre-amplified receiver by combining PDM-QPSK and 16-PPM with pilot-assisted digital coherent detection. Optical Fiber Communication Conference; 2011. p PDPB.1.
72. Selmy H, Shalaby HMH, Kawasaki Z-I. Proposal and performance evaluation of a hybrid BPSK-modified MPPM technique for optical fiber communications systems. *J Lightwave Technol* 2013;31(22):3535–3545.
73. Eriksson TA, Johannisson P, Puttnam BJ, Agrell E, Andrekson PA, Karlsson M. K-over-L Multidimensional position modulation. *J Lightwave Technol* 2014;32(12):2254–2262.

74. Lau APT, Kahn JM. Signal design and detection in presence of nonlinear phase noise. *J Lightwave Technol* 2007;25(10):3008–3016.
75. Agrell E, Karlsson M. Satellite constellations: towards the nonlinear channel capacity. 25th IEEE Photonics Conference, IPC 2012; 2012. p 316–317.
76. Häger C, Graell i Amat A, Alvarado A, Agrell E. Constellation optimization for coherent optical channels distorted by nonlinear phase noise. IEEE Global Communications Conference (GLOBECOM); 2012. p 2870–2875.
77. Kayhan F, Montorsi G. Constellation design for memoryless phase noise channels. *IEEE Trans Wireless Commun* 2014;13(5):2874–2883.
78. Steiner M. The strong simplex conjecture is false. *IEEE Trans Inf Theory* 1994;40(3):721–731.
79. Chang F, Onohara K, Mizuochi TK-A. Forward error correction for 100 G transport networks. *IEEE Commun Mag* 2010;48(3):S48–S55.
80. Djordjevic IB, Arabaci M, Minkov LL. Next generation FEC for high-capacity communication in optical transport networks. *J Lightwave Technol* 2009;27(16):3518–3530.
81. Smith BP, Kschischang FR. Future prospects for fec in fiber-optic communications. *IEEE J Sel Top Quantum Electron* 2010;16(5):1245–1257.
82. Viterbi AJ. On coded phase-coherent communications. *IRE Trans Space Electron Telem* 1961;SET-7(1):3–14.
83. Lin S, Costello DJ Jr. Error Control Coding. 2nd ed. Upper Saddle River (NJ): Prentice Hall; 2004.
84. ITU-T. Forward error correction for submarine systems. Recommendation G.975, International Telecommunication Union; 2000.
85. Shannon CE. A mathematical theory of communication. *Bell Syst Tech J* 1948;27:379–423, 623–656.
86. Alvarado A, Agrell E. Achievable rates for four-dimensional coded modulation with a bit-wise receiver. Optical Fiber Communication Conference; Optical Society of America; 2014. p M2C.1.
87. Yamazaki H, Hashizume Y, Saida T. Simple three-dimensional simplex modulator. Optical Fiber Communication Conference; 2014. p W1I.3.
88. Sjödin M, Johannisson P, Wymeersch H, Andrekson PA, Karlsson M. Comparison of polarization-switched QPSK and polarization-multiplexed QPSK at 30 Gbit/s. *Opt Express* 2011;19(8):7839–7846.
89. Millar DS, Lavery D, Makovejs S, Behrens C, Thomsen BC, Bayvel P, Savory SJ. Generation and long-haul transmission of polarization-switched QPSK at 42.9 Gb/s. *Opt Express* 2011;19(10):9296–9302.
90. Nelson LE, Zhou X, Mac Suibhne N, Ellis AD, Magill P. Experimental comparison of coherent polarization-switched QPSK to polarization-multiplexed QPSK for 10 x 100 km WDM transmission. *Opt Express* 2011;19(11):10849–10856.
91. Fischer JK, Molle L, Nölle M, Gross D-D, Schubert C. Experimental investigation of 28-GBd polarization-switched quadrature phase-shift keying signals. European Conference on Optical Communications; Optical Society of America; 2011. p Mo.2.B.1.
92. Renaudier J, Serena P, Bononi A, Salsi M, Bertran-Pardo O, Mardoyan H, Tran P, Dutisseuil E, Charlet G, Bigo S. Generation and detection of 28 Gbaud polarization switched-QPSK in WDM long-haul transmission systems. *J Lightwave Technol* 2012;30(9):1312–1318.

93. Yamazaki H, Goh T, Saida T, Hashizume Y, Mino S. IQ-coupling-loss-free polarization-switched QPSK modulator. Optical Fiber Communication Conference; 2012. p PDP5A.8.

94. Behrens C, Lavery D, Millar DS, Makovejs S, Thomsen BC, Killey RI, Savory SJ, Bayvel P. Ultra-long-haul transmission of 7x42.9Gbit/s PS-QPSK and PM-BPSK. European Conference on Optical Communication; 2011. p Mo.2.B.2.

95. Lavery D, Behrens C, Makovejs S, Millar DS, Killey RI, Savory SJ, Bayvel P. Long-haul transmission of PS-QPSK at 100 Gb/s using digital backpropagation. *IEEE Photonics Technol Lett* 2012;24(3):176–178.

96. Sjödin M, Puttnam BJ, Johannisson P, Shinada S, Wada N, Andrekson PA, Karlsson M. Transmission of PM-QPSK and PS-QPSK with different fiber span lengths. *Opt Express* 2012;20(7):7544–7554.

97. Nölle M, Fischer JKK, Molle L, Schmidt-Langhorst C, Schubert C. Influence of channel spacing on ultra long-haul WDM transmission of 8× 112 Gb/s PS-QPSK signals. Optical Fiber Communication Conference; 2012. p OTh3A.4.

98. Sjödin M, Puttnam BJ, Johannisson P, Shinada S, Wada N, Andrekson PA, Karlsson M. Comparison of PS-QPSK and PM-QPSK at different data rates in a 25 GHz-spaced WDM system. Optical Fiber Communication Conference; Optical Society of America; 2012. p OTu2A.3.

99. Masalkina E, Dischler R, Bülow H. Experimental study of polarization-switched-QPSK subcarrier modulation and iterative demapping on optical OFDM systems. Optical Fiber Communication Conference; 2011. p OThO.6.

100. Buchali F, Bülow H. Experimental transmission with POLQAM and PS-QPSK modulation format using a 28-Gbaud 4-D transmitter. European Conference on Optical Communication; Optical Society of America; 2012. p We.3.A.1.

101. Fischer JK, Alreesh S, Elschner R, Frey F, Meuer C, Schmidt-Langhorst LMC, Tanimura T, Schubert C. Experimental investigation of 126-Gb/s 6PolSK-QPSK signals. *Opt Express* 2012;20(26):B232–B237.

102. Ding D, Zhang Y, Yuan X, Zhang J, Li Y, Huang Y. Generation of 6polsk-qpsk using dual-drive mach-zehnder modulators. Asia Communications and Photonics Conference; 2014. p ATh3A.130.

103. Tanimura T, Alreesh S, Fischer JK, Schmidt-Langhorst C, Frey F, Meuer C, Elschner R, Molle L, Schubert C. Nonlinear transmission of 6PolSK-QPSK signals using coded modulation and digital back propagation. Optical Fiber Communication Conference; 2013. p OTu3B.3.

104. Eriksson TA, Sjödin M, Andrekson PA, Karlsson M. Experimental demonstration of 128-SP-QAM in uncompensated long-haul transmission. Optical Fiber Communication Conference; 2013. p OTu3B.2.

105. Renaudier J, Bertran-Pardo O, Ghazisaeidi A, Tran P, Mardoyan H, Brindel P, Voicila A, Charlet G, Bigo S. Experimental transmission of Nyquist pulse shaped 4-D coded modulation using dual polarization 16QAM set-partitioning schemes at 28 Gbaud. Optical Fiber Communication Conference; 2013. p OTu3B.

106. Zhang H, Batshon HG, Foursa DG, Mazurczyk M, Cai JX, Davidson CR, Pilipetskii A, Mohs G, Bergano NS. 30.58 Tb/s transmission over 7,230 km using PDM half 4D-16QAM coded modulation with 6.1 b/s/Hz spectral efficiency. Optical Fiber Communication Conference; 2013. p OTu2B.3.

107. Eriksson TA, Sjödin M, Johannisson P, Andrekson PA, Karlsson M. Comparison of 128-SP-QAM and PM-16QAM in long-haul WDM transmission. *Opt Express* 2013;21(16):19269–19279.

108. Zhang H, Cai J-X, Batshon HG, Mazurczyk M, Sinkin O, Foursa D, Pilipetskii A, Mohs G, Bergano N. 200 Gb/s and dual wavelength 400 Gb/s transmission over transpacific distance at 6 b/s/Hz spectral efficiency. Optical Fiber Communication Conference; 2013. p PDP5A.6.

109. Sun H, Egorov R, Basch BE, McNicol J, Wu K-T. Comparison of two modulation formats at spectral efficiency of 5 bits/dual-pol symbol. European Conference on Optical Communication; 2013. p Th.2.D.3.

110. Rios-Muller R, Renaudier J, Bertran-Pardo O, Ghazisaeidi A, Tran P, Charlet G, Bigo S. Experimental comparison between hybrid-QPSK/8QAM and 4D-32SP-16QAM formats at 31.2 GBaud using Nyquist pulse shaping. European Conference on Optical Communication; 2013. p Th.2.D.2.

111. van Uden R, Okonkwo C, Chen H, de Waardt H, Koonen A. 6×28GBaud 128-SP-QAM transmission over 41.7 km few-mode fiber with a 6×6 MIMO FDE. Optical Fiber Communication Conference; 2014. p W4J.4.

112. Fischer JK, Schmidt-Langhorst C, Alreesh S, Elschner R, Frey F, Berenguer PW, Molle L, Nölle M, Schubert C. Generation, transmission and detection of 4d set-partitioning qam signals. *J Lightwave Technol* 2015;33:1445–1451.

113. Karout J, Liu X, Chandrasekhar S, Agrell E, Karlsson M, Essiambre R-J. Experimental demonstration of an optimized 16-ary four-dimensional modulation format using optical OFDM. Optical Fiber Communication Conference; 2013. p OW3B.4.

114. Bülow H, Rahman T, Buchali F, Idler W, Kuebart W. Transmission of 4-D modulation formats at 28-Gbaud. Optical Fiber Communication Conference; 2013. p JW2A.39.

115. Chagnon M, Osman M, Zhuge Q, Xu X, Plant DV. Analysis and experimental demonstration of novel 8PolSK-QPSK modulation at 5 bits/symbol for passive mitigation of nonlinear impairments. *Opt Express* 2013;21(25):30204–30220.

116. Bülow H, Lu X, Schmalen L, Klekamp A, Buchali F. Experimental performance of 4d optimized constellation alternatives for PM-8QAM and PM-16QAM. Optical Fiber Communication Conference; 2014. p M2A.6.

117. Eriksson TA, Johannisson P, Sjödin M, Agrell E, Andrekson PA, Karlsson M. Frequency and polarization switched QPSK. European Conference on Optical Communication; 2013. p Th.2.D.4.

118. Eriksson TA, Johannisson P, Agrell E, Andrekson PA, Karlsson M. Biorthogonal modulation in 8 dimensions experimentally implemented as 2PPM-PS-QPSK. Optical Fiber Communication Conference; 2014. p W1A.5.

119. Shiner AD, Reimer M, Borowiec A, Oveis Gharan S, Gaudette J, Mehta P, Charlton D, Roberts K, O'Sullivan M. Demonstration of an 8-dimensional modulation format with reduced inter-channel nonlinearities in a polarization multiplexed coherent system. *Opt Express* 2014;22(17):20366–20374.

120. Puttnam BJ, Eriksson TA, Delgado Mendinueta J-M, Luís RS, Awaji Y, Wada N, Karlsson M, Agrell E. Modulation formats for multi-core fiber transmission. *Opt Express* 2014;22(26):32457–32469.

121. Liu X, Chandrasekhar S, Wood TH, Tkach RW, Winzer PJ, Burrows EC, Chraplyvy AR. M-ary pulse-position modulation and frequency-shift keying with additional

polarization/phase modulation for high-sensitivity optical transmission. *Opt Express* 2011;19(26):B868–B881.

122. Liu X, Wood TH, Tkach RW, Chandrasekhar S. Demonstration of record sensitivities in optically preamplified receivers by combining PDM-QPSK and M-ary pulse-position modulation. *J Lightwave Technol* 2012;30(4):406–413.
123. Ludwig A, Schulz M-L, Schindler P, Schmogrow R, Mustafa A, Moos B, Brunsch S, Dippon T, Malsam D, Hillerkuss D, Roos F, Freude W, Koos C, Leuthold J. Stacking PS-QPSK and 64PPM for long-range free-space transmission. Advanced Photonics Congress; 2013. p NW2C.2.
124. Sjödin M, Eriksson TA, Andrekson PA, Karlsson M. Long-haul transmission of PM-2PPM-QPSK at 42.8 Gbit/s. Optical Fiber Communication Conference; 2013.
125. Savory SJ. Digital filters for coherent optical receivers. *Opt Express* 2008;16(2):804–817.
126. Kuschnerov M, Chouayakh M, Piyawanno K, Spinnler B, De Man E, Kainzmaier P, Alfiad MS, Napoli A, Lankl B. Data-aided versus blind single-carrier coherent receivers. *IEEE Photonics J* 2010;2(3):387–403.
127. Elschner R, Frey F, Meuer C, Fischer JK, Alreesh S, Schmidt-Langhorst C, Molle L, Takahito T, Schubert C. Experimental demonstration of a format-flexible single-carrier coherent receiver using data-aided digital signal processing. *Opt Express* 2012;20(27):28786–28791.
128. Yan M, Tao Z, Zhang H, Yan W, Hoshida T, Rasmussen J. Adaptive blind equalization for coherent optical BPSK system. European Conference on Optical Communication; 2010. p Th.9.A.4.
129. Johannisson P, Sjödin M, Karlsson M, Wymeersch H, Agrell E, Andrekson PA. Modified constant modulus algorithm for polarization-switched QPSK. *Opt Express* 2011;19(8):7734–7741.
130. Millar DS, Savory SJ. Blind adaptive equalization of polarization-switched QPSK modulation. *Opt Express* 2011;19(9):8533–8538.
131. Alreesh S, Fischer JK, Nolle M, Schubert C. Joint-polarization carrier phase estimation for PS-QPSK signals. *IEEE Photonics Technol Lett* 2012;24(15):1282–1284.
132. Bülow H. Ideal POL-QAM modulation for coherent detection schemes. Signal Processing in Photonic Communications; 2011. p SPTuB.1.
133. Bülow H, Masalkina ES. Coded modulation in optical communications. Optical Fiber Communication Conference; 2011. p OThO1.
134. Chen C, Li C, Zamani M, Zhang Z. Coherent detection of a 32-point 6PolSK-QPSK modulation format. Optical Fiber Communication Conference; 2013. p OTh3C.4.
135. Conway JH, Sloane NJA. Soft decoding techniques for codes and lattices, including the Golay code and the Leech lattice. *IEEE Trans Inf Theory* 1986;32(1):41–50.
136. Shannon CE. Communication in the presence of noise. *Proc IRE* 1949;37(1):10–21.
137. Shannon CE. Probability of error for optimal codes in a Gaussian channel. *Bell Syst Tech J* 1959;38(3):611–656.
138. Sjödin M, Agrell E, Johannisson P, Lu G-W, Andrekson PA, Karlsson M. Filter optimization for self-homodyne coherent WDM systems using interleaved polarization division multiplexing. *J Lightwave Technol* 2011;29(9):1219–1226.
139. Benedetto S, Poggiolini P. Theory of polarization shift keying modulation. *IEEE Trans Commun* 1992;40(4):708–721.

140. Karlsson M. Four-dimensional rotations in coherent optical communications. *J Lightwave Technol* 2014;32(6):1246–1257.

141. Chandrasekhar S, Liu X, Zhu B, Peckham DW. Transmission of a 1.2-Tb/s 24-carrier no-guard-interval coherent OFDM superchannel over 7200-km of ultra-large-area fiber. European Conference on Optical Communications; 2009. p Th.13.B.1.

142. Edfors O, Johansson AJ. Is orbital angular momentum (OAM) based radio communication an unexploited area? *IEEE Trans Antennas Propag* 2012;60(2):1126–1131.

143. Bülow H, Al-Hashimi H, Schmauss B. Coherent multimode-fiber MIMO transmission with spatial constellation modulation. European Conference on Optical Communications; 2011. p Tu.5.B.3.

144. Puttnam BJ, Delgado Mendinueta J-M, Luis RS, Klaus W, Sakaguchi J, Awaji Y, Wada N, Eriksson TA, Agrell E, Andrekson PA, Karlsson M. Energy-efficient modulation formats for multi-core fibers. Opticoelectronics and Communications Conference, OECC/ACOFT; 2014. p We9B.1.

145. Kolner B. Space-time duality and the theory of temporal imaging. *IEEE J Quantum Electron* 1994;30(8):1951–1963.

3

ADVANCES IN DETECTION AND ERROR CORRECTION FOR COHERENT OPTICAL COMMUNICATIONS: REGULAR, IRREGULAR, AND SPATIALLY COUPLED LDPC CODE DESIGNS

LAURENT SCHMALEN[1], STEPHAN TEN BRINK[2], AND ANDREAS LEVEN[1]

[1] *Bell Labs, Nokia, Stuttgart, Germany*
[2] *Institute of Telecommunications, University of Stuttgart, Stuttgart, Germany*

3.1 INTRODUCTION

Forward error correction (FEC) in optical communications has been first demonstrated in 1988 [1]. Since then, coding technology has evolved significantly. This pertains not only to the codes but also to encoder and decoder architectures. Modern high-speed optical communication systems require high-performance FEC engines that support throughputs of 100 Gbit/s or multiples thereof, that have low power consumption, that realize *net coding gains* (NCGs) close to the theoretical limits at a target *bit error rate* (BER) of less than 10^{-15}, and that are preferably adapted to the peculiarities of the optical channel.

FEC coding is based on deterministically adding redundant bits to a source information bit sequence. After transmission over a noisy channel, a decoding system tries to exploit the redundant information for fully recovering the source information. Several methods for generating the redundant bit sequence from the

Enabling Technologies for High Spectral-efficiency Coherent Optical Communication Networks, First Edition. Edited by Xiang Zhou and Chongjin Xie.
© 2016 John Wiley & Sons, Inc. Published 2016 by John Wiley & Sons, Inc.

source information bits are known. Transmission systems with 100 Gbit/s and 400 Gbit/s today typically use one of two coding schemes to generate the redundant information: *block-turbo codes* (BTCs) or *low-density parity-check* (LDPC) codes. In coherent systems, so-called soft information is usually readily available and can be used in high-performance systems within a soft-decision decoder architecture. Soft-decision information means that no binary 0/1 decision is made before entering the FEC decoder. Instead, the (quantized) samples are used together with their statistics to get improved estimates of the original bit sequence. This chapter focuses on soft-decision decoding of LDPC codes and the evolving spatially coupled (SC) LDPC codes.

In coherent optical communications, the signal received after carrier recovery may be affected by distortions that are different from those that commonly occur in wireless communications. For instance, the signal at the input of the signal space demapper may be affected by phase slips (also called *cycle slips* [2]), with a probability depending on the nonlinear phase noise introduced by the optical transmission link [3]. The phase slips are *not* an effect of the physical waveform channel but, rather, an artifact of coarse blind phase recovery algorithms with massive parallelization at the initial digital signal processing (DSP) receiver steps [4]. If such a phase slip is ignored, error propagation will occur at the receiver and all data following the phase slip cannot be properly recovered. Several approaches to mitigate phase slips have been proposed. Of these, the most common is differential coding, rendering a phase slip into a single error event. In order to alleviate the penalty caused by differential coding, iterative decoding between an FEC decoder and a differential decoder can be beneficial [5]. This solution leads, however, to an increased receiver complexity, as several executions of a soft-input soft-output differential decoder (usually based on the BCJR algorithm[1]) have to be carried out.

In this chapter, we first show how the use of differential coding and the presence of phase slips in the transmission channel affect the total achievable information rates and capacity of a system. By means of the commonly used *quadrature phase-shift keying* (QPSK) modulation, we show that the use of differential coding *does not decrease the capacity*, that is, the total amount of reliably conveyable information over the channel remains the same. It is a common misconception that the use of differential coding introduces an unavoidable "differential loss." This *perceived* differential loss is rather a consequence of simplified differential detection and decoding at the receiver. Afterwards, we show how capacity-approaching coding schemes based on LDPC and spatially coupled LDPC codes can be constructed by combining iterative demodulation and decoding. For this, we first show how to modify the differential decoder to account for phase slips and then how to use this modified differential decoder to construct good LDPC codes. This construction method can serve as a blueprint to construct good and practical LDPC codes for other applications with iterative detection, such as higher-order modulation formats with nonsquare constellations [7], multidimensional optimized modulation formats [8], turbo equalization

[1] Termed after the initial letters of its inventors Bahl, Cocke, Jelinek, and Raviv [6].

to mitigate ISI (e.g., due to nonlinearities) [9, 10], and many more. Finally, we introduce the class of *spatially coupled* (SC)-LDPC codes, which are a specialization of LDPC codes with some outstanding properties and which can be decoded with a very simple windowed decoder. We show that the universal behavior of spatially coupled codes makes them an ideal candidate for iterative differential demodulation/detection and decoding.

This chapter is structured as follows: In Section 3.2, we formally introduce the notation, system model, and differential coding. We highlight some pitfalls that one may encounter when phase slips occur on the equivalent channel. We propose a modified differential decoder that is necessary to construct a capacity-approaching system with differential coding. In Section 3.3, we introduce LDPC codes and iterative detection. We highlight several possibilities of realizing the interface between the LDPC decoder and the detector, and give design guidelines for finding good degree distributions of the LDPC code. We show that with iterative detection and LDPC codes, the differential loss can be recovered to a great extent. Finally, in Section 3.4, we introduce SC-LDPC codes and show how a very simple construction can be used to realize codes that outperform LDPC codes while having similar decoding complexity.

3.2 DIFFERENTIAL CODING FOR OPTICAL COMMUNICATIONS

In this section, we describe and study the effect of differential coding on coherent optical communication systems and especially on the maximum conveyable information rate (the so-called *capacity*). We assume a simple, yet accurate channel model based on *additive white Gaussian noise* (AWGN) and random phase slips. We start by giving a rigorous description of higher-order modulation schemes frequently used in coherent communications and then introduce in Section 3.2.2 the channel model taking into account phase slips that are due to imperfect phase estimation in the coherent receiver. We then introduce differential coding and show how the differential decoder has to be modified in order to properly take into account phase slips. We show that differential coding as such does not limit the capacity of a communication system, provided that an adequate receiver is used.

3.2.1 Higher-Order Modulation Formats

In this section, the interplay of coding and modulation is discussed in detail. We only take on an IQ-perspective of digital modulation, representing digital modulation symbols as complex numbers. The sequence of complex numbers (where I denotes the real part and Q the imaginary part) is then used to generate the actual waveform (taking into account pulse shaping and eventually electronic predistortion), that is, to drive the optical modulators generating the I and Q components. For a thorough overview of coding and modulation in the context of coherent communications, we refer the interested reader to [11, 12].

When talking about digital modulation, especially in the context of coded modulation, we are mostly interested in the *mapping function*, which is that part of the

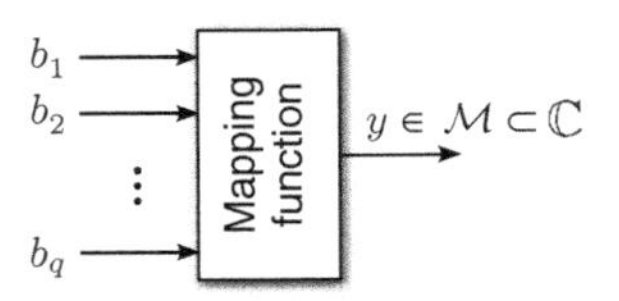

FIGURE 3.1 Mapping of a group of q bits $(b_1, b_2, \ldots, b_q)$ to a modulation symbol $y \in \mathbb{C}$.

modulator that assigns (complex) modulation symbols to bit patterns. We introduce in what follows the notation necessary for describing the mapping function. Let q denote the number of bits that are assigned to one complex modulation symbol $y \in \mathbb{C}$, and let $\boldsymbol{b} = (b_1, b_2, \ldots, b_q) \in \mathbb{F}_2^q$ be a binary q-tuple with $\mathbb{F}_2 = \{0, 1\}$ denoting the field of binary numbers. The one-to-one modulation mapping function $y = \phi(\boldsymbol{b})$ maps the q-tuple $\boldsymbol{b}$ to the (complex) modulation symbol $y \in \mathbb{C}$, where y is chosen from the set of $Q = 2^q$ modulation symbols $\mathcal{M} = \{M_0, M_1, \ldots, M_{Q-1}\}$. The set $\mathcal{M}$ is also commonly referred to as *constellation*. The mapping function is illustrated in Figure 3.1. In this chapter, we only consider one-to-one mappings. One such mapping is $\phi_{\text{Nat}}(\boldsymbol{b}) = \phi_{\text{Nat}}(b_1, b_2, \ldots, b_q) = M_{(b_1 b_2 \ldots b_q)_{10}}$, where $(b_1 b_2 \ldots b_q)_{10}$ denotes the decimal expansion of the binary q-digit number $b_1 b_2 \ldots b_q$.

In the context of differential coding of higher-order modulation formats, it is advantageous if the constellation $\mathcal{M}$ fulfills certain properties. One such property is the *rotational invariance* of the constellation.

Definition 1 (Rotational Invariance of Constellation) *We say that a constellation $\mathcal{M} = \{M_0, M_1, \ldots, M_{Q-1}\}$ exhibits a V-fold rotational invariance if we recover the original constellation $\mathcal{M}$ after rotating each modulation symbol M_i by an amount $\frac{2\pi}{V}k$, $\forall k \in \{1, \ldots, V\}$ in the complex plane. Formally, we say that a constellation exhibits a V-fold rotational invariance if (with $\imath = \sqrt{-1}$)*

$$\{M_i \cdot e^{\imath \frac{2\pi}{V} k} : M_i \in \mathcal{M}\} = \mathcal{M} \qquad \textit{for all} \quad k \in \{1, \ldots, V\}$$

EXAMPLE 3.1

Consider the two constellations with 8 and 16 points shown in Figure 3.2. The rectangular 8QAM (*quadrature amplitude modulation*) constellation of Figure 3.2(a) has a $V = 2$-fold rotational invariance as any rotation of the constellation by π leads again to the same constellation. The 16QAM constellation shown in Figure 3.2(b) exhibits a $V = 4$-fold rotational invariance as any rotation of the constellation by $\frac{\pi}{2}$ leads again to the same constellation.

Before introducing differential coding and modulation, we first describe the channel model including phase slips.

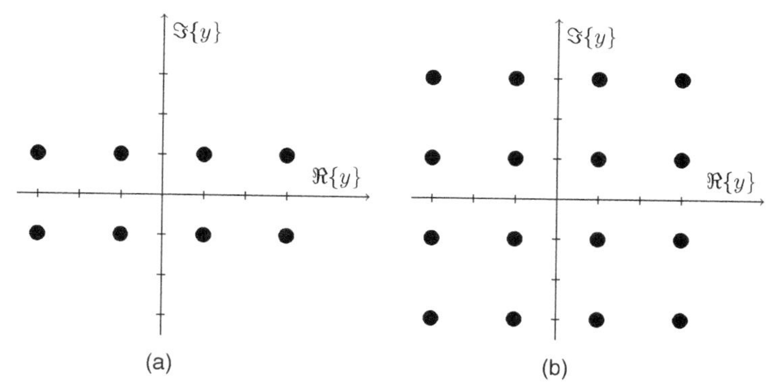

FIGURE 3.2 Two common higher-order constellations: (a) Rectangular 8QAM with $Q = 8$ and $V = 2$, and (b) 16QAM with $Q = 16$ and $V = 4$.

3.2.2 The Phase-Slip Channel Model

In coherent receivers for high-speed optical communications, it is usually not feasible to employ decision-directed blind phase recovery [4] so that usually, feed-forward phase recovery algorithms have to be employed. Feed-forward carrier recovery algorithms exploit the rotational invariance of the constellation to remove the modulation before estimating the phase. However, due to the necessary phase unwrapping algorithm in the feed-forward phase estimator, a phenomenon called *phase slip* occurs.[2] These are mostly due to coarse blind phase recovery algorithms with massive parallelization, including preliminary hard decisions and phase unwrapping at the initial *DSP* receiver steps [4].

Figure 3.3 displays the phase-slip channel model we employ in the following. The channel input is a complex modulation symbol $y[t] \in \mathcal{M} \subset C$. The first noise contribution is complex-valued AWGN. In the field of coding and in the broad body of literature on FEC, the terms E_s/N_0 and E_b/N_0 are frequently used to characterize AWGN channels. Therein, $E_s = \mathsf{E}\{|y|^2\}$ denotes the energy per modulation symbol.[3] The noise $n[t] = n_I[t] + \imath n_Q[t]$ (where $\imath = \sqrt{-1}$) is characterized by the two-sided noise power spectral density $N_0 = 2\sigma_n^2$, where σ_n^2 is the variance of both noise components $n_I[t]$ and $n_Q[t]$, that is, $\sigma_n^2 = \mathsf{var}(n_I[t]) = \mathsf{var}(n_Q[t])$. The received symbol $z[t]$ in our model is obtained by $z[t] = (y[t] + n[t]) \cdot p[t]$, where $p[t]$ describes the phase slips. Phase slips and $p[t]$ are discussed in detail later.

Frequently, especially for comparing different coding schemes, E_b/N_0 is used instead of E_s/N_0. Herein, E_b denotes the energy per *information bit*, whereas E_s denotes the energy per *transmit symbol*. For example, if a code of rate $r = 4/5$, corresponding to an overhead of $\Omega = \frac{1}{r} - 1 \doteq 25\%$, is used, the ratio of n code bits

[2]Sometimes, phase slips are also denoted as *cycle slips*; however, we employ the term phase slip in this chapter.

[3]Note that in this chapter we use lowercase letters to denote random variables as well as their realizations to avoid confusion, unless it is not clear from the context.

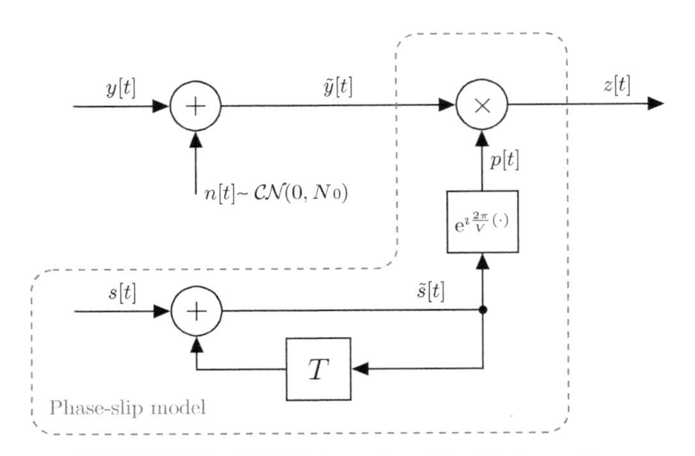

FIGURE 3.3 AWGN channel model with phase slips.

versus k information bits amounts to $n/k = 5/4 = 1.25$, that is, $1.25 = 1/r$ code bits are transmitted for each information bit. Thereof, q code bits are assigned to one modulation symbol $y[t]$. This means that if the modulation symbols are transmitted each with energy E_s, the amount of energy conveyed by each information bit amounts to

$$E_s = E_b \cdot q \cdot r \quad \Longleftrightarrow \quad E_b = \frac{E_s}{q \cdot r}$$

As E_b is normalized with respect to the information bits of the transmission system, it allows us to immediately evaluate the NCG. The NCG is frequently used to assess the performance of a coding scheme and is defined as the difference (in decibels) of required E_b/N_0 values between coded and uncoded transmission for a given output BER. Note that the NCG takes into account the coding rate r and the number of bits assigned to each modulation symbol, which are included in E_b.

In optical communications, the *optical signal-to-noise ratio* (OSNR) is also frequently employed. The OSNR is the signal-to-noise ratio (SNR) measured in a reference optical bandwidth, where frequently a bandwidth B_{ref} of 12.5 GHz is used corresponding to 0.1 nm wavelength. The OSNR relates to the E_s/N_0 and E_b/N_0 as

$$\text{OSNR}\Big|_{\text{dB}} = \frac{E_s}{N_0}\Big|_{\text{dB}} + 10\log_{10}\frac{R_S}{B_{\text{ref}}} = \frac{E_b}{N_0}\Big|_{\text{dB}} + 10\log_{10}\frac{q \cdot r \cdot R_S}{B_{\text{ref}}}$$

where B_{ref} is the previously introduced reference bandwidth, R_S corresponds to the symbol rate of the transmission, r is the aforementioned rate of the code with $r = k/n$, and q corresponds to the number of bits mapped to each modulation symbol.

Returning to the description of the channel model of Figure 3.3, we see that the noisy signal $\tilde{y}[t]$ additionally undergoes a potential phase rotation yielding $z[t]$. If the constellation shows a V-fold rotational invariance with V even (which is the case for most of the practically relevant constellations), we introduce the following probabilistic phase-slip model

$$P(s[t] = \pm 1) = \xi$$
$$P(s[t] = \pm 2) = \xi^2$$
$$\vdots$$
$$P\left(s[t] = \pm \frac{V}{2}\right) = \xi^{V/2}$$

The probability that a phase slip occurs is thus

$$P_{\text{slip}} = 2\sum_{i=1}^{V/2} \xi^i = 2\left(\frac{1-\xi^{\frac{V}{2}+1}}{1-\xi} - 1\right) = \frac{2\xi\left(1-\xi^{V/2}\right)}{1-\xi} \tag{3.1}$$

For a given phase-slip probability, which may be obtained from measurements [2], and which may also depend on the nonlinear phase noise introduced by the optical transmission link and the variance of the additive Gaussian noise due to amplification, we obtain the value ξ by solving (3.1) for ξ. For the practically most important cases with $V = 2$, and $V = 4$, we get

$$\xi = \begin{cases} \frac{P_{\text{slip}}}{2}, & \text{if } V = 2 \\ \frac{\sqrt{2P_{\text{slip}}+1}}{2} - \frac{1}{2}, & \text{if } V = 4 \end{cases} \tag{3.2}$$

Experimental measurements [13] suggest that the phase-slip probability depends on the equivalent BER before the FEC decoder. Such a dependency was also suggested in Ref. [3]. We may thus model P_{slip} empirically as

$$P_{\text{slip}} = \begin{cases} \min\left(1, \frac{\gamma}{2}\text{erfc}\left(\sqrt{\frac{E_s}{N_0}}\right)\right), & \text{for BPSK} \\ \min\left(1, \frac{\gamma}{2}\text{erfc}\left(\sqrt{\frac{E_s}{2N_0}}\right)\right), & \text{for QPSK} \end{cases} \tag{3.3}$$

where γ is the factor between slip rate and pre-FEC BER for the equivalent BPSK channel. Given E_s/N_0 and γ, we can compute P_{slip} from (3.3) and subsequently ξ from (3.2) or (3.1). Using ξ, we can use a pseudo-random number generator to generate a sequence of $s[t]$ with the probability mass function defined earlier.

3.2.3 Differential Coding and Decoding

Several approaches to mitigate phase slips have been proposed in the literature. Probably the most common is differential coding, rendering a phase slip into a single error event. In this section, we restrict ourselves for simplicity to constellations with a V-fold rotational invariance where $V = 2^v$, $v \in \mathbb{N}$, that is, $V \in \{2, 4, 8, 16, \ldots\}$.

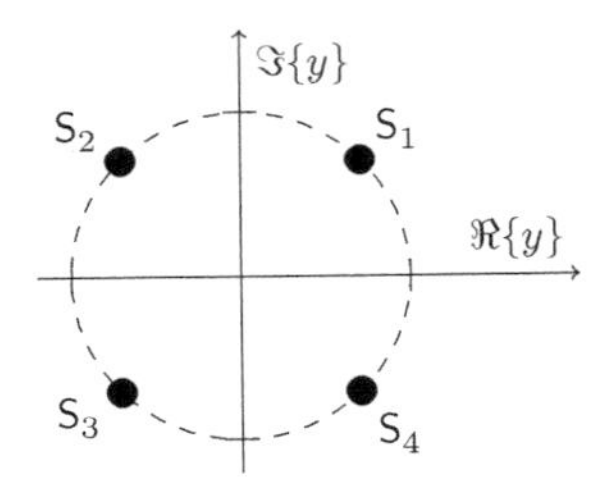

FIGURE 3.4 Example of a QPSK constellation with state assignment in rotational order.

We consider two different cases:

1. In the first case, we have $V = Q$. To each constellation point, we assign a state S_i, $i \in \{1, \ldots, V\}$. An example of such a constellation is the widely used QPSK constellation with $V = 4$, which is shown in Figure 3.4 together with its state assignment.
2. In the second case, we have $Q > V$. We restrict ourselves to the practical case with $Q = J \cdot V$, where J is an integer number. In this case, we employ differential coding as described in Ref. [14]: The constellation is divided into V disjoint regions such that these regions are preserved when rotating the constellation by $\pm\frac{2\pi}{V}$. We assign a state label S_i to each disjoint region. The regions are selected such that each region contains exactly $J = \frac{Q}{V} = 2^j$ constellation points and such that a rotation of the constellation by an angle $\kappa \cdot \frac{2\pi}{V}$, $\kappa \in \{0, \pm 1, \pm 2, \ldots\}$ does neither change the regions nor the assignment of points to a region. For the constellation points within each region, we employ a *rotationally invariant* bit mapping, which means that the bit mapping of points inside a region is not changed by a rotation of the constellation by an angle $\kappa \cdot \frac{2\pi}{V}$. The popular 16QAM constellation is an example of such a constellation with $Q = 16$, $V = 4$, and $J = 4$. The state assignment and rotationally invariant mapping are exemplarily discussed in Example 3.2 and shown in Figure 3.5.

EXAMPLE 3.2

We consider the transmission of the popular 16QAM constellation [15]. It can be easily verified that the 16QAM constellation shows a $V = 4$-fold rotational invariance. As shown in Figure 3.5, we label the four quadrants of the complex plane by *states* S_1, S_2, S_3, and S_4. Inside the first quadrant S_1, we employ a Gray labeling (also denoted by *mapping*) to assign the bits b_3 and b_4 to the four points. The mapping of the bits b_3 and b_4 in the three remaining quadrants is obtained by applying a rotational invariant mapping, that is, by rotating the Gray mapping of S_1 by multiples of $\frac{\pi}{2}$. In this case, even by rotating the constellation by multiples of $\frac{\pi}{2}$, the bits b_3 and b_4 can always be recovered unambiguously.

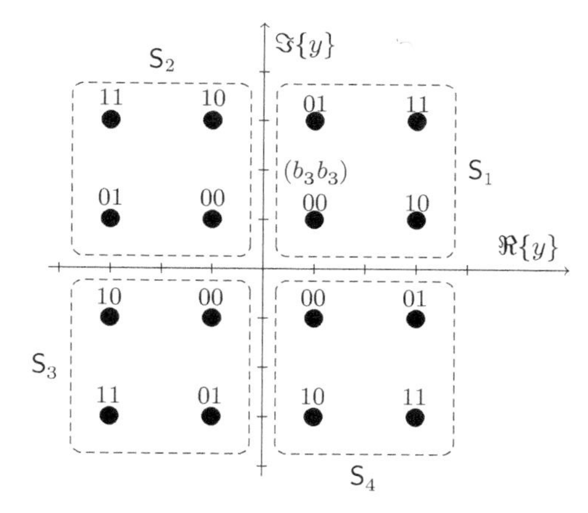

FIGURE 3.5 Differential coding for the 16QAM constellation ($V = 4$) with rotational invariant bit mapping in each quadrant.

We employ differential coding with $v = \log_2(V)$ bits to encode and reliably transmit the region, that is, the state. Within each of these regions, exactly Q/V constellation points are placed, to which a *rotationally invariant* bit mapping is assigned. This means that whenever the constellation is rotated by an angle that is a multiple of $\frac{2\pi}{V}$, the bit patterns assigned to constellation points within the region can still be uniquely identified. Note that we restrict ourselves to state-region assignments such that the rotation of a complete region gives another valid region, that is, $\forall i \in \{1, \ldots, V\}$, there exists a $j \in \{1, \ldots, V\}$, such that

$$\left\{ z \cdot \mathrm{e}^{\imath\kappa\frac{2\pi}{V}} : z \in \mathsf{S}_i \right\} = \mathsf{S}_j, \quad \forall \kappa \in \{0, \pm 1, \pm 2, \ldots\}$$

Note that this restriction does not impose any problems for practical systems as most of the practically relevant constellations can be described in this form. In what follows, we impose another, slightly more stringent condition on the states. We assume that the states S_i are assigned in what we denote as *rotational order*. Formally,

Definition 2 *We define a sequence of states* S_i, $i \in \{1, \ldots, V\}$ *that are assigned to a region of the complex plane, to be in* rotational order, *if and only if the following condition*

$$\left\{ z \cdot \mathrm{e}^{\imath\kappa\frac{2\pi}{V}} : z \in \mathsf{S}_i \right\} = \mathsf{S}_{((i+\kappa-1) \bmod V)+1}, \quad \forall \kappa \in \mathbb{N}$$

is fulfilled.

We can easily verify that the state assignments of the constellations given in Figures 3.4 and 3.5 are in rotational order. Again, note that the restriction of the states to be in *rotational order* does not yet impose any major constraint, as we have not yet defined an encoding map. We group the V states into the set $\mathcal{S} := \{\mathsf{S}_1, \mathsf{S}_2, \ldots, \mathsf{S}_V\}$.

The main step in differential coding is to impose memory on the modulation. We assume that the transmission starts at time instant $t = 1$. We introduce the differential memory $\mathsf{d}_t^{[\mathrm{mem}]} \in \mathcal{S}$ and set $\mathsf{d}_0^{[\mathrm{mem}]} = \mathsf{S}_1$. The differential encoder can be considered to be the function

$$
\begin{aligned}
f_{\mathrm{diff}} &: \mathcal{S} \times \mathbb{F}_2^v \rightarrow \mathcal{S} \\
&\left(\mathsf{d}_{t-1}^{[\mathrm{mem}]}, b_{t,1}, b_{t,2}, \ldots, b_{t,v}\right) \mapsto f_{\mathrm{diff}}\left(\mathsf{d}_{t-1}^{[\mathrm{mem}]}, b_{t,1}, b_{t,2}, \ldots, b_{t,v}\right) = \mathsf{d}_t^{[\mathrm{mem}]}
\end{aligned}
$$

which takes as input the bits $b_{t,1}, \ldots, b_{t,v}$ and the differential memory $\mathsf{d}_{t-1}^{[\mathrm{mem}]}$ and generates a new state that is saved in the differential memory $\mathsf{d}_t^{[\mathrm{mem}]}$. This new state $\mathsf{d}_t^{[\mathrm{mem}]}$ selects the symbol to be transmitted (if $V = Q$) or the region from which the symbol is selected using the bits $b_{t,v+1}, \ldots, b_{t,q}$. Note that the differential function is not unique but depends on the assignment of bit patterns to state transitions. Consider the example of the QPSK constellation shown in Figure 3.4. We can give two distinct differential encoding maps. The first differential encoding function is the *natural differential code*. The state transition diagram of the natural differential code is visualized in Figure 3.6 and is also given in Table 3.1. The second encoding function, baptized *Gray differential code* is given in Table 3.2. Note that all other differential coding maps for the QPSK constellation can be transformed into one of these two forms by elementary transformations of the constellation and the state assignment.

As the differential code can be understood as a Markov process, we can employ the BCJR algorithm [6] to carry out *bit-wise Maximum A Posteriori (MAP)* decoding of

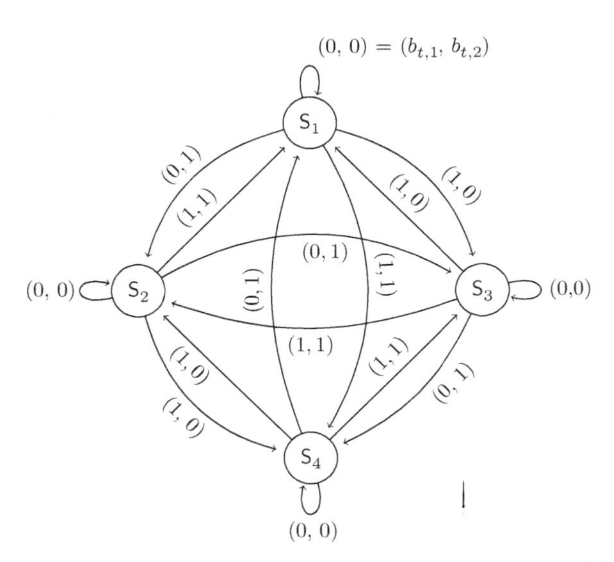

FIGURE 3.6 Differential encoding state transition diagram for the natural differential code. Arrow annotations are binary labels $(b_{t,1}, b_{t,2})$.

TABLE 3.1 Differential encoding map f_{diff} for the natural differential code

$(b_{t,1}, b_{t,2})$	$\mathsf{d}_{t-1}^{[\text{mem}]} = \mathsf{S}_1$	$\mathsf{d}_{t-1}^{[\text{mem}]} = \mathsf{S}_2$	$\mathsf{d}_{t-1}^{[\text{mem}]} = \mathsf{S}_3$	$\mathsf{d}_{t-1}^{[\text{mem}]} = \mathsf{S}_4$
(0, 0)	S_1	S_2	S_3	S_4
(0, 1)	S_2	S_3	S_4	S_1
(1, 0)	S_3	S_4	S_1	S_2
(1, 1)	S_4	S_1	S_2	S_3

TABLE 3.2 Differential encoding map f_{diff} for the Gray differential code

$(b_{t,1}, b_{t,2})$	$\mathsf{d}_{t-1}^{[\text{mem}]} = \mathsf{S}_1$	$\mathsf{d}_{t-1}^{[\text{mem}]} = \mathsf{S}_2$	$\mathsf{d}_{t-1}^{[\text{mem}]} = \mathsf{S}_3$	$\mathsf{d}_{t-1}^{[\text{mem}]} = \mathsf{S}_4$
(0, 0)	S_1	S_2	S_3	S_4
(0, 1)	S_2	S_3	S_4	S_1
(1, 0)	S_4	S_1	S_2	S_3
(1, 1)	S_3	S_4	S_1	S_2

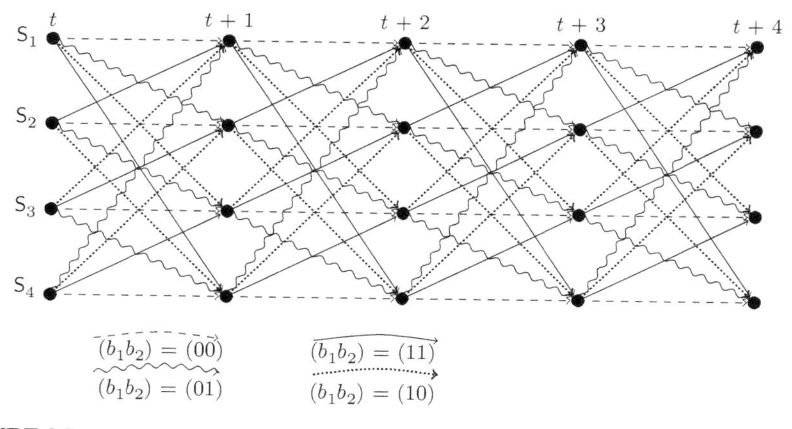

FIGURE 3.7 Trellis diagram for *natural* differential encoding of a constellation with $V = 4$ as given in Table 3.1.

the differential code. For this, we may represent the differential code using a so-called trellis diagram. The trellis diagram is an "unrolled" version of the state diagram of Figure 3.6. Figure 3.7 shows four segments of a trellis diagram for the natural differential encoding map. Four segments of the trellis diagram of the Gray differential encoding map are given in Figure 3.8. The different input bit patterns (b_1, b_2) can be distinguished by different line styles (dashed, dotted, solid, and "waved").

If phase slips occur on the channel, memory is imposed on the channel as well. If this additional memory is not properly accounted for in the BCJR decoder of the

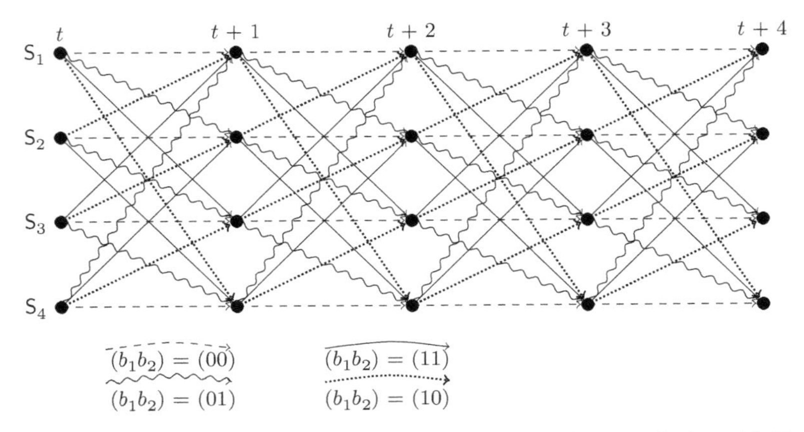

FIGURE 3.8 Trellis diagram for *Gray* differential encoding of a constellation with $V = 4$ as given in Table 3.2.

differential code, the performance of the decoder will rapidly decrease, due to the decoder not being properly adapted to the channel model, as has been observed in Ref. [16]. We therefore need to extend the trellis to properly *take into account the phase slips*. One such extension introduces additional states that correspond to the memory of the phase-slip channel [17]. We introduce states $\mathsf{S}_{i,\tilde{s}}$ where the second index $\tilde{s}$ tracks the current phase-slip state $\tilde{s}[t]$ mod 4 (see Figure 3.3), while the first index i is still responsible for describing the differential code. The occurrence of a phase slip ($s[t] \neq 0$) leads to a different $\tilde{s}[t]$. For the running example of a differential code for $V = 4$, we have no longer a trellis diagram (or a state transition diagram) with 4 states and $4 \cdot 4 = 16$ state transitions, but instead a trellis diagram with $4 \cdot 4 = 16$ states and $16 \cdot 16 = 256$ state transitions. One segment of this extended trellis diagram is shown in Figure 3.9 for the Gray differential encoding map. In order to distinguish the additional state transitions corresponding to phase slips, we use Gray scales corresponding to the probability of occurrence of the phase slips. The original trellis is obtained by utilizing only those state transitions that correspond to $s[t] = 0$, which correspond to the black lines. The state transitions corresponding to $s[t] = 1$ and $s[t] = 3$ are given by gray lines while the state transitions corresponding to $s[t] = 2$ are given by light gray lines, as these have the lowest probability of occurrence.

As the trellis diagram of Fig. 3.9 may be challenging to implement, we seek for a way to reduce its complexity. By observing that the memory of the phase-slip channel collapses with the memory of the differential encoder, we may get a more compact representation of the trellis and only need V states. This is possible as a phase slip does not introduce a new state, but only leads to a different state transition to one of the V existing states. In fact we have

$$\mathsf{d}_t^{[\mathrm{mem}]\prime} = \mathsf{S}_{((i+s[t]-1) \bmod V)+1} \quad \text{with } \mathsf{S}_i = \mathsf{d}_t^{[\mathrm{mem}]}$$

The state transitions are given exemplarily for the case of the Gray differential encoder in Table 3.3. This means that we can still use a trellis diagram with V states

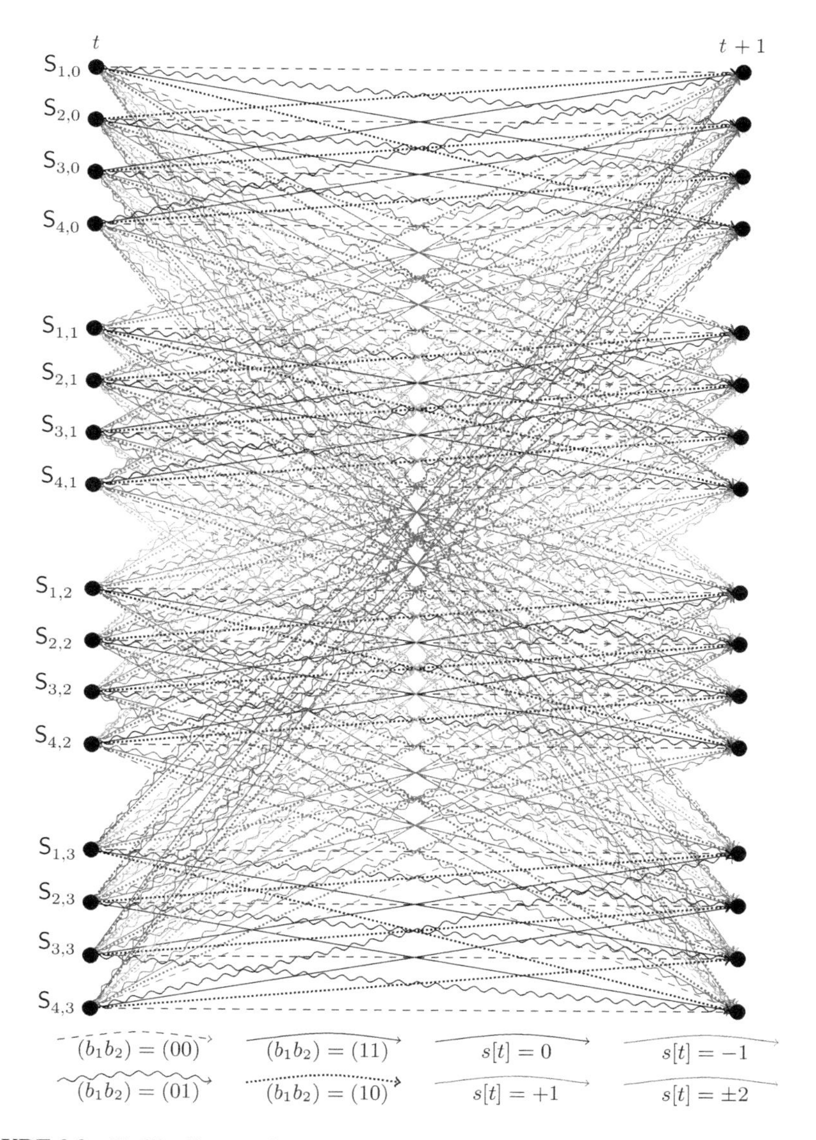

FIGURE 3.9 Trellis diagram for *Gray* differential encoding of a constellation with $V = 4$ taking into account the possible phase slips and tracking the phase-slip state (indicated by four distinct line types and gray scales).

but have to insert additional state transitions taking into account all possible values of $s[t]$. Figure 3.10 shows the simplified extended trellis diagram taking into account the possible slips, indicated by the slip value $s[t] \in \{0, 1, 2, 3\}$. Again, we use differential gray scales to represent the state transitions corresponding to different values of $s[t]$. The trellis diagram of Figure 3.10 is a simplification of the extended

TABLE 3.3 Differential encoding map f_{diff} for the Gray differential code taking into account the phase-slip variable $s[t]$

$s[t]$	$(b_{t,1}, b_{t,2})$	$\mathsf{d}_{t-1}^{[\text{mem}]} = \mathsf{S}_1$	$\mathsf{d}_{t-1}^{[\text{mem}]} = \mathsf{S}_2$	$\mathsf{d}_{t-1}^{[\text{mem}]} = \mathsf{S}_3$	$\mathsf{d}_{t-1}^{[\text{mem}]} = \mathsf{S}_4$
0	(0, 0)	S_1	S_2	S_3	S_4
0	(0, 1)	S_2	S_3	S_4	S_1
0	(1, 1)	S_3	S_4	S_1	S_2
0	(1, 0)	S_4	S_1	S_2	S_3
1	(0, 0)	S_2	S_3	S_4	S_1
1	(0, 1)	S_3	S_4	S_1	S_2
1	(1, 1)	S_4	S_1	S_2	S_3
1	(1, 0)	S_1	S_2	S_3	S_4
2	(0, 0)	S_3	S_4	S_1	S_2
2	(0, 1)	S_4	S_1	S_2	S_3
2	(1, 1)	S_1	S_2	S_3	S_4
2	(1, 0)	S_2	S_3	S_4	S_1
3	(0, 0)	S_4	S_1	S_2	S_3
3	(0, 1)	S_1	S_2	S_3	S_4
3	(1, 1)	S_2	S_3	S_4	S_1
3	(1, 0)	S_3	S_4	S_5	S_2

trellis diagram with only $V = 4$ states (instead of 16) and $4 \cdot 16 = 64$ state transitions (instead of 256). Another approach to take into account phase slips into an extended trellis has been presented in Ref. [13].

3.2.4 Maximum a Posteriori Differential Decoding

In what follows, we use the BCJR decoder [6] to carry out bit-wise *maximum a posteriori* differential decoding. The BCJR decoder makes a decision on the transmitted symbol (equivalent to a state) based on the maximization

$$\hat{\mathsf{S}}[t] = \arg \max_{\mathsf{s} \in \{\mathsf{S}_1, \ldots, \mathsf{S}_V\}} P(\mathsf{S}[t] = \mathsf{s} | z_1^{\tilde{n}})$$

At each time instant t, the most probable state S_t is computed given the complete received sequence $z_1^{\tilde{n}} = (z[1], z[2], \ldots, z[\tilde{n}])$. We do not give a complete derivation of the BCJR algorithm and refer the interested reader to the literature, for example, [6, 18]. We merely summarize the equations in the Appendix.

We use the technique of *EXtrinsic Information Transfer* (EXIT) charts [19] to characterize the behavior of the differential decoder based on the BCJR algorithm. EXIT charts plot the extrinsic output mutual information as a function of the input mutual information and are a tool to characterize single components in iterative decoders. Bit interleavers statistically decouple the respective encoding/decoding components

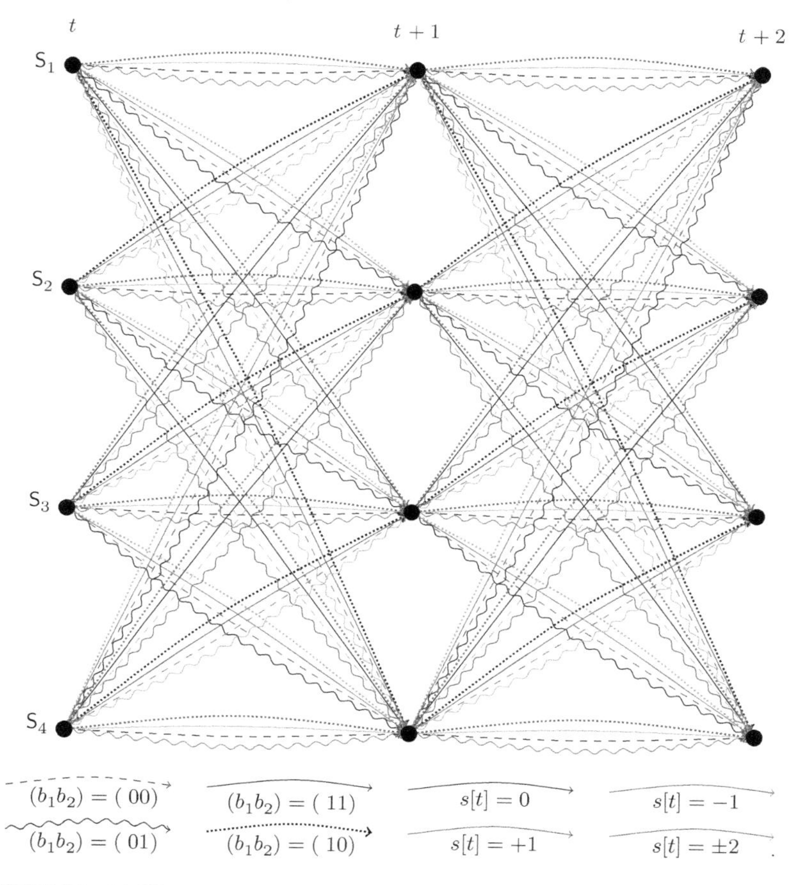

FIGURE 3.10 Trellis diagram for *Gray* differential encoding of a constellation with $V = 4$ taking into account the possible phase slip (indicated by four distinct line types and gray scales).

such that a single parameter is sufficient to track their input/output relations. This parameter may be the SNR at the output of a processing block, or, as is the case for EXIT charts, the mutual information between transmitted bits and the received and processed soft bit *log-likelihood ratio* (LLR) values. For some channels and some codes, the individual *transfer characteristics* (or EXIT curves) can be obtained analytically, while for most cases, one has to resort to Monte Carlo simulation for computing the mutual information. EXIT curves can be defined not only for channel encoders/decoders such as convolutional codes or parity-check codes, but also for components of many serially or parallel concatenated detection and decoding schemes: For example, EXIT curves have been used for describing channel interfaces such as mappers/demappers (detectors) for spectrally efficient modulation, or equalizers of multipath channels; even the decoder of an LPDC code can be viewed as a serial concatenation, with a variable node decoder and a check node decoder that, both, can be described by EXIT curves, respectively.

The main advantage of the EXIT chart technique is that the individual component processing blocks can be studied and characterized separately using EXIT curves, and that the interaction of two (or more) such processing blocks can be graphically predicted in the EXIT chart without performing a complex simulation of the actual fully fletched concatenated coding scheme itself. As it turns out, the EXIT curves must not intersect to allow convergence to low BERs, and thus, code design reduces to finding good pairs of EXIT curves that match well, or, more constructively as in the case of LDPC codes, to apply curve-fitting algorithms to determine variable and check node degree profiles that match well. A decoding trajectory visualizes the iterative exchange of information between the processing blocks, and shows the progress of the decoding.

While the EXIT chart is exact on the *binary erasure channel* (BEC) for sufficiently long/infinite sequence lengths, the reduction to single parameter tracking of the involved distributions is just an approximation for other channels. It has been observed, however, that the predicted and actually simulated decoding trajectories match quite well, proving the usefulness of the method, with many successful code designs performed in practice up to date.

Figure 3.11 shows the EXIT characteristics of the differential decoder for a QPSK constellation and both differential encoding maps. We can clearly see that the characteristic of the detector employing the nonmatched trellis diagram has a nonincreasing shape, which is an indicator of a *mismatched model* used within the decoder: the decoder trellis does not leave the possibility open for phase slips to occur, but *forces* the result to a simply differentially encoded target sequence, which is, however, not

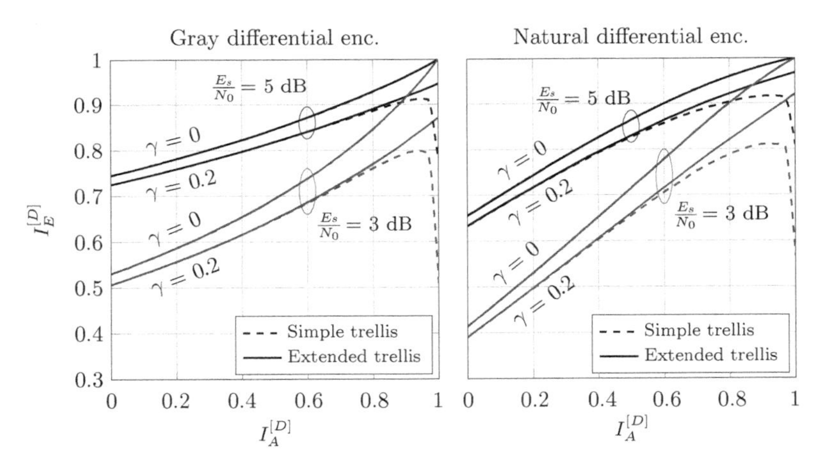

FIGURE 3.11 EXIT characteristics of differential detectors using the model-matched decoder with the trellis diagram of Figure 3.10 (solid lines, ———) and using the conventional unmatched decoder based on the trellis diagram of Figure 3.8 (or Figure 3.7, respectively) (dashed lines, – – –).

the case after the phase-slip channel. This nonincreasing shape is the reason for the error floor that has been observed in Ref. [16]. The decreasing EXIT characteristic means that during iterative decoding, the overall system performance actually decreases, which can lead to a severe error floor. In Reference [20], the authors propose to employ *hybrid turbo differential decoding* (HTDD): by a careful execution of the differential decoder only in those iterations where the extrinsic information is low enough, the operating point in the EXIT chart is kept in the range of an increasing characteristic. This approach allows the authors of [20] to mitigate the detrimental effect of phase slips on iterative differential decoding and to realize codes with relatively low error floors, which can be combated using a high-rate outer code.

If we employ the trellis diagram of Figure 3.10 incorporating the phase-slip model instead of the nonmatched trellis diagram, we can see that the EXIT characteristics are monotonically increasing, which is a prerequisite for successful decoding with low error floors. In the next section, we use the EXIT characteristics to compute the information theoretic achievable rates of the differentially encoded system. Further note that for $\gamma > 0$ (see Section 3.2.2), the value of $I_E^{[D]} < 1$, even for $I_A^{[D]} = 1$, which may entail an error floor unless the channel code is properly designed.

3.2.5 Achievable Rates of the Differentially Coded Phase-Slip Channel

According to Shannon's information theory [21, 22], the *capacity* of a communication channel is the maximum amount of information (usually expressed in terms of *bits per channel use*) that can be reliably conveyed over the channel. In information theory, the capacity is usually maximized over the input distribution of the channel. In this chapter, we are only interested in the maximum achievable information rate for uniform channel inputs y, as we do not wish to impose any constraints on the data sequence. One possibility to achieve a nonuniform channel input is the use of *constellation shaping* [23, 24], which is, however, beyond the scope of this chapter. The comparison between the achievable rate of the channel affected by phase slips and the achievable rate of the original AWGN channel shows how much performance may be lost in the presence of phase slips. In order to compute the achievable rates of the differentially encoded channel affected by phase slips, we employ the EXIT chart technique.

By utilizing a slightly modified way of computing EXIT curves of the BCJR decoder, we can also compute the achievable rates of the coded modulation schemes [25]. For this, we make use of the chain-rule of mutual information [26, 27] and compute the mutual information of the equivalent bit channel experienced by the channel decoder *after* differential detection. This can be done by (numerically, simulation-based) computing the EXIT curve $\tilde{I}_E^{[D]}$ of the differential detector using *a priori* knowledge that is modeled as coming from a BEC, and integrating over such curves. Specifically, EXIT curves such as those depicted in Figure 3.11 are determined for many different E_s/N_0-values (and several different phase-slip probability factors γ) but now with *a priori* knowledge based on a BEC model: By integration, we determine the area $q \int_0^1 \tilde{I}_E^{[D]}(I_A)dI_A$ under these curves [25–27] and obtain the respective mutual information limits that are plotted into Figures 3.12

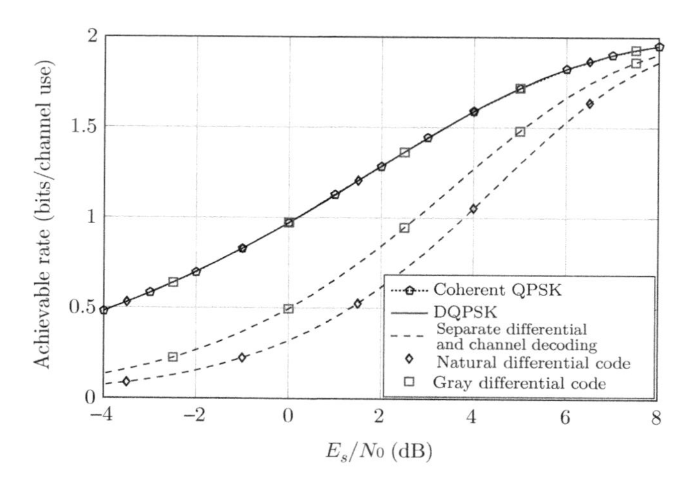

FIGURE 3.12 Achievable rates of the DQPSK channel (solid lines, ———) and of conventional separate differential and channel decoding (dashed lines, – – –) for an AWGN channel without phase slips ($\gamma = 0$).

and 3.13 at the corresponding E_s/N_0-values and phase-slip probabilities factors γ, respectively. Note that this mutual information is available to the channel decoder provided that *perfect* iterative decoding over inner differential detector and outer LDPC decoder is performed. Thus, we still need to design an appropriate LDPC code and iterative decoding scheme to actually approach these promised rates as closely as possible. Indeed, the subsequent sections explain how to construct such codes and coding schemes in more detail. The achievable rate of the noniterative system with separate differential decoding and channel decoding is obtained from $q\tilde{I}_E^{[D]}(0) = qI_E^{[D]}(0)$.

Figures 3.12 and 3.13 show the numerically computed achievable rates for the QPSK constellation *without* differential coding on an AWGN channel that is *not affected* by phase slips (dotted lines, marker "o") as a reference and additionally the achievable rates for differentially encoded QPSK for a channel affected by phase slips (solid lines) with $P_{\text{slip}} = \min\left(1, \frac{\gamma}{2}\,\text{erfc}\left(\sqrt{\frac{E_s}{2N_0}}\right)\right)$. In Figure 3.12, we set $\gamma = 0$, and we observe that the achievable rate of the differential QPSK transmission equals the achievable rate of a conventional coherent QPSK transmission, *independent* of the differential encoding map. In addition, we plot the achievable rates for a simplified system that carries out differential decoding (leading to the well-known effect of error doubling) followed by error correction decoding (dashed lines). We see that at a spectral efficiency of 1.6 (corresponding to system with $\Omega = 25\%$ overhead for coding), the simplified non-iterative system leads to an unavoidable loss in E_s/N_0 of 1.5 dB (Gray differential encoding map) or 2.5 dB (natural differential encoding map) respectively. This performance difference becomes even more severe if low spectral efficiencies (i.e., high coding overheads) are targeted.

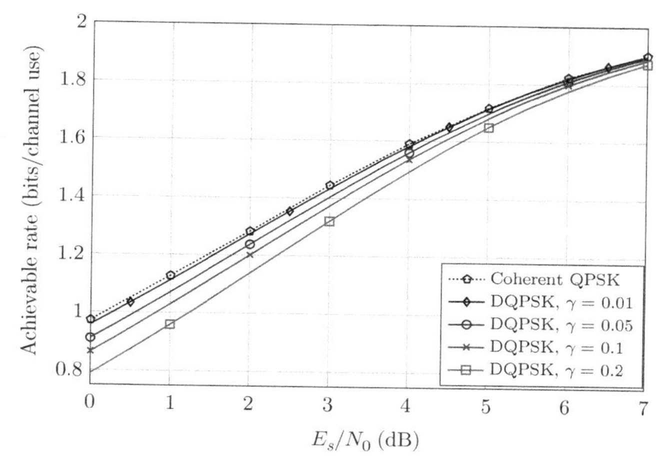

FIGURE 3.13 Capacity of the differential DQPSK system for transmission over an AWGN channel affected by phase slips with probability of occurrence depending on pre-FEC BER, given by (3.3)

If phase slips occur on the channel ($\gamma > 0$), we can observe in Figure 3.13 that for high spectral-efficiencies (above 1.5 bits/channel use), the loss in information rate due to the phase slips is not severe, unless γ becomes large. For example, for $\gamma = 0.2$, the capacity loss at a spectral efficiency of 1.5 bit/channel use is only approximately 0.7 dB. The transmission at very low spectral efficiencies, requiring codes with very large overheads, is however seriously affected by the phase-slip channel.

3.3 LDPC-CODED DIFFERENTIAL MODULATION

In the previous section, we have compared the achievable rates of various systems for an AWGN channel ($\gamma = 0$) and we have found that differential coding can be used without entailing a decrease of the communication system's achievable rate. This means that at least from an information theoretic perspective, we can employ differential coding to combat phase slips without introducing any decoding penalty. Information theory, however, does not tell us what constructive method we may use to achieve this capacity.

One particularly promising way to approach the capacity with differential coding is the use of coded differential modulation with iterative decoding, as proposed first in Ref. [5] with convolutional codes and in Ref. [28] with LDPC codes. This scheme extends the *bit-interleaved coded modulation* (BICM) [29] method to account for differential encoding and employs iterative decoding and detection [30, 31] to improve the overall system performance. The adaptation of this scheme to optical communications has been considered in Ref. [32] for the channel not affected by phase slips and in Refs [13, 16, 17, 20] for the channel affected by phase slips. Note that

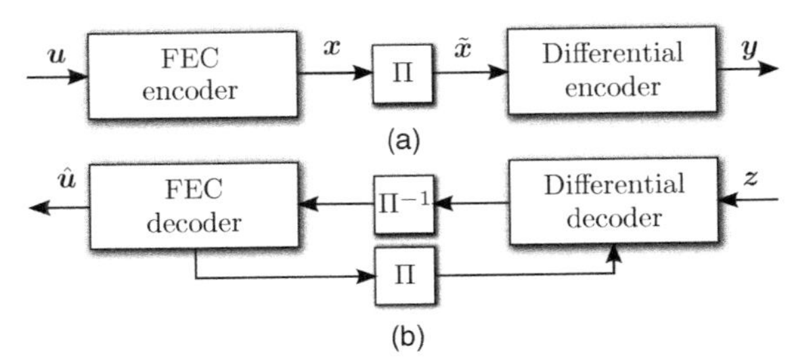

FIGURE 3.14 Block diagram of LDPC-coded differential modulation transmitter (a) with iterative detector (b).

other schemes have been proposed that do not rely on iterative differential decoding, including the slip resilient code presented in Refs [33, 34] and block differential modulation [35].

Figure 3.14 shows the general transmitter (a) and iterative receiver (b) of the coded differential modulation system with iterative decoding and detection. In this general block diagram, a block FEC encoder takes as input a binary length-k vector of input bits $\boldsymbol{u} = (u_1, u_2, \dots, u_k)$, where $u_i \in \mathbb{F}_2 = \{0, 1\}$ and generates a binary length-n vector of code bits $\boldsymbol{x} = (x_1, x_2, \dots, x_n)$. Almost all of the popular channel codes that are used in optical communications are such block codes. The amount $n - k$ of redundant bits that are added by the FEC encoder is commonly expressed in terms of the code rate r, which is defined as the ratio of the information block length k and the code dimension n, that is,

$$r := \frac{k}{n}$$

In optical communications, often the overhead is used to quantify the amount of redundant information. The overhead Ω of the code and its rate are interrelated by

$$\Omega := \frac{n}{k} - 1 = \frac{n-k}{k} = \frac{1}{r} - 1 = \frac{1-r}{r}$$

The block $\boldsymbol{x}$ of code bits is interleaved by a permutation Π to yield a permuted version $\tilde{\boldsymbol{x}}$. Ideally, a random permutation is employed, but sometimes, a structure in the permutation is necessary to facilitate implementation (parallelization) or to improve the error correction capabilities of the code. Note that the permutation Π is sometimes implicitly included in the FEC encoder and does not need to be explicitly implemented. The interleaved block $\tilde{\boldsymbol{x}}$ is differentially encoded (as discussed in Section 3.2.3) yielding a block of $\tilde{n} = \lceil \frac{n}{q} \rceil$ modulation symbols (where $\lceil \tau \rceil$ denotes the smallest integer larger or equal than τ).

At the receiver, the differential decoder and the FEC decoder iteratively decode the signal, where the output of the FEC decoder is used to yield an improved differential decoding result in a subsequent iteration by sharing so-called *extrinsic information* between the decoder components. For a thorough description and introduction to the concept of iterative detection and decoding, we refer the interested reader to [18, 36]. In the remainder of this section, we assume that the employed FEC scheme is an LDPC [18, 37] code. We first give an introduction to LDPC codes and then show how irregular LDPC codes can be designed to be well-adapted to differential coding. We do not show explicitly how decoding is performed, as we intend to take on a more code design-oriented perspective. We only give equations for performing differential decoding and LDPC decoding in the Appendix.

We restrict ourselves in the remainder of this chapter to the case where $V = Q = 2^q$, that is, every state S_i is assigned to the modulation symbol M_i. We, however, give hints on how to deal with the case $V < Q$ in Section 3.3.3.

3.3.1 Low-Density Parity-Check (LDPC) Codes

LDPC codes were developed in the 1960s by Gallager in his landmark Ph.D. thesis [37]. These codes were not further investigated for a long time due to the perceived complexity of long codes. With the discovery of turbo codes in 1993 [38] and the sudden interest in iteratively decodable codes, LDPC codes were rediscovered soon afterwards [39, 40]. In the years that followed, numerous publications from various researchers paved the way for a thorough understanding of this class of codes leading to numerous applications in various communication standards, such as WLAN (IEEE 802.11) [41], DVB-S2 [42], and 10G Ethernet (IEEE 802.3) [43]. LDPC codes for soft-decision decoding in optical communications were studied in Ref. [44]. Modern high-performance FEC systems are sometimes constructed using a soft-decision LDPC inner code that reduces the BER to a level of 10^{-3} to 10^{-5} and a hard-decision outer code that pushes the system BER to levels below 10^{-12} [44]. An outer cleanup code is used, as most LDPC codes exhibit a phenomenon called *error floor*: above a certain *SNR*, the BER does not drop rapidly anymore but follows a curve with a small slope. This effect is mainly due to the presence of *trapping sets* or *absorbing sets* [45, 46]. The implementation of a coding system with an outer cleanup code requires a thorough understanding of the LDPC code and a properly designed interleaver between the LDPC and outer code for avoiding that the errors at the output of the LDPC decoder—which typically occur in clusters—cause uncorrectable blocks after outer decoding. With increasing computing resources, it is now also feasible to evaluate very low target BERs of LDPC codes and optimize the codes to have very low error floors below the system's target BER [47]. A plethora of LDPC code design methodologies exist, each with its own advantages and disadvantages. The goal of an LDPC code designer is to find a code that yields high coding gains and possesses some structure facilitating the implementation of the encoder and decoder. We point the interested reader to numerous articles published on this topic, for example, [48–50] and references therein. An introduction to LDPC codes in the context of optical communications is given in Ref. [51]. An overview of coding schemes for optical communications is also provided in Ref. [12] and the references therein. For a thorough reference to LDPC codes together with an overview of decoding algorithms and construction methods, we refer the interested reader to [18].

An LDPC code is defined by a sparse binary parity check matrix $\boldsymbol{H}$ of dimension $m \times n$, where n is the code word length (in bits) of the code and m denotes the number of parity check equations defining the code. Usually,[4] the number of information bits equals $n - m$. The overhead of the code is defined as $\Omega = \frac{m}{n-m}$. A related measure is the *rate* of the code, which is defined as $r = \frac{n-m}{n}$. Sparse means that the number of "1"s in $\boldsymbol{H}$ is small compared with the number of zero entries. Practical codes usually have a fraction of "1"s that is below 1% by several orders of magnitude. We start by introducing some notation and terminology related to LDPC codes. Each column of the parity check matrix $\boldsymbol{H}$ corresponds to 1 bit of the FEC frame. The n single bits of the code are also often denoted as *variables*. Similarly, each row of $\boldsymbol{H}$ corresponds to a parity check equation and ideally defines a single parity bit (if $\boldsymbol{H}$ has full rank).

3.3.1.1 Regular and Irregular LDPC Codes LDPC codes are often classified into two categories: regular and irregular LDPC codes. In this chapter, we consider the latter, which also constitutes the more general, broader class of codes. The parity check matrix of regular codes has the property that the number of "1"s in each column is constant and amounts to $\mathtt{v}_{\text{reg.}}$ (called *variable degree*) and that the number of "1"s in each row is constant and amounts to $\mathtt{c}_{\text{reg.}}$ (called *check degree*). Clearly, $n \cdot \mathtt{v}_{\text{reg.}} = m \cdot \mathtt{c}_{\text{reg.}}$ has to hold and we furthermore have $r = 1 - \frac{\mathtt{v}_{\text{reg.}}}{\mathtt{c}_{\text{reg.}}}$. Irregular LDPC codes [52] have the property that the number of "1"s in the different columns of $\boldsymbol{H}$ is not constant. In this chapter, we mainly consider *column-irregular* codes, which means that only the number of "1"s in the columns is not constant but the number of "1"s in each row remains constant. The irregularity of the parity-check matrix is often characterized by the degree profile of the parity check matrix $\boldsymbol{H}$ [50].

We denote the number of columns of the parity-check matrix $\boldsymbol{H}$ with i ones by Λ_i. We say that these columns have *degree i*. Normalizing this value to the number of total bits n per codewords yields

$$L_i = \frac{\Lambda_i}{n}$$

which is the *fraction* of columns with degree i, that is, with i ones (e.g., if $L_3 = \frac{1}{2}$, half the columns of $\boldsymbol{H}$ have three "1"s).

Similarly, we can define the *check degree profile* by defining that P_j denotes the number of rows of $\boldsymbol{H}$ with exactly j "1"s. The normalized check profile is given by R_j, the *fraction* of rows with j "1"s. We have the $R_j = \frac{P_j}{m}$. In most of the codes we consider, however, *all* rows of $\boldsymbol{H}$ have the same number of $\mathtt{c}_{\text{reg.}}$ "1"s. In that case, we have $R_{\mathtt{c}_{\text{reg.}}} = 1$ and $R_1 = R_2 = \cdots = R_{\mathtt{c}_{\text{reg.}}-1} = R_{\mathtt{c}_{\text{reg.}}+1} = \cdots = R_\infty = 0$. Example 3.3 illustrates the degree distribution of such an irregular LDPC code.

[4]Provided that the parity-check matrix has full row rank, that is, $\operatorname{rank} \boldsymbol{H} = m$. If the parity-check matrix $\boldsymbol{H}$ is rank-deficient, the number of information bits $k \geq n - m$.

EXAMPLE 3.3

Consider the following LDPC code of size $n = 32$ with parity-check matrix of size $\dim \boldsymbol{H} = m \times n = 8 \times 32$, that is, of rate $r = \frac{32-8}{32} = 0.75$, corresponding to an overhead of $33.\overline{3}\%$. Note that the zeros in $\boldsymbol{H}$ are not shown for clarity.

$$\boldsymbol{H} = \left(\begin{array}{cccccccc|cccccccccccccccc|cccccccc}
1 & & & 1 & & & & & & & 1 & 1 & & & 1 & & & & & 1 & & 1 & & 1 & 1 & & & 1 & 1 & 1 & 1 & \\
1 & & & 1 & & & & & & & 1 & & & 1 & & 1 & & 1 & & & & & 1 & 1 & 1 & 1 & 1 & & 1 & & & 1 \\
& 1 & & & & & & 1 & & 1 & & 1 & & & & 1 & 1 & & 1 & & & & & 1 & & 1 & & & 1 & 1 & 1 & 1 \\
& & 1 & & 1 & & & & & 1 & & & 1 & & & 1 & & & 1 & & & 1 & 1 & & & 1 & 1 & & 1 & 1 & 1 & \\
& 1 & & & & & & 1 & 1 & & & & 1 & 1 & & & & 1 & 1 & & 1 & & & & 1 & & & 1 & 1 & 1 & & 1 \\
& & & & & 1 & 1 & & 1 & & & & 1 & & 1 & & 1 & 1 & & & & & 1 & & 1 & 1 & 1 & 1 & & & & 1 \\
& & 1 & & 1 & & & & 1 & & & & & 1 & 1 & & 1 & & & 1 & 1 & & & & 1 & & 1 & 1 & & & 1 & 1 \\
& & & & & 1 & 1 & & & 1 & 1 & 1 & & & & & & & & 1 & 1 & 1 & & & & 1 & 1 & 1 & & 1 & 1 &
\end{array}\right)$$

The first eight columns of $\boldsymbol{H}$ have two "1"s per column, that is, $\Lambda_2 = 8$. Furthermore, the middle 16 columns each contain three "1"s, that is, $\Lambda_3 = 16$. Finally, the last eight columns contain five "1"s, that is, $\Lambda_5 = 8$. Normalizing leads to

$$L_2 = \frac{\Lambda_2}{n} = \frac{1}{4}, \quad L_3 = \frac{\Lambda_3}{n} = \frac{1}{2}, \quad L_5 = \frac{\Lambda_5}{n} = \frac{1}{4}$$

Note that $L_1 = L_4 = L_6 = L_7 = \cdots = 0$. The number of "1"s in each row of $\boldsymbol{H}$ is constant and amounts to $c_{\text{reg.}} = 13$.

3.3.1.2 Graph Representation of LDPC Codes LDPC codes are often represented by a so-called *Tanner* graph [50]. This graph is an undirected bipartite graph in which the nodes can be partitioned into two disjoint sets and each edge connects a node from the first set to a node from the second set. The Tanner graph allows for an easy description of the decoding algorithm of LDPC codes, which we do not detail here. We give a summary of the iterative decoding algorithm in the Appendix.

Figure 3.15 shows the graph representation of the toy code given in Example 3.3. The circular nodes on the bottom of the graph represent the *variable nodes*, which correspond to the bits in the codeword. As each codeword contains n bits, there are n variable nodes $x_1, x_2, \ldots, x_n$. The variable node x_i has one connection to the transmission channel (arrow from the bottom) and j additional connections toward the top where j equals the number of "1"s in the ith column of $\boldsymbol{H}$. For instance, the first Λ_2 variables $x_1, \ldots, x_{\Lambda_2}$ (where $\Lambda_2 = 8$) of the code have two connections toward the graph part of the code and an additional connection from the transmission channel. As shown in Example 3.3, the variable nodes can be divided into three groups, corresponding to the degree of these variables.

The rectangular nodes on the top of the graph are the so-called *check nodes*. Each check node c_i corresponds to one of the m rows of the parity-check matrix $\boldsymbol{H}$ of the

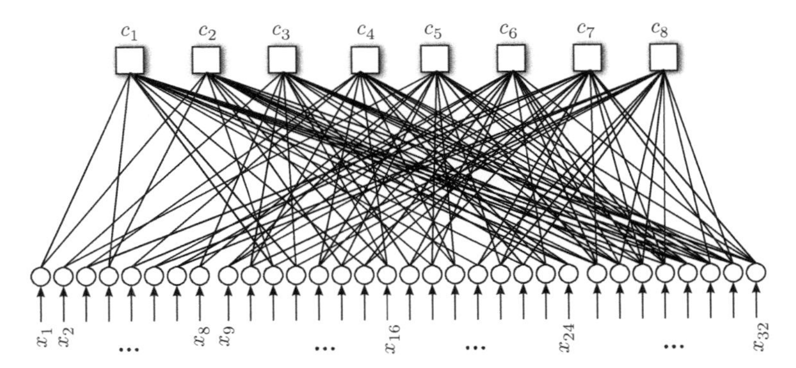

FIGURE 3.15 Graph of the code defined by the parity check matrix given in Example 3.3.

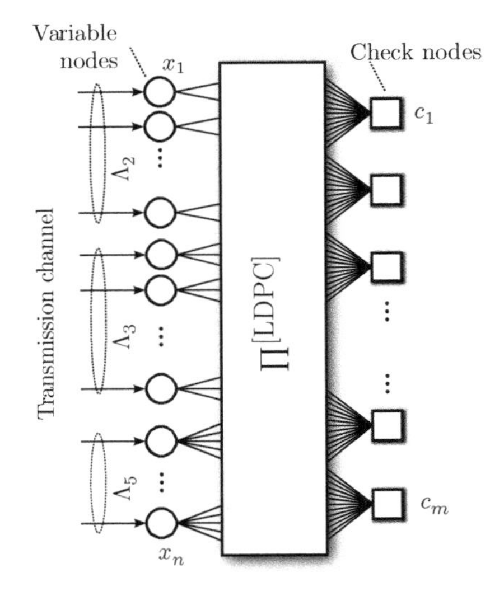

FIGURE 3.16 Simplified graph representation of an irregular LDPC code with $\mathtt{v} \in \{2,3,8\}$ and $\mathtt{c}_{\text{reg.}} = 8$.

code and defines a code *constraint*. The number of connections of the check nodes with the graph corresponds to the number of "1"s in the respective row of $\boldsymbol{H}$. In the above-mentioned example, every row has $\mathtt{c}_{\text{reg.}} = 13$ "1"s, so that each of the check nodes has exactly $\mathtt{c}_{\text{reg.}} = 13$ connected edges. If $\boldsymbol{H}$ has a nonzero entry at row i and column j, that is, $H_{i,j} = 1$, then an edge connects variable node x_j to check node c_i.

As drawing the graph of the code in this way quickly becomes cumbersome and confusing due to the large number of edges, we resort to a simplified (and rotated) representation shown in Figure 3.16. In this figure, we do not draw all the edges, but only the beginning and end of each edge and assume that the permutation of the edges is managed by an interleaver $\Pi^{[\text{LDPC}]}$. The interleaver $\Pi^{[\text{LDPC}]}$ thus ensures that the connections between the different nodes corresponds to the one given by the parity-check matrix $\boldsymbol{H}$.

3.3.1.3 Design of Irregular LDPC Codes The design of irregular LDPC codes consists of finding good degree distributions, that is, good values Λ_i and P_i (or $\mathtt{c}_{\text{reg.}}$) such that the rate of the code has the desired value (given by the system designer) and such that the NCG achievable by this code is maximized, that is, the code is able to successfully recover the bit stream at the lowest possible E_s/N_0 value. A comprehensive body of literature on the design of irregular codes exists (see [18] and references therein) and we only introduce the basics to describe the optimization of codes tailored to slip-tolerant differential decoding in Section 3.3.2.

The optimization of irregular LDPC codes requires the use of *edge-perspective* degree distributions [50].

Definition 3 (Edge-Perspective Degree Distribution) *In the Tanner graph representation of the code, we denote by λ_i the fraction of* edges *that are connected to variable nodes of degree i. We have*

$$\lambda_i = \frac{i \cdot L_i}{\sum_{j=1}^{\infty} j \cdot L_j} \tag{3.4}$$

Similarly, ρ_i denotes the fraction of edges that are connected to check nodes of degree i. Again, we have

$$\rho_i = \frac{i \cdot R_i}{\sum_{j=1}^{\infty} j \cdot R_j}$$

Using the technique of EXIT charts [19, 27, 53], good values of λ_i and potentially ρ_i may be found that can then be used to design a parity-check matrix $\boldsymbol{H}$ fulfilling these degree distributions. We constrain the maximum possible variable node degree to be $\mathtt{v}_{\max}$ and the maximum possible check node degree to be $\mathtt{c}_{\max}$.

The inverse relationship between λ_i and L_i, or between ρ_i and R_i, respectively, reads

$$L_i = \frac{\frac{\lambda_i}{i}}{\sum_{j=1}^{\mathtt{v}_{\max}} \frac{\lambda_j}{j}} \quad \text{and} \quad R_i = \frac{\frac{\rho_i}{i}}{\sum_{j=1}^{\mathtt{c}_{\max}} \frac{\rho_j}{j}} \tag{3.5}$$

The (iterative) LDPC decoding process may be understood as a process where two decoders pass information between each other. The first decoder is the *variable node decoder* (VND), which processes each of the n variable nodes of the code. The second decoder is the *check node decoder* (CND), which processes each of the m check nodes. Each of these decoders has a certain *information transfer* (EXIT) characteristic. Before describing the transfer characteristics, we introduce the J-function that interrelates mean μ (and variance, which amounts 2μ in the case of symmetric messages, for details, see [19] and [50]) and mutual information for the Gaussian random variable describing the messages that are exchanged in the iterative decoder, with

$$J(\mu) = 1 - \int_{-\infty}^{\infty} \frac{e^{-(\tau-\mu)^2/(4\mu)}}{\sqrt{4\pi\mu}} \log_2(1 + e^{-\tau})\, d\tau$$

which can be conveniently approximated [54] by

$$I = J(\mu) \approx \left(1 - 2^{-H_1 (2\mu)^{H_2}}\right)^{H_3}$$

$$\mu = J^{-1}(I) \approx \frac{1}{2}\left(-\frac{1}{H_1}\log_2\left(1 - I^{\frac{1}{H_3}}\right)\right)^{\frac{1}{H_2}}$$

with $H_1 = 0.3073$, $H_2 = 0.8935$, and $H_3 = 1.1064$.

In the case of LDPC codes and transmission over an AWGN channel, the information transfer characteristics are obtained as [55]

$$I_E^{[V]} = f_V\left(I_A^{[V]}, \frac{E_s}{N_0}\right) := \sum_{i=1}^{v_{\max}} \lambda_i J\left(4\frac{E_s}{N_0} + (i-1)J^{-1}(I_A^{[V]})\right) \tag{3.6}$$

$$I_E^{[C]} = f_C(I_A^{[C]}) := \sum_{i=1}^{c_{\max}} \frac{\rho_i}{\log(2)} \sum_{j=1}^{\infty} \frac{\left(\Phi_j\left(J^{-1}(I_A^{[C]})\right)\right)^{i-1}}{2j(2j-1)} \tag{3.7}$$

where

$$\Phi_i(\mu) = \int_{-1}^{1} \frac{2\tau^{2i}}{(1-\tau^2)\sqrt{4\pi\mu}} \exp\left(-\frac{\left(\mu - \log\frac{1+\tau}{1-\tau}\right)^2}{4\mu}\right) d\tau$$

Equation (3.6) describes the characteristic of the VND while (3.7) describes the characteristic of the CND. For codes with regular check node degree, (3.7) can be simplified to

$$I_E^{[C]} = f_C(I_A^{[C]}) := \frac{1}{\log(2)} \sum_{j=1}^{\infty} \frac{\left(\Phi_j\left(J^{-1}(I_A^{[C]})\right)\right)^{c_{\text{reg.}}-1}}{2j(2j-1)}$$

As $I_A^{[V]} = I_E^{[C]}$ holds in the context of iterative decoding, a condition for successful decoding is that

$$f_V\left(I, \frac{E_s}{N_0}\right) > f_C^{-1}(I), \forall I \in [0; 1) \tag{3.8}$$

where the inverse function $f_C^{-1}(I)$ of the strictly monotonically increasing function f_C given in (3.7) can be found using numerical methods. The task of the code designer is to find a degree distribution minimizing E_s/N_0 such that (3.8) is fulfilled. Usually, the condition (3.8) is evaluated at discrete values of I only, simplifying the optimization.

Some more conditions usually apply to the degree distributions. One of these is the so-called *stability condition* [50], which, in the case of an AWGN channel ensures that

$$\lambda_2 \leq \frac{\exp\left(\frac{E_s}{N_0}\right)}{\sum_{i=1}^{c_{\max}} \rho_i(i-1)}$$

3.3.2 Code Design for Iterative Differential Decoding

As described in Section 3.2.5, the differential decoder based on the BCJR algorithm can be characterized by an EXIT characteristic $I_E^{[D]} = f_D(I_A^{[D]}, E_s/N_0)$. Before optimizing the LDPC code toward the interworking with the differential decoding, we first have to define the *decoder scheduling* as we are concerned with a threefold iterative decoder loop: decoding iterations are carried out within the LDPC decoder and between LDPC decoder and differential decoder. In this chapter, we restrict ourselves to the following scheduling:

(a) In a first initial step, the differential decoder is executed and generates initial channel-related information.
(b) Using this initial channel-related information, a *single* LDPC iteration is carried out, that is, a single execution of the check node and variable node computing processors.
(c) Using the accumulated variable node information from the LDPC graph, *excluding* the intrinsic channel-related information from the initial differential decoding execution (step (a)), the differential decoder is executed again, yielding improved channel-related information.
(d) With the improved information from step (c), another *single* LDPC iteration is carried out. If the maximum number of allowed iterations is not yet reached, we continue with step (c).
(e) If the maximum number of iterations is reached, the accumulated variable node information is used to get an *a posteriori* estimate of each bit.

In what follows, we now describe in detail how to find good degree distributions for iterative differential decoding. In References [56] and [57], conditions for degree distributions were derived and it was analyzed if it is possible to construct codes that work equally well for differential coding and conventional nondifferential transmission. In this work, we solely consider the case of differential coding and we aim at showing different possibilities of degree distribution optimization with the goal to find the best possibility for LDPC-coded differential modulation with the above-mentioned decoder scheduling.

We only consider *column irregular* codes in the remainder of this chapter, that is, the number of "1"s in each row of the parity-check matrix $\boldsymbol{H}$ is constant and amounts to $c_{\text{reg.}}$. Such a constraint is often imposed as it simplifies the hardware that is needed to implement the check node decoding operation, which is the most difficult operation in the LDPC decoder. The complexity of this operation scales roughly linearly with the check node degree (i.e., the number of "1"s per row) and having a constant degree allows the hardware designer to implement a fixed and optimized check node computation engine. The second constraint that we impose is that we only have three different variable node degrees, namely Λ_2 variable nodes of degree 2, Λ_3 variable nodes of degree 3, and $\Lambda_{\mathrm{v}_{\max}}$ variable nodes of degree $\mathrm{v}_{\max}$. This is in line with the findings given in Ref. [58] that show that the degree distributions are often sparse and that only a few different values are often sufficient. Having only three different variable node degrees simplifies the hardware implementation, especially the design of the required bit widths in a fixed point implementation.

Contrary to many degree distribution approaches proposed in the literature [50, 56, 59], we first fix the rate r of the final code as the rate is usually constrained by the system design parameters (e.g., speed of analog-to-digital and digital-to-analog converters, pulse shape, channel bandwidth, framing overhead). With fixed rate r, we remove the dependencies [58] of the degree distribution. We further assume that no nodes of degree 1 are present in the code, that is, $\Lambda_1 = 0$ and thus $\lambda_1 = 0$. As $\sum_i \lambda = 1$, we can uniquely determine λ_2 as

$$\lambda_2 = 1 - \sum_{i=3}^{\mathrm{v}_{\max}} \lambda_i \tag{3.9}$$

As the rate of the code is given by [50]

$$r = 1 - \frac{\sum_{i=1}^{\mathrm{c}_{\max}} \frac{\rho_i}{i}}{\sum_{i=1}^{\mathrm{v}_{\max}} \frac{\lambda_i}{i}} \tag{3.10}$$

we can eliminate another dependency and by combining (3.10) with (3.9), we get

$$\lambda_3 = 3 + 6 \sum_{i=4}^{\mathrm{v}_{\max}} \lambda_i \left(\frac{1}{i} - \frac{1}{2} \right) - \frac{6}{1-r} \sum_{i=1}^{\mathrm{c}_{\max}} \frac{\rho_i}{i} \tag{3.11}$$

For check-regular codes with regular check node degree $\mathrm{c}_{\mathrm{reg.}}$ (i.e., $\rho_{\mathrm{c}_{\mathrm{reg.}}} = 1$), (3.11) can be simplified to

$$\lambda_3 = 3 - 6 \left(\frac{1}{\mathrm{c}_{\mathrm{reg.}}(1-r)} - \sum_{i=4}^{\mathrm{v}_{\max}} \lambda_i \left(\frac{1}{i} - \frac{1}{2} \right) \right) \tag{3.12}$$

This means that λ_2 and λ_3 are uniquely determined by $\lambda_4, \lambda_5, \ldots, \lambda_{\mathrm{v}_{\max}}$. If we only allow λ_2, λ_3 and $\lambda_{\mathrm{v}_{\max}}$ to be nonzero, then λ_2 and λ_3 are uniquely determined by $\lambda_{\mathrm{v}_{\max}}$ and we have

$$\lambda_3 = 3 - 6 \left(\frac{1}{\mathrm{c}_{\mathrm{reg.}}(1-r)} - \lambda_{\mathrm{v}_{\max}} \left(\frac{1}{\mathrm{v}_{\max}} - \frac{1}{2} \right) \right) \tag{3.13}$$

$$\lambda_2 = -2 - \lambda_{\mathrm{v}_{\max}} + 6 \left(\frac{1}{\mathrm{c}_{\mathrm{reg.}}(1-r)} - \lambda_{\mathrm{v}_{\max}} \left(\frac{1}{\mathrm{v}_{\max}} - \frac{1}{2} \right) \right) \tag{3.14}$$

For determining the degree distribution, the choice of the interleaving scheme between LDPC code and differential encoder/decoder is crucial. In fact, this choice

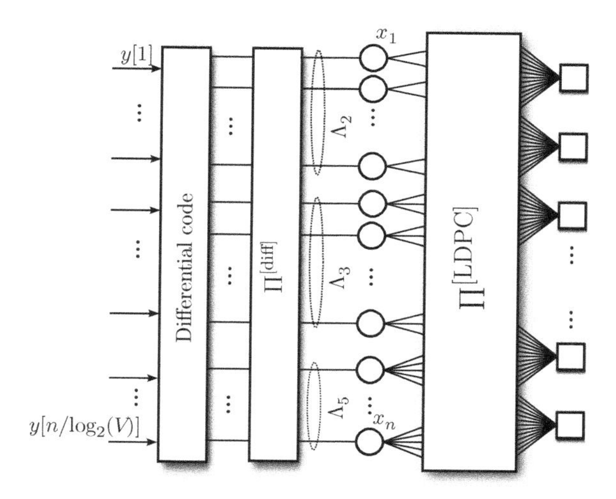

FIGURE 3.17 Schematic of the LDPC code with full interleaving between LDPC code and differential decoder.

determines how to select the degree distribution and finally has an influence on the overall system performance.

3.3.2.1 Design of LDPC Codes—Full Interleaving The first way of interleaving consists of placing a full interleaver $\Pi^{[\text{diff}]}$ of size n between differential code and LDPC code, as depicted in Figure 3.17. The interleaver $\Pi^{[\text{diff}]}$ is placed between the differential decoder and the variable nodes of the LDPC code, such that the interleaved output of the differential decoder mimics the transmission channel output. This is the approach that has been followed in Refs [59] and [28].

As the transmission channel is in this case the combination of differential decoder and interleaver, we need to modify the convergence condition (3.8). Instead of having a function describing the information transfer of the VND, we introduce a function $f_{V,D}$ that describes the information transfer of the combined differential decoder and VND. This combined information transfer function is given by

$$I_E^{[V,D]} = \sum_{i=1}^{\text{v}_{\max}} \lambda_i J\left(\mu_c + (i-1)J^{-1}(I_A^{[V,D]})\right)$$

The value μ_c is the mean of the message that is sent from the differential decoder towards the LDPC code. Using the EXIT characteristic of the differential decoder $f_D(I, E_s/N_0)$, which can be prerecorded and potentially represented by a polynomial [53], we can express μ_c as [53]

$$\mu_c = J^{-1}\left(f_D\left(\sum_{i=1}^{\text{v}_{\max}} L_i J\left(i \cdot J^{-1}(I_A^{[V,D]})\right), \frac{E_s}{N_0}\right)\right)$$

which leads to the overall EXIT characteristic of the combined VND and differential decoder

$$f_{V,D}\left(I, \frac{E_s}{N_0}\right) = \sum_{i=2}^{\mathrm{v}_{\max}} \lambda_i J\left(J^{-1}\left(f_D\left(\sum_{j=2}^{\mathrm{v}_{\max}} L_j J\left(j \cdot J^{-1}\left(I_A^{[V,D]}\right)\right), \frac{E_s}{N_0}\right)\right) + (i-1)J^{-1}(I_A^{[V,D]})\right)$$

$$= \sum_{i=2}^{\mathrm{v}_{\max}} \lambda_i J\left(J^{-1}\left(f_D\left(\sum_{j=2}^{\mathrm{v}_{\max}} \frac{\lambda_j J\left(j \cdot J^{-1}(I_A^{[V,D]})\right)}{j \sum_{\kappa=2}^{\mathrm{v}_{\max}} \frac{\lambda_\kappa}{\kappa}}, \frac{E_s}{N_0}\right)\right) + (i-1)J^{-1}(I_A^{[V,D]})\right) \tag{3.15}$$

where we have used (3.5) in the second line of the equation. This leads to the condition for successful decoding

$$f_{V,D}\left(I, \frac{E_s}{N_0}\right) > f_C^{-1}(I), \quad \forall I \in [0; 1) \tag{3.16}$$

In this case, the stability condition reads [59]

$$\lambda_2 < \frac{1}{\mathrm{c}_{\mathrm{reg.}} - 1} \exp\left(\frac{J^{-1}\left(f_D\left(1, \frac{E_s}{N_0}\right)\right)}{4}\right) \tag{3.17}$$

As the function $f_{V,D}(\cdot, \cdot)$ is not linear in λ_i (and even not necessarily convex), the elegant linear programming-based optimization [50] cannot be applied. We have to resort to heuristic optimization methods such as *differential evolution* [58] or *simulated annealing*. If we assume, however, that the degree distribution only consists of three degrees 2, 3, and $\mathrm{v}_{\max}$, then we have seen before that by fixing $\lambda_{\mathrm{v}_{\max}}$, the values of λ_2 and λ_3 are immediately given. Thus, the problem of finding the optimal degree distribution reduces to a one-dimensional problem. By sweeping $\lambda_{\mathrm{v}_{\max}}$ between the extremes of the admissible interval [0, 1], we can find the best possible degree distribution.

We use the following binary search to find the best possible degree distribution. We first assume that the differential decoder EXIT characteristic is available for any E_s/N_0 value between $\frac{E_s}{N_0}|_{\min}$ and $\frac{E_s}{N_0}|_{\max}$. We fix a minimum step size $\Delta_{\min}$ and use Algorithm 3.1, which outputs the optimum $\mathrm{c}_{\mathrm{reg.}}$, λ_2, λ_3, and $\lambda_{\mathrm{v}_{\max}}$.

We have found that using only three different variable node degrees and only a fixed check node degree does not impose a noteworthy limitation and that the performance of the obtained codes is very close to the performance of codes designed with less constraints, provided that $\mathrm{v}_{\max}$ is chosen large enough.

Algorithm 3.1

```
Input {
    Maximum variable node degree $\mathbf{v}_{\max}$
    Minimum and maximum check node degrees $\mathbf{c}_{\min}$ and $\mathbf{c}_{\max}$
    Discretization steps $D$ (mutual information) and $D_\lambda$
    Minimum stepsize $\Delta_{\min}$
}
Output {
    Optimum variable node degree distribution $\boldsymbol{\lambda}_{\text{best}} = (\lambda_2, \lambda_3, \lambda_{\mathbf{v}_{\max}})$
    Optimum check node degree $\mathbf{c}_{\text{best}}$
    SNR threshold $E_{\text{best}}$
}
Initialization {
  $E \leftarrow \frac{1}{2}\left(\frac{E_s}{N_0}\big|_{\min} + \frac{E_s}{N_0}\big|_{\max}\right)$, $E_{\text{best}} \leftarrow \infty$, $\Delta \leftarrow \frac{1}{2}\left(\frac{E_s}{N_0}\big|_{\max} - \frac{E_s}{N_0}\big|_{\min}\right)$
}
Optimization Algorithm {
  while ($\Delta > \Delta_{\min}$)
  {
      Success $\leftarrow$ false
      for ($\mathbf{c}_{\text{reg.}} = \mathbf{c}_{\min} \cdots \mathbf{c}_{\max}$)
      {
          compute $f_C^{-1}\left(\frac{i}{D}, \mathbf{c}_{\text{reg.}}\right)$, $\forall i \in \{0, 1, \ldots, D-1\}$
          for ($j \in \{0, 1, \ldots, D_\lambda\}$)
          {
              $\lambda_{\mathbf{v}_{\max}} \leftarrow \frac{j}{D_\lambda}$
              determine $\lambda_2$ and $\lambda_3$ via (3.14) and (3.13)
              compute $f_{V,D}\left(\frac{i}{D}, E\right)$, $\forall i \in \{0, \ldots, D-1\}$ using (3.15)
              if $f_{V,D}\left(\frac{i}{D}, E\right) > f_C^{-1}\left(\frac{i}{D}\right)$, $\forall i$ and (3.17) fulfilled
              {
                  if ($E < E_{\text{best}}$)
                  {
                      $E_{\text{best}} \leftarrow E$
                      $\boldsymbol{\lambda}_{\text{best}} = (\lambda_2, \lambda_3, \lambda_{\mathbf{v}_{\max}})$
                      $\mathbf{c}_{\text{best}} \leftarrow \mathbf{c}_{\text{reg.}}$
                  }
                  Success $\leftarrow$ true
              }
          }
      }
      if (Success = True)
      {
          $E \leftarrow E - \frac{\Delta}{2}$
      } else {
          $E \leftarrow E + \frac{\Delta}{2}$
      }
      $\Delta \leftarrow \frac{\Delta}{2}$
  }
}
```

The full interleaving scheme has several limitations. For instance, the EXIT chart-based optimization assumes that the messages exchanged between the different decoder components are Gaussian distributed. This is, however, not the case when interleaving the outputs of all different variables; in this case, the messages show rather a distribution that can be described by a *Gaussian mixture*, leading to inaccuracies of the model. Even though the messages may be conveniently approximated by Gaussian distributions (if the variances of the different parts of the mixture do not vary much), the codes designed according to this model may not yield the best possible performance.

3.3.2.2 Design of LDPC Codes—Partial Interleaving In order to mitigate the limitations of the full interleaving approach of the previous section, we replace the single interleaver $\Pi^{[\text{diff}]}$ by multiple *partial* interleavers, as described in Ref. [60] and inspired by the analysis for BICM-ID with convolutional codes in Ref. [61]. In the partial interleaving case, we group all Λ_i variable nodes of degree i, assign an interleaver of size Λ_i to these nodes, and employ a separate differential decoder/encoder for this group of variable nodes. The graph-based model with partial interleaving is shown in Figure 3.18 (where we assume that each Λ_i is a multiple of V).

If partial interleaving is used, the equations for analyzing the convergence and finding good degree distributions have to be modified as well. In this case, every variable node group (of degree i) has to be treated separately and is assigned its own differential decoder output $\mu_{c,i}$ and we can write

$$I_E^{[V,D]} = \sum_{i=1}^{v_{\max}} \lambda_i J\left(\mu_{c,i} + (i-1)J^{-1}(I_A^{[V,D]})\right) \tag{3.18}$$

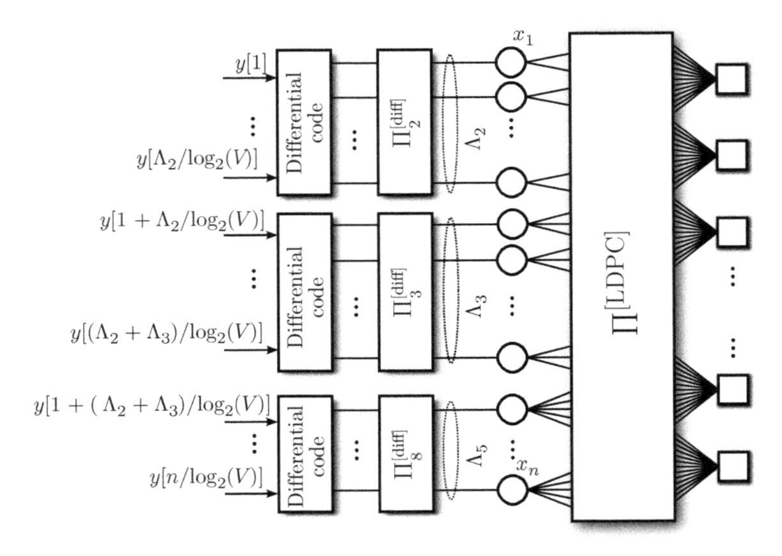

FIGURE 3.18 Schematic of the LDPC code with partial interleaving between LDPC code and differential decoder.

where $\mu_{c,i}$ can be computed as

$$\mu_{c,i} = J^{-1}\left(f_D\left(J\left(i \cdot J^{-1}(I_A^{[V]})\right), \frac{E_s}{N_0}\right)\right)$$

This leads to the overall EXIT characteristic of the combined variable node differential decoder

$$f_{V,D}\left(I, \frac{E_s}{N_0}\right) := \sum_{i=1}^{\mathrm{v}_{\max}} \lambda_i J\left(J^{-1}\left(f_D\left(J\left(i \cdot J^{-1}(I)\right), \frac{E_s}{N_0}\right)\right) + (i-1)J^{-1}(I)\right)$$

which is a linear function in λ. Due to the linearity of $f_{V,D}(\cdot,\cdot)$, we can employ a simple linear programming optimization to find good values of λ_i for a given check node degree distribution, as described in Ref. [50]. However, if we only allow three different variable node degrees (and thus three partial interleavers), the problem reduces to a one-dimensional problem again, which we can solve in a similar way (using Algorithm 3.1) as for the case with full interleaving. However, we would like to point out that due to the linearity of $f_{V,D}(\cdot,\cdot)$, it is much easier to find *optimal* degree distributions. Numerical methods such as differential evolution cannot guarantee to find the global optimum of the problem.

3.3.2.3 Comparison of Interleaving Schemes—Results Using both interleaving schemes and differential QPSK transmission, we design degree distributions. We impose the constraint that the maximum variable node degree shall be $\mathrm{v}_{\max} = 12$. For each interleaving scheme, we either use Gray differential encoding or natural differential encoding. For both options, we have optimized codes using Algorithm 3.1 with variable degrees $\in \{2, 3, 12\}$ and regular check node degree. We additionally design codes for the constraint where $L_2 \leq 1 - r$. This constraint is necessary to avoid a high number of degree-2 nodes. It can be shown [62] that a potential error floor can occur if the fraction of degree-2 nodes is larger than $1 - r$. If we impose the constraint $L_2 \leq 1 - r$, then we can design a code that avoids—in the graph description—cycles containing only degree-2 nodes and with no information bits assigned to degree-2 nodes [63]. The condition $L_2 \leq 1 - r$ translates in the general case into

$$\lambda_2 \leq 2\left(\frac{1}{r} - 1\right)\sum_{j=3}^{\mathrm{v}_{\max}} \frac{\lambda_j}{j}$$

which can be added to Algorithm 3.1.

We generate codes of target rate $r = \frac{4}{5} = 0.8$, a rate typically used in optical communications. A rate of $r = 0.8$ is a viable selection for current and future 100 Gbit/s (with QPSK) or 200 Gbit/s systems operating in a *dense wavelength division multiplex* (DWDM) setting with 50 GHz channel spacing and an exploitable bandwidth of roughly 37.5 GHz due to frequent filtering with non-flat frequency characteristic. The best possible achievable values of E_s/N_0 (corresponding to E_{best} in Algorithm 3.1) for the different code designs are shown in Table 3.4 and the degree distributions of the

TABLE 3.4 Theoretical thresholds values of E_s/N_0 (in dB) for the designed codes

	No constraint on L_2		$L_2 \leq 1 - r$	
	Full interleavers	Partial interleavers	Full interleavers	Partial interleavers
Gray Differential	4.536	4.43	4.946	4.907
Natural Differential	4.905	4.839	5.382	5.33

TABLE 3.5 Degree distributions of all considered codes

Code	L_2	L_3	L_{12}	$c_{\text{reg.}}$
Gray differential, full interleaving	0.959	0.002	0.039	12
Gray differential, partial interleaving	0.919	0.002	0.079	14
Natural differential, full interleaving	0.979	0.002	0.019	11
Natural differential, partial interleaving	0.959	0.002	0.039	12
Gray differential, full interleaving, $L_2 \leq 1 - r$	0.198	0.78	0.022	15
Gray differential, partial interleaving, $L_2 \leq 1 - r$	0.198	0.78	0.022	15
Natural differential, full interleaving, $L_2 \leq 1 - r$	0.198	0.78	0.022	15
Natural differential, partial interleaving, $L_2 \leq 1 - r$	0.198	0.78	0.022	15

resulting codes are summarized in Table 3.5. We can see that Gray differential coding with partial interleaving leads to the best coded transmission schemes operating at the lowest possible E_s/N_0 values.

Figure 3.19 shows a first simulation results for the case of a QPSK modulation. We have constructed codes of size $n = 32,000$ having the degree distributions from Table 3.5. The utilized channel model is the phase-slip channel model from Figure 3.3

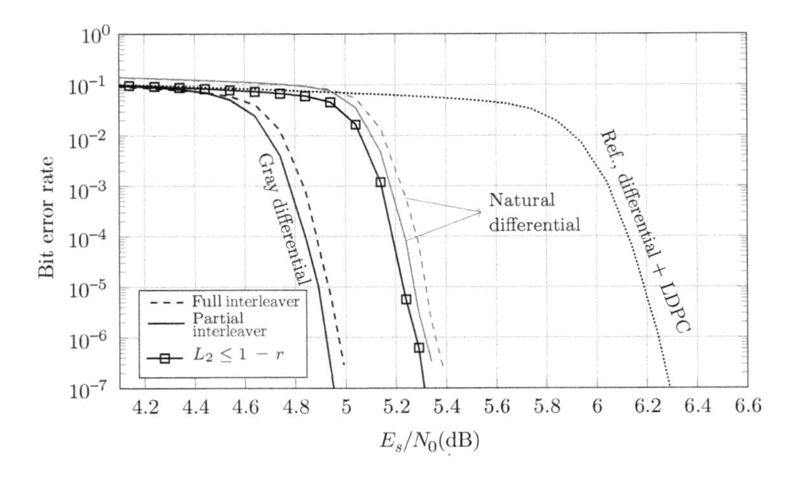

FIGURE 3.19 Simulation example, QPSK with $\gamma = 0$.

with $\gamma = 0$. In this example, we wish to confirm the results of Table 3.4. As a reference scheme, we optimized an LDPC code for a simple AWGN channel with $v_{max} = 12$ and regular check node degree $c_{reg.}$, which is used with *noniterative* differential decoding. We can see that the results of Table 3.4 are indeed confirmed and the code with Gray differential coding and partial interleaving yields the best performance. As decoder, we use a conventional decoder as described in the Appendix of this chapter with 18 decoding iterations, where in each iteration, we invoke the differential decoder. Note that with the use of a layered decoder [64], the convergence speed can be increased and the same performance can be obtained by using only approximately 12 layered iterations. For this reason, the choice of 18 iterations is practical, as 12 layered LDPC iterations are deemed to be implementable [20].

In the second example, we increase the phase-slip probability on the channel by choosing $\gamma = 0.2$. The simulation results for this case are shown in Figure 3.20. We can observe that the formerly best case with Gray differential coding now shows a significant error floor. With natural differential coding, an error floor is observed as well, however, at several orders of magnitude smaller. This floor is mainly due to the large number of degree-2 nodes and the fact that $I_E^{[D]}(I_A^{[D]} = 1) < 1$ if $\gamma > 0$. Indeed, for this optimization, most of the variable nodes are of degree-2, that is, λ_2 and consequently L_2 becomes very large. It has also been observed [56] that LDPC codes designed for differentially coded modulation require many degree-2 nodes. Degree-2 variable nodes are, however, a non-negligible contributor to the error floor, especially if there are cycles in the graph that connect only degree-2 variable nodes. It has been shown that cycles containing only degree-2 variable nodes can be avoided [62] if $\Lambda_2 \leq m = n(1-r)$, that is, if $L_2 \leq 1 - r$, which is why we have included that constraint into the optimization. In this case, we may design a systematic code and assign only parity bits to the degree-2 variable nodes. This further reduces the error floor as the BER is calculated purely based on the systematic bits and the higher the variable node degree, the more reliable a bit is after decoding.

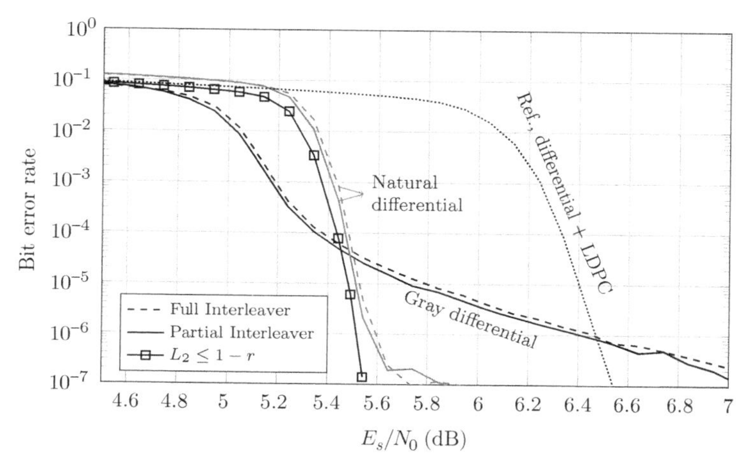

FIGURE 3.20 Simulation example, QPSK with $\gamma = 0.2$.

We have added the constraint $L_2 \leq 1 - r$ to the optimization routine and have found the corresponding degree distribution (summarized in Table 3.5). In the simulation results shown in Figure 3.19, the performance of the code obtained with the $L_2 \leq 1 - r$ condition is shown by the curve with square markers for Gray differential coding. We see that in this case, we get a slightly better performance than natural differential coding, but have advantages with respect to the residual BER, at the expense of an increased required E_s/N_0 to allow for successful decoding in the case where $\gamma = 0$ (see Figure 3.19). Thus, depending on the required target BER, we may either use the condition $L_2 \leq 1 - r$ or not. If an outer code is used that can correct up to an input BER of 4.2×10^{-3} (e.g., the staircase code [65]) we may use the code designed without the constraint on L_2, but if we use a higher rate outer code as mentioned in Refs [20, 63] that requires a very low input BER, we may select the code designed with $L_2 \leq 1 - r$.

We can thus summarize that there are somewhat conflicting code design strategies. If $\gamma = 0$, we can use the code optimized for Gray differential coding with partial interleaving. This code will, however, lead to an elevated error floor if phase slips occur on the channel. This error floor has to be combated either with a properly designed outer code[5] or by using only the code with the constraint $L_2 \leq 1 - r$, which however leads to a suboptimal performance in the phase-slip-free case ($\gamma = 0$). Another solution is the implementation of two codes, one for each case ($\gamma = 0$ and large γ). This latter method can guarantee best performance depending on the channel, but requires a feedback loop from the receiver to the transmitter, which may not be available in the network and of course it requires the implementation of two different codes, which may be too complex on an *application specific integrated circuit* (ASIC). In Section 3.4, we show a solution that requires only a single code and shows a more universal, channel-agnostic behavior.

3.3.3 Higher-Order Modulation Formats with $V < Q$

In practical systems, we often have to deal with the case where $V < Q$, for example, if 16QAM is used, where we have $V = 4$ and $Q = 16$. In this case, we may use different techniques to optimize the code. We propose to refine the method of partial interleaving and to use only differential coding on a fraction of $\frac{\log_2 V}{\log_2 Q} = \frac{v}{q}$ of the bits. This is shown in Figure 3.21. In this case, the value $\mu_{c,i}$ required in (3.18) is computed as

$$\mu_{c,i} = \frac{v}{q} J^{-1}\left(f_D\left(J\left(i \cdot J^{-1}(I_A^{[V]})\right)\right)\right) + \left(1 - \frac{v}{q}\right)\overline{\mu}_c$$

where $\overline{\mu}_c$ denotes the mean of the LLRs of the bits that are not differentially encoded, obtained using a conventional bit-wise decoder [25] (see (3.25)) and averaged over all these bits. These bits correspond to the part of the constellation encoded with the rotationally invariant mapping (symbols within a region associated with a state S_i).

[5]Note that the implementation of a coding system with an outer cleanup code requires a thorough understanding of the LDPC code and a properly designed interleaver between the LDPC code and the outer code.

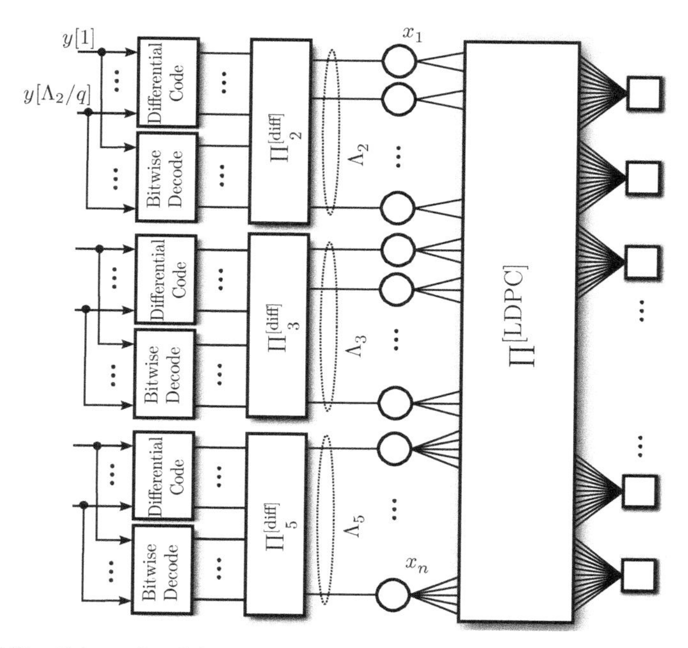

FIGURE 3.21 Schematic of the LDPC code with partial interleaving between LDPC code and differential decoder for the case where $V < Q$.

Besides this simple and direct approach, we can also use more involved methods, for example, using the technique of *multi-edge-type* (MET) codes, as described in Refs [50, 66, Chapter 7], but this is outside the scope of this chapter.

3.4 CODED DIFFERENTIAL MODULATION WITH SPATIALLY COUPLED LDPC CODES

In the 1960s and the following decades, most coding research focused on block coding techniques, but many practical coding schemes were based upon convolutional codes [67]. With the advent of turbo codes [38], the rediscovery of LDPC codes, and advances in semiconductor technology, this suddenly changed so that today most new coding schemes are, again, block codes. The trend is, however, to return to convolutional-like structures [68] that can be efficiently encoded and decoded using sliding-window techniques.

In the last few years, the class of spatially coupled (SC) code ensembles has emerged [69, 70]. Spatially coupled codes were originally introduced more than a decode ago [71] and were then called *LDPC convolutional codes*. The appealing properties of SC codes were only recently noticed, when it was found that the performance of terminated SC-LDPC codes with simple belief propagation decoding approaches the *maximum a posteriori* (MAP) thresholds of the underlying ensemble [72, 73].

Thus, contrary to a common belief, introducing structure into LDPC codes leads to a class of (degenerated) realizations of LDPC codes that demonstrate superior performance under belief propagation decoding. This effect of threshold saturation has been analyzed for the BEC in Ref. [69] and it has been shown that spatially coupled LDPC codes can asymptotically achieve the MAP threshold of the underlying ensemble under belief propagation decoding. Recently, this result has been extended to more general channels and it has been shown in Ref. [70] that spatially coupled regular LDPC ensembles *universally* achieve capacity over binary-input memoryless output-symmetric channels: most codes in this ensemble are good for each channel realization in this class of channels.

Spatially coupled codes are now emerging in various applications. Two examples in the context of optical communications are the staircase code [65] and the braided BCH codes of [74], which are both rate $R = 239/255$ codes targeted for 100 Gbit/s applications with hard-decision decoding. Both codes are spatially coupled BCH product codes that allow for a natural windowed decoder implementation. These codes can be interpreted as being generalized spatially coupled LDPC codes with variable node degree $d_v = 2$ and every bit participating in two BCH component codes, where the component BCH codes are able to correct up to 4 errors each. Another example is the IEEE 1901 power line communications standard, where an LDPC convolutional code is specified for the wavelet physical layer [75]. For a basic introduction to spatially coupled codes, we refer the interested reader to [51, 76].

3.4.1 Protograph-Based Spatially Coupled LDPC Codes

In this chapter, we restrict ourselves to the class of protograph-based construction of SC-LDPC codes as introduced in, for example, [72, 73, 77]. A protograph [78] is a convenient way of describing LDPC codes. Protograph codes are constructed from the P-cover of a relatively small graph that conveys the main properties of the code. In contrast to the graph representation of the LDPC code, the protograph may contain multiple edges. The code itself is constructed by placing P copies of the protograph next to each other (note that these have no interconnecting edges) and permuting the edges between the different copies of the protograph, such that the relation between the group of edges is respected. The construction of a small toy code is illustrated in Example 3.4.

EXAMPLE 3.4

We illustrate the construction of larger codes based on protographs using a simple toy example. Starting with a prototype matrix, also called *protomatrix*

$$\boldsymbol{B} = \begin{pmatrix} 1 & 2 & 3 & 0 \\ 1 & 2 & 0 & 2 \end{pmatrix}$$

we show how a P-cover is constructed. First, we construct an equivalent graph of the base matrix in the same way as we constructed the graph of the LDPC code in Section 3.3.1.2. The difference is that the nonzero entries in $\boldsymbol{B}$ indicate the number of *parallel edges* connecting the variables with the checks. The graph representation of the protomatrix $\boldsymbol{B}$ is given by

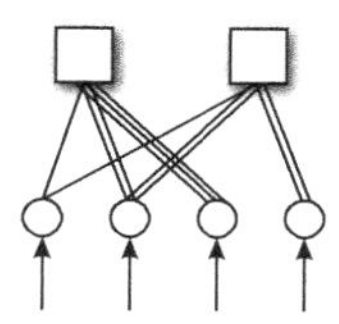

In the next step, we construct the P-cover of this graph, which means that we simply place P copies of this graph next to each other:

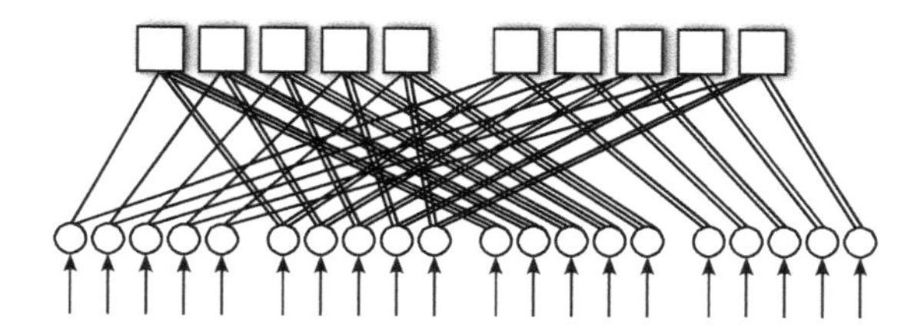

This graph is still not a valid LDPC code as it contains parallel edges and the P subgraphs are not connected. In order to remove the parallel edges and to construct a more randomized code, we permute in a next step all edges that are within one *edge group*, that is, that correspond to a single entry $B_{i,j}$ of $\boldsymbol{B}$. This permutation is performed in such a way that no parallel edges persist. The final code graph is then obtained by

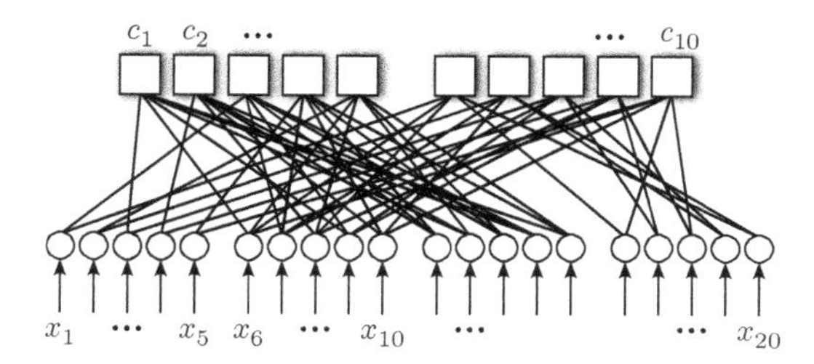

Note that it is *not* possible to draw this code with a single interleaver $\Pi^{[\mathrm{LDPC}]}$ as the code in Figure 3.16, because it is actually a *multi-edge-type* ensemble [50, Section 7.1] and for every single entry $B_{i,j}$ of $\boldsymbol{B}$, an individual interleaver is required. The parity-check matrix $\boldsymbol{H}$ can be constructed from $\boldsymbol{B}$ by replacing each entry $B_{i,j}$ (row i, column j) of $\boldsymbol{B}$ by the superposition of $B_{i,j}$ permutation matrices,[6] chosen such that no two "1"s are in the same position (avoiding parallel edges in the final graph). For example, a parity-check matrix corresponding to the above-mentioned $\boldsymbol{B}$ with $P = 5$

[6] A permutation matrix $\boldsymbol{P}$ is a square binary matrix (i.e., a matrix containing only "0" and "1") where each row contains exactly one "1" *and* where each column contains exactly one "1."

is given by

$$
\boldsymbol{H} = \left(\begin{array}{ccccc|ccccc|ccccc|ccccc}
 & & 1 & & & 1 & & & 1 & & & 1 & & 1 & 1 & & & & & \\
 & & & 1 & & & & 1 & & 1 & 1 & 1 & & 1 & & & & & & \\
1 & & & & & & 1 & & 1 & & 1 & & 1 & 1 & & & & & & \\
 & & & & 1 & & 1 & & & 1 & 1 & & 1 & & 1 & & & & & \\
 & 1 & & & & 1 & & 1 & & & & 1 & 1 & & 1 & & & & & \\
\hline
1 & & & & & 1 & & 1 & & & & & & & & 1 & & & & 1 \\
 & 1 & & & & & 1 & & & 1 & & & & & & & & & 1 & 1 \\
 & & 1 & & & & 1 & & 1 & & & & & & & & 1 & & 1 & \\
 & & & 1 & & & & 1 & 1 & & & & & & & & 1 & 1 & & \\
 & & & & 1 & 1 & & & & 1 & & & & & & 1 & & 1 & &
\end{array}\right).
$$

We follow the approach given in Ref. [79] to describe protograph-based SC-LDPC codes. The protograph of a *time-invariant, terminated* spatially coupled LDPC with syndrome former memory m_s and replication factor L is obtained from a collection of $(m_s + 1)$ distinct protomatrices $\boldsymbol{B}_i$, $i \in \{0, 1, \dots, m_s\}$ each of size dim $\boldsymbol{B}_i = m' \times n'$. The protomatrix of the spatially coupled code is then given by

$$
\boldsymbol{B}^{[\text{conv}]}(L) = \begin{pmatrix}
\boldsymbol{B}_0 & & & \\
\boldsymbol{B}_1 & \boldsymbol{B}_0 & & \\
\vdots & \boldsymbol{B}_1 & \ddots & \\
\boldsymbol{B}_{m_s} & \vdots & \ddots & \boldsymbol{B}_0 \\
 & \boldsymbol{B}_{m_s} & \ddots & \boldsymbol{B}_1 \\
 & & \ddots & \vdots \\
 & & & \boldsymbol{B}_{m_s}
\end{pmatrix}_{(L+m_s)m' \times Ln'} \tag{3.19}
$$

$\boldsymbol{B}^{[\text{conv}]}(L)$ can also be viewed as being composed by a stack of $L + m_s$ shifted (by n') and overlapping versions of

$$
\boldsymbol{B}_r = \begin{pmatrix} \boldsymbol{B}_{m_s} & \cdots & \boldsymbol{B}_0 \end{pmatrix}
$$

Note that the termination, which cuts the boundaries of the stacked matrix, leads to an inevitable rate loss, which becomes, however, negligible for increasing L. The rate of the code amounts to [79]

$$
r_L = 1 - \left(\frac{L + m_s}{L}\right)\frac{m'}{n'}
$$

If we are allowed to choose L large enough, we immediately see that $\lim_{L\to\infty} r_L = 1 - \frac{m'}{n'}$, which corresponds to the rate of the original protomatrices $\boldsymbol{B}_i$.

One can say that the convergence of spatially coupled codes is well understood meanwhile. A thorough analysis for the BEC is given in Ref. [69] and extended to general binary input memoryless channels in Ref. [70]. The convergence behavior of the iterative decoder can be subdivided into two convergence regions

- the region of *macro-convergence*, where convergence is dominated by the code's degree distribution and convergence prediction by conventional EXIT charts is possible.
- the region of *micro-convergence*, where the convergence is dominated by spatial coupling and termination effects.

The region of *macro-convergence* is observed in the first decoding iterations and at high channel SNRs, whereas the region of *micro-convergence* is observed at low SNRs close to the thresholds of the code. In the region of micro-convergence, the decoding process can be visualized as a decoding wave [80] that slowly progresses through the graph from the boundaries onward. This wave-like behavior allows the efficient design of windowed decoders [81] where the decoding window follows the decoding wave. The understanding of the dynamics of the decoding wave, especially its speed [82], is essential for designing high-performance codes and effective windowed decoders.

3.4.2 Spatially Coupled LDPC Codes with Iterative Demodulation

In this section, we combine spatially coupled codes with a differential decoder and use common analysis techniques to show how the detector front-end influences the performance of the codes. As we have seen, in conventional LDPC code design, usually the code needs to be "matched" to the transfer curve of the detection front-end. If the code is not well matched to the front-end, a performance loss occurs. If the detector front-end has highly varying characteristics, due to, for example, varying channels or varying phase-slip probabilities, several codes need to be implemented and always the right code needs to be chosen for maximum performance, which appears impractical in optical networks where a feedback channel from the receiver to the transmitter can potentially be difficult to realize.

In contrast to a random ensemble of the same degree profile, spatially coupled LDPC codes can converge below the pinch-off in the EXIT chart, so even if the code is not well matched to the differential decoder we can hope to successfully decode due to the micro-convergence effect. So, even with a varying channel and detector characteristics, we can use a single code that is *universally* good in all scenarios. This means that the code design can stay *agnostic* to the channel/detector behavior.

We determine the thresholds of the protograph-based spatially coupled codes combined with demodulation and detection by an extension of the PEXIT technique [83] with the Gaussian approximation of the check node operation [84]. A refined version of the PEXIT technique taking into account the windowed decoder of spatially coupled codes has been presented in Ref. [85]. The mutual information analysis used for the design of degree distributions in LDPC codes has to be modified slightly to account for the protograph structure. Instead of analyzing and tracking a single mutual information value I, we now have to track an individual mutual information value for each nonzero entry at row i and column j of the protomatrix $\boldsymbol{B}^{[\mathrm{conv}]}$. We denote the respective outgoing (incoming) edge mutual information by $I_{E,i,j}^{[V,D]}$ ($I_{A,i,j}^{[V,D]}$) or $I_{E,i,j}^{[C]}$ ($I_{A,i,j}^{[C]}$), depending on whether the message is computed by the combined variable node detector engine ("V, D") or by the check node engine ("C"). Note that we assume that the messages are Gaussian distributed and can described by a single parameter: their mean μ (with the variance 2μ, see [19] and [50] for details).

Similar to the previous section, we assume that the demodulator/detector properties can be described by means of an EXIT characteristic [27, 53], which we denote by $f_D(\cdot,\cdot)$. If the message at the input of the detector is Gaussian distributed with mean μ, then the detector output mean for (protograph) variable j is obtained by

$$\mu_{c,j} = J^{-1}\left(f_D\left(J\left(\sum_{i=1}^{(L+m_s)m'} B_{i,j}^{[\mathrm{conv}]} J^{-1}(I_{A,i,j}^{[V,D]})\right), \frac{E_s}{N_0}\right)\right) \tag{3.20}$$

leading to the combined variable node and detector update characteristic

$$I_{E,i,j}^{[V,D]} = J\left(\mu_{c,j} + \left(B_{i,j}^{[\mathrm{conv}]} - 1\right) J^{-1}\left(I_{A,i,j}^{[V,D]}\right) + \sum_{\substack{k=1\\k\neq i}}^{(L+m_s)m'} B_{k,j}^{[\mathrm{conv}]} J^{-1}\left(I_{A,k,j}^{[V,D]}\right)\right) \tag{3.21}$$

which has to be evaluated for all $(i,j) \in [1,\ldots,(L+m_s)m'] \times [1,\ldots,Ln']$ where $B_{i,j}^{[\mathrm{conv}]} \neq 0$. The check node information is computed according to [84]

$$I_{E,i,j}^{[C]} = J\left(\phi^{-1}\left(1 - \left[1 - \phi\left(J^{-1}(I_{A,i,j}^{[C]})\right)\right]^{B_{i,j}^{[\mathrm{conv}]}-1} \prod_{\substack{k=1\\k\neq j}}^{Ln'} \left[1 - \phi\left(J^{-1}(I_{A,i,k}^{[C]})\right)\right]^{B_{i,k}^{[\mathrm{conv}]}}\right)\right) \tag{3.22}$$

which again has to be evaluated for all combinations of (i,j) such that $B_{i,j}^{[\mathrm{conv}]} \neq 0$. The function $\phi(\mu)$, which is used to compute the evolution of the mean of the Gaussian messages in the check node update is given by

$$\phi(x) = \begin{cases} 1 - \frac{1}{\sqrt{4\pi x}} \int_{-\infty}^{\infty} \tanh\left(\frac{u}{2}\right) \exp\left(-\frac{(u-x)^2}{4x}\right) du, & \text{if} \quad x > 0 \\ 1, & \text{if} \quad x = 0 \end{cases} \tag{3.23}$$

Numerical approximations for (3.23) and its inverse function are given in Ref. [84].

The evaluation of the information is carried out in an iterative way: First, we initialize the process by setting $I_{A,i,j}^{[V,D]}(1) = 0$ for all possible (i,j) where the "(1)" denotes the first iteration. Using (3.20), we first compute $\mu_{c,j}(1)$, $\forall j \in [1, Ln']$ and use $\mu_{c,j}(1)$ to compute $I_{E,i,j}^{[V,D]}(1)$ by evaluating (3.21) for all $(i,j) \in [1, \ldots, (L+m_s)m'] \times [1, \ldots, Ln']$. We then set $I_{A,i,j}^{[C]}(1) = I_{E,i,j}^{[V,D]}(1)$ and evaluate (3.22) yielding $I_{E,i,j}^{[C]}(1)$. By setting $I_{A,i,j}^{[V,D]}(2) = I_{E,i,j}^{[C]}(1)$ we may proceed to the second iteration and compute $\mu_{c,j}(2)$, $I_{E,i,j}^{[V,D]}(2)$ and $I_{E,i,j}^{[C]}(2)$ in this sequence. Finally, after $\mathcal{I}$ iterations, we may—for each variable node in the protograph—determine the *a posteriori* reliability by

$$I_{\text{ap},j}^{[V]} = J\left(\mu_{c,j} + \sum_{k=1}^{(L+m_s)m'} B_{k,j}^{[\text{conv}]} J^{-1}\left(I_{A,k,j}^{[V,D]}(\mathcal{I})\right)\right)$$

We illustrate the behavior of $I_{\text{ap},j}^{[V]}$, which gives an indication of the reliability of the P bits that will be assigned to position j in the protograph by means of an example. We consider a spatially coupled code of rate $r = 0.8$ with $m_s = 2$ and with $\boldsymbol{B}_0 = \boldsymbol{B}_1 = \boldsymbol{B}_2 = (1\ \ 1\ \ 1\ \ 1\ \ 1)$. We use QPSK with Gray differential coding and set $\gamma = 0.2$ at $E_s/N_0 = 4.8$ dB; that is, according to (3.3), phase slips occur on the channel with a probability $P_{\text{slip}} \approx 0.0082$. Figure 3.22 shows the behavior of the *a posteriori* mutual information $I_{\text{ap},j}^{[V]}(\mathcal{I})$ as a function of the decoding iterations $\mathcal{I}$. We can see that the mutual information $I_{\text{ap},j}^{[V]}$ increases in a wave-like manner. Starting from the boundaries of the codeword, the mutual information converges toward 1 with an increasing number of iterations from the outside toward the inside until both waves meet and the whole codeword has been successfully decoded.

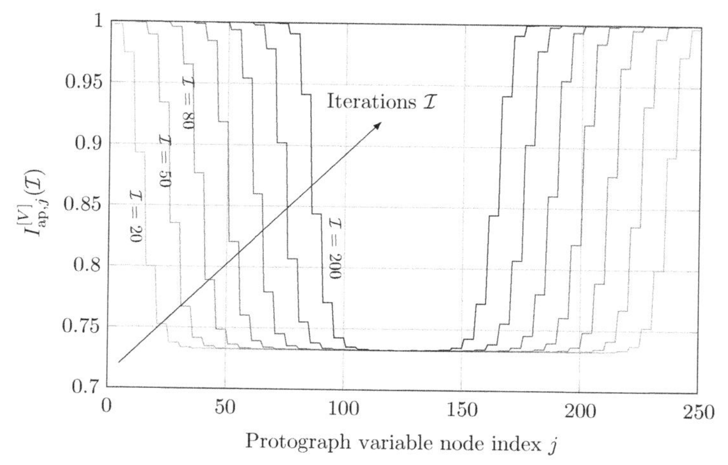

FIGURE 3.22 Wave-like decoding behavior of the spatially coupled, protograph-based LDPC code ($L = 50$) with Gray differential coding and $\gamma = 0.2$ for $E_s/N_0 = 4.8$ dB.

3.4.3 Windowed Differential Decoding of SC-LDPC Codes

By observing the wave-like decoding behavior [80] in Figure 3.22, we note that only parts of the protograph get updated when carrying out iterations. For instance, the reliability of the protograph variable node indices 110–140 stay at an almost constant value during the first 200 iterations. Likewise, the protograph variable node indices 1–30 have already converged to 1 after 110 iterations and do not benefit anymore from additional decoding iterations. This wave-like behavior thus leads to an efficient *windowed decoder* [81, 86], which follows the decoding wave and only carries out operations on the part of the protograph that benefits from further decoding iterations. The windowed decoder works in principle as an LDPC decoder, just with the difference that it operates on a fraction of the parity-check matrix. The windowed decoder is characterized by a window length w and operates on a portion of the protomatrix containing w vertically stacked copies of $\boldsymbol{B}_r$

$$\tilde{\boldsymbol{B}}^{[\text{conv}]} = \begin{pmatrix} \boldsymbol{B}_{m_s} & \cdots & \boldsymbol{B}_0 & & & \\ & \boldsymbol{B}_{m_s} & \cdots & \boldsymbol{B}_0 & & \\ & & \ddots & \cdots & \ddots & \\ & & & \boldsymbol{B}_{m_s} & \cdots & \boldsymbol{B}_0 \end{pmatrix}_{wm' \times (w+m_s)n'} \tag{3.24}$$

The decoder takes a block of $(w + m_s)n'$ protograph variables (i.e., $(w+ m_s)Pn'$ code bits), and carries out $\mathcal{I}_w$ decoding iterations on this block (using the conventional decoding scheme described in the Appendix). After having carried out $\mathcal{I}_w$ iterations, the window is shifted by n' protograph variables (Pn' code bits) and the left-most portion of Pn' code bits are considered to be decoded. Then, the process starts again. At the beginning of the process, the decoding window is initialized with perfect knowledge of the boundary values. For details, we refer the interested reader to [81] and [87].

3.4.4 Design of Protograph-Based SC-LDPC Codes for Differential-Coded Modulation

We show by means of an example how to construct good protographs for SC-LDPC codes in the context of differential-coded modulation [88]. For simplicity, we restrict ourselves to SC-LDPC codes with $m_s = 2$ leading to a protomatrix given by

$$\boldsymbol{B}_c = \begin{pmatrix} \boldsymbol{B}_0 & & & \\ \boldsymbol{B}_1 & \boldsymbol{B}_0 & & \\ \boldsymbol{B}_2 & \boldsymbol{B}_1 & \ddots & \\ & \boldsymbol{B}_2 & \ddots & \boldsymbol{B}_0 \\ & & \ddots & \boldsymbol{B}_1 \\ & & & \boldsymbol{B}_2 \end{pmatrix}$$

We select $m_s = 2$, as this choice leads to windowed decoders with a relatively compact decoding window. Note that the minimum required decoding window size grows almost linearly with m_s. Another viable choice would be $m_s = 1$; however, simulation results that are not shown here have indicated that a better performance can be expected with $m_s = 2$.

We consider iterative differential decoding as described in the previous section using the modified slip-resilient BCJR algorithm based on the trellis of Figure 3.10 and wish to design coding schemes of rate $r = 0.8$ (25% OH). We use protographs leading to regular SC-LDPC codes with variable degree $\mathtt{v}_{\text{reg.}} = 3$ and check degree $\mathtt{c}_{\text{reg.}} = 15$. The reason for using regular codes is that it has been shown in Ref. [69] that regular SC-LDPC are sufficient to achieve capacity and this particular code has a MAP threshold very close to capacity. Furthermore, due to the regularity, the implementation of the decoder can be simplified and the error floor is expected to be very low (increasing $\mathtt{v}_{\text{reg.}}$ further may even lead to lower error floors).

Although $\mathtt{v}_{\text{reg.}}$, $\mathtt{c}_{\text{reg.}}$, and m_s are fixed, we still need to find good protomatrices $\boldsymbol{B}_0$, $\boldsymbol{B}_1$, and $\boldsymbol{B}_2$. In order to have a very simple structure, we fix $m' = 1$ and $n' = 5$, leading to the smallest possible protomatrices. We have constructed all 1837 possible such combinations of protographs (unique up to column permutation) and computed decoding thresholds for all these protographs using the above-described method. We selected the protographs with the 50 best thresholds and carried out Monte Carlo simulations with three different windowed decoders:

- The first setup uses a window size of $w = 4$ and carries out $\mathcal{I}_w = 3$ iterations per decoding step.
- The second setup uses a window size of $w = 7$ and carries out $\mathcal{I}_w = 2$ iterations per decoding step.
- The third setup uses a window size of $w = 16$ and carries out a single iteration per decoding step, that is, $\mathcal{I}_w = 1$.

Note that with all three setups, every coded bit undergoes and equivalent number of 18 iterations, leading to the same complexity of all decoders and the same complexity as the LDPC-coded schemes presented in Section 3.3. Furthermore, note that the number of differential decoder executions per coded bit amounts to $w + m_s$ and thus depends on the setup. In the case of LDPC-coded differential demodulation, we have executed the differential decoder for each iteration. The required E_s/N_0 (in dB) to achieve a target BER of 10^{-6} for the 50 selected protographs is shown in Figure 3.23 for $L = 100$ and $P_{\text{slip}} \in \{0, 0.01\}$. Based on the results of Figure 3.23, we select the protograph with index 3, which has the best performance compromise at $P_{\text{slip}} = 0$ and $P_{\text{slip}} = 0.01$, and which is given by $\boldsymbol{B}_0 = (1\ \ 1\ \ 1\ \ 1\ \ 1)$, $\boldsymbol{B}_1 = (1\ \ 0\ \ 0\ \ 0\ \ 0)$, and $\boldsymbol{B}_2 = (1\ \ 2\ \ 2\ \ 2\ \ 2)$.

Finally, we use this protograph to construct a code by first lifting the protographs $\boldsymbol{B}_i$ with $P = 40$ and using the intermediate result in a second step to generate a *quasi-cyclic* (QC) code with circulant permutation matrices of size 50×50. The resulting parity-check submatrices $\boldsymbol{H}_i$ associated with $\boldsymbol{B}_i$ have size $\dim\ \boldsymbol{H}_i = 2000 \times 10,000$. As reference, we use QPSK with Gray differential coding and pick the two best codes found in Section 3.3.2: partial interleaving with no restriction on L_2 and partial interleaving with $L_2 \leq 1 - r$. As in Figures 3.19

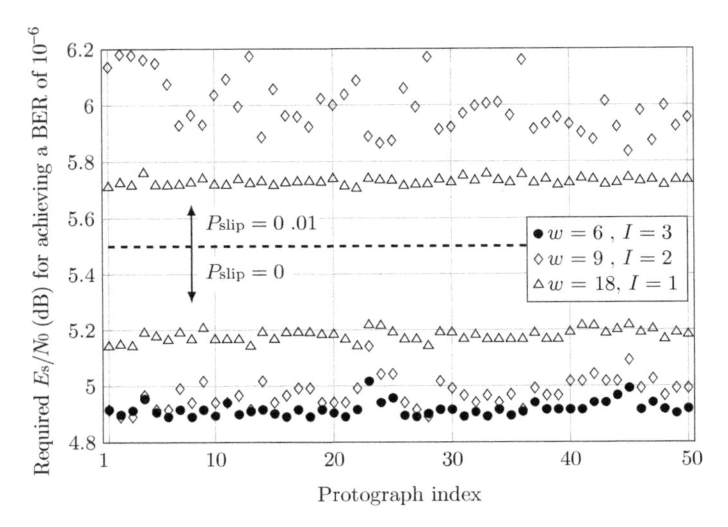

FIGURE 3.23 Required E_s/N_0 for 50 different protographs.

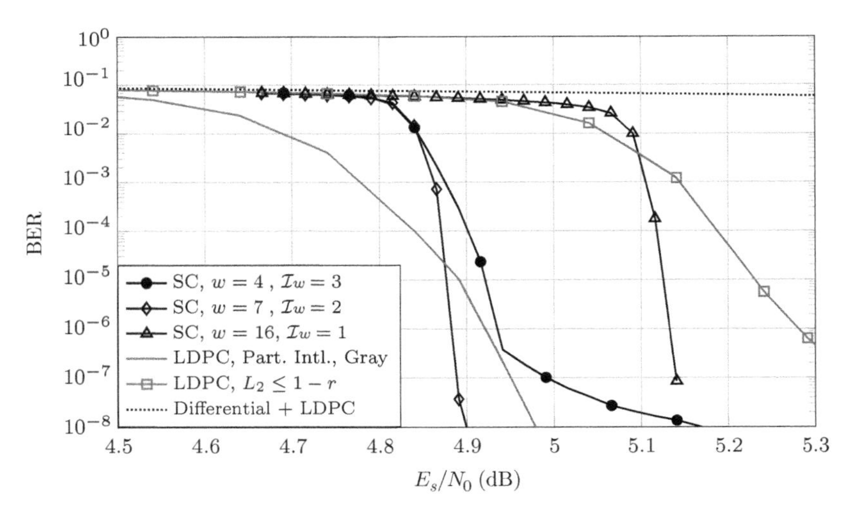

FIGURE 3.24 Simulation results of proposed and reference schemes for QPSK with $\gamma = 0$.

and 3.20, we use $I = 18$ decoding iterations in all cases. The results are shown in Figure 3.24 for $\gamma = 0$ (i.e., no phase slips) and in Figure 3.25 for $\gamma = 0.2$.

We can see from the results that for $\gamma = 0$, the LDPC code with partial interleaving (no restriction on L_2) already yields a very good performance within 1 dB of the theoretically minimum E_s/N_0, while the code with the constraint on L_2 entails a performance loss of about 0.4 dB (see also Figures 3.19 and 3.20). The SC-LDPC code with the second decoder setup ($w = 7$) outperforms the LDPC code for low BERs

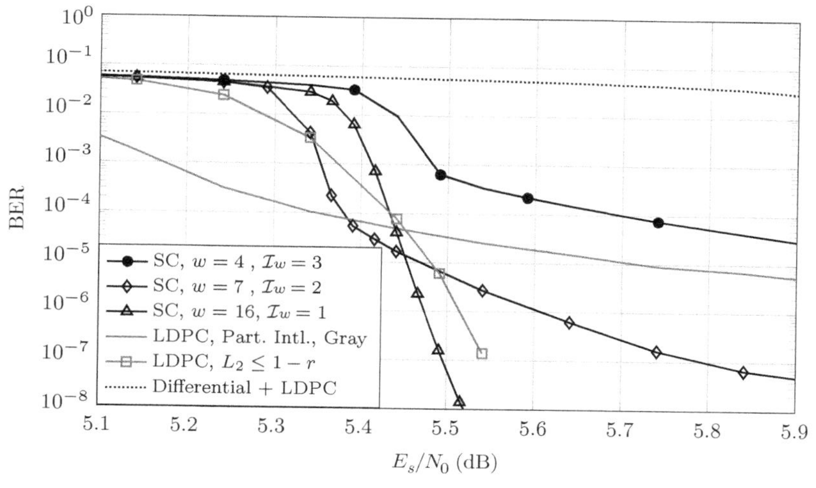

FIGURE 3.25 Simulation results of proposed and reference schemes for QPSK with $\gamma = 0.2$.

due to the steepness of the waterfall curve. If the phase-slip probability is nonzero with $\gamma = 0.2$, which may occur in a highly nonlinear DWDM transmission with OOK neighbors, we observe from Figures 3.19 and 3.25 that the LDPC code optimized without any constraints no longer performs well and has a severe error floor requiring strong outer coding. The error floor can be reduced by designing a code with $L_2 \leq 1 - r$. If $\gamma = 0.2$, the SC-LDPC code with decoder setup 3 almost shows the same performance as the LDPC code with the constraint $L_2 \leq 1 - r$ and outperforms it at low error rates (we did not observe any error floor in our simulations).

The proposed SC-LDPC code has the important advantage of *universality* [77]: A single SC code is powerful in all transmission scenarios requiring only a single *channel agnostic transmitter* and a receiver selecting the best decoding setup depending on γ and P_{slip}. The transmission scheme can thus be kept simple and only a *single* code needs to be implemented. In the setup with conventional LDPC codes, two different codes have to be implemented, depending on the setup: one code for small γ and another code, possibly with $L_2 \leq 1 - r$, for larger values of γ. The transmitter has to know the expected γ on the channel and adapt the coding scheme accordingly, which requires an undesired feedback channel and changes in the network control plane. Another possibility is to deliberately design an LDPC code that shall perform well in both cases [56]: this leads, however, to a compromise in the construction and codes that are not able to compete with codes optimized for a specific scenario. With the SC-LDPC code, we can use a single code and use receiver processing to estimate γ and setup the windowed decoder (w and $\mathcal{I}_w$) accordingly. Note that for large γ, it is advantageous to use a long window w, which means that the number of differential decoding executions per bit shall be maximized. This is in contrast to the HTDD approach [20], which however uses a differential decoder not adapted to the channel. We conclude that executing the differential detector does not degrade the performance as long as it is well adapted to the channel model.

3.5 CONCLUSIONS

In this chapter, we have described some important aspects of soft-decision FEC with iterative decoding. We have shown that the phenomenon of phase slips, which occurs frequently in the case of coherent long-haul optical communications, can be combated effectively using iterative differential decoding. We have shown that the achievable information rate is not affected by iterative differential decoding and that differential decoding only leads to an unavoidable performance loss if not properly decoded. We have further shown how phase slips affect the achievable rate and how to design a trellis diagram describing the differential code affected by phase slips. In order to achieve the best possible performance, this differential decoder needs some well-adapted code design. We have proposed different design guidelines for LDPC codes and have shown that, depending on the channel quality, there is a different code design that may lead to the desired performance, especially if very low residual BERs are targeted. Finally, we have shown that spatially coupled codes offer a more universal code design and lead to codes that are more agnostic to the channel and thus enable the implementation of a single code that performs equally well in all channel conditions and that even outperforms conventional LDPC codes.

APPENDIX: LDPC-CODED DIFFERENTIAL MODULATION—DECODING ALGORITHMS

The decoders we consider (which can include simplified versions such as binary message passing decoders [89]) rely on the knowledge of the channel.[7] The communication channel, or an equivalent communication channel comprising the physical channel as well as various inner receiver and signal processing stages, can be characterized by its *conditional probability* $P(z|y)$, that is, the probability of observing z at the receiver assuming a transmitted modulation symbol $y \in \mathcal{M}$. Note that we restrict ourselves to memoryless channels, which can be achieved in practice by sufficiently long interleaving of the channel input. Notable examples are the *binary symmetric channel* (BSC) with $(y = x)$

$$
\begin{aligned}
P(z = 0|x = 0) &= 1 - \epsilon \quad & P(z = 1|x = 0) &= \epsilon \\
P(z = 0|x = 1) &= \epsilon \quad & P(z = 1|x = 1) &= 1 - \epsilon
\end{aligned}
$$

[7] In optical coherent receivers, either channel models such as the AWGN channel are assumed together or histogram-based methods may be employed [12, Section 6.2]. Sometimes, the worst-case channel for which the system is designed may be assumed.

which is frequently used to model the hard-decision channel. Another famous example is the real-valued *AWGN* channel with *binary phase-shift keying* (BPSK) modulation with $y = (-1)^x$ and (real-valued) noise variance $\sigma_n^2 = N_0/2$, leading to

$$p(z|y) = \frac{1}{\sqrt{\pi N_0}} \exp\left(\frac{-(z-y)^2}{N_0}\right)$$

As the computation with probabilities tends to be numerically unstable and hinders potential hardware implementations, frequently LLRs [90] are employed. The LLR $L(z[t])$ for received symbol $z[t]$ at time instant t is defined for BPSK as

$$L(z[t]) = \log \frac{p(z[t]|x[t]=0)}{p(z[t]|x[t]=1)}$$

where $\log(\cdot)$ is the natural logarithm. It turns out that for the AWGN channel with BPSK modulation, we have

$$L_{\text{AWGN}}(z[t]) = \frac{2}{\sigma_n^2} z[t] = 4\frac{E_s}{N_0} z[t] = 4r\frac{E_b}{N_0} z[t] =: L_c \cdot z[t]$$

This last equation means that the LLR $L_{\text{AWGN}}(z[t])$ is obtained by multiplication of $z[t]$ with a constant $L_c := \frac{2}{\sigma_n^2}$, which depends on the noise variance only. Usually, the noise variance is assumed to be constant, and the constant L_c is predetermined and set to a value suitable for implementation. The noise variance may also be estimated at the receiver [91].

If higher-order modulation formats based on a constellation $\mathcal{M}$ are employed, we have to use a more involved computation rule. Starting from the mapping function $\phi(\boldsymbol{b})$, we first define the ith inverse mapping function

$$\phi_i^{-1}(b) =$$
$$\{y = \phi(\tilde{b}_1, \ldots, \tilde{b}_{i-1}, b, \tilde{b}_{i+1}, \ldots, \tilde{b}_q) : y \in \mathcal{M}, (\tilde{b}_1, \ldots, \tilde{b}_{i-1}, \tilde{b}_{i+1}, \ldots, \tilde{b}_q) \in \mathbb{F}_2^{q-1}\}$$

Thus, $\phi_i^{-1}(b)$ returns the set of all modulation symbols to which a bit pattern whereof the ith bit takes on the value b is assigned. The LLR for the ubiquitous *bit-wise* decoder [25] is then given by

$$L(b_i[t]) = \log\left(\frac{\sum_{y\in\phi_i^{-1}(0)} p(z[t]|y)}{\sum_{y\in\phi_i^{-1}(1)} p(z[t]|y)}\right) \tag{3.25}$$

Before we step ahead and describe the decoding algorithms, we first introduce the max* operation, which simplifies the description of the BCJR decoder [92]

Definition 4 *The* max* *operations is defined as*

$$\max{}^*(\delta_1, \delta_2) := \max(\delta_1, \delta_2) + \log\left(1 + e^{-|\delta_1 - \delta_2|}\right) = \log\left(e^{\delta_1} + e^{\delta_2}\right)$$

and can be conveniently approximated by

$$\max{}^*(\delta_1, \delta_2) \approx \max(\delta_1, \delta_2)$$

The max* operation has several properties, namely

$$\begin{aligned}
\max{}^*(\delta_1, \delta_2) &= \max{}^*(\delta_2, \delta_1) \\
\lim_{\delta_1 \to -\infty} \max{}^*(\delta_1, \delta_2) &= \delta_2 \\
\max{}^*(\delta_1, \delta_2, \delta_3) &= \max{}^*(\delta_1, \max{}^*(\delta_2, \delta_3))
\end{aligned}$$

The latter property allows us to define

$$\operatorname*{max^*}_{j=1}^{\chi} \delta_j = \max{}^*(\delta_1, \delta_2, \ldots, \delta_\chi) = \max{}^*(\delta_1, \max{}^*(\delta_2, \cdots \max{}^*(\delta_{\chi-1}, \delta_\chi) \cdots))$$

and with the trivial case

$$\operatorname*{max^*}_{j=1}^{1} \delta_j = \delta_1$$

Differential Decoding

The soft-input soft-output differential decoding is carried out using the BCJR algorithm [6]. We just summarize the operations of the BCJR algorithm in the LLR domain, such that it can be immediately applied. We give the equations for the case $V = 4$, which we have used in our simulations in this chapter. We use the trellis diagram of Figure 3.10 to describe the BCJR algorithm. The algorithm consists of computing a forward and a backward recursion. In the forward recursion, the variables $\tilde{\alpha}_t(\mathsf{S}_i)$ are updated for $i \in \{1, 2, 3, 4\}$. The initialization, which describes the initial differential memory, is usually carried out as

$$\begin{aligned}
&\tilde{\alpha}_0(\mathsf{S}_1) = 0 \\
&\tilde{\alpha}_0(\mathsf{S}_i) = -\infty \qquad \text{for} \quad i \in \{2, 3, 4\}
\end{aligned}$$

The recursive update is given by (for all $j \in \{1, 2, 3, 4\}$)

$$\tilde{\alpha}_t(\mathsf{S}_j) = \operatorname*{max^*}_{i=1}^{4} \operatorname*{max^*}_{s=0}^{3} \left(\tilde{\alpha}_{t-1}(\mathsf{S}_i) + \tilde{\gamma}_t(i, j, s)\right)$$

Similarly, the backward recursion is carried out with the initialization $\tilde{\beta}_{\tilde{n}}(\mathsf{S}_j) = 0$, for $j \in \{1, 2, 3, 4\}$, where $\tilde{n}$ denotes the length of the sequence of modulation symbols to be decoded. We have

$$\tilde{\beta}_{t-1}(\mathsf{S}_i) = \operatorname*{max^*}_{i=1}^{4} \operatorname*{max^*}_{s=0}^{3} \left(\tilde{\beta}_t(\mathsf{S}_j) + \tilde{\gamma}(i, j, s)\right)$$

Before giving the equation to compute $\tilde{\gamma}(i,j,s)$, we first introduce the sets $\mathcal{M}_{\mathsf{S}_i} \subset \mathcal{M}$, which contain all modulation symbols that are associated with state S_i (that are within the region associated with state S_i). For the example of the QPSK constellation of Figure 3.4, $\mathcal{M}_{\mathsf{S}_1} = \{\frac{1+\imath}{\sqrt{2}}\}$ and for the example of the 16QAM constellation of Figure 3.5, we have

$$\mathcal{M}_{\mathsf{S}_1} = \left\{ \frac{1+\imath 1}{\sqrt{10}}, \frac{3+\imath 1}{\sqrt{10}}, \frac{1+\imath 3}{\sqrt{10}}, \frac{3+\imath 3}{\sqrt{10}} \right\}$$

The variable $\tilde{\gamma}_t(i,j,\zeta)$ describes the (logarithmic) probability of a state transition from state S_i at time $t-1$ to state S_j at time t provided that the phase-slip occurrence descriptor $s[t]$ takes on the value ζ. We have

$$\begin{aligned}\tilde{\gamma}_t(i,j,\zeta) = &\frac{1}{N_0} \sum_{\chi \in \mathcal{M}_{\mathsf{S}_j}} |z[t] - \chi|^2 + \frac{1}{2} \sum_{\kappa=1}^{v} \left(1 - 2\check{f}^{-1}_{\mathrm{diff},\kappa}(\mathsf{S}_i, \mathsf{S}_j, \zeta)\right) L^{[\mathrm{apriori},\Pi]}_{(t-1)v+\kappa} \\ &+ \log P(s = \zeta)\end{aligned}$$

Note that with the phase-slip model introduced in Section 3.2.2, we may abbreviate $\log P(s = \zeta) = |\zeta| \log \xi$. The function $\check{f}^{-1}_{\mathrm{diff},\kappa}(\mathsf{S}_i, \mathsf{S}_j, s)$ returns the κth bit b_κ of the differential encoding map that causes a state transition from S_i to $\mathsf{S}_{j'}$, where j' is an intermediate state leading to the final state $j = ((j' + s - 1) \bmod V) + 1$ after taking into account the phase slip. $L_i^{[\mathrm{apriori},\Pi]}$ contains the input LLR values $L_i^{[\mathrm{apriori}]}$ that are provided by the LDPC decoder after full or partial interleaving. In the initial execution (first iteration) of the differential decoder, we may set $L_i^{[\mathrm{apriori},\Pi]} = 0$.

Finally, we obtain for each $t \in \{1, \ldots, \tilde{n}\}$ and $\kappa \in \{1, \ldots, v\}$

$$\begin{aligned}L^{[\mathrm{diff},\Pi]}_{(t-1)v+\kappa} = &\max^*_{\substack{(i,j,s)\\ \check{f}^{-1}_{\mathrm{diff},\kappa}(\mathsf{S}_i,\mathsf{S}_j,s)=0}} \left(\tilde{\alpha}_{t-1}(\mathsf{S}_i) + \tilde{\gamma}_t(i,j,s) + \tilde{\beta}_t(\mathsf{S}_j)\right) - \\ &- \max^*_{\substack{(i,j,s)\\ \check{f}^{-1}_{\mathrm{diff},\kappa}(\mathsf{S}_i,\mathsf{S}_j,s)=0}} \left(\tilde{\alpha}_{t-1}(\mathsf{S}_i) + \tilde{\gamma}_t(i,j,s) + \tilde{\beta}_t(\mathsf{S}_j)\right) - L^{[\mathrm{apriori},\Pi]}_{(t-1)v+\kappa}\end{aligned}$$

After (partial or full) deinterleaving of $L_i^{[\mathrm{diff},\Pi]}$, we obtain $L_i^{[\mathrm{diff}]}$, which is used as input of the LDPC decoder.

LDPC Decoding

A vast collection of various decoding algorithms for LDPC codes exist and we refer to the broad body of literature for a good introduction (see, for example, [18, 93]). Most of these decoders are message passing decoders, where the most prominent is probably the sum–product decoder [18]. In what follows, we describe the sum–product decoder and the closely related min-sum decoder, which can be interpreted as being an approximation of the sum–product decoder.

In order to describe the sum–product and min-sum decoders, we introduce the set $\mathcal{N}(m) := \{j : H_{m,j} \neq 0\}$, which contains the positions (columns) of nonzero entries at row m of the parity-check matrix $\boldsymbol{H}$. For the matrix given in Example 3.3, we have $\mathcal{N}(1) = \{1; 4; 11; 12; 15; 22; 24; 25; 28; 29; 30; 31\}$. Similarly, the set $\mathcal{M}(n) := \{i : H_{i,n} \neq 0\}$ contains the positions (rows) of nonzero entries at column n of the parity-check matrix $\boldsymbol{H}$. Again, for the exemplary matrix of Example 3.3, we have $\mathcal{M}(1) = \{1; 2\}$, $\mathcal{M}(2) = \{3; 5\}$ and so on.

Within the sum–product decoder, messages are exchanged between the variable nodes and the check nodes, thus the name *message passing decoder*. We denote the message that is passed from variable node i toward check node j by $L_{i,j}^{[v\to c]}$. Similarly, the message that is passed from check node j toward variable node i is denoted by $L_{i,j}^{[v\leftarrow c]}$. Before first executing the LDPC decoder with a new frame of data, all messages are set to zero, that is, $L_{i,j}^{[v\to c]} = L_{i,j}^{[v\leftarrow c]} = 0$ for all combinations of $(j, i) \in [1, \dots, m] \times [1, \dots, n]$ such that $H_{j,i} \neq 0$.

The sum–product LDPC coder computes for each of the n variables, that is, for each transmitted bit, the total sum

$$L_i^{[\text{tot}]} = L_i^{[\text{diff}]} + \sum_{j\in\mathcal{M}(i)} L_{i,j}^{[v\leftarrow c]}, \quad \forall i \in \{1, \dots, n\}$$

Using this total sum, the variable-to-check messages may be computed as

$$L_{i,j}^{[v\to c]} = L_i^{[\text{tot}]} - L_{i,j}^{[v\leftarrow c]}, \quad \forall i \in \{1, \dots, n\}, \forall j \in \mathcal{M}(i)$$

In the second step, the check node update rule is carried out to compute new check-to-variable messages

$$L_{i,j}^{[v\leftarrow c]} = 2\tanh^{-1}\left(\prod_{i'\in\mathcal{N}(j)\backslash\{i\}} \tanh\left(\frac{L_{i',j}^{[v\to c]}}{2}\right)\right), \quad \forall j \in \{1, \dots, m\}, \forall i \in \mathcal{N}(j)$$

where the inner product is taken over all entries in $\mathcal{N}(j)$ except the one under consideration i. This is indicated by the notation $\mathcal{N}(j)\backslash\{i\}$. Usually, in practical implementations, simplified approximations to this update rule are implemented, for example, the scaled min-sum rule [94]

$$L_{i,j}^{[v\leftarrow c]} = \nu\left(\prod_{i'\in\mathcal{N}(j)\backslash\{i\}} \operatorname{sign}\left(L_{i',j}^{[v\to c]}\right)\right) \min_{i'\in\mathcal{N}(j)\backslash\{i\}} |L_{i',j}^{[v\to c]}|$$

where ν is an appropriately chosen scaling factor. See [95] for other simplified variants of the sum–product algorithm.

With the updated check-to-variable node messages, a new *a priori* message that is transmitted to the differential decoder may be computed as

$$L_i^{[\text{apriori}]} = \sum_{j \in \mathcal{M}(i)} L_{i,j}^{[\text{v}\leftarrow\text{c}]}, \quad \forall i \in \{1, \ldots, n\}$$

Note that the convergence of the sum–product decoder as described here can be considerable improved by using the so-called row-layered decoder [64], which allows to roughly halve the number of decoding iterations. Many additional possible decoder variants, which may be better suited for an ASIC implementation than the message passing decoder described here, are discussed in Ref. [18].

REFERENCES

1. Grover W. Error correction in dispersion-limited lightwave systems. *J Lightwave Technol* 1988;6:643–654.
2. Fludger C, Nuss D, Kupfer T. Cycle-slips in 100G DP-QPSK transmission systems. Proceedings OFC/NFOEC; 2012. Paper OTu2G.1.
3. Leong MY, Larsen KJ, Jacobsen G, Popov S, Zibar D, Sergeyev S. Dimensioning BCH codes for coherent DQPSK systems with laser phase noise and cycle slips. *J Lightwave Technol* 2014;32(21):3446–3450.
4. Pfau T. Carrier recovery algorithms and real-time DSP implementation for coherent receivers. Proceedings of OFC, Mar 2014. Paper W4K.1.
5. Hoeher P, Lodge J. "Turbo DQPSK": iterative differential PSK demodulation and channel decoding. *IEEE Trans. Commun.* 1999;47:837–843.
6. Bahl L, Cocke J, Jelinek F, Raviv J. Optimal decoding of linear codes for minimizing symbol error rate. *IEEE Trans. Inf. Theory* 1974;20:284–287.
7. Schmalen L. Energy efficient FEC for optical transmission systems. Proceedings of the Optical Fiber Communications Conference (OFC); San Francisco (CA); Mar 2014.
8. Bülow H, Lu X, Schmalen L. Experimental analysis of transmission and soft decoding of optimized 4D constellations. Asia Communications and Photonics Conference (ACP); Beijing, China; 2013.
9. Koetter R, Singer AC, Tüchler M. Turbo equalization. *IEEE Signal Process Mag* 2004;21:67–80.
10. Fujimori T, Koike-Akino T, Sugihara T, Kubo K, Koguchi K, Mizuochi T, Ohshima C, Nakashima H, Hoshida T. A study on the effectiveness of turbo equalization with FEC for nonlinearity compensation in coherent WDM transmissions. Proceedings of Optoelectronics and Communications Conference (OECC); June 2013.
11. Beygi L, Agrell E, Kahn J, Karlsson M. Coded modulation for fiber-optic networks. *IEEE Signal Process Mag* 2014;31:93–103.
12. Djordjevic IB, Ryan W, Vasic B. Coding for Optical Channels. New York, Dordrecht, Heidelberg, London: Springer-Verlag; 2010.
13. Koike-Akino T, Kojima K, Millar DS, Parsons K, Miyata Y, Matsumoto W, Sugihara T, Mizuochi T. Cycle slip-mitigating turbo demodulation in LDPC-coded coherent optical communications. Proceedings of OFC, 2014. Paper M3A.3.

14. Weber W. Differential encoding for multiple amplitude and phase shift keying systems. *IEEE Commun. Lett.* 1978;26:385–391.
15. Pfau T, Hoffmann S, Noe R. Hardware-efficient coherent digital receiver concept with feedforward carrier recovery for *M*-QAM constellations. *J Lightwave Technol* 2009;27:989–999.
16. Bisplinghoff A, Langenbach S, Kupfer T, Schmauss B. Turbo differential decoding failure for a coherent phase slip channel. Proceedings of European Conference on Optical Communications (ECOC), Sept 2012. Paper Mo.1.A.5.
17. Leven A, ten Brink S. Method of decoding optical data signals. European Patent Application EP2506516; Oct 2012.
18. Ryan WE, Lin S. Channel Codes: Classical and Modern. Cambridge: Cambridge University Press; 2009.
19. ten Brink S. Convergence behavior of iteratively decoded parallel concatenated codes. *IEEE Trans Commun* 2001;49:1727–1737.
20. Bisplinghoff A, Langenbach S, Beck N, Fludger C, Schulien C. Cycle slip tolerant hybrid turbo differential decoding. Proceedings of European Conference on Optical Communications (ECOC); Cannes, France; Sept 2014. Paper Th.2.3.3.
21. Shannon CE. A mathematical theory of communication. *Bell Syst Tech J* 1948;27:379–423, 623–656.
22. Cover TM, Thomas JA. Elements of Information Theory. Hoboken (NJ): John Wiley & Sons, Inc.; 2006.
23. Forney GD Jr. Trellis shaping. *IEEE Trans. Inf. Theory* 1992;38:281–300.
24. Smith BP, Kschischang FR. A pragmatic coded modulation scheme for high-spectral-efficiency fiber-optic communications. *J Lightwave Technol* 2012;30:2047–2053.
25. Guillén i Fàbregas A, Martinez A, Caire G. Bit-interleaved coded modulation. *Found Trends Commun Inf Theory* 2008;5:1–153.
26. ten Brink S. Exploiting the chain rule of mutual information for the design of iterative decoding schemes. Allerton Annual Conference on Communication, Control and Computing; Monticello (IL); Oct 2001.
27. Ashikhmin A, Kramer G, ten Brink S. Extrinsic information transfer functions: model and erasure channel properties. *IEEE Trans Inf Theory* 2004;50:2657–2673.
28. Franceschini M, Ferrari G, Raheli R, Curtoni A. Serial concatenation of LDPC codes and differential modulations. *IEEE J Sel Areas Commun* 2005;23:1758–1768.
29. Caire G, Taricco G, Biglieri E. Bit-interleaved coded modulation. *IEEE Trans Inf Theory* 1998;44:927–946.
30. ten Brink S, Speidel J, Yan R-H. Iterative demapping and decoding for multilevel modulation. Proceedings of GLOBECOM; Sydney; 1998.
31. Li X, Ritcey JA. Bit-interleaved coded modulation with iterative decoding. IEEE International Conference on Communications (ICC); Vancouver, BC, Canada; June 1999.
32. Yu F, Stojanovic N, Hauske F, Chang D, Xiao Z, Bauch G, Pflüger D, Xie C, Zhao Y, Jin L, Li Y, Li L, Xu X, Xiong Q. Soft decision LDPC turbo decoding for DQPSK modulation in coherent optical receivers. Proceedings of ECOC, Geneva (CH), Sept 2011. Paper We.10.P1.70.
33. Schmalen L. Low-complexity phase slip tolerant LDPC-based FEC scheme. Proceedings of ECOC; Cannes, France, Sept 2014. Paper Th.2.3.2.

34. Schmalen L. A low-complexity LDPC coding scheme for channels with phase slips. *J Lightwave Technol* 2015;33(7):1319–1325.
35. Bellini S, Ferrari M, Tomasoni A, Costantini C, Razzetti L, Gavioli G. LDPC design for block differential modulation in optical communications. *J Lightwave Technol* 2015;33(1):78–88.
36. Hanzo L, Liew TH, Yeap BL. Turbo Coding, Turbo Equalisation and Space-Time Coding for Transmission over Fading Channels. Chichester: John Wiley & Sons, Ltd.; 2002.
37. Gallager RG. Low-Density Parity-Check Codes. Cambridge (MA): M.I.T. Press; 1963.
38. Berrou C, Glavieux A, Thitimajshima P. Near Shannon limit error-correcting coding and decoding: Turbo-codes (1). IEEE International Conference on Communications (ICC); May 1993. p 1064–1070.
39. MacKay DC, Neal R. Near Shannon limit performance of low density parity check codes. *Electron Lett* 1996;32:1645–1646.
40. Luby M, Mitzenmacher M, Shokrollahi A, Spielman DA, Stemann V. Practical loss-resilient codes. Proceedings of the ACM Symposium on Theory of Computing; 1997. p 150–159.
41. IEEE. 802.11, wireless LAN medium access control (MAC) and physical layer (PHY) specifications; 2012.
42. ETSI. DVB-S2, ETSI standard EN 302 307 v.1.2.1; 2009.
43. IEEE. 802.3an, local and metropolitan area networks-specific requirements part 3: carrier sense multiple access with collision detection (CSMA/CD): Access method and physical layer specifications; 2006.
44. Miyata Y, Kubo K, Yoshida H, Mizuochi T. Proposal for frame structure of optical channel transport unit employing LDPC codes for 100 Gb/s FEC. Proceedings of OFC/NFOEC, paper NThB2; 2009.
45. Richardson T. Error floors of LDPC codes. Proceedings of the Annual Allerton Conference on Communications, Control and Computing; 2003.
46. Dolecek L, Lee P, Zhang Z, Anantharam V, Nikolic B, Wainwright M. Predicting error floors of structured LDPC codes: deterministic bounds and estimates. *IEEE J Sel Areas Commun* 2009;27:908–917.
47. Morero DA, Castrillon MA, Ramos FA, Goette TA, Agazzi OE, Hueda MR. Non-concatenated FEC codes for ultra-high speed optical transport networks. Proceedings of GLOBECOM; 2011. p 1–5.
48. Bonello N, Chen S, Hanzo L. Design of low-density parity-check codes. *IEEE Veh Technol Mag* 2011;6(4):16–23.
49. Liva G, Song S, Lan L, Zhang Y, Lin S, Ryan W. Design of LDPC codes: a survey and new results. *J Commun Softw Syst* 2006;2(3):191–211.
50. Richardson T, Urbanke R. Modern Coding Theory. New York: Cambridge University Press; 2008.
51. Leven A, Schmalen L. Status and recent advances on forward error correction technologies for lightwave systems. *J Lightwave Technol* 2014;32:2735–2750.
52. Richardson TJ, Shokrollahi MA, Urbanke RL. Design of capacity-approaching irregular low-density parity-check codes. *IEEE Trans Inf Theory* 2001;47:619–637.
53. ten Brink S, Kramer G, Ashikhmin A. Design of low-density parity-check codes for modulation and detection. *IEEE Trans Commun* 2004;52(4):670–678.

54. Schreckenbach F. Iterative decoding of bit-interleaved coded modulation [PhD thesis]. Munich: TU Munich; 2007.
55. Sharon E, Ashikhmin A, Litsyn S. Analysis of low-density parity-check codes based on EXIT functions. *IEEE Trans Commun* 2006;54:1407–1414.
56. Pflüger D, Bauch G, Hauske F, Zhao Y. Design of LDPC codes for hybrid 10 Gbps/100 Gbps optical systems with optional differential modulation. Proceedings of ITG SCC; 2013.
57. Pflüger D, Bauch G, Zhao Y, Hauske FN. Conditions on degree distributions to compensate differential penalty by LDPC turbo decoding. Proceedings of IEEE VTC; 2013.
58. Shokrollahi A. and Storn R. Design of Efficient Erasure Codes with Differential Evolution. In: Price KV, Storn R, Lampinen JA, editors. Differential evolution—a practical approach to global optimization, Chapter 7.7. Berlin, Heidelberg: Springer–Verlag; 2005.
59. Lechner G, Sayir J, Land I. Optimization of LDPC codes for receiver frontends. Proceedings ISIT; Seattle (WA); 2006.
60. Benammar B, Thomas N, Pouillat C, Boucheret M, Dervin M. Asymptotic analysis and design of iterative receivers for non linear ISI channels. Proceedings of the International Symposium on Turbo Codes & Iterative Information Processing, (Bremen, Germany); Aug 2014.
61. Alvarado A, Szczecinski L, Agrell E, Svensson A. On BICM-ID with multiple interleavers. *IEEE Commun Lett* 2010;14:785–787.
62. Tillich J, Zemor G. On the minimum distance of structured LDPC codes with two variable nodes of degree 2 per parity-check equation. Proceedings of IEEE ISIT; Seattle (WA); July 2006. p 1549–1553.
63. Sugihara K, Miyata Y, Sugihara T, Kubo K, Yoshida H, Matsumoto W, Mizuochi T. A spatially-coupled type LDPC code with an NCG of 12 dB for optical transmission beyond 100 Gb/s. Proceedings of OFC; 2013. Paper OM2B.4.
64. Hocevar D. A reduced complexity decoder architecture via layered decoding of LDPC codes. Proceedings of IEEE Workshop on Signal Processing Systems (SiPS); 2004.
65. Smith BP, Farhood A, Hunt A, Kschischang FR, Lodge J. Staircase codes: FEC for 100 Gb/s OTN. *J Lightwave Technol* 2012;30(1):110–117.
66. Zhang L, Kschischang FR. Multi-edge-type low-density parity-check codes for bandwidth-efficient modulation. *IEEE Trans Commun* 2013;61:43–52.
67. Costello DJ Jr., Forney GD Jr. Channel coding: the road to channel capacity. *Proc IEEE* 2007;95(6):1150–1177.
68. Tavares M. On low-density parity-check convolutional codes: constructions, analysis and VLSI implementations [PhD thesis]. Dresden: TU Dresden; 2010.
69. Kudekar S, Richardson T, Urbanke R. Threshold saturation via spatial coupling: why convolutional LDPC ensembles perform so well over the BEC. *IEEE Trans Inf Theory* 2011;57:803–834.
70. Kudekar S, Richardson T, Urbanke R. Spatially coupled ensembles universally achieve capacity under belief propagation. *IEEE Trans Inf Theory* 2013;59 (12):7761–7813.
71. Felström AJ, Zigangirov KS. Time-varying periodic convolutional codes with low-density parity-check matrix. *IEEE Trans Inf Theory* 1999;45:2181–2191.
72. Lentmaier M, Fettweis GP, Zigangirov KS, Costello DJ Jr. Approaching capacity with asymptotically regular LDPC codes. Proceedings of Information Theory and Applications Workshop (ITA); Feb 2009.

73. Lentmaier M, Mitchell DGM, Fettweis G, Costello DJ Jr. Asymptotically good LDPC convolutional codes with AWGN channel thresholds close to the Shannon limit. Proceedings of the International Symposium on Turbo Codes & Iterative Information Proceedings (ISTC); Brest, France; Sept 2010.
74. Jian Y-Y, Pfister HD, Narayanan KR, Rao R, Mazareh R. Iterative hard-decision decoding of braided BCH codes for high-speed optical communication. Proceedings of GLOBE-COM; Atlanta (GA); Dec 2013. p 2398–2403.
75. IEEE 1901. IEEE standard for broadband over power line networks: medium access control and physical layer specifications. New York: IEEE; 2010.
76. Costello DJ, Dolecek L, Fuja TE, Kliewer J, Mitchell DGM, Smarandache R Jr. Spatially coupled sparse codes on graphs: theory and practice. *IEEE Commun Mag* 2014;52:168–176.
77. Schmalen L, ten Brink S. Combining spatially coupled LDPC codes with modulation and detection. Proceedings of ITG SCC; Munich; Jan 2013.
78. Thorpe J. Low-Density Parity-Check (LDPC) codes constructed from protographs. Tech. Rep., IPN Progress Report 42-154. Pasadena (CA): JEP Propulsion Laboratory, California Institute of Technology; 2003.
79. Mitchell DGM, Lentmaier M, Costello DJ Jr. AWGN channel analysis of terminated LDPC convolutional codes. Proceedings of Information Theory and Applications Workshop (ITA); 2011.
80. Kudekar S, Richardson T, Urbanke R. Wave-like solutions of general one-dimensional spatially coupled systems. Tech. Rep. arXiv:1208:5273. CoRR; 2012, submitted to Trans Inform Theory.
81. Iyengar AR, Siegel PH, Urbanke R, Wolf JK. Windowed decoding of spatially coupled codes. Proceedings of ISIT; July 2011.
82. Aref V, Schmalen L, ten Brink S. On the convergence speed of spatially coupled LDPC ensembles. Proc. Allerton Conference on Communications, Control, and Computing; Oct 2013. arXiv:1307.3780.
83. Liva G, Chiani M. Protograph LDPC codes design based on EXIT analysis. Proceedings of GLOBECOM; 2007.
84. Chung S, Richardson TJ, Urbanke RL. Analysis of sum-product decoding of low-density parity-check codes using a Gaussian approximation. *IEEE Trans Inf Theory* 2001;47:657–670.
85. Häger C, Graell i Amat A, Brännström F, Alvarado A, Agrell E. Improving soft FEC performance for higher-order modulations via optimized bit channel mappings. *Opt. Express* 2014;22:14544–14558.
86. Lentmaier M, Prenda MM, Fettweis G. Efficient message passing scheduling for terminated LDPC convolutional codes. Proceedings of ISIT; St. Petersburg, Russia; 2011.
87. Iyengar AR, Papaleo M, Siegel PH, Wolf JK, Vanelli-Coralli A, Corazza GE. Windowed decoding of protograph-based LDPC convolutional codes over erasure channels. *IEEE Trans Inf Theory* 2012;58:2303–2320.
88. Schmalen L, ten Brink S, Leven A. Spatially-coupled LDPC protograph codes for universal phase slip-tolerant differential decoding. Proceedings of OFC; Los Angeles (CA); 2015.
89. Lechner G, Pedersen T, Kramer G. Analysis and design of binary message passing decoders. *IEEE Trans Commun* 2012;60:601–607.

90. Hagenauer J, Offer E, Papke L. Iterative decoding of binary block and convolutional codes. *IEEE Trans Inf Theory* 1996;42:429–445.

91. Mecklenbräuker CF, Paul S. On estimating the signal to noise ratio from BPSK signals. Proceedings of ICASSP; Philadelphia (PA); Mar 2005.

92. Robertson P, Villebrun E, Hoeher P. A comparison of optimal and sub-optimal MAP decoding algorithms operating in the log domain. IEEE International Conference on Communications (ICC); Seattle (WA); June 1995. p 1009–1013.

93. Moon TK. Error Correction Coding - Mathematical Methods and Algorithms. Hoboken (NJ): John Wiley & Sons, Inc.; 2005.

94. Chen J, Fossorier MPC. Near optimum universal belief propagation based decoding of low-density parity check codes. *IEEE Trans Commun* 2002;50:406–414.

95. Zhao J, Zarkeshvari F, Banihashemi AH. On implementation of min-sum algorithm and its modifications for decoding low-density parity-check (LDPC) codes. *IEEE Trans Commun* 2005;53:549–554.

4

SPECTRALLY EFFICIENT MULTIPLEXING: NYQUIST-WDM

GABRIELLA BOSCO
Dipartimento di Elettronica e Telecomunicazioni (DET), Politecnico di Torino, Torino, Italy

This chapter discusses the basics of the Nyquist signaling theory and various aspects regarding its application in coherent optical communication systems to improve the transport spectral efficiency (SE). An alternative, and dual, approach to achieve a high SE is based on the use of orthogonal frequency-division multiplexing (OFDM), which is described in the following chapter.

4.1 INTRODUCTION

After the advent of digital coherent technology [1], there has been a significant interest in using advanced multilevel modulation formats [2], in combination with advanced multiplexing techniques [3] and efficient digital signal processing (DSP) algorithms [4], to increase the spectral efficiency (SE) in fiber-optic communication systems.

This chapter focuses on wavelength-division multiplexing (WDM) optical systems, for which the SE is defined as the information capacity of a single channel (in bit/s) divided by the frequency spacing Δf (in Hz) between the carriers of the WDM comb:

$$\mathrm{SE} = \frac{R_s}{\Delta f}\frac{\log_2(M)}{(1+r)} \tag{4.1}$$

Enabling Technologies for High Spectral-efficiency Coherent Optical Communication Networks, First Edition. Edited by Xiang Zhou and Chongjin Xie.
© 2016 John Wiley & Sons, Inc. Published 2016 by John Wiley & Sons, Inc.

where R_s is the symbol rate, M is the number of constellation points of the modulation format, and r is the redundancy of the forward-error correction (FEC) code, for example $r = 0.07$ for an FEC with overhead (OH) equal to 7%. For more details on the basics of digital communications and coding, the interested reader is referred to classical textbooks such as [5, 6]. The total system capacity (defined as the maximum information in bit/s that can be transmitted by the WDM comb) is obtained as the product between the SE and the available bandwidth. The maximization of the SE thus plays an important role in the maximization of the overall system capacity.

In the past years, the SE of optical systems has significantly increased, mainly due to the advent of coherent-detection technologies, which enabled the use of high-order modulation formats based on polarization-division multiplexing (PDM) [2], such as PDM-QPSK (quadrature phase-shift keying), with $M = 4$, PDM-16QAM (quadrature-amplitude modulation), with $M = 8$, and PDM-64QAM, with $M = 12$. However, the use of high-order modulation formats requires a higher optical signal-to-noise ratio (OSNR), which may result in a significantly reduced achievable transmission distance [2, 7].

An alternative way of increasing the SE, and consequently the overall system capacity, involves reducing the frequency spacing Δf between the WDM sub-carriers. In this chapter, the focus is on the minimization of the *normalized* frequency spacing δf, defined as

$$\delta f = \frac{\Delta f}{R_s} \tag{4.2}$$

To achieve ultimate spectral efficiency, WDM channel spacings are reduced until the optical spectra of neighboring channels start to noticeably overlap. In this limit of ultra-dense WDM systems, linear crosstalk between adjacent WDM channels becomes a main source of degradation [8]. An efficient countermeasure to limit the crosstalk is based on an accurate spectral shaping of each subchannel of the WDM comb [9–12]: this technique has been widely used in radio links for decades and has been lately proposed and demonstrated for optical links too. Employing this technique, known in optical communications as "Nyquist-WDM," the transmission of PDM-QPSK WDM signals with channel spacing equal to the symbol rate has been demonstrated over transpacific distances [13–15]. The technique has also been successfully applied to the generation and transmission of higher-order modulation formats, such as PDM-8QAM [16], PDM-16QAM [17–19], PDM-32QAM [20–22], and PDM-64QAM [23–25], with frequency spacing values equal or very close to the symbol rate.

The ideal shape of the transmitted spectra, which allows to achieve a channel spacing equal to the symbol rate R_s with no crosstalk and no intersymbol interference (ISI), is rectangular with bandwidth equal to R_s [26]. In such an ideal scenario, Nyquist-WDM can achieve the optimum matched filter performance in additive white Gaussian noise (AWGN) systems [5]. In practice, penalties are to be expected when the ideal constraints on Nyquist WDM implementation are relaxed [27], for instance with the transmission of channels with not perfectly rectangular spectra.

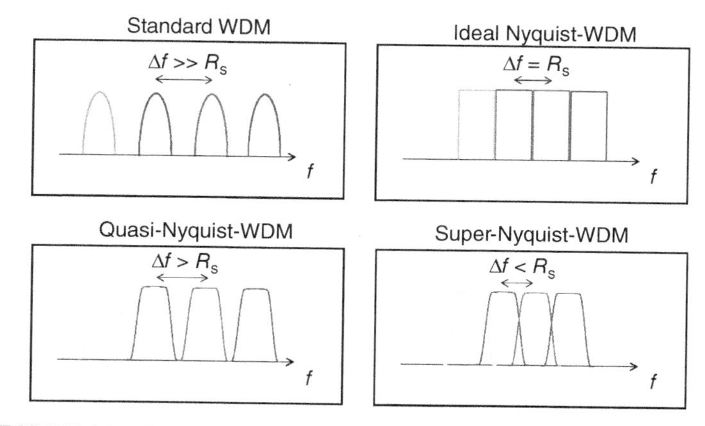

FIGURE 4.1 Spectrum of the WDM comb in four different configurations.

In this chapter, the fundamental Nyquist-WDM signaling theory is presented and various aspects regarding its implementation and application in coherent optical communication systems to improve the transport spectral efficiency are discussed. Depending on the normalized frequency spacing δf among the WDM channels, three different categories of Nyquist-WDM signaling, which are described in detail in Section 4.2, can be identified:

- $\delta f = 1$ (i.e., $\Delta f = R_s$): Ideal Nyquist-WDM (Section 4.2.1).
- $\delta f > 1$ (i.e., $\Delta f > R_s$): Quasi-Nyquist-WDM (Section 4.2.2).
- $\delta f < 1$ (i.e., $\Delta f < R_s$): Super-Nyquist-WDM (Section 4.2.3).

Figure 4.1 shows an example of the WDM spectrum in the three cases described here, as well as in a more standard WDM configuration.

Section 4.3 discusses the receiver (Rx) structure used to detect Nyquist-WDM signals. Section 4.4 is devoted to a general discussion of practical implementations of the Nyquist-WDM transmitter (Tx), focusing on the different available options to perform spectral shaping in the optical and digital domains. In Section 4.5, a theoretical analysis of the trade-offs between capacity and reach is presented, followed by a review of the most relevant experimental demonstrations of Nyquist-WDM transmission over multispan optical links. Finally, Section 4.6 concludes the chapter.

4.2 NYQUIST SIGNALING SCHEMES

This section reports a review of the fundamental results of the digital communications theory, highlighting the properties of the transmitted signal spectrum that avoids both linear crosstalk between subcarriers (in the frequency domain) and ISI between adjacent pulses (in the time domain). The section is organized in three subsections, for each of the categories of Nyquist signaling identified in Section 4.1.

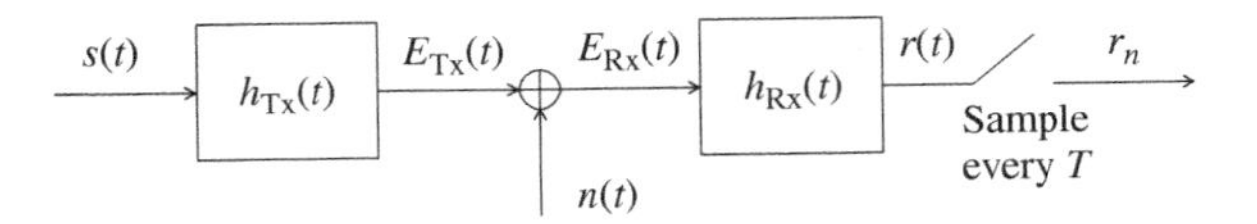

FIGURE 4.2 Baseband model of a classical digital modulation system.

4.2.1 Ideal Nyquist-WDM ($\Delta f = R_s$)

Figure 4.2 shows the baseband model of a classical digital modulation system, which is a good approximation of the behavior of an optical system with coherent detection [12]. The model assumes propagation in the linear regime, as well as ideal modulation onto the optical carrier and perfect carrier recovery. Complex symbols α_k are transmitted with a continuous-time pulse shape $h_{\text{Tx}}(t)$, resulting in a transmit electrical field that can be written as

$$E_{\text{Tx}}(t) = \sum_k \alpha_k h_{\text{Tx}}(t - kT) \tag{4.3}$$

where T is the signaling period, equal to the inverse of the symbol rate ($T = 1/R_s$). This can alternatively be viewed as applying a filter with transfer function $H_{\text{Tx}}(f)$ (equal to the Fourier transform of the impulse response $h_{\text{Tx}}(t)$) to a signal $s(t)$, defined as

$$s(t) = \sum_k \alpha_k \delta(t - kT) \tag{4.4}$$

The transfer function $H_{\text{Tx}}(f)$ takes into account all filtering effects at the transmitter side. Under the assumption of zero-mean and uncorrelated modulation symbols α_k, the power spectral density (PSD) of the baseband transmitted electrical field of Equation 4.3 is proportional to $|H_{\text{Tx}}(f)|^2$ [5]; that is, it depends exclusively on the transmit pulse shape.

The optical channel is modeled as an AWGN channel, with a flat frequency response. In an optical transmission system, this corresponds to assuming that the effects of chromatic dispersion (CD) and other propagation effects are perfectly compensated for. The signal at the input of the receiver can then be written as

$$E_{\text{Rx}}(t) = E_{\text{Tx}}(t) + n(t) = \sum_k \alpha_k h_{\text{Tx}}(t - kT) + n(t) \tag{4.5}$$

where $n(t)$ is a Gaussian random process with two-sided PSD equal to $N_0/2$.

The coherent receiver (Rx) is modeled as a linear filter with transfer function $H_{\text{Rx}}(f)$, followed by a sampler, corresponding to an analog-to-digital conversion device, operating at the symbol rate R_s. The analog waveform at the decision point can be written as

$$r(t) = E_{\text{Rx}}(t) * h_{\text{Rx}} = \sum_k \alpha_k h(t - kT) + \nu(t) \tag{4.6}$$

with $h(t) = h_{\mathrm{Tx}}(t) * h_{\mathrm{Rx}}(t)$ and $v(t) = n(t) * h_{\mathrm{Rx}}(t)$. After sampling,

$$r_n = r(t_0 + nT) = \alpha_n h(t_0) + \sum_{k \neq n} \alpha_k h_{n-k} + v_k \tag{4.7}$$

where $h_{n-k} = h(t_0 + (n-k)T)$, $v_k = v(t_0 + nT)$ and t_0 is the optimum sampling point. The condition for no ISI is

$$h_i = \begin{cases} 1 & (i = 0) \\ 0 & (i \neq 0) \end{cases} \tag{4.8}$$

The Nyquist theorem [5] states that the necessary and sufficient condition for $h(t)$ to satisfy Equation 4.8 is that its Fourier transform $H(f)$ satisfies

$$B(f) = \sum_k H\left(f + \frac{k}{T}\right) = T \tag{4.9}$$

Three cases can be distinguished, depending on the spectral width B_H of the overall transfer function $H(f)$:

- $R_s < B_H$: the left term of Equation 4.9 consists of overlapping replicas of $H(f)$, separated by $R_s = 1/T$, and there exist numerous choices for $H(t)$ that satisfy Equation 4.9 . This case is discussed in Section 4.2.2.
- $R_s > B_H$: since the left term of Equation 4.9 consists of nonoverlapping replicas of H(f), separated by $R_s = 1/T$, there is no choice of $H(f)$ that satisfies Equation 4.9 ; that is, there is no way for a system to be designed without ISI. This case is discussed in Section 4.2.3.
- $R_s = B_H$: this case corresponds to the so-called "Nyquist limit," which is the subject of this section.

In the "Nyquist limit" case, there exists only one $H(f)$ satisfying Equation 4.9 , namely

$$H(f) = \begin{cases} T & (|f| < B_H/2) \\ 0 & (\text{otherwise}) \end{cases} \tag{4.10}$$

which corresponds to the pulse in time

$$h(t) = \frac{\sin\left(\frac{\pi t}{T}\right)}{\frac{\pi t}{T}} = \text{sinc}\left(\frac{\pi t}{T}\right) \tag{4.11}$$

The difficulty with this choice of $h(t)$ is that it is noncausal, and therefore nonrealizable. In order to make it realizable, usually a delayed version of it is used: $h(t - T_d)$, with $h(t - T_d) \simeq 0$ if $t < 0$, which also implies a shift in the sampling instant. Another problem with this pulse is that it decays to zero very slowly and, consequently, a small error in the sampling time results in a large ISI.

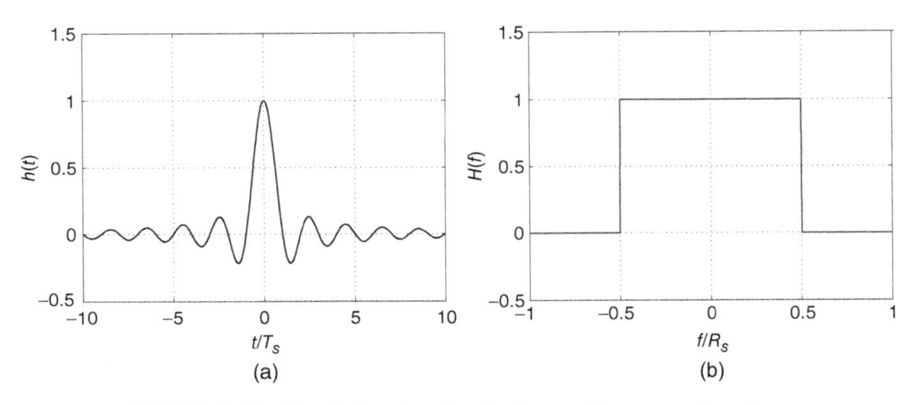

FIGURE 4.3 Ideal Nyquist pulse in time and frequency domains.

Figure 4.3 shows the time- and frequency-domain pulse shapes corresponding to the ideal Nyquist-WDM signal. From the optimum detection theory in AWGN channel [5], the receiver filter yielding the best performance is matched to the transmit pulse shape; that is, $|H_{\mathrm{Rx}}(f)| = |H_{\mathrm{Tx}}(f)| = \sqrt{H(f)}$. In the ideal Nyquist case, this would in theory require an infinite length digital filter at the receiver, which is not physically realizable, as discussed in Section 4.3. One way of limiting this problem involves increasing the bandwidth of the spectrum (which corresponds to the case $R_s < B_H$). Since an increase of the bandwidth of the spectrum of the WDM channels would induce linear crosstalk between them, the frequency spacing needs to be increased, as well, yielding the "quasi-Nyquist-WDM" transmission described in the following section.

4.2.2 Quasi-Nyquist-WDM ($\Delta f > R_s$)

A particular class of pulse spectra that satisfy Equation 4.9 is characterized by a raised-cosine (RC) shape, whose frequency characteristic is given by

$$H_{\mathrm{RC}}(f) = \begin{cases} T & \left(0 \le |f| \le \frac{1-\rho}{2T}\right) \\ \frac{T}{2}\left\{1+\cos\left[\frac{\pi T}{\rho}\left(|f| - \frac{1-\rho}{2T}\right)\right]\right\} & \left(\frac{1-\rho}{2T} \le |f| \le \frac{1+\rho}{2T}\right) \\ 0 & \left(|f| > \frac{1+\rho}{2T}\right) \end{cases} \tag{4.12}$$

where ρ is called the *roll-off factor* and takes values in the range [0, 1]. In the time domain, the shape of the pulse having an RC spectrum is

$$h_{\mathrm{RC}}(t) = \frac{\sin\left(\frac{\pi t}{T}\right)}{\frac{\pi t}{T}} \frac{\cos\left(\frac{\pi \rho t}{T}\right)}{1 - \frac{4\rho^2 t^2}{T^2}} \tag{4.13}$$

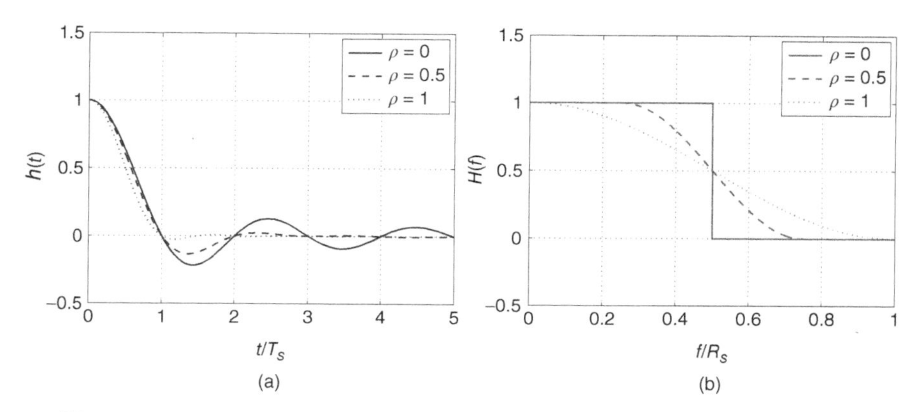

FIGURE 4.4 Raised-cosine pulses in time domain (a) and frequency domain (b).

The pulse shape in the time and frequency domains for different values of ρ is shown in Figure 4.4. The case $\rho = 0$ corresponds to the rectangular spectrum with bandwidth R_s. Due to the smooth characteristics of the raised cosine spectrum, it is possible to design practical filters for both the transmitter and the receiver that approximate the overall desired frequency response. In the case in which the channel has a flat frequency response, the optimum receiver filter is matched to the transmit one, thus both assume a square-root raised cosine (SRRC) shape. Note that an additional delay is required to ensure the physical realizability of the filters [5].

Figure 4.5 shows the eye diagrams of the signal after the receiver filter for both two- and four-level amplitude modulations, when SRRC filters are used both at the Tx and at the Rx side, for different values of roll-off. The lowest the roll-off, the more sensitive is the system to error in the sampling instant due to a nonperfect timing recovery.

When a WDM signal is generated by multiplexing subchannels with nonrectangular spectra, depending on the relationship between the frequency spacing and the symbol rate linear crosstalk among subchannels can be generated, potentially degrading performance. The impact of crosstalk as a function of the normalized frequency spacing is shown in Figure 4.6(a), in terms of signal-to-noise ratio (SNR) needed to achieve a target bit error rate (BER) of 10^{-3}, for three different modulation formats (QPSK, 16QAM, and 64QAM). The SNR is defined as

$$\mathrm{SNR} = \frac{P_{\mathrm{Rx}}}{p\, N_0\, R_s} \tag{4.14}$$

where P_{Rx} is the power of the useful signal at the input of the Rx filter, N_0 is the one-sided PSD of the additive Gaussian noise (GN) and p is a parameter that is equal to 1 in the single-polarization case and is equal to 2 in the dual-polarization case. SRRC filters with roll-off equal to 0.1 were used both at Tx and Rx sides. Ideally, no crosstalk is present when $\delta f > (1 + \rho)$. In fact, if $\delta f > 1.1$, the performance of all three formats converges to the theoretical value, whilst for lower spacings an SNR penalty is incurred, which is higher for higher cardinality modulation formats.

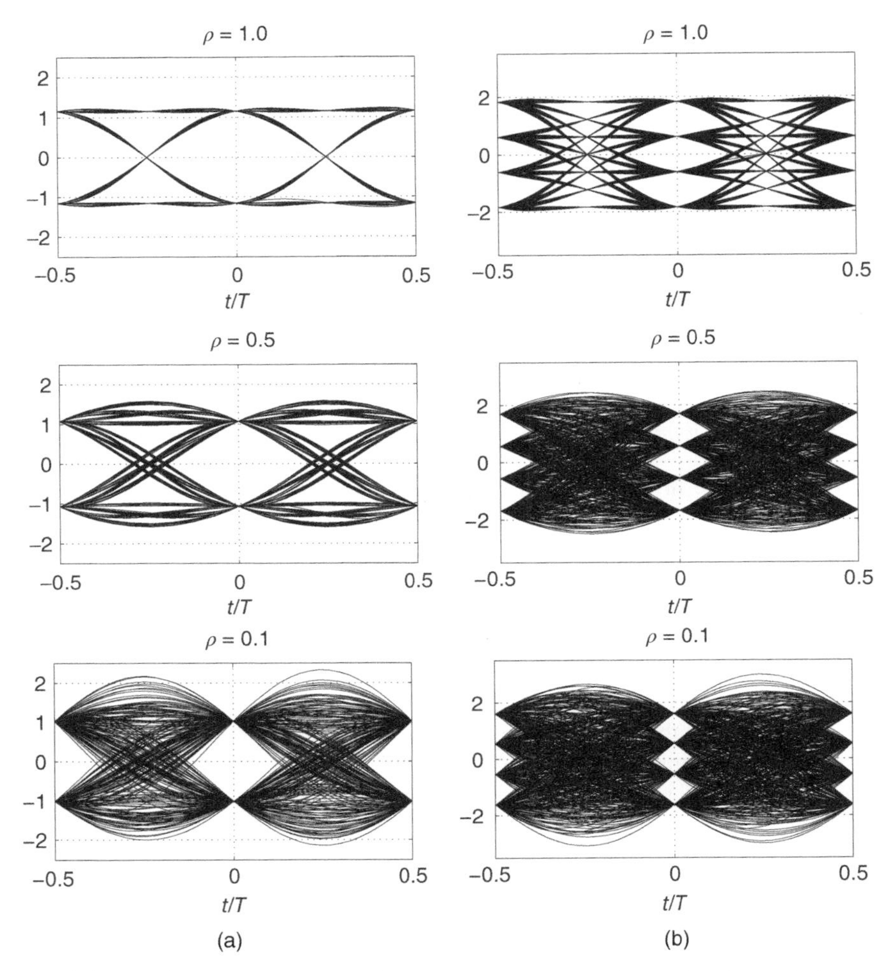

FIGURE 4.5 Eye diagram for a two-level (a) and a four-level (b) modulation with RC spectrum for different values of ρ.

Figure 4.6(b) shows the impact of crosstalk at a fixed frequency spacing equal to $1.1 \cdot R_s$ as function of the roll-off. For values of ρ lesser than 0.1, no penalty is present, whilst for higher values the SNR penalty due to crosstalk increases, with a larger impact for higher-order modulation formats.

4.2.3 Super-Nyquist-WDM ($\Delta f < R_s$)

As observed in Section 4.2.1, it is not possible to achieve zero-ISI if $\Delta f < R_s$. However, it is possible to achieve a channel spacing lower than the symbol rate by exploiting the so-called *partial response signaling*. The basic idea of partial response

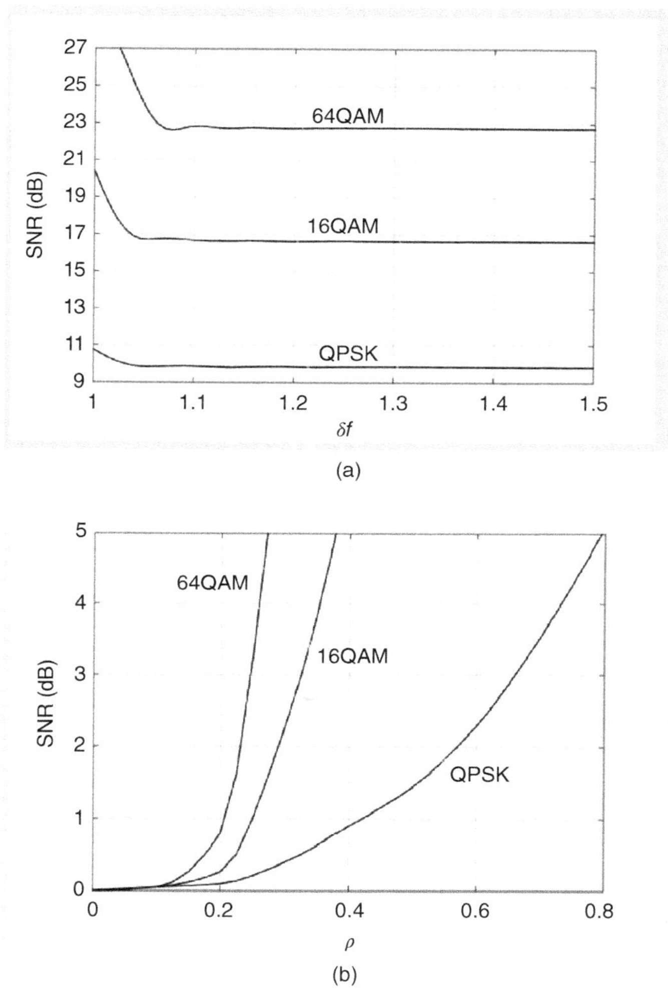

FIGURE 4.6 (a) SNR needed to achieve a target bit error rate of 10^{-3} as a function of the normalized frequency spacing $\delta f = \Delta f / R_s$ with $\rho = 0.1$. (b) SNR penalty at a target bit error rate of 10^{-3} as a function of the roll-off factor ρ at $\delta f = 1.1$. SNR is defined over a bandwidth equal to the symbol rate R_s.

signaling involves introducing some amount of controlled ISI, which can be removed at the Rx side: this process allows to reduce the spectral width of the signal, and thus to increase the SE. The counterpart is some power penalty and/or additional complexity in the transponder.

Duobinary (DB) line-coding is an example of partial-response transmission format. It is based on the introduction of correlation among symbols through some amount of controlled ISI. The transmission based on DB was first proposed in the

1960s by A. Lender [28], for radio frequency (RF) communications. In the 1990s, DB has re-emerged in the field of optical communications thanks to its high tolerance against fiber chromatic dispersion (CD) [29–31]. Later, it was overcome by multilevel modulation schemes that could reach even higher spectral efficiencies, combined with coherent detection, which allows to compensate for huge amounts of CD in the digital domain. Recently, the use of the DB concept has been proposed in order to further increase the SE in coherent optical transmission systems based on WDM multilevel modulation schemes [32–37]: since the optical signals have a narrower spectrum bandwidth than conventional signals, a tighter frequency spacing can be used, thus potentially increasing the SE and the overall capacity.

The condition for zero ISI (see Eq. 4.8) is that $h(iT) = 0$ for $i \neq 0$. The DB coding allows one additional nonzero value in the samples $h(iT)$:

$$h_i = \begin{cases} 1 & (i = 0, 1) \\ 0 & (i \neq 0, 1) \end{cases} \tag{4.15}$$

The introduced ISI is deterministic, thus it can be taken into account and compensated for at the Rx.

The effect of partial response coding can be viewed as a filtering of the signal, that is, as the multiplication of the global transfer function $H(f)$ by a periodic transfer function $C(f)$ with period R_s. In the case of DB modulation scheme, $C(f)$ is given by Proakis and Salehi [5]

$$C(f) = 1 + \exp\{-j2\pi fT\} = 2\cos(\pi fT)\exp\{-j\pi fT\} \tag{4.16}$$

It corresponds in the time domain to the addition of a signal and a replica delayed by T. Figure 4.7 shows the shape of the duobinary pulse in frequency domain

$$H_{\mathrm{DB}}(f) = \begin{cases} 2\cos(\pi fT)\exp\{-j\pi fT\} & |f| < \frac{1}{2T} \\ 0 & \text{otherwise} \end{cases} \tag{4.17}$$

and in time domain

$$h_{\mathrm{DB}}(t) = \mathrm{sinc}\left(\pi\frac{t}{T}\right) + \mathrm{sinc}\left(\pi\frac{(t-T)}{T}\right) \tag{4.18}$$

where the *sinc* function is defined as in Equation 4.11 . Since the spectrum decays to zero smoothly, physically realizable filters that closely approximate this spectrum can be designed.

The effect of a DB coding on the scattering diagram of a QPSK modulation is shown in Figure 4.8: while the standard constellation (a) is composed of four points, the number of constellation points is increased to nine for the DB QPSK signal (b), due to the ISI introduced by the DB coding.

Two methods can be used to detect signals that contain a controlled amount of ISI. One is the symbol-by-symbol method, which is easy to implement, but gives an asymptotic SNR penalty in the order of 2 dB for DB line coding

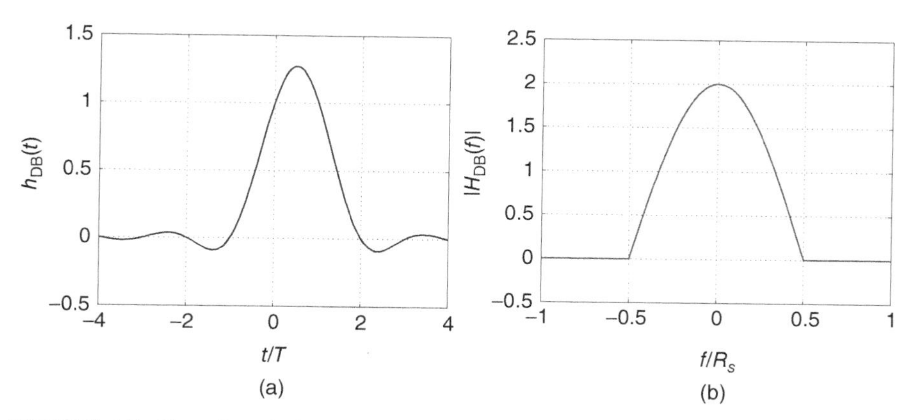

FIGURE 4.7 Time domain (a) and frequency domain (b) characteristic of a duobinary signal.

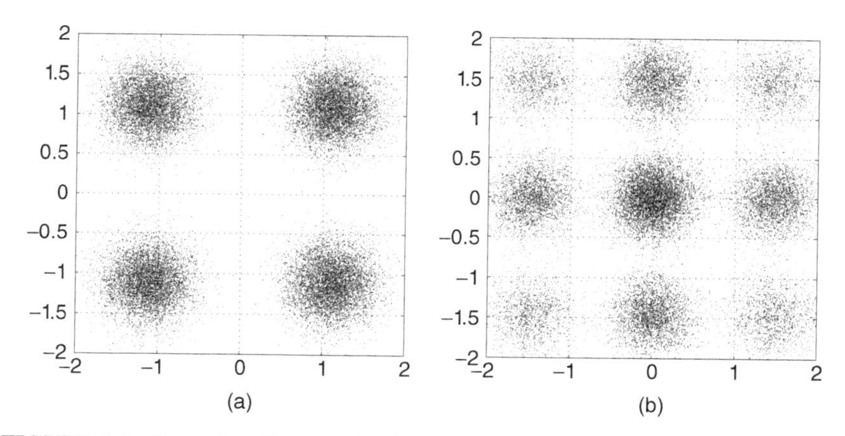

FIGURE 4.8 Scattering diagram of noisy QPSK (a) and DB-QPSK (b) constellations.

modulations [5]. The other is based on multisymbol detection techniques, such as maximum-likelihood sequence estimation (MLSE) and maximum-a-posteriori probability (MAP) algorithms, which minimize the error probability, at the expenses of a higher computational complexity. In Reference [34], the detection strategies for quadrature duobinary (i.e., DB PDM-QPSK), coherent optical systems are analyzed in detail. As an example, Figure 4.9 shows the back-to-back performance, in terms of Q-factor versus OSNR, for both the symbol-by-symbol detection scheme and the maximum-likelihood sequence detection (MLSD) algorithm. A Q-factor gain in the order of 1.7 dB was achieved thanks to the use of the more complex MLSD scheme.

In addition to the results based on the standard DB technique, several simulative and experimental demonstrations of long-haul transmission in coherent optical systems at a spacing lower than the symbol rate (also termed as “Faster-than-Nyquist” signaling) can be found in Refs [14, 38–42].

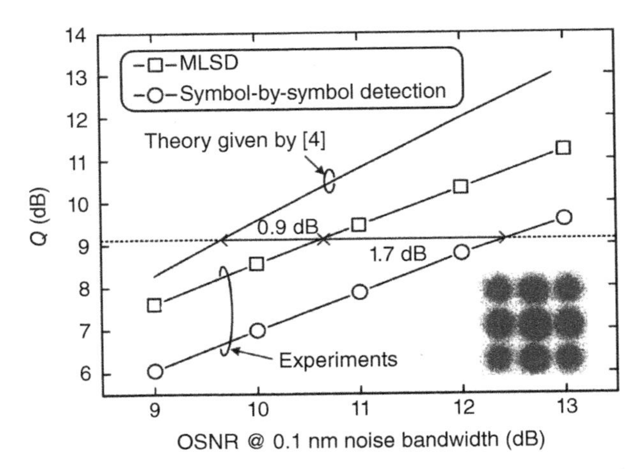

FIGURE 4.9 *Q*-factor versus OSNR for different detection schemes. The inset depicts the quadrature duobinary constellation before data detection at OSNR = 12 dB. After Ref. [34]; ©2011 IEEE.

4.3 DETECTION OF A NYQUIST-WDM SIGNAL

In conventional WDM systems, subchannels are first demultiplexed in the optical domain and then separately detected. This approach cannot be applied to Nyquist-WDM signals, in which the subchannels are too closely spaced to be separated through optical filtering without incurring in any penalty. However, sharp filter shapes can be efficiently implemented at the Rx in the digital domain, enabling detection of single subchannels without the need of any tight optical filtering at the Rx.

The schematic of a coherent Rx, which detects a single PDM subchannel in a Nyquist-WDM comb, is shown in Figure 4.10. It is composed of a 90-° hybrid, followed by four balanced photodetectors (BPDs), whose functionality is to map the optical field into four electrical signals, corresponding to the in-phase and quadrature field components for the two polarizations [4]. The single subchannel is selected by tuning the local oscillator (LO) to the center frequency of the subchannel. The electrical low-pass filters (LPFs) in Figure 4.10 represents the cascade of all band-limiting components in the Rx. The four analog electrical signals are sampled by four

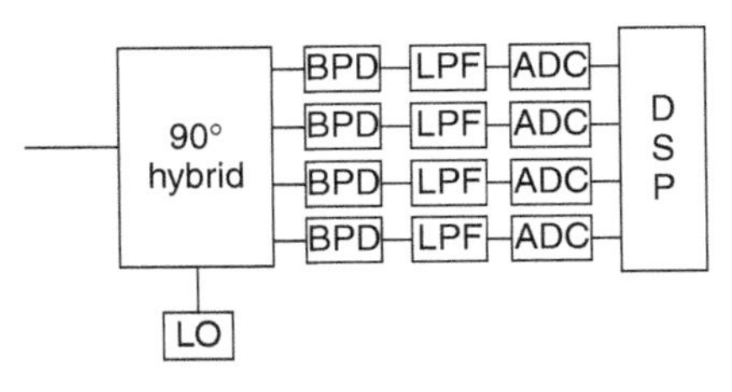

FIGURE 4.10 Schematics of a coherent receiver for the detection of a single subchannel in a Nyquist-WDM comb. LO: local oscillator, BPD: balanced photodetector, LPF: low-pass filter, ADC: analog-to-digital converter, DSP: digital signal processing.

analog-to-digital converters (ADCs) with sampling speed f_{ADC} and the signal samples are elaborated by *ad hoc* DSP algorithms [4], which perform polarization recovery and compensation of propagation linear (and possibly nonlinear) impairments.

Note that antialiasing electrical filters can be placed before the ADCs in order to reduce the bandwidth of the input signal, thus relaxing the requirements for the ADC sampling frequency and enabling more efficient DSP for polarization recovery and impairments compensation. The Nyquist sampling theorem [5] states that a sufficient and necessary condition to avoid the generation of aliasing replica of the analog signal is that the sampling frequency is greater than twice the bandwidth occupation of the signal. In the case of ideal (rectangular) Nyquist shaping with ideal (rectangular) antialiasing filters, the sampling frequency needs to be greater than R_s in order to avoid aliasing. In practical cases (quasi-Nyquist-WDM with realistic antialiasing filters), a higher sampling speed is needed. Typically, $f_{ADC} = 2 \cdot R_s$, corresponding to a number of samples per symbol N_{SpS} equal to 2, but lower values can be used without incurring in substantial penalties, as shown in the experiments reported in Refs [13] and [43], where a value of N_{SpS} equal to 1.67 was used.

As an example, Figure 4.11(a) shows the results of a simulative analysis of the robustness of the system to the limited speed of the ADC (which translates into a limited number of samples per symbol) in terms of the back-to-back SNR penalty, at a reference BER of 4×10^{-3}, with respect to the 2 SpS case. A Nyquist-WDM comb was simulated, with an SRRC shape with $\rho = 0.1$. The Rx electrical low-pass characteristic was modeled as a fifth-order Bessel filter with bandwidth B_{Rx}. The value of B_{Rx} was optimized for each value of N_{SpS}, obtaining the optimum values shown in Figure 4.11(b) (normalized with respect to the symbol rate). A low value of N_{SpS} corresponds to a low value of f_{ADC}/R_s: to avoid aliasing, the Rx bandwidth has to be decreased accordingly. This, however, causes a distortion on the useful signal, which in turn induces an SNR penalty, which is higher for higher-order modulation formats.

A relevant characteristics of Nyquist-WDM signaling is its good performance even at very low values of the Rx bandwidth, as also demonstrated in Ref. [27]. In fact, thanks to the compact spectral shape of the Nyquist-WDM subchannels, the bandwidth of the electrical Rx filter can be kept low with respect to the symbol rate (close to $R_s/2$), without introducing filtering penalty on the subchannel and without breaking the orthogonality between subcarriers [26].

In Reference [44], the matched filter design for SRRC spectrally shaped Nyquist-WDM systems has been addressed. Typically, two equalizers are implemented in the coherent Rx DSP: one CD equalizer and one adaptive butterfly blind equalizer. The CD equalizer is used to compensate for the large amount of accumulated CD in the fiber link and usually is a static frequency domain equalizer (FDE) [45]. The adaptive equalizer is typically implemented using finite-impulse-response (FIR) digital filters, which perform polarization demultiplexing, polarization-mode dispersion (PMD) compensation, residual CD compensation, and ISI mitigation. As the butterfly equalizer is dynamically adjusted, it is harder to implement and usually much smaller than the CD equalizer in terms of the FIR tap numbers or FDE overlap lengths. Optimum performance can be achieved if the adaptive equalizer can converge to a matched filter. For a Nyquist signal with a smaller roll-off factor,

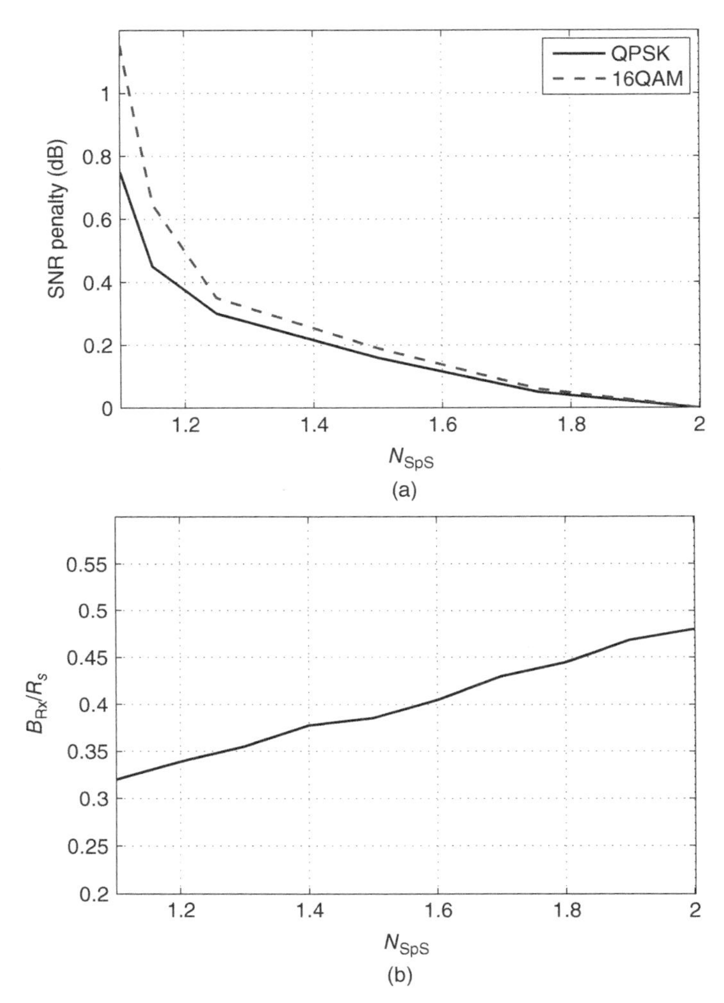

FIGURE 4.11 SNR penalty (a) and optimum Rx bandwidth (b) as a function of the number of samples per symbol (N_{SpS}) at the output of the ADC.

a larger number of taps are required for an FIR filter-based adaptive equalizer, as shown in Figure 4.12, which reports the plots of the impulse response of SRRC filters for different values of roll-off. The length of the impulse response in time increases when the roll-off decreases.

In Reference [44], it is shown that incorporating a matched filter in the bulk CD equalizer for an SRRC-shaped signal can significantly reduce the complexity of the blind equalizer, with no additional complexity added to the CD equalizer. It is also shown that Nyquist-WDM systems with matched filtering are sensitive to the frequency offset between the Tx laser and the LO, and that the induced penalty decreases with increased SRRC roll-off factor.

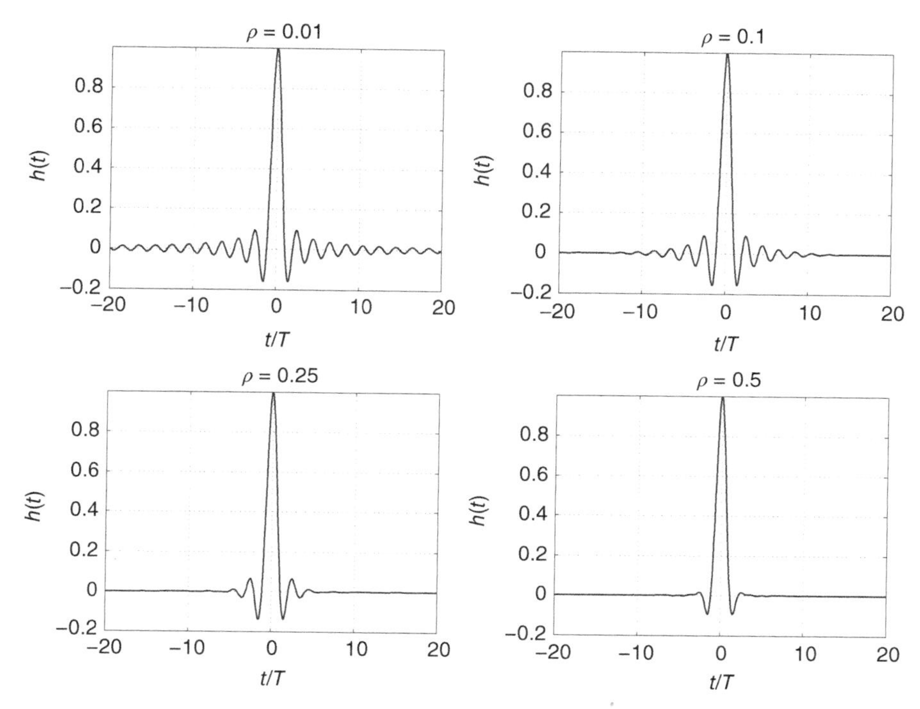

FIGURE 4.12 SRRC filter shape for different values of roll-off.

4.4 PRACTICAL NYQUIST-WDM TRANSMITTER IMPLEMENTATIONS

As shown in Section 4.2.1, the ideal shape of the transmitted spectra, which allows to achieve a channel spacing equal to the symbol rate R_s, is rectangular with bandwidth equal to R_s. In such an ideal scenario, Nyquist-WDM can achieve the optimum matched filter performance in AWGN systems [5]. In practice, penalties are to be expected when the ideal constraints on Nyquist WDM implementation are relaxed [26], like for instance with the transmission of channels with not perfectly rectangular spectra.

The key-component in the generation of Nyquist-WDM signals is the "Nyquist filter", which performs a tight spectral shaping on the generated signals at the Tx side in order to obtain an almost rectangular spectrum. Figure 4.13 shows the shape of the Nyquist filter needed to transform an ideal nonreturn-to-zero (NRZ) pulse, characterized by a rectangular pulse in the time domain with length equal to T, into a Nyquist spectrum with SRRC shape, for different values of roll-off. If $H_{\mathrm{NRZ}}(f)$ is the Fourier transform of the NRZ pulse, the transfer function of the Nyquist filter is obtained as the product between the SRRC transfer function $\sqrt{H_{\mathrm{RC}}(f)}$ (see Eq. 4.12) and the inverse of $H_{\mathrm{NRZ}}(f)$:

$$H(f) = \sqrt{H_{\mathrm{RC}}(f)}\frac{\pi fT}{\sin(\pi fT)} \tag{4.19}$$

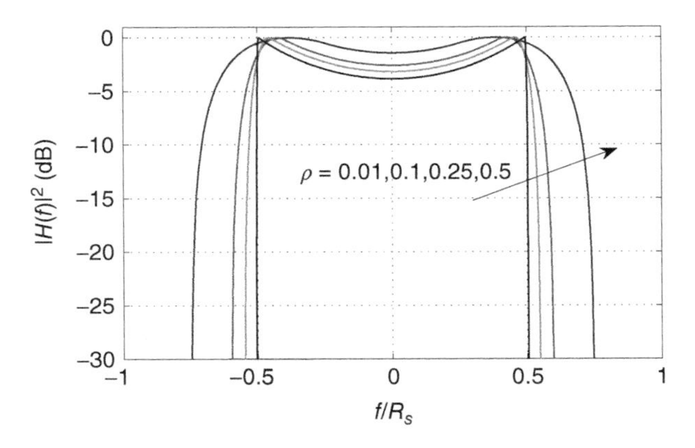

FIGURE 4.13 Nyquist-filter shape for different values of roll-off.

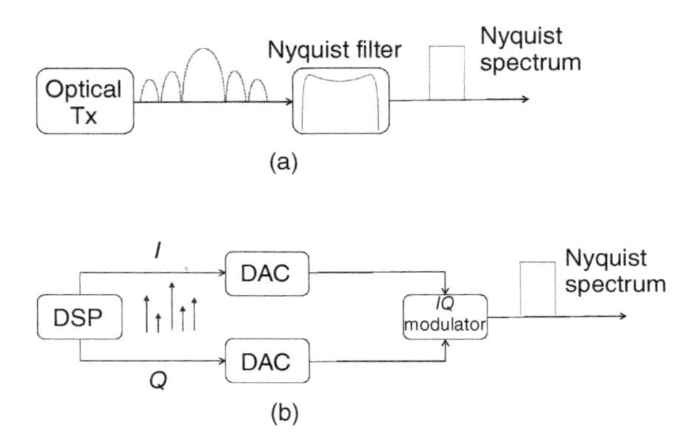

FIGURE 4.14 Spectral shaping in the optical (a) and digital (b) domains.

The quasi-rectangular subchannel spectral shaping needed for Nyquist-WDM can be obtained either by band-limiting the signal coming out of each transmitter through an optical filter, as schematically shown in Figure 4.14(a), or by driving the electro-optical modulator with suitable electrical signals so that the optical modulated signal takes on the desired spectral shape, as shown in Figure 4.14(b). The two techniques are known as *optical Nyquist-WDM* and *digital Nyquist-WDM*, respectively. The latter technique requires digital-to-analog-converters (DACs) to generate the electrical driving signals. Early demonstrations of near-Nyquist channel spacing were performed by approximating the desired rectangular spectral shape through the use of optical filters (see Section 4.4.1). More recently, several experimental demonstrations appeared based on digital Nyquist-WDM generation and transmission (see Section 4.4.2). In the following, the characteristics of Nyquist-WDM signals generated through spectral shaping in either optical or electrical/digital domain are analyzed and discussed, taking into account the implementation nonidealities of state-of-the-art components.

4.4.1 Optical Nyquist-WDM

The Tx setup for Nyquist-WDM signal generation using spectral shaping in the optical domain is shown in Figure 4.15. Each of the N transmitters generates a modulated optical signal with NRZ spectral shape. An array of optical filters is then used to transform the NRZ shape into a quasi-rectangular one. The N Nyquist signals are then wavelength multiplexed, generating the overall Nyquist-WDM spectrum.

Successful ultra-long-haul experiments exploiting Nyquist-WDM, based on BPSK, QPSK, and 8QAM modulation format (with polarization domain multiplexing) using tight optical filtering at the Tx side, have been performed (see Section 4.5.1). In all experiments, the ideal optical filter shape was approximated using state-of-the art components, such as the Finisar Waveshaper filter in Refs [13, 16, 46]. In addition to standard filtering, the Waveshaper allows a high-frequency pre-emphasis to be introduced in the optical spectrum of each channel in order to better approximate the Nyquist filter transfer function.

Figure 4.16 shows the transfer function of the optical shaping filter in four different cases: Finisar filter w/o pre-enhancement, super-Gaussian (SG) filter, which better approximates the Finisar filter transfer function, Finisar filter with pre-enhancement, and SRRC filter shape with $\rho = 0.1$. Clearly, the main limitation of the Waveshaper component is the fact that its profile is not particularly steep (approximately second-order super-Gaussian) with respect to the ideal one. This introduces linear crosstalk between adjacent subchannels, inducing a penalty in the back-to-back transceiver performance. The impact of such a nonideal filtering on system performance was investigated by simulations in Ref. [27], showing that the constraints on the steep spectral shaping requested to satisfy the orthogonality condition and to minimize crosstalk between the subcarriers can be relaxed by increasing the channel spacing, at the expenses of a little loss in spectral efficiency.

As an example, in Figure 4.17 the SNR needed to achieve a target BER equal to 4×10^{-3} is shown as a function of the normalized frequency spacing. The modulation format is PDM-QPSK and the simulation setup is the one described in Ref. [27]. Since the profile of the realistic filters is less steep than the ideal filter (see Figure 4.16),

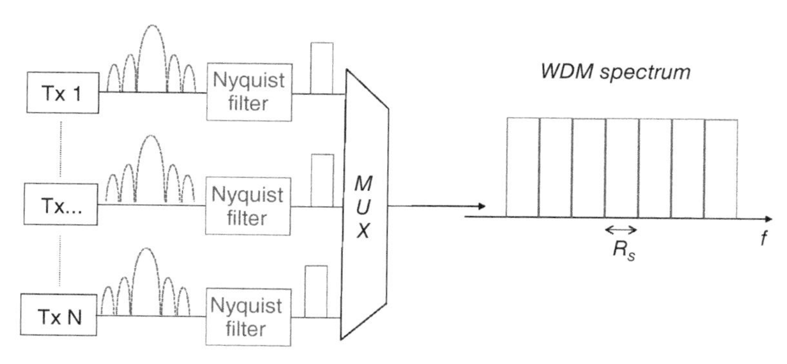

FIGURE 4.15 Setup for optical Nyquist-WDM generation.

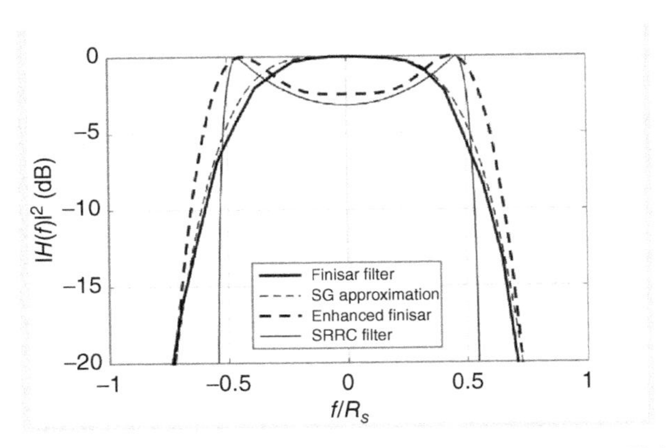

FIGURE 4.16 Examples of optical shaping filter transfer functions. The Finisar filter shapes shown in the figure were measured from the experimental setup in Ref. [13]. The shape of the SG filter is second-order super-Gaussian with bandwidth $0.9{\cdot}R_s$. The SRRC filter has a roll-off equal to 0.1.

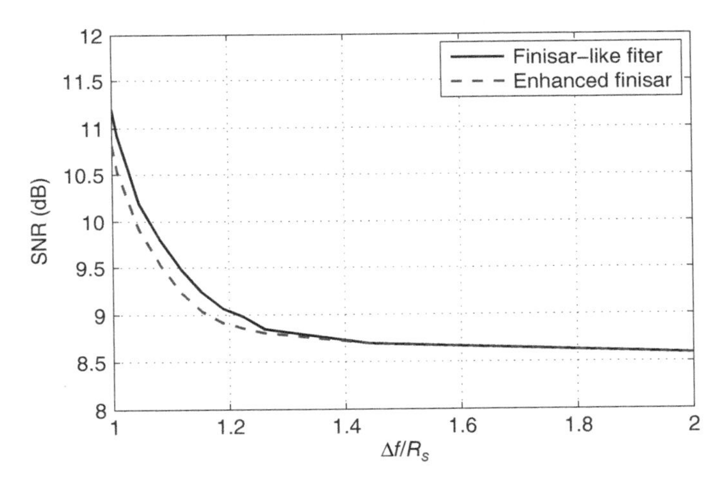

FIGURE 4.17 SNR (defined over a bandwidth equal to the symbol rate) versus normalized frequency spacing with realistic optical filters.

substantial crosstalk would occur at symbol rate spacing. Penalties can, however, be canceled by increasing the subchannel spacing, as shown in Ref. [27].

In Reference [47], a transmission of a 96 × 128 Gbit/s PDM-QPSK quasi-Nyquist WDM comb over 11,680 km was demonstrated, with a channel spacing equal to $1.17{\cdot}R_S$. A programmable optical filter was used, engineering the top of the intensity response in order to follow a quadratic intensity profile of variable depth in decibels. The narrow profile of the standard flat-top filter was found to be responsible for 0.5-dB penalty with respect to the single-channel case without optical filtering. However, engineering the spectral response according to a quadratic profile with variable depth, which enhances the power of the spectral components that are farther from the carrier frequency, yielded a significant performance improvement, with the Q-factor becoming larger than in the case of single channel without optical filter. The optimum depth (6 dB) was found to be identical in the back-to-back configuration and after transmission. The improvement with respect to the single-channel case without filtering was enabled by the fact that, in addition to narrow spectral shaping, the programmable filtering device allows a high-frequency pre-emphasis to be introduced in the optical spectrum of each channel in order to precompensate for Tx and Rx electrical bandwidth limitations. Using a similar technique, the transmission of a 10 × 120 Gbit/s PDM-QPSK Nyquist-WDM signal over 9000 km was demonstrated in Ref. [13], with a channel spacing equal to the symbol rate.

Even though several record experiments have been performed in the past years using optical spectral shaping on PDM-QPSK signals, the difficulty in applying the technique to higher-order modulation formats was evident. In Reference [16], quasi-Nyquist-WDM signals were generated using a PDM-8QAM modulation, but the normalized frequency spacing had to be kept as high as $1.22 \cdot R_s$ in order not to incur in substantial crosstalk penalty. In practice, the main drawback in performing spectral shaping in the optical domain is the need of optical filters with very steep transfer functions, with the requirements on tight filtering becoming more stringent with the increase of the modulation cardinality. A possible solution is to perform the spectral shaping in the electrical/digital domain by using DACs. In such a way, it is possible to accurately "design" through DSP the signal spectrum at the output of the DAC and ideally obtain a perfectly square spectrum, as shown in the following section.

4.4.2 Digital Nyquist-WDM

In the following, it is shown that very accurate spectral shaping can potentially be performed in the digital domain, with the main limitations being the sampling speed of the DAC and the availability of suitable analog antialiasing filters. The fundamental results from the signal theory are first reviewed and then applied to the case of optical transmission.

The "Nyquist sampling theorem" [5] states that any analog signal $x(t)$, band-limited in $[-W, W]$, can be perfectly reconstructed from its samples provided that the sampling frequency f_s is greater than $2 \cdot W$. Figure 4.18 shows the schematics

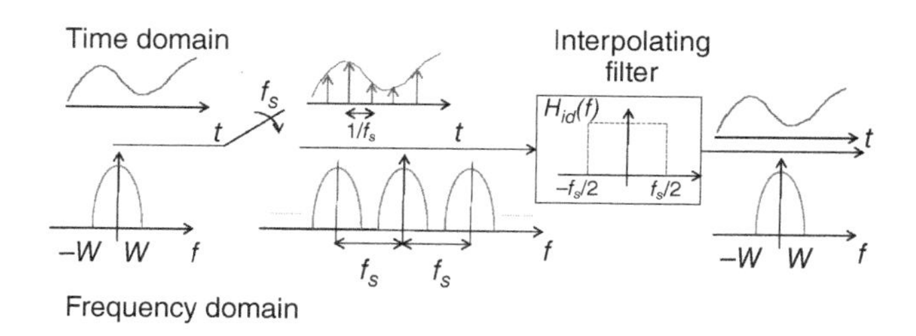

FIGURE 4.18 Schematics of ideal D/A conversion.

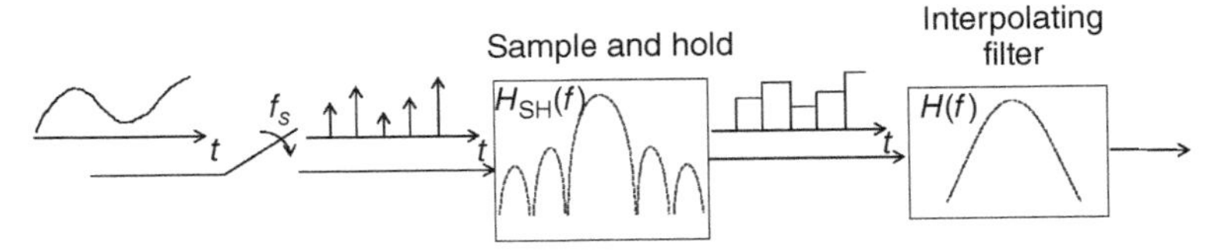

FIGURE 4.19 Schematics of the generation of pulses for Nyquist-WDM with realistic D/A conversion.

of an ideal digital-to-analog (D/A) conversion process [5]. Ideally, to generate a perfectly rectangular Nyquist-spectrum with bandwidth equal to the symbol rate R_s, a DAC is needed operating at a speed equal to R_s samples/s and with a perfectly rectangular transfer function with bandwidth $0.5 \cdot R_s$. On the other hand, today commercial DACs are characterized by a transfer function which is far from rectangular [48, 49], which makes it not possible to perform Nyquist spectral shaping to generate signals with a symbol rate equal to the DAC sampling speed.

The scheme of a realistic D/A conversion process used to generate each quadrature of a Nyquist-WDM signal is shown in Figure 4.19. The DAC can be modeled as a sample&hold (S&H) device, which generates a "step" function whose levels correspond to the samples of the ideal signal, followed by an interpolating filter, used to reconstruct the original signal. The evolution of the signal spectrum is shown in Figure 4.20(a)–(d) for the case $f_s = 2 \cdot R_s$ and in Figure 4.20(e) and (f) for the case $f_s = 1.5 \cdot R_s$. The presence in the generated spectrum of spurious frequencies is due to non-perfectly-rectangular antialiasing filtering (the transfer function of the DAC is shown in plot (d) and (f) as dashed line). A certain amount of ISI, introduced by both the interpolating filter and the S&H device, is also present, resulting in a non-flat spectrum in Figures 4.20(c)–(f). ISI can be partially mitigated by performing a pre-enhancement on the signal samples generated in DSP, while crosstalk can be canceled by adding a steep analog antialias filter at the output of the DAC.

A realistic DAC device is characterized by two main parameters:

- The sampling speed f_{DAC}, which limits the achievable symbol rate $R_s = f_{\mathrm{DAC}}/N_{\mathrm{SpS}}$, where N_{SpS} is the number of samples per symbol (also indicated as "oversampling factor"). State-of-the art DACs are characterized by a maximum sampling speed around 34 Gsamples/s [48].
- The number of resolution bits N_{DAC}, which limits the cardinality of the modulation format. In fact, the higher is the order of the modulation format, the higher is the required value of N_{DAC}.

Typically if f_{DAC} increases, N_{DAC} decreases. The achievable symbol rate can clearly be increased by decreasing the oversampling factor. In doing so, penalties could be incurred due to interference produced by spectral replica of the useful spectrum in the DAC process. In Reference [50], a 1.5 SpS DAC-supported Nyquist-WDM PDM-16QAM experiment was reported, using a DAC with $f_{\mathrm{DAC}} = 23.4$ GHz

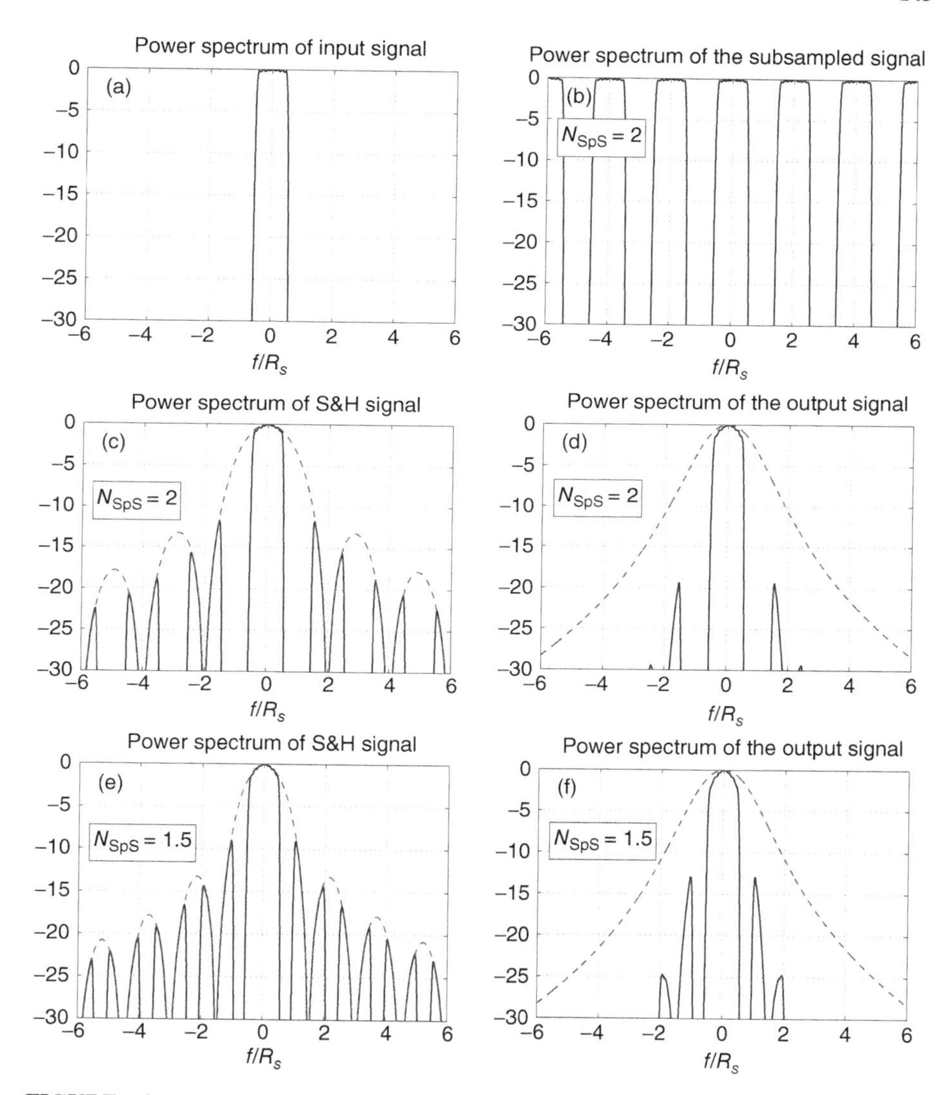

FIGURE 4.20 Evolution of digital spectra in the digital-to-analog conversion process of Figure4.19 for $f_s = 2 \cdot R_s$ and $f_s = 1.5 \cdot R_s$. In (c) and (e), the transfer function of the S&H process is shown as a dashed line. In (d) and (f), the transfer function of the DAC is shown as a dashed line.

and thus achieving a symbol rate $R_s = 15.6$ Gbaud. In Reference [51], 1.33 SpS were employed in a 100-km PDM-64QAM single-channel transmission at 252 Gbit/s.

In Reference [25], an oversampling factor as low as 1.15 was used, limiting the penalty due to spectrum replica thanks to the use of *ad hoc* antialiasing

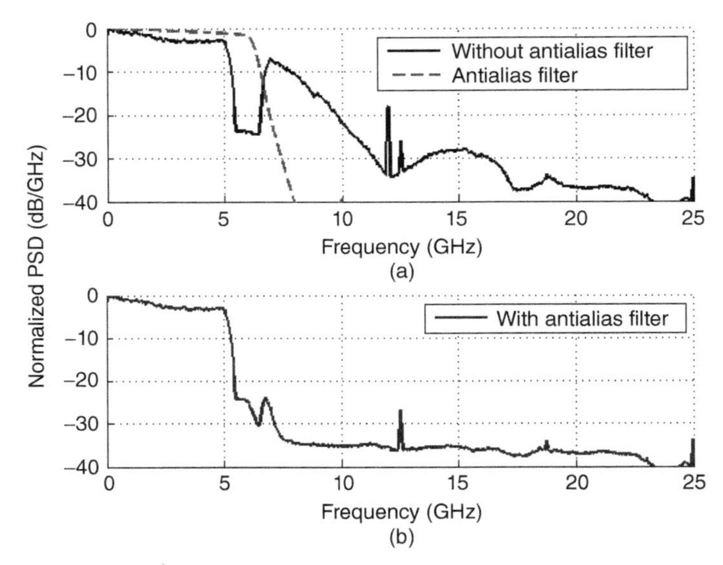

FIGURE 4.21 Spectrum of the modulator driving signal (eight-level NRZ PAM) without (a) and with (b) antialias filter. In (a), the dashed line is the antialias filter transfer function. After Ref. [25]; © 2014 OSA.

electrical filters. The modulation format was 10.4-Gbaud PDM-64QAM. The filtered eight-level signals driving the IQ modulator were generated by the DACs operating at 11.96 GS/s. As a result, an alias signal replica was also generated, centered at 11.96 GHz, as shown in Figure 4.21(a). The replica was partially filtered out by the low-pass frequency response of the DAC, but to suppress it to the extent of making it negligible, a specifically designed antialiasing filter, with steep cutoff, was interposed between the DAC and the modulators. The antialiasing filter frequency response and its output are shown in Figure 4.21(a) and (b), respectively.

In the following, a set of simulation results is reported showing a comparison between digital and optical spectral shaping. Both PDM-QPSK and PDM-16QAM signaling are considered. The use of a DAC working at 24 Gsamples/s with bandwidth equal to 9.6 GHz [49] is assumed. An oversampling factor equal to 2 was used, yielding a value of symbol rate equal to 12 Gbaud. The DAC transfer function was modeled as a two-pole filter with dumping factor equal to $\sqrt{2}/2$, which is a good approximation of the profile of state-of-the-art commercial devices [49]. The modulator driving voltages were optimized and a proper pre-enhancement was applied to the digital samples in order to compensate for both the interpolating filter and the S&H process [52]. For the Nyquist filter generation, instead of an ideal square shape in $[0, R_s/2]$, a more realistic raised-cosine shape with roll-off 0.15 was used. In case of optical shaping, a fourth-order super-Gaussian shaping filter was assumed. The back-to-back performance is shown in Figure 4.22. The channel spacing, equal to R_s for PDM-QPSK, was increased to $1.1 \cdot R_s$ for PDM-16QAM in order

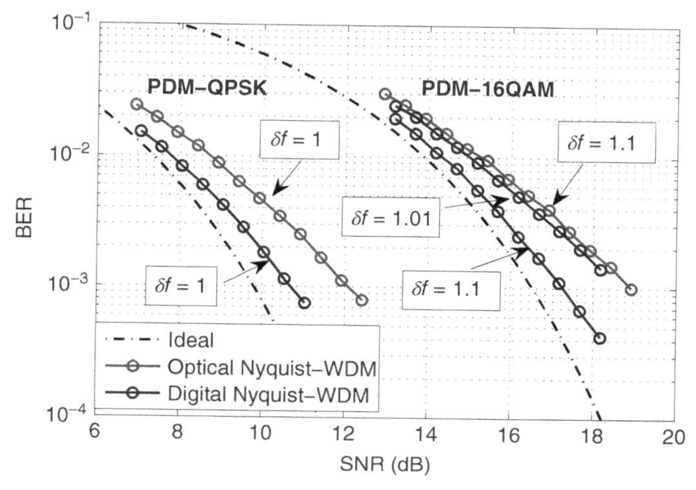

FIGURE 4.22 BER versus SNR for PDM-QPSK and PDM-16QAM with digital and optical spectral shaping. SNR is defined over a bandwidth equal to the symbol rate.

to limit the crosstalk penalty, which for this format can heavily affect the performance. The advantage of digital over optical spectral shaping highlighted by the results of Figure 4.22 is mainly due to the ability to precisely control the Tx spectrum to a degree far beyond what it possible with practical analog optical filters.

An additional advantage of a DAC-based Tx is the ability to compensate for the linear transfer function of the analog components of the Tx and Rx, thus relaxing the requirements for these components. Compensation is typically performed by first measuring the transfer function between the desired waveform and the optical output. This transfer function is then inverted and included in the digital filtering performed at the Tx. This is very attractive since the linear transfer function of the required equalization can be combined with the spectral shaping filter without increasing the ASIC resources needed to perform the filtering [53].

Nyquist pulse shaping in the digital domain is typically performed using FIR filters. The required length of the digital filter needed to generate SRRC pulses increases when the roll-off decreases (see Figure 4.12). In References [54–56], complexity and performance of digital pulse shaping has been investigated, showing that FIR filters with 17 taps allow for a reduction in channel spacing to 1.1 the symbol rate within a 1-dB penalty [56]. A higher number of FIR filter taps are needed for tighter channel spacing. In Reference [57], using a 600-tap FIR filter to perform digital preshaping, the generation of 20 Gbit/s QPSK WDM signals with $\Delta f = R_s = 10$ GHz was demonstrated without back-to-back penalty. A twofold oversampling was used in the DAC. In Reference [58], a real-time demonstration of generation of Nyquist-like pulses with 14-Gbaud PDM-16QAM was reported, showing that the use of a 32-tap FIR filter to shape the signals would allow a channel spacing equal to $1.06{\cdot}R_s$ without substantial crosstalk penalty.

4.5 NYQUIST-WDM TRANSMISSION

The transmission performance of Nyquist-WDM systems has been extensively studied both experimentally (see Sections 4.5.1 and 4.5.2) and numerically [7]. Using the theory described in Ref. [59], based on the GN-model for nonlinear propagation in uncompensated coherent optical systems [60], the relationship between the SE and total link length for Nyquist-WDM or quasi-Nyquist-WDM signals in arbitrary transmission scenarios can be derived. In this section, this possibility is exemplified by analyzing PDM-QAM formats in two different multispan link scenarios with erbium-doped fiber amplifier (EDFA) amplification (noise figure $F = 5$ dB): a terrestrial link over standard single-mode fiber (SSMF) with 100-km span length and a submarine link over pure silica-core fiber (PSCF) with 60-km span length. The parameters of the fibers are shown in Table 4.1.

The WDM signal is assumed to occupy the entire C-band ($B_{WDM} = 5$ THz) with a spacing among WDM subchannels equal to 1.1 R_s, that is, the lowest value assuring the absence of any linear crosstalk for RC spectra with roll-off 0.1. In order to analyze a realistic scenario, it is assumed to operate with a conservative 2-dB margin with respect to the ideal BER-versus-SNR performance and with a realistic soft FEC 1.5-dB penalty with respect to ideal soft FEC performance.

The dependence of the SE on total link length is plotted in Figure 4.23 for PDM-QAM modulation formats with cardinality ranging from 4 to 64. The results of Figure 4.23 clearly highlight the trade-off between distance and SE, in relation to the different modulation formats: increasing the cardinality of the constellation, a higher SE can be achieved, but typically over a shorter transmission distance and/or at a higher required FEC overhead. State-of-the-art soft FEC with 20% overhead can now operate at pre-FEC BER of 2.7×10^{-2} [21]: so, for all modulation formats, the points corresponding to BER $= 2.7 \times 10^{-2}$ are marked in the figures. The section of the SE lines to the left of the dots is, therefore, the "practicable" section, whereas moving to the right will be possible only if better FECs become available.

Considering the currently possible systems, in a terrestrial link with EDFA-only amplification, PDM-QPSK is the best choice for ultra-long-haul transmissions beyond 5000 km, while PDM-8QAM can be used to achieve an SE around 5–5.5 bit/s/Hz in 3000 km links. PDM-16QAM allows to reach 2000 km with an SE of 6 bit/s/Hz. The reach of higher-order modulation formats, such as PDM-32QAM and PDM-64QAM, is very limited in this kind of systems, but can be significantly

TABLE 4.1 Parameters of the SSMF and PSCF

Fiber type	Dispersion (ps/nm/km)	Loss (dB/km)	Non-linearity coeff. (W^{-1} km^{-1})	Span length (km)
SSMF	16.7	0.2	1.3	100
PSCF	20.1	0.17	0.8	60

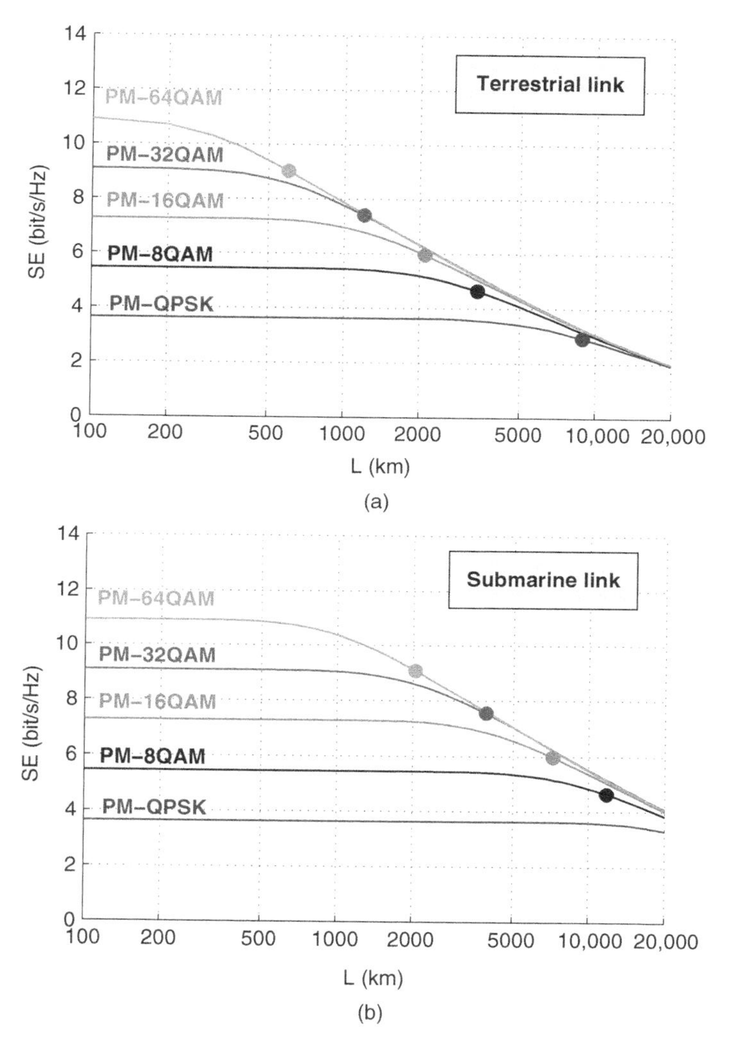

FIGURE 4.23 Spectral efficiency (SE) versus total link length in two different transmission scenarios described in the text. B_{WDM}=5 THz, $\Delta f = 1.1 \cdot R_s$. Assumptions: 2-dB SNR margin from ideal performance and 1.5-dB penalty of soft FEC with respect to infinite-length codes ideal performance. Dots correspond to a pre-FEC BER = 2.7×10^{-2}.

increased by using new generation fibers and shorter span length, similar to that in the analyzed submarine-like system over PSCF, where they reach 4000 and 2000 km, respectively.

The plot also shows that PDM-16QAM could reach ultra-long-haul distances over submarine-like links. Note that this reach can be further increased by using better-performing fibers, higher-performance FEC, and nonlinearity compensation techniques at the Rx, as done in Ref. [61], where 10,000 km could be achieved at an SE of 6 bit/(s Hz).

4.5.1 Optical Nyquist-WDM Transmission Experiments

In Table 4.2, experimental demonstrations of long-haul transmission based on super-Nyquist-WDM, Nyquist-WDM, and quasi-Nyquist-WDM using optical shaping at the Tx are listed. The most commonly used modulation format was PDM-QPSK, with a symbol rate in the order of 30 Gbaud. Only one demonstration is present with a higher-order modulation (PDM-8QAM), achieving the highest value of SE (4.1 bit/s/Hz), but with a limited reach (4000 km).

In References [41] and [64], the transmission of 96×100 Gbit/s PDM-QPSK super-Nyquist-WDM signal was demonstrated over 4730 km with a sub-Nyquist channel spacing equal to $0.89 \cdot R_s$. The individual subchannels were bandwidth constrained by applying an aggressive prefiltering at the Tx: this created a significant ISI that produced a complex signal constellation (composed of a total of 36 points). Using a 5-tap MAP algorithm at the Rx side, ISI was efficiently removed and the constellation was recovered back to that of a typical QPSK signal.

A similar technique to mitigate ISI was used in Ref. [14] to reach a record length of 11,860 km. With the use of symbol-by-symbol detection, the maximum system length at the same BER was around 9000 km. The reduced performance was due to the fact that the adaptive digital equalizer, while compensating for the ISI introduced by strong filtering, simultaneously enhanced the high-frequency noise, thus inducing a significant SNR penalty. With a sub-Nyquist spacing equal to $1.2 \cdot R_s$, more than 15,000 km could be reached without the need of multisymbol detection.

4.5.2 Digital Nyquist-WDM Transmission Experiments

In Table 4.3, experimental demonstrations of long-haul transmission based on Nyquist-WDM and quasi-Nyquist-WDM using digital shaping at the Tx are listed.

TABLE 4.2 Experimental demonstrations of optical Nyquist-WDM

References	Format (PDM)	R_s (Gbaud)	$\delta f = \Delta f / R_s$	Net SE (bit/s/Hz)	FEC OH (%)	Capacity (Tbit/s)	Reach (km)
[41]	QPSK	28	0.89	3	7	9.6	4370
[42]	QPSK	28	0.89	3	7	20	6860
[13]	QPSK	30	1	3.3	20	1	9000
[14]	QPSK	28	1	3.6	7	20	11,860
[13]	QPSK	30	1.1	3	20	1	10,000
[62]	QPSK	28	1.1	3.4	7	1	2300
[46]	QPSK	25	1.1	3.4	7	0.93	2226
[63]	BPSK	43	1.16	1.6	7	0.25	440
[47]	QPSK	32	1.17	2.7	23	9.6	11,680
[41]	QPSK	28	1.18	4	7	9.6	10,610
[16]	8QAM	23	1.22	4.1	20	0.92	4072

TABLE 4.3 Experimental demonstrations of digital Nyquist-WDM

References	Format (PDM)	R_s (Gbaud)	$\delta f = \Delta f / R_s$	Net SE (bit/s/Hz)	FEC OH (%)	Capacity (Tbit/s)	Reach (km)
[18]	16QAM	12.5	1	6.4	25	26	227
[17]	16QAM	20	1.015	4.93	60	44.1	9100
[21]	Hybrid	9.7	1.02	8.25	20	4.95	3200
[20]	32QAM	9	1.022	8.4	7	3.36	800
[50]	16QAM	15.625	1.024	6.48	20.5	1.66	3590
[19]	16QAM	32	1.04	6	28	21.2	4370
[43]	16QAM	48	1.042	5	47	38.75	6600
[61]	16QAM	32	1.042	6	28	21.2	10,220
[65]	16QAM	14	1.05	6.32	20.5	1.5	3700
[23]	64QAM	5.7	1.1	9.1	20	102.3	240
[66]	64QAM	5.6	1.116	10	7	42.25	240
[24]	64QAM	10.4	1.154	8.67	20	2.1	1306
[67]	64QAM	10.7	1.168	7.9	28	30	320

The modulation formats range from PDM-16QAM to PDM-64QAM. Thanks to digital spectral shaping, values of frequency spacing very close to the symbol rate could be used in all experimental demonstrations. Typically, SRRC shapes were generated, with roll-off values ranging from 0.001 in Refs [17, 19] to 0.05 in Refs [16, 24].

In Reference [21], the Nyquist-WDM approach was used together with hybrid time-domain QAM modulation [68]. A net SE of 8.25 bit/s/Hz was achieved interleaving PDM-32QAM and PDM-64QAM symbols in the time domain. Digital spectral shaping was used to achieve a frequency spacing as low as $1.02 \cdot R_S$. In References [21] and [69], it is shown that time-domain hybrid QAM, together with digital spectral shaping, provides a new degree of design freedom to optimize the transmission performance by fine tuning the SE of the modulation format for a specific channel bandwidth and FEC redundancy requirement. In Reference [70], Nyquist-WDM based on time-domain hybrid QAM is indeed proposed as an enabling technology for future elastic optical networks [71], thanks to its improved ability to optimize system SE as a function of optical channel conditions compared with conventional individual format-based transceivers.

4.6 CONCLUSIONS

The principles and recent developments of Nyquist signaling in coherent optical transmission systems have been reviewed. The generation and detection of Nyquist-WDM signals have been discussed, together with the main experimental demonstrations of long-haul transmission. It was shown that, with either optical or electrical spectral shaping, WDM channel spacing equal or close to the symbol rate has been achieved, significantly outperforming unfiltered NRZ in terms of spectral efficiency.

The main drawback when performing spectral shaping in the optical domain is the need for analog filters with very steep profile, which are critical to design and build using state-of-the art technology. This limits the achievable spectral efficiency, especially when using higher-order modulation formats. On the contrary, one of the main advantages in performing spectral shaping in the digital domain is the ability to precisely control the Tx spectrum to a degree far beyond what it possible with practical analog filters, which enables the generation of closely spaced Nyquist-WDM combs with channel spacing equal (or very close) to the symbol rate, even with high-cardinality constellations.

The main limitation of digital spectral shaping is the finite sampling speed of state-of-the-art DACs, which limits the achievable symbol rate. Also, an advantage of optical spectral shaping compared with digital pulse shaping is the fact that costly DACs are not required and power consumption can be significantly reduced.

Finally, a major advantage of digital spectral shaping through DSP and DACs over analog optical shaping is its high flexibility: in fact, keeping the symbol rate fixed, the same Tx hardware can be used to generate different modulation formats, and thus achieve different line rates. This will enable software-defined optical transmission, based on the optimization of the channel throughput depending on the link conditions.

In conclusion, recent experimental demonstrations confirmed that the Nyquist-WDM concept is a promising technology for ultrahigh spectral density long-haul links and that, exploiting state-of-the-art DAC and DSP technology for digital spectral shaping, it is a good candidate for next generation flexible optical networks.

REFERENCES

1. Taylor MG. Coherent detection method using DSP for demodulation of signal and subsequent equalization of propagation impairments. *IEEE Photonics Technol Lett* 2010;16(2):674–676.
2. Winzer PJ. High-spectral-efficiency optical modulation formats. *J Lightwave Technol* 2012;30(24):3824–3835.
3. Chandrasekhar S, Liu X. Advances in Tb/s superchannels. In: Kaminov I, Li T, Willner A, editors. Optical Fiber Telecommunications Volume VIB: Systems and Networks. Oxford: Elsevier; 2013.
4. Savory SJ. Digital filters for coherent optical receivers. *Opt Express* 2008;16(2):804–817.
5. Proakis JG, Salehi M. Digital Communications. 5th ed. New York: McGraw-Hill; 2007.
6. Benedetto S, Biglieri E. Principles of Digital Transmission: With Wireless Applications. New York: Kluwer Academic Publishers; 1999.
7. Bosco G, Curri V, Carena A, Poggiolini P, Forghieri F. On the performance of Nyquist-WDM terabit superchannels based on PM-BPSK, PM-QPSK,PM-8QAM or PM-16QAM subcarriers. *J Lightwave Technol* 2011;29(1):53–61.
8. Winzer PJ, Pfennigbauer M, Essiambre R-J. Coherent crosstalk in ultradense WDM systems. *J Lightwave Technol* 2005;23(4):1734–1744.

9. Bosco G. Spectral shaping in ultra-dense WDM systems: optical vs. electrical approaches. Proceedings of Optical Fiber Communication Conference; Los Angeles (CA); 2012, Paper OM3H.1.
10. Schmogrow R, Ben-Ezra S, Schindler PC, Nebendahl B, Koos C, Freude W, Leuthold J. Pulse-shaping with digital, electrical, and optical filters—a comparison. *J Lightwave Technol* 2013;31(15):2570–2577.
11. Schmogrow R, Schindler PC, Freude W, Leuthold J. Digital pulse-shaping for spectrally efficient and flexible coherent optical networks. Proceedings Advanced Photonics for Communications; San Diego (CA); 2014, Paper NM4D.3.
12. Mazurczyk M. Spectral shaping in long haul optical coherent systems with high spectral efficiency. *J Lightwave Technol* 2014;32(16):2915–2924.
13. Torrengo E, Cigliutti R, Bosco G, Gavioli G, Alaimo A, Carena A, Curri V, Forghieri F, Piciaccia S, Belmonte M, Brinciotti A, La Porta A, Abrate S, Poggiolini P. Transoceanic PM-QPSK Terabit superchannel transmission experiments at Baud-rate subcarrier spacing. Proceedings, European Conference on Optical Communication; Torino, Italy; 2010, Paper We.7.C.2.
14. Cai J-X, Sinkin O, Zhang H, Sun Y, Pilipetskii A, Mohs G, Bergano NS. ISI Compensation up to Nyquist Channel Spacing for Strongly Filtered PDM RZ-QPSK using Multi-Tap CMA. Proceedings of Optical Fiber Communication Conference; Los Angeles (CA); 2012, Paper JW2A.47.
15. Cai J-X. 100G transmission over transoceanic distance with high spectral efficiency and large capacity. *J Lightwave Technol* 2012;30(24):3845–3856.
16. Cigliutti R, Torrengo E, Bosco G, Caponio NP, Carena A, Curri V, Poggiolini P, Yamamoto Y, Sasaki T, Forghieri F. Transmission of 9 138 Gb/s prefiltered PM-8QAM signals Over 4000 km of pure silica-core fiber. *J Lightwave Technol* 2011;29(15):2310–2318.
17. Foursa DG, Batshon HG, Zhang H, Mazurczyk M, Cai J-X, Sinkin O, Pilipetskii A, Mohs G, Bergano NS. 44.1 Tb/s transmission over 9,100 km using coded modulation based on 16QAM signals at 4.9 bits/s/Hz spectral efficiency. Proceedings, European Conference on Optical Communication; London; 2013, Paper PD3.E.1.
18. Hillerkuss D, Schmogrow R, Meyer M, Wolf S, Jordan M, Kleinow P, Lindenmann N, Schindler PC, Melikyan A, Yang X, Ben-Ezra S, Nebendahl B, Dreschmann M, Meyer J, Parmigiani F, Petropoulos P, Resan B, Oehler A, Weingarten K, Altenhain L, Ellermeyer T, Moeller M, Huebner M, Becker J, Koos C, Freude W, Leuthold J. Single-laser 32.5 Tbit/s Nyquist WDM transmission. *J Opt Commun Netw* 2012;4(10):715–723.
19. Cai J-X, Zhang H, Batshon HG, Mazurczyk M, Sinkin OV, Foursa DG, Pilipetskii AN, Mohs G, Bergano NS. 200 Gb/s and dual wavelength 400 Gb/s transmission over transpacific distance at 6.0 bit/s/Hz spectral efficiency. *J Lightwave Technol* 2014;32(4):832–939.
20. Zhou X, Nelson LE, Magill P, Isaac R, Zhu B, Peckham DW, Borel PI, Carlson K. PDM-Nyquist-32QAM for 450-Gb/s per-channel WDM transmission on the 50 GHz ITU-T grid. *J Lightwave Technol* 2012;30(4):553–559.
21. Zhou X, Nelson LE, Magill P, Isaac R, Zhu B, Peckham DW, Borel PI, Carlson K. High spectral efficiency 400 Gb/s transmission using PDM time-domain hybrid 32-64 QAM and training-assisted carrier recovery. *J Lightwave Technol* 2013;31(7):999–1005.
22. Zhou X, Nelson L, Isaac R, Magill P, Zhu B, Peckham DW, Borel P, Carlson K. 800 km transmission of 5x450Gb/s PDM-32QAM on the 50GHz grid using electrical and optical

spectral shaping. Proceedings, European Conference on Optical Communication; Geneva, Switzerland, 2011, Paper We.8.B.2.

23. Sano A, Kobayashi T, Yamanaka S, Matsuura A, Kawakami H, Miyamoto Y, Ishihara K, Masuda H. 102.3-Tb/s (224 x 548-Gb/s) C- and extended L-band all-Raman transmission over 240 km using PDM-64QAM single carrier FDM with digital pilot tone. Proceedings of Optical Fiber Communication Conference; Los Angeles (CA); 2012, Paper PDP5C.3.
24. Nespola A, Straullu S, Bosco G, Carena A, Jiang Y, Poggiolini P, Forghieri F, Yamamoto Y, Hirano M, Sasaki T, Bauwelinck J, Verheyen K. 1306-km 20x124.8-Gb/s PM-64QAM transmission over PSCF with Net SEDP 11,300 (b·km)/s/Hz using 1.15 samp/symb DAC. Proceedings, European Conference on Optical Communication; London; 2013, Paper Th.2.D.1.
25. Nespola A, Straullu S, Bosco G, Carena A, Yanchao J, Poggiolini P, Forghieri F, Yamamoto Y, Hirano M, Sasaki T, Bauwelinck J, Verheyen K. 1306-km 20x124.8-Gb/s PM-64QAM transmission over PSCF with Net SEDP 11,300 (b·km)/s/Hz using 1.15 samp/symb DAC. *Opt Express* 2014;22(1):1796–1805.
26. Bosco G, Carena A, Curri V, Poggiolini P, Forghieri F. Performance limits of Nyquist-WDM and CO-OFDM in high-speed PM-QPSK systems. *IEEE Photonics Technol Lett* 2010;22(15):1129–1131.
27. Bosco G, Carena A, Curri V, Poggiolini P, Torrengo E, Forghieri F. Investigation on the robustness of a Nyquist-WDM Terabit superchannel to transmitter and receiver non-idealities. Proceedings, European Conference on Optical Communication; Torino, Italy, 2010, Paper Tu.3.A.4.
28. Lender A. The duobinary technique for high-speed data transmission. *IEEE Trans Commun Electron* 1963;82:214–218.
29. Price AJ, Le Mercier N. Reduced bandwidth optical digital intensity modulation with improved chromatic dispersion tolerance. *Electron Lett* 1995;31(1):58–59.
30. Yonenaga K, Kuwano S, Norimatsu S, Shibata N. Optical duobinary transmission system with no receiver sensitivity degradation. *Electron Lett* 1995;31(4):302–304.
31. Penninckx D, Chbat M, Pierre L, Thierry J-P. The Phase-Shaped Binary Transmission (PSBT): a new technique to transmit far beyond the chromatic dispersion limit. *IEEE Photonics Technol Lett* 1997;9:259–261.
32. Li J, Sjödin M, Karlsson M, Andrekson PA. Building up low-complexity spectrally-efficient Terabit superchannels by receiver-side duobinary shaping. *Opt Express* 2012;20(9):10271–10282.
33. Igarashi K, Tsuritani T, Morita I. Bit-error rate performance of super-Nyquist-WDM DPQPSK signals with duobinary-pulse shaping. Proceedings of Optical Fiber Communication Conference; San Francisco (CA); 2014, Paper Th2A.18.
34. Li J, Tao Z, Zhang H, Yan W, Hoshida T, Rasmussen JC. Spectrally efficient quadrature duobinary coherent systems with symbol-rate digital signal processing. *J Lightwave Technol* 2011;29(8):1098–1104.
35. Zhang J, Huang B, Li X. Improved quadrature duobinary system performance using multi-modulus equalization. *IEEE Photonics Technol Lett* 2013;25(16):1630–1633.
36. Igarashi K, Tsuritani T, Morita I, Tsuchida Y, Maeda K, Tadakuma M, Saito T, Watanabe K, Imamura K, Sugizaki R, Suzuki M. Super-Nyquist-WDM transmission over 7,326-km seven-core fiber with capacity-distance product of 1.03 Exabit/s·km. *Opt Express* 2014;22(2):1220–1228.

37. Li J, Tipsuwannakul E, Eriksson T, Karlsson M, Andrekson PA. Approaching Nyquist limit in WDM systems by low-complexity receiver-side duobinary shaping. *J Lightwave Technol* 2012;30(11):1664–1675.

38. Colavolpe G, Foggi T, Modenini A, Piemontese A. Faster-than-Nyquist and beyond: how to improve spectral efficiency by accepting interference. *Opt Express* 2011;19(27):26600–26609.

39. Li L, Lu Y, Liu L, Chang D, Xiao Z, Wei Y. 20*times*224 Gbps (56 Gbaud) PDM-QPSK transmission in 50 GHz grid over 3040 km G.652 fiber and EDFA only link using soft output faster than Nyquist technology. Proceedings of Optical Fiber Communication Conference; Anaheim (CA); 2013, Paper W3J.2.

40. Cai Y, Cai JX, Davidson CR, Foursa D, Lucero A, Sinkin O, Pilipetskii A, Mohs G, Bergano NS. High spectral efficiency long-haul transmission with pre-filtering and maximum a posteriori probability detection. Proceedings, European Conference on Optical Communication; Torino, Italy; 2010, Paper We.7.C.4.

41. Cai J-X, Cai Y, Davidson CR, Foursa DG, Lucero AJ, Sinkin OV, Patterson WW, Pilipetskii AN, Mohs G, Bergano NS. Transmission of 96 100-Gb/s bandwidth-constrained PDM-RZ-QPSK channels with 300% spectral efficiency over 10610 km and 400% spectral efficiency over 4370 km. *J Lightwave Technol* 2011;29(4):491–498.

42. Cai J-X, Davidson CR, Lucero A, Zhang H, Foursa DG, Sinkin OV, Patterson WW, Pilipetskii AN, Mohs G, Bergano NS. 20 Tbit/s transmission Over 6860 km With sub-Nyquist channel spacing. *J Lightwave Technol* 2012;30(4):651–657.

43. Salsi M, Rios-Muller R, Renaudier J, Tran P, Schmalen L, Ghazisaeidi A, Mardoyan H, Brindel P, Charlet G, Bigo S. 38.75 Tb/s transmission experiment over transoceanic distance. Proceedings, European Conference on Optical Communication; London; 2013, Paper PD3.E.2.

44. Wang J, Xie C, Pan Z. Matched filter design for RRC spectrally shaped Nyquist-WDM systems. *IEEE Photonics Technol Lett* 2013;25(23):2263–2266.

45. Kudo R, Kobayashi T, Ishihara K, Takatori Y, Sano A, Miyamoto Y. Coherent optical single carrier transmission using overlap frequency domain equalization for long-haul optical systems. *J Lightwave Technol* 2009;27(16):3721–3728.

46. Gavioli G, Torrengo E, Bosco G, Carena A, Curri V, Miot V, Poggiolini P, Belmonte M, Forghieri F, Muzio C, Piciaccia S, Brinciotti A, La Porta A, Lezzi C, Savory S, Abrate S. Investigation of the impact of ultra-narrow carrier spacing on the transmission of a 10-carrier 1Tb/s superchannel. Proceedings of Optical Fiber Communication Conference; San Diego (CA); 2010, Paper OThD3.

47. Salsi M, Koebele C, Tran P, Mardoyan H, Dutisseuil E, Renaudier J, Bigot-Astruc M, Provost L, Richard S, Sillard P, Bigo S, Charlet G. Transmission of 96×100 Gb/s with 23% super-FEC overhead over 11,680 km, using optical spectral engineering. Proceedings of Optical Fiber Communication Conference; Los Angeles (CA); 2011, Paper OMR2.

48. http://www.micram.de/index.php/products/vega?start=1. Accessed 2015 Sep 22.

49. http://www.tek.com/products/signal-generator/awg7000/. Accessed 2015 Sep 22.

50. Cigliutti R, Nespola A, Zeolla D, Bosco G, Carena A, Curri V, Forghieri F, Yamamoto Y, Sasaki T, Poggiolini P. 16 × 125 Gb/s Quasi-Nyquist DAC-generated PM-16QAM transmission over 3590 km of PSCF. *IEEE Photonics Technol Lett* 2012;24(23):2143–2146.

51. Schmogrow R, Meyer M, Schindler PC, Josten A, Ben-Ezra S, Koos C, Freude W, Leuthold J. 252 Gbit/s real-time Nyquist pulse generation by reducing the oversampling

factor to 1.33. Proceedings of Optical Fiber Communication Conference; Anaheim (CA); 2013, Paper OTu2I.1.

52. Bosco G, Curri V, Carena A, Poggiolini P, Forghieri F. Performance of digital Nyquist-WDM. Proceedings of 2011 OSA Summer Topical Meeting on Signal Processing in Photonics Communications (SPPCom); Toronto, Canada; 2011, Paper SPMA4.
53. Mazurczyk M. Optical spectral shaping and high spectral efficiency in long haul systems. Proceedings of Optical Fiber Communication Conference; San Francisco (CA); 2013, Paper Tu3J.4.
54. Schmogrow R, Winter M, Meyer M, Hillerkuss D, Wolf S, Baeuerle B, Ludwig A, Nebendahl B, Ben-Ezra S, Meyer J, Dreschmann M, Huebner M, Becker J, Koos C, Freude W, Leuthold J. Real-time Nyquist pulse generation beyond 100 Gbit/s and its relation to OFDM. *Opt Express* 2011;20(1):317–337.
55. Schmogrow R, Bouziane R, Meyer M, Milder PA, Schindler PC, Bayvel P, Killey RI, Freude W, Leuthold J. Real-time digital Nyquist-WDM and OFDM signal generation: spectral efficiency versus DSP complexity. Proceedings, European Conference on Optical Communication; Amsterdam, The Netherlands, 2012, Paper Mo.2.A.4.
56. Wang J, Xie C, Pan Z. Generation of spectrally efficient Nyquist-WDM QPSK signals using digital FIR or FDE filters at transmitters. *J Lightwave Technol* 2012;30(23):3679–3686.
57. Igarashi K, Mori Y, Katoh K, Kikuchi K. Bit-error rate performance of Nyquist wavelength-division multiplexed quadrature phase-shift keying optical signals. Proceedings of Optical Fiber Communication Conference; Los Angeles (CA); 2011, Paper OMR6.
58. Schmogrow R, Winter M, Meyer M, Hillerkuss D, Nebendahl B, Meyer J, Dreschmann M, Huebner M, Becker J, Koos C, Freude W, Leuthold J. Real-time Nyquist pulse modulation transmitter generating rectangular shaped spectra of 112 Gbit/s 16QAM signals. Proceedings of 2011 OSA Summer Topical Meeting on Signal Processing in Photonics Communications (SPPCom); Toronto, Canada; 2011, Paper SPMA5.
59. Bosco G, Poggiolini P, Carena A, Curri V, Forghieri F. Analytical results on channel capacity in uncompensated optical links with coherent detection. *Opt Express* 2011;19(26):B440–B451.
60. Poggiolini P. The GN model of non-linear propagation in uncompensated coherent optical systems. *J Lightwave Technol* 2012;30(24):3857–3879.
61. Zhang H, Cai J-X, Batshon HG, Mazurczyk M, Sinkin OV, Foursa DG, Pilipetskii A, Mohs G, Bergano NS. 200 Gb/s and dual-wavelength 400 Gb/s transmission over transpacific distance at 6 bit/s/Hz spectral efficiency. Proceedings of Optical Fiber Communication Conference; Anaheim (CA); 2013, Paper PDP5A.6.
62. Gavioli G, Torrengo E, Bosco G, Carena A, Savory SJ, Forghieri F, Poggiolini P. Ultra-narrow-spacing 10-channel 1.12 Tb/s DWDM long-haul transmission over uncompensated SMF and NZDSF. *IEEE Photonics Technol Lett* 2010;22(19):1419–1421.
63. Bousselet P, Bissessur H, Lestrade J, Salsi M, Pierre L, Mongardien D. High capacity (64 x 43 Gb/s) unrepeated transmission over 440 km. Proceedings of Optical Fiber Communication Conference; Los Angeles (CA); 2011, Paper OMI2.
64. Cai J-X, Cai Y, Davidson CR, Foursa DG, Lucero A, Sinkin O, Patterson W, Pilipetskii A, Mohs G, Bergano NS. Transmission of 96x100 G pre-filtered PDM-RZ-QPSK channels with 300% spectral efficiency over 10,608 km and 400% spectral efficiency over 4,368 km. Proceedings of Optical Fiber Communication Conference; San Diego (CA); 2010, Paper PDPB10.

65. Cigliutti R, Nespola A, Zeolla D, Bosco G, Carena A, Curri V, Forghieri F, Yamamoto Y, Sasaki T, Poggiolini P. Ultra-long-haul transmission of 16x112 Gb/s spectrally-engineered DAC-generated Nyquist-WDM PM-16QAM channels with 1.05x(Symbol-Rate) frequency spacing. Proceedings of Optical Fiber Communication Conference; Los Angeles (CA); 2012, Paper OTh3A.3.
66. Kobayashi T, Sano A, Matsuura A, Yoshida1 M, Sakano T, Kubota H, Miyamoto Y, Ishihara K, Mizoguchi M, Nagatani M. 45.2 Tb/s C-band WDM transmission over 240 km using 538 Gb/s PDM-64QAM single carrier FDM signal with digital pilot tone. Proceedings, European Conference on Optical Communication; Geneva, Switzerland; 2011, Paper Th.13.C.6.
67. Yu J, Dong1 Z, Chien H-C, Xiao X, Jia Z, Chi N. 430-Tb/s (3×12.84-Tb/s) signal transmission over 320 km using PDM 64-QAM modulation. Proceedings of Optical Fiber Communication Conference; Los Angeles (CA); 2012, Paper OM2A.4.
68. Peng W-R, Morita I, Tanaka H. Hybrid QAM transmission techniques for single-carrier ultra-dense WDM systems. Proceedings of the OptoElectronics and Communication Conference; Taiwan; 2011, p 824–825.
69. Zhuge Q, Xu X, Morsy-Osman M, Chagnon M, Qiu M, Plant DV. Time domain hybrid QAM based rate-adaptive optical transmissions using high speed DACs. Proceedings of Optical Fiber Communication Conference; Anaheim (CA); 2013, Paper OTh4E.6.
70. Zhuge Q, Morsy-Osman M, Xu X, Chagnon M, Qiu M, Plant DV. Spectral efficiency-adaptive optical transmission using time domain hybrid QAM for agile optical networks. *J Lightwave Technol* 2013;31(15):2013–2621.
71. Gerstel O, Jinno M, Lord A, Yoo SJB. Elastic optical networking: a new dawn for the optical layer? *IEEE Commun Mag* 2012;50(2):s12–s20.

5

SPECTRALLY EFFICIENT MULTIPLEXING – OFDM

AN LI, DI CHE, QIAN HU, XI CHEN, AND WILLIAM SHIEH

Department of Electrical and Electronic Engineering, The University of Melbourne, Melbourne, VIC, Australia

Single-carrier modulation (SCM) has been the *de facto* modulation choice and has long been implemented in commercialized products for both long- and short-reach optical fiber communications. The SCM gains its popularity due to low hardware complexity for relatively low-speed (10 Gb/s or less) communication systems. In the recent decade, multicarrier modulation (MCM), which has higher spectral efficiency (SE) than the conventional SCM, has been argued as a promising alternative to satisfy the exponential growth of the Internet traffic. Several schemes that utilize MCM have been demonstrated since 2006. For instance, direct-detection optical orthogonal frequency-division multiplexing (DDO-OFDM) is proposed for short-reach applications [1, 2], and coherent optical OFDM (CO-OFDM) for long-haul transmission [3]. While the implementation (hardware and software) of DDO-OFDM is simpler [1, 2], CO-OFDM promises higher SE, better receiver sensitivity, and polarization-dispersion resilience [4, 5]. Lab demonstrations confirm CO-OFDM with a data rate higher than 1-Terabit per second (Tb/s) can be delivered over 600-km standard single-mode fiber (SSMF) [6–10].

OFDM is one of the variations of MCM in which the data information is carried over multiple low-rate subcarriers. It has been widely understood that OFDM has the robustness against channel dispersion and has its ease of phase and channel estimation in a time-varying environment. However, due to the nature of MCM, OFDM has the disadvantages of high peak-to-average power ratio (PAPR) and high sensitivity to frequency offset and phase noise. In order to fully discuss the applications of OFDM in optical communication, we present an introduction to the fundamentals

Enabling Technologies for High Spectral-efficiency Coherent Optical Communication Networks, First Edition. Edited by Xiang Zhou and Chongjin Xie.
© 2016 John Wiley & Sons, Inc. Published 2016 by John Wiley & Sons, Inc.

of OFDM, including its basic mathematical formulation, discrete Fourier transform (DFT) implementation, cyclic prefix (CP), spectral efficiency (SE), and PAPR characteristics. We review two major schemes of OFDM: CO-OFDM and DDO-OFDM that are popular and have been widely adopted by the optical communications. Furthermore, we show a novel variant of OFDM, which is called DFT-spread OFDM system (DFT-S OFDM), with many interesting features that can significantly improve the system performance and SE. Finally, we show a few OFDM-based superchannel transmission technologies to achieve high SE for high-speed optical transports.

5.1 OFDM BASICS

OFDM is a special form of a broader class of MCM. The principle of OFDM is to transmit the information through a large number of orthogonal subcarriers. The OFDM signal in time domain consists of a continuous stream of OFDM symbols with a regular period T_s. The OFDM baseband signal $s(t)$ is written as [11, 12]

$$s(t) = \sum_{i=-\infty}^{+\infty} \sum_{k=-N_{sc}/2+1}^{k=N_{sc}/2} c_{ki} \exp(j2\pi f_k(t - iT_s)) f(t - iT_s) \tag{5.1}$$

$$f_k = \frac{k-1}{t_s}, \quad \Delta f = \frac{1}{t_s} \tag{5.2}$$

$$f(t) = \begin{cases} 1, & (-\Delta_G < t \le t_s) \\ 0, & (t \le -\Delta_G, t > t_s) \end{cases} \tag{5.3}$$

where c_{ki} is the ith information symbol at the kth subcarrier, $f(t)$ is the pulse waveform of the symbol, f_k is the frequency of the subcarrier, and Δf is the subcarrier spacing, and N_{sc}, Δ_G, and t_s are the number of OFDM subcarriers, guard interval (GI) length, and observation period, respectively.

The optimum detector for each subcarrier could use a filter that matches the subcarrier waveform, or a correlator matched to the subcarrier as shown in Figure 5.1. Therefore, the detected information symbol $\vec{c}_{ki}$ at the output of the correlator is given by

$$\vec{c}_{ki} = \int_0^{T_s} r(t - iT_s) s_k^* \, dt = \int_0^{T_s} r(t - iT_s) \exp(-j2\pi f_k t) dt \tag{5.4}$$

where $r(t)$ is the received time-domain signal, s_k is the kth subcarrier waveform, and the * stands for complex conjugate. The classical MCM uses nonoverlapped band-limited signals, and can be implemented with a bank of large number of oscillators and filters at both transmit and receive ends. The major disadvantage of nonoverlapped MCM is that it requires excessive bandwidth. This is because in order to design the filters and oscillators cost-efficiently, the channel spacing has to be multiples of the symbol rate, greatly reducing the spectral efficiency. On the

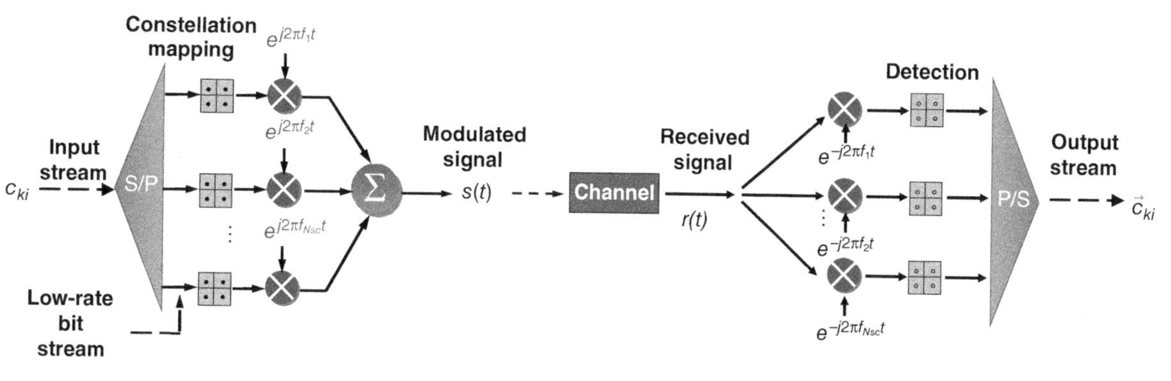

FIGURE 5.1 Conceptual diagram for a generic multicarrier modulation (MCM) system.

contrary, OFDM employs overlapped yet orthogonal signal set [13, 14]. This orthogonality originates from the straightforward correlation between any two subcarriers, given by

$$\delta_{kl} = \frac{1}{T_s}\int_0^{T_s} s_k s_l^* \, dt = \frac{1}{T_s}\int_0^{T_s} \exp(j2\pi(f_k - f_l)t)dt$$
$$= \exp(j\pi(f_k - f_l)T_s)\frac{\sin(\pi(f_k - f_l)T_s)}{\pi(f_k - f_l)T_s} \quad (5.5)$$

If the following condition

$$f_k - f_l = m\frac{1}{T_s} \quad (5.6)$$

is satisfied, the two subcarriers are orthogonal to each other. This signifies that these orthogonal subcarrier sets, with their frequency spaced at multiple of inverse of the symbol rate can be recovered with the matched filters (Eq. 5.4) without intercarrier interference (ICI), in spite of strong signal spectral overlapping.

One of the enabling techniques for OFDM is the insertion of cyclic prefix (CP), which is also known as guard interval (GI) [11, 12]. Cyclic prefix was proposed to resolve the channel dispersion-induced intersymbol interference (ISI) and ICI [11, 12, 15]. Figure 5.2 shows insertion of a cyclic prefix by cyclic extension of the OFDM waveform into the guard interval, Δ_G. As shown in Figure 5.2, the waveform in the guard interval is essentially an identical copy of that in the DFT window, with time-shifted by "t_s" behind.

It can be seen that, if the maximum delay spread of multipath fading is smaller than the guard time Δ_G, the CP can perfectly accommodate the ISI. In the context of optical transmission, the delay spread due to the chromatic dispersion (CD) among

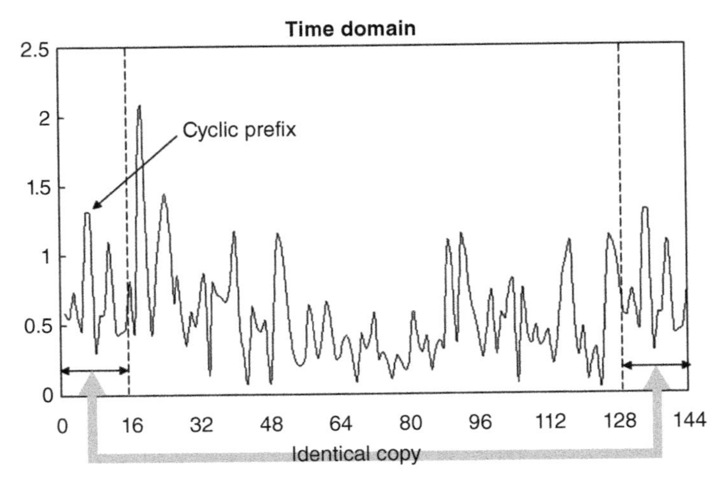

FIGURE 5.2 OFDM symbol in time domain.

the subcarriers should not exceed the guard time, and the fundamental condition for the complete elimination of ISI in optical medium is thus given by

$$\frac{c}{f^2}|D_t| \cdot N_{sc} \cdot \Delta f \leq \Delta_G \tag{5.7}$$

where f is the frequency of the optical carrier, c the speed of light, D_t the total accumulated chromatic dispersion in units of ps/km, and N_{sc} is the number of subcarriers.

5.2 COHERENT OPTICAL OFDM (CO-OFDM)

CO-OFDM offers good performance in the sense of receiver sensitivity, spectral efficiency, and robustness against polarization dispersion, but requires high complexity in transceiver design. In the open literature, CO-OFDM was first proposed by Shieh and Athaudage [3], and the concept of the coherent optical multiple-input-multiple-output (MIMO)-OFDM was formalized by Shieh et al. [4]. The early CO-OFDM experiments were carried out by Shieh et al. for a 1000 km SSMF transmission at 8 Gb/s [16], and by Jansen et al. for 4160 km SSMF transmission at 20 Gb/s [17]. The principle and transmitter/receiver design for CO-OFDM are given in the following sections.

5.2.1 Principle of CO-OFDM

5.2.1.1 Coherent-Detection and Optical OFDM The synergies between coherent optical communications and OFDM are twofold. OFDM enables channel and phase estimation for coherent detection in a computationally efficient way. Coherent detection provides linearity in radio frequency (RF)-to-optical (RTO) up-conversion and optical-to-RF (OTR) down-conversion, much needed for OFDM. Consequently, CO-OFDM is a natural choice for optical transmission in the linear regime. A generic CO-OFDM system is depicted in Figure 5.3. In general, a CO-OFDM system can be divided into five functional blocks including (i) RF OFDM transmitter, (ii) RTO up-converter, (iii) the optical channel, (iv) the OTR down-converter, and (v) the RF OFDM receiver. The detailed architecture for RF OFDM transmitter/receiver has already been shown in Figure 5.3, which generates/recovers OFDM signals either in baseband or an RF band. Let us assume for now a linear channel where optical fiber nonlinearity is not considered. It is apparent that the challenges for CO-OFDM implementation are to obtain a linear RTO up-converter and linear OTR down-converter. It has been proposed and analyzed that by biasing the Mach–Zehnder modulators (MZMs) at null point, a linear conversion between the RF signal and optical field signal can be achieved [3, 18]. Meanwhile, by using coherent detection, a linear transformation from "optical signal" to RF (or baseband electrical) signal can be achieved [3, 18–20]. Therefore, combining such a composite system cross RF and optical domain [3, 16, 17], a linear channel can be constructed where OFDM can perform its best role of mitigating channel dispersion impairment in both RF and optical domains. In this section, we use the term "RF domain" and "electrical domain" interchangeably.

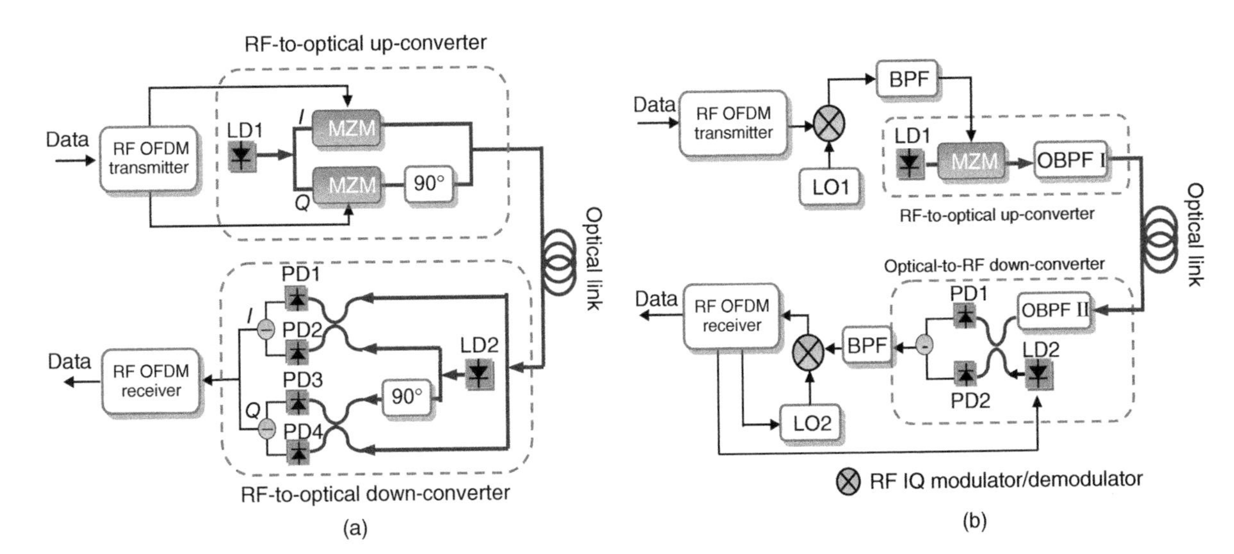

FIGURE 5.3 A CO-OFDM system in (a) direct up/down-conversion architecture, and (b) intermediate frequency (IF) architecture.

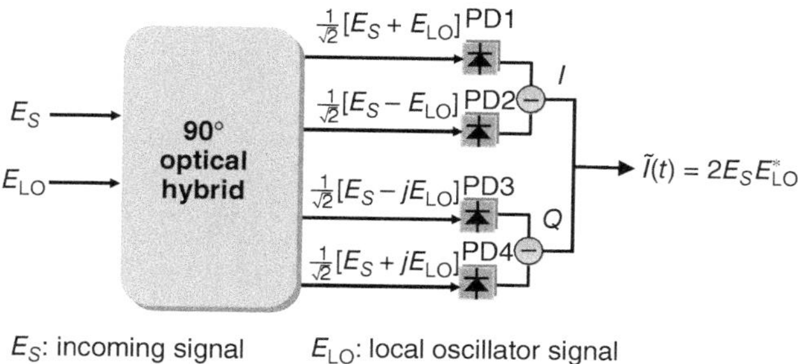

FIGURE 5.4 Coherent detection using an optical hybrid and balanced photodetection.

As shown in Figure 5.4, coherent detection uses a 2×4 90° optical "hybrid mixer" and a pair of balanced photodetectors. The main purposes of coherent detection are (i) to linearly recover the *In-Phase* (I) and *Quadrature* (Q) components of the incoming signal, and (ii) to suppress or cancel the common mode noise. Using a six-port 90° hybrid, signal detection and analysis have been realized in RF domain for decades [21, 22], and its application to single-carrier coherent optical systems can be found also in [19, 20]. In what follows, in order to illustrate its working principle, we perform an analysis of down-conversion via coherent detection assuming ideal condition for each component shown in Figure 5.4.

The purpose of the four output ports of the 90° optical hybrid is to generate a 90° phase shift between I and Q components, and 180° phase shift between balanced detectors. Ignoring imbalance and loss of the optical hybrid, the output signals E_{1-4} can be expressed as

$$E_1 = \frac{1}{\sqrt{2}}[E_s + E_{\text{LO}}], \quad E_2 = \frac{1}{\sqrt{2}}[E_s - E_{\text{LO}}]$$
$$E_3 = \frac{1}{\sqrt{2}}[E_s - jE_{\text{LO}}], \quad E_4 = \frac{1}{\sqrt{2}}[E_s + jE_{\text{LO}}] \tag{5.8}$$

where E_s and E_{LO} are, respectively, the electric field of the incoming signal and local oscillator (LO) signal. We further decompose the incoming signal into two components: (i) the received signal free from the amplified spontaneous noise (ASE), $E_r(t)$, and (ii) the ASE noise, $n_o(t)$, namely

$$E_s = E_r + n_o \tag{5.9}$$

We first study how the I component of the photodetected current is generated, and the Q component can be derived accordingly. The I component is obtained by using a pair of the photodetectors, PD1 and PD2 in Figure 5.4, whose photocurrent I_{1-2} can be described as

$$I_1 = |E_1|^2 = \frac{1}{2}\{|E_s|^2 + |E_{LO}|^2 + 2\mathrm{Re}\{E_s E_{LO}^*\}\} \tag{5.10}$$

$$I_2 = |E_2|^2 = \frac{1}{2}\{|E_s|^2 + |E_{LO}|^2 - 2\mathrm{Re}\{E_s E_{LO}^*\}\} \tag{5.11}$$

$$|E_s|^2 = |E_r|^2 + |n_o|^2 + 2\mathrm{Re}\{E_r n_o^*\} \tag{5.12}$$

$$|E_{LO}|^2 = I_{LO}(1 + I_{RIN}(t)) \tag{5.13}$$

where I_{LO} and $I_{RIN}(t)$ are the average power and relative intensity noise (RIN) of the LO laser, and "Re" or "Im" denotes the real or imaginary part of a complex signal. For simplicity, the photodetection responsivity is set to unity. The three terms at the right-hand side of (5.12) represent the signal-to-signal beat noise (SSBN), signal-to-ASE beat noise, and ASE-to-ASE beat noise. Because of the balanced detection, using (5.10) and (5.11), the I component of the photocurrent becomes

$$I_I(t) = I_1 - I_2 = 2\mathrm{Re}\{E_s E_{LO}^*\} \tag{5.14}$$

Now, the noise suppression mechanism is completely revealed: the three noise terms in (5.12) and the RIN noise in (5.13) from a single detector are completely cancelled via balanced detection. Meanwhile, it has been shown that coherent detection can be performed by using a single photodetector, but at the cost of reduced dynamic range [23].

In a similar fashion, the Q component from the other pair of balanced detectors can be derived as

$$I_Q(t) = I_3 - I_4 = 2\mathrm{Im}\{E_s E_{LO}^*\} \tag{5.15}$$

Using the results of (5.14) and (5.15), the complex detected signal $\tilde{I}(t)$ consisting of both I and Q components becomes

$$\tilde{I}(t) = I_I(t) + jI_Q(t) = 2E_s E_{LO}^* \tag{5.16}$$

From (5.16), the linear down-conversion process via coherent detection becomes quite clear; the complex photocurrent $\tilde{I}(t)$ is in essence a linear replica of the incoming complex signal that is frequency down-converted by a local oscillator frequency. Thus, with linear coherent detection at receiver and linear generation at transmitter, complex OFDM signals can be readily transmitted over the optical fiber channel.

5.2.1.2 Digital Signal Processing (DSP) of CO-OFDM In Section 5.2.1.1, we have introduced the concept of CO-OFDM with focus on the principle of coherent detection. Here, we revisit the principle of CO-OFDM from experimental and signal processing point of view. Figure 5.5 shows a conceptual diagram of a complete CO-OFDM system [3, 16, 17, 24, 25]. The function of the OFDM transmitter is to

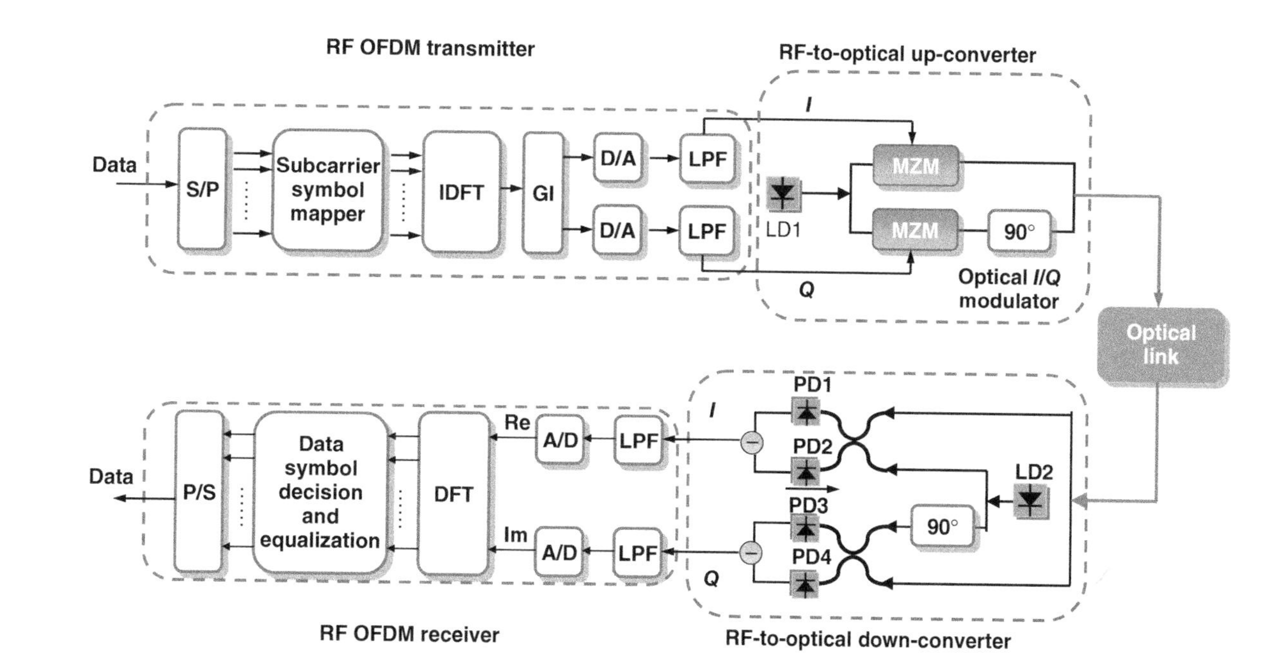

FIGURE 5.5 Conceptual diagram of CO-OFDM system. S/P: Serial-to-parallel, GI: Guard Interval, (I)DFT: (Inverse) Discrete Fourier Transform, LPF: Low-Pass Filter, MZM: Mach–Zehnder Modulator, PD: Photodiode.

map the data bits into each OFDM symbol, and generate the time series by inverse discrete Fourier transform (IDFT) expressed in (Eq. 5.1), including insertion of the guard interval. The digital signal is then converted to analog signal through the digital-to-analog converter (DAC), and filtered with a low-pass filter (LPF) to remove the alias signal. In Figure 5.5, direct up-conversion architecture is used where RF OFDM transmitter outputs a baseband OFDM signal. The subsequent RTO up-converter transforms the baseband signal to the optical domain using an optical IQ modulator comprising a pair of MZMs with a 90° phase offset. The baseband OFDM signal is directly up-converted to the optical domain given by

$$E(t) = \exp(j\omega_{\mathrm{LD1}}t + \phi_{\mathrm{LD1}}) \cdot s_B(t) \tag{5.17}$$

where ω_{LD1} and ϕ_{LD1}, respectively, are the angular frequency and the phase of the transmitter laser. The up-converted signal $E(t)$ traverses the optical medium with the impulse response $h(t)$, and the received optical signal becomes

$$E(t) = \exp(j\omega_{\mathrm{LD1}}t + \phi_{\mathrm{LD1}})s_B(t) \otimes h(t) \tag{5.18}$$

where "⊗" stands for convolution. The optical OFDM signal is then fed into the OTR down-converter where the optical OFDM signal is converted to RF OFDM signal. There are two ways to do the down-conversion. One is direct down-conversion architecture where the intermediate frequency (IF) is near DC. The other is to first down-convert the signal to RF domain with an IF and then down-convert to baseband. The IF signal can be expressed as

$$r(t) = \exp(j\omega_{\mathrm{off}}t + \Delta\phi)r_0(t), \quad r_0(t) = s_B(t) \otimes h(t) \tag{5.19}$$

$$\omega_{\mathrm{off}} = \omega_{\mathrm{LD1}} - \omega_{\mathrm{LD2}}, \quad \Delta\phi = \phi_{\mathrm{LD1}} - \phi_{\mathrm{LD2}} \tag{5.20}$$

where $\Delta\omega_{\mathrm{off}}$ and $\Delta\phi$ are, respectively, the angular frequency offset and phase offset between transmit and receive lasers. In the RF OFDM receiver, the down-converted OFDM signal is first sampled with an analog-to-digital converter (ADC). Then, the signal needs to go through three levels of synchronizations before the symbol decision can be made. The three levels of synchronizations are (i) FFT window synchronization where OFDM symbol is properly delineated to avoid ISI, (ii) frequency synchronization, namely, frequency offset ω_{off} needs to be estimated and compensated, and (iii) the subcarrier recovery, where each subcarrier channel is estimated and compensated. Assuming successful completion of DFT window synchronization and frequency synchronization, the RF OFDM signal through DFT of the sampled value of Eq. (5.19) becomes

$$r_{ki} = e^{\phi_i}h_{ki}c_{ki} + n_{ki} \tag{5.21}$$

where r_{ki} is the received information symbol, ϕ_i is the OFDM symbol phase (OSP) or common phase error (CPE), h_{ki} is the frequency domain channel transfer function, and n_{ki} is the noise. The third synchronization of subcarrier recovery involves estimation of OSP ϕ_i and the channel transfer function h_{ki}. Once they are known, an estimated value of c_{ki}, $\hat{c}_{ki}$ is given by zero-forcing method as

$$\hat{c}_{ki} = \frac{h^*_{ki}}{|h_{ki}|^2} e^{-i\phi_i} r_{ki} \tag{5.22}$$

$\hat{c}_{ki}$ is used for symbol decision or to recover the transmitter value c_{ki}, which can subsequently be mapped back to the original transmitted digital bits.

This description of CO-OFDM processing has so far left out the pilot-subcarrier or training-symbol insertion where a proportion of the subcarriers or all the subcarriers in one OFDM symbol are known values to the receiver. The purpose of these pilot subcarrier or training symbol is to assist the above-mentioned three-level synchronization. Another important aspect of the CO-OFDM signal processing not involved is the error-correction coding involving the error-correction encoder/decoder and the interleaver/de-interleaver.

5.2.1.3 Polarization-Mode Dispersion (PMD) Supported CO-OFDM It is well-known that optical fiber can support two polarization modes. The propagation of an optical signal is influenced by the polarization effects including polarization coupling, polarization-dependent loss (PDL), and polarization-mode dispersion (PMD). By utilizing the MIMO algorithm in digital signal processing (DSP), the capacity of CO-OFDM system can be doubled by using polarization-division multiplexing (PDM), and the impact of PMD can be digitally removed from the signal.

As shown in Figure 5.6, a two-input-two-output (TITO) scheme of CO-OFDM is usually applied to support polarization-multiplexed transmission in the presence of PMD [4, 26]. It consists of two sets of CO-OFDM transmitter and receiver, each

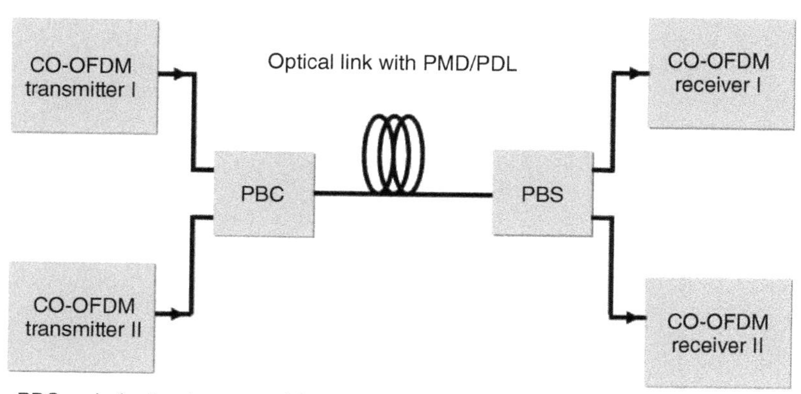

FIGURE 5.6 A variation of CO-MIMO-OFDM system: two-input two-output (TITO) [25].

pair for one polarization. In such a scheme, because the transmitted OFDM information symbol $\vec{c}_{ki}$ can be considered as polarization modulation or polarization multiplexing, the capacity is thus doubled compared with single-input-single-output (SISO) scheme. As the impact of the PMD is to simply rotate the subcarrier polarization, it can be treated by channel estimation and constellation reconstruction. Therefore, the doubling of the channel capacity will not be affected by PMD. Due to the polarization-diversity receiver employed at the receive end, TITO scheme does not need polarization tracking at the receiver.

Similar to the single-polarization OFDM signal model described in Section 5.2, the OFDM time-domain signal, $\boldsymbol{s}(t)$ is described using Jones vector given by Shieh et al. [24]

$$\boldsymbol{s}(t) = \sum_{i=-\infty}^{+\infty} \sum_{k=-N_{sc}/2+1}^{N_{sc}/2} \boldsymbol{c}_{ik} \prod (t - iT_s) \exp(j2pf_k(t - iT_s)) \tag{5.23}$$

$$\boldsymbol{s}(t) = \begin{bmatrix} s_x \\ s_y \end{bmatrix}, \quad \boldsymbol{c}_{ik} = \begin{bmatrix} c_{ik}^x \\ c_{ik}^y \end{bmatrix} \tag{5.24}$$

$$f_k = \frac{k-1}{t_s} \tag{5.25}$$

$$\prod(t) = \begin{cases} 1, & (-\Delta_G < t < t_s) \\ 0, & (t \le -\Delta_G, t > t_s) \end{cases} \tag{5.26}$$

where s_x and s_y are the two polarization components in the time domain, $\boldsymbol{c}_{ik}$ is the transmitted OFDM symbol in the form of Jones vector for the kth subcarrier in the ith OFDM symbol, c_{ik}^x and c_{ik}^y are the two polarization elements for $\boldsymbol{c}_{ik}$, f_k is the frequency for the kth subcarrier, N_{sc} is the number of OFDM subcarriers, and T_s, Δ_G, and t_s are the OFDM symbol period, guard interval length, and observation period, respectively. The Jones vector $\boldsymbol{c}_{ik}$ is employed to describe generic OFDM information symbol regardless of the methods of the OFDM transmitter polarization configuration. In particular, the $\boldsymbol{c}_{ik}$ encompasses various modes of the polarization generation including single polarization, polarization multiplexing, and polarization modulation, as they all can be represented by the two-element Jones vector $\boldsymbol{c}_{ik}$. The transmitted Jones vectors can be recovered from the received Jones vectors by using training symbols.

We select a guard interval long enough to handle the fiber dispersion including PMD and CD. This time margin condition is given by

$$\frac{c}{f^2} |D_t| \cdot N_{sc} \cdot \Delta f + \text{DGD}_{max} \le \Delta_G \tag{5.27}$$

where f is the frequency of the optical carrier, c is the speed of light, D_t is the total accumulated chromatic dispersion in units of ps/pm, N_{sc} is the number of the subcarriers, Δf is the subcarrier channel spacing, and DGD_{max} is the maximum budgeted differential-group-delay (DGD), which is about 3.5 times of mean PMD to have sufficient margin.

Following the same procedure as in [3], assuming long-enough symbol period, we arrive at the received symbol given by

$$\vec{c}\,'_{ik} = e^{j\phi_i} \cdot e^{j\Phi_D(f_k)} \cdot T_k \cdot \vec{c}_{ik} + \vec{n}_{ik} \tag{5.28}$$

$$T_k = \prod_{i=1}^{N} \exp\left\{\left(-\frac{1}{2}j\vec{\beta}_i f_k - \frac{1}{2}\vec{\alpha}_l\right)\vec{\sigma}\right\} \tag{5.29}$$

$$\Phi_D(f_k) = \pi \cdot c \cdot D_t \cdot f_k^2 / f_{\mathrm{LD1}}^2$$

where $\boldsymbol{c}'_{ik} = \left[c_{ik}^{\prime x}\ c_{ik}^{\prime y}\right]^T$ is the received information symbol in the form of the Jones vector for the kth subcarrier in the ith OFDM symbol, $\boldsymbol{n}_{ik} = \left[n_{ik}^x\ n_{ik}^y\right]^T$ is the noise including two polarization components, $\boldsymbol{T}_k$ is the Jones matrix for the fiber link, $\Phi_D(f_k)$ is the phase dispersion owing to the fiber chromatic dispersion, and ϕ_i is the OSP noise owing to the phase noises from the lasers and RF local oscillators (LOs) at both the transmitter and receiver [3]. ϕ_i is usually dominated by the laser phase noise.

5.3 DIRECT-DETECTION OPTICAL OFDM (DDO-OFDM)

Direct-detection optical OFDM (DDO-OFDM) usually has simpler transmitter/receiver than CO-OFDM, thus lower system cost. It has many variants that trade-off between the spectral efficiency and the system cost from a broad range of applications. For instance, the first report of the DDO-OFDM [27] takes advantage of the fact that the OFDM signal is more immune to the impulse clipping noise seen in CATV networks. Another example is single-sideband (SSB)-OFDM, which has been recently proposed by Djordjevic and Vasic [2] and Lowery and Armstrong [28] for long-haul transmission. Tang et al. have proposed an adaptively modulated optical OFDM (AMO-OFDM) that uses bit and power loading showing promising results for both multimode fiber and short-reach SMF fiber links [29–31]. The common feature for DDO-OFDM is the use of a simple square-law photodiode at the receiver. DDO-OFDM can be divided into two categories according to how the optical OFDM signal is generated: (i) linearly mapped DDO-OFDM (LM-DDO-OFDM) where the optical OFDM spectrum is a replica of baseband OFDM, and (ii) nonlinearly mapped DDO-OFDM (NLM-DDO-OFDM) where the optical OFDM spectrum does not display a replica of baseband OFDM. In what follows, we discuss the principles and design choices for these two categories of direct-detection OFDM systems.

5.3.1 Linearly Mapped DDO-OFDM

As shown in Figure 5.7, the optical spectrum of an LM-DDO-OFDM signal at the output of the optical OFDM (O-OFDM) transmitter is a linear copy of the RF OFDM

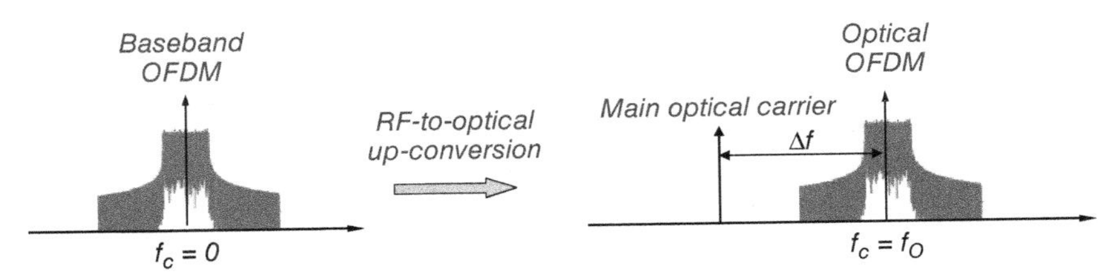

FIGURE 5.7 Illustration of linearly mapped DDO-OFDM (LM-DDO-OFDM) where the optical OFDM spectrum is a replica of the baseband OFDM spectrum.

spectrum plus an optical carrier, which usually occupies 50% of the overall power. The position of the main optical carrier can be one OFDM spectrum bandwidth away [28, 32] or right at the end of the OFDM spectrum [33, 34]. Formally, such type of DDO-OFDM can be described as

$$s(t) = e^{j2\pi f_0 t} + \alpha e^{j2\pi (f_0+\Delta f)t} \cdot s_B(t) \tag{5.30}$$

where $s(t)$ is the optical OFDM signal, f_0 is the main optical carrier frequency, Δf is guard band between the main optical carrier and the OFDM band (Figure 5.7), and α is the scaling coefficient that describes the OFDM band strength related to the main carrier. $s_B(t)$ is the baseband OFDM signal given by

$$s_B = \sum_{k=-\frac{1}{2}N_{\rm sc}+1}^{\frac{1}{2}N_{\rm sc}} c_k e^{j2\pi f_k t} \tag{5.31}$$

where c_k and f_k are, respectively, the OFDM information symbol and the frequency for the kth subcarrier. For explanatory simplicity, only one OFDM symbol is shown in (5.31). After the signal passing through fiber link with chromatic dispersion, the OFDM signal can be approximated as

$$r(t) = e^{j(2\pi f_0 t + \Phi_D(-\Delta f) + \phi(t))} + \alpha e^{j(2\pi (f_0+\Delta f)t + \phi(t))} \cdot \sum_{k=-\frac{1}{2}N_{\rm sc}+1}^{\frac{1}{2}N_{\rm sc}} c_{ik} e^{(j2\pi f_k t + \Phi_D(f_k))} \tag{5.32}$$

$$\Phi_D(f_k) = \pi \cdot c \cdot D_t \cdot f_k^2 / f_0^2 \tag{5.33}$$

where $\Phi_D(f_k)$ is the phase delay due to chromatic dispersion for the kth subcarrier. D_t is the accumulated chromatic dispersion in the unit of ps/pm, f_0 is the center frequency of optical OFDM spectrum, and c is the speed of light. At the receiver, the photodetector can be modeled as a square-law detector and the resultant photocurrent signal is

$$I(t) \propto |r(t)|^2 = 1 + 2\alpha \mathrm{Re}\left\{ e^{j2\pi\Delta ft} \sum_{k=-\frac{1}{2}N_{sc}+1}^{\frac{1}{2}N_{sc}} c_{ik} e^{(j2\pi f_k t + \Phi_D(f_k) - \Phi_D(-\Delta f))} \right\}$$

$$+ |\alpha^2| \sum_{k_1=-\frac{1}{2}N_{sc}+1}^{\frac{1}{2}N_{sc}} \sum_{k_2=-\frac{1}{2}N_{sc}+1}^{\frac{1}{2}N_{sc}} c_{k_2}^{*} c_{k_1} e^{(j2\pi(f_{k1}-f_{k_2})t + \Phi_D(f_{k_1}) - \Phi_D(f_{k_2}))} \tag{5.34}$$

The first term is a DC component that can be easily filtered out. The second term is the fundamental term consisting of linear OFDM subcarriers that are to be retrieved. The third term is the second-order nonlinearity term that needs to be removed.

There are several approaches to minimize the penalty due to the second-order nonlinearity term:

(A) Offset SSB-OFDM. Sufficient guard band is allocated such that the linear term and the second-order nonlinearity of the RF spectra are nonoverlapping. As such, the third term in Equation 5.34 can be easily removed using an RF or DSP filter, as proposed by Lowery and Armstrong [28].

(B) Baseband Optical SSB-OFDM. α coefficient is reduced as much as possible such that the distortion as result of the third-term is reduced to an acceptable level. This approach has been adopted by Djordjevic and Vasic [2] and Hewitt [33].

(C) Subcarrier interleaving. From Equation 5.34, it follows that if only odd subcarriers are filled, that is, c_k is nonzero only for the odd subcarriers, the second-order intermodulation will be at even subcarriers, which are orthogonal to the original signal at the odd subcarrier frequencies. Subsequently, the third-term does not produce any interference. This approach has been proposed by Peng et al. [35].

(D) Iterative distortion reduction. The basic idea is to go through a number of iterations of estimation of the linear term, and compute the second-order term using the estimated linear term, and removing the second-order term from the right side of Equation 5.34. This approach has been proposed by Peng et al. [34].

There are advantages and disadvantages among all these four approaches. For instance, Approach B has the advantage of better spectral efficiency, while sacrificing receiver sensitivity. Approach D has both good spectral efficiency and receiver sensitivity, but has a burden of computational complexity.

Figure 5.8 shows one offset SSB-OFDM proposed by Lowery et al. [36]. They show that such DDO-OFDM can mitigate enormous amount of chromatic dispersion up to 5000 km standard SMF (SSMF) fiber. The proof-of-concept experiment was demonstrated by Schmidt et al. from the same group for 400 km DDO-OFDM

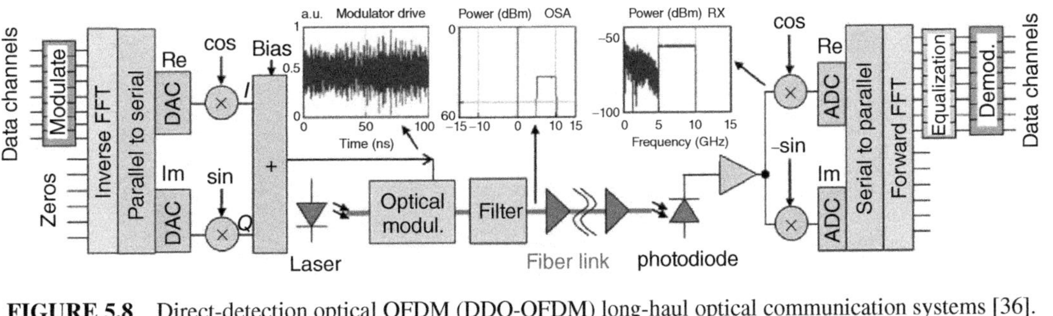

FIGURE 5.8 Direct-detection optical OFDM (DDO-OFDM) long-haul optical communication systems [36].

transmission at 20 Gb/s [32].The simulated system is 10 Gb/s with 4QAM modulation with the baud rate around 5 GHz [36]. In the electrical OFDM transmitter, the OFDM signal is up-converted to an RF carrier at 7.5 GHz generating an OFDM band spanning from 5 to 10 GHz. The RF OFDM signal is fed into an optical modulator. The output optical spectrum has the two OFDM sidebands that are symmetric across the main optical subcarrier. An optical filter is then used to filter out one OFDM sideband. This SSB is crucial to ensure that there is one-to-one mapping between the RF OFDM signal and the optical OFDM signal. The power of the main optical carrier is optimized to maximize the sensitivity. At the receiver, only one photodetector is used. The RF spectrum of the photocurrent is depicted as an inset in Figure 5.8. It can be seen that the second-order intermodulation, the third-term in Equation 5.34 is from DC to 5 GHz, whereas the OFDM spectrum, the second term in Equation 5.34, spans from 5 to 10 GHz. As such, the RF spectrum of the intermodulation does not overlap with the OFDM signal, signifying that the intermodulation does not cause detrimental effects after proper electrical filtering.

5.3.2 Nonlinearly Mapped DDO-OFDM (NLM-DDO-OFDM)

The second class of DDO-OFDM is nonlinearly mapped OFDM. Instead of linearly mapping the electric field (baseband OFDM) to the optical field, NLM-DD-OFDM aims to obtain a linear mapping between baseband OFDM and optical intensity. For simplicity, we assume the generation of NLM-DDO-OFDM using the direct modulation, the waveform after the direct modulation can be expressed as [37]

$$E(t) = e^{j2\pi f_o t} A(t)^{1+jC} \tag{5.35}$$

$$A(t) \equiv \sqrt{P(t)} = A_0 \sqrt{1 + \alpha \mathrm{Re}(e^{j(2\pi f_{IF} t)} \cdot s_B(t))} \tag{5.36}$$

$$s_B(t) = \sum_{k=-\frac{1}{2}N_{sc}+1}^{\frac{1}{2}N_{sc}} c_k e^{j2\pi f_k t} \tag{5.37}$$

$$m \equiv \alpha \sqrt{\sum_{k=-\frac{1}{2}N_{sc}+1}^{\frac{1}{2}N_{sc}} |c_k|^2} \tag{5.38}$$

where $E(t)$ is the optical OFDM signal, $A(t)$ and $P(t)$ are the instantaneous amplitude and power of the optical OFDM signal, c_k is the transmitted information symbol for the kth subcarrier, C is the chirp constant for the direct-modulated DFB laser [37], f_{IF} is the IF for the electrical OFDM signal for modulation, m is the optical modulation index, α is a scaling constant to set an appropriate modulation index m to minimize the clipping noise, and $s_B(t)$ is the baseband OFDM signal. Assuming the chromatic dispersion is negligible, the detected current is

$$I(t) = |E(t)|^2 = |A|^2 = A_0(1 + \alpha \mathrm{Re}(e^{j(2\pi f_{\mathrm{IF}} t)} \cdot s_B(t))) \tag{5.39}$$

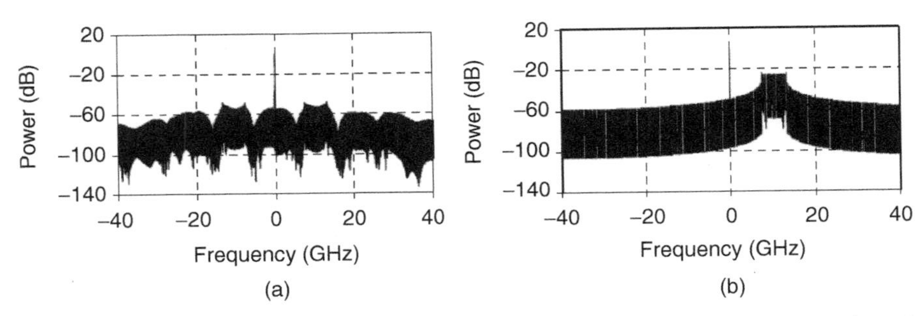

FIGURE 5.9 Comparison of optical spectra between (a) NLM-DDO-OFDM through direct-modulation of DFB laser, and (b) externally modulated offset SSB DDO-OFDM. The chirp constant C of 1 and the modulation index m of 0.3 are assumed for direct-modulation in (a). Both OFDM spectrum bandwidths are 5 GHz comprising 256 subcarriers.

Equation 5.39 shows that the photocurrent contains a perfect replica of the OFDM signal $s_B(t)$ with a DC current. We also assume that the modulation index m is small enough so that the clipping effect is not significant. Equation 5.39 shows that by using NLM-DDO-OFDM without chromatic dispersion, the OFDM signal can be perfectly recovered. The fundamental difference between the NLM- and LM-DDO-OFDM can be gained by studying their respective optical spectra. Figure 5.9 shows the optical spectra of NLM-DDO-OFDM using (i) direct modulation with the chirp coefficient C of 1 in (5.35) and modulation index m of 0.3 in (5.38) and (ii) offset SSB-OFDM. It can be seen that, in sharp contrast to SSB-OFDM, NLM-DDO-OFDM has a multiple of OFDM bands with significant spectral distortion, indicating the nonlinear mapping from the baseband OFDM to the optical OFDM. The consequence of this nonlinear mapping is fundamental. Any type of dispersion (such as chromatic dispersion, polarization dispersion, and modal dispersion) occurs in the link result in the fact that the linear baseband OFDM signal can no longer be recovered; namely, any dispersion will cause the nonlinearity for NLM-DD-OFDM systems. In particular, unlike SSB-OFDM, the channel model for direct-modulated OFDM is no longer linear under any form of optical dispersion. Subsequently, NLM-DD-OFDM only fits short-haul application such as multimode fiber for local area networks (LAN), or short-reach single-mode fiber (SMF) transmission. This class of optical OFDM has attracted attention recently due to its low cost. Some notable works of NLM-DD-OFDM are experimental demonstrations and analysis of optical OFDM over multimode fibers [29, 31, 38] and compatible SSB-OFDM (CompSSB) proposed by Schuster et al. to achieve higher spectral efficiency than offset SSB-OFDM [39].

5.4 SELF-COHERENT OPTICAL OFDM

While long-haul networks have witnessed a capacity evolution to multi-Terabit during the last decade with the revival of coherent communications [5–10], short-reach networks [40–45] within distance of hundreds of kilometers need to increase their capacity per wavelength beyond 40 or even 100 Gb/s to meet the ever-increasing traffic demand. Different from long-haul communications, these short-reach networks require massive number of transceivers across diverse geographic areas. Direct detection (DD) [38–45] can significantly lower the expense compared with the coherent counterpart, making it suitable for short-reach applications. However, the conventional single-ended PD-based direct-detection systems cannot undertake the task due to two fundamental bottlenecks: (i) chromatic dispersion (CD)-induced signal fading due to the lack of receiver phase diversity [38, 41], which limits the transmission distance and (ii) second-order nonlinearity due to photodetection [42, 43], which limits the system capacity. A multitude of solutions have been proposed to overcome the problems, among which a host of self-coherent (SCOH) systems have attracted most attention. The SCOH systems send both the modulated signal (S) and the carrier (C) at the transmitter; at the receiver, the carrier serves as a reference and beats with the signal during photodetection, namely the SCOH detection. SCOH provides the following advantages compared with conventional DD: (i) The receiver sensitivity increases dramatically with the help of carrier; and it is even not surprising that several SCOH receivers achieve the phase diversity [43–45]. (ii) The signal and the carrier are generated with the same laser source, which guarantees that the phase between S and C are coherent with each other at the transmitter; therefore, SCOH naturally mitigates the systems vulnerability (especially the coherent optical OFDM) to the laser frequency offset and phase noise [43–45]. (iii) The SCOH system is capable of being compatible with the nowadays powerful DSP technology, which provides the flexibility when deployed to the short-reach applications. In this chapter, we review the current status of the SCOH systems. Depending on the receiver structure, we divide them into two categories: (i) DD based on single-ended photodetector (PD) and (ii) DD based on balanced receiver.

5.4.1 Single-Ended Photodetector-Based SCOH

Traditional direct-detection (DD) scheme uses intensity modulation at the transmitter [1, 41]. The modulator is driven by a real-valued RF signal, leading to a Hermitian symmetric optical spectrum. This method wastes half of the optical spectral efficiency; more importantly, it gives rise to the problem of CD-induced power fading, which limits the transmission distance. The most straightforward approach to avoid the problem is to use SSB field modulation, which is adopted by a set of OFDM-based SCOH systems as illustrated in Figure 5.10 [42]. The offset-SSB [1] scheme shown in Figure 5.10(a) first up-converts the RF signal with an electrical IQ mixer; then uses the output to drive an MZM, which is biased off the null point to provide an optical carrier. The frequency gap between the signal (S) and the carrier (C) is reserved for the second-order SSBN, whose bandwidth is equal to that of the signal. The electrical spectral efficiency (E-SE) is sacrificed by half in this case. To increase the E-SE, the virtual-SSB OFDM [42] is proposed shown by Figure 5.10(b). Virtual-SSB arranges one inserted RF tone at the leftmost OFDM subcarrier; then transfers the signal along with the RF tone to a complex signal by the IFFT; the real and imaginary parts of the signal drive an optical I/Q modulator biased at the null point. The original optical

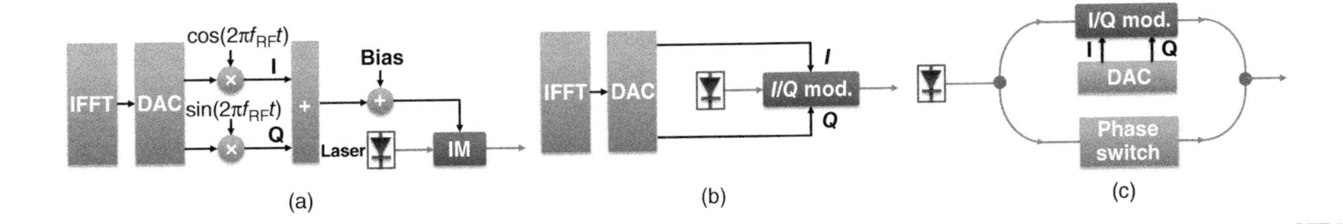

FIGURE 5.10 The transmitter structures for the single-ended PD-based direct detection. (a) Offset SSB OFDM; (b) Virtual SSB OFDM; (c) BPS-DD. IFFT: inverse fast Fourier transform; DAC: digital-to-analog converter; IM: intensity modulator; *I*/*Q* mod.: *I*/*Q* modulator; *I*: in-phase; *Q*: quadrature.

carrier is thus suppressed, and the new optical carrier is provided by the inserted RF tone located at the left edge of the signal spectrum. The RF transmitter structure is simplified compared with offset-SSB; moreover, virtual-SSB does not reserve the frequency gap between the *S* and *C*, which effectively doubles the E-SE. To eliminate the SSBN, an iterative SSBN estimation and cancellation technique is introduced at the receiver using DSP, which sacrifices the system computational complexity mainly induced by the iterative FFT operation inside the cancellation algorithm. As the electrical bandwidth is the dictating factor for the transponder cost, it is necessary to apply the double sideband (DSB) modulation to further increase the E-SE. The block-wise phase switching (BPS) DD [43] follows the idea. In Figure 5.10(c), for two identical consecutive signal blocks, the phase of the main carrier is switched by 90° or 180° at the transmitter; while the receiver recovers the DSB signal with phase diversity. It is noted that BPS is suitable for both OFDM and single-carrier systems.

5.4.2 Balanced Receiver-Based SCOH

After field modulation is introduced into SCOH schemes, it becomes meaningful for DD receivers to achieve the phase diversity. Balanced receiver-based DD provides the following advantages at the sacrifice of system expense: (i) phase-diverse signals can be recovered; (ii) the SSBN can be eliminated spontaneously by the balanced photodetector instead of the iterative SSBN cancellation, which significantly increases the computational complexity. A classical SCOH enabling coherent-like detection at the receiver without LO is as follows: (i) transmitter sends the carrier (*C*) along with the signal (*S*); (ii) at reception, *C* and *S* are first separated into two paths, which are then served as the two inputs of the standard coherent receiver [40]. The polarization multiplexing (POL-MUX) DD [40] first realizes this by separating the *S* and *C* in frequency domain by a narrow bandwidth LPF, as shown in Figure 5.11(a). A frequency gap between the *S* and *C* needs to be reserved for (i) laser wavelength drift and (ii) filter bandwidth. In fact, the laser may have a frequency drift of 10 GHz; moreover, it is quite expensive for current commercial filter to achieve a low-pass bandwidth within 10 GHz. Therefore, the POL-MUX DD has a huge system cost while still wastes the SE conspicuously. The signal carrier interleaved (SCI) DD [44] avoids this problem by separating the *S* and *C* in time domain. Two signal blocks are followed by one carrier block at the transmitter. At reception, the stream is divided into two paths as shown in Figure 5.11(b), while the lower path is delayed by one block length. Therefore, the *S* and *C* from different paths can be mixed with each other at the coherent receiver. SCI-DD sacrifices the SE by one-third.

5.4.3 Stokes Vector Direct Detection

Recently, the Stokes vector (SV) DD [45] is proposed, which at the first time achieves 100% SE with reference to the single polarization-modulated coherent detection. Compared with POL-MUX coherent systems, SV-DD has the following advantages: the transmitter only requires one-polarization modulation; the receiver does not need an LO; the DSP is much simpler due to using less number of FFTs, and the laser

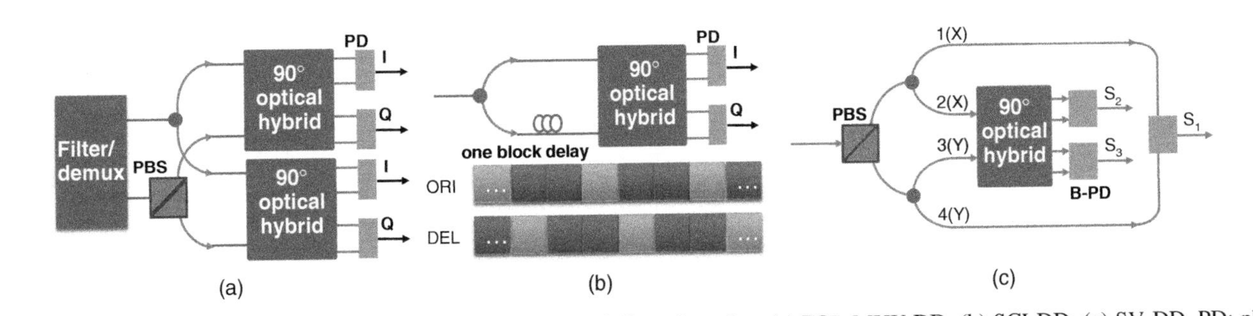

FIGURE 5.11 The receiver structure for the balanced receiver-based direct detection. (a) POL-MUX DD; (b) SCI-DD; (c) SV-DD. PD: photodetector; B-PD: balanced PD; ORI: original data stream; DEL: delayed data stream; PBS: polarization beam splitter.

frequency offset and phase noise need not be tracked. SV-DD achieves 100% spectral efficiency with reference to the single-polarization coherent detection. SV-DD receiver is polarization independent. Although several prior DD schemes [40] have demonstrated dual-polarization modulation and reception, the transmitter complexity approaches that of the POL-MUX coherent systems. Using the SV-DD scheme, Ref. [45] has experimentally demonstrated the first successful direct detection of 160 Gb/s single-wavelength single-polarization-modulated signal after transmission over 160-km SSMF.

In the SV-DD scheme, the signal (S) and carrier (C) are, respectively, placed onto two orthogonal polarizations at the transmitter. The signal can be expressed by a Jones vector $\boldsymbol{J} = [S, C]^T$, where a vector or matrix is represented in bond font and superscript "T" stands for transpose. Converting the Jones vector to the Stokes space, we arrive at the Stokes vector $\boldsymbol{S} = [s_1, s_2, s_3]^T = [|s|^2 - |c|^2,\ \mathrm{Re}(s \cdot c^*),\ \mathrm{Im}(s \cdot c^*)]^T$, where Re( ) and Im( ) represent real and imaginary parts of a complex number. This Stokes vector $\boldsymbol{S}$ can be detected as follows: the signal is split with a polarization beam splitter (PBS) into two outputs of X and Y, respectively. Since the signal polarization has been randomly rotated in the fiber, the received signals X and Y are the mixture of the transmitted signals of S and C. One of the critical tasks of SV-DD is to acquire this polarization rotation (PR) between the input and output signals. Compared with the coherent detection for which the PR is accomplished in the Jones space, SV-DD is done in the Stokes space by the subsystem shown in Figure 5.11(c). Both X and Y are split with 3 dB couplers, which are identified as ports 1 and 4, respectively. Ports 1 and 4 are fed into a balanced PD directly, resulting in the output of $|X|^2 - |Y|^2$, which is the first component of SV. Ports 2 and 3 are fed into a standard balanced receiver consisting of an optical hybrid and two balanced PDs. The output of the balanced receiver is $\mathrm{Re}(X \cdot Y^*)$ and $\mathrm{Im}(X \cdot Y^*)$, respectively, which are the second and third components of SV. For simplicity, the above-mentioned analysis omits some simple scaling constants for coupler and photodiode outputs.

To recover the signal, the remaining task is to acquire the 3×3 SV rotation matrix (RM) of the channel and rotate the Stokes vectors at the receiver back to those at the transmitter. The effectiveness of the SV-DD algorithm is fully revealed: (i) combining the second and third components of SV, we have the final output of $S \cdot C^*$ that has full phase diversity of signal S, from which the input signal S is fully recovered without being affected by the CD induced fading and (ii) the nonlinearity term is completely lumped into the first SV component without affecting the recovered signals, which are derived from second and third SV components. SV-DD signal is complex modulated and no frequency gap is required. Therefore, SV-DD has an electrical SE of four times of offset-OFDM [1, 28] and two times of virtual single-sideband (VSSB) [42] or block-wise phase switching (BPS) [43]. This makes SV-DD an ideal DD format for achieving high data rate with reasonable electrical bandwidth.

5.5 DISCRETE FOURIER TRANSFORM SPREAD OFDM SYSTEM (DFT-S OFDM)

Optical communication has rapidly advanced toward 1-Tb/s and beyond transport. As the available bandwidth of SSMF is limited, high spectral efficiency (SE) becomes an important issue. CO-OFDM has become one of the promising candidates due to its high SE and resilience to linear channel impairments such as CD. Experimental demonstration at data rate of 1-Tb/s [6–10] and beyond [46–53] has been achieved using either Nyquist wavelength-division multiplexing (WDM) or CO-OFDM. Despite many promising features, CO-OFDM system suffers from high PAPR, which leads to inferior tolerance to fiber nonlinearity compared with SC system, and has become an obstacle to its practical implementation in long-haul transmission systems. Although specialty fibers such as ultra-large area fiber (ULAF) [7, 49, 50, 54] or low-loss low-nonlinearity pure silica core fiber (PSCF) [55] with Raman amplification has been suggested to further extend the reach of transmission systems, it could be either more expensive or not compatible with the deployed links. To solve the nonlinear tolerance problem, DFT-S OFDM has recently been proposed with an attractive feature of much reduced PAPR [56–58]. DFT-S OFDM is called single-carrier frequency-division multiplexing (SC-FDM) that has been incorporated into the 3GPP-LTE standard in uplink for the next generation mobile system with many interesting features [58]. Furthermore, benefited from the sub-band or subwavelength accessibility of CO-OFDM, properly designed multiband DFT-S OFDM (MB-DFT-S OFDM) potentially have better nonlinearity tolerance over either conventional CO-OFDM or SC system for ultra-high-speed transmission [59, 60]. The nonlinearity advantage of MB-DFT-S OFDM has been verified through simulation in [59, 60]. In addition, optical transmission experiments utilizing the DFT-S OFDM or SC-FDM have been demonstrated very recently by several groups [55, 61–65], which shows a potential advantage of better nonlinear tolerance and high SE. The reconfigurable optical add/drop multiplexer (ROADM) functionally has also been demonstrated on SC-FDM superchannel [64].

5.5.1 Principle of DFT-S OFDM

The DSP at the transmitter and receiver of a DFT-S OFDM system is shown in Figure 5.12. For comparison, the signal processing of conventional OFDM is also illustrated.

Similar to the conventional OFDM, signal processing in DFT-S OFDM is repetitive in a few different time intervals called blocks. At the input to the transmitter, a baseband modulator transforms the binary serial input data to a multilevel modulation formats such as M-ary phase-shift keying (M-PSK) or M-ary quadrature amplitude modulation (M-QAM). The most commonly used modulation formats in OFDM system include binary phase-shift keying (BPSK), QPSK, 16QAM, and 64QAM. The modulation format can be made adaptive by the system to match the current channel conditions, and thereby the transmission data rate. The transmitter next groups the

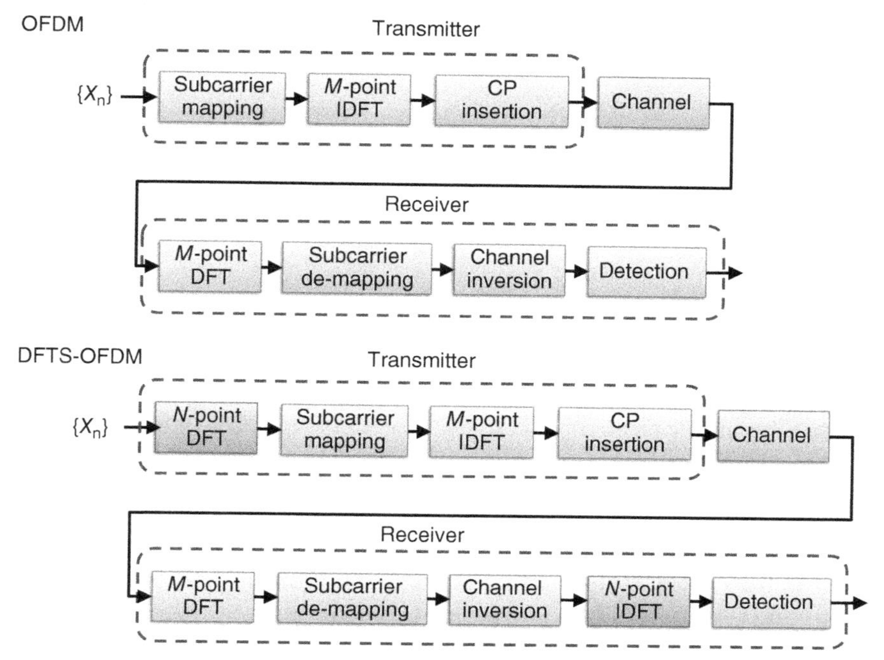

FIGURE 5.12 Signal processing of conventional OFDM and DFT-S OFDM.

modulation symbols X_n into many OFDM blocks (serial to parallel), each containing N symbols. After that, the first unique step in DFT-S OFDM is an N-point DFT before the subcarrier mapping operation to produce a frequency-domain representation X_k of the input symbols. Then, each of the N-point DFT outputs is mapped to one of the M (>N) subcarriers in a conventional OFDM that can be transmitted. As in conventional OFDM, the typical value of M must be a power of 2 (e.g., 64, 128, or 256). The choice of N in DFT-S OFDM must follow the relationship $N = M/Q$, which means N must be an integer submultiple of M. Q is defined as the bandwidth expansion factor of the symbol sequence. DFT-S OFDM can handle Q simultaneous transmissions without co-channel interference (CCI) when each terminal is allocated N symbol per block. The result of the subcarrier mapping is the set $\widetilde{X}_l$ ($l = 0, 1, 2, \ldots, M-1$) of complex subcarrier amplitudes, where the N of amplitudes are nonzero. As in conventional OFDM, an M-point IDFT transforms the subcarrier amplitudes to a complex time-domain signal $\widetilde{X}_m$. Each $\widetilde{X}_m$ then modulates a single frequency carrier and all the modulated symbols are transmitted sequentially. The transmitter then inserts CP in order to provide a guard time to prevent ISI. The modulated DFT-S OFDM signal is then launched into a wireless or fiber-optic channel for transmission. After transmission, the receiver first transforms the time-domain received signal into the frequency domain via DFT, de-maps the subcarriers, and then performs frequency-domain equalization to remove the channel distortion. Minimum mean square error (MMSE) frequency-domain equalization method is generally preferred over zero forcing (ZF) due to the robustness against noise. Subsequently, the equalized symbols are transformed back from frequency- to time-domain via IDFT, and finally the detection and decoding is performed.

In DFT-S OFDM, since the DFT size $M > N$, several approaches have been proposed to the mapping of transmission symbols X_k to DFT-S OFDM subcarriers. These approaches can be divided into two categories: distributed and localized. Distributed subcarrier mapping means the DFT outputs of the input data are allocated over the entire bandwidth with the unused subcarriers filled with zeros, resulting in a noncontinuous comb-shaped spectrum. The well-known interleaved DFT-S OFDM (IDFT-S OFDM), or so-called interleaved SC-FDMA (IFDMA) [66] is at special case of distributed DFT-S OFDM [56, 57]. On the contrary, localized subcarrier mapping means consecutive subcarriers are occupied by the DFT outputs of the input data, resulting in a continuous spectrum that occupies a fraction of the total available bandwidth. For IDFT-S OFDM, time symbols are simply a repetition of the original input symbols with a systematic phase rotation applied to each symbol in the time domain [66]. Therefore, the PAPR of IDFT-S OFDM signal is the same as in the case of a conventional single carrier signal. In the case of localized DFT-S OFDM (LDFT-S OFDM), or so-called localized SC-FDMA (LFDMA), the time signal has exact copies of input time symbols in N sample positions. The other $M - N$ time samples are weighted sums of all the symbols in the input block [67]. As we can see from Figure 5.12, the first obvious difference between conventional OFDM and DFT-S OFDM is the additional pair of N-point DFT/IDFT in the DFT-S OFDM, with DFT in the transmitter and IDFT in the receiver. The second fundamental difference between DFT-S OFDM and conventional OFDM is in the receiver equalization and detection processes [56]. In conventional OFDM, since the data symbol is carried by individual subcarriers, channel equalization, channel inversion, and data detection is performed individually on each subcarrier [12]. Channel coding or power/rate adaptation is required for OFDM to protect individual subcarriers if there are nulls in the channel spectrum, which would severely degrade the system performance since there is essentially no way to recover the data affected by the null. In the case of DFT-S OFDM, channel equalization and inversion is done similarly in the frequency domain but data detection is performed after the frequency-domain-equalized data are reverted back to time domain by IDFT [56]. Hence, it is more robust to spectral nulls compared with conventional OFDM since the noise is averaged out over the entire bandwidth. Additional advantages of DFT-S OFDM include less sensitivity to the carrier frequency offset (CFO), less sensitivity to the laser phase noise, and less nonlinear distortion due to the much reduced PAPR [67], while conventional OFDM suffers due to the multicarrier nature of OFDM [68].

5.5.2 Unique-Word-Assisted DFT-S OFDM (UW-DFT-S OFDM)

Unique-word (UW) was first proposed for single-carrier frequency-domain equalization (SC-FDE) systems and has been extensively studied in wireless communications [66, 68–71]. The data pattern structure of UW-DFT-S OFDM for two polarizations is illustrated in Figure 5.13. The unique-words (UWs), normally comprising a Zadoff–Chu (ZC) sequence [72] and an optional GI is inserted periodically at

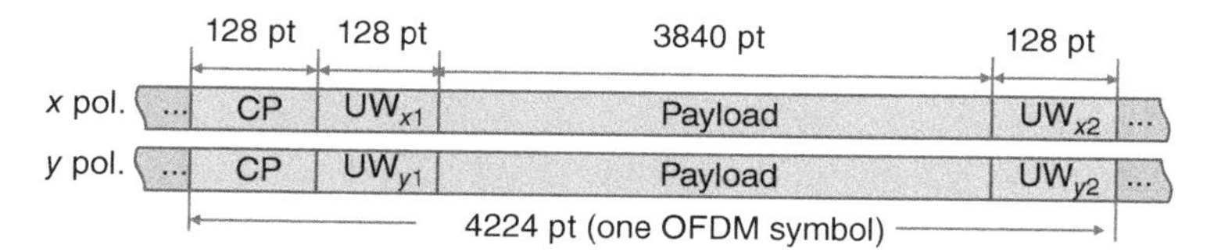

FIGURE 5.13 Structure of UW-DFT-S OFDM data symbol. UW: unique-word; CP: cyclic prefix.

both ends of the payload. The Zadoff–Chu sequence is generated with the following equation:

$$x_u(n) = \begin{cases} e^{-j\frac{\pi un(n+1)}{N_{ZC}}} & \text{if } N_{ZC} \text{ is odd} \\ e^{-j\frac{\pi un^2}{N_{ZC}}} & \text{if } N_{ZC} \text{ is even} \end{cases} \quad (0 \le n \le N_{ZC} - 1) \tag{5.40}$$

where $x_u(n)$ is the Zadoff–Chu sequence, N_{zc} is sequence length, and u is an integer relatively prime of N_{ZC}. A similar OFDM symbol structure can be drawn for the second polarization by using different UWs. The reason to use two UWs within one OFDM symbol is the compatibility with polarization diversity. The first and second UWs are orthogonal to each other when combining the two OFDM symbols for two polarizations in a Jones vector form. The two UWs for the two polarizations (see Figure 5.13), $\begin{pmatrix} \mathrm{UW}_{x1} \\ \mathrm{UW}_{y1} \end{pmatrix}$ and $\begin{pmatrix} \mathrm{UW}_{x2} \\ \mathrm{UW}_{y2} \end{pmatrix}$ are given by

$$\begin{pmatrix} \mathrm{UW}_{x1} & \mathrm{UW}_{x2} \\ \mathrm{UW}_{y1} & \mathrm{UW}_{y2} \end{pmatrix} = \begin{pmatrix} \mathrm{UW} & -\mathrm{cshift}(\mathrm{UW})^* \\ \mathrm{cshift}(\mathrm{UW}) & \mathrm{UW}^* \end{pmatrix} \tag{5.41}$$

where cshift(.) denotes a circular shift of the sequence by half of the sequence length and "*" denotes complex conjugate. The circular shift ensures that UWs for two polarizations, for example, UW_{x1} and UW_{y1} are uncorrelated so long as the channel length is shorter than half of the unique-word length. The short UWs in each OFDM symbol can be used for multiple purposes: timing synchronization, channel estimation, phase estimation, and so on.

5.6 OFDM-BASED SUPERCHANNEL TRANSMISSIONS

In the past few decades, optical communication systems have evolved rapidly thanks to the advance of electronics such as high-speed DSP and fiber-optic technology. For high-speed optical transmission systems, multiplexing technique for the optical signals has been extensively studied and implemented. One of the most significant schemes is the WDM, in which the signals are multiplexed in wavelength (or frequency) domain as wavelength channels. The WDM transmission technique can be classified by the frequency spacing between the adjacent channels as the coarse WDM (CWDM), dense WDM (DWDM), or even high-density WDM (e.g., superchannel transmission). These patterns are characterized by the ratio between the channel spacing Δf and the modulation symbol rate of the channel B. Coherent detection and DSP enable signals with extremely narrow guard band (or no guard band) can be recovered with minimum sacrifice of receiver sensitivity. The associated technologies in single-carrier cases are termed as "quasi-Nyquist" WDM representing the scenario that $1 \leq \Delta f/B \leq 1.2$, "Nyquist" WDM for $\Delta f/B = 1$, and "super-Nyquist" WDM for $\Delta f/B < 1$. Given that signals are densely packed, spectral overlapping may possibly affect the system performance. Thus, usually prefiltering, either done optically or electronically, is required to mitigate crosstalk. For instance, optical prefiltering has been utilized in "quasi-Nyquist" WDM [14] and "super-Nyquist" WDM [15], and electronic prefiltering has been used in [16, 17].

One of the very elegant ways to fundamentally eliminate the crosstalk among different wavelength channels and achieve the "Nyquist" condition is to perform OFDM modulation. The OFDM modulation guarantees crosstalk-free reception of symbol-rate-spaced channels without any prefiltering [18–23]. The "OFDM conditions" essentially mean the following [20]: First, the carrier spacing should be equal to the symbol rate, which requires the carriers on which the modulation is imprinted to be frequency locked. Second, to prevent ISI, the time window for modulation and demultiplexing should be aligned (see Figure 2 of [20]). Failing to do so results in large crosstalk and destroys the orthogonality condition. It is worth noting that after long distance transmission while the fiber dispersion becomes significant, the symbols of neighboring carriers can be severely displaced. In this scenario, dispersion compensation can be adopted to rewind the orthogonality condition. Third, sufficient bandwidth is needed at the transmitter and the receiver to modulate each subcarrier, due to the fact that the spectral representative of the modulated symbols is usually a sinc function. In other words, sufficient oversampling is required to capture most of the sinc function for each of the modulated subcarriers.

5.6.1 No-Guard-Interval CO-OFDM (NGI-CO-OFDM) Superchannel

In the conventional CO-OFDM, the freedom to select the OFDM symbol length (namely the size of FFT, N_{FFT}) is constrained by the fiber dispersion, particularly the CD and PMD. The CP inserted between OFDM symbols offers an effective approach to remove the ISI induced by the CD and PMD. We use the percentage of GI (GI%) to characterize the SE penalty caused by the GI, defined as the ratio between the GI length N_{GI} and N_{FFT}. A given GI% limits the minimum value of N_{FFT} for a fixed dispersion value. Although the GI% can be simply reduced by increasing the OFDM symbol length, this will inevitably increase the CO-OFDM vulnerability to the fiber nonlinearity, laser phase noise, and CFO. In short, there is a trade-off between choosing a small N_{FFT} to achieve better receiver sensitivity and a small GI% to achieve larger SE. For the conventional CO-OFDM demonstrated in previous works, the GI% was typically kept in the range of 10–25% [73]. The length of GI becomes larger for systems with higher baud rate and longer transmission distance, making GI one of the dominant factors to limit the system SE. To increase the SE as large as possible, the no-guard-interval (NGI) CO-OFDM [73–75] is proposed in 2008. Instead of using the GI to remove the ISI induced by the dispersion, NGI-CO-OFDM compensates the CD and PMD at the receiver by enabling the DSP, namely the electronic dispersion compensation (EDC). The transmitter configuration maintains the same as the conventional CO-OFDM, in which the PDM is applied to double the spectral efficiency. The receiver front-end consists of a polarization-diversity optical hybrid, four balanced photodetectors (B-PDs), and four ADCs. The outputs of the ADCs are the in-phase and quadrature part of the dual polarizations. The receiver back-end contains several unique DSP procedures for NGI-CO-OFDM, as shown in Figure 5.14.

The back-end first applies the EDC to the signals of the two polarizations, respectively, by using the overlap frequency domain equalization [76], which includes several FFT and IFFT pairs. Then, carrier separation is conducted by shifting each subcarrier to the baseband. The data sequence of each subcarrier is obtained by the FFT. The polarization demultiplexing as well as the channel equalization is fulfilled by the adaptive equalizer, such as the least mean square (LMS) algorithm and the constant modulus algorithm (CMA), which is widely applied in the single-carrier systems [77]. The remaining DSP is the same as the conventional CO-OFDM. The carrier recovery includes the frequency offset compensation and the phase estimation. The final symbol decision can be made after all the DSP procedure and the BER can be calculated.

The DSP of NGI-CO-OFDM has two major differences compared with the conventional CO-OFDM. First, the CD compensation is added before any further DSP; second, the adaptive equalization is adopted for the polarization demultiplexing and channel estimation, instead of the classic training symbol (TS)-aided equalization in CO-OFDM. Alternatively, a modified NGI-CO-OFDM [73], called zero GI (ZGI) CO-OFDM, uses the training symbols with CP, followed by the data symbols without CP. Therefore, the channel estimation can maintain its accuracy without being affected by the dispersion. The channel estimation is conducted (in the form of a 2×2 matrix for each subcarrier) from the OFDM demodulator once the TSs have been processed, and then the PMD compensation can be achieved by applying the inverse of the channel matrix.

The effectiveness of NGI-CO-OFDM has been verified by several remarkable experiments. Sano et al. [74] demonstrate the 13.4-Tb/s (134×111-Gb/s/ch) NGI-CO-OFDM transmission over 3600 km of SMF in 2008; while Liu et al.

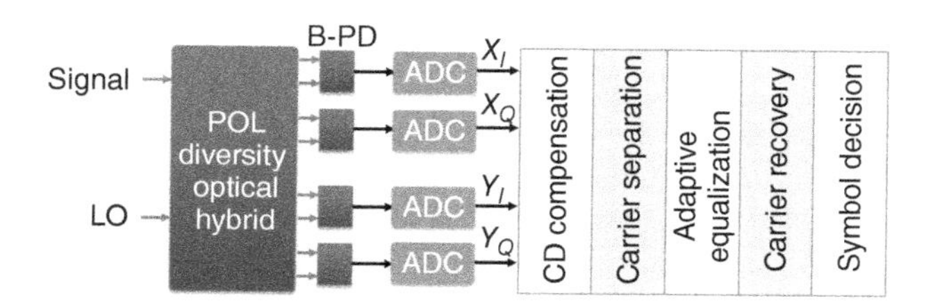

FIGURE 5.14 Block diagram for the NGI-CO-OFDM receiver.

[75] demonstrate the transmission of a 1.2-Tb/s 24-Carrier NGI-CO-OFDM super-channel over 7200-km of ultra-large-area fiber (ULAF) in 2009. NGI-CO-OFDM provides an effective approach to maximize the system SE, without any need to consider the large dispersion induced by the ultra-long transmission distance and wide optical bandwidth. However, NGI-CO-OFDM significantly sacrifices the receiver computational complexity. Therefore, we would rather conclude that there is always a trade-off between the SE and the computational complexity, and NGI-CO-OFDM offers one extreme scheme in terms of SE.

5.6.2 Reduced-Guard-Interval CO-OFDM (RGI-CO-OFDM) Superchannel

Section 5.5.1 described the implementation of NGI-CO-OFDM scheme for the improvement of the SE. NGI-CO-OFDM removes the GI completely and compensates the fiber dispersion at the receiver using blind channel equalization [74]. However, since no GI is added to the OFMD symbol, NGI-CO-OFDM suffers from the ISI caused by the transmitter bandwidth limitation, and complex equalization algorithm is needed at the receiver to compensate the effect of PMD.

For a more cost-effective DSP at the receiver, RGI-CO-OFDM scheme is proposed [78, 79]. In this scheme, the ISI with short memory, such as the transmitter bandwidth limitations or PMD-induced ISI, is accommodated by a short length of GI between the adjacent OFDM symbols, while the ISI with long memory, such as CD-induced ISI, is compensated at the receiver using EDC-based on DFT, IDFT and overlap-add [76]. By using RGI-CO-OFDM, the DSP complexity can get greatly reduced compared with NGI-CO-OFDM, and the GI length and OFDM size can be much reduced compared with the conventional OFDM.

Consider a 112-Gb/s PMD-OFDM system over 1500-km SSMF (D = 17 ps/nm/km) with baud rate of 56-GHz [80]. For conventional OFDM, the DFT size is chosen to be 2048, and the GI length is chosen to be 512 samples to accommodate the CD, leading to the GI% of 25% (512/2048). For RGI-CO-OFDM, four samples of GI (71.4 ps) is used to accommodate the instantaneous DGD, and the DFT size can be shortened to 128. The GI% is dramatically reduced to 3.13% (4/128) and the subcarrier spacing is increased by a factor of 16, relaxing the requirements on frequency locking between transmitter laser and receiver OLO.

A data rate of 448-Gb/s with SE of 7 b/s/Hz has been demonstrated in [78, 79] using RGI-CO-OFDM with 16QAM modulation. The experimental setup in [78, 79] is shown in Figure 5.15. The DFT of size is 128 with 75 filled with data. Four samples of GI are added into each symbol, resulting in a symbol length of 132. Training symbols (TSs) were inserted at the beginning of each OFDM frame. Inset (a) of Figure 5.15 shows the TSs after the PDM. The first two TSs are used for frame synchronization, and the last two are for channel estimation. The sampling rate 10 GS/s is used, resulting in 22.4 Gb/s OFDM signal with a spectral bandwidth of 6.016 GHz. The generated OFDM signal is then shifted by 6.016 GHz and combined with a time-delayed OFDM signal in the original frequency, forming a 44.8-Gb/s signal consisting of two decorrelated bands, as illustrated in inset (b) of Figure 5.15. The 2-band signal is then expanded by a 5-comb generator to form a 10-band 224-Gb/s

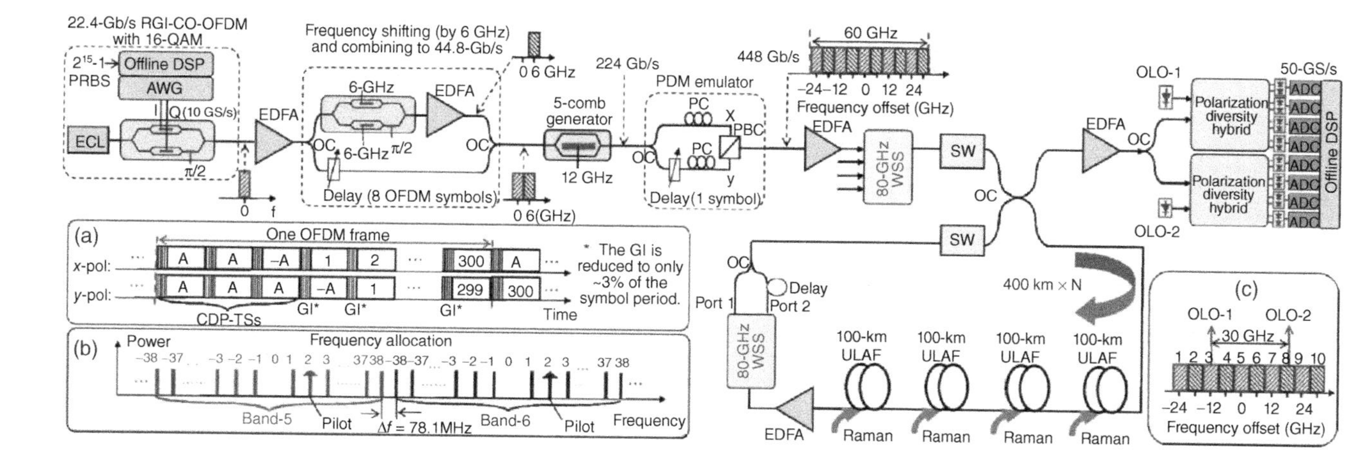

FIGURE 5.15 Schematic of RGI-CO-OFDM transmission experimental setup. Inserts: (a) OFDM frame arrangement; (b) frequency allocation of the OFDM subcarriers; and (c) configuration of the banded digital coherent detection with 2 OLOs. OC: optical coupler; PC: polarization controller; EDFA: erbium-doped fiber amplifier; SW: optical switch.

signal. A PDM emulator is used to enable the PDM, achieving the bit rate of 448-Gb/s signal within a bandwidth of 60.16 GHz. The signal is launched into a transmission loop, consisting of four Raman-amplified 100-km ULAF spans. At the receiver, two optical local oscillators (OLOs) have to be used to recover the entire 448-Gb/s signal due to the ADC bandwidth limitation as shown in the inset (c) of Figure 5.15; 50-GS/s ADCs are used to collect the data. Most of the DSP modules were similar to those described in [81]. New DSP module is added to compensate self-phase modulation (SPM) and CD through a multistep FFT-based algorithm, similar to that for single-carrier transmission [82].

Figure 5.16(a) shows the measured bit error ratio (BER) as a function of the optical signal-to-noise ratio. At BER $= 3.8 \times 10^{-3}$, which is the threshold for 7% forward error correction (FEC), the required OSNR for the 448-Gb/s signal is 25 dB. Figure 5.16(b) shows the Q^2 factor as a function of transmission distance. The reach distance is improved by 25% using the fiber nonlinearity compensation (NLC). With NLC, the mean BER of the 448-Gb/s signal is below 3×10^{-3} after 2000-km transmission.

5.6.3 DFT-S OFDM Superchannel

DFT-S OFDM superchannel is another potential candidate for high SE optical transports. The nonlinear advantage of multiband DFT-S OFDM for fiber transmission has been theoretically predicted and analyzed in [59, 60], enabling longer reach or higher SE (by using higher-order modulation) compared with conventional optical OFDM system at the same FEC threshold. In multiband DFT-S OFDM systems, each sub-band is essentially filled with a digitally generated single-carrier signal [59, 60]. It has been numerically studied that DFT-S OFDM signal processes lower PAPR compared with conventional OFDM signal. For instance, the PAPR value of

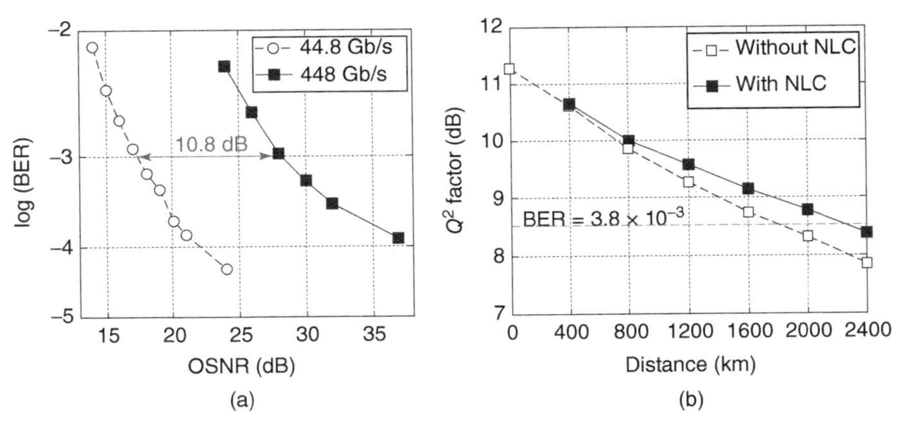

FIGURE 5.16 (a) Measured BER performance of the 448-Gb/s RGI-CO-OFDM signal as compared with the original signal band 44.8-Gb/s signal; (b) measured Q^2 factor as a function of transmission distance.

7.5 dB occupies probability higher than 99.9%, and this PAPR is 3.2 dB lower than the value in conventional OFDM with the same probability [83]. Furthermore, one of the important findings of DFT-S OFDM for optical transmission is that there exists an optimal bandwidth within which the sub-bands should be partitioned. The insertion of UWs and the partitioned subcarrier mapping can slightly alter the performance of the DFT-S OFDM, but nevertheless the advantage of reduced PAPR remains significant. In the following section, we show experimental demonstration of 1-Tb/s PDM-QPSK UW-DFT-S OFDM transmission over 80-km span engineering SSMF and EDFA-only amplification that is compatible with most of the deployed links.

The experimental setup of 1-Tb/s UW-DFT-S OFDM system is shown in Figure 5.17. Our laser sources are 16 external-cavity lasers (ECLs) with low laser linewidth (<100 kHz) combined together and fed into an optical intensity modulator to impress three tones on each wavelength. The tone spacing is set at 6.5625 GHz driven by a synthesizer. The wavelength spacing of all the ECLs is carefully controlled and stabilized at ~20.1875 GHz (1.615 nm). Inset (i) of Figure 5.17 shows the generated densely spaced 48 tones monitored at point (i) using a high-resolution (0.01 nm) optical spectrum analyzer (OSA). After tone generation, the optical carrier is split into two equal branches by a 3-dB polarization-maintaining (PM) coupler and two arbitrary waveform generators (AWGs) are used to drive two IQ modulators to modulate different data patterns on the two polarizations. The baseband spectra for the data pattern are shown in the inset (ii) of Figure 5.17. After IQ modulation, the optical outputs on the two polarizations are multiplexed with a polarization beam combiner. The optical spectrum of generated 16-channel PDM-OFDM signal occupying a bandwidth of 323 GHz is monitored at point (iii), shown as inset (iii) of Figure 5.17. The 16-channel OFDM signal is then launched into a recirculating loop, which consists of two spans of 80-km SSMF with loss compensated by EDFAs. The received OFDM signal after transmission is shown in inset (iv) of Figure 5.17. At the receiver, a 10-GHz optical filter is used to filter out one band each time, and the optical signal is converted to the electrical domain by an optical coherent receiver. The baseband signal is then received by a four-channel Tektronix oscilloscope at 50-GSa/s sampling rate. The DSP at the transmitter and receiver is shown in Figure 5.18.

The transmitted data pattern is generated following the procedure as shown in the "transmitter" block of Figure 5.18. In order to facilitate a stable comparison, time-domain signals of DFT-S OFDM and conventional OFDM are cascaded digitally in MATLAB before loading onto the AWGs. The parameters of DFT-S OFDM and conventional OFDM are for DFT-S, middle 2625/4096 subcarriers are filled with data while the center 65 subcarriers around DC are nullified to avoid performance degradation due to DC leakage, occupying a bandwidth of 6.409 GHz/band. For conventional OFDM the middle 83/128 subcarriers are filled with data while 3 subcarriers around DC are nullified, occupying a bandwidth of 6.484 GHz/band. The difference between UW-DFT-S OFDM and conventional OFDM is that the 2560 data subcarriers in UW-DFT-S OFDM are mapped from DFT precoded UW-assisted data pattern, which will be described later, while in conventional OFDM the 80 data subcarriers are directly mapped with QPSK data pattern. After IFFT to convert data from frequency to time domain, a 128-point cyclic prefix (CP) is appended before

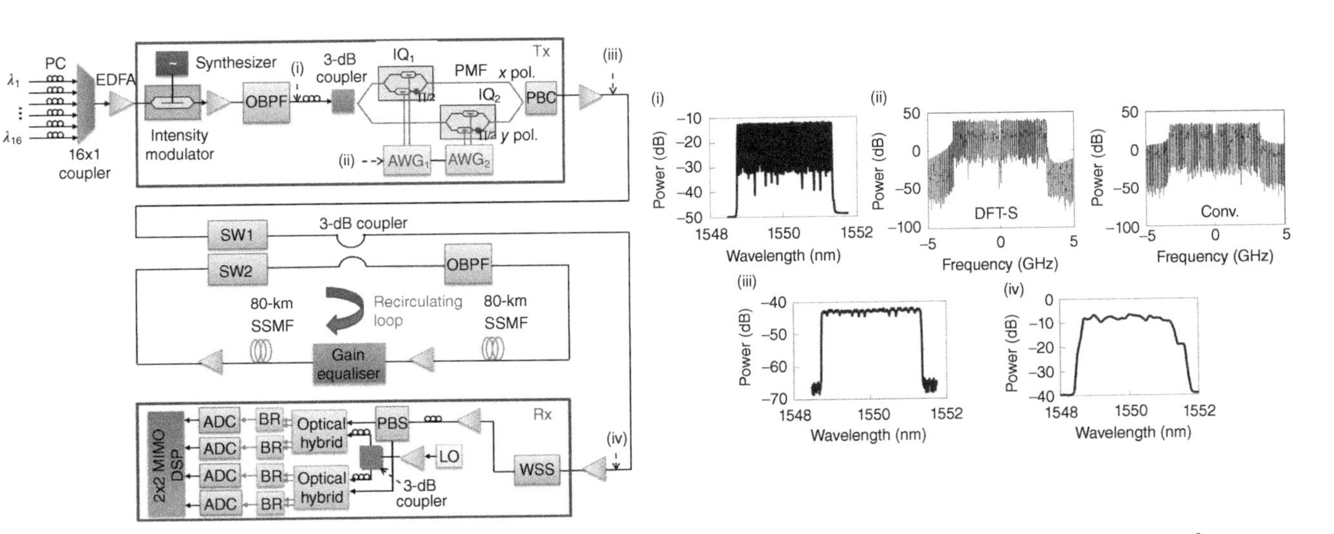

FIGURE 5.17 Experimental setup of 1-Tb/s UW-DFT-S OFDM system. OBPF: optical band-pass filter; AWG: arbitrary waveform generator; PMF: polarization maintaining fiber; PBC/PBS: polarization beam combiner/splitter; SSMF: standard single-mode fiber; SW: (optical) switch; WSS: wavelength selective switch; LO: local oscillator; BR: balanced receiver; ADC: analog-to-digital converter. Insets: measured optical and electrical spectra of (i) 48-tone source; (ii) data pattern loaded onto AWG; (iii) transmitted OFDM signal; (iv) received OFDM signal.

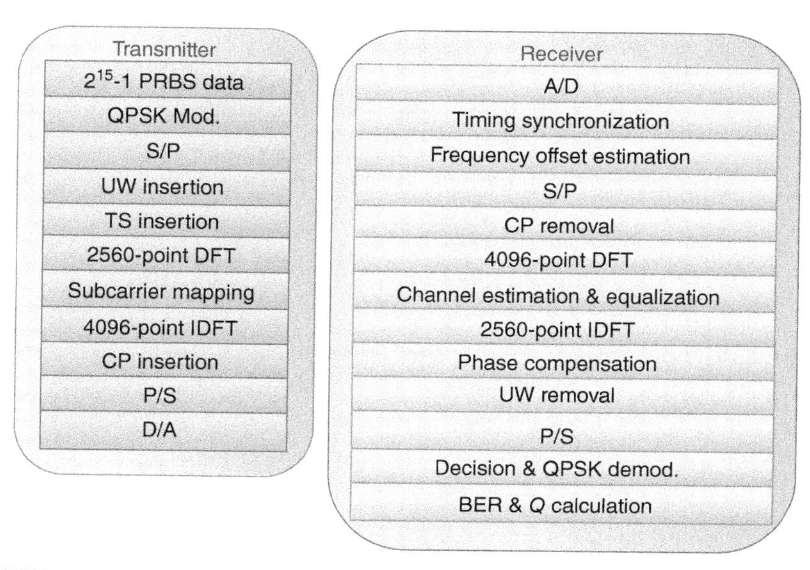

FIGURE 5.18 Digital signal processing at the transmitter and receiver of DFT-S OFDM.

each symbol. The dissimilar number of subcarriers used in conventional OFDM and UW-DFT-S OFDM is because that conventional OFDM can only compensate CPE within one OFDM symbol and thus imposes a constraint to the use of long symbol unless other complicated phase noise compensation methods such as RF-pilot tone are used [17]. However for ultra-long-haul transmission, a large number of subcarriers are preferred or else the CP overhead is too much (in conventional OFDM the overhead of CP is more than 50%, while in DFT-S OFDM the overhead of CP is reduced to only 3%). At receiver, the received four data streams I_x, Q_x, I_y, and Q_y, are first converted from analog to digital by ADCs, and then combined to complex signal and timing synchronized using pilot symbols [84]. The frequency offset is then estimated and compensated, also using the pilot symbols [84]. The time domain signal is then converted from serial to parallel followed by the removal of CP. Subsequently, the data are transformed from time to frequency domain by a 4096-point DFT. Channel estimation and equalization are first performed with the assistance of short- and long-UW pattern training sequence. Then, phase noise compensation is realized with the assistance of short UWs in each OFDM symbol using a novel channel estimation and phase estimation method, which is discussed later. The UWs are then removed followed by the payload data decision, QPSK demodulation and finally BER calculation for the performance evaluation.

The raw data rate of our UW-DFT-S OFDM signal is 1.2-Tb/s (6.25 GHz × 48 band × 2 bit/s × 2 pol), and the net data rate is 1.0-Tb/s after excluding all the overheads. The net spectral efficiency is 3.1 bit/s/Hz. Figure 5.19 shows the BER sensitivity for optical back-to-back of 1-band (with 1 laser), 3-band (with 1 laser), and 48-band (with 16 lasers) PDM-QPSK UW-DFT-S OFDM system corresponding to

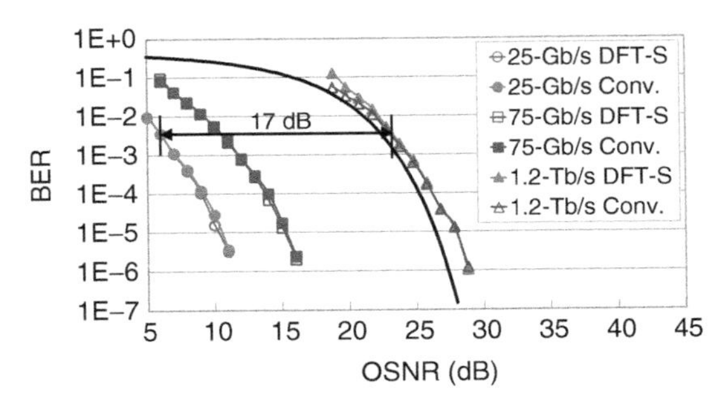

FIGURE 5.19 Measured optical back-to-back BER performance. The data rates shown are raw data rates. DFT-S: DFT-S OFDM; Conv.: conventional OFDM.

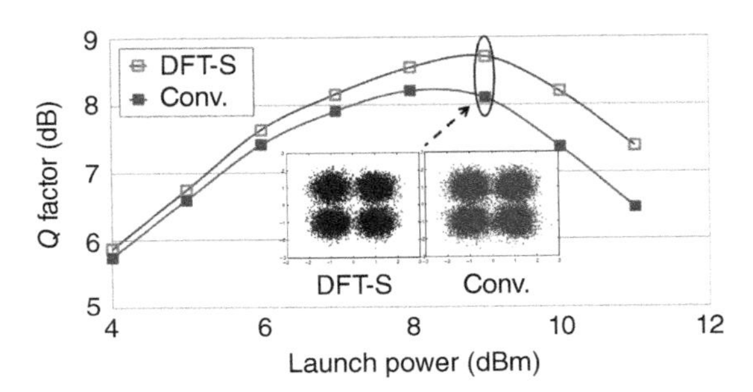

FIGURE 5.20 Measured Q factor versus launch power after 8000-km transmission. Inset: recovered constellations at the launch power of 9 dBm.

a raw data rate of 25-Gb/s, 75-Gb/s, and 1.2 Tb/s. The required OSNR for DFT-S OFDM system is similar to conventional OFDM system measured at 22.9 dB for a BER of 4.6×10^{-3} (7% FEC). This is 17 dB more than that of a single-band system and only 0.9 dB away from the theoretical value. Figure 5.20 shows the Q factor performance against launch power. The optimum launch power is 8 dBm for conventional OFDM and 9 dBm for DFT-S OFDM, which agrees well with our simulated value. A noticeable 0.6 dB improvement in Q factor is observed for DFT-S OFDM compared with conventional OFDM. Figure 5.21 shows the measured BER against the transmission distance for DFT-S OFDM and conventional OFDM at launch powers of 8 and 9 dBm. It can be seen that the maximum possible transmission distance at a BER of 4.6×10^{-3} (7% FEC) [85] is 8300 and 7300 km for DFT-S OFDM and conventional OFDM, respectively, which shows a 20% increase in reach for DFT-S OFDM. If BER of 2×10^{-2} (20% FEC) [86] is used, the reach of DFT-S OFDM can be extended to more than 10,000 km. Finally, the BER performance of all 48 bands is measured at the launch power of 9 dBm and transmission distance of 8000 km (80 km $\times$ 100) as shown in Figure 5.22. The BERs of all bands in DFT-S OFDM are below

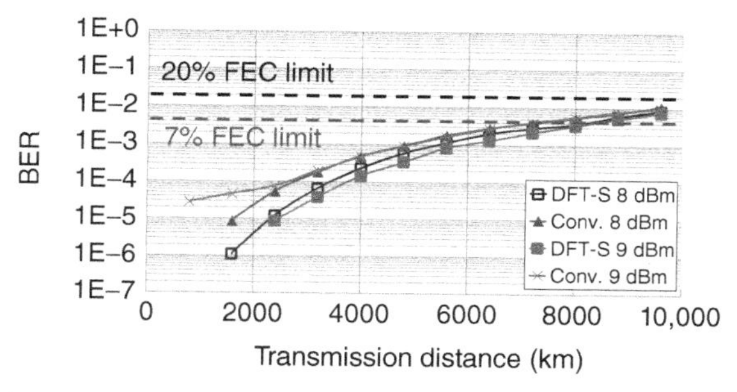

FIGURE 5.21 Measured BER performance of 1.0-Tb/s UW-DFT-S OFDM system at different transmission distances.

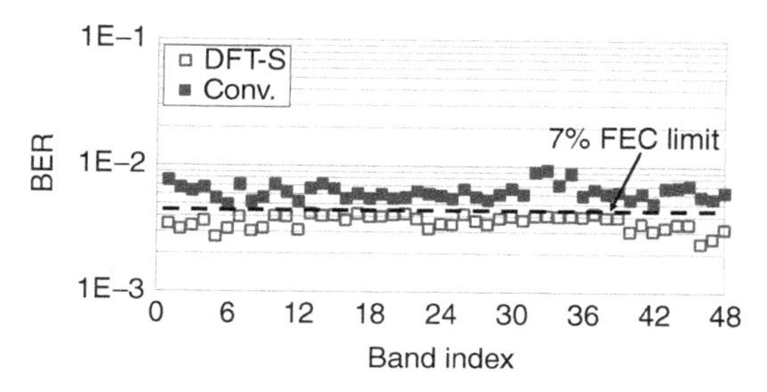

FIGURE 5.22 Measured BER performance for all 48 bands for 8000-km transmission at a launch power of 9 dBm.

the 7% FEC threshold, whereas for conventional OFDM all bands have crossed the BER limit.

5.7 SUMMARY

This chapter describes the basic concept of OFDM theory and its application in optical transmission systems. In Sections 5.2 and 5.3, principles of coherent optical OFDM and direct-detection optical OFDM are provided. The system configuration, BER and Q performance, and tolerance to channel impairments including the fiber attenuation, chromatic dispersion, PMD, and fiber nonlinearities are provided. In Section 5.4, a novel variant of OFDM called DFT-S OFDM is introduced. The basic figures of merit of DFT-S OFDM including the PAPR advantage and unique-word-assisted DSP for multiple purposes (timing synchronization, channel estimation, phase estimation, etc.) are discussed. In Section 5.5, OFDM-based superchannel transmission technologies are reviewed. The concept of superchannel transmission is introduced and how it can increase the system spectral efficiency is explained. The most recent process in OFDM superchannel transmission, including the reduced-guard-interval OFDM superchannel, no-guard-interval OFDM superchannel, and DFT-S OFDM superchannel is provided. Experimental results are given and their performances are compared with the conventional OFDM system.

REFERENCES

1. Lowery AJ , Du LB , Armstrong J. Orthogonal frequency division multiplexing for adaptive dispersion compensation in long haul WDM systems. Optical Fiber Communication Conference, paper no. PDP 39, Anaheim, CA; 2006.
2. Djordjevic IB, Vasic B. Orthogonal frequency division multiplexing for high-speed optical transmission. *Opt Express* 2006;14:3767–3775.
3. Shieh W, Athaudage C. Coherent optical orthogonal frequency division multiplexing. *Electron Lett* 2006;42:587–589.
4. Shieh W, Yi X, Ma Y, Tang Y. Theoretical and experimental study on PMD-supported transmission using polarization diversity in coherent optical OFDM systems. *Opt Express* 2007;15:9936–9947.
5. Jansen SL, Morita I, Tanaka H. 16 × 52.5-Gb/s, 50-GHz spaced, POLMUX-CO-OFDM transmission over 4160 km of SSMF enabled by MIMO processing. European Conference on Optical Communication, paper no. PD1.3. Berlin, Germany; 2007.
6. Ma Y, Yang Q, Tang Y, Chen S, Shieh W. 1-Tb/s single-channel coherent optical OFDM transmission over 600-km SSMF fiber with subwavelength bandwidth access. *Opt Express* 2009;17:9421–9427.
7. Yamada E, Sano A, Masuda H, Yamazaki E, Kobayashi T, Yoshida E, Yonenaga K, Miyamoto Y, Ishihara K, Takatori Y, Yamada T, Yamazaki H. 1 Tbit/s (111 Gbit/s/ch × 10 ch) no-guard-interval CO-OFDM transmission over 2100 km DSF. *Electron Lett* 2008;44:1417–1419.
8. Chandrasekhar S, Liu X, Zhu B, Peckham DW. Transmission of a 1.2-Tb/s 24-carrier no-guard-interval coherent OFDM superchannel over 7200-km of ultra-large-area fiber. ECOC; 2009. p 1–2.
9. Yu J, Dong Z, Chi N. 1.96 Tb/s (21 × 100 Gb/s) OFDM optical signal generation and transmission over 3200-km fiber. *IEEE Photon Technol Lett* 2011;23:1061 –1063.
10. Zhao C, Chen Y, Zhang S, Li J, Zhang F, Zhu L, Chen Z. Experimental demonstration of 1.08 Tb/s PDM CO-SCFDM transmission over 3170 km SSMF. *Opt Express* 2012;20:787–793.
11. Nee R, Prasad R. OFDM for Wireless Multimedia Communications. 1st ed. Norwood, MA, USA: Artech House, Inc.; 2000.
12. Hara S, Prasad R. Multicarrier Techniques for 4G Mobile Communications. Norwood, MA, USA: Artech House, Inc.; 2003.
13. Chang RW. Synthesis of band-limited orthogonal signals for multichannel data transmission. *Bell Sys Tech J* 1966;45:1775–1796.
14. Saltzberg BR. Performance of an efficient parallel data transmission system. *IEEE Trans Commun* 1967;15:805–813.
15. Shieh W, Djordjevic I. Orthogonal Frequency Division Multiplexing for Optical Communications. Academic Press; 2009.

16. Shieh W, Yi X, Tang Y. Transmission experiment of multi-gigabit coherent optical OFDM systems over 1000 km SSMF fiber. *Electron Lett* 2007;43:183–185.
17. Jansen SL, Morita I, Takeda N, Tanaka H. 20-Gb/s OFDM transmission over 4,160-km SSMF enabled by RF-Pilot tone phase noise compensation. Optical Fiber Communication Conference (OFC), paper PDP15; 2007.
18. Tang Y, Shieh W, Yi X, Evans R. Optimum design for RF-to-optical up-converter in coherent optical OFDM systems. *IEEE Photon Technol Lett* 2007;19:483–485.
19. Ly-Gagnon DS, Tsukarnoto S, Katoh K, Kikuchi K. Coherent detection of optical quadrature phase-shift keying signals with carrier phase estimation. *J Lightwave Technol* 2006;24:12–21.
20. Savory SJ, Gavioli G, Killey RI, Bayvel P. Electronic compensation of chromatic dispersion using a digital coherent receiver. *Opt Express* 2007;15:2120–2126.
21. Cohn SB, Weinhouse NP. An automatic microwave phase measurement system. *Microwave J* 1964;7:49–56.
22. Hoer CA, Roe KC. Using an arbitrary six-port junction to measure complex voltage ratios. *IEEE Trans MTT* 1975;MTT-23:978–984.
23. Tang Y, Chen W, Shieh W. Study of nonlinearity and dynamic range of coherent optical OFDM receivers. Optical Fiber Communication Conference (OFC), paper JWA65; 2008.
24. Shieh W, Bao H, Tang Y. Coherent optical OFDM: theory and design. *Opt Express* 2008;16:841–859.
25. Shieh W, Yi X, Ma Y, Yang Q. Coherent optical OFDM: has its time come? [Invited]. *J Opt Netw* 2008;7:234–255.
26. Shieh W. PMD-supported coherent optical OFDM systems. *IEEE Photon Technol Lett* 2007;19:134–136.
27. Pan Q, Green RJ. Bit-error-rate performance of lightwave hybrid AM/OFDM systems with comparison with AM/QAM systems in the presence of clipping impulse noise. *IEEE Photon Technol Lett* 1996;8:278–280.
28. Lowery AJ, Armstrong J. Orthogonal-frequency-division multiplexing for dispersion compensation of long-haul optical systems. *Opt Express* 2006;14(6):2079–2084.
29. Tang JM, Lane PM, Shore KA. 30 Gb/s transmission over 40 km directly modulated DFB laser-based SMF links without optical amplification and dispersion compensation for VSR and metro applications. Optical Fiber Communication Conference (OFC), paper JThB8; 2006.
30. Tang JM, Shore KA. Maximizing the transmission performance of adaptively modulated optical OFDM signals in multimode-fiber links by optimizing analog-to-digital converters. *J Lightwave Technol* 2007;25:787–798.
31. Jin XQ, Tang JM, Spencer PS, Shore KA. Optimization of adaptively modulated optical OFDM modems for multimode fiber-based local area networks. *J Opt Networking* 2008;7:198–214.
32. Schmidt BJC, Lowery AJ, Armstrong J. Experimental demonstrations of 20 Gbit/s direct-detection optical OFDM and 12 Gbit/s with a colorless transmitter. Optical Fiber Communication Conference (OFC), paper PDP18; 2007.
33. Hewitt DF. Orthogonal frequency division multiplexing using baseband optical single sideband for simpler adaptive dispersion compensation. Optical Fiber Communication Conference (OFC), paper OME7; 2007.

34. Peng WR, Wu X, Arbab VR, Shamee B, Yang JY, Christen LC, Feng KM, Willner AE, Chi S. Experimental demonstration of 340 km SSMF transmission using a virtual single sideband OFDM signal that employs carrier suppressed and iterative detection techniques. Optical Fiber Communication Conference (OFC), paper OMU1; 2008.
35. Peng WR, Wu X, Arbab VR, Shamee B, Christen LC, Yang JY, Feng KM, Willner AE, Chi S. Experimental demonstration of a coherently modulated and directly detected optical OFDM system using an RF-Tone insertion. Optical Fiber Communication Conference (OFC), paper OMU2; 2008.
36. Lowery AJ, Du LB, Armstrong J. Performance of optical OFDM in ultralong-haul WDM lightwave systems. *J Lightwave Technol* 2007;25:131–138.
37. Agrawal GP. Fiber-Optic Communication Systems. Fourth ed. Hoboken, NJ, USA: John Wiley & Sons, Inc.; 2010.
38. Jolley NE, Kee H, Pickard P, Tang J, Cordina K. Generation and propagation of a 1550 nm 10 Gbit/s optical orthogonal frequency division multiplexed signal over 1000m of multimode fibre using a directly modulated DFB. Optical Fiber Communication Conference (OFC), paper OFP3; 2005.
39. Schuster M, Randel S, Bunge CA, Lee SCJ, Breyer F, Spinnler B, Petermann K. Spectrally efficient compatible single-sideband modulation for OFDM transmission with direct detection. *IEEE Photon Technol Lett* 2008;20:670–672.
40. Brendon JC, Schmidt ZZ, Du LB, Lowery AJ. 120 Gbit/s over 500-km using single-band polarization-multiplexed self-coherent optical OFDM. *J Lightwave Technol* 2010;28(4):328–335.
41. Cartledge JC, Karar AS. 100 Gbit/s using intensity modulation and direct detection. Proceeding of ECOC, Paper We. 4. C. 3; 2013.
42. Peng W-R, Wu X, Feng K-M, Arbab VR, Shamee B, Yang J-Y, Christen LC, Willner AE, Chi S. Spectrally efficient direct-detected OFDM transmission employing an iterative estimation and cancellation technique. *Opt Express* 2009;17(11):9099–9111.
43. Chen X, Li A , Che D , Hu Q, Wang Y, He J, Shieh W. High-speed Fading-free Direct Detection for Double-Sideband OFDM Signal via Block-wise Phase Switching. OFC'2013, PDP5B.7; 2013.
44. Chen X, Che D, Li A, He J, Shieh W. Signal-carrier interleaved optical OFDM for direct detection optical communication. *Opt Express* 2013;21(26):32501–32507.
45. Che D, Li A, Chen X, Hu Q, Wang Y, Shieh W. 160-Gb/s stokes vector direct detection for short reach optical communication. OFC'2014, PDP5C.7; 2014.
46. Yu J, Zhou X, Huang MF, Shao Y, Qian D, Wang T, Cvijetic M, Magill P, Nelson L, Birk M, Ten S, Matthew HB, Mishra SK. 17 Tb/s (161x 114 Gb/s) PolMux-RZ-8PSK transmission over 662 km of ultra-low loss fiber using C-band EDFA amplification and digital coherent detection. 34th European Conference on Optical Communication (ECOC), paper Th.3.E.2; 2008.
47. Sano A, Masuda H, Kobayashi T, Fujiwara M, Horikoshi K, Yoshida E, Miyamoto Y, Matsui M, Mizoguchi M, Yamazaki H, Sakamaki Y, Ishii H. 69.1-Tb/s (432 × 171-Gb/s) C- and extended L-band transmission over 240 km using PDM-16-QAM modulation and digital coherent detection. Optical Fiber Communication Conference (OFC), paper PDPB7; 2010.

48. Gnauck AH, Winzer PJ, Chandrasekhar S, Liu X, Zhu B, Peckham DW. 10 x 224-Gb/s WDM transmission of 28-Gbaud PDM 16-QAM on a 50-GHz grid over 1,200 km of fiber. Optical Fiber Communication Conference (OFC), paper PDPB8; 2012.
49. Zhou X, Yu J, Huang MF, Shao Y, Wang T, Nelson L, Magill PD, Birk M, Borel PI, Peckham DW, Lingle R. 64-Tb/s (640x 107-Gb/s) PDM-36QAM transmission over 320 km using both pre- and post-transmission digital equalization. Optical Fiber Communication Conference (OFC), paper PDPB9; 2010.
50. Huang Y-K, Ip E, Huang M-F, Zhu B, Ji PN, Shao Y, Peckham DW, Lingle R, Aono Y, Tajima T, Wang T. 10x456-Gb/s DP-16QAM transmission over 8 x 100 km of ULAF using coherent detection with a 30-GHz analog-to-digital converter. 15th OptoElectronics and Communications Conference (OECC); Sapporo, Japan, paper PD3; July 2010.
51. Alfiad MS, Kuschnerov M, Jansen SL, Wuth T, van den Borne D, de Waardt H. 11 x 224-Gb/s POLMUX-RZ-16QAM transmission over 670 km of SSMF with 50-GHz channel spacing. *IEEE Photon Technol Lett* 2010;22(15):1150–1152.
52. Cai J-X, Cai Y, Sun Y, Davidson CR, Foursa DG, Lucero A, Sinkin O, Patterson W, Pilipetskii A, Mohs G, Bergano NS. 112x112 Gb/s transmission over 9,360 km with channel spacing set to the baud rate (360% spectral efficiency). 36th European Conference and Exposition on Optical Communication (ECOC), paper PD2.1; 2010.
53. Nölle M, Hilt J, Molle L, Seimetz M, Freund R. 8x224 Gbit/s PDM 16QAM WDM transmission with real-time signal processing at the transmitter. 36th European Conference and Exposition on Optical Communication (ECOC), paper We.8.C.4; 2010.
54. Liu X, Chandrasekhar S, Zhu B, Winzer P, Gnauck A, Peckham D. Transmission of a 448-Gb/s reduced-guard-interval CO-OFDM signal with a 60-GHz optical bandwidth over 2000 km of ULAF and five 80-GHz-grid ROADMs. Optical Fiber Communication Conference (OFC), paper PDPC2; 2010.
55. Kobayashi T, Sano A, Matsuura A, Yoshida M, Sakano T, Kubota H, Miyamoto Y, Ishihara K, Mizoguchi M, Nagatani M. 45.2Tb/s C-band WDM transmission over 240km using 538Gb/s PDM-64QAM single carrier FDM signal with digital pilot tone. 37th European Conference and Exposition on Optical Communication (ECOC), paper Th.13.C6; 2011.
56. Myung HG, Lim J, Goodman DJ. Single carrier FDMA for uplink wireless transmission. *IEEE Veh Technol Mag* 2006;1(3):30–38.
57. Fazel K, Kaiser S. Multi-Carrier and Spread Spectrum Systems: From OFDM and MC-CDMA to LTE and WiMAX. 2nd ed. Chichester, UK: John Wiley & Sons Ltd.; 2008.
58. 3rd Generation Partnership Project (3GPP). http://www.3gpp.org/LTE.
59. Shieh W, Tang Y. Ultrahigh-speed signal transmission over nonlinear and dispersive fiber optic channel: the multicarrier advantage. *IEEE Photon J* 2010;2(3):276–283.
60. Tang Y, Shieh W, Krongold BS. DFT-spread OFDM for fiber nonlinearity mitigation. *IEEE Photon Technol Lett* 2010;22(16):1250–1252.
61. Chen X, Li A, Gao G, Shieh W. Experimental demonstration of improved fiber nonlinearity tolerance for unique-word DFT-spread OFDM systems. *Opt Express* 2011;19:26198–26207.
62. Li A, Chen X, Gao G, Shieh W. Transmission of 1 Tb/s unique-word DFT-spread OFDM superchannel over 8000 km EDFA-only SSMF link (invited). *J Lightwave Technol* 2012;30(24):3931–3937.

63. Li A, Chen X, Gao G, Shieh W, Krongold B. Transmission of 1.63-Tb/s PDM-16QAM Unique-word DFT-Spread OFDM Signal over 1,010-km SSMF. Optical Fiber Communication Conference (OFC), paper OW4C.1; 2012.
64. Li J, Zhao C, Zhu L, Zhang F, He Y, Chen Z. Experimental demonstration of ROADM functionality on an optical SCFDM superchannel. *IEEE Photon Technol Lett* 2012;24(3):215–217.
65. Ishihara K, Kobayashi T, Kudo R, Takatori Y, Sano A, Miyamoto Y. Frequency-domain equalization for coherent optical single-carrier transmission systems. *IEICE Trans Commun* 2009;E92-B(12).
66. Sorger U, De Broeck I, Schnell M. Interleaved FDMA – a new spread- spectrum multiple-access scheme. Proceedings of IEEE ICC '98; Atlanta, GA; June 1998. p 1013–1017.
67. Myung HG, Lim J, Goodman DJ. Peak-to-average power ratio of single carrier FDMA signals with pulse shaping. The 17th Annual IEEE International Symposium on Personal, Indoor and Mobile Radio Communications (PIMRC '06), Helsinki, Finland, September 2006.
68. Falconer D, Ariyavisitakul SL, Benyamin-Seeyar A, Eidson B. Frequency domain equalization for single-carrier broadband wireless systems. *IEEE Commun Mag* 2002;40(4):58–66.
69. Witschnig H, Mayer T, Springer A, Koppler A, Maurer L, Huemer M, Weigel R. A different look on cyclic prefix for SC/FDE. The 13th IEEE International Symposium on Personal, Indoor and Mobile Radio Communications, 2002. vol. 2; 2002. p 824–828.
70. Huemer M, Witschnig H, Hausner J. Unique word based phase tracking algorithms for SC/FDE-systems. Global Telecommunications Conference, 2003. GLOBECOM '03. IEEE, vol. 1; 2003. p 70–74.
71. Coon J, Sandell M, Beach M, McGeehan J. Channel and noise variance estimation and tracking algorithms for unique-word based single-carrier systems. *IEEE Trans Wireless Commun* 2006;5(6):1488–1496.
72. Chu D. Polyphase codes with good periodic correlation properties (Corresp.). *IEEE Trans Inf Theor* 1972;18(4):531–532.
73. Chen C, Zhuge Q, Plant DV. Zero-guard-interval coherent optical OFDM with overlapped frequency-domain CD and PMD equalization. *Opt Express* 2011;19(8):7451–7467.
74. Sano A, Yamada E, Masuda H, Yamazaki E, Kobayashi T, Yoshida E, Miyamoto Y, Kudo R, Ishihara K, Takatori Y. No-Guard-Interval Coherent Optical OFDM for 100-Gb/s Long-Haul WDM Transmission. *J Lightwave Technol* 2009;27(16):3705–3713.
75. Liu X, Chandrasekhar S, Zhu B, Peckham DW. Efficient digital coherent detection of a 1.2-Tb/s 24-carrier no-guard-interval CO-OFDM signal by simultaneously detecting multiple carriers per sampling. OFC'2010, paper OWO2; 2010.
76. Ishihara K, Kobayashi T, Kudo R, Takatori Y, Sano A, Yamada E, Masuda H, Matsui M, Mizoguchi M, Miyamoto Y. Frequency-domain equalisation without guard interval for optical transmission systems. *Electron Lett* 2008;44(25).
77. Savory SJ. Digital filters for coherent optical receivers. *Opt Express* 2008;16:804–817.
78. Liu X, Chandrasekhar S, Zhu B, Winzer PJ, Gnauck AH, Peckham DW. 448-Gb/s reduced-guard-interval CO-OFDM transmission over 2000 km of ultra-large-area fiber and five 80-GHz-Grid ROADMs. *J Lightwave Technol* 2011;29(4):483–490.

79. Liu X, Chandrasekhar S, Zhu B, Winzer PJ, Gnauck AH, Peckham DW. Transmission of a 448-Gb/s reduced-guard-interval CO-OFDM signal with a 60-GHz optical bandwidth over 2000 km of ULAF and five 80-GHz-Grid ROADMs. OFC/NFOEC 2010, paper PDPC2; 2010.
80. X. Liu and F. Buchali, Intra-symbol frequency-domain averaging based channel estimation for coherent optical OFDM. *Opt Exp* , 16, 21944-21957 (2008).
81. Liu X, Buchali F, Tkach RW. Improving the nonlinear tolerance of polarization-division-multiplexed CO-OFDM in long-haul fiber transmission. *J Lightwave Technol* 2009;27(16):3632–3640.
82. Millar DS, Makovejs S, Mikhailov V, Killey RI, Bayvel P, Savory SJ. Experimental compensation of nonlinear compensation in long-haul PDM-QPSK transmission at 42.7 and 85.4 Gb/s. ECOC 2009, paper 9.4.4; 2009.
83. Myung HG, Lim J, Goodman DJ. Peak-to-average power ratio of single carrier FDMA signals with pulse shaping. The 17th Annual IEEE International Symposium on Personal, Indoor and Mobile Radio Communications (PIMRC '06), Helsinki, Finland; 2006.
84. Schmidl TM, Cox DC. Robust frequency and timing synchronization for OFDM. *IEEE Trans Commun* 1997;45:1613–1621.
85. Chang F, Onohara K, Mizuochi T. Forward error correction for 100 G transport networks. *Communications Magazine, IEEE* 2010;48(3):S48–S55.
86. Mizuochi T, Miyata Y, Kubo K, Sugihara T, Onohara K, Yoshida H. Progress in soft-decision FEC. OFC/NFOEC 2011, paper NWC2; 2011.

6

POLARIZATION AND NONLINEAR IMPAIRMENTS IN FIBER COMMUNICATION SYSTEMS

CHONGJIN XIE

R&D Lab, Ali Infrastructure Service, Alibaba Group, San Mateo, CA, USA

6.1 INTRODUCTION

Polarization is a property of waves that can oscillate with more than one orientation. Electromagnetic waves such as light exhibit polarization [1]. Although polarization can be used as another dimension to carry information, in practice, this feature is not utilized in direct-detection optical communication systems due to the difficulties of polarization demultiplexing in the optical domain [2–4]. It was only until recently, with the advent of digital coherent detection, polarization-division multiplexing (PDM), which transmits signals on two orthogonal states of polarization (SOPs) at an identical wavelength, started to be massively used in optical communication systems [5–10]. As the full optical field information is preserved and accessible after coherent detection, optical phase and polarization can be used to encode data, which significantly increases the spectral efficiency and capacity of an optical communication system. With advances in high-speed electronics, in today's coherent optical communication systems, signals are first converted to digital format through analog-to-digital-convertors (ADCs) after mixing with local oscillators (LOs) in photodetectors at coherent receivers, and digital signal processing (DSP) is used to recover signals. For example, carrier phase recovery and polarization alignment and separation, the main obstacles for analog implementations of coherent receivers in earlier years, can be realized in the electrical domain using sophisticated DSP.

Enabling Technologies for High Spectral-efficiency Coherent Optical Communication Networks, First Edition. Edited by Xiang Zhou and Chongjin Xie.
© 2016 John Wiley & Sons, Inc. Published 2016 by John Wiley & Sons, Inc.

Digital coherent detection revolutionizes the ways to design optical communication systems. First, linear impairments such as chromatic dispersion (CD) and polarization-mode dispersion (PMD), used to be compensated with optical methods in direct-detection systems, in principle can be completely compensated in the electrical domain using DSP in digital coherent-detection systems [5, 6]. It does not mean that they do not cause any performance degradations. For example, the interaction between electronic equalizers and LO phase noise induces amplitude noise and additional phase noise, which cause penalties in coherent systems without optical dispersion compensators [11, 12]. It has been shown that PMD can still degrade the performance of an optical communication system using coherent detection due to limited complexity of DSP in a real system [13–15]. Polarization-dependent loss (PDL) generates power and optical signal-to-noise-ratio (OSNR) fluctuations and repolarizes amplified spontaneous emission (ASE) noise in optical communication systems, and these effects cannot be compensated in coherent receivers either [16–18].

Second, digital coherent detection significantly changes the ways to manage nonlinear impairments in optical communication systems. Many nonlinearity mitigation techniques developed for direct-detection systems such as dispersion management are not effective anymore and become suboptimal in coherent-detection systems [19–25]. PMD can be helpful to reduce fiber nonlinearities [26–30], and fiber nonlinear effects can be mitigated with DSP in both transmitters and receivers [31–34].

In direct-detection on-off-keyed (OOK) systems, fiber nonlinearities manifest themselves as timing and amplitude jitter induced by intra- and interchannel nonlinearities [35, 36]. In coherent-detection systems, as information is also coded in phase and polarization, depending on system parameters such as bit rates and modulation formats, phase and polarization distortions caused by fiber nonlinearities can be the dominant nonlinear effects [20, 21, 25, 37–39], and these make fiber nonlinearity impacts on coherent-detection systems significantly different from direct-detection systems and largely change the ways to manage nonlinearities in optical communication systems.

In this chapter, polarization and nonlinear impairments in single-carrier coherent optical communication systems are discussed. To make the chapter self-contained, first basics on polarization of light are given in Section 6.2. The phenomena of PMD and PDL in optical communication systems are presented in Section 6.3. Section 6.4 shows the modeling of nonlinear transmission in optical fibers. Digital coherent optical communication systems and some electrical equalization techniques are discussed in Section 6.5. In Section 6.6, the PMD and PDL impairments in digital coherent optical communication systems are presented. Section 6.7 focuses on nonlinear impairments in coherent-detection optical communication systems, and nonlinearities in three different dispersion-managed systems are discussed. The chapter is summarized in Section 6.8.

6.2 POLARIZATION OF LIGHT

Light is electromagnetic waves composed of electrical field vector $\vec{E}$ and magnetic field vector $\vec{H}$. In free space or homogeneous isotropic nonattenuating medium, light is properly described as transverse waves, meaning that the electric field vector $\vec{E}$ and the magnetic field vector $\vec{H}$ are in directions perpendicular to (or "transverse" to) the direction of wave propagation, and $\vec{E}$ and $\vec{H}$ are also perpendicular to each other. In optics, the electrical field vector is usually chosen to represent the electromagnetic field and monochromatic light is mathematically described as

$$\vec{E}(z,t) = [\hat{x}E_x + \hat{y}E_y e^{j\varphi}]e^{-j(\omega t - \beta z + \theta)} \tag{6.1}$$

where $\vec{E}(z,t)$ is the electrical field vector, $\hat{x}$ and $\hat{y}$ are unit vectors of x and y axes, ω is the angular frequency of the light, β is the wave number, θ is the initial phase of the light, $j = \sqrt{-1}$, z and t are the distance in the propagation direction of the light and time. The amplitudes of the electrical field in x and y axes and the phase difference, φ, between the electrical field in x and y axes determine the polarization of light, which describes the temporal evolution of the electrical field vector at a certain distance. Figure 6.1 shows the temporal evolution of the electrical field vector for light with linear, circular, and elliptical polarization.

Polarization of light can be described by many representations, and two widely used representations are Jones vector representation and Stokes vector representation. As shown in Equation 6.1, the amplitudes and phases of the x and y components of an electrical field vector provide the full information on completely polarized light. The amplitude and phase information can be conveniently represented as a two-dimensional complex vector, which is called Jones vector [40]

$$|s\rangle = \frac{1}{|E|}\begin{pmatrix} E_x e^{j\varphi_x} \\ E_x e^{j\varphi_x} \end{pmatrix} = \begin{pmatrix} s_x \\ s_y \end{pmatrix} \tag{6.2}$$

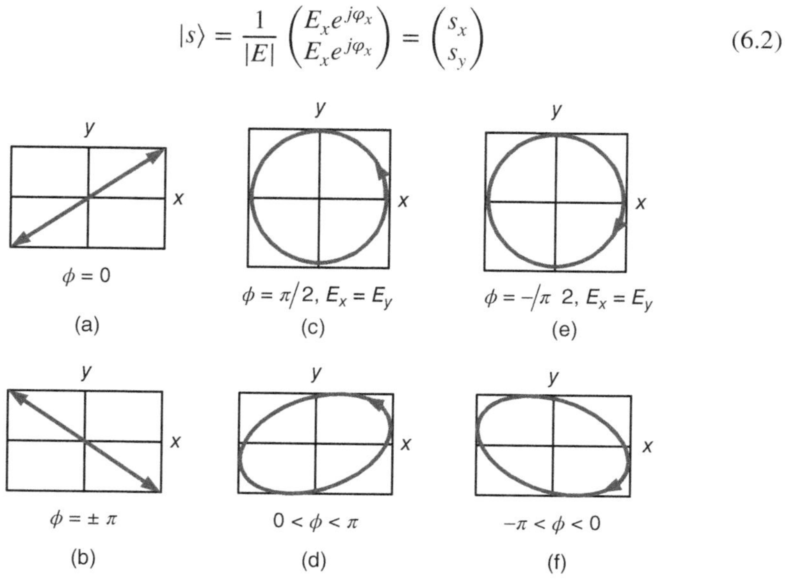

FIGURE 6.1 Temporal evolution of the electrical field vector for linear polarization (a) (b), right-hand circular polarization (c), right-hand elliptical polarization (d), left-hand circular polarization (e), and left-hand elliptical polarization (f).

where $|E| = \sqrt{E_x^2 + E_y^2}$, φ_x, and φ_y are the phases of x and y components. The Jones vector has unit magnitude and is usually written as

$$|s\rangle = \begin{pmatrix} \cos\theta \\ \sin\theta\, e^{j\varphi} \end{pmatrix} \tag{6.3}$$

$\varphi = 0$ and $\pm\pi$ mean linear polarization, for $\varphi = \pm\pi/2$ and $\theta = \pi/4$, it is circular polarization and all the other values represent elliptical polarization.

Another representation to describe polarization of light is a four-dimensional real vector called Stokes vector. Unlike Jones vector, which can only be used to describe completely polarized light, where all the frequency components of the light have the same polarization, Stokes vector can be used to describe partially polarized light. A Stokes vector is defined as

$$\vec{S} = \begin{pmatrix} S_0 \\ S_1 \\ S_2 \\ S_3 \end{pmatrix} = \begin{pmatrix} I_x + I_y \\ I_x - I_y \\ I_{\pi/4} - I_{-\pi/4} \\ I_{\mathrm{RHC}} - I_{\mathrm{LHC}} \end{pmatrix} \tag{6.4}$$

where I_x, I_y, $I_{\pi/4}$, $I_{-\pi/4}$, I_{RHC}, and I_{LHC} are the intensities of light in the states of horizontal, vertical, 45° linear, −45° linear, right-hand circular, and left-hand circular polarization, respectively.

The four components of a Stokes vector have the relation of $S_1^2 + S_2^2 + S_3^2 \le S_0^2$. For completely polarized light, $S_1^2 + S_2^2 + S_3^2 = S_0^2$. The unpolarization of a signal is quantified by degree of polarization (DOP), which is the ratio of the power of polarized component to the total power of a signal, defined as

$$\mathrm{DOP} = \frac{\sqrt{S_1^2 + S_2^2 + S_3^2}}{S_0} \tag{6.5}$$

DOP = 1 means light is completely polarized. Monochromatic light is totally polarized, but for a signal with a certain bandwidth, it can be completely polarized, completely unpolarized, or partially polarized.

In most cases, instead of using a four-dimensional Stokes vector, a three-dimensional normalized standard Stokes vector is used, which is defined as

$$\hat{s} = \frac{1}{S_0}\begin{pmatrix} S_1 \\ S_2 \\ S_3 \end{pmatrix} = \begin{pmatrix} s_1 \\ s_2 \\ s_3 \end{pmatrix} \tag{6.6}$$

For completely polarized light, a three-dimensional Stokes vector can be obtained from the corresponding Jones vector

$$s_1 = s_x s_x^* - s_y s_y^*,\ s_2 = s_x s_y^* + s_x^* s_y,\ s_3 = j(s_x s_y^* - s_x^* s_y). \tag{6.7}$$

A three-dimensional Stokes vector can be virtualized with the Poincaré sphere in Stokes space, as shown in Figure 6.2. The Poincaré sphere has a radius of 1. On the surface of the Poincaré sphere, DOP = 1, and inside the sphere, DOP < 1. The north and south poles on the Poincaré sphere represent right- and left-hand circular polarization, respectively, and linear polarization resides on the equator of the sphere.

Polarization of light can be measured with a polarimeter. As shown in Figure 6.3, it measures the total power and the powers of three perpendicular polarization states in Stokes space, I_x, $I_{\pi/4}$, and $I_{\rm RHC}$, from which the Stokes vector and DOP of light can be obtained.

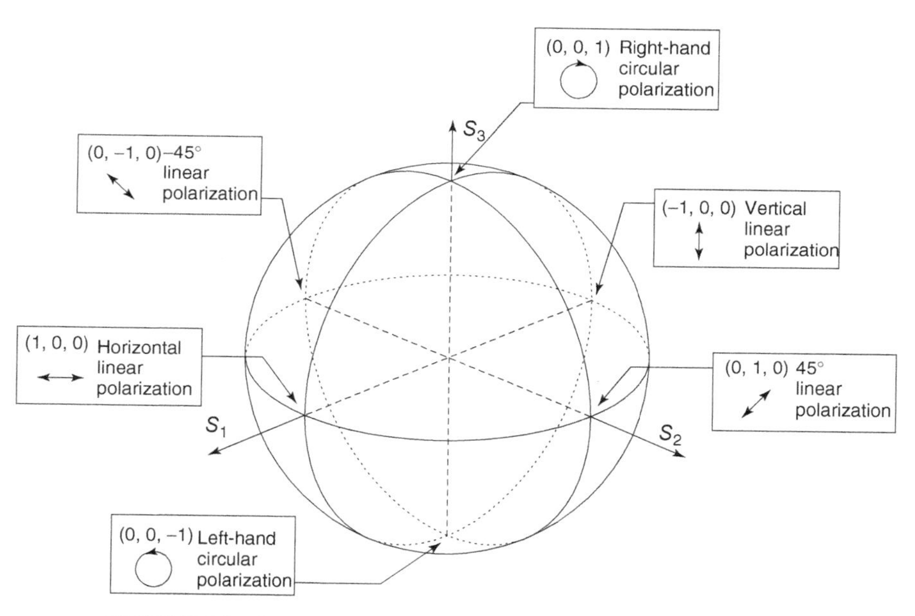

FIGURE 6.2 Representation of Stokes vectors on the Poincaré sphere.

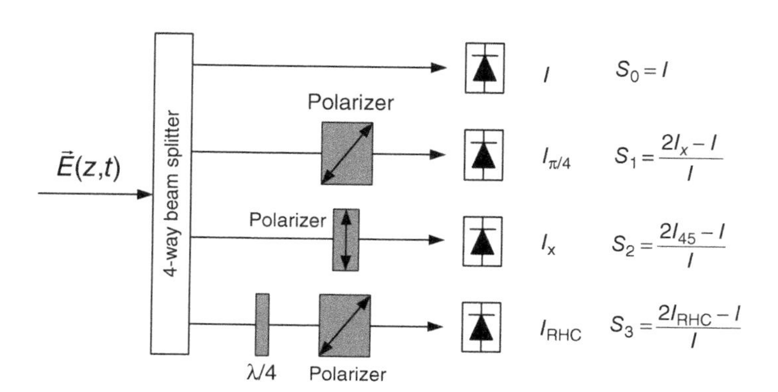

FIGURE 6.3 The block diagram of a polarimeter.

6.3 PMD AND PDL IN OPTICAL COMMUNICATION SYSTEMS

6.3.1 PMD

In an ideal optical fiber, the fiber core has a perfectly circular cross-section. In this case, the fundamental mode has two orthogonal polarization states that travel at the same speed. The signal that is transmitted over the fiber is randomly polarized, but that would not matter in an ideal fiber because the two polarization states would propagate identically. In a realistic fiber, however, there are random imperfections induced during the manufacturing process and by external environment such as stress, which break the circular symmetry. This asymmetry causes local birefringence, which manifests as a difference in the refractive indexes and thus the propagation constants for the two orthogonal polarization states

$$\Delta\beta = \beta_s - \beta_f = \frac{\omega}{c}(n_s - n_f) = \frac{\omega \Delta n}{c} \tag{6.8}$$

where β_s and β_f are the propagation constants in slow and fast axes and $\Delta n = n_s - n_f$ is the refractive index difference between the two axes, which causes light on the two polarization states to propagate with different speeds. For standard telecommunication-type fibers, $\Delta n \sim 10^{-7}$, and for polarization-maintaining fibers (PMF), Δn is much larger and is $\sim 10^{-3}$.

In long transmission fibers, due to localized stress during spooling/cabling/deployment, there are random variations in the axes of the birefringence along the fiber length, causing polarization-mode coupling where the fast and slow polarization modes from one segment split into both the fast and slow modes in the next segment [41]. PMD in fibers is the combined effect of local birefringence and random polarization-mode coupling along fibers, which causes random spreading of optical pulses, similar to the effects of other kinds of dispersion. Due to random mode coupling and pulse splitting, the propagation of a pulse along a long fiber is very complicated but, surprisingly, it can be described by the principal states model. The principal states model was originally developed by Poole and Wagner [42], which states that even for a long fiber, there exist two special orthogonal polarization states at the fiber input that result in an output pulse that is undistorted to the first order. These two special orthogonal polarization states are called the principal states of polarization (PSPs).

PMD can be characterized in Stokes space by the PMD vector $\vec{\Omega}$ [41]

$$\vec{\Omega} = \Delta\tau \hat{p} \tag{6.9}$$

where the magnitude, $\Delta\tau$, is differential group delay (DGD), the group-delay difference between the slow and fast principal state modes, and the unit vector, $\hat{p}$, points in the direction of the slower PSP, whereas the vector, $-\hat{p}$, is in the direction of the orthogonal faster PSP. In the time domain, PMD manifests as pulse splitting and broadening, which induces intersymbol interference (ISI) in optical communication systems. In the frequency domain, PMD causes the output SOP of a signal changes with frequency described as

$$\frac{d\hat{t}}{d\omega} = \vec{\Omega} \times \hat{t} \tag{6.10}$$

where vector $\hat{t}$ is the output SOP. Equation 6.10 indicates when a signal with all its frequency components having the same SOP propagates in a fiber with PMD, at the output, different frequency components of the signal will have different SOPs, which means that the signal is depolarized by PMD.

PMD vector is not constant but varies randomly with optical angular frequency. A Taylor-series expansion of $\vec{\Omega}(\omega)$ is typically used for a signal with a large bandwidth [43]

$$\vec{\Omega}(\omega_0 + \Delta\omega) = \vec{\Omega}(\omega_0) + \vec{\Omega}_{\omega}(\omega_0)\Delta\omega + \frac{1}{2}\vec{\Omega}_{\omega\omega}(\omega_0)\Delta\omega^2 + \cdots \tag{6.11}$$

where $\vec{\Omega}(\omega_0)$, $\vec{\Omega}_{\omega}(\omega_0)$, and $\vec{\Omega}_{\omega\omega}(\omega_0)$ are first-, second-, and third-order PMD, and the subscript ω indicates differentiation with angular frequency. Second-order PMD is described by the derivative of the PMD vector

$$\vec{\Omega}_{\omega} = \frac{d\vec{\Omega}}{d\omega} = \Delta\tau_{\omega}\hat{p} + \Delta\tau\,\hat{p}_{\omega} \tag{6.12}$$

Equation 6.12 shows that second-order PMD has two components. The first term on the right-hand side of Equation 6.12 is $\vec{\Omega}_{\omega||}$, the component that is parallel to $\vec{\Omega}(\omega_0)$, and is called polarization-dependent chromatic dispersion (PCD). The second term $\vec{\Omega}_{\omega\perp}$ is the component that is perpendicular to $\vec{\Omega}(\omega_0)$, which describes the change of PSP with frequency and is called depolarization. These two components cause different impairments in an optical communication system. Figure 6.4 shows a vector diagram of first- and second-order PMD.

The characteristics of first- and second-order PMD have been extensively studied and well understood, including its statistical properties. For a sufficiently long fiber,

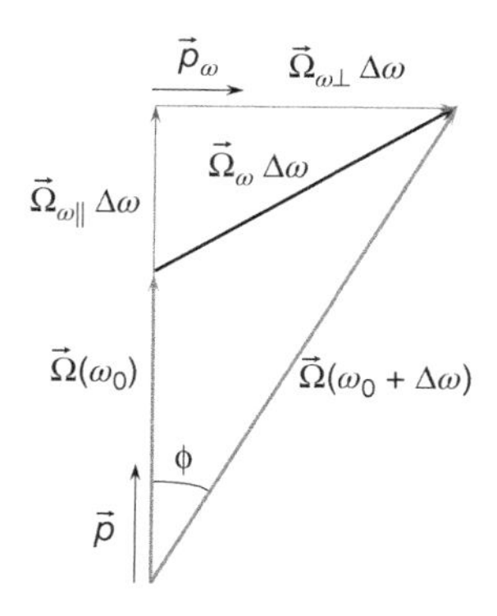

FIGURE 6.4 Schematic diagram of PMD vector, first- and second-order PMD.

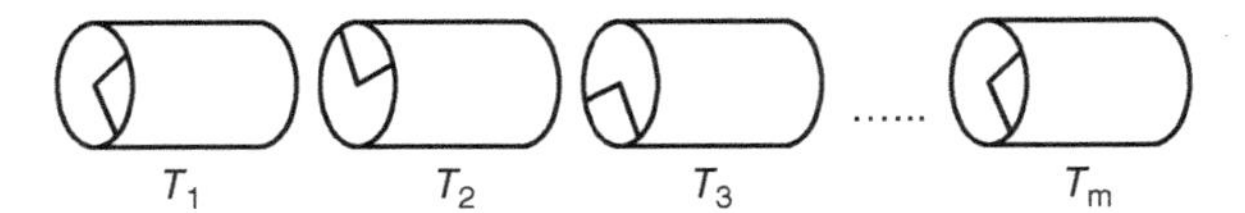

FIGURE 6.5 Modeling of PMD.

DGD is Maxwellian distributed. Its probability density function (PDF) is given by Kogelnik et al. [41]

$$p_{\Delta\tau}(x) = \frac{8}{\pi^2\langle\Delta\tau\rangle}\left(\frac{2x}{\langle\Delta\tau\rangle}\right)^2 e^{-(2x/\langle\Delta\tau\rangle)^2/\pi}; \quad x \geq 0 \tag{6.13}$$

with mean and root mean square (RMS) values being $E(x) = \langle\Delta\tau\rangle$ and $E(x^2) = 3\pi\langle\Delta\tau\rangle^2/8$, respectively. The statistical properties of second-order PMD can be found in Refs. [41] and [43].

PMD in fiber can be modeled as a concatenation of randomly rotated waveplates, each having a certain birefringence and certain length, as shown in Figure 6.5. Simulations are typically performed in Jones space and the overall transmission matrix of Figure 6.5 is given by

$$T = T_m T_{m-1} \cdots T_2 T_1 \tag{6.14}$$

where the transmission matrix of ith waveplate is

$$T_i = R(-\theta_i, -\varphi_i)\begin{pmatrix} \exp\left(-j\omega\,\Delta\tau_i/2\right) & 0 \\ 0 & \exp(j\omega\,\Delta\tau_i/2) \end{pmatrix} R(\theta_i, \varphi_i) \tag{6.15}$$

where $\Delta\tau_i$ is the DGD of the ith waveplate, and

$$R(\theta, \varphi) = \begin{pmatrix} \cos\theta & \sin\theta\exp(j\varphi) \\ -\sin\theta\exp(-j\varphi) & \cos\theta \end{pmatrix} \tag{6.16}$$

There are some variations when one simulates PMD. For example, one can choose the angle between waveplates to be totally random or have some correlation, and the DGD of each waveplate can be constant or random [44].

6.3.2 PDL

PDL usually occurs in optical components, such as isolators and couplers, whose insertion loss varies with the SOPs of input signals. Besides common loss, the polarization-dependent component of the power gain of an optical component with PDL can be described as $1 + \vec{\alpha} \cdot \hat{s}$, where $\hat{s}$ is a unit Stokes vector corresponding to the SOP of the incident optical signal and $\vec{\alpha}$ is the vector of PDL [45]. The highest and lowest gains are $1 \pm \alpha$ with $\alpha = |\vec{\alpha}|$, and they are achieved when the SOP of the input signal is either parallel to $\vec{\alpha}$ in Stokes space (highest gain) or antiparallel to it (lowest gain). In most common definition, PDL refers to the ratio of the lowest and highest losses of a PDL component in decibels

$$\Gamma = 10\log_{10}\frac{1+\alpha}{1-\alpha} \tag{6.17}$$

In an optical communication system with many PDL components distributed in the system, the statistical distribution of the overall PDL expressed in decibels is Maxwellian [45, 46].

Unlike PMD, PDL does not cause signal distortions, but it generates power variations along the optical link and changes of OSNR at the receiver, and repolarizes ASE noise and PDM signals as well [18, 47–51]. These effects can significantly degrade the performance of an optical communication system. PDL can also be modeled as the concatenation of many PDL elements with randomly oriented axis. It is found that to accurately simulate the noise variations induced by PDL, ASE noise has to be distributed along the link [18].

6.4 MODELING OF NONLINEAR EFFECTS IN OPTICAL FIBERS

When polarization effects can be neglected and the signal is launched in a single polarization state, the scalar nonlinear Schrödinger equation (NLSE) is a fairly good model to study transmission impairments in optical fibers including nonlinear effects. However, to consider polarization effects such as PMD and cross-polarization modulation (XPolM) and to study the nonlinear transmission of PDM signals in optical fibers, the coupled nonlinear Schrödinger equation (CNLSE) has to be used [44, 52, 53]

$$\frac{\partial \vec{E}}{\partial z} - j\Delta\beta\Sigma\vec{E} + \Delta\tau\Sigma\frac{\partial \vec{E}}{\partial t} + \frac{j}{2}\beta_2\frac{\partial^2 \vec{E}}{\partial t^2} = j\gamma\left[\left|\vec{E}\right|^2\vec{E} - \frac{1}{3}(\vec{E}^{+}\sigma_3\vec{E})\sigma_3\vec{E}\right] \tag{6.18}$$

where $\vec{E} = [E_x, E_y]^T$ is the electrical field column vector, $\Delta\beta$ is the birefringence parameter, $\Delta\tau$ is the DGD parameter related to PMD coefficient, Σ is the local Jones matrix describing polarization changes, β_2 is the group velocity dispersion (GVD), γ is the fiber nonlinear coefficient, $\vec{E}^{+} = [E_x^*, E_y^*]$ is the transpose conjugate of $\vec{E}$, and σ_3 is one of the Pauli spin matrices [40]

$$\sigma_3 = \begin{pmatrix} 0 & -j \\ j & 0 \end{pmatrix} \tag{6.19}$$

In Equation 6.18, z is the distance along the fiber axis and t is the retarded time moving at group velocity of the carrier frequency of the signal. By averaging the nonlinear effects over the Poincaré sphere under the assumption of complete mixing (averaging over the random polarization changes that uniformly cover the Poincaré sphere) and neglecting PMD, the CNLSE can be transformed to the Manakov equation [44, 52, 53]

$$\frac{\partial \vec{E}}{\partial z} + \frac{j}{2}\beta_2\frac{\partial^2 \vec{E}}{\partial t^2} - j\frac{8}{9}\gamma|\vec{E}|^2\vec{E} = 0 \tag{6.20}$$

Assume that we have a WDM system with two channels, channels a and b, and the two channels have no overlapping spectra. By neglecting four-wave mixing (FWM) between the two channels, one can separate the equations for channels a and b from the Manakov equation as [54–58]

$$\frac{\partial \vec{E}_a}{\partial z} + \frac{j}{2}\beta_2 \frac{\partial^2 \vec{E}_a}{\partial t^2} - j\frac{8}{9}\gamma(|\vec{E}_a|^2\vec{E}_a + |\vec{E}_b|^2\vec{E}_a + \vec{E}_b^{+}\vec{E}_a\vec{E}_b) = 0 \tag{6.21}$$

$$\frac{\partial \vec{E}_b}{\partial z} + \frac{j}{2}\beta_2 \frac{\partial^2 \vec{E}_b}{\partial t^2} - j\frac{8}{9}\gamma(|\vec{E}_b|^2\vec{E}_b + |\vec{E}_a|^2\vec{E}_b + \vec{E}_a^{+}\vec{E}_b\vec{E}_a) = 0 \tag{6.22}$$

Within the parentheses of the two equations, the first term is self-phase modulation (SPM), the second term is polarization-independent cross-phase modulation (XPM), and the third term is polarization-dependent XPM. SPM does not depend on the polarization, but XPM is polarization dependent. The third nonlinear term is the same as the second nonlinear term when the two channels have the same polarization and it is zero when they are orthogonally polarized, which means that the XPM between two channels with parallel polarizations is two times that with orthogonal polarizations.

The last two terms in each of Equations 6.21 and 6.22 show that XPM between channels also causes XPolM. Due to XPolM, the SOP of one channel can be changed by other channels. Neglecting CD, one can derive the SOP evolution of channel a induced by channel b due to XPolM as

$$\frac{d\vec{S}_a}{dz} = \frac{8}{9}\gamma(\vec{S}_a \times \vec{S}_b) = \frac{8}{9}\gamma(\vec{S}_a \times \vec{S}_{\text{sum}}) \tag{6.23}$$

where $\vec{S}_a = (S_{a1}, S_{a2}, S_{a3})$ and $\vec{S}_b = (S_{b1}, S_{b2}, S_{b3})$ are Stokes vectors for channels a and b, respectively, and $\vec{S}_{\text{sum}} = \vec{S}_a + \vec{S}_b$ is the sum of the two Stokes vectors. The evolution of $\vec{S}_b$ can be obtained by exchanging the subscripts in Equation 6.23.

To model nonlinear polarization effects in fiber-optic communication systems, one can directly solve the CNLSE given in Equation 6.18 with the split-step Fourier method [59]. To increase the speed of the simulations, the CNLSE can be solved with the approach proposed by Marcuse et al. by integrating with steps small enough to follow the detailed polarization evolution and using larger steps for CD and nonlinear effects [52]. The other widely used method is the coarse-step method, which assumes that within each step the polarization does not change and the signal propagation is described by the following CNLSE [52, 60]

$$\frac{\partial E_x}{\partial z} - \frac{1}{2}\Delta\beta_1 \frac{\partial E_x}{\partial t} + \frac{j}{2}\beta_2 \frac{\partial^2 E_x}{\partial t^2} = j\gamma\left(|E_x|^2 + \frac{2}{3}|E_y|^2\right)E_x \tag{6.24}$$

$$\frac{\partial E_y}{\partial z} + \frac{1}{2}\Delta\beta_1 \frac{\partial E_y}{\partial t} + \frac{j}{2}\beta_2 \frac{\partial^2 E_y}{\partial t^2} = j\gamma\left(\left|E_y\right|^2 + \frac{2}{3}|E_x|^2\right)E_y \tag{6.25}$$

At the interval of the fiber coupling length, which is typically one or a few step sizes, the polarization of the field is randomly rotated to generate complete mixing over the Poincaré sphere. Two scattering matrices have been used to rotate signal polarizations. One scattering matrix is [2]

$$\begin{pmatrix} \cos\alpha \exp(i\phi) & \sin\alpha \exp(i\phi) \\ -\sin\alpha & \cos\alpha \end{pmatrix} \tag{6.26}$$

and the other one is [60]

$$\begin{pmatrix} \cos\alpha & \sin\alpha \exp(i\phi) \\ -\sin\alpha \exp(-i\phi) & \cos\alpha \end{pmatrix} \tag{6.27}$$

where $\cos 2\alpha$ and ϕ are randomly chosen from uniform distributions in Equation 6.26, and α and ϕ are randomly chosen from uniform distributions in Equation 6.27. As shown by Marcuse et al. [52], although neither matrix introduces a uniform scattering on the Poincaré sphere, concatenating several of these matrices does lead to rapid uniform mixing on the Poincaré sphere.

6.5 COHERENT OPTICAL COMMUNICATION SYSTEMS AND SIGNAL EQUALIZATION

6.5.1 Coherent Optical Communication Systems

The block diagram of a digital coherent optical communication system is illustrated in Figure 6.6. For simplicity, only one channel of a WDM system is shown in the figure. Note that DSP is massively used in such system, not only in the receiver, but in the transmitter as well. Typically, both polarization and phase of lightwave are used to carry information in a coherent optical communication system to increase spectral efficiency and system capacity. In the transmitter, a continuous wave (CW) from a low linewidth laser such as an external cavity laser (ECL) is split into two parts, one for each polarization, and each part is modulated with an in-phase/quadrature (I/Q) modulator by electrical signals from digital-to-analog converters (DACs) after DSP. Due to the use of DAC and DSP, lots of functions can be performed in the transmitter, for example, to perform predistortion and generate signals with specific waveforms and spectral shapes for purposes such as improving nonlinear tolerance and/or spectral efficiency [61–64].

During the propagation in fibers, signal polarizations are not maintained but randomly rotated. At the polarization and phase diversity receiver, the received signal is split by a polarization beam splitter (PBS). Each polarization of the signal after the PBS is combined with a LO in a 90° hybrid. The four tributaries (x and y polarizations, I/Q branches) of the combined signal and LO after the hybrids are detected by four detectors (or four pairs of balanced detectors). After antialias filtering, the signals are sampled and converted to digital form by ADCs. The signals are then processed by DSP to recover the transmitted data, including retiming and resampling,

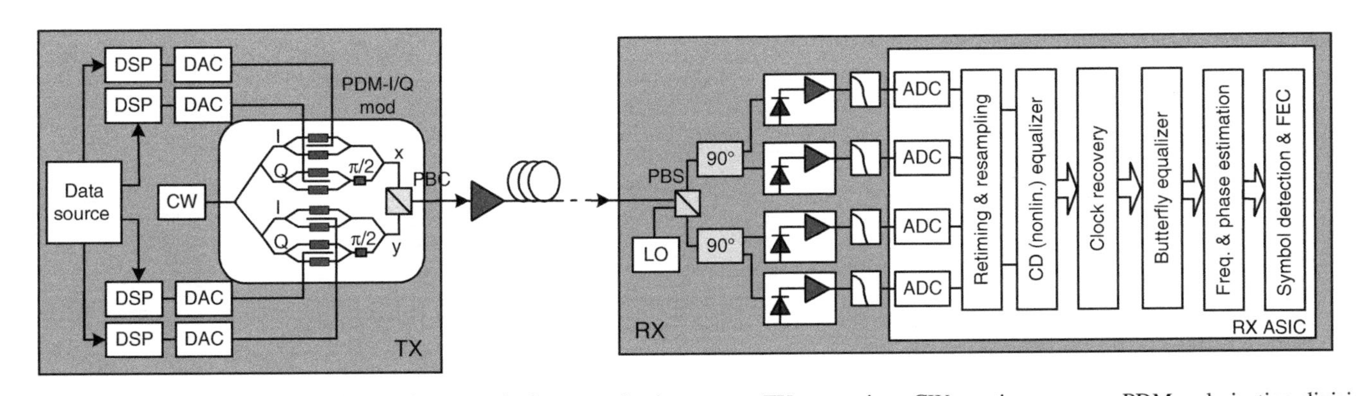

FIGURE 6.6 Block diagram of a digital coherent optical communication system. TX: transmitter, CW: continuous wave, PDM: polarization-division multiplexed, *I*/*Q*: in-phase/quadrature, Mod: modulator, RX: receiver, LO: local oscillator, ADC: analog-to-digital converter, PBC/S: polarization-beam combiner/splitter, ASIC: application-specific integrated circuit, FEC: forward-error correction. From Ref. [24].

CD compensation (nonlinearity compensation if allowed by the complexity of the DSP), clock recovery, polarization demultiplexing, PMD compensation, carrier frequency and phase estimation, symbol detection, and forward-error correction (FEC). One of the key distinguishing features of a digital coherent optical communication system is its ability to compensate for most transmission impairments in the electrical domain with DSP.

As CD can be completely compensated with DSP in coherent receivers, it is not necessary to have optical dispersion compensators in green field systems. But most existing systems have optical dispersion compensators. It has been shown that fiber nonlinear effects are significantly different in coherent optical communication systems with optical dispersion compensators from those in the systems without optical dispersion compensators.

6.5.2 Signal Equalization

In principle, distortion equalization in a coherent receiver can be realized in one equalizer, but it is in general beneficial to perform the equalization with one static equalizer and one dynamic equalizer. As CD usually changes very slowly, a static equalizer is typically used for CD, which requires large filters and is bulky and virtually static. A dynamic equalizer typically requires much shorter adaptive filters and is used to deal with time-varying effects such as polarization demultiplexing and PMD compensation.

According to Equation 6.18, the effect of CD on an optical signal propagating in fibers can be modeled as

$$\frac{\partial \vec{E}}{\partial z} = -\frac{j}{2}\beta_2 \frac{\partial^2 \vec{E}}{\partial t^2} \tag{6.28}$$

The conventional approach is to solve this equation in the frequency domain. By taking Fourier transform of Equation 6.28, one can obtain the frequency-domain transfer function of CD as

$$F_{\mathrm{CD}}(z, \omega) = \exp\left(\frac{j}{2}\beta_2 \omega^2 z\right) \tag{6.29}$$

To compensate for CD effect, CD equalizer in a coherent receiver has to provide a transfer function opposite to Equation 6.29. There are two approaches to realize CD compensation in a coherent receiver. One is to use finite impulse response (FIR) filters. The CD-induced impulse response can be obtained by inverse Fourier transform of Equation 6.29, which is used to set the tap coefficients of the FIR filters [65, 66]. The other approach is to use frequency-domain equalizers [67]. For a short impulse response, the time-domain FIR filter approach is preferred, but for systems with large CD, the frequency-domain equalizers are more efficient.

Due to the existence of two polarizations, the dynamic equalizer has a butterfly structure, which consists of four subequalizers. Each subequalizer can be an FIR filter, as shown in Figure 6.7, where τ is the sampling interval, which is half the symbol period for two times oversampling. The butterfly equalizer is mainly used for polarization demultiplexing, PMD compensation, and PDL mitigation. Apart from these,

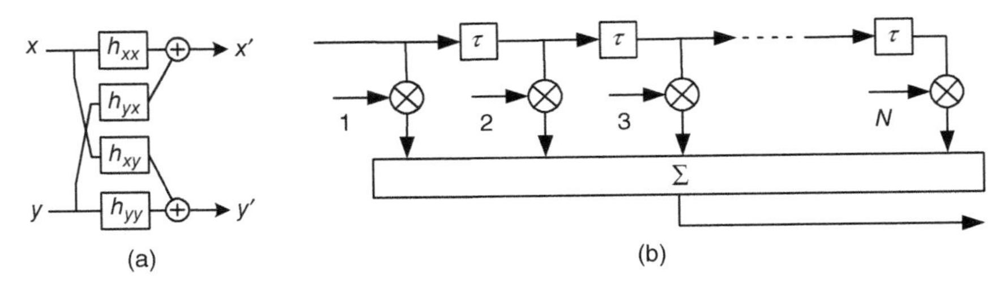

FIGURE 6.7 The structure of a butterfly equalizer (a) and FIR filter for each subequalizer (b).

the butterfly equalizer also mitigates ISI caused by other factors such as residual CD after the CD compensator, filtering, or nonlinearities in a system.

The butterfly equalizer performs multi-input-multi-output (MIMO) processing. The output of the equalizer is given by

$$x' = \vec{h}_{xx} \cdot \vec{x}^T + \vec{h}_{xy} \cdot \vec{y}^T \tag{6.30a}$$

$$y' = \vec{h}_{yx} \cdot \vec{x}^T + \vec{h}_{yy} \cdot \vec{y}^T \tag{6.30b}$$

where $\vec{h}_{xx}$, $\vec{h}_{xy}$, $\vec{h}_{yx}$, and $\vec{h}_{yy}$ are coefficient vectors for the four adaptive FIR filters, each of which has a length N taps, as shown in Figure 6.7(b). $\vec{h}_{xx} = (h_{xx}^1, h_{xx}^2, \ldots, h_{xx}^N)$ is an FIR filter coefficient vector, and $\vec{x} = (x^1, x^2, \ldots, x^N)$ and $\vec{y} = (y^1, y^2, \ldots, y^N)$ are input signal vectors. Superscript T means transpose. The equalizer is dynamically adjusted according to some criteria, with the goal to achieve the polarization demultiplexing and minimum ISI for the output signal.

Many algorithms can be used to control the butterfly equalizer, including blind equalization algorithms and data-aided algorithm [68]. One widely used blind equalization algorithm is the constant modulus algorithm (CMA) proposed by Godard [69]. The CMA aims to minimize the mean square errors of the output signal deviation from a unit amplitude $\varepsilon_{x'}^2 = (1 - |x'|^2)^2$ and $\varepsilon_{y'}^2 = (1 - |y'|^2)^2$. Using stochastic gradient algorithm, the butterfly equalizer tap coefficients are recursively updated in the following way:

$$\vec{h}_{xx}(n+1) = \vec{h}_{xx}(n) - \mu \frac{\partial(\varepsilon_{x'}^2)}{\partial \vec{h}_{xx}} \tag{6.31a}$$

$$\vec{h}_{xy}(n+1) = \vec{h}_{xy}(n) - \mu \frac{\partial(\varepsilon_{x'}^2)}{\partial \vec{h}_{xy}} \tag{6.31b}$$

$$\vec{h}_{yx}(n+1) = \vec{h}_{yx}(n) - \mu \frac{\partial(\varepsilon_{y'}^2)}{\partial \vec{h}_{yx}} \tag{6.31c}$$

$$\overrightarrow{h}_{yy}(n+1) = \overrightarrow{h}_{yy}(n) - \mu \frac{\partial(\varepsilon_{y'}^2)}{\partial \overrightarrow{h}_{yy}} \tag{6.31d}$$

where u is a convergence parameter. After some algebraic manipulation, one can obtain the following update rule for CMA:

$$\overrightarrow{h}_{xx}(n+1) = \overrightarrow{h}_{xx}(n) + \mu\varepsilon_{x'}\overrightarrow{x}^* x' \tag{6.32a}$$

$$\overrightarrow{h}_{xy}(n+1) = \overrightarrow{h}_{xy}(n) + \mu\varepsilon_{x'}\overrightarrow{y}^* x' \tag{6.32b}$$

$$\overrightarrow{h}_{yx}(n+1) = \overrightarrow{h}_{yx}(n) + \mu\varepsilon_{y'}\overrightarrow{x}^* y' \tag{6.32c}$$

$$\overrightarrow{h}_{yy}(n+1) = \overrightarrow{h}_{yy}(n) + \mu\varepsilon_{y'}\overrightarrow{y}^* y' \tag{6.32d}$$

where the superscript * means complex conjugate.

CMA works well for quadrature-phase-shift-keying (QPSK) signals, and it can also be used for initial convergence for higher-order modulation formats such as 16-ary quadrature-amplitude modulation (16QAM) and 64QAM. For these higher-order modulation formats, after initial convergence, some other control algorithms such as a multimodulus algorithm (MMA) or a decision-directed least mean square (DD-LMS) algorithm can be used to finely tune the equalizer for further performance improvement [70]. For both MMA and DD-LMS, the butterfly equalizer coefficients are updated with the same rule as Equations 6.32a–6.32d describe, the only difference is that the error functions $\varepsilon_{x'}$ and $\varepsilon_{y'}$ are different. For MMA, $\varepsilon_{x'} = R_{ix} - |x'|^2$ and $\varepsilon_{y'} = R_{iy} - |y'|^2$, where R_{ix} and R_{iy} are the moduli closest to $|x'|^2$ and $|y'|^2$, respectively. For 16QAM, there are three moduli. For DD-LMS, $\varepsilon_{x'} = d_x - x'$ and $\varepsilon_{y'} = d_y - y'$ with d_x and d_y being the symbols that are closest to x' and y', respectively.

6.6 PMD AND PDL IMPAIRMENTS IN COHERENT SYSTEMS

In principle, the dynamic butterfly equalizer in a digital coherent receiver can generate a Jones matrix inverse to that of the optical channel and all signal distortions induced by PMD and PDL can be completely compensated. However, PMD and PDL can still degrade the performance of a digital coherent optical communication system due to the following reasons. First, in a real system, the complexity of the butterfly equalizer is limited and it is challenging to implement a butterfly equalizer with a long length of taps; second, the butterfly is not only used for PMD compensation but mitigates any other ISI as well, irrespective of its origin; third, PDL causes signal power and OSNR fluctuations, which result in SNR variations for the received electrical signal, and degraded SNR cannot be brought back in the receiver. Therefore, it is important for a system designer to understand the PMD and PDL impairments in a digital coherent system and know the amount of PMD and PDL that a system can tolerate. In this section, the PMD- and PDL-induced impairments in a 112-Gb/s

PDM-QPSK coherent system are presented. Note that the results can be extended to other modulation formats.

6.6.1 PMD Impairment

For single-polarization (SP) signals, PMD penalties are caused by ISI [71], but for PDM signals, PMD penalties are mainly caused by crosstalk between two polarization tributaries [72, 73]. It has been shown that, in a direct-detection system in the absence of PMD compensation, PMD impairments can be well evaluated with only first-order PMD, but in a system with PMD compensation, higher-order PMD has to be considered [71, 74, 75]. As a coherent receiver in general has a PMD equalizer, to have a clear picture of PMD impairments in a coherent optical communication system, not only first-order, but also second-order and all-order PMDs have to be considered.

Here, a 112-Gb/s PDM-QPSK coherent optical communication system is used as an example to show the PMD impairments. As shown in Figure 6.8, a transmitter generates a 112-Gb/s non-return-to-zero (NRZ) PDM-QPSK signal. A polarization controller is used to adjust the input SOP of the signal to a PMD emulator (PMDE), the PMDE can be a first-, second-, or all-order PMDEs. ASE noise is loaded to a coherent receiver to generate required OSNR. In the coherent receiver, two times oversampling is used and the butterfly equalizer is adjusted with CMA. Bit error ratios (BERs) are calculated using a direct error counting method.

6.6.1.1 First- and Second-Order PMD Impairments To assess first- and second-order PMD-induced penalties in the coherent system, the PMDE is built in the following way: First-order PMD is in S_1 direction and depolarization is in S_2 direction in Stokes space. The Jones matrix including first- and second-order PMD can be expressed as [75, 76]

$$U = \begin{pmatrix} \exp\left(-j\frac{\phi}{2}\right) & -\frac{\hat{p}_\omega \Delta\omega}{2}\sin\left(\frac{\phi}{2}\right) \\ \frac{\hat{p}_\omega \Delta\omega}{2}\sin\left(\frac{\phi}{2}\right) & \exp\left(j\frac{\phi}{2}\right) \end{pmatrix} \tag{6.33}$$

where $\phi = \Delta\tau\,\Delta\omega + \Delta\tau_\omega \Delta\omega^2/2$, including first-order PMD and PCD term, and $\hat{p}_\omega$ is depolarization term. Equation 6.33 shows that depolarization causes linear changes of the Jones matrix with frequency and PCD induces phase changes similar to CD.

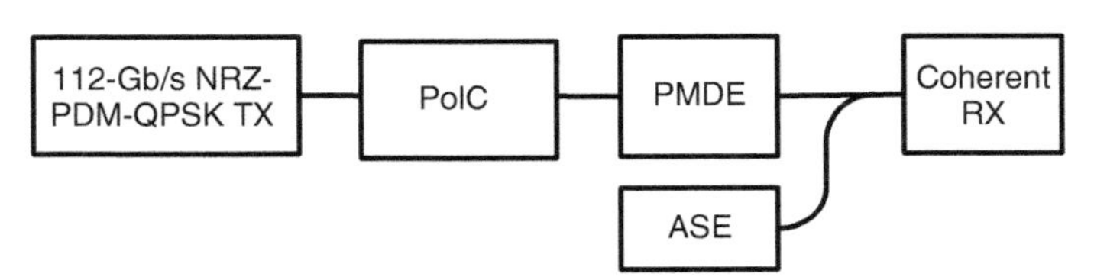

FIGURE 6.8 System model to evaluate PMD impairments. PolC: Polarization controller, PMDE: PMD emulator.

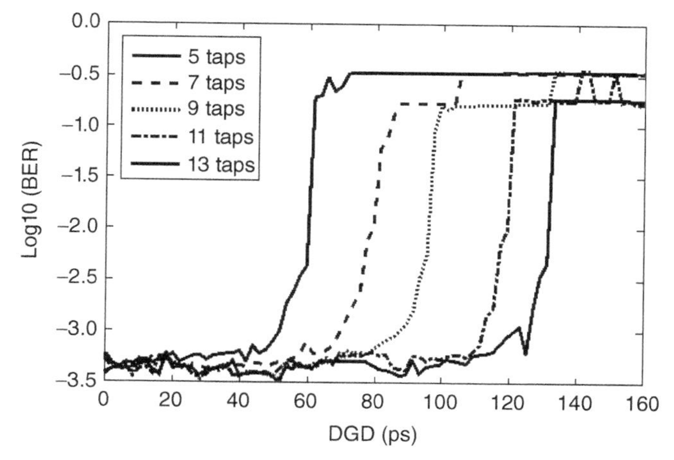

FIGURE 6.9 BER vs. DGD in the worst case at 14.9-dB OSNR for the 112-Gb/s NRZ-PDM-QPSK coherent system with the butterfly equalizer of different number of taps.

Figure 6.9 shows first-order PMD-induced impairments in the 112-Gb/s PDM-QPSK coherent system [14], which depicts the dependence of BER on DGD with the butterfly equalizer in the receiver having different numbers of taps. In the figure, the received OSNR is fixed at 14.9 dB, which is 0.5-dB higher than the required OSNR at BER $= 10^{-3}$, and the input SOP of the signal (one polarization of the PDM signal) is set at 45° to the PSP of the PMDE.

In a direct-detection system using SP signals, the PMD penalties are caused by pulse-broadening-induced ISI and typically gradually increase with the increasing DGD. Figure 6.9 shows that for a coherent system using PDM signals, PMD penalties have some different features [14]. There is a DGD threshold for the impairments. When DGD is less than the threshold, there is little penalty from PMD, but when DGD is larger than the threshold, BER increases sharply. The reason is that the impairments are mainly caused by crosstalk between the two polarizations, and once DGD is larger than the threshold that the equalizer can handle, PMD-induced crosstalk rapidly degrades the system performance. Note that the "threshold" is dependent on the length of the butterfly equalizer and signal launch SOP. A larger "threshold" is anticipated for the equalizer with a larger number of taps, as shown in Figure 6.9, and the signal with SOP not aligned in the worst case. This indicates that unlike that in a direct-detection system, a larger OSNR margin may not significantly increase the system tolerance to PMD in a PDM coherent system.

The effect of PCD and depolarization on the performance of the coherent system is given in Figure 6.10 [14]. In this figure, 7-tap butterfly equalizer is used. The penalty caused by PCD does not depend on the input SOP of the signal and is similar to that caused by CD of half the value, as indicated in Equation 6.33. For example, the penalty induced by 1500-ps^2 PCD is similar to that caused by 750-ps^2 CD. Figure 6.10(b) shows that depolarization has little impact on the coherent system.

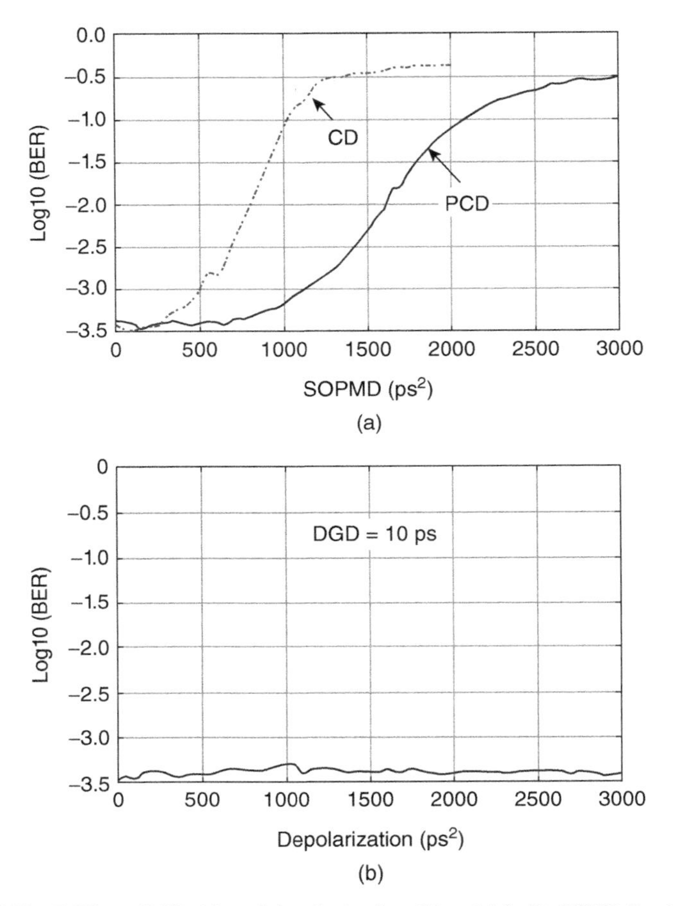

FIGURE 6.10 BER vs. PCD (a) and depolarization (b) at 14.9-dB OSNR for the 112-Gb/s NRZ-PDM-QPSK coherent system with a 7-tap butterfly equalizer. DGD is fixed at 0 and 10 ps in (a) and (b), respectively. From Ref. [14].

This can be explained by Equation 6.31. Depolarization generates a linear change of the Jones matrix with frequency, and FIR filter equalizers can compensate for linear effects efficiently, independent of the filter length. It is also found that the signal SOP has little impact on the results in Figure 6.10(b). As the average PCD is only about 1/9 the average second-order PMD [77], Figure 6.10 indicates that using PCD to emulate second-order PMD effects could significantly overestimate PMD impairments.

6.6.1.2 PMD-Induced Outage Probabilities As PMD is a stochastic phenomenon, PMD penalties at specific PMD values and launch SOPs are not sufficient to quantify PMD impairments in an optical communication system and PMD-induced outage probabilities (OPs), the probabilities that a system is out of service caused by PMD, are often used by system designers and engineers. When designing an optical communications system, a certain OSNR margin, which is an additional OSNR on top of the required OSNR for the system to work properly, is allocated to PMD. When the PMD-induced penalty is larger than the margin, an outage occurs. PMD tolerance is typically quantified at an OP of 10^{-5}, which means that if actual PMD value in a system is less than the PMD value that the system can tolerate, PMD will not cause larger than 10^{-5} OPs.

To accurately evaluate PMD-induced OPs, one needs to choose an appropriate PMDE. Figure 6.11 depicts the PMD-induced OPs in the 112-Gb/s NRZ-PDM-QPSK coherent system with three different PMDEs, a first-order PMDE, a PMDE including first-order PMD and PCD, and an all-order PMDE [14]. The all-order PMDE is the concatenation of 100 waveplates, and the first- and second-order PMDEs are built based on Equation 6.33. In order to have a fair comparison among the three PMDEs, first a Monte Carlo simulation for the all-order PMDE is performed, and for each realization of PMD, the instantaneous PMD values including both first- and second-order PMD are obtained. Then, the parameters of the first- and second-order PMDEs based on Equation 6.33 are adapted such that they generate the same amount of first- and second-order PMD values and same PSPs. An OSNR margin of 0.5 dB is set at BER of 10^{-3} in Figure 6.11. The figure shows that only considering first-order PMD underestimates PMD impairments, and the PMD tolerance will be overestimated by 10–20%. As stated earlier, modeling second-order PMD as PCD significantly overestimates PMD impairments. If one considers first-order PMD and takes second-order PMD as PCD, the tolerable PMD will be underestimated by about 30%.

One way to increase the PMD tolerance of a coherent system is to increase the length of the butterfly equalizer, as indicated in Figure 6.12 [14]. Figure 6.12(a) plots

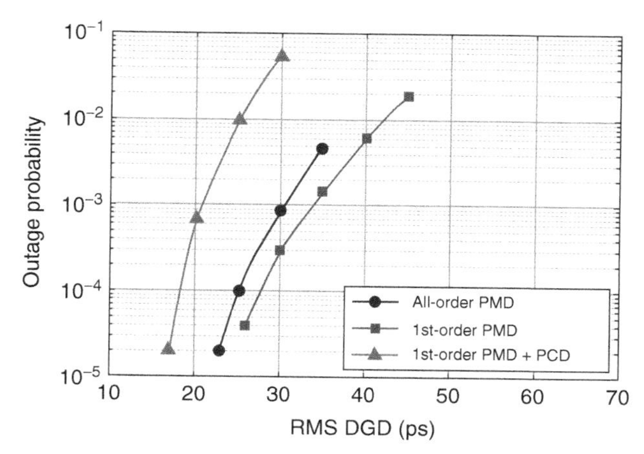

FIGURE 6.11 PMD-induced OPs at a 0.5-dB OSNR margin for the 112-Gb/s NRZ-PDM-QPSK system with a 7-tap butterfly equalizer for different PMDEs. From Ref. [14].

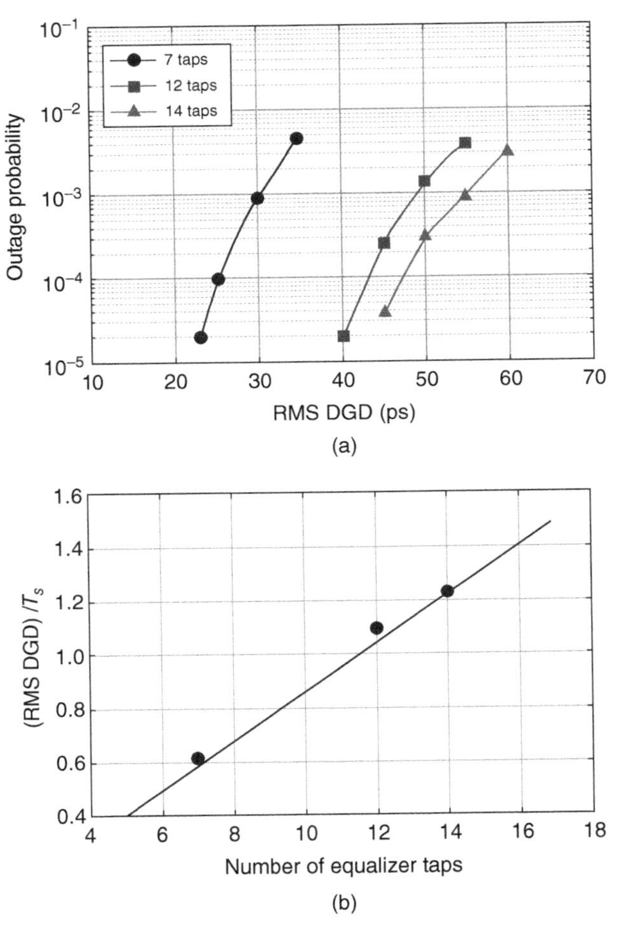

FIGURE 6.12 (a) PMD-induced OPs at for the 112-Gb/s NRZ-PDM-QPSK system with a butterfly equalizer of different tap numbers; (b) Dependence of normalized tolerable PMD of the NRZ-PDM-QPSK system at an OP of 10^{-5}. All-order PMDE and a 0.5-dB OSNR margin are used. From Ref. [14].

the PMD-induced OPs in the 112-Gb/s NRZ-PDM-QPSK system with a butterfly equalizer of three different numbers of taps in the coherent receiver. It shows that increasing the butterfly equalizer complexity increases the PMD tolerance of the system, as expected. In Figure 6.12(b), the dependence of the tolerable PMD in the PDM-QPSK system, measured in RMS DGD normalized with the symbol period, on the length of butterfly equalizer filter is given. It shows the tolerable PMD increases linearly with the number of the butterfly equalizer taps. With a 12-tap butterfly equalizer, larger than symbol period PMD can be tolerated in the system.

6.6.1.3 Joint Optimization of Static and Dynamic Equalizers The above-described discussions on PMD impairments assume that the dynamic butterfly equalizer in a coherent receiver is solely used for PMD compensation, but in fact the equalizer is also used to counteract other ISI impairments such as residual CD, tight optical filtering, transmitter and receiver hardware bandwidth limitations, and fiber nonlinearities. This is because that most control algorithms used to adjust the equalizer filter coefficients such as CMA and DD-LMS do not differentiate between physical sources of ISI. Therefore, any ISI impairments in the system consume the resource of the butterfly equalizer and can reduce the capability of the butterfly equalizer to compensate PMD.

Figure 6.13 gives an example of PMD tolerance degradation caused by residual CD, where the residual CD is the CD after the CD equalizer due to inaccurate CD estimation, resulting in a slight CD compensation error [15]. In the figure, an all-order PMDE composed of 100 concatenated waveplates is used, and a 0.5-dB OSNR margin at a BER of 10^{-3} is allocated to PMD. It shows that the PMD tolerance of the coherent system can be significantly reduced if CD in the system is not accurately compensated by the static CD equalizer in the receiver. For the 112-Gb/s PDM-QPSK coherent system with a 7-tap butterfly equalizer in the receiver, a 300-ps/nm residual CD can more than halve the PMD tolerance of the system, from about 22 ps to less than 10 ps at an OP of 10^{-5}.

One way to increase the PMD tolerance of a coherent receiver in the presence of other ISI impairments is to increase the length of the butterfly equalizer, but this increases the complexity and power consumption of the DSP. Another technique is to jointly optimize the CD and butterfly equalizers. The block diagram of the technique is shown in Figure 6.14 [15]. As most ISI impairments such as residual CD and filtering are the same for x and y polarizations, one can monitor the

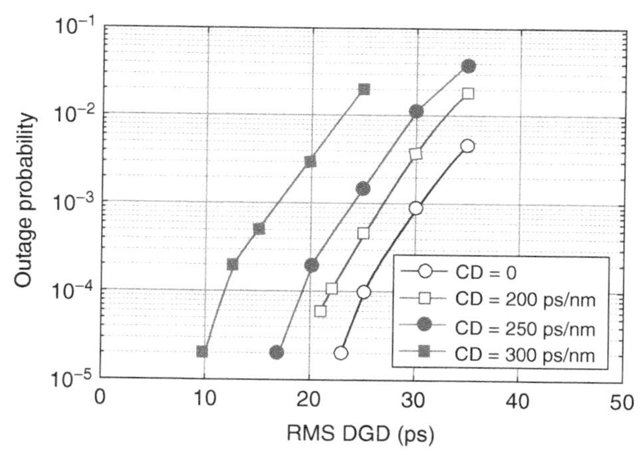

FIGURE 6.13 PMD-induced OPs for the 112-Gb/s PDM-QPSK system with a 7-tap butterfly equalizer and different values of residual CD. A 0.5-dB OSNR margin is used. From Ref. [15]. Reproduced with permission of OSA.

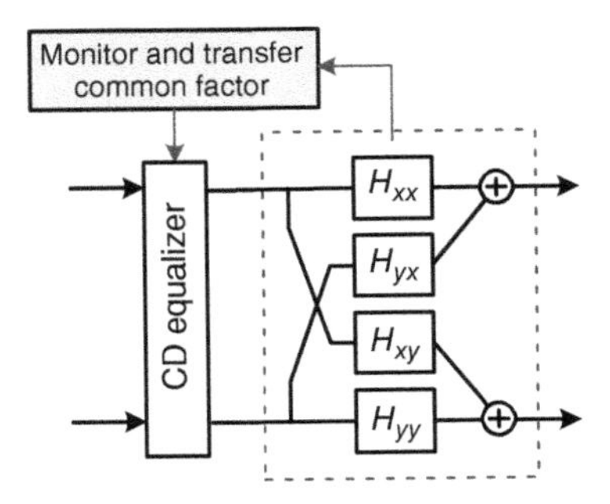

FIGURE 6.14 Block diagram of joint optimization of CD and butterfly equalizers. From Ref. [15]. Reproduced with permission of OSA.

common factor of the four subequalizers of the butterfly equalizer and move it into the CD equalizer. By doing this, the butterfly equalizer is only used to compensate for polarization-dependent impairments such as PMD and PDL, and the PMD compensation capability of the butterfly equalizer is almost fully restored. Since the CD equalizer is much larger than the butterfly equalizer, moving some functions of the butterfly equalizer to the CD equalizer has little impact on the ability of the CD equalizer to compensate for CD.

There are two methods to extract the common factor from the butterfly equalizer [15]. The first method is to use the transfer function of any of the four subequalizers as the common factor. For example, if H_{xx} is used for the common factor, one can write the transfer function of the butterfly equalizer as

$$H_{\mathrm{be}}(f) = \begin{pmatrix} H_{xx}(f) & H_{xy}(f) \\ H_{yx}(f) & H_{yy}(f) \end{pmatrix} = H_{xx}(f) \begin{pmatrix} 1 & H_{xy}(f)/H_{xx}(f) \\ H_{yx}(f)/H_{xx}(f) & H_{yy}(f)/H_{xx}(f) \end{pmatrix} \tag{6.34}$$

The second method assumes that the butterfly equalizer is only used to compensate PMD, and thus its transfer function can be treated as a unitary matrix. One then has

$$H_{\mathrm{be}}(f) = \begin{pmatrix} H_{xx}(f) & H_{xy}(f) \\ H_{yx}(f) & H_{yy}(f) \end{pmatrix} = H_0(f) \begin{pmatrix} u(f) & v(f) \\ -v^*(f) & u^*(f) \end{pmatrix} \tag{6.35}$$

where $H_0(f) = \sqrt{H_{xx}(f) * H_{yy}(f) - H_{xy}(f) * H_{yx}(f)}$ is the common factor. In a real implementation, there is no need to do division operation. One only needs to update the CD equalizer with the common factor and reoptimize the butterfly equalizer with its control algorithms.

6.6.2 PDL Impairment

6.6.2.1 PDL Effects on SP Signals For SP signals, PDL impairments manifest as the fluctuations of signal power and OSNR. Consider an SP signal with an arbitrary polarization state and two noise modes passing through a PDL element, as shown in Figure 6.15. The PDL value of the PDL element in dB is $\Gamma = 10\log 10[(1+\alpha)/$

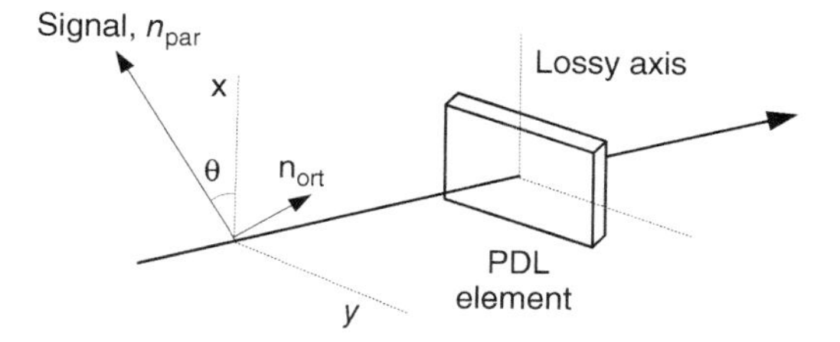

FIGURE 6.15 Schematic diagram of an SP signal and noise with a PDL element.

$(1-\alpha)]$, and its lossy axis is x-axis. The input signal and the noise fields can be described as [50]

$$\vec{s}_{\text{in}}(t) = \sqrt{P_{\text{in}}(t)}(\hat{x}\cos\theta + \hat{y}\sin\theta) \tag{6.36a}$$

$$\vec{n}_{\text{in}}^{\text{par}}(t) = \sqrt{N_{\text{in}}^{\text{par}}(t)}(\hat{x}\cos\theta + \hat{y}\sin\theta) \tag{6.36b}$$

$$\vec{n}_{\text{in}}^{\text{ort}}(t) = \sqrt{N_{\text{in}}^{\text{ort}}(t)}(\hat{x}\sin\theta - \hat{y}\cos\theta) \tag{6.36c}$$

where $\vec{s}_{\text{in}}(t)$, $\vec{n}_{\text{in}}^{\text{par}}(t)$, and $\vec{n}_{\text{in}}^{\text{ort}}(t)$ are the signal field, the noise field polarized parallel to the signal, and the noise field polarized orthogonally to the signal, respectively, $\hat{x}$ and $\hat{y}$ are unit vectors, and $P_{\text{in}}(t)$, $N_{\text{in}}^{\text{par}}(t)$, and $N_{\text{in}}^{\text{ort}}(t)$ are powers of signal, parallel noise, and orthogonal noise, respectively. After passing through the PDL element, the output signal and noise fields are

$$\vec{s}_{\text{out}}(t) = \sqrt{P_{\text{in}}(t)}\sqrt{1-\alpha\cos 2\theta}\,\hat{p} \tag{6.37a}$$

$$\vec{n}_{\text{out}}^{\text{par}}(t) = \sqrt{N_{\text{in}}^{\text{par}}(t)}\sqrt{1-\alpha\cos 2\theta}\,\hat{p} \tag{6.37b}$$

$$\vec{n}_{\text{out}}^{\text{ort}}(t) = \sqrt{N_{\text{in}}^{\text{ort}}}(a_p\hat{p} + a_q\hat{q}) \tag{6.37c}$$

where $\hat{p} = (\hat{x}\sqrt{1-\alpha}\cos\theta + \hat{y}\sqrt{1+\alpha}\sin\theta)/\sqrt{1-\alpha\cos 2\theta}$ is the polarization direction of the signal after the PDL element, and $\hat{q}$ is the polarization direction that is orthogonal to the signal after the PDL element, $a_p = -\alpha\sin 2\theta/\sqrt{1-\alpha\cos 2\theta}$ and $a_q = \sqrt{1-\alpha^2}/\sqrt{1-\alpha\cos 2\theta}$. The signal and noise powers after the PDL element are

$$P_{\text{out}}(t) = P_{\text{in}}(t)(1-\alpha\cos 2\theta) \tag{6.38a}$$

$$N_{\text{out}}^{\text{par}}(t) = N_{\text{in}}^{\text{par}}(t)(1-\alpha\cos 2\theta) + N_{\text{in}}^{\text{ort}}(t)\frac{\alpha^2\sin^2 2\theta}{1-\alpha\cos 2\theta} \tag{6.38b}$$

$$N_{\text{out}}^{\text{ort}}(t) = N_{\text{in}}^{\text{ort}}(t)\frac{1-\alpha^2}{1-\alpha\cos 2\theta} \tag{6.38c}$$

This analysis shows that PDL tends to change the polarization state of the signal and causes the orthogonal noise to couple to the parallel noise.

*6.6.2.2 **PDL Effects on PDM Signals*** Except for the power and OSNR variations, PDL has two additional effects on a PDM signal. First, PDL induces nonorthogonality between two originally orthogonal polarizations for a PDM signal, and second, PDL causes power imbalance between two polarization components of a PDL signal, as illustrated in Figures 6.16(a) and (b), respectively [73].

Using A and B to denote the two components of a PDM signal, the normalized input signal can be expressed as

$$\begin{cases} \vec{A}_{\text{in}} = \sin\theta\hat{x} + \cos\theta\hat{y} \\ \vec{B}_{\text{in}} = -\cos\theta\hat{x} + \sin\theta\hat{y} \end{cases} \tag{6.39}$$

where θ is the angle between the SOP of $\vec{A}_{\text{in}}$ and the axis of the PDL element in Jones space. After the PDL element with $\hat{x}$ being lossy axis, the normalized output signal is

$$\begin{cases} \vec{A}_{\text{out}} = \sin\theta\sqrt{1-\alpha}\hat{x} + \cos\theta\sqrt{1+\alpha}\hat{y} \\ \vec{B}_{\text{out}} = -\cos\theta\sqrt{1-\alpha}\hat{x} + \sin\theta\sqrt{1+\alpha}\hat{y} \end{cases} \tag{6.40}$$

where is α related to PDL value in dB as we stated above. From Equation 6.40, one can derive the angle between the components of the output as [73]

$$\gamma = \arctan\left(10^{-\frac{\Gamma}{20}}\tan\theta\right) + \arctan\left(10^{-\frac{\Gamma}{20}}\cot\theta\right) \tag{6.41}$$

The change of output angle with input SOP and PDL value is plotted in Figure 6.17. It shows that the output angle decreases with the increase of PDL value and changes with the launch SOP. The largest nonorthogonality (minimum γ) is reached when the launched SOP is at 45°.

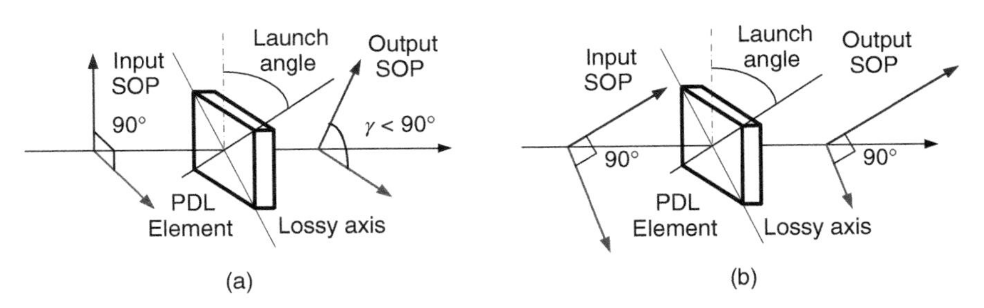

FIGURE 6.16 Schematic diagram of a PDM signal with a PDL element. The input SOP is 45° (a) and 0° (b). From Ref. [73]. Reproduced with permission of OSA.

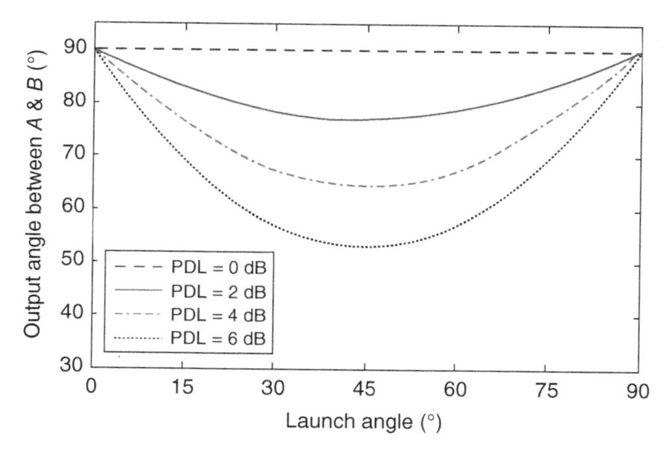

FIGURE 6.17 Output angle versus the launch SOP for different PDL values. From Ref. [73]. Reproduced with permission of OSA.

6.6.2.3 PDL-Induced Penalties in PDM Coherent-Detection Systems Unlike CD and PMD, PDL effects cannot be well compensated in a coherent receiver due to non-unitary nature of PDL [16–18]. As shown earlier, PDL causes signal power and OSNR fluctuations and repolarizes ASE noise. In addition, it induces loss of orthogonality and power/OSNR imbalance between the two polarizations of a PDM signal. Although the nonorthogonality between the two polarizations can be corrected by electronic equalizers, some penalties will be introduced. When the polarization of a PDM signal is aligned with the PDL axis at the input, one polarization is improved, but the other polarization is degraded and the overall performance is mainly determined by the degraded polarization tributary.

There are two PDL models to study PDL effects in a coherent system. One is a lumped model and the other is a distributed model, as shown in Figure 6.18. In the lumped model, there is one PDL emulator. ASE noise is loaded at the receiver and a polarization controller or polarization scrambler is inserted before the PDL emulator in the lumped model to get the PDL penalty at a particular input SOP or an average PDL penalty. For a PDM signal, the worst and best performance degradations occur when a PDM signal is 0° and 45° aligned to the axes of a PDL element, respectively. In the worst case, the performance of one polarization tributary

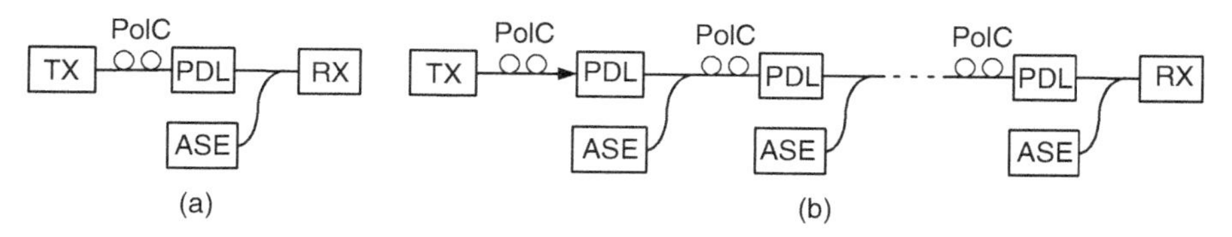

FIGURE 6.18 (a) lumped PDL model, (b) distributed PDL model.

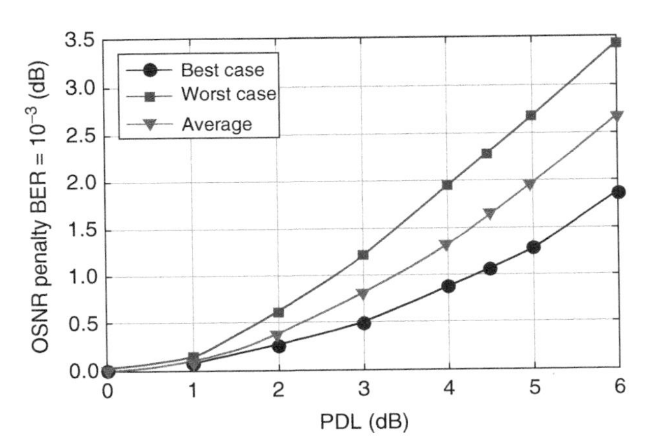

FIGURE 6.19 PDL-induced OSNR penalty versus instant PDL value at BER = 10^{-3} for a 112-Gb/s PDM-QPSK coherent detection system in different cases using the lumped model. From Ref. [18].

is degraded while that of the other tributary is improved; therefore, the overall performance degradation induced by PDL in the worst case for a PDM signal is smaller than that for an SP signal. In the best case, PDL induces the largest nonorthogonality between the two polarization tributaries and the two tributaries have the same PDL penalty.

Figure 6.19 gives PDL induced at BER = 10^{-3} in the worst and best cases for a 112-Gb/s PDM-QPSK coherent system [18]. The average OSNR penalty is calculated assuming that a polarization scrambler generates a uniform SOP distribution on the Poincaré sphere. There is a large difference of the penalties among different cases. With 4-dB PDL value, the OSNR penalty is about 0.8 dB and 2.0 dB in the best and worst cases, respectively, while the average penalty is about 1.3 dB.

The lumped model, usually used in lab tests, is simple and helpful to understand some PDL effects, but it does not include all PDL effects such as repolarization of ASE noise. In the distributed model, many PDL emulators and ASE noise sources are distributed along a link, with random polarization rotations between the PDL emulators. The distributed model is similar to a real system and automatically takes into account all PDL effects. Figure 6.20 depicts the probability distribution of OSNR variations in one polarization calculated with the lumped model and distributed model at 3-dB RMS PDL value [18]. For the distributed model, 20 PDL elements are used, and PDL and ASE noise are equally distributed along the link. These results show that the lumped model generates much larger OSNR variations than the distributed model, which means that the lumped model will significantly overestimate the PDL penalties in an optical communication system.

As PDL is a statistical phenomenon similar to PMD, its impact on an optical communication system also needs to be quantified statistically. Figure 6.21 gives PDL-induced OPs at BER = 10^{-3} in a 112-Gb/s NRZ-PDM-QPSK system with

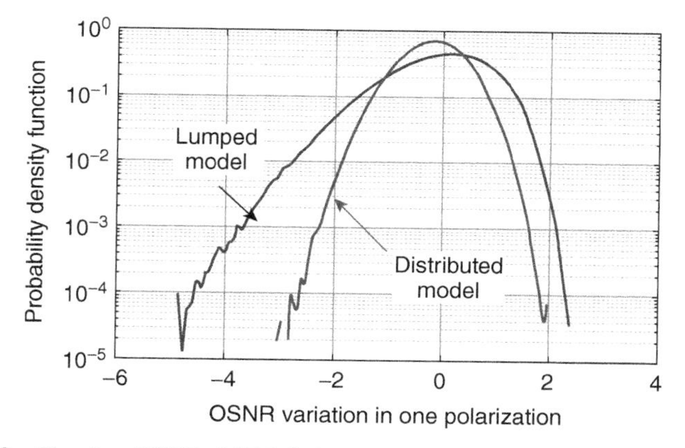

FIGURE 6.20 Simulated PDF of PDL-induced OSNR variations in one polarization using the lumped model and the distributed model at the RMS PDL value of 3 dB. From Ref. [18].

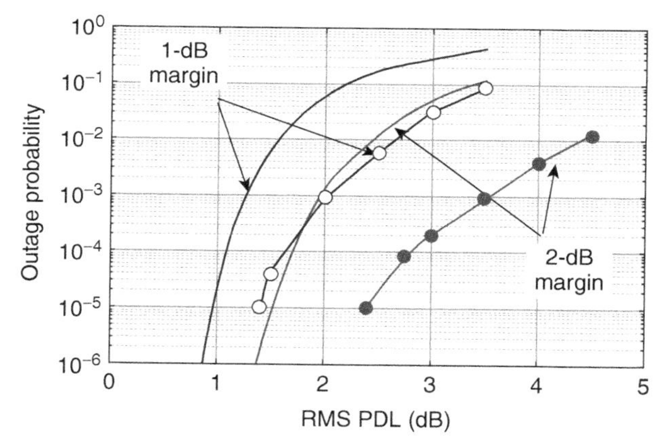

FIGURE 6.21 PDL-induced OPs at BER = 10^{-3} versus RMS PDL in a 112-Gb/s PDM-QPSK system using the distributed model (lines with symbols) and lumped model (lines without symbols). From Ref. [18].

1- and 2-dB OSNR margins. The results of the distributed model are obtained with Monte Carlo simulations and the lumped model results are obtained with the following equation [18]:

$$\mathrm{OP} = \int_0^\infty \int_0^\pi I(\mathrm{PDL}, \theta) \cdot f(\mathrm{PDL}) \cdot f(\theta) \cdot d\theta \cdot d\mathrm{PDL} \tag{6.42}$$

where $f(\mathrm{PDL})$ is the PDF of PDL, which is a Maxwellian distribution. Assuming uniform distribution of SOP over the Poincaré sphere, one has $f(\theta) = \sin\theta/2$,

$0 < \theta \leq \pi$. $I(\mathrm{PDL}, \theta)$ is an outage function

$$I(\mathrm{PDL}, \theta) = \begin{cases} 1, & \mathrm{BER}\,(\mathrm{PDL}, \theta) > \mathrm{BER}_0 \\ 0, & \text{else} \end{cases} \tag{6.43}$$

where BER_0 is the designated BER threshold. Figure 6.21 clearly shows that the lumped model significantly overestimates PDL penalties.

6.7 NONLINEAR IMPAIRMENTS IN COHERENT SYSTEMS

Signals propagating in optical fibers experience rich nonlinear effects. In general, fiber nonlinear effects can be categorized into intra- and inter-channel nonlinearities. Intra-channel nonlinearities can be further categorized into SPM, intra-channel FWM (IFWM), and intra-channel XPM (IXPM) [35, 78]. Inter-channel nonlinearities include FWM, XPM, and XPolM. Depending on system parameters such as CD values, bit rates, and modulation formats, one or a few nonlinear effects are more dominant than the others. For example, in a system with a low bit rate and CD value, FWM is the dominant nonlinear effect [79, 80], and in a dispersion-managed homogeneous WDM PDM-QPSK system, XPolM can be the most detrimental nonlinear effect [20–25].

In most direct-detection systems, information is encoded in amplitude with OOK modulation and polarization is not used to carry any information. In such systems, it is the amplitude and timing jitter caused by inter- and intra-channel FWM and XPM that severely degrades system performance; phase noise and XPolM do not have significant impacts on the systems. Dispersion management [79–81], which distributes optical dispersion compensators such as dispersion compensation fibers (DCF) along a link with a certain dispersion map, can significantly reduce FWM- and XPM-induced amplitude and timing jitter and is an effective technology to mitigate nonlinearities in a direct-detection system.

With the advent of digital coherent detection, it is found that dispersion management, which has been successfully used in direct-detection optical communication systems to reduce fiber nonlinear impairments, becomes suboptimal in coherent systems [19–22]. Fiber nonlinearities are the same for signals propagating in fibers, regardless of direct detection or coherent detection, and the difference is caused by the way that information is carried. In a coherent-detection system, in addition to amplitude, information is also carried by phase, and PDM is generally used to double the spectral efficiency. As a result, phase and polarization distortions caused by XPM and XPolM, which are neglected in a direct-detection system, can cause severe impairments in a coherent-detection system [23–25].

In a coherent system without any optical dispersion compensators, as signals are rapidly spread in time due to large accumulated CD, the dominant nonlinear effects are intra-channel nonlinearities such as IFWM and IXPM. For dispersion-compensated coherent optical communication systems with inline optical dispersion compensators, the dominant nonlinear effects are inter-channel nonlinearities, including inter-channel XPolM and inter-channel XPM, but whether XPolM or XPM is dominant depends on the actual system configurations. Nonlinearities in non-dispersion-compensated systems are covered in another chapter in the book. This chapter focuses on nonlinearities in dispersion-managed coherent systems, including homogeneous PDM-QPSK, hybrid PDM-QPSK, and OOK, and homogeneous PDM-16QAM [24]. Fiber nonlinearities in these systems are described with numerical simulations.

6.7.1 System Model

The transmission system model is shown in Figure 6.22. The system has seven channels with a channel spacing of 50 GHz. Depending on the system to study, the transmitters can generate either 28-Gbaud QPSK, 28-Gbaud 16QAM, or 10-Gb/s OOK signals. The transmission line consists of 10 spans of standard single-mode fiber (SSMF) with a CD coefficient of 17.0 ps/(nm km), a nonlinear coefficient of 1.17 $(\text{km W})^{-1}$ and a loss coefficient of 0.21 dB/km. The span length is 100 km. Although it has been shown that more than 20 channels are needed to accurately assess the performance of a coherent system with fibers of low CD and at high nonlinear penalties [82, 83], seven channels are sufficient to show the difference in nonlinearities in these three systems. In the homogenous PDM-QPSK and in the hybrid PDM-QPSK and OOK systems, an erbium-doped-fiber amplifier (EDFA) after each span is used to compensate for the transmission loss, while in the homogeneous PDM-16QAM system, transmission loss is compensated for by hybrid Raman/EDFA to improve the delivered OSNR. Two different dispersion maps are studied and compared, one with legacy dispersion management supporting the 10-Gb/s OOK channels, and the another without optical dispersion compensators. In the dispersion-managed system, the CD in each span is compensated by DCF with a residual dispersion per span (RDPS) of 30 ps/nm and dispersion precompensation is optimized for coherent channels, which is about −400 ps/nm. The net residual CD after transmission is compensated in the electrical domain by DSP in the coherent receiver. The dispersion map for the dispersion-managed system used here is a typical map for a direct-detection optical communication system. In the system without any optical dispersion compensators, the CD is entirely compensated with electrical equalizers in the coherent receiver. The dotted line modules in Figure 6.22 are used in some system configurations.

The nonlinear transmission effects in these systems are studied using the coarse-step method, that is, numerically solving Equations 6.24 and 6.25 with

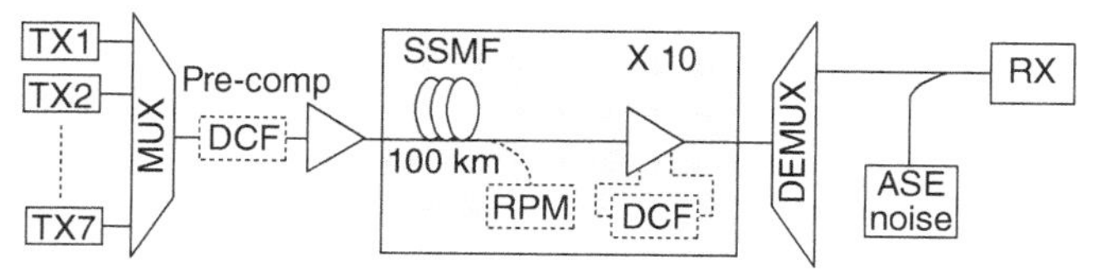

FIGURE 6.22 System model. TX: transmitter, RX: receiver, MUX: multiplexer, DEMUX: demultiplexer, RPM: Raman pump module. The modules in dotted lines are used for some configurations, as indicated in the text.

split-step Fourier method [59] and randomly rotating the field every coupling length, which is set as 500 m. Note that in the simulations, SOPs of all the channels are set in the same direction unless explicitly stated. It has been shown that system performance varies with the SOP changes of channels, but for systems with SSMF and penalties less than 2 dB, the performance variation is small [83].

6.7.2 Homogeneous PDM-QPSK System

For a homogeneous PDM-QPSK system, all the channels carry PDM-QPSK signals with the same bit rate. The coherent receiver has two times oversampling and the butterfly equalizer has 13 taps and is optimized with CMA. Carrier phase recovery is performed using the Viterbi-Viterbi phase estimation method with block length of 10 [84], and the BER is calculated with direct error counting. Differential coding and decoding is employed to avoid cycle-slip-induced error propagation [85]. In a homogeneous PDM-QPSK system with DCF, the amplitudes for all the channels are almost constant (neglect transient between symbols) and the same after each span and the dominant nonlinear effect is XPolM-induced nonlinear polarization scattering. Figure 6.23 shows the required OSNR at a BER of 10^{-3} after 1000-km transmission versus launch power per channel for both 42.8-Gb/s and 112-Gb/s NRZ-PDM-QPSK coherent systems with and without inline DCF. To separate the penalty caused by XPolM from that by XPM, we also plot the results of one 42.8-Gb/s and 112-Gb/s NRZ-PDM-QPSK channel surrounded by six 21.4-Gb/s and 56-Gb/s NRZ-SP-QPSK channels with the same symbol rate as the PDM-QPSK channel. As shown in Figure 6.24, the SOPs of NRZ-PDM-QPSK in different symbols are at S_2, S_3, $-S_2$, and $-S_3$ on the Poincaré sphere (in x and y polarizations in the Jones space). We set the SOPs of the six surrounding NRZ-SP-QPSK channels at S_1 (in x polarization in the Jones space). By doing this, at the same launch power, the average XPM on the center NRZ-PDM-QPSK channel from the surrounding NRZ-SP-QPSK and NRZ-PDM-QPSK channels are the same. As NRZ-SP-QPSK has constant amplitude (not considering the transient between symbols) and its SOP is the same in different symbols, XPolM from the surrounding NRZ-SP-QPSK channels cause little nonlinear polarization scattering in the middle PDM-QPSK channel. Therefore, there is almost no XPolM-induced penalty for the PDM-QPSK channel when it is surrounded by SP-QPSK channels, as shown in Figure 6.23.

Figure 6.23 shows that when all the channels carry NRZ-PDM-QPSK signals, the system with inline DCF performs worse than that without DCF [21]. For 42.8-Gb/s and 112-Gb/s NRZ-PDM-QPSK, the maximum launch powers at 1-dB OSNR penalty for the systems with DCF are about 2-dB and 1.5-dB lower than those without DCF, respectively. However, when the surrounding channels carry NRZ-SP-QPSK signals, the systems with inline DCF perform better than those without DCF. At 1-dB OSNR penalty, the maximum launch powers for the systems with DCF are about 2.0-dB and 1.0-dB higher than those without DCF for the 42.8-Gb/s and 112-Gb/s NRZ-PDM-QPSK, respectively. In addition, Figure 6.23 also shows that when the surrounding channels are changed from NRZ-SP-QPSK to NRZ-PDM-QPSK, the allowed launch power is reduced by about 3 dB and 2 dB for

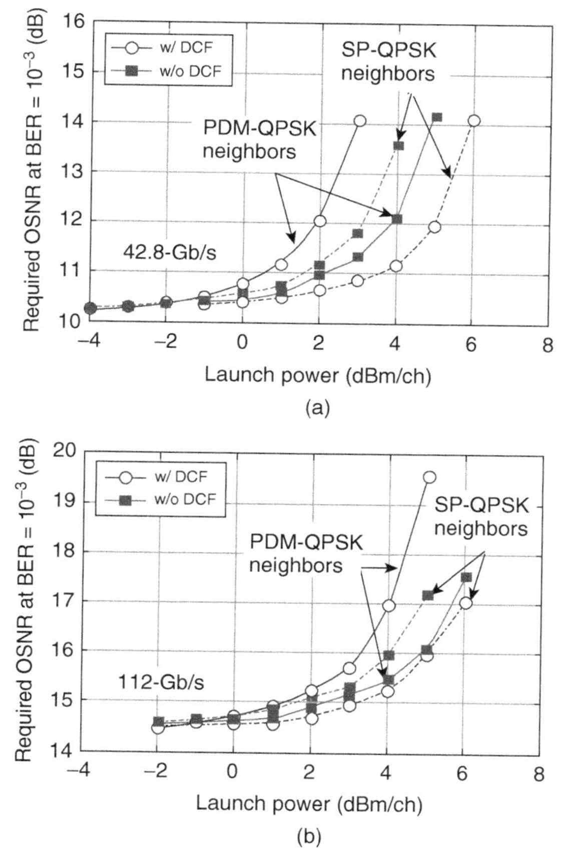

FIGURE 6.23 Required OSNR at a BER of 10^{-3} after 1000-km transmission versus launch power per channel for the 42.8-Gb/s (a) and 112-Gb/s NRZ-PDM-QPSK (b) coherent systems with and without inline DCF. Solid lines are for the results when all the channels are PDM-QPSK channels and dash-dotted lines are the results when the center PDM-QPSK channel is surrounded by six SP-QPSK channels. From Ref. [21].

the 42.8-Gb/s and 112-Gb/s systems with DCF, respectively, whereas it is increased by about 1 dB for both 42.8-Gb/s and 112-Gb/s systems without DCF. This indicates that XPolM is the dominant nonlinear effect in the homogeneous NRZ-PDM-QPSK system with DCF, and it is XPolM that makes homogeneous PDM-QPSK systems with DCF perform worse than those without DCF. The reason why the PDM-QPSK channels cause less interchannel penalty than SP-QPSK channels in the systems without DCF is because that the impact of the interchannel XPM is much larger than XPolM in the systems without DCF and the peak powers of PDM signals are smaller than those of SP signals for a given average power due to different data in the two polarizations for PDM signals. Note that the performance difference between the

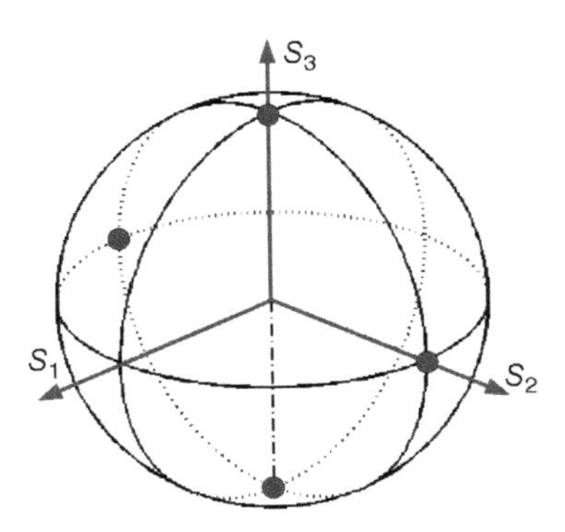

FIGURE 6.24 SOP of PDM-QPSK on the Poincaré sphere. The symbols on x and y polarizations are aligned.

112-Gb/s PDM-QPSK systems with DCF and without DCF is smaller than that between the 42.8-Gb/s systems because of the increased symbol rate.

This conclusion is further confirmed in Figure 6.25, which shows a reduction in the DOP of 21.4-Gb/s and 56-Gb/s SP-QPSK reference channels caused by XPolM-induced depolarization from six surrounding 42.8-Gb/s and 112-Gb/s PDM-QPSK channels [21]. For the systems with inline DCF, the DOP decreases more rapidly with the launch power than for those without DCF, indicating that the nonlinear polarization scattering is much larger in systems with DCF than without DCF. We also note that the depolarization in the 112-Gb/s PDM-QPSK systems is smaller than that in the 42.8-Gb/s systems due to the increase of the symbol rate.

One technique to suppress XPolM in a PDM-QPSK system is to use iRZ-PDM modulation format, which can reduce or eliminate the dependence of SOP on the data carried by the two polarizations [20, 21]. This modulation format uses RZ pulses and time interleaves the two polarizations by half a symbol period, as shown in Figure 6.26. One can see that at the center of each symbol, the SOP is either at S_1 or $-S_1$ on the Poincaré sphere, and it does not depend on data carried by the two polarizations. In addition, for an iRZ-PDM signal, the SOP at each symbol alternates between S_1 and $-S_1$ on the Poincaré sphere, which causes opposite nonlinear polarization rotation according to Equation 6.23, and its signal peak power is also reduced compared with that of a time-aligned signal, leading to reduced XPolM between channels.

The transmission performance of 42.8-Gb/s and 112-Gb/s iRZ-PDM-QPSK WDM systems is given in Figure 6.27. The RZ pulses have 50% duty cycle. For the 42.8-Gb/s system with inline DCF, using iRZ-PDM-QPSK can increase the allowed launch power by 7 dB at 1-dB OSNR penalty compared with NRZ-PDM-QPSK (Figure 6.23a), from about 1-dBm per channel launch power to about 8 dBm, and perform better than the system without DCF. For the 112-Gb/s system with DCF, the improvement obtained by using iRZ-PDM-QPSK is smaller than that for the 42.8-Gb/s system due to the symbol rate increase, but it can still increase the launch power tolerance by about 3 dB and achieve similar performance as the system without DCF.

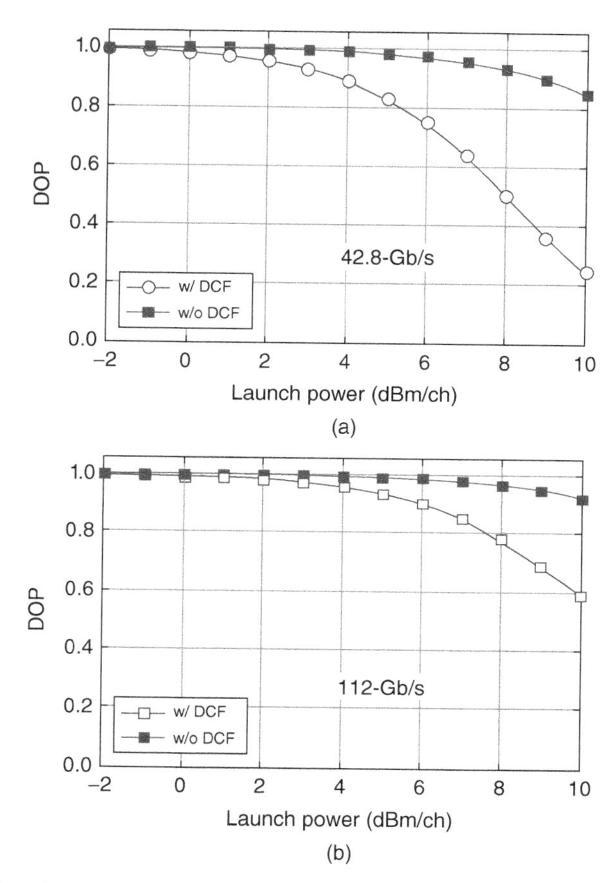

FIGURE 6.25 XPolM-induced depolarization in the 42.8-Gb/s (a) and 112-Gb/s (b) PDM-QPSK systems with and without inline DCF after 1000-km transmission. From Ref. [21].

6.7.3 Hybrid PDM-QPSK and 10-Gb/s OOK System

When upgrading 10-Gb/s OOK systems to 100-Gb/s PDM-QPSK and higher-bit rate QAM, PDM-QPSK and QAM channels may copropagate with 10-Gb/s OOK channels. In such hybrid systems, PDM-QPSK and QAM channels can be severely degraded by copropagating 10-Gb/s OOK-channels [37, 38]. In these systems, the penalty is mainly caused by interchannel XPM, not XPolM, because of the non-constant amplitude of 10-Gb/s OOK signals. Figure 6.28 gives the performance of 112-Gb/s PDM-QPSK in a hybrid system, where one 112-Gb/s NRZ-PDM-QPSK channel is surrounded by six 10-Gb/s NRZ-OOK channels at 50-GHz channel spacing [37]. The SOPs of the 10-Gb/s OOK channels are set at S_1 in Stokes space (x polarization in Jones space) to maximum XPolM effects [37, 54, 55], and the SOP of the PDM-QPSK channel is the same as that shown in Figure 6.24. As shown by Equation 6.23, the XPolM between two channels is the largest when their SOPs are

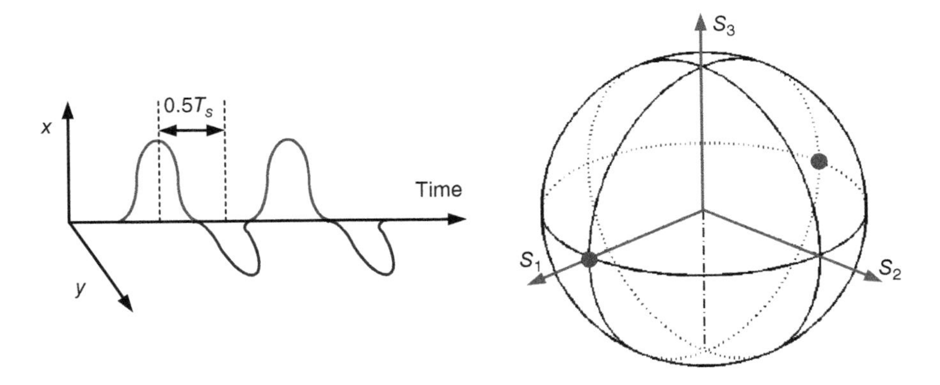

FIGURE 6.26 Waveform and SOP diagram of iRZ-PDM-QPSK. T_s: symbol period.

perpendicular to each other in Stokes space. For comparison, we also plot the performance of 112-Gb/s PDM-QPSK in a homogeneous PDM-QPSK system. Due to large interchannel nonlinearities from the OOK channels, the maximum launch powers for the systems with DCF and without DCF at 1-dB penalty are reduced by 5 and 3 dB, respectively.

Figure 6.29 depicts the DOP reduction of a 56-Gb/s SP-QPSK reference channel caused by XPolM-induced signal depolarization from the surrounding six 10-Gb/s OOK channels [37]. Comparing the result with that in Figure 6.25(b) (note the different scale of the x-axes) shows that XPolM is larger in the hybrid system than in the homogeneous system due to the lower symbol rate (and hence the slower waveform evolution) of the 10-Gb/s OOK channels. The fact that XPolM is not the main degrading factor in the hybrid system can be seen from Figure 6.28, where in the system with DCF at −1-dBm per-channel launch power, the OOK channels already induce more than a 3-dB penalty on the PDM-QPSK channel, while the depolarization caused by the 10-Gb/s OOK channels at this power level (Figure 6.29) is still very small (DOP above 0.98), which by itself would not cause any noticeable performance degradation for the 112-Gb/s PDM-QPSK channel. Along the same lines, we see from Figures 6.28 and 6.29 that the penalty caused by XPolM-induced depolarization in the hybrid system without DCF is also small.

XPM effects are further illustrated in Figure 6.30. No ASE noise and laser phase noise are added. The larger phase spread of the PDM-QPSK channel is caused by XPM from the neighboring 10-Gb/s OOK channels. As the OOK channels are aligned with the x-polarization, XPM on the x-polarization of the PDM-QPSK signal is twice that on the y-polarization.

6.7.4 Homogeneous PDM-16QAM System

To get sufficient delivered OSNR for PDM-16QAM transmission, hybrid Raman/EDFA amplification is used to compensate for the transmission loss,

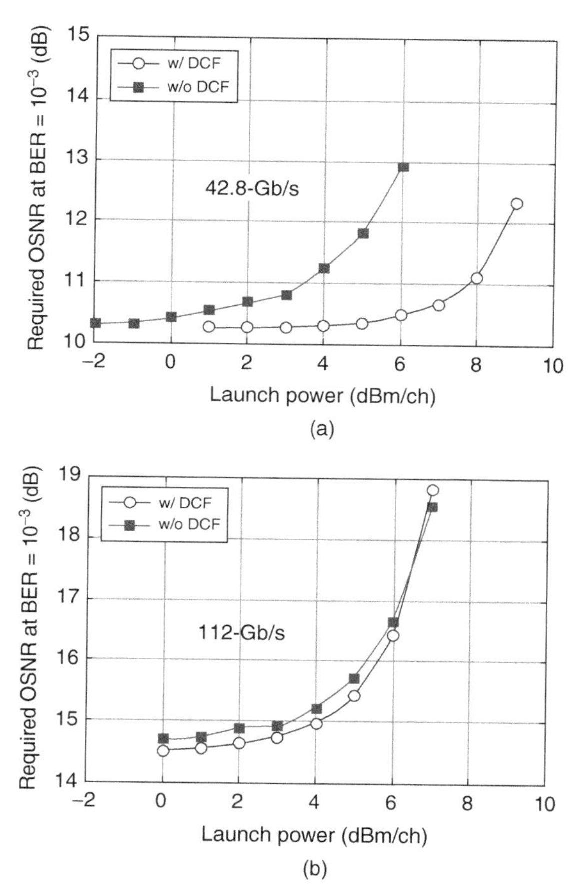

FIGURE 6.27 Required OSNR at a BER of 10^{-3} after 1000-km transmission versus launch power per channel for the 42.8-Gb/s (a) and 112-Gb/s iRZ-PDM-QPSK (b) coherent systems with and without inline DCF.

with 15-dB on/off gain provided by the Raman amplifier. As iRZ has been shown to have better nonlinear tolerance than NRZ, iRZ-PDM-16QAM is used here. In the PDM-16QAM receiver, polarization demultiplexing and residual distortion equalization is performed with a butterfly equalizer consisting of four 13-tap FIR filters. These are first optimized with the CMA for preconvergence and then finely tuned with the DD-LMS algorithm. Carrier phase recovery is performed with a decision-directed phase estimation method [39, 70]. The BER is evaluated by direct error counting with Gray coding.

To separate SPM, XPM and XPolM, a single-channel, 112-Gb/s RZ-SP-16QAM transmission is performed, with the RDPS varying from 30 to 1700 ps/nm by changing the DCF length in each span (RDPS of 1700 ps/nm means no inline DCF). Figure 6.31 depicts the impact of the RDPS on the transmission system

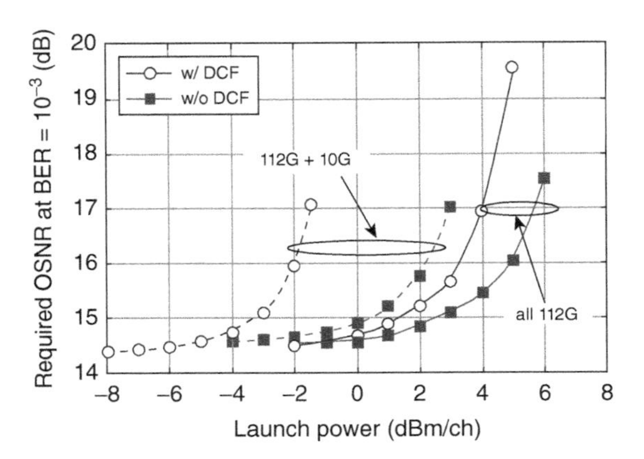

FIGURE 6.28 Required OSNR at a BER of 10^{-3} after 1000-km transmission versus launch power per channel for 112-Gb/s NRZ-PDM-QPSK co-propagating with six 10-Gb/s OOK channels. The results of all 112-Gb/s channels are plotted for comparison. From Ref. [37].

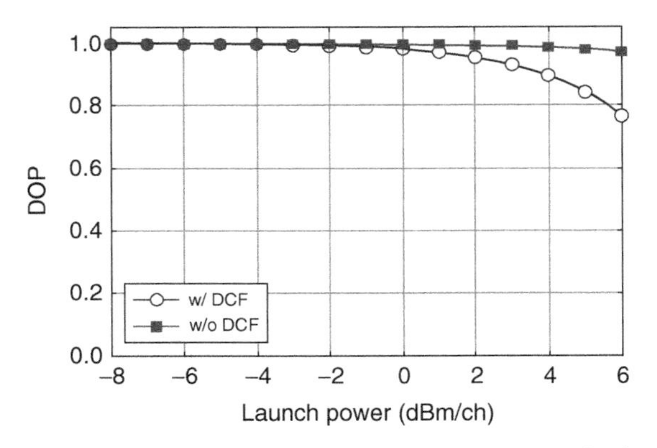

FIGURE 6.29 XPolM from 10-Gb/s OOK channels induced depolarization in the 112-Gb/s PDM-QPSK channel with and without inline DCF after 1000-km transmission. From Ref. [37].

performance [39]. It shows that the penalty peaks at an RDPS of 90 ps/nm, and then decreases with the increase of RDPS. Moreover, it shows that even for single-channel SP-16QAM, the system with DCF cannot perform better than that without DCF. This is completely different from QPSK systems, where systems with DCF can achieve better performance than those without DCF for SP-QPSK and iRZ-PDM-QPSK [19–22].

To exclude XPolM effects, SP signals are then used. Figure 6.32(a) shows the OSNR penalty at a BER of 10^{-3} after 1000-km transmission versus launch power per channel for the 112-Gb/s RZ-SP-16QAM coherent systems with and without DCF,

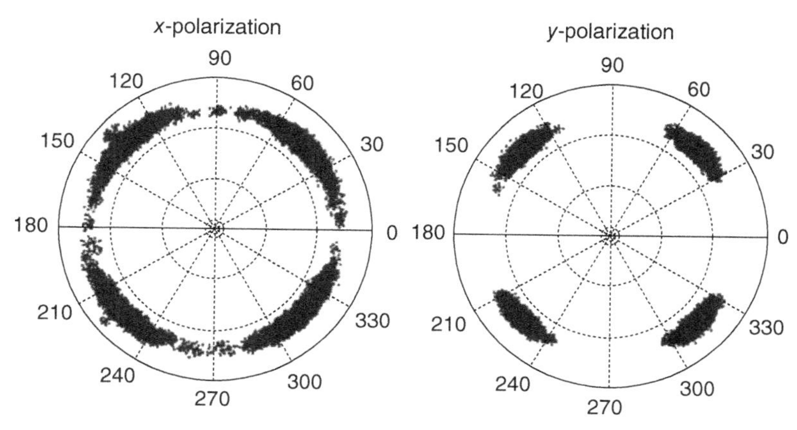

FIGURE 6.30 Signal constellations of 112-Gb/s PDM-QPSK after co-propagating with 10-Gb/s OOK channels over 1000-km SSMF with DCF at −1-dBm per channel launch power. From Ref. [24].

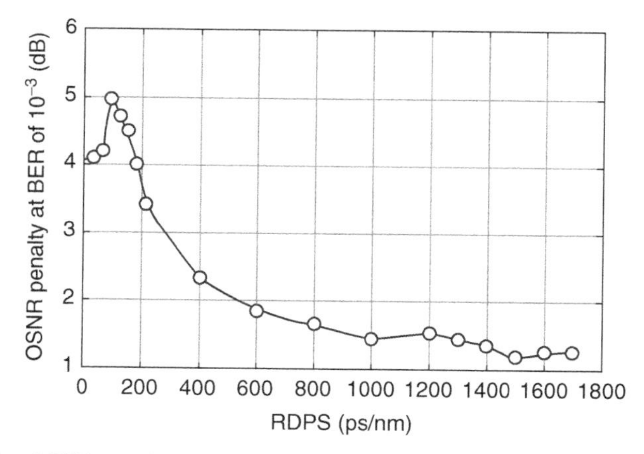

FIGURE 6.31 OSNR penalty after 1000 km transmission versus RDPS for single-channel, 112-Gb/s RZ-SP-16QAM with 1-dBm launch power and optimized dispersion precompensation. From Ref. [39].

where the RDPS is set as 30 ps/nm. For the system with DCF, dispersion precompensation is optimized, whereas there is no dispersion precompensation for that without DCF. It shows that for both single-channel and WDM transmission, the system with DCF has less tolerance to fiber nonlinearities than that without DCF. Constellations in Figure 6.32(b)–(e) shows that in the system with DCF, SPM and XPM cause large nonlinear phase distortions, and it is this effect that significantly degrades the system performance, whereas in the system without DCF, there is almost no nonlinear phase distortion.

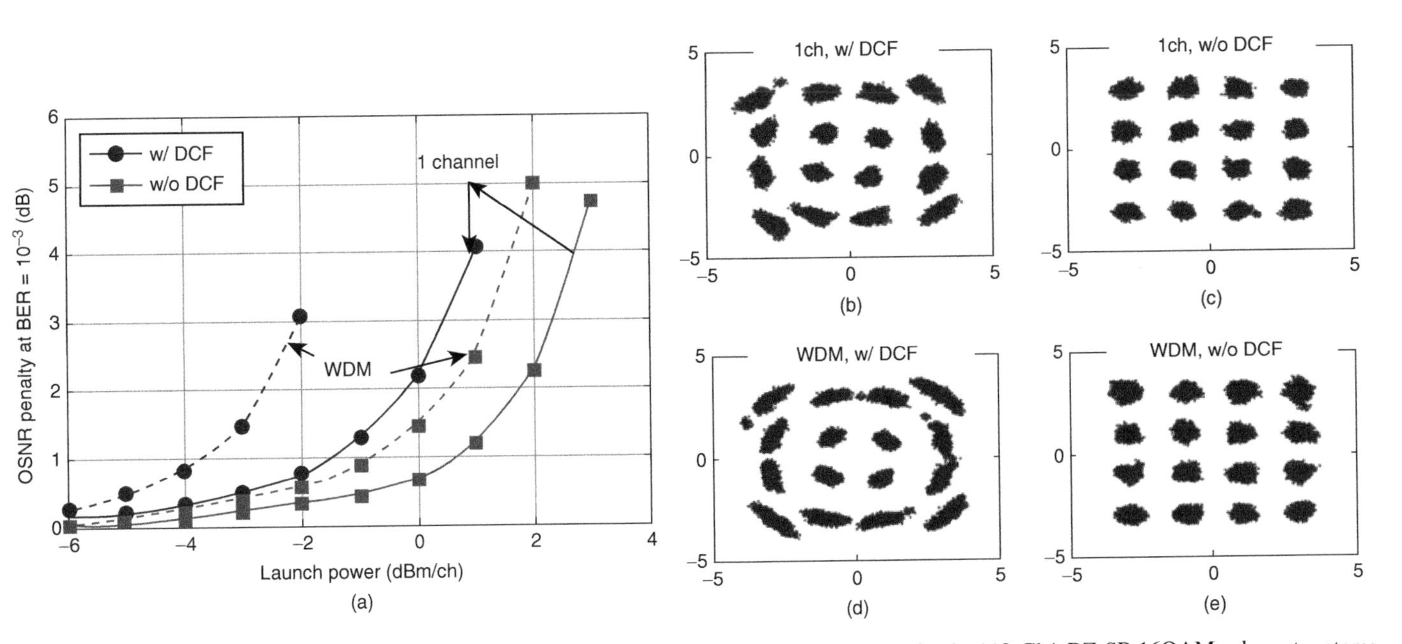

FIGURE 6.32 (a) OSNR penalty at BER of 10^{-3} after 1000-km transmission versus launch power for the 112-Gb/s RZ-SP-16QAM coherent systems with and without DCF; (b)–(e): signal constellations for single-channel and WDM systems at 1-dBm/ch and −1-dBm/ch launch powers, respectively.

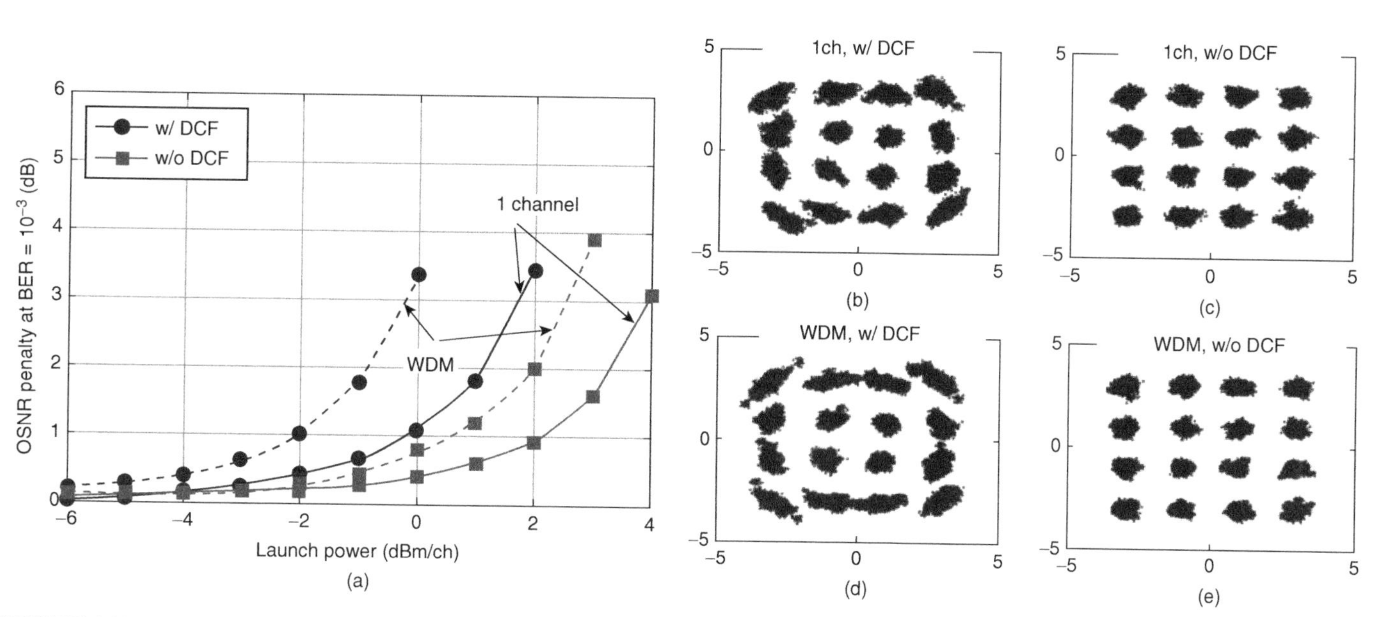

FIGURE 6.33 (a) OSNR penalty at BER of 10^{-3} after 1000-km transmission versus launch power for the 224-Gb/s iRZ-PDM-16QAM coherent systems with and without DCF; (b)–(e): signal constellations for single-channel and WDM systems at 3-dBm/ch and 1-dBm/ch launch powers, respectively. From Ref. [24].

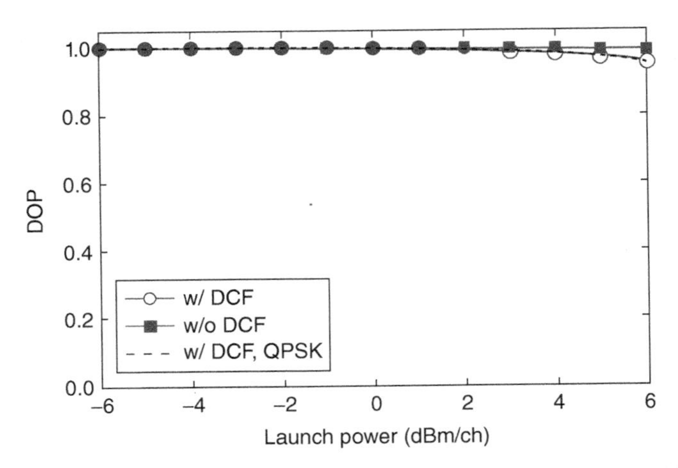

FIGURE 6.34 XPolM-induced depolarization in the 224-Gb/s iRZ-PDM-16QAM system with and without inline DCF after 1000-km transmission, and in the 112-Gb/s iRZ-PDM-QPSK system with DCF. From Ref. [24].

Nonlinear transmission performance of 224-Gb/s iRZ-PDM-16QAM is given in Figure 6.33. Similar to the SP signals, the PDM signals perform worse in the system with DCF than that without DCF for both single-channel and WDM transmission. Compared with 112-Gb/s RZ-SP-16QAM, the launch powers per channel of 224-Gb/s iRZ-SP-16QAM are improved by 1.5–2 dB for all the cases (note that power per polarization is lower). Constellations in Figure 6.33(b)–(e) show that it is SPM- and XPM-induced nonlinear phase distortions that significantly degrade the performance of the 224-Gb/s iRZ-PDM-16QAM system with DCF.

Figure 6.34 gives the XPolM-induced depolarization in the iRZ-PDM-16QAM transmission systems, which is measured by the DOP of a 112-Gb/s RZ-SP-16QAM channel surrounded by six 224-Gb/s iRZ-PDM-16QAM channels. For comparison, the result of the 112-Gb/s iRZ-PDM-QPSK system with DCF is also plotted in the figure. The XPolM-induced depolarization is similar in the PDM-16QAM system and in the PDM-QPSK system. The figure again indicates that the dominant nonlinear effect in the WDM PDM-16QAM system with DCF is not XPolM, but XPM, as the OSNR penalty at 1-dBm per channel launch power in the WDM system with DCF is more than 3 dB, but the XPolM-induced depolarization is very small at this launch power (DOP is 9.93) and will not cause any noticeable penalties.

6.8 SUMMARY

This chapter reviews recent advances in understanding polarization effects and fiber nonlinearities in coherent optical communication systems. To make the chapter self-contained, at the beginning of the chapter, basics of polarization of light are presented.

First, polarization effects are described. Although in principle, PMD can be entirely compensated in a coherent receiver with DSP, due to the limited complexity of DSP in the coherent receiver, the PMD tolerance of a coherent receiver is limited, and is proportional to the complexity of the butterfly equalizer. A technique to jointly optimize CD and butterfly equalizer is discussed. Due to the non-unitary nature of PDL, it cannot be well compensated in a coherent receiver. Two models to evaluate PDL impairments are presented and it is shown that to accurately evaluate the PDL penalties in coherent optical communication systems, the distributed PDL model has to be used.

Then, the nonlinear effects in coherent optical communication systems are discussed, including nonlinear transmission modeling techniques. While the dominant nonlinear effects in coherent optical communication systems without optical dispersion compensators are mostly intra-channel nonlinearities, the dominant nonlinear effects are inter-channel nonlinearities in coherent optical systems with inline DCF. In coherent optical communication systems with DCF, when modulation formats of constant amplitude such as QPSK are used, the dominant nonlinear effect is XPolM, which generates nonlinear polarization scattering and induces severe crosstalk between polarization tributaries, whereas when the channels carry nonconstant amplitude modulation formats such as hybrid QPSK and OOK or 16QAM, XPM is the dominant nonlinear effect.

REFERENCES

1. Huard S. Polarization of Light. New York: John Wiley & Sons; 1997.
2. Evangelides SG, Mollenauer LF, Gordon JP, Bergano NS. Polarization multiplexing with solitons. *J Lightwave Technol* 1992;10:28–35.
3. Chraplyvy AR, Gnauck AH, Tkach RW, Zyskind JL, Sulhoff JW, Lucero AJ, Sun Y, Jopson RM, Forghieri F, Derosier RM, Wolf C, McCormick AR. 1 Tb/s transmission experiment. *IEEE Photon Technol Lett* 1996;8:1264–1266.
4. Gnauck AH, Charlet G, Tran P, Winzer PJ, Doerr CR, Centanni JC, Burrows EC, Kawanishi T, Sakamoto T, Higuma K. 25.6-Tb/s WDM transmission of polarization-multiplexed RZ-DQPSK signals. *J Lightwave Technol* 2008;26:79–84.
5. Savory SJ, Stewart AD, Wood S, Gavioli G, Taylor MG, Killey RI, Bayvel P. Digital equalisation of 40 Gbit/s per wavelength transmission over 2480 km of standard fibre without optical dispersion compensation. Proceedings of ECOC'2006; Cannes, France, paper Th2.5.5; Sep 2006.
6. Sun H, Wu KT, Roberts K. Real-time measurements of a 40 Gb/s coherent systems. *Opt Express* 2008;16:873–879.
7. Shieh W, Bao H, Tang Y. Coherent optical OFDM: theory and design. *Opt Express* 2008;16:841–859.
8. Gnauck AH, Winzer PJ, Konczykowska A, Jorge F, Dupuy J-Y, Riet M, Charlet G, Zhu B, Peckham DW. Generation and transmission of 21.4-Gbaud PDM 64-QAM using a novel high-power DAC driving a single I/Q modulator. *IEEE/OSA J Lightwave Technol* 2012;30:532–536.

9. Huang Y-K, Qian D, Yaman F, Wang T, Mateo E, Inoue T, Inada Y, Toyoda Y, Ogata T, Sato M. et al. Real-time 400G superchannel transmission using 100-GbE based 37.5-GHz spaced subcarriers with optical Nyquist shaping over 3,600-km DMF link. Proceedings of OFC/NFOEC'2013; Anaheim, CA, USA, paper NW4E.1; 2013.
10. Ghazisaeidi A, Schmalen L, de Jauregui IF, Tran P, Simonneau C, Brindel P, Charlet G. 52.9 Tb/s transmission over transoceanic distances using adaptive multi-rate FEC. Proceedings of ECOC'2014; Cannes, France, paper PD.3.4; Sep 2014.
11. Shieh W, Ho K-P. Equalization-enhanced phase noise for coherent-detection systems using electronics digital signal processing. *Opt Express* 2008;16:15718–15727.
12. Xie C. Local oscillator phase noise induced penalties in optical coherent detection systems using electronic chromatic dispersion compensation. Proceedings of OFC'09; San Diego, CA, paper OMT4; Mar 2009.
13. Mantzoukis N, Vgenis A, Petrou CS, Roudas I, Kamalakis T, Raptis L. Design guidelines for electronic PMD equalizers used in coherent PDM QPSK systems. Proceedings of ECOC'2010; Torino, Italy, paper P4.16; Sep 2010.
14. Xie C. Polarization-mode-dispersion impairments in 112-Gb/s PDM-QPSK coherent systems. Proceedings of ECOC'2010; Torino, Italy, paper Th.10.E.6; Sep 2010.
15. Xie C, Winzer P. Increasing polarization-mode dispersion tolerance of coherent receivers by joint optimization of chromatic dispersion and butterfly equalizers. OSA Summer Topicals, SPPCom'2013; Puerto Rico, paper SPM2E.5; Jul 2013.
16. Duthel T, Fludger CRS, Geyer J, Schulien S. Impact of polarisation dependent loss on coherent POLMUX-NRZ-DQPSK. Proceedings of OFC/NFOEC'2008; San Diego, CA, USA, paper OThU5; Feb 2008.
17. Shtaif M. Performance degradation in coherent polarization multiplexed systems as a result of polarization dependent loss. *Opt Express* 2008;16:13918–13932.
18. Xie C. Polarization-dependent loss induced penalties in PDM-QPSK coherent optical communication systems. Proceedings of OFC/NFOEC'2010; San Diego, CA, USA, paper OWE6; Mar 2010.
19. Curri V, Poggiolini P, Carena A, Forghieri F. Dispersion compensation and mitigation of nonlinear effects in 111-Gb/s WDM coherent PM-QPSK systems. *IEEE Photon Technol Lett* 2008;20:1473–1475.
20. Xie C. Inter-channel nonlinearities in coherent polarization-division-multiplexed quadrature-phase-shift-keying systems. *IEEE Photon Technol Lett* 2009;21:274–276.
21. Xie C. WDM coherent PDM-QPSK systems with and without inline optical dispersion compensation. *Opt Express* 2009;17:4815–4823.
22. Xie C. Dispersion management in WDM coherent PDM-QPSK systems. Proceedings of ECOC'2009; Vienna, Austria, paper 9.4.3, Sep 2009.
23. Xie C. Nonlinear polarization effects and mitigation in polarization multiplexed transmission. Proceedings of OFC/NFOEC'2010; San Diego, CA, USA, paper OWE1; Mar 2010.
24. Xie C. Impact of nonlinear and polarization effects in coherent systems. *Opt Express* 2011;19(26):B915–B930.
25. Bononi A, Rossi N, Serena P. Transmission limitations due to fiber nonlinearity. Proceedings of OFC/NFOEC'2011; Los Angeles, CA, USA, paper OWO7; Mar 2011.
26. Charlet G, Renaudier J, Salsi M, Mardoyan H, Tran P, Bigo, S. Efficient mitigation of fiber impairments in an ultra-long haul transmission of 40Gbit/s polarization-multiplexed data

by digital processing in a coherent receiver. Proceedings of OFC/NFOEC'2007; Anaheim, CA, USA, paper PDP17; 2007.

27. Kaneda N, Leven A. Coherent polarization-division-multiplexed QPSK receiver with fractionally spaced CMA for PMD compensation. *IEEE Photon Technol Lett* 2009;21:203–205.
28. Serena P, Rossi N, Bononi A. Nonlinear penalty reduction induced by PMD in 112 Gbit/s WDM PDM-QPSK coherent systems. Proceedings of ECOC'2009; Vienna, Austria, paper 10.4.3; Sep 2009.
29. Bertran-Pardo O, Renaudier J, Charlet G, Tran P, Mardoyan H, Bertolini M, Salsi M, Bigo S. Demonstration of the benefits brought by PMD in polarization-multiplexed systems. Proceedings of ECOC'2010; Torino, Italy, paper Th.10.E.4.; Sep 2010.
30. Xia C, da Silva Pina JF, Striegler A, van den Borne D. PMD-induced nonlinear penalty reduction in coherent polarization-multiplexed QPSK transmission. Proceedings of ECOC'2010; Torino, Italy, paper Th.10.E.5; Sep 2010.
31. Xie C, Essiambre R-J. Electronic nonlinearity compensation in 112-Gb/s PDM-QPSK optical coherent transmission systems. Proceedings of ECOC'2010; Torino, Italy, paper Mo.1.C.1; Sep 2010.
32. Li G, Mateo E, Zhu L. Compensation of nonlinear effects using digital coherent receivers. Proceedings of OFC/NFOEC'2011; Los Angeles, CA, USA, paper OWW1; Mar 2011.
33. Tao Z, Dou L, Yan W, Li L, Hoshida T, Rasmussen JC. Multiplier-free intrachannel nonlinearity compensating algorithm operating at symbol rate. *IEEE/OSA J Lightwave Technol* 2011;29:2570–2576.
34. Li L, Tao Z, Dou L, Yan W, Oda S, Tanimura T., Hoshida T, Rasmussen JC. Implementation efficient nonlinear equalizer based on correlated digital backpropagation. Proceedings of OFC/NFOEC'2011; Los Angeles, CA, USA, paper OWW3; Mar 2011.
35. Essiambre R-J, Mikkelsen B, Raybon G. Intrachannel cross-phase modulation and four-wave mixing in high-speed TDM systems. *Electron Lett* 1999;35:1576–1578.
36. Mamyshev PV, Mollenauer LF. Soliton collisions in wavelength-division-multiplexed dispersion-managed systems. *Opt Lett* 1999;24:448–450.
37. Xie C. Inter-channel nonlinearities in hybrid OOK and coherent PDM-QPSK transmission systems with dispersion management. Proceedings of IEEE Photonic Society Summer Topicals; Cancun, Mexico, paper TuA.3.2; Jul 2010.
38. van den Borne D, Fludger CRS, Duthel T, Wuth T, Schmidt ED, Schulien C, Gottwald E, Khoe GD, de Waardt H. Carrier phase estimation for coherent equalization of 43-Gb/s POLMUXNRZ-DQPSK transmission with 10.7-Gb/s NRZ neighbours. Proceedings of ECOC'2007; Berlin, Germany, paper 7.2.3; Sep 2007.
39. Xie C. Fiber nonlinearities in 16QAM transmission systems. Proceedings of ECOC'2011; Geneva, Switzerland, paper We.7.B.6; Sep 2011.
40. Gordon JP, Kogelnik H. PMD fundamentals: polarization mode dispersion in optical fibers. *Proc Natl Acad Sci U S A* 2000;97:4541–4550.
41. Kogelnik H, Jopson RM, Nelson LE. Polarization-mode dispersion. In: Kaminov I, Li T, editors. Optical Fiber Telecommunications, IVB. San Diego, USA: Academic Press; 2002. (Chapter 15).
42. Poole CD, Wagner RE. Phenomenological approach to polarization dispersion in long single-mode fibers. *Electron Lett* 1986;22:1029–1030.

43. Foschini GJ, Poole CD. Statistical theory of polarization dispersion in single mode fibers. *IEEE/OSA J Lightwave Technol* 1991;9:1439–1456.
44. Wai PKA, Menyuk CR. Polarization mode dispersion, decorrelation, and diffusion in optical fibers with randomly varying birefringence. *IEEE/OSA J Lightwave Technol* 1996;14:148–157.
45. Mecozzi A, Shtaif M. The statistics of polarization-dependent loss in optical communication systems. *IEEE Photon Technol Lett* 2002;14(3):313–315.
46. Lu P, Chen L, Bao X. Statistical distribution of polarization dependent loss in the presence of polarization mode dispersion in single mode fibers. *IEEE Photon Technol Lett* 2001;13:451–453.
47. Bruyère F, Audouin O. Penalties in long-haul optical amplifier systems due to polarization dependent loss and gain. *IEEE Photon Technol Lett* 1994;6(5):654–656.
48. Shtaif M, Mecozzi A, Tkach RW. Noise enhancement caused by polarization dependent loss and the effect of dynamic gain equalizers. Proceedings of OFC/NFOEC'2002; Anaheim, CA, USA, paper TuI2; Mar 2002.
49. Yu M, Kan C, Lewis M, Sizmann A. Statistics of signal-to-noise ratio and path-accumulated power due to concatenation of polarization-dependent loss. *IEEE Photon Technol Lett* 2002;14(10):1418–1420.
50. Xie C, Mollenauer LF. Performance degradation induced by polarization dependent loss in optical fiber transmission systems with and without polarization mode dispersion. *J Lightwave Technol* 2003;21:1953–1957.
51. Xie C. Polarization-dependent loss-induced outage probabilities in optical communication systems. *IEEE Photon Technol Lett* 2008;20:1091–1093.
52. Marcuse D, Menyuk CR, Wai PKA. Application of the Manakov-PMD equation to studies of signal propagation in optical fibers with randomly varying birefringence. *IEEE/OSA J Lightwave Technol* 1997;15:1753–1746.
53. Menyuk CR, Marks BS. Interaction of polarization mode dispersion and nonlinearity in optical fiber transmission systems. *IEEE/OSA J Lightwave Technol* 2006;24:2806–2826.
54. Mollenauer LF, Gordon JP, Heismann F. Polarization scattering by soliton-soliton collisions. *Opt Lett* 1995;20:2060–2062.
55. Collings BC, Boivin L. Nonlinear polarization evolution induced by cross-phase modulation and its impact on transmission systems. *IEEE Photon Technol Lett* 2000;12:1582–1584.
56. Wang D, Menyuk CR. Polarization evolution due to the Kerr nonlinearity and chromatic dispersion. *IEEE/OSA J Lightwave Technol* 1999;17:2520–2529.
57. Bononi A, Vannucci A, Orlandini A, Corbel E, Lanne S, Bigo S. Degree of polarization degradation due to cross-phase modulation and its impact on polarization-mode dispersion compensators. *IEEE/OSA J Lightwave Technol* 2003;21:1903–1913.
58. Karlsson M, Sunnerud H. Effects of nonlinearities on PMD-induced system impairments. *IEEE/OSA J lightwave Technol* 2006;24:4127–4137.
59. Agrawal GP. Nonlinear Fiber Optics. San Diego, CA, USA: Academic; 2001.
60. Wai PKA, Menyuk CR, Chen HH. Stability of solitons in randomly varying birefringent fibers. *Opt Lett* 1991;16:1231–1233.
61. Dou L, Tao Z, Li L, Yan W, Tanimura T, Hoshida T, Rasmussen JC. A low complexity pre-distortion method for intra-channel nonlinearity. Proceedings OFC/NFOEC'2011; Los Angeles, CA, USA, paper OThF5; Sep 2011.

62. Châtelain B, Laperle C, Roberts K, Xu X, Chagnon M, Borowiec A, Gagnon F, Cartledge J, Plant DV. Optimized pulse shaping for intra-channel nonlinearities mitigation in a 10 Gbaud dual-polarization 16-QAM system. Proceedings of OFC/NFOEC'2011; Los Angeles, CA, USA, paper OWO5; Sep 2011.
63. Wang J, Xie C, Pan Z. Generation of spectrally efficient Nyquist-WDM QPSK signals using digital FIR or FDE filters at transmitters. *IEEE/OSA J Lightwave Technol* 2012;30:3679–3686.
64. Wang J, Xie C, Pan Z. Optimization of DSP to generate spectrally efficient 16QAM Nyquist-WDM signals. *IEEE Photon Technol Lett* 2013;25:772–775.
65. Savory S. Digital coherent optical receivers: algorithms and subsystems. *IEEE J Selected Topics Quan Electron* 2010;16:1164–1179.
66. Savory S. Digital filters for coherent optical receivers. *Opt Express* 2008;16:804–817.
67. Kuschnerov M, Hauske F, Piyawanno K, Spinnler B, Alfiad M, Napoli A, Lankl B. DSP for coherent single-carrier receivers. *IEEE/OSA J Lightwave Technol* 2009;27:3614–3622.
68. Kuschnerov M, Chouayakh M, Piyawanno K, Spinnler B, de Man E, Kainzmaier P, Alfiad MS, Napoli A, Lankl B. Data-aided versus blind single-carrier coherent receivers. *IEEE Photon J* 2010;2:387–403.
69. Godard D. Self-recovering equalization and carrier tracking in two dimensional data communication systems. *IEEE Trans Commun* 1980;28:1867–1875.
70. Fatadin I, Savory SJ. Blind equalization and carrier phase recovery in a 16-QAM optical coherent system. *IEEE/OSA J Lightwave Technology* 2009;27:3042–3049.
71. Sunnerud H, Xie C, Karlsson M, Samuelsson R, Andrekson PA. A comparison between different PMD compensation techniques. *IEEE/OSA J Lightwave Technol* 2002;20:368–378.
72. Nelson LE, Nielsen TN, Kogelnik H. Observation of PMD induced coherent crosstalk in polarization-multiplexed transmission. *IEEE Photon Technol Lett* 2001;13:738–740.
73. Wang Z, Xie C. PMD and PDL impairments in polarization division multiplexing signals with direct detection. *Opt Express* 2009;17:7993–8004.
74. Bülow H, Xie C, Klekamp A, Liu X, Franz B. PMD compensation/mitigation techniques for high-speed optical transport. *Bell Labs Tech J* 2009;14(1):105–124.
75. Xie C, Möller L. The accuracy assessment of different polarization mode dispersion models. *Opt Fiber Technol* 2006;12(2):101–109.
76. Kogelnik H, Nelson LE, Gordon JP. Emulation and inversion of polarization-mode dispersion. *IEEE/OSA J Lightwave Technol* 2003;21:482–495.
77. Foschini GJ, Jopson RM, Nelson LE, Kogelnik H. The statistics of PMD-induced chromatic fiber dispersion. *IEEE/OSA J Lightwave Technol* 1999;17:1560–1565.
78. Mamyshev PV, Mamysheva NA. Pulse-overlapped dispersion-managed data transmission and intra-channel four-wave mixing. *Opt Lett* 1999;24:1454–1456.
79. Tkach RW, Chraplyvy AR, Forghieri F, Gnauck AH. Four-photon mixing and high-speed WDM systems. *IEEE/OSA J Lightwave Technol* 1995;13:841–849.
80. Winzer PJ, Essiambre R-J. Advanced modulation formats for high-capacity optical transport networks. *IEEE/OSA J Lightwave Technol* 2006;24:4711–4728.
81. Killey RI, Thiele HJ, Mikhailov V, Bayvel P. Reduction of intrachannel nonlinear distortion in 40-Gb/s-based WDM transmission over standard fiber. *IEEE Photon Technol Lett* 2000;12:1624–1626.

82. Renaudier J, Bertran-Pardo O, Charlet G, Salsi M, Bertolini M, Tran P, Mardoyan H, Bigo S. On the required number of WDM channels when assessing performance of 100Gb/s coherent PDM-QPSK overlaying legacy systems. Proceedings of ECOC'2009; Vienna, Austria, paper 3.4.5; Sep 2009.
83. Xia C, van den Borne D. Impact of the channel count on the nonlinear tolerance in coherently-detected POLMUX-QPSK modulation. Proceedings of OFC/NFOEC'2011; Los Angeles, CA, USA, paper OWO1; 2011.
84. Viterbi A. Nonlinear estimation of PSK-modulated carrier phase with application to burst digital transmission. *IEEE Trans Inf Theory* 1983;29:543–551.
85. Noé R. PLL-free synchronous QPSK polarization multiplex/diversity receiver concept with digital I&Q baseband processing. *IEEE Photon Technol Lett* 2005;17:887–889.

7

ANALYTICAL MODELING OF THE IMPACT OF FIBER NON-LINEAR PROPAGATION ON COHERENT SYSTEMS AND NETWORKS

PIERLUIGI POGGIOLINI[1], YANCHAO JIANG[1,2], ANDREA CARENA[1], AND FABRIZIO FORGHIERI[3]

[1]*Dipartimento di Elettronica, Politecnico di Torino, Torino, Italy*
[2]*College of Information Engineering, Dalian University, Dalian, China*
[3]*Cisco Photonics, Vimercate, Italy*

7.1 WHY ARE ANALYTICAL MODELS IMPORTANT?

Analytical models of the impact of non-linear effects on system and network performance are important for several reasons: in the context of point-to-point (PTP) systems, they allow to explore design strategies efficiently, without resorting to lengthy computer simulations; in the context of networks, they can help in the optimization of the network architecture and layout, and can provide physical layer awareness for real-time control-plane tasks such as channel routing. In all contexts, they can be used for research purposes, to devise and theoretically test new and disruptive technologies.

7.1.1 What Do Professionals Need?

Ideally, system and network engineers, as well as researchers, would like to have analytical models that are:

1. easy to set-up and parameterize,

Enabling Technologies for High Spectral-efficiency Coherent Optical Communication Networks, First Edition. Edited by Xiang Zhou and Chongjin Xie.

© 2016 John Wiley & Sons, Inc. Published 2016 by John Wiley & Sons, Inc.

2. fast to compute,
3. accurate.

Unfortunately, no single model currently fully complies with all three of these requirements simultaneously. As a result, there are trade-offs that must be managed to provide the best compromise for each specific application.

On the other hand, recent progress in modeling has made it possible to work out acceptable solutions for a broad range of utilization scenarios, from PTP link design to real-time network optimization. In the "model selection" part of this chapter (Section 7.4) we will adopt a rather practical approach towards the goal of helping the reader find a suitable trade-off among the three modeling features listed above, for the different utilization scenarios.

As an important disclaimer that we make upfront, Section 7.4 focuses on the GN–EGN class of models, partly because it has been the subject of substantial research by the authors of this chapter, partly because it has recently enjoyed considerable attention by the technical community, and partly because of the practical difficulty of encompassing many different classes of models within a single book chapter. No claim of intrinsic superiority versus other models is implied.

As a counterbalance, Section 7.2 aims at providing some general theoretical background that should prove useful for the readers to carry out an informed model selection themselves. The idea is that, apart from the indications that we propose, the readers should be able to autonomously make or refine their choices regarding the identification of the most suitable models for their own applications, possibly outside of the GN–EGN model class.

As a whole, the field of research on modeling is currently extremely active. Which of the various model classes will emerge as a front-runner or eventual winner, it will probably become clear as the field progresses and new modeling efforts come to fruition. It may also be the case that different model classes are best suited for different types of problems or contexts of utilization.

This chapter uses several acronyms. They are comprehensively listed in Section A.5. For convenience, most of them are also defined where they occur for the first time.

7.2 BACKGROUND

In this section we provide some background information on modeling approaches.

In general, one could call "non-linear fiber propagation model" any form of analytical description of the non-linear behavior of the optical fiber. In this respect, the well-known dual-polarization non-linear Schroedinger equation (DP-NLSE, [1–3]) is one such "model," and a quite successful one. Numerical integration of the DP-NLSE, or of the related Manakov equation (see Section 7.2.1.1), typically within a Monte-Carlo simulation environment, has been and still is one of the most powerful tools for the study and design of optical systems in the presence of non-linear effects.

However, in this chapter we are interested in "derived models." They all assume that the DP-NLSE accurately represents the underlying physics of the fiber and try to obtain from it simpler results, if possible closed-form, using various approximations, mostly for the purpose of quantifying the *system impact* of the fiber non-linear behavior.

Many such models have been proposed over time. An extensive bibliography, encompassing several of the modeling efforts carried out in the last 25 years, is available at the end of this chapter. Some of the listed papers [4–14] contain, in turn, further extensive referencing, so that the interested reader can directly or indirectly find rather exhaustive orientation across this complex field. A comprehensive bibliography on non-linearity modeling can also be found in the influential papers [15, 16], even though the main focus of those papers was different.

Describing, classifying and discussing in detail all, or even just the majority, of non-linearity modeling efforts is far beyond the scope of a single book chapter. Instead, we collect in the following a list of the main approximations which have been used to derive such models from the DP-NLSE. Discussing the underlying approximation is key to the understanding of the various models because, in essence, it is the employed approximations that define them and characterize their behavior and effectiveness. We briefly discuss the impact and implications of the most common approximations and provide practical recommendations regarding their viability, pros and cons.

Those readers who are chiefly interested in indications about practical modeling solutions for specific system scenarios may skip this section and go to Section 7.4. Those readers who would like to know more and possibly wish to investigate the matter themselves, may instead want to read it, keeping in mind that this is an introductory overview which is neither exhaustive nor comprehensive. Its goal is that of providing some orientation and pointers towards the relevant research material.

7.2.1 Modeling Approximations

The non-linear propagation models derived from the DP-NLSE typically exploit one or more of the following approximations:

1. The Manakov equation approximation
2. The single-polarization approximation
3. The perturbation approximation
4. The signal Gaussianity approximation
5. The NLI additive-Gaussian-noise (AGN) approximation
6. The locally-white NLI noise approximation
7. The lossless fiber approximation
8. The incoherent NLI accumulation approximation
9. The noiseless propagation approximation
10. The frequency-domain XPM approximation.

In the following we introduce them one by one and discuss their implications. Before proceeding, we define the term "NLI." The acronym stands for "non-linear interference." Calling $s_{\mathrm{WDM}}(t)$ and $s_{\mathrm{WDM}}^{\mathrm{NL}}(t)$ the received WDM signal in the *absence* and in the *presence* of fiber non-linearity, respectively, then NLI is the difference signal:

$$s_{\mathrm{NLI}}(t) = s_{\mathrm{WDM}}^{\mathrm{NL}}(t) - s_{\mathrm{WDM}}(t) \tag{7.1}$$

Rearranging the above formula, NLI can also be viewed as a disturbance created by non-linear effects which gets formally added to the WDM signal and degrades it.

7.2.1.1 The Manakov Equation Approximation The DP-NLSE is deemed to accurately account for the effect of the evolution of the state of polarization (SOP) of the signal on the generation of NLI. However, SOP variations along the fiber are a random process and this makes the DP-NLSE difficult to use, also because the characteristic lengths related to SOP evolution may be quite short (from a hundred meters down to a fraction of a meter [4]).

The Manakov equation (ME) [4, 17, 18] is an approximation of the DP-NLSE based on analytically averaging over the random evolution of the SOP along the fiber. As a result, the ME is a deterministic differential equation, which *does* capture the non-linear effect of one polarization onto the other, but averages over the fast dynamic of SOP variations. More in detail, the SOP-evolution averaging procedure generates the Manakov-PMD (polarization-mode dispersion) equation, as Equation 12 in Ref. [4]. A *simplified version of it*, consisting of only the left-hand side of said equation, is the ME that is widely employed in analytical modeling and computer simulations. It entails the further approximation of neglecting both the linear and non-linear effects of PMD [4]. On the other hand, *linear* PMD is no longer a factor in modern coherent systems thanks to receiver digital signal processing (DSP), whereas in Ref. [4] *non-linear* PMD was assessed to be virtually negligible in typical transmission links.[1] At present, the ME is generally regarded as quite accurate and effective.

Note that another polarization-related effect, polarization dependent loss (PDL), is indeed a source of penalty even in current DSP-assisted systems, both because of its linear and non-linear effects [19]. However, its linear impact is regarded as prevalent, especially in dispersion uncompensated systems, and is typically studied separately [20, 21]. Therefore, we consider PDL beyond our target level-of-detail for system-oriented non-linearity modeling and we disregard it henceforth.

In conclusion, the ME, that is, the left-hand side of Equation 12 in Ref. [4], adequately accounts for the main cross-polarization non-linear effects impacting practical coherent systems. For this reason, the GN–EGN models have been proposed based on it.

[1] Some subtle effects on the statistical features of the NLI produced on a specific channel may be lost when neglecting PMD. The reason is that PMD scrambles the channels' relative SOPs, over a large-enough optical bandwidth and propagation distance, and this may in principle affect both the quantity and quality of NLI generation. Whether this effect is significant is a matter of current investigation.

Practical recommendation: For system-oriented studies, the use of the Manakov equation for the modeling of fiber non-linear propagation provides a good compromise between accuracy and complexity and is therefore recommended.

7.2.1.2 The Single-Polarization Approximation This approximation assumes that propagation obeys the *single-polarization* (SP) NLSE. This is a scalar equation which completely neglects all polarization-related effects, either linear or non-linear. The transmitted signal must be assumed single-polarization or, otherwise, it must be assumed that the two multiplexed signal polarizations propagate in a completely separate and independent way.

This approximation has been very popular over the years, especially prior to the coherent systems revolution, for various reasons. First, dealing with a scalar equation eases calculations and may lead to simpler final results. Secondly, before the advent of coherent systems, polarization multiplexing was not used and the transmitted intensity-modulation direct-detection (IM/DD) signals of the time were fundamentally scalar signals, detected through polarization-insensitive photodiodes. As a result, the impact of polarization-mediated non-linear effects was minor and was typically neglected.[2] PMD was a problem, but it was mostly studied separately as a linear effect.

On the other hand, today's coherent systems are almost exclusively dual-polarization, and therefore cross-polarization effects should be properly accounted for.[3] The sensible practical recommendation is then to use the Manakov equation to suitably account for non-linear polarization interactions. In principle, this might lead to increased model complexity. However, at least in the case of the GN–EGN models, the final equations have identical complexity whether the single or dual-polarization derivation is used.

Practical recommendation: Do not use the single-polarization approximation for the modeling of the impact of NLI on dual-polarization coherent systems.

7.2.1.3 The Perturbation Approximation The vast majority of non-linear propagation models makes the assumption that non-linearity is relatively small, i.e., that it

[2]The main consequence of assuming that different IM/DD WDM channels may travel on different polarizations, with random SOPs, is that inter-channel non-linearity actually decreases somewhat (by a factor 5/6 or 8/9, depending on assumptions [4]). This effect is relatively minor and was often disregarded.

[3]In principle, it might be possible to use the SP-NLSE for dual-polarization systems. One could propagate the two polarizations independently and then *approximately* account a posteriori for their interaction, for instance based on pre-computed results on the increase in NLI when adding a cross-polarized signal, as discussed in Ref. [22]. However, the cross-polarized interaction has statistical features that are somewhat different than that of same-polarization [22], which makes this procedure non-straightforward. In particular, with reference to the GN–EGN model and to Equation 7.10, it can be shown that the relative strength of the so-called "non-Gaussianity correction term" $G_{\mathrm{NLI}}^{\mathrm{corr}}(f)$ versus the GN model term $G_{\mathrm{NLI}}^{\mathrm{GN}}(f)$ is smaller in the dual-polarization case, depending in a complex way on signal features. As a final remark, the ME takes into account both the so-called XPolM (cross-polarization modulation) effect and the effect of inter-polarization four-wave-mixing (FWM). For more details on XPolM, see for instance the very recent papers [23, 24], published after this manuscript was drafted. In the remainder of this chapter, for simplicity, we choose not to explicitly single-out XPolM among the various non-linear effects, but to account for it tacitly as the dual-polarization component of other effects, such as XPM.

is a *perturbation* as compared to the useful signal. Thanks to this assumption, model derivation can exploit *perturbation* techniques, which allow to find approximate analytical solutions to the Manakov (or other) propagation equations.

One possible perturbation approach consists of assuming that the signal propagates linearly from input to output, subject only to the action of fiber chromatic dispersion (CD) and loss/amplification; at each point along the fiber, this linearly-propagating signal excites fiber non-linearity and creates the NLI disturbance, which is calculated based on the employed propagation equation (the ME, the SP-NLSE or others) but is kept separate from the signal itself. At the end of the link, the sum of the linearly propagated signal and the NLI produced by it constitutes the overall non-linear fiber output signal (see Section A.4 for an example of this procedure).

The key feature of this method is that it leads to NLI computation formulas which are remarkably simple. This popular approach belongs to the class of *regular perturbation (RP) methods* [5] and the one described above is its *first-order* version. Higher-order versions are possible whereby not only the linearly propagated signal but the produced NLI itself cooperate to create further NLI. To the purpose of obtaining system-impact models, first-order versions have however been used, with few exceptions such as [25].

Another perturbation method, based on truncated Volterra series, was proposed in Ref. [26]. Interestingly, in Ref. [5] it was shown that the RP method and the VS-method are equivalent, so in this context they can be unified as RP–VS methods. Other first-order perturbation method, which can be re-conduced or bear substantial similarities to the RP–VS methods, were proposed in Refs [11, 27, 28]. Further perturbation methods have also been proposed, such as the *logarithmic perturbation (LP) method* [29], a combination of the RP and LP method [30], the frequency-resolved LP (FRLP) method [14, 31], the enhanced RP method [13], and still others.

Many simulations and experiments have shown the perturbation approximation to produce rather accurate results within the typical range of optimal system launch powers. Irrespective of the specific method, however, all perturbation techniques can be expected to break down at highly non-linear regimes. For instance, first-order methods typically do not take into account that NLI noise is created at the expense of the WDM signal power, that is, the signal is assumed "undepleted." In those systems where the power of the generated NLI is not small as compared to that of the WDM signal, neglecting signal power depletion may induce substantial model inaccuracy. This effect was recently studied in Ref. [32]. It was found that a simple semi-heuristic correction, similar to the factor c in Equation 8 of [33], can approximately account for such depletion, extending the range of usability of first-order perturbation methods. Caution should anyway be used in those cases where substantial signal-power depletion may occur.

Specifically regarding the GN–EGN model class, the derivation of these models is based on a first-order RP method.

Practical recommendation: The use of the perturbation approximation, in particular of first-order perturbation methods, to solve the fiber non-linear propagation equations appears adequate for typical system operating conditions.

7.2.1.4 *The Signal Gaussianity Approximation*

According to this approximation, the transmitted signal is modeled as stationary circular Gaussian noise, whose power spectrum (or power spectral density, PSD) is shaped as the PSD of the actually transmitted WDM channels. It allows to drastically simplify model derivation and strongly decreases the model final analytical complexity. One of its implications is that the model results are independent of the used transmission format, since this information is completely removed from the signal itself.

This approximation has been repeatedly used over the years. It is also one of the main approximations employed by the GN model. It is already found in what can be considered the first GN-model-class paper [34], dating back to 1993. It is instead removed in the EGN model [23, 35], generalizing a procedure proposed in Ref. [36] for the XPM component of NLI (see Section 7.2.1.10). The EGN model is however quite substantially more complex than the GN.

One key feature of the error incurred by using the signal Gaussianity approximation is that the impact of NLI is *never underestimated* for QAM transmission formats. Rather, it is overestimated, to an extent that depends on several aspects, among which fiber type, span length, amplification scheme, transmission format, symbol rate, and others. In typical EDFA-amplified PTP links, the NLI overestimation can lead to between 5% and 15% system reach underestimation [37, 38], for 32 GBaud systems. Interestingly, the error due to the Gaussianity assumption tends to vanish when going towards large symbol rates (>32 GBaud). It *also* tends to vanish for massively multi-carrier systems, such as OFDM, Nyquist-FDM or similar. See Section 7.4.4 for more details on this aspect. Nonetheless, the practical recommendation for PTP links is to avoid using the signal Gaussianity approximation, if possible. Either the very accurate, but more complex, EGN model, or an approximation of it should be used, as extensively discussed in Sections 7.4.2 and 7.4.3.

In the context of dynamically reconfigurable networks (DRNs), however, the situation is less clear-cut (see Section 7.4.5). In that environment neighboring channels can have different symbol rate, format, and accumulated dispersion. In addition, the re-routing process across the network may change the neighbors of a channel many times along a lightpath. All these factors produce a non-linear behavior which is varied and on average less distant from the signal-Gaussianity approximation. In addition, its use makes it possible to obtain very simple and powerful performance prediction tools, well-suited for real-time physical-layer awareness. The combination of these aspects suggests that in DRNs the signal Gaussianity approximation may actually be recommended, in which case the GN model may be a possible option.

Practical recommendation: Avoid the signal Gaussianity assumption for PTP links, if practical. For DRNs, it may instead be an effective option, provided that one is fully aware of its limitations.

7.2.1.5 *The NLI Circular-Additive Gaussian-Noise Approximation*

This approximation consists of assuming that the disturbance generated by non-linearity, that is, NLI, manifests itself as AGN at the output of the link, *circular* and *independent* of either the signal or ASE noise (for short CIAGN). One of the key implications of this assumption is that the analytical assessment of the impact of non-linearity on system performance can be carried out by calculating the variance of NLI noise on the signal constellation, and then simply adding it to the noise variance due to ASE. In particular, this approximation makes it possible to characterize a channel BER based on a modified "non-linear" OSNR:

$$\mathrm{OSNR}_{\mathrm{NL}} = \frac{P_{\mathrm{ch}}}{P_{\mathrm{ASE}} + P_{\mathrm{NLI}}} \tag{7.2}$$

where P_{NLI} is a suitably calculated power of NLI (see Eq. 7.8). Due to its simplicity, Equation 7.2 has been used in many non-linearity modeling studies to assess the system impact of NLI. It has been extensively validated in many practical system scenarios, where BER estimates based on it have been shown to come very close to Monte-Carlo simulation results [10, 39–41], also when using the very accurate EGN model [35, 42]. Experimental results too, within the limitations of their error margins, have confirmed its validity [41, 43–49].

In reality, NLI noise is only approximately CIAGN. Specifically, it has recently been pointed out that even in uncompensated systems NLI contains a "phase noise" component, which is non-additive [11, 14, 31, 36, 50]. Higher-order formats, such as PM-16QAM, tend to create more of it than lower-order formats, such as PM-QPSK. In addition, it has also been shown that a variable fraction of NLI may have a long correlation time, on the order of many tens, to hundreds of symbols. On the other hand, the non-CIAGN and long-correlation components of NLI tend to be quite large only in special conditions. Specifically, they get larger the more the "ideal distributed amplification" condition is approached[4] (see "lossless fiber approximation," Section 7.2.1.7), whereas they are much smaller in conventional long-haul lumped-amplification systems [51–53]. In such systems, they can typically be neglected and Equation 7.2 can be used to predict system performance with satisfactory accuracy.

This matter is however not settled. New results are steadily appearing, which should be closely monitored for potentially disruptive innovation.

Another aspect related to Eq. 7.2 is that, as already pointed out in Section 7.2.1.3, at high non-linear regimes WDM signal depletion may occur. A heuristic correction of Equation 7.2 is possible to approximately account for this phenomenon [32, 33].

Practical recommendation: Equation 7.2 can typically be trusted, with caution to be exercised for system scenarios that depart substantially from conventional (see Footnote 4). We also advise the reader to monitor the technical literature for possible new results on this topic.

7.2.1.6 The Locally-White NLI Noise Approximation Provided that the previous approximation is accepted, then the NLI noise is fully characterized by its PSD. Such PSD is, in general, non-flat, even if looked at locally, over any single WDM channel. On the other hand, it is generally not far from flat either. The typical shapes of NLI

[4] A potentially greater impact of long-correlated phase and even polarization noise appears to be possible in all-Raman systems using simultaneous co- and counter-propagating pumping, as well as in systems using very short spans (<50 km) [50, 51].

PSDs, found using the GN model [12, 54], show that the error incurred assuming a NLI locally-white-noise (LWN) approximation, with constant PSD value equal to that of the channel center frequency, is modest. In addition, it typically leads to over-estimating noise slightly rather than underestimating it, that is, it is a conservative approximation. An indicative value of such overestimation was found to be 0.3 dB of NLI power for various typical system configurations [38]. On the other hand, a similar investigation is not yet available regarding the EGN model, so this result should currently be taken with some caution.

The LWN approximation has the obvious substantial advantage of requiring the estimate of the NLI PSD at one specific frequency only, typically the center frequency of the channel-under-test (CUT). Besides reducing the computational burden, it also makes it easier to obtain closed-form formulas for the overall NLI power impinging on a channel (i.e., P_{NLI} in Eq. 7.2). On the other hand, if very accurate predictions are needed, this approximation should not be used. In this case, the PSD of NLI must be evaluated at multiple frequencies within a CUT, as many as necessary to make the calculation of P_{NLI} accurate (see Section 7.4, Eq. 7.8).

Practical recommendation: The LWN approximation is acceptable for approximate system performance assessment. It should be removed if high-accuracy predictions are needed.

7.2.1.7 ***The Lossless Fiber Approximation*** One approximation which has been used both in modeling efforts and in theoretical studies consists of assuming that the fiber is lossless or, equivalently, that ideal distributed amplification is present in the link, exactly canceling out loss, so that the signal power stays constant throughout the fiber. This approximation is often found together with the single-polarization approximation (see Section 7.2.1.2).

The reasons behind the use of this approximation have been varied. In certain cases it was that model derivation becomes easier, while it was perhaps assumed that the essential features of non-linearity generation would be preserved. In other contexts, it was viewed as a limiting case of distributed Raman amplification, and hence considered a sufficiently plausible scenario where to estimate, or find bounds related to, the "ultimate" fiber capacity.

Very recent studies have however shown that the lossless fiber approximation creates propagation regimes whose features are markedly different from those of typical practical systems [11, 14, 31, 36, 50, 53]. In particular, a lossless fiber leads to the creation of a large phase noise component within the NLI noise, which may exhibit long correlation time (tens to hundreds of symbols). In addition, it exacerbates the format-dependence of phase-noise generation. These features are much less evident in conventional lumped-amplifications systems [51–53]. In particular the strength of phase noise is less [52] and its long-correlated component tends to decrease substantially [51–53].

Certain simultaneously forward- and backward-amplified all-Raman systems may exhibit a behavior which is, to some extent, similar to that of lossless systems [51] (see also Footnote 4). The features of completely lossless fiber systems are however substantially more extreme than even these special Raman systems [51].

Practical recommendation: The lossless fiber approximation should be avoided as it may produce results that diverge quite considerably, both quantitatively and qualitatively, from the actual behavior of realistic transmission systems.

7.2.1.8 The Incoherent NLI Accumulation Approximation This approximation assumes that the NLI produced in each span adds up incoherently, that is in power, at the receiver site. Specifically, defining as $G_{\text{NLI}}^{(n)}(f)$ the PSD of NLI generated in the nth span, and assuming that it is linearly propagated till the end of the link, the incoherent accumulation approximation implies that the total PSD of NLI at the end of the link[5] is:

$$G_{\text{NLI}}(f) \approx \sum_{n=1}^{N_{\text{span}}} G_{\text{NLI}}^{(n)}(f) \tag{7.3}$$

In reality, the NLI contributions generated in each span should be added together coherently, at the *field* level, keeping both their amplitude and phase into account. This approximation, however, allows to greatly simplify the computation of the accumulation of NLI along a link. Its use is especially beneficial in the context of physical-layer aware DRNs, for various reasons that are explained in Section 7.4.5.2.

The accuracy issue for the incoherent approximation is complex. It is discussed in detail in Sections 7.4.3.4 and 7.4.3.5. Different behaviors can be obtained depending on the model it is used with (such as for instance the GN model, EGN model or others). When applied to the GN model, it produces rather accurate estimates of system maximum reach for typical lumped-amplification PM-QAM systems [38, 40], taking advantage of an error cancelation circumstance. In other contexts it can yield less favorable results. As a rule of thumb, its accuracy is poor at very low span count and at very low, and especially single, channel count.

In essence, the incoherent accumulation approximation should be viewed as a practical heuristic tool for achieving drastic complexity reduction for certain specific modeling needs and application scenarios. As such, it needs targeted ad hoc validation. If properly tailored and used, it may prove very effective.

Practical recommendation: The incoherent accumulation approximation should not be employed for high-accuracy link design/analysis purposes, or for system research. It can however be a quite effective solution for specific applications (such as real-time management of DRNs), provided that its limitations are understood and pre-assessed in those scenarios.

7.2.1.9 The Noiseless Propagation Approximation This approximation consists of neglecting the NLI produced by ASE noise or by the interplay of ASE noise with the WDM signal. It can be equivalently stated by saying that NLI is generated only by the WDM signal propagating along the link. This approximation relies on the

[5] Equation 7.3 assumes link transparency, that is, that the loss of each span is exactly compensated for by the span amplifier (lumped or distributed). Otherwise, the balance of all loss and amplification, from each span all the way to the receiver, should be taken into account. This can be done but, for simplicity, we assume transparency throughout this chapter, unless otherwise stated.

observation that the OSNR of a coherent optical system cannot be too low, to ensure sufficiently low BER at the receiver. If the WDM signal is substantially stronger than ASE noise, then indeed NLI generation is almost exclusively due to the WDM signal itself and ASE noise can be neglected.

The noiseless approximation is widely used in modeling efforts and system investigations. It makes modeling easier and allows certain useful results to be obtained in closed-form, such as the optimum channel launch power of Equation 29 in Ref. [38]. On the other hand, a recent study [32] has pointed out that non-negligible loss of predictive accuracy may be incurred when neglecting ASE-generated NLI, even in systems that do not operate at extremely low OSNRs. For instance, in a PM-QPSK system with 15 channels (32 GBaud, spacing 33.6 GHz, non-zero dispersion-shifted fiber as in Table 7.1, operating at BER $= 2 \times 10^{-2}$), neglecting ASE-generated NLI caused an overestimation of the maximum reach by over 6%. In this system the OSNR due to ASE alone was about 8 dB at the receiver, at maximum reach, over a bandwidth equal to the symbol rate. This result shows that even apparently sizeable OSNRs do not guarantee that NLI generation is completely unaffected by ASE.

The reach overestimation error further increases for systems capable of operating at even lower OSNRs, which can be the case either because their FECs can tolerate higher BERs or because intrinsically more robust, such as those using PM-BPSK. In these cases, ASE-generated NLI must necessarily be accounted for to avoid large reach overestimation errors. Both rigorous and heuristic methods have been proposed, for instance [11, 22, 23, 32, 55].

Practical recommendation: The noiseless propagation approximation is sufficiently accurate as long as the ASE-only OSNR at the receiver (over a bandwidth equal to the symbol rate) is larger than about 9–10 dB. For lower OSNRs, ASE noise should be accounted for in NLI generation.

7.2.1.10 The XPM Approximation The XPM approximation consists of taking into account only the "cross-phase-modulation" [3] contribution among all the possible NLI-generating processes. Specifically, XPM accounts exclusively for the non-linear distortion induced on the CUT by the power fluctuations of each single interfering (INT) channel in the WDM comb, individually. Neglected are all effects involving the mutual non-linear interaction of two different INTs or three INTs, affecting the CUT, often referred to as FWM [3]. Note that in the absence of dispersion, XPM would manifest itself as pure phase noise on the CUT, which is why the XPM acronym refers to "phase."

TABLE 7.1 Parameters of the fiber types addressed in this chapter

Fiber type	α_{dB} (dB/km)	D (ps/nm/km)	γ (1/W/km)
PSCF	0.17	20.1	0.8
SMF	0.2 / 0.22	16.7	1.3
NZDSF	0.2 / 0.22	3.8	1.5
LS	0.2 / 0.22	−1.8	2.2

Note that XPM *does not include* single-channel non-linearity, which is the non-linear effect of a channel onto itself (also called self-channel interference, SCI, or SPM, self-phase modulation [3]). For the purpose of this discussion, though, we intend the "XPM approximation" as neglecting certain *inter-channel* non-linear effects, whereas we assume that single-channel ones are separately and correctly accounted for.

In the context of the GN–EGN model class, the XPM approximation was proposed in Refs [36, 50]. To distinguish it from XPM approximations applied to other model classes, we call it frequency-domain (FD) XPM. This denomination appears appropriate, because in Ref. [36], Section 3, the proposed model based on the XPM approximation is introduced as "frequency-domain analysis."

A visually intuitive way of depicting the FD-XPM approximation is in relation to Figure 7.1. This figure will be exhaustively explained in Section 7.3.2, so the reader interested in the theoretical details should refer to that section. Here we simply point out that Figure 7.1 represents the two-dimensional frequency plane where the integral equations of the GN and EGN model must be evaluated. In particular each closed lozenge or triangle is a specific integration sub-domain, which contributes some of the NLI power falling on the center channel of a WDM comb. To account for all contributions, all such sub-domains (or "islands") should be considered. Note that the actual analytical form of the integrals also changes according to the specific islands.

The FD-XPM approximation consists of considering the contribution of the X1 islands only, while neglecting all others (see also Figure 7 in Ref. [35]). Specifically, the X1 regions are those straddling the f_1 and f_2 axes [35, 54]. This approach clearly simplifies the model calculations considerably. On the other hand, the majority of the NLI contribution islands is neglected. Yet, despite this circumstance, the FD-XPM approximation may be sufficiently accurate because the X1 islands, although a minority in number, often contribute most of the total NLI.

For certain system scenarios, though, the FD-XPM approximation may underestimate NLI considerably. This happens specifically when fiber dispersion is small, channel spacing is Nyquist or quasi-Nyquist, and the symbol rate is low, these three conditions re-enforcing one another [35, 42]. As an indicative rule for Nyquist or quasi-Nyquist systems, derived from considerations on the relative strength of the XPM and FWM contributions present in the GN–EGN model, a substantial NLI underestimation error may be incurred by the FD-XPM approximation when $\beta_2 R^2 < 1/100$ (in units km^{-1}), with R the symbol rate and β_2 the fiber dispersion coefficient (see Section 7.4.2 for more detailed symbol definitions).

For instance, in Ref. [42] (Figure 2(b) and (c)) two examples of a 15-channel, 32 GBaud PM-QPSK quasi-Nyquist system are shown, operating over either NZDSF or LS fiber (see Table 7.1 for fiber parameters). The resulting values of $\beta_2 R^2$ are $1/200$ and $1/400$. After 20 spans (span length 100 km) the FD-XPM approximation underestimates NLI by 1.7 and 2.7 dB, respectively.

According to the $\beta_2 R^2$ rule above, the FD-XPM approximation should be increasingly critical for decreasing symbol rates. Provided that the symbol rate is low enough, inaccuracy should be observable over any fiber, including high-dispersion ones. This is indeed what is found when analyzing the problem of assessing the

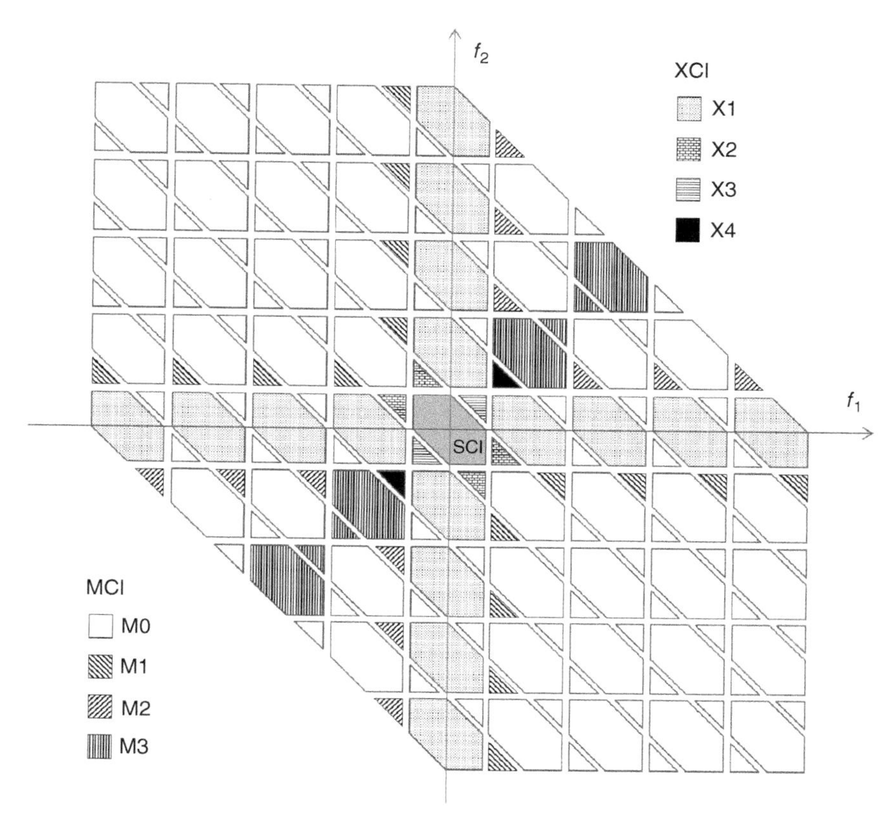

FIGURE 7.1 GN–EGN model overall integration domain for a 9-channel system with spacing 1.05 times the symbol rate and ideal rectangular channel spectra. This example assumes that the PSD of NLI is being evaluated at the center frequency of the center channel, formally $f = 0$ in Equation 7.10. The white "islands" have a GN model term only, whereas the other islands have both a GN model term and a non-Gaussianity correction term (or "EGN model term"). The FD-XPM approximation (see Section 7.2.1.10) consists of considering the X1 islands only, while neglecting all others. MCI is multi-channel interference and XCI is cross-channel interference, see Section 7.4.3 for details.

symbol rate minimizing the generation of NLI, for a constant spectrally efficiency. This topic is dealt with in detail in Section 7.4.4. When trying to address it using the FD-XPM approximation, an increasing NLI estimation inaccuracy is found when decreasing the symbol rate. Incidentally, this feature of the FD-XPM approximation makes it also inadequate for dealing with those OFDM or Nyquist-FDM systems whose per-subcarrier rate is low.

Another form of XPM approximation is used by the FRLP (frequency-resolved logarithmic-perturbation) model [14, 31]. This model bears some similarities with the EGN model both because it adopts a first-order perturbation method and because it assumes that the CUT is decomposed into elementary frequency components, as it is done in the EGN model and in general in the SpS methods (see Section 7.3). However, it departs substantially from the EGN model because it uses a logarithmic-perturbation (LP) approach rather than a RP one, and because the effect of non-linearity on the CUT is expressed in the form of a time-dependent transfer function. Yet another version of the XPM approximation is used in a time-domain (TD) first-order RP model proposed in Ref. [11] and used also in Refs [36, 51] (TD-XPM). These further XPM models are quite effective in depicting certain specific features of NLI, and in particular non-linear phase noise (see Section 7.2.1.5), for which they even allow to find simple closed-form results. Contrary to the FD-XPM approximation, there is currently no specific targeted investigation of the possible errors incurred by these further XPM models, when going towards low $\beta_2 R^2$ products and low channel spacing. Since they explicitly neglect FWM, they may suffer from similar limitations, but this topic has not been explicitly explored as yet.

Practical recommendation: The XPM approximation should be used with caution, as it may substantially underestimate NLI at low symbol rates and/or low dispersion. In general, when used within any model, the accuracy of the XPM approximation should be verified when $\beta_2 R^2 < 1/100$ (km^{-1}). In addition, if the overall NLI is needed, single-channel effects must be somehow re-introduced, since XPM does not include them.

7.3 INTRODUCING THE GN–EGN MODEL CLASS

In this section, we provide background information on the GN–EGN model class. We start out with a brief overview of the literature on the GN model and on various prior modeling efforts which are related to the GN model. We then address its early validation and the subsequent observation of its limitations that have eventually led to the formalization of an enhanced-GN (EGN) model.

Those readers who are chiefly interested in indications about practical modeling solutions for specific system scenarios may skip this section and go to Section 7.4.

7.3.1 Getting to the GN Model

Numerous models were proposed prior to the GN model, which bear substantial similarities to the GN model itself. They all rely on first-order perturbation approaches.

To the best of our knowledge, the earliest of these models dates back to 1993 [34]. It was based on directly postulating that all non-linearity was produced by FWM acting among the WDM signal spectral components, assumed "incoherent." This latter assumption is equivalent to the signal-Gaussianity approximation of Section 7.2.1.4. Though limited to single-polarization, ideal-distributed amplification and a rectangular overall WDM spectrum, the derived equations were essentially the same as those of the GN model for such idealized system scenario. In 2003, it was shown in Ref. [33] that results similar to [34] could also be derived using a different perturbation approach, earlier proposed in Ref. [28]. Equations similar to the single-polarization GN model were also independently derived in Ref. [55] using the truncated VS approach in frequency domain, introduced in Ref. [26].

More recently (2008– 2010), a derivation approach analogous to that used in Ref. [34] was taken up again, based on ideally slicing up the signal spectrum into discrete spectral components. This "spectral slicing" (SpS) approach naturally lends itself to describing OFDM systems, and in fact it was first used to model NLI limited to OFDM [56, 57]. There, the generated NLI PSD was found through first-order perturbation dual-polarization analytical FWM formulas, applied to the OFDM subcarriers. This modeling effort obtained what can be viewed as a specialized version of the GN model for OFDM.

The SpS approach was independently exploited in 2010–2011 to address generic WDM systems [39, 40, 58]. In these papers, SpS was used early in the derivation and then it was removed through a transition to continuous spectra. A first-order regular perturbation method applied to the Manakov equation was used. The dual-polarization general form of the GN model was obtained as a result. The name "GN model" was first introduced in Ref. [40].

Two further papers proposed detailed re-derivations of the GN model [12, 13]. Specifically, [12] used a variation on the SpS approach while [13] was based on a modified version of the first-order RP method, called enhanced-RP. Both independently confirmed the GN model equations and provided insightful extensions and generalizations. Various follow-up papers have been published on the GN model, providing further generalizations and numerous approximate closed-form solutions to the GN model reference integral equation formula (GNRF, see [38] Eq. 1) [38, 54, 59–61].

7.3.1.1 GN Model Analytical Derivation Having introduced the GN model theoretical and bibliographical background, it would be in order to provide the model formulas and an outline of the model derivation analytical steps. Since, however, the GN model has recently evolved into the more accurate EGN model, of which it is one of the constituents and with which it shares most of the analytical derivation, we elect to address these aspect in a unified way, for both the GN and EGN models. The model formulas will be introduced in Section 7.4.3, whereas an outline of the main derivation steps is provided in Section A.4. We therefore go directly to the topic of the accuracy tests on the GN model, since the observation of some of the GN model accuracy limitations was essential for motivating its evolution into the EGN model.

7.3.1.2 GN Model Simulative and Experimental Tests The GN model was proposed specifically to address uncompensated transmission (UT) multi-channel systems, operating at typical commercial system symbol rates (>10 GBaud), over fibers whose dispersion was not too low. It was deemed inaccurate for DM systems or for UT systems operating near zero-dispersion, and caveats were also put forth regarding its use for single-channel systems [39, 40, 54]. For this reason, its initial simulative validation was targeted at *UT multi-channel* conditions, at a minimum dispersion value of $D = 3.8$ ps/(*nm km*), corresponding to that of a typical long-haul non-zero dispersion-shifted fiber (NZDSF) [39, 40].

Figure 7.2 shows a comprehensive set of maximum reach results at 32 GBaud (see the figure caption for more system details), where markers are simulations and lines are predictions based on the GN model. The solid lines refer to the "incoherent" GN model, which makes use of the incoherent NLI accumulation approximation (Section 7.2.1.8), and reproduce the results of [40], whereas the dashed ones do not use such approximation. In both cases the GN model performance is rather good. It holds up well from 1 bit/(sHz) spectral efficiency (SE) and 20,000 km maximum reach to 6 bit/(sHz) SE and 200 km maximum reach, across four transmission formats, three different fibers and six channel spacings. Similar good agreement was found in Ref. [39] also when changing the symbol rate and various other parameters, at Nyquist spacing. Simulative tests were run independently by groups not involved in the development of the GN model, who also found good model accuracy [10, 41].

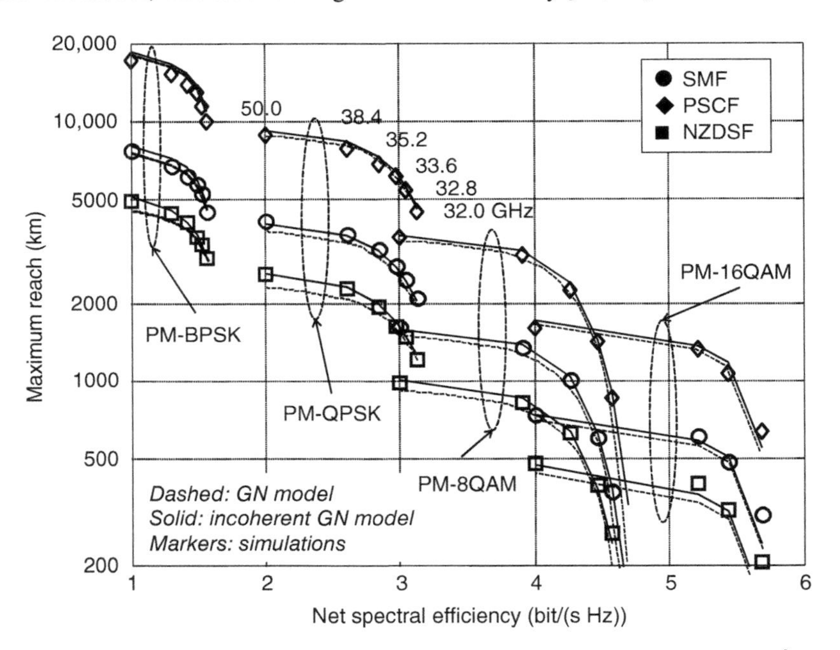

FIGURE 7.2 Maximum system reach at optimum launch power, for BER = 10^{-3}, versus net spectral efficiency. Markers are simulations, lines are analytical predictions using the GN model (solid: with the incoherent NLI accumulation approximation of Section 7.2.1.8; dashed: without). The shape of the markers identifies three different fiber types, SMF, NZDSF and PSCF, whose parameters are collected in Table 7.1. The span length is 100 km in all cases, with EDFA amplification (noise-figure 5 dB) and UT. Transmission employed the four formats indicated in figure, at 32 GBaud. The labeling in GHz indicates channel spacing, which starts at 50 GHz for all formats. The total combined framing and FEC overhead is 28%, resulting for PM-QPSK in a net bit rate of 100 Gb/s. Spectral shaping is performed using fourth-order super-Gaussian optical filters with bandwidth 32 GHz. The drop in system performance observed at low channel spacing is mostly due to linear inter-channel crosstalk, caused by the assumption of the use of the super-Gaussian optical filters, instead of more ideal raised-cosine pulse-spectrum shaping based on transmitter DSP.

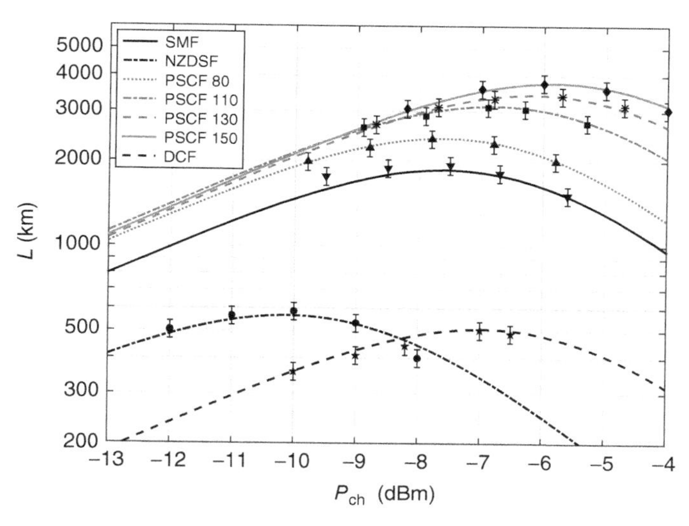

FIGURE 7.3 System reach versus launch power per channel P_{ch}, at BER = 1.5×10^{-2}. Markers are experimental results, lines are analytical predictions using the GN model. Error bars show a total 15% error interval (±7.5%) over the reach. The parameters of the fibers can be found in Ref. [47]. The numbers in the legend refer to the fiber effective area, in μm². The system had 22 channels, modulated PM-16QAM at 15.625 GBaud (125 Gb/s), quasi-Nyquist spaced (16 GHz). The span length was between 50 and 55 km for all fibers, except for the DCF, whose span length was 20 km.

The GN model has also enjoyed extensive experimental confirmation, among which [41, 43–48, 49]. Two of these experiments were specifically designed to test the model over different fiber types [43, 47]. In particular, the latter addressed seven fiber types, including a dispersion-compensating fiber (DCF) used as transmission fiber. It was based on a 22-channel WDM PM-16QAM system, running at 15.625 GBaud. Overall, a good match between predictions and experiments was found (see Figure 7.3).

Other quite extensive independent experiments were carried out in [45, 46, 48]. In particular, in Ref. [48] a 152-channel WDM signal using quasi-Nyquist transmission (32 GBaud, 33 GHz spacing) with PM-QPSK, PM-16QAM or PM-64QAM was transmitted over 60-km spans of high-performance PSCF, and subjected to numerous tests. Figure 7.4 shows the measured and GN-model predicted Q factor on the center channel (ch. 77) versus distance, for various formats.[6] The results of this experiment also agreed with some of the GN-model predicted general system features, at least within the accuracy of the obtained experimental measurements. For instance, the prediction that the optimum launch power should be format-independent agreed with the experimental results. In Reference [45], in a similar set-up, the GN model

[6]To reduce experimental uncertainties, the method used in Ref. [48] consisted of best-fitting the model parameters for PM-QPSK and then use such parameters and the theoretical GN model equations to extrapolate the PM-16QAM and PM-64QAM predictions (no further best-fitting).

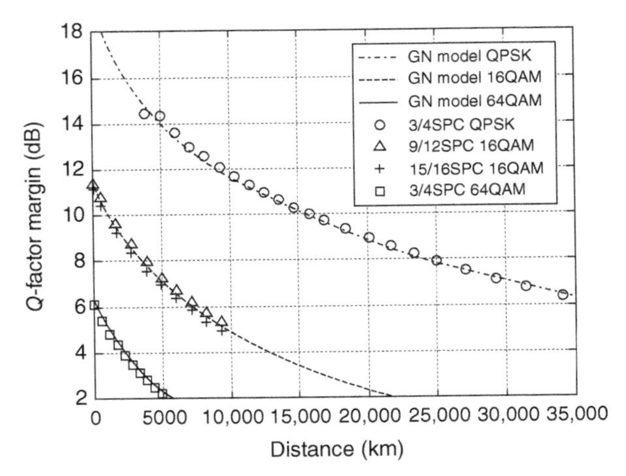

FIGURE 7.4 Measured and GN-model-predicted Q factor on the center channel (ch. 77) versus transmission distance for various formats. Data taken from Ref. [48]. The system consisted of 152-channels, quasi-Nyquist, at 32 GBaud, with 33 GHz spacing. The employed raw formats were PM-QPSK, PM-16QAM and PM-64QAM. Coded modulation (SPC, single parity-check) was used to improve sensitivity, see [48] and Footnote 6 for details.

prediction regarding the extent of the benefit obtainable through single-channel digital BP (backward-propagation) were experimentally confirmed to be rather accurate.

Despite the ample validation, certain discrepancies between the GN model predictions and certain simulative results were observed. For instance, a paradox is present in Figure 7.2: the GN model curves appear somewhat less accurate than the incoherent GN model ones. This is puzzling, since the GN model is a less approximate model that the incoherent GN model. It has later been understood that the incoherent GN model benefits from two approximations canceling each other error's out [38]. In any case, the paradox clearly signaled the problem of some inaccuracy being intrinsic to the GN model.

Issues with the GN model accuracy were first pointed out, at the same time, in Refs [37, 62]. They were simulatively investigated down to the level of span-by-span NLI accumulation along the link in Ref. [37] (see Section 7.3.2). Another group then independently found large discrepancies (several dBs) between the GN model prediction and the simulated NLI accumulation along the link, when assuming single-polarization lossless fibers [36] or, but to a lesser extent, single-polarization lossy fiber with short spans [50]. These discrepancies were attributed in all papers to the signal-Gaussianity approximation (see Section 7.2.1.4). The first paper where a suitable procedure for removing it was outlined was [36], limited to XPM. The full removal of this assumption from all NLI components has eventually led to the EGN model.

One question that needs to be addressed is: why were these GN model issues not detected in experiments? The likely answer is that experiments were mostly run on set-ups having typical realistic parameters, where the difference between the GN model predictions and the actual system performance is modest. Note that in typical systems the inaccuracy of the GN model can be expected to be similar to what is shown in Figure 7.2, that is, quite small. Given the extent of the possible experimental uncertainty, such differences may well have gone undetected.

Another plausible explanation is the following. As it will be shown in detail in the next section, the GN model always *overestimates* non-linearity, leading to somewhat pessimistic maximum reach predictions. In experiments, a number of small impairments often add up causing some penalty versus the expected performance, which may have brought the experimental results close to the GN model predictions. Yet another aspect is that in most experimental papers the GN model predictions were made neglecting the non-linear effects produced by the co-propagating ASE noise, as well as neglecting channel power depletion. These are reasonable approximations (see Sections 7.2.1.3 and 7.2.1.9) whose impact may however make up, in certain system configurations, for a substantial portion of the distance between the GN model predictions and the actual system performance.

7.3.2 Towards the EGN Model

Figure 7.2 indicates that the GN model returns a somewhat pessimistic maximum reach prediction, especially over lower-dispersion fibers. While the extent of the error has been found to be limited for typical system configurations at 28–32 GBaud (5–15%), it would obviously be desirable to obtain a model that would avoid such error and could therefore be used reliably in less-typical system configurations too. Such model would be useful not only in the analysis and design of current systems, but also in the exploration of possible innovative non-standard system solutions.

Research towards this goal started by analyzing not just the predicted system maximum reach, but the detailed prediction of the generation and accumulation of NLI, span by span, along the link. The first paper that performed such an analysis, by comparing accurate simulation results with the GN model prediction, as mentioned in the previous section, was [37]. Some of those results are shown in Figure 7.5. The plotted non-linearity parameter is:

$$\eta_{\text{NLI}} = P_{\text{NLI}} / P_{\text{ch}}^3 \tag{7.4}$$

that has dimensions of (1/W^2). The advantage of looking at the parameter η_{NLI} is that it is independent of the signal launch power used for testing, since $P_{\text{NLI}} \propto P_{\text{ch}}^3$. Figure 7.5 shows that over the first few spans, where the signal is certainly farthest from Gaussian-distributed, the GN model (dashed-dotted line) substantially overestimates NLI noise power (black solid line), up to several dB's. Such overestimation then abates considerably along the link and, in the examples of Figure 7.5, it drops to about 1.3 and 0.9 dB for PM-QPSK and PM-16QAM, respectively.

Note that these NLI errors translate into comparatively smaller maximum reach errors because the relationship between such quantities is approximately [54]:

$$\Delta L^{\max}\ [\text{dB}] \approx -\frac{1}{3} \Delta\eta_{\text{NLI}}\ [\text{dB}] \tag{7.5}$$

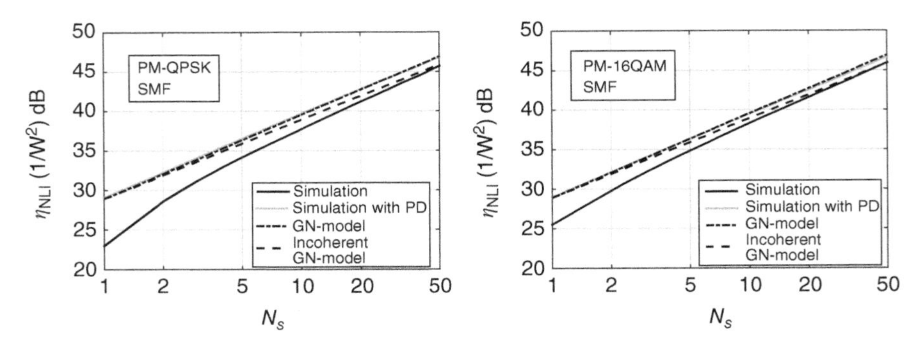

FIGURE 7.5 Normalized NLI noise power η_{NLI} (see Eq. 7.4) over the center channel, versus the number of spans N_s, for quasi-Nyquist, 9-WDM-channel systems (32 GBaud, 33.6 GHz channel spacing), with 100 km SMF span length and EDFA amplification. "PD" means that the simulated signal was thoroughly pre-dispersed so that it essentially behaved as Gaussian noise.

where $L^{\max}$ is the maximum reach (i.e., the system reach at the optimum launch power) and the symbol Δ means the ratio of two values of the same quantity. So a 1 dB overestimation of NLI power leads to only 1/3 dB (or 7%) maximum reach underestimation. Equation 7.5 is one of the main reasons why detailed NLI accumulation studies are necessary for in-depth model investigations: maximum reach alone is inadequate because it has too weak a sensitivity versus NLI estimation errors.

In Reference [37] it was conjectured that NLI overestimation was due to the signal-Gaussianity approximation (see Section 7.2.1.4) used by the GN model. To test this hypothesis, system simulations were run where a considerable amount of pre-dispersion (200,000 ps/nm in figure, though similar results were obtained with 100,000 ps/nm) was applied, so that the signal started out very dispersed and did essentially behave as Gaussian noise.[7] In this case (Figure 7.5, light gray solid curves) there was an excellent coincidence between simulations and GN model. So it was surmised that it had to be the non-Gaussian nature of the signal that caused the actual NLI curve to depart from the GN model one, since a Gaussian-distributed signal behaved exactly as the GN model prediction.

[7]In Reference [37] it was conjectured that the gradual convergence of the actual NLI curve (non-PD) towards the GN model, shown in Figure 7.5, was due to the accumulated dispersion increasingly making the signal more Gaussian-distributed along the link. This interpretation however does not explain why the distance between actual NLI and the GN model tends to stabilize to an asymptotic value, rather than gradually going to zero (such asymptote has been recently calculated [42] and is discussed in Section 7.4.3.3). A different interpretation has been proposed [50], whereby the notable abatement of the error between the GN model and the actual generated NLI along the link is due to the dynamics of the XPM collisions among the pulses carrying the symbols in the CUT and INTs. In particular, the onset of incomplete collisions, as the pulses considerably spread out due to dispersion, would be responsible for such an abatement. A more detailed analysis along the same lines is presented in Ref. [24], a paper appeared after this chapter manuscript was finalized. This pulse collision interpretation becomes however more involved in dual-polarization conditions and does not account for FWM effects. Research is ongoing on the topic and it is possible that new results may emerge, perhaps providing a more unified picture.

Remarkable progress towards overcoming this GN model limitation was made in Ref. [36], which succeeded in analytically removing the GN model signal Gaussianity approximation from one of the main contributions to NLI, the cross-phase-modulation one (XPM, see Section 7.2.1.10). This result also confirmed the dependence of XPM generation on the fourth moment of the signal constellation, a result that other groups also found within different modeling approaches [11, 14]. Note that the GN model, due to the signal Gaussianity approximation, takes into account the signal constellation second moment only, both for the XPM contribution and for all other non-linearity contributions.

The approach of [36] was then extended and generalized, to derive a complete 'enhanced' GN model, the 'EGN' model [35], which addresses not just XPM (i.e., the X1 islands in Figure 7.1) but all NLI components (all the islands in Figure 7.1), including SCI.[8] Interestingly, SCI turned out to involve not just the fourth, but also the sixth moment of the signal constellation. Note that in this chapter the XPM model proposed in Ref. [36], which includes the X1 regions exclusively, is called "FD-XPM" (frequency-domain XPM, see also Sect 7.2.1.10), to distinguish it from other models that were defined as XPM-based, such as [11, 14].

The accuracy of the EGN model has been since tested simulatively in various configurations. It has, so far, always been found excellent, with maximum reach errors on the order of few percent, at the limit of the accuracy of the computer simulations run for comparison [23, 35, 42, 63, 64]. In-depth tests carried out at the more error-sensitive level of NLI accumulation (the η_{NLI} parameter of Eq. 7.4) have shown the EGN model to supersede the limitations of both the GN model and the FD-XPM model (see Section 7.2.1.10 and [35, 42]). In particular, very low dispersion (such as LS fibers) is handled successfully, as well as single or low span count, single-channel transmission, very low symbol rates (see Section 7.4.4), links with mixed fiber types (SMF and LS were tested in Ref. [64]), and special conditions found in dynamic reconfigurable networks (see Section 7.4.5).

A few examples of these NLI accumulation tests are shown in Figure 7.6. The EGN model curve is always much closer (or virtually superimposed) to the simulation results than either the GN model or the FD-XPM approximation.

On the other hand, the price to pay for such wide-scope predictive accuracy is increased complexity. This is not just measured in terms of computing time, but also in terms of the more complex management of the model. This aspect will be made clearer in Section 7.4.5, where it will be shown that the EGN model, to be ideally capable of accurately predicting the NLI induced on any CUT propagating through a network, requires to be supplied with the detailed trace of the propagation history of all other INT channels. This includes their format, where they started interacting with the CUT and where they stopped doing so, as well as how much dispersion they had accumulated when interacting with the CUT. Specifically, the latter is key information, which impacts NLI generation on the CUT in a substantial way.

[8] After this manuscript was completed, an interesting EGN model re-derivation using a different methodology, with significant extensions towards analyzing NLI correlation features, has been published [23].

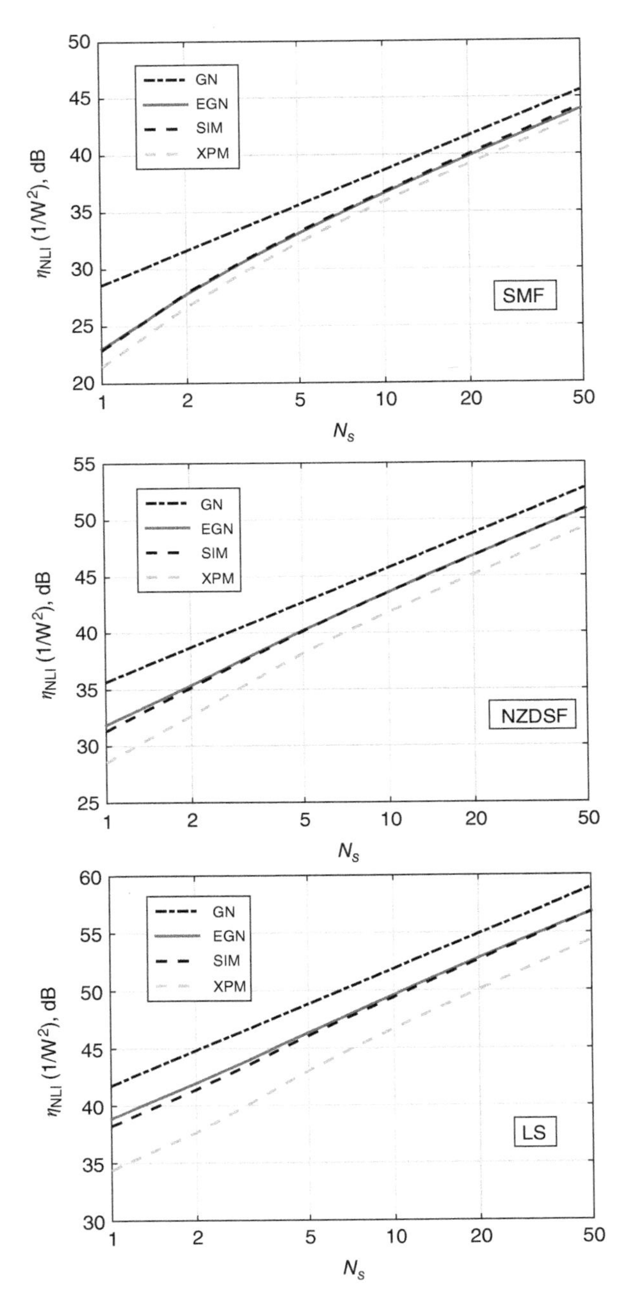

FIGURE 7.6 Normalized NLI noise power η_{NLI} (see Eq. 7.4) over the center channel, versus the number of spans N_{span}, for quasi-Nyquist WDM 15-channel PM-QPSK systems (32 GBaud, 33.6 GHz channel spacing), with 100 km span length and EDFA amplification. *Single-channel non-linearity is removed from all curves* (including simulation results) to make a comparison possible with the FD-XPM approximation. The legend labels mean the following. GN: GN model; EGN: EGN model; SIM: split-step simulations; XPM: the FD-XPM approximation (Section 7.2.1.10).

We should also mention that, at present, the EGN model validation is mostly simulative, with the exception of [65]. There certainly is the need for more targeted experiments, specifically designed for the purpose of confirming the GN–EGN model dichotomy. As for [65], the EGN model predictions regarding the existence of an optimum transmission symbol rate were experimentally confirmed. For more details on this topic, see Section 7.4.4.

At any rate, the field of non-linearity modeling is quite effervescent and though we tried to include the latest results in this section, we encourage the reader to check for possible further developments,[9] which are likely to appear in the near future.

7.4 MODEL SELECTION GUIDE

In this section we provide practical guidance as to which model to use, within the GN–EGN model class, to best tackle certain specific analysis and design needs.

We make a first splitting distinction between PTP links and DRNs. The main difference is that in PTP links it is assumed that the channels travel together from source to destination, whereas in DRNs a certain CUT can change its INTs several times during propagation. This latter circumstance alters the non-linearity picture quite substantially.

Both for the PTP link and DRN cases we will provide a significant "case study." In PTP links it consists of the determination of the symbol rate minimizing the generation of NLI (Section 7.4.4). In DRNs it is the analysis of the widely different NLI profile that spectrally identical WDM signals may generate depending on routing assumptions (Section 7.4.5.1).

7.4.1 From Model to System Performance

Before we discuss NLI modeling, it is necessary to clarify how system performance is calculated and what the model needs to deliver to make it possible to calculate it. For both PTP links and DRNs, we will accept the CIAGN approximation for NLI (see Section 7.2.1.5). Then, the system BER can be estimated by inserting the non-linear OSNR of Equation 7.2 into a suitable BER formula, which depends on the transmission format. For instance, for PM-QPSK, the BER formula is:

$$\mathrm{BER}_{\mathrm{PM-QPSK}} = \frac{1}{2}\mathrm{erfc}\left(\sqrt{\frac{\mathrm{OSNR}_{\mathrm{NL}}}{2}}\right) \tag{7.6}$$

[9] One example of ongoing research is the following. As discussed in Section 7.2.1.5, the NLI at the receiver contains a certain amount of long-correlated non-linear phase noise (LC-NLPN). The EGN model includes LC-NLPN in the estimate of the NLI variance that it provides. However, recent results show that LC-NLPN gets typically suppressed by the receiver carrier-phase estimation circuitry and hence should not affect reception. This in turn means that the current EGN model, which includes it, may overestimate the actual amount of NLI that affects reception. Correcting the EGN model for this aspect is a current topic of investigation. LC-NLPN may be significant in links with very-low-loss spans or close-to-ideal distributed amplification, especially with high-order modulation formats. See Section 7.2.1.5 for more details.

The BER formulas for the other main QAM formats are provided in Section A.1.

The non-linear OSNR of Equation 7.2 must be computed as follows. Note that all PSDs are assumed to be *unilateral*. P_{ASE} in Equation 7.2 is:

$$P_{\text{ASE}} = G_{\text{ASE}}(f_{\text{CUT}}) \cdot R_{\text{CUT}} \tag{7.7}$$

where $G_{\text{ASE}}(f)$ is the PSD of ASE noise, f_{CUT} is the center frequency of the CUT and R_{CUT} is the CUT symbol rate. P_{NLI} in Equation 7.2 is given by:

$$P_{\text{NLI}} = \frac{R_{\text{CUT}}}{B_H} \int_{-\infty}^{\infty} G_{\text{NLI}}(f + f_{\text{CUT}})\, |H_{\text{Rx}}(f)|^2 df \tag{7.8}$$

where B_H is a normalization factor, defined as:

$$B_H = \int_{-\infty}^{\infty} |H_{\text{Rx}}(f)|^2 df \tag{7.9}$$

and $H_{\text{Rx}}(f)$ is the receiver overall baseband transfer function, which is assumed to be *matched* to the transmitted signal *baseband* pulse. Note that, in coherent systems, the DSP adaptive equalizer tends to make $H_{\text{Rx}}(f)$ matched, so this assumption appears reasonable. The above formulas also assume that inter-symbol interference (ISI) be absent. Otherwise, a penalty can be expected with respect to their predictions.

Equation 7.8 clearly shows that the key quantity that must be estimated through NLI models is the PSD of the NLI noise, that is, $G_{\text{NLI}}(f)$, at least over the bandwidth of the CUT.[10]

7.4.2 Point-to-Point Links

As shown in previous sections, the level of accuracy of the EGN model in characterizing the power of NLI interfering with a certain CUT has been found to be very good in all system situations tested so far, including some non-typical ones. The EGN model would therefore seem to be the natural first pick for the analysis and design of PTP links. However, such good and reliable performance comes at the price of substantial model complexity. In addition, from its integral equations it is not possible to glean

[10]Note that Equations 7.7–7.8 assume an "OSNR noise bandwidth" equal to the CUT symbol rate R_{CUT}. Other common choices are 0.1 or 0.5 nm. However, choosing R_{CUT} makes the OSNR_{NL} versus BER laws, such as Equation 7.6, invariant versus the symbol rate. That is, a given value of OSNR_{NL} always corresponds to the same BER, independently of R_{CUT}. This greatly simplifies the comparison of systems operating at different symbol rates. Note that the conversion from a generic noise bandwidth B_{N} to R_{CUT} is simply:

$$\text{OSNR}_{\text{NL}}\,|_{R_{\text{CUT}}} = \text{OSNR}_{NL}\,|_{B_{\text{N}}} \cdot \frac{B_{\text{N}}}{R_{\text{CUT}}}$$

Under the mentioned assumptions of a matched $H_{\text{Rx}}(f)$, no ISI, and a noise bandwidth equal to R_{CUT}, OSNR_{NL} also corresponds to the signal-to-noise ratio that can be measured on the Rx electrical signal constellation, at the input of the decision stage. For more details on these aspects see [38], Section IV.

the actual dependence of NLI on many of the key system parameters, something that is of great interest from both a theoretical and a practical viewpoint.

So, we first introduce and discuss the full EGN model. Then, we provide a number of alternatives, found by gradually relaxing the accuracy constraints while gaining in either ease of use, speed of computation or parameter-dependence readability.

A comprehensive list of the symbols used in the following is provided here for convenience. The indicated units make them dimensionally consistent. All PSDs are assumed to be *unilateral*.

- z is the longitudinal spatial coordinate, along the link (km).
- α is the fiber *field* loss coefficient (km^{-1}), such that the signal *power* is attenuated as $\exp(-2\alpha z)$.
- β_2 is the dispersion coefficient (ps^2·km^{-1}). The relationship between β_2 and the widely used dispersion parameter D in ps/(nm·km) is: $D = -(2\pi c/\lambda^2)\beta_2$, with c the speed of light in km/s and λ the light wavelength in nm.
- γ is the fiber non-linearity coefficient (W^{-1}·km^{-1}).
- L_s is the span length (km).
- L_{eff} is the span effective length defined as: $[1 - \exp(-2\alpha L_s)]/2\alpha$ (km).
- N_s is the total number of spans in a link, sometimes written N_{span} when necessary for clarity.
- $G_{\text{WDM}}(f)$ is the PSD of the overall WDM transmitted signal (W/Hz).
- $G_{\text{NLI}}(f)$ is the PSD of NLI noise (W/Hz).
- N_{ch} is the total number of channels present in the WDM comb.
- P_n is the launch power of the nth channel in the WDM comb (W). The power of a single channel is also sometimes written P_{ch} when necessary for clarity.
- R_n is the symbol rate of the nth channel (TBaud). The symbol rate of a single channel is also written R, or R_{ch} when necessary for clarity.
- $T_n = R_n^{-1}$ is the symbol time of the nth channel (ps).
- Δf is the channel spacing, used for systems where it is uniform (THz).
- $s_n(t)$ is the pulse used by the nth channel, in time domain. Its Fourier transform is $s_n(f)$. The pulse is assumed to be normalized[11] so that the integral of its absolute value squared is T_n. If any pre-distortion or dispersion pre-compensation is applied at the transmitter, this should be taken into account in $s_n(t)$ and $s_n(f)$.
- B_n is the full bandwidth of the nth channel (THz). If the channel is Nyquist then $B_n = R_n$.
- f_n is the center frequency of the nth channel (THz).
- $a_{x,n}^k, a_{y,n}^k$ are random variables corresponding to the symbols sent on the nth channel at the kth signaling time, on either the polarization $\hat{x}$ or $\hat{y}$; we will assume henceforth that they are all statistically independent of one another

[11]Note that according to such normalization a channel with an ideal rectangular spectrum and bandwidth R_n would have the flat-top value of its Fourier transform $s_n(f)$ equal to R_n^{-1}.

and that within each channel they are all equally distributed. They can have different distributions in different channels. Due to the units assumed for the pulses $s_n(t)$, and the way the overall WDM signal is written in Equation 7.12, then $|a_{x,n}^k|^2, |a_{y,n}^k|^2$ must have dimensions of power (W). See also Equation 7.13.

7.4.3 The Complete EGN Model

In the following we provide the complete set of analytical formulas expressing the GN and EGN model. A summary of the derivation procedure of these models is provided separately, in Section A.4.

According to the EGN model, the PSD of NLI generated at a certain frequency f in the optical spectrum is made up of two contributions:

$$G_{\rm NLI}^{\rm EGN}(f) = G_{\rm NLI}^{\rm GN}(f) - G_{\rm NLI}^{\rm corr}(f) \tag{7.10}$$

where the $G_{\rm NLI}^{\rm GN}(f)$ contribution is calculated according to the GN model, that is, according to the signal Gaussianity assumption of Section 7.2.1.4, and the contribution $G_{\rm NLI}^{\rm corr}(f)$ accounts for the actual non-Gaussian statistical features of the signal. Note that for PM-QAM systems of any order, $G_{\rm NLI}^{\rm corr}(f) > 0$. This means that the correction term always *detracts* from the value of the PSD of NLI found through the GN model.[12] This shows the GN model to be a guaranteed *upper bound* to NLI for all PM-QAM systems.

The GN model contribution is expressed by the GN model reference formula (GNRF):

$$G_{\rm NLI}^{\rm GN}(f) = \frac{16}{27}\int_{-\infty}^{\infty}\int_{-\infty}^{\infty} G_{\rm WDM}(f_1)\, G_{\rm WDM}(f_2)\, G_{\rm WDM}(f_1+f_2-f)\cdot|\mu(f_1,f_2,f)|^2\, df_2\, df_1 \tag{7.11}$$

where $\mu(f_1,f_2,f)$ is the "link function" which depends only on the physical link parameters, that is, fiber parameters and amplification features, but not on the characteristics of the launched signal. The link function is detailed and thoroughly discussed in Section A.2.

Equation 7.11 formally integrates over all frequencies in f_1 and f_2 but in practice its integration domain is shaped by the WDM signal PSD, $G_{\rm WDM}(f)$, whose presence in the integrand function induces a segmentation of the integration domain into many "islands." Figure 7.1 shows an example of such islands for a 9-channel WDM system with rectangular spectra and quasi-Nyquist spacing.

Regarding the correction term $G_{\rm NLI}^{\rm corr}(f)$ in Equation 7.10, again with reference to Figure 7.1, such term is zero over the white-filled islands (the M0 islands). It then

[12]It is possible to conceive signal constellations for which $G_{\rm corr}(f)$ is negative [62]. However no constellation in practical use has this feature. The ideal zero-mean two-dimensional Gaussian constellation produces $G_{\rm NLI}^{\rm corr}(f) = 0$, since in that case the signal Gaussianity assumption is exact.

takes *different formal expressions* over the eight other types of islands (SCI, X1-X4 and M1-M3) and therefore $G_{\text{NLI}}^{\text{corr}}(f)$ cannot be expressed through a single simple formula as for the GN model contribution. Note that the acronyms used in Figure 7.1 refer to the type of non-linear interaction, according to the taxonomy[13] proposed in Ref. [54]. Specifically:

- SCI: It is NLI caused by the CUT onto itself.
- XCI: It is NLI affecting the CUT caused by the non-linear interaction of the CUT with any *single* interfering (INT) channel.
- Multiple-channel interference (MCI): It is NLI affecting the CUT, caused either by the non-linear interaction of the CUT with two INTs or by the non-linear interaction of three INT channels.

A more rigorous set of such definitions, based on the actual spectral position of the WDM signal components beating together, can be found in Ref. [54], Section VI.

Before we proceed, we have to introduce some notation, different from that used in Equation 7.11. The reason is that when dealing with $G_{\text{NLI}}^{\text{corr}}(f)$ it is not possible to just look at the WDM signal PSD. Rather, the Fourier transforms of the individual channel pulses are called into play.

The overall WDM transmitted signal is written in time-domain as:

$$s_{\text{WDM}}(t) = \sum_{n=1}^{N_{\text{ch}}} \sum_{k} (a_{x,n}^{k} \hat{x} + a_{y,n}^{k} \hat{y})\, s_n(t - kT_n)\, e^{j2\pi f_n t} \tag{7.12}$$

where the index k identifies the signaling time-slot, and the index n the WDM channel. According to the symbol definitions established in Section 7.4.2, the power carried by each channel is given by:

$$P_n = \mathrm{E}\{|a_{x,n}^{k}|^2 + |a_{y,n}^{k}|^2\} \tag{7.13}$$

where $\mathrm{E}\{\cdot\}$ is the statistical average (or "expectation") operator.

We also define the following quantities related to the fourth and sixth moments of the channel constellations:

$$\Phi_n = 2 - \frac{\mathrm{E}\{|a_n|^4\}}{\mathrm{E}^2\{|a_n|^2\}}, \quad \Psi_n = -\frac{\mathrm{E}\{|a_n|^6\}}{\mathrm{E}^3\{|a_n|^2\}} + 9\frac{\mathrm{E}\{|a_n|^4\}}{\mathrm{E}^2\{|a_n|^2\}} - 12 \tag{7.14}$$

[13]The islands of Figure 7.1 can also be classified using the traditional non-linear effect taxonomy. Specifically, XPM is the X1 islands only. All other islands can be attributed to FWM, with the exception of the center island, which corresponds to SPM. For the latter, the alternative denomination SCI, was proposed for uniformity with XCI and MCI, and to stress the fact that in UT systems the nature of the disturbance generated by single-channel non-linearity is very different from the well-known pulse-shape distortions occurring in DM systems and typically associated with the acronym SPM.

TABLE 7.2 Values of the Φ and Ψ parameters

Format	Φ	Ψ
PM-BPSK	1	−4
PM-QPSK	1	−4
PM-16QAM	17/25	−52/25
PM-64QAM	13/21	−1161/646
PM-∞-QAM	3/5	−12/7
PM-Gaussian	0	0

where a_n is any of the $a_{x,n}^k$ or of the $a_{y,n}^k$, which are assumed to be all identically distributed. The values of Φ and Ψ for some relevant constellations are shown in Table 7.2. The values of both Φ and Ψ steadily decrease in absolute value when going from a simpler to a more complex PM-QAM constellation. We report their limit values for a QAM constellation made up of infinitely many signal points uniformly distributed within a square region whose center is the origin (the PM-∞-QAM entry in Table 7.2). Note that the values for PM-64QAM are very close to such limit. Also, if the distribution of the transmitted symbols $a_{x,n}^k$ and $a_{y,n}^k$ is assumed to be Gaussian and zero-mean, that is, a "Gaussian constellation" is used, then Φ and Ψ vanish and $G_{\rm NLI}^{\rm corr}(f) = 0$.

We can now return to Equation 7.10. To compute the GN model contribution according to the signal notation introduced above, Equation 7.11 can still be used, with the substitution:

$$G_{\rm WDM}(f) = \sum_{n=1}^{N_{\rm ch}} P_n R_n |s_n(f - f_n)|^2 \tag{7.15}$$

Regarding the correction term $G_{\rm NLI}^{\rm corr}(f)$ in Equation 7.10, in the following we show the equations related to the contributions of two specific types of islands, SCI and X1. The complete set of formulas for all types of islands of Figure 7.1 is reported in Section A.3. The detailed derivation of all these formulas for the case of identical WDM channels can be found in the arXiv version of [35]. Here we show them in a more general form, which can handle arbitrarily different WDM channels.

- The SCI island

We assume the CUT to be the mth channel, *not necessarily the center channel in the WDM comb.* The SCI island is the center integration region in Figure 7.1 accounting for the effect of non-linearity due to the CUT onto itself. Therefore, there is only one SCI island contributing to the NLI PSD correction term at a given frequency f within the CUT band, that is, for:

$$f \in [f_m - B_m/2, f_m + B_m/2] \tag{7.16}$$

The formulas for the SCI island contribution to $G_{\text{NLI}}^{\text{corr}}(f)$ are as follows:

$$G_{\text{SCI}}^{\text{corr}}(f) = P_m^3\ [\Phi_m\ \kappa_{\text{SCI}}^m(f) + \Psi_m\ \varsigma_{\text{SCI}}^m(f)] \tag{7.17}$$

$$\begin{aligned}\kappa_{\text{SCI}}^m(f) = {} & \frac{80}{81}R_m^2 \int_{f_m-B_m/2}^{f_m+B_m/2} df_1 \int_{f_m-B_m/2}^{f_m+B_m/2} df_2 \int_{f_m-B_m/2}^{f_m+B_m/2} df_2' \cdot \\ & |s_m(f_1)|^2 s_m(f_2) s_m^*(f_2') s_m^*(f_1+f_2-f) s_m(f_1+f_2'-f) \cdot \\ & \mu\,(f_1,f_2,f)\mu^*(f_1,f_2',f) \\ & + \frac{16}{81}R_m^2 \int_{f_m-B_m/2}^{f_m+B_m/2} df_1 \int_{f_m-B_m/2}^{f_m+B_m/2} df_2 \int_{f_m-B_m/2}^{f_m+B_m/2} df_2' \cdot \\ & |s_m(f_1+f_2-f)|^2 s_m(f_1) s_m(f_2) s_m^*(f_1+f_2-f_2') s_m^*(f_2') \cdot \\ & \mu\,(f_1,f_2,f)\mu^*(f_1+f_2-f_2',f_2',f)\end{aligned} \tag{7.18}$$

$$\begin{aligned}\varsigma_{\text{SCI}}^m(f) = {} & \frac{16}{81}R_m \int_{f_m-B_m/2}^{f_m+B_m/2} df_1 \int_{f_m-B_m/2}^{f_m+B_m/2} df_2 \int_{f_m-B_m/2}^{f_m+B_m/2} df_1' \int_{f_m-B_m/2}^{f_m+B_m/2} df_2' \cdot \\ & s_m(f_1) s_m(f_2) s_m^*(f_1+f_2-f) s_m^*(f_1') s_m^*(f_2') s_m(f_1'+f_2'-f) \cdot \\ & \mu\,(f_1,f_2,f)\mu^*(f_1',f_2',f)\end{aligned} \tag{7.19}$$

- The X1 islands

Given a link with N_{ch} channels, there are $2(N_{\text{ch}} - 1)$ islands of type X1 contributing to $G_{\text{NLI}}^{\text{corr}}(f)$, two for each INT channel. However, due to symmetries, the two islands related to a single INT produce exactly the same result, so that there are only $(N_{\text{ch}} - 1)$ different contributions to be summed. Assuming again that the CUT is the mth channel, that is, that f obeys Equation 7.16, then:

$$G_{\text{X1}}^{\text{corr}}(f) = P_m \sum_{\substack{n=1 \\ n \neq m}}^{N_{\text{ch}}} P_n^2\ \Phi_n\ \kappa_{\text{X1}}^n(f) \tag{7.20}$$

$$\begin{aligned}\kappa_{\text{X1}}^n(f) = {} & \frac{80}{81}R_m R_n \int_{f_m-B_m/2}^{f_m+B_m/2} df_1 \int_{f_n-B_n/2}^{f_n+B_n/2} df_2 \int_{f_n-B_n/2}^{f_n+B_n/2} df_2' \cdot \\ & |s_m(f_1)|^2 s_n(f_2) s_n^*(f_2') s_n^*(f_1+f_2-f) s_n(f_1+f_2'-f) \cdot \\ & \mu(f_1,f_2,f)\mu^*(f_1,f_2',f)\end{aligned} \tag{7.21}$$

7.4.3.1 Practical Remarks The inspection of the above formulas for the SCI and X1 islands, together with those for the other six island types in Section A.3, clearly demonstrates the challenge posed by the EGN model. Although it can be shown that the triple and quadruple integrals in the formulas can always be reduced to double integrals[14] ([35], Appendix C), the number and diversity of all these contributions make the use of the full EGN model rather difficult.

Note that most of the complexity comes from the correction term. The GN model term, especially if cast in the form of Equation 7.11, is relatively easier to tackle, because it consists of a single comprehensive formula. In addition, various analytical and semi-analytical results are available for the GN model, either to speed up its integration, or to obtain simpler and even closed-form approximate solutions [13, 54, 59–61]. So, if the full EGN model is too complex for a specific application, then it is mostly the correction term that should be targeted for substantial reduction of complexity, through suitable approximations. This will be done in the next sections.

However, it should be mentioned that any approximations to either the GN model or the correction term must be very carefully validated. The reason is that the EGN model PSD of Equation 7.10 is found by *subtracting* two terms which can be of comparable value. This circumstance greatly amplifies in the final result the possible errors incurred when approximating either one of them.

If instead high accuracy is the target, especially in new untested system configurations, then of course the full EGN model should be used with no approximations.

7.4.3.2 The EGN-SCI-X1 and EGN-X1 Approximate Models One reasonable simplifying approximation consists of retaining only the SCI and X1 islands in the calculation of $G_{\mathrm{NLI}}^{\mathrm{corr}}(f)$. This is suggested by the observation, substantiated below, that for typical systems these islands contribute the majority of the *correction* term. The caveat is that this is a typical occurrence but not a general rule, and it should be kept in mind that especially for research-type applications where new non-standard systems are explored, the other islands contributing to $G_{\mathrm{NLI}}^{\mathrm{corr}}(f)$ may be significant too. We call this approximate model EGN-SCI-X1.

As an even more drastic approximation, only the X1 islands can be considered for $G_{\mathrm{NLI}}^{\mathrm{corr}}(f)$. This means that the signal non-Gaussianity correction related to SCI is neglected, that is, SCI is overestimated. Clearly, this approximation is better suited for systems with a large number of channels, where the impact of SCI is smaller. Its inaccuracy also decreases as the channel spacing decreases, since the relative strength of SCI versus the other NLI contributions (XCI and MCI) tends to go down with the spacing. We call this model EGN-X1.

Both these approximations are conservative, in the sense that with PM-QAM systems they deliver a result that is *guaranteed* to be an upper bound to the NLI PSD delivered by the full EGN model, though a substantially tighter bound than the GN model. In Fig 7.7 we show an example of the use of the EGN-SCI-X1 and EGN-X1 models. The system data are in the figure caption. The curves plot the NLI power on the center channel, normalized as shown in Equation 7.4, versus the number of spans. The picture shows that the EGN-SCI-X1 model incurs no practically relevant error. As expected, the EGN-X1 model is somewhat pessimistic. Its NLI overestimation

[14] It is this circumstance that makes it possible to state that the integration regions for the overall EGN model can be depicted over the two-dimensional plane of Figure 7.1, despite the fact that the correction terms contain either triple or quadruple integrals. For more details, see [35].

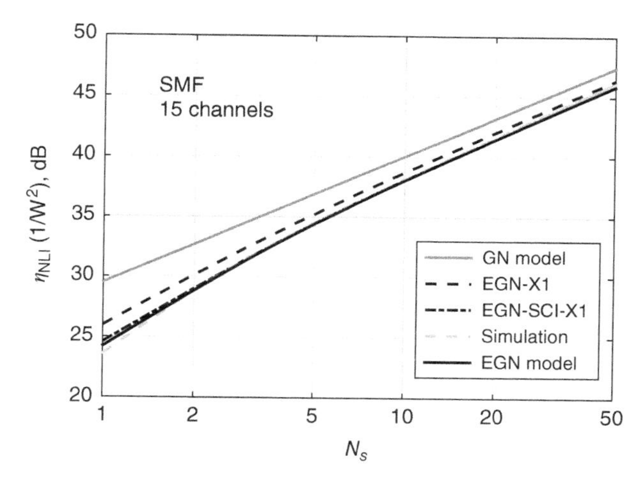

FIGURE 7.7 Normalized NLI noise power η_{NLI} (see Eq. 7.4) over the center channel, versus the number of spans N_s, for a 15-channel PM-QPSK system (32 GBaud, $\Delta f = 33.6$ GHz), with 100 km SMF identical spans and EDFA (lumped) amplification. The channels have rectangular spectra with bandwidth equal to the symbol rate.

error is 0.5 dB at large span count. This is a very contained error, especially in view of Equation 7.5, which indicates that it would cause less than 4% max reach underestimation. Over NZDSF (not shown) a very similar results is found, with again virtually no appreciable error for the EGN-SCI-X1 model and a modest 0.65 dB error at large span count for the EGN-X1. In passing, we remark that the simulation curve in Figure 7.7 matches very well the full EGN model, as it has been consistently found in Refs [35, 42] over a very wide range of system scenarios.

An intermediate-complexity approximation between the EGN-SCI-X1 and EGN-X1 models is also possible. It consists of the EGN-X1 model with a SCI correction including only the first term of $\kappa^{m}_{\text{SCI}}(f)$ in Equation 7.18, namely the term whose leading factor is 80/81, while discarding the term whose leading factor is 16/81. Also, the SCI correction contribution Equation 7.19 is neglected entirely. Preliminary results show the retained term to provide by far most of the SCI correction, for symbol rates of 28 GBaud or higher. In Figure 7.7, in particular, this approximation would be indistinguishable from the EGN-SCI-X1 curve.

In summary, when using the approximations proposed in this section, the overall model complexity decreases sharply, versus the full EGN model, and is no longer much larger than that of the GN model. The practical usability of these models is hence much improved. Of course the EGN-SCI-X1 and EGN-X1 are approximate models and due caution must be exerted not to step out of their boundaries of validity.[15]

[15]The EGN-X1 model considers only the X1 islands in the correction term $G^{\text{corr}}_{\text{NLI}}(f)$, but includes *all* islands in the GN model term $G^{\text{GN}}_{\text{NLI}}(f)$. Much larger errors can be incurred if the GN model term, too, considers just the X1 islands, while discarding all others (except, of course SCI). This would amount to the XPM approximation, whose limitations are discussed in Section 7.2.1.10.

A further substantial simplification that leads to a complexity which is comparable to that of the GN model is introduced in the next section. It consists of an *approximate closed-form* formula for $G_{\text{NLI}}^{\text{corr}}(f)$.

7.4.3.3 The Asymptotic EGN-X1 Approximate Model Recently a *closed-form* approximate formula for the correction term $G_{\text{NLI}}^{\text{corr}}(f)$ has been proposed. Its accuracy improves as the number of spans goes up, achieving an approximate asymptotic convergence versus N_s. The derivation[16] is reported in Ref. [42]. The formula is:

$$G_{\text{NLI}}^{\text{corr}}(f) \approx \underset{\rightarrow}{G}_{\text{NLI}}^{\text{corr}} = \frac{40}{81} \frac{\gamma^2 P_m N_s \overline{L}_{\text{eff}}^{\,2}}{R_m \pi \beta_2 \overline{L}_s} \left(\sum_{\substack{n=1 \\ n \neq m}}^{N_{\text{ch}}} \Phi_n \frac{P_n^2}{R_n \left| f_n - f_m \right|} + \Phi_m \frac{2P_m^2}{R_m^2} \right) \tag{7.22}$$

with the mth channel being the CUT. The arrow underneath $\underset{\rightarrow}{G}_{\text{NLI}}^{\text{corr}}$ is a reminder of the asymptotic behavior of the formula. If the channels are all identical and equally spaced, and the CUT is the center channel, then the formula can be re-written as:

$$\underset{\rightarrow}{G}_{\text{NLI}}^{\text{corr}} = \frac{80}{81} \Phi \frac{\gamma^2 \overline{L}_{\text{eff}}^{\,2} P_{\text{ch}}^3 N_s}{R^2 \Delta f \pi \beta_2 \overline{L}_s} \left[\text{HN}\left(\left[N_{\text{ch}} - 1 \right] / 2 \right) + \frac{\Delta f}{R} \right] \tag{7.23}$$

where HN stands for harmonic number series, defined as: $\text{HN}(N) = \sum_{n=1}^{N} (1/n)$.

Equation 7.22 can be used to correct the GN model term and therefore obtain an approximate EGN model, as follows:

$$G_{\text{NLI}}^{\text{EGN}}(f) \approx G_{\text{NLI}}^{\text{GN}}(f) - \underset{\rightarrow}{G}_{\text{NLI}}^{\text{corr}} \tag{7.24}$$

Because of the assumptions involved in the derivation of Equation 7.22, Equation 7.24 should be considered an asymptotic approximation, versus N_s, of the EGN-SCI-X1 model discussed in Section 7.4.3.2.

Equation 7.22 has limitations, which impact Equation 7.24 as well. It assumes that the same type of fiber is used in all spans. Spans can be of different length, though: Equation 7.22 uses the average span length $\overline{L}_s$ and the average span effective length $\overline{L}_{\text{eff}}$. Accuracy is quite good for links having all individual span lengths within $\pm 15\%$ of the average. Caution should be used for larger deviations. Equation 7.22 also assumes that lumped amplification is used, exactly compensating for span loss. Raman amplification can be present, provided that it contributes negligibly to NLI generation. As a rule of thumb, this is the case if Raman is backward-pumped and the signal power at the end of the span is at least 6 dB lower than it is at the beginning

[16]More precisely, in Ref. [42] the detailed derivation of Equation 7.23 is actually shown, rather than that of Equation 7.22. However, the latter is a rather straightforward generalization of the former. Also, the SCI correction term, which is the second term in brackets in Equation 7.22, was not dealt with in Ref. [42]. Its addition is discussed in Ref. [63].

of the span. It is possible that for very low number of spans (1–3, i.e., completely outside of the range of validity of the formula), in conjunction with very low number of channels (1–3), Equation 7.24 returns negative values. Equation 7.22 is derived assuming ideally rectangular channel spectra. If spectra have a significantly different shape (such as sinc-shaped), some error may be incurred. Finally, Equation 7.22 neglects the frequency-dependence of $G_{\text{NLI}}^{\text{corr}}(f)$, that is, it assumes that the correction is flat over the CUT bandwidth.

In addition, one of the approximations used in deriving Equation 7.22 assumes that the symbol rate of the CUT channel (which we labeled as the mth) is not too low. For accuracy to be preserved, the following relation should be satisfied:

$$R_m \geq \left| \frac{1}{\pi \beta_2 N_s L_s \left(f_n - R_n/2 \right)} \right|, \quad n = m \pm 1 \tag{7.25}$$

Note that this constraint becomes less and less stringent for larger span count, span length and dispersion. At ideal Nyquist WDM, with identical channels, the constraint simplifies to: $R \geq \sqrt{2/(\pi \mid \beta_2 \mid L_s N_s)}$. Using SMF fiber parameters, 80 km spans and $N_s = 20$, the lower limit for the symbol rate R is about 4.4 GBaud. Going to NZDSF, it is 8.9 GBaud. In practice, single-carrier type systems never pose any problems, whereas Equation 7.22 should not be used with either OFDM or multi-subcarrier channels with very low subcarrier symbol rate. Loss of accuracy is quite gradual when the symbol rate approaches the limit.

Incidentally, Equation 7.25, as well as Equation 7.22, have been found to have an important meaning in the context of the study of the dependence of NLI generation on the channel symbol rate. For more details see Section 7.4.4 and [66].

The accuracy of Equation 7.22 in estimating $G_{\text{NLI}}^{\text{corr}}(f)$ for large N_s was thoroughly tested in Refs [42, 63]. Despite all the above limitations and approximations, it was found to be excellent over a very broad range of system scenarios. Its effectiveness can be appreciated in Figure 7.8, plotted for a 31-channel, 32-GBaud PM-QPSK Nyquist WDM system, over three different fibers. Again, the NLI normalized power η_{NLI} defined in Equation 7.4 is represented, versus N_s. The dashed curves are generated according to Equation 7.24. The residual error at large N_s is negligible.

In summary, the closed-form approximation for $G_{\text{NLI}}^{\text{corr}}(f)$ provided by Equation 7.22 can be a quite effective tool in correcting for most of the GN model overestimation error in most cases of practical interest, adding essentially negligible complexity to the GN model itself. However, the GN model computational burden may still be excessive for certain applications and it may be desirable to reduce it further.

7.4.3.4 Incoherent NLI Accumulation Models A substantial problem in the numerical integration of both the GN and the EGN model equations is posed by the link function $\mu(f_1,f_2,f)$, and in particular by its $\nu(f_1,f_2,f)$ factor (see Eq. A.5). This latter factor is 1 for a single span but, as the number of spans grows, it displays increasingly sharp peaks (see [54], Figure 21) which are difficult to integrate numerically. These peaks may get smoothed somewhat if the spans are not all identical, but they are not eliminated. The question is then whether some suitable

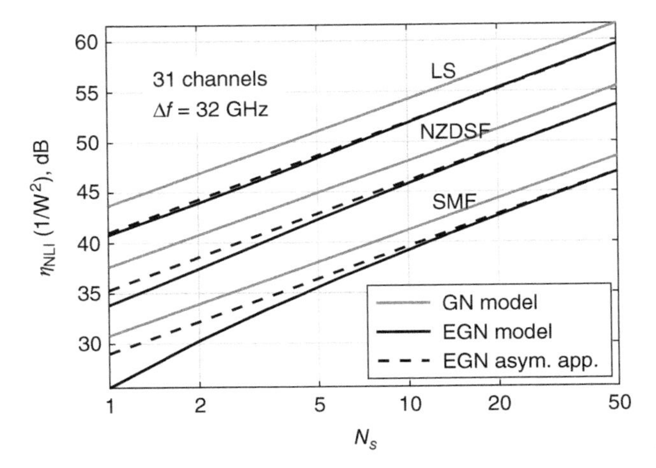

FIGURE 7.8 Normalized NLI noise power η_{NLI} (see Eq. 7.4) over the center channel, versus the number of spans N_s, for a 31-channel Nyquist-WDM PM-QPSK system (32 GBaud, channel spacing $\Delta f = 32$ GHz), with 100 km identical spans and EDFA (lumped) amplification. The fiber parameters are reported in Table 7.1, attenuation was 0.22 dB/km. The dashed curve (label "EGN asym. app.") is the approximate asymptotic model of Equation 7.24.

approximations may help in dealing with this problem. In particular, if NLI could be estimated for any value of N_s based on calculations carried out for $N_s = 1$, then $\nu(f_1,f_2,f) = 1$ and such calculations would drastically simplify. We first discuss cross- and multi-channel effects (XCI and MCI), and then deal with SCI.

In identical-span systems, assuming transparency (as defined in Footnote 5) and conventional symbol rates (≥28 GBaud), XCI and MCI (called XMCI for brevity) turn out to accumulate very close to linearly as the number of spans. In particular, the GN model term for XMCI, which we call $G^{\text{GN}}_{\text{XMCI}}(f)$, grows as N_s already from the first span, whereas the correction term for XMCI does so asymptotically. In fact, removing the SCI correction from Equation 7.22, we get an asymptotic correction formula for XMCI:

$$\underset{\rightarrow}{G}^{\text{corr}}_{\text{XMCI}} = N_s \frac{40}{81} \frac{\gamma^2 P_m \overline{L}_{\text{eff}}^2}{R_m \pi \beta_2 \overline{L}_s} \sum_{\substack{n=1 \\ n \neq m}}^{N_{\text{ch}}} \Phi_n \frac{P_n^2}{R_n |f_n - f_m|} \tag{7.26}$$

which grows as N_s . This circumstance suggests that, regarding XMCI, one could compute the GN model contribution $G^{\text{GN}}_{\text{XMCI}}(f)$ in the first span and then scale it linearly versus N_s. As for the correction $G^{\text{corr}}_{\text{XMCI}}(f)$, its formula inherently scales as N_s.

Regarding SCI, a possible coarse approximation is to calculate the GN model SCI contribution for $N_s = 1$, and then scale it linearly versus N_s, while completely neglecting the SCI non-Gaussianity correction. In reality, the GN model SCI contribution $G^{\text{GN}}_{\text{SCI}}(f)$ scales *super-linearly* versus N_s [54], so that scaling it just as N_s underestimates it. On the other hand, neglecting the SCI non-Gaussianity correction tends to lead to SCI overestimation, so that these two approximations tend to cancel each other out. At any rate, when the number of channels is large, the impact of these SCI approximations tends to be relatively minor (though non-negligible) because SCI tends to be a minority contribution to the overall NLI.

Pulling all these approximations together, we finally have:

$$G_{\text{NLI}}(f) \approx N_s \cdot G_{\text{NLI}}^{\text{GN}}(f)|_{1\ \text{span}} - \underset{\rightarrow\text{XMCI}}{G^{\text{corr}}} \quad (7.27)$$

This formula is well suited for a quick appraisal of identical span systems with large WDM bandwidths and large number of channels. Note that Equation 7.27, as opposed to the EGN-X1 approximation, is not guaranteed to either over- or underestimate the NLI PSD. As a rule of thumb, it can be expected to somewhat underestimate it for low channel count, whereas for high channel count nothing can be said.

One example of the use of Equation 7.27 is shown in Figure 7.9. A quasi-Nyquist-WDM system is considered, made up of 41 PM-QPSK rectangular channels at 32-GBaud, spaced either 33.6 or 50 GHz, over SMF. Once more, the NLI normalized power η_{NLI} defined in Equation 7.4 is represented, versus N_s. The dashed line is Equation (7.27), which incurs a 0.5 and 0.65 dB error at $N_s = 50$, for $\Delta f = 33.6$ and 50 GHz, respectively. The error is towards underestimation, which confirms that now there is no guarantee of getting a conservative result. On the other hand the error is modest.

Equation 7.27 can be re-written as:

$$G_{\text{NLI}}(f) \approx N_s \cdot \left[G_{\text{NLI}}^{\text{GN}}(f) - \underset{\rightarrow\text{XMCI}}{G^{\text{corr}}}\right]_{1\ \text{span}} \quad (7.28)$$

where the quantities within square brackets are calculated for one span *only*. The formula shows that this approximation inherently implies that the total PSD of NLI at the end of the system is given by the sum *in power* of the contribution of each single span. This in turn suggests that such assumption could be generalized towards non-identical span links, using the *incoherent accumulation approximation* of Section 7.2.1.8, Equation 7.3, resulting into:

$$G_{\text{NLI}}(f) \approx \sum_{n_s=1}^{N_s} \left[G_{\text{NLI}}^{\text{GN}}(f) - \underset{\rightarrow\text{XMCI}}{G^{\text{corr}}}\right]_{n_s\text{th span}} \quad (7.29)$$

where the quantities within square brackets are calculated for the nth span *only*. that is, as if the nth span was the only one in the link.

An important caveat is that this heuristic formula is expected to be reasonably accurate if the spans are not very different in either fiber type or length. At present no extensive testing or error appraisal has been carried out for Equation 7.29 in more general cases where spans are substantially different. Its use in such contexts should be subjected to proper pre-validation for the specific application of interest. Nonetheless, Equation 7.29 could potentially be a possible practical tool to be used in DRNs, as discussed in Section 7.4.5.

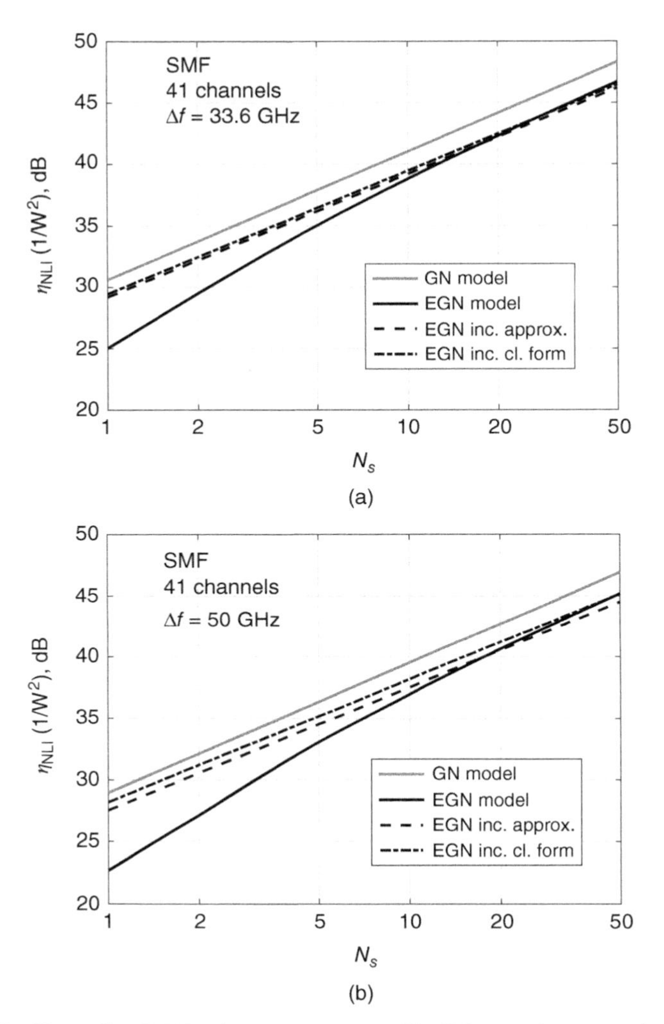

FIGURE 7.9 Normalized NLI noise power η_{NLI} (see Eq. 7.4) over the center channel, versus the number of spans N_s, for 41-channel PM-QPSK systems at 32 GBaud, with 100 km SMF identical spans and EDFA amplification. The channels have rectangular spectra with bandwidth equal to the symbol rate. The dashed curve is Equation 7.27, the dashed-dotted curve is Equation 7.32. (a) Channel spacing $\Delta f = 33.6$ GHz. (b) Channel spacing $\Delta f = 50$ GHz.

Finally, an even simpler and theoretically much coarser simplifying approach consists of entirely neglecting $G_{\text{NLI}}^{\text{corr}}(f)$ in Equation 7.29, obtaining:

$$G_{\text{NLI}}(f) \approx \sum_{n_s=1}^{N_s} \left[G_{\text{NLI}}^{\text{GN}}(f) \right]_{n_s\text{th span}} \tag{7.30}$$

This formula combines the incoherent accumulation approximation of Section 7.2.1.8 with the signal Gaussianity approximation of Section 7.2.1.4. The result is an extremely simple model that has been called the "incoherent GN model." It was the first proposed version of the GN model [39, 40]. Despite the drastic approximations, it actually delivers rather accurate maximum reach predictions for lumped amplification long-haul systems with moderate number of channels, operating in the 28–32 GBaud range, as extensive simulative tests showed (see for instance Figure 5 in Ref. [39] and Figure 7.2). Good accuracy is also found at the NLI accumulation level for large span count, as shown in Figure 7.5, dashed-dotted line.

The reason for such good predictivity is that the signal Gaussianity approximation and the incoherent accumulation approximation used by Equation 7.30 tend to balance each other's error out, when the number of channels is moderate. For very large and very low channel count, this compensation is less accurate, leading to NLI over- and underestimation, respectively. Also, some loss of accuracy can be expected when operating at lower symbol rates than 28–32 GBaud, a regime which appears to be of possible future interest according to the results discussed in Section 7.4.4.

In summary, both Equations 7.29 and 7.30 are rather coarse models, which may incur substantial errors. They cannot be recommended for highly accurate PTP link analysis and design. On the other hand, for certain applications they may be quite attractive, due to their simplicity and specific features. One such application can be DRNs, a topic which will be dealt with in Section 7.4.4.

7.4.3.5 Closed-Form Analytical NLI Formulas With reference to Equation 7.10, various approximate closed-form formulas[17] for $G_{\mathrm{NLI}}^{\mathrm{GN}}(f)$, or for quantities that can be traced back to it, have been proposed over the years, for instance in Refs [9, 13, 33, 34, 54, 59–61, 68]. Regarding $G_{\mathrm{NLI}}^{\mathrm{corr}}(f)$ instead, only Equations 7.22 and 7.23 are currently available, to the best of the authors' knowledge, due to the more recent introduction of the EGN model. It may be foreseen that more closed-form formulas will emerge in the near future for both $G_{\mathrm{NLI}}^{\mathrm{GN}}(f)$ and $G_{\mathrm{NLI}}^{\mathrm{corr}}(f)$.

Due to the limited space available, only two closed-form EGN model approximate formulas will be presented here. The first one is significant as it addresses the limiting case of ideal Nyquist WDM. The second one allows to implement a generic-system quick approximate evaluator.

- The Nyquist-WDM case with identical spans

We remind the reader that by "Nyquist WDM" we mean a system where the WDM channels have a rectangular spectrum with bandwidth equal to the symbol rate R, and the channel spacing is equal to R as well. In practice, the system spectrally looks like a compact rectangle with overall WDM bandwidth $B_{\mathrm{WDM}} = N_{\mathrm{ch}} R$.

For the GN model contribution, Equations 13 and 23 of [54] are used. For the correction term Equation 7.23 is used. As a result, the following formula can be viewed

[17] The term "closed-form" is used with different meanings in the literature. We intend it here in the sense attributed to it by Wikipedia [67].

as a fully closed-form approximation of the EGN-SCI-X1 model of Section 7.4.3.2, specific for Nyquist-WDM systems with identical spans:

$$G_{\mathrm{NLI}}(f) \approx \frac{8}{27}\frac{\gamma^2 P_{\mathrm{ch}}^3 N_s}{\pi|\beta_2|R^3} \cdot \left\{\frac{N_s^{\epsilon}}{2\alpha}\,\mathrm{asinh}\left(\frac{|\beta_2|}{4\alpha}\pi^2 B_{\mathrm{WDM}}^2\right) - \Phi\frac{10}{3}\frac{L_{\mathrm{eff}}^2}{L_s}\left[\mathrm{HN}\left(\frac{N_{\mathrm{ch}}-1}{2}\right)+1\right]\right\}$$

$$\epsilon \approx \frac{3}{10}\log_e\left(1+\frac{3}{\alpha L_s}/\mathrm{asinh}\left(\frac{|\beta_2|}{4\alpha}\pi^2 B_{\mathrm{WDM}}^2\right)\right) \tag{7.31}$$

The exponent $\epsilon > 0$ accounts for coherent NLI accumulation. Note that for large overall WDM bandwidths B_{WDM}, then $\epsilon \approx 0$ (see [54], Section IX).

Equation 7.31 has the compounded limitations of Equation 7.23, listed in Section 7.4.3.3, and of Equations 13 and 23 from [54]. The latter is essentially that the span loss L_s should not be too small. In Reference [54] a value of 7 dB was indicated but a more conservative figure of 10 dB is advised. Within such constraints, the main source of inaccuracy of Equation 7.31 is the asymptotic convergence of Equation 7.23 versus N_s. The inaccuracy due to the approximations involved in the reduction to closed-form is instead minor.

Despite its limitations, for large-enough number of spans, Equation 7.31 provides a very accurate estimate of the NLI PSD at the center of a Nyquist-WDM comb. In Figure 7.10, we show an example of a 31-channel Nyquist-WDM PM-QPSK system,

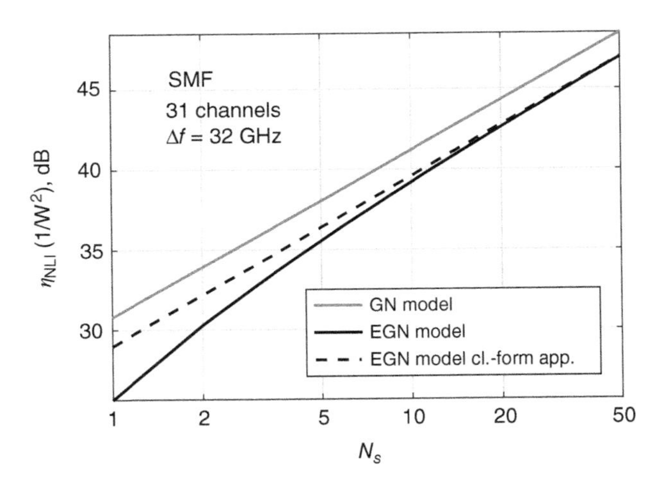

FIGURE 7.10 Normalized NLI noise power η_{NLI} (see Eq. 7.4) over the center channel, versus the number of spans N_s, for a 31-channel Nyquist-WDM PM-QPSK system (32 GBaud, Δf = 32 GHz), with 100 km SMF identical spans and EDFA (lumped) amplification. The dashed curve is the EGN-model closed-form approximation Equation 7.31.

at 32 GBaud, over SMF ($L_s = 100$ km). The asymptotic error for large-enough N_s is essentially negligible. Similar very good accuracy is found for NZDSF and LS.

- The generic WDM comb case

We adopt the *incoherent accumulation approximation* Equation 7.29 and then we use the closed-form expression for the GN model contribution proposed in Ref. [38], Equations 41–43. For the non-Gaussianity correction term we use Equation 7.26, as prescribed by Equation 7.29. We assume that the same type of fiber is used in all spans. The resulting formula provides $G_{\text{NLI}}(f)$ at the center frequency of the mth channel in the comb:

$$G_{\text{NLI}}(f_m) \approx \frac{8}{27\pi} G_m \frac{N_s \gamma^2 \overline{L}_{\text{eff}}}{\beta_2} \sum_{n=1}^{N_{\text{ch}}} G_n^2\, \psi_{n,m} \tag{7.32}$$

$$\psi_{n,m} \approx \operatorname{asinh}\left(\frac{\pi^2 |\beta_2|}{2\alpha} [f_n - f_m + B_n/2] B_m \right)$$

$$- \operatorname{asinh}\left(\frac{\pi^2 |\beta_2|}{2\alpha} [f_n - f_m - B_n/2] B_m \right) - \Phi_n \frac{R_n}{|f_n - f_m|} \frac{5}{3} \frac{\overline{L}_{\text{eff}}}{\overline{L}_s}, \quad n \neq m$$

$$\psi_{m,m} \approx \operatorname{asinh}\left(\frac{\pi^2 |\beta_2|}{4\alpha} B_m^2 \right)$$

The limitations and constraints are similar to those of Equation 7.31, compounded by those induced by the use of the incoherent accumulation approximation,[18] discussed in Section 7.4.3.4

In Figure 7.9, we test Equation 7.32 on two 41-channel PM-QPSK 32-GBaud systems over SMF, with different channel spacing $\Delta f = 33.6$ or 50 GHz. The error versus the full EGN model at large span count is small: at 50 spans it is −0.2 and negligible for $\Delta f = 33.6$ and 50 GHz, respectively. Good results are found over NZDSF (not shown) too: −0.7 and negligible, respectively. However, it must be understood that the high accuracy found in these cases is due in part to error cancellation. In general, possibly more substantial errors of either under- or overestimation could occur. On the other hand, the above results address typical and relevant scenarios. A small error in these cases is a significant result, especially considering that Equation 7.32 is a fully closed-form model.

Assuming that all channels are identical, a much more compact closed-form formula can be written based on [54], Equation 15, but the additional constraint of $R \geq 1/(16\sqrt{\beta_2})$ must be satisfied for sufficient accuracy.

[18] The incoherent accumulation approximation could be removed from Equation 7.32 through the introduction of an ϵ exponent, similar to Equation 7.31, see [54] Section IX, or through alternative approaches such as shown in Ref. [57]. In that case, though, the SCI non-Gaussianity correction should be introduced as well, by resorting to Equation 7.22 rather than Equation 7.26. However, given the less strict accuracy goals of Equation 7.32, we consider the incoherent accumulation approximation an acceptable compromise.

7.4.4 Case Study: Determining the Optimum System Symbol Rate

A recent experiment [70] has shown a rather strong maximum reach gain (20%) in long-haul transmission when a single serial-channel (SC) was broken up into either FDM quasi-Nyquist subcarriers [70]. Simulative evidence of a dependence of performance on the per-subcarrier symbol rate has also been found in [69, 71–73]. Investigating the behavior of NLI when changing the symbol-rate of WDM channels then appears to be an interesting case-study.

This case study is very well suited to show the different behavior of three of the modeling approaches mentioned in this chapter: the GN model, the EGN model and the FD-XPM model (or FD-XPM approximation [36], see Section 7.2.1.10). It addresses a key question for PTP links, which is simply formulated as follows:

given pre-determined total WDM bandwidth, spectral efficiency, spectrum roll-off and modulation format, what is the symbol rate which minimizes NLI generation?

Note that the above assumptions make the total raw bit rate, conveyed by the overall WDM signal, a fixed constant, too.

We chose to address this case study using the following link parameters: SMF and NZDSF fibers (Table 7.1, att. 0.22 dB/km), span length 100 km, lumped amplification, roll-off 0.05, total WDM bandwidth 504 GHz, total raw bit rate 1920 Gb/s. The channel spacing is 1.05 times the symbol rate, corresponding to a raw spectral efficiency of 3.81 b/(s Hz). The only free parameter is the number of channels N_{ch} that the overall WDM bandwidth is split into. We look at the accumulated NLI at 30 and 50 spans, for NZDSF and SMF, respectively.

The NLI-related quantity chosen for the study is $\tilde{G}_{\text{NLI}}$, defined as the PSD of the non-linear noise NLI falling over the center channel and averaged over it. It is also normalized versus the transmission signal PSD cube, G_{ch}^3. In math:

$$\tilde{G}_{\text{NLI}} = \frac{P_{\text{NLI}}}{R_{\text{ch}} \cdot G_{\text{ch}}^3} \tag{7.33}$$

where P_{NLI} is the total NLI power affecting the center channel, as defined in Equation 7.8. The convenient features of $\tilde{G}_{\text{NLI}}$, are: it is independent of the power per channel launched into the link; the *same value* of $\tilde{G}_{\text{NLI}}$ for different symbol rates means that the corresponding systems would achieve the *same maximum reach*.

The results are shown in Figure 7.11. The GN model line is essentially flat, that is, it predicts no change of performance versus the number of channels the total WDM bandwidth is split into.

The EGN model, on the contrary, shows a change, and in particular it shows a *minimum*, which for SMF and NZDSF is located at about 200 and 70 channels, that is, at about 2.4 and 6.8 GBaud, respectively. These results agree very well with the computer simulations (markers) shown in Figure 7.11. Interestingly, Figure 7.11 also shows that the GN and EGN model tend to come together both at very large and very small symbol rates.

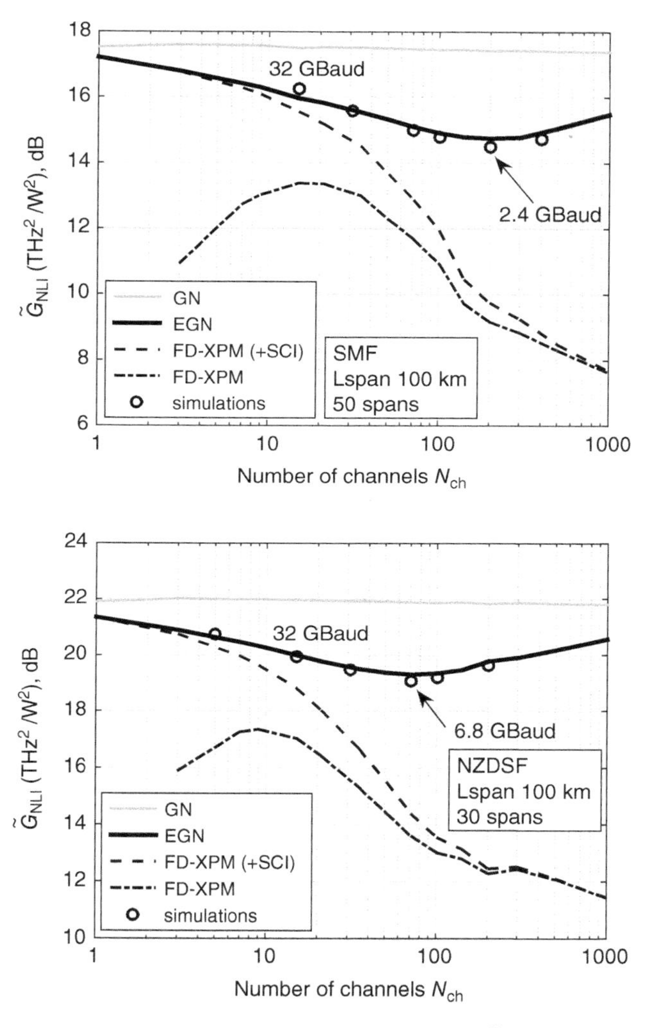

FIGURE 7.11 Normalized NLI noise power spectral density $\tilde{G}_{NLI}$ over the center channel (see Eq. (7.33)), versus the number of channels N_{ch}, measured at 50 spans of SMF or 30 spans of NZDSF. Span length is 100 km. The modulation format is PM-QPSK, with roll-off 0.05 and channel spacing 1.05 times the symbol rate. The total WDM bandwidth is 504 GHz. Lumped amplification is assumed.

The FD-XPM model does not include single-channel non-linearity. When plotted by itself it generates the curve marked "FD-XPM" in figure. We supplemented it with the SCI contribution calculated through the EGN model, so that a comparison could be carried out. The plot shows that, as pointed out in Section 7.2.1.10, the FD-XPM (with SCI) model is accurate at large symbol rates. However, it departs from the EGN model when moving towards low symbol rates. At the optimum $N_{ch,}$ the FD-XPM (with SCI) model underestimates NLI by about 5 dB, both for SMF and NZDSF. Note also that its prediction appears to decrease steadily for $N_{ch} \rightarrow \infty$. The reason for this behavior is explained in Section 7.2.1.10 and has to do with the neglect by FD-XPM of all integration islands in Figure 7.1, except X1. This can be alternatively stated by saying that FD-XPM takes only XPM into account but neglects all of FWM. Depending chiefly on symbol rate and dispersion, this approximation may be accurate or not.

By using results from the derivation of the asymptotic closed-form correction formula of Section 7.4.3.3, which can be found in Ref. [42], a closed-form expression of the optimum symbol rate can be derived. Interestingly, it coincides with Equation 7.25. For quasi-Nyquist systems, with all identical spans, Equation 7.25 can be re-written as:

$$R_{opt} = \sqrt{2/(\pi|\beta_2|L_{span}\, N_{span})} \tag{7.34}$$

This formula indicates that the optimum rate is a function not only of the accumulated dispersion per span $|\beta_2| \cdot L_{span}$ but also of the link length through N_{span}. Owing to the square root in Equation 7.34, the range of optimum rates is relatively narrow. It is difficult to push R_{opt} outside of the interval 2–10 GBaud, unless rather extreme scenarios are assumed. Equation 7.34 was tested using the EGN model and found to be very accurate over a wide range of dispersions and span numbers (at 100 km span length) [66]. Another significant result has to do with the approximate asymptotic expression Equation 7.24. It was indicated in Section 7.4.3.3 that it starts losing accuracy *beyond* the symbol rate given by Equation 7.25, which coincides with R_{opt}. At that rate, Equation 7.24 is still rather accurate and can therefore be used to carry out approximate NLI calculations with reduced overall complexity.

Regarding the possible practical impact of the results of Figure 7.11, the NLI reduction between the optimum rates and the current industry-standard (32 GBaud, 15 channels in the plots) is 1.2 and 0.7 dB, respectively. According to Equation 7.5, this leads to about 0.4 and 0.27 dB (or 10% and 6%) max-reach increases for SMF and NZDSF, respectively. These numbers are significant but not disruptive.

Nonetheless, these results might still influence future trends. In particular, the general industry push towards higher symbol rates must be weighed versus the greater penalties that are incurred there. In Figure 7.11, the NLI gap for SMF grows to 2 dB between R_{opt} and 100 GBaud. It is a full 3 dB between $R_{opt} = 2.8$ and 100 GBaud over 60-km-spans PSCF (plot not shown). This means that, apart from the obvious technological hurdles towards higher rates, there are also fundamental disadvantages to straightforward serial-rate increase, which are going to add up.

As for trying to move to lower rates, to take advantage of lower NLI, implementing this technique by increasing the number of *optical* carriers is clearly impractical. One solution could be subcarrier multiplexing over a single carrier by means of DSP-DAC enabled transmitters, as it was done in Ref. [70].

It should also be mentioned that for more complex formats than PM-QPSK, the potential NLI mitigation gets reduced. This is due to the correction term $G_{NLI}^{corr}(f)$, which is responsible for the appearance of the NLI minimum, getting smaller because of the smaller Φ and Ψ coefficients (see Table 7.2). Qualitatively similar plots to Figure 7.11 are found for instance for PM-16QAM. The general curve shapes are identical and the optimum symbol rates are the same, but the NLI mitigation is smaller: for SMF the drop from 32 to 2.4 GBaud is only 0.66 dB, resulting in about 5% potential max reach increase. Further investigation is, however, in order, since PM-16QAM is affected more than PM-QPSK by long-correlated non-linear phase and polarization noise (see Section 7.2.1.5), whose thorough removal might improve the effectiveness of symbol-rate optimization.

Several other aspects of this topic need further investigation. One of them is the variation of the NLI mitigation versus the total WDM bandwidth. For NLI mitigation to be of interest, it must still be significant at C band and, in prospect, even at larger WDM bandwidths. The dependence of the NLI mitigation on the span number where NLI is assessed, as well as on span length and dispersion, also need to be carefully evaluated.[19]

7.4.4.1 Summary on PTP Links In the subsections of Section 7.4 we have presented a range of alternative approaches that can be used to assess NLI in the context of PTP links. The ordering has roughly gone from greater accuracy and complexity towards simplicity and speed, at the cost of loss of accuracy.

For research purposes, the full EGN model is highly recommended. For routine accurate design applications, the EGN-SCI-X1 model is an efficient and still very accurate alternative. For fast preliminary assessment, the asymptotic EGN-SCI-X1 model is attractive, whereas all other presented solutions can be considered when even faster results are needed and accuracy can be relaxed.

7.4.5 NLI Modeling for Dynamically Reconfigurable Networks

As mentioned at the beginning of Section 7.4, in DRNs each optical transmission channel (or "lightpath") can be re-routed at each network node and hence, contrary to PTP links, it can change its spectrally neighboring channels, possibly many times. Such neighboring channels can have a different symbol rate, format, and accumulated dispersion.

This complicates drastically the non-linearity modeling problem, since the final amount of NLI impacting any given channel (assumed as the CUT) depends on the detailed overall "propagation history" of the CUT itself and all of its INTs, from source to destination. The EGN model can be extended to take such propagation history fully into account and deliver a very accurate end result. On the other hand, its complexity, already substantial for the PTP case, is further exacerbated.

In practice, in DRNs there is a need for a fast assessment of physical layer impairments, so that the control plane can enact "physical-layer aware" routing and traffic allocation decisions, essentially "real-time." Given this requirement, it is hard to picture the EGN model, made more complex by the need to take into account the propagation history of each CUT and INT, as a practical real-time solution for DRNs.

[19]After the manuscript of this chapter was completed, several papers on symbol rate optimization have appeared, including a targeted experimental study [65] which confirms the EGN model predictions.

Besides practicality, there are other reasons why a full-fledged EGN model approach may not be the right option for DRNs. These reasons will be discussed after an initial set of examples is provided, in the next section.

7.4.5.1 CUT Performance Dependence on INT History We look at five different link scenarios, that have the following features in common:

- 50 spans of NZDSF (see Table 7.1), 100 km each. Transparency is assumed (see Footnote 5).
- 41 channels are transmitted, with symbol rate 32 GBaud and 33.6 GHz spacing; all spectra are raised-cosine with roll-off 0.05, all channels are launched with the same power, the total WDM bandwidth is 1.377 THz.
- The spectrum of the WDM signal launched is *the same* across the five scenarios at every point along the link.

Assuming that the CUT is the center channel in the WDM comb, the scenarios 1– 5 have the specific features:

1. The CUT and the INTs are all PM-QPSK, and they propagate together from source to destination.
2. The CUT and the INTs are all PM-16QAM, and they propagate together from source to destination.
3. The CUT is PM-QPSK and the INTs are all PM-16QAM, and they propagate together from source to destination.
4. The CUT and INTs are all PM-QPSK. The INTs are completely replaced every 10 spans with others with identical features but independent data. This mimicks a situation where the CUT is re-routed every 10 spans, changing all of its INTs. The new INTs are assumed not to originate at the CUT routing nodes. For simplicity, it is assumed that all of them have already travelled 10 spans before the CUT joins them.
5. Same as #4 but all channels (CUT and INTs) are PM-16QAM.

The NLI accumulation curves for the five scenarios are shown in Figure 7.12. The GN model approximation is shown as a gray solid line. There is only one such line because the GN model prediction is the *same* for all these scenarios, since the GN model only looks at the PSD of the WDM signal, which is identical. Note also that the GN model curve is pessimistic, that is, it predicts more NLI, in all cases.

Regarding the EGN results, the lowest curve is that of scenario #1, that is, a PTP link with all PM-QPSK channels. The other curves are comprised between this curve and the GN model. In particular, scenario #4 shows that it is important to take the detailed INT history into account. A comparison of scenarios #1 and #2 shows the impact of changing the format of the INTs. A comparison of scenarios #2 and #3 shows that the format of the INTs is more important in the generation of NLI than that

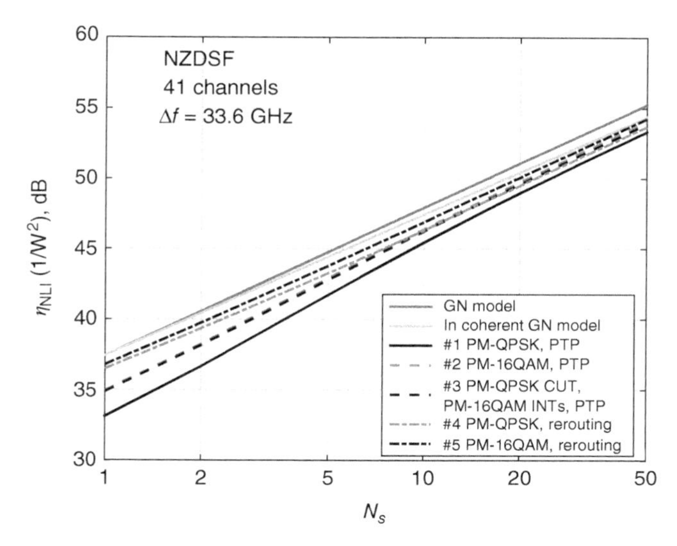

FIGURE 7.12 Normalized NLI noise power η_{NLI} (see Equation 7.4) over the center channel, versus the number of spans N_s, for 41-channel systems at 32 GBaud with channel spacing $\Delta f = 33.6$ GHz, 100 km NZDSF identical spans and EDFA (lumped) amplification. A similar plot for the SMF case can be found in Ref. [74].

of the CUT itself. Overall, Figure 7.12 shows that various scenarios whose spectrum is everywhere identical along the link may produce rather different NLI curves.

We also did the evaluation of scenarios #1–5 over SMF, and for other channel spacings (37.5 and 50 GHz). With SMF we also tested 60 and 80 km spans. The qualitative appearance of Figure 7.12 is maintained in all these links, including the incoherent GN model curve essentially merging into scenario #5 at large span count. The specific plot for SMF, 100 km spans, 41 channels and 33.6 GHz spacing can be found in Ref. [74].

In an actual DRN, many more situations that are also spectrally identical to these could show up, where the INTs could change more or less frequently and could come into the link with any amount of accumulated dispersions. INTs and CUTs could have any mix of different formats. Remarkably, it turns out that all the corresponding NLI curves would fall within the relatively narrow region, comprised between the curve of the PTP-like scenario using the lowest-cardinality format and the GN model curve, such as shown in Figure 7.12. The GN model curve is in fact an upper bound for all possible different situations that may present themselves.[20]

[20]It may instead be shown that the EGN model curve for the PTP-like scenario using the lowest-cardinality format is not a rigorous lower-bound. However, from the practical viewpoint of managing a DRN, the upper bound appears to be more significant, so the discussion of a rigorous lower bound will not be dealt with here. It should also be mentioned that the EGN-model NLI results in Figure 7.12 include long-correlated non-linear phase and polarization noise. Their amount is modest in lumped-amplification 100-km span links (see Section 7.2.1.5) but, depending on how thoroughly these components are removed

Based on this circumstance, an approximate but conservative modeling approach could be that of adopting the GN model. This means that performance prediction would be pessimistic, to some variable degree. On the other hand, the added complexity required to obtain the accurate EGN model curves shown in Figure 7.12 would be extremely large. Also, it would typically gain a relatively modest improvement in accuracy versus the GN model, considering what Equation 7.5 shows: an improvement of 1 dB in NLI noise estimation accuracy only improves the maximum reach estimation accuracy by about 1/3 dB.

As a result of the above remarks, the approach followed in the remainder of this section on DRNs will be that of taking as reference the GN model, on the basis that it appears to be a reasonable and conservative starting point, which represents a compromise between accuracy and complexity and possibly one of the few practicable alternatives to begin approaching real-time operation. However, not even the GN model itself is sufficiently lean to directly allow it.

7.4.5.2 Real-Time DRN Physical-Layer Awareness The GN model still requires to keep track of some of the propagation history of the CUT, namely the features of all the spans traversed by it, as well as the full WDM spectra present in such spans, though the format and propagation history of the INTs is no longer needed. In addition, it still requires numerical integration, which may be particularly hard to perform due to the presence of the ν factor in the link function, as discussed in Section 7.4.3.4. These requirements place the pure GN model approach still far away from handling real-time. Hence, further approximations are necessary.

An effective simplification strategy is that of combining the incoherent accumulation approximation of Section 7.2.1.8 with the GN model, resulting in the incoherent GN model of Equation 7.30. Per se, it does not remove the need for numerical integration, but it removes the problematic ν factor from the link function. If numerical integration is nonetheless too heavy, an approximate closed-form formula can be used. In particular, Equation 7.32, with Φ_n set to zero, is an effective solution, though others are possible.

As for accuracy, it can be appreciated from Figure 7.12, where the incoherent GN model is shown as a light gray solid line. The interesting result is that the incoherent GN model appears to act as a tighter upper bound than the GN model to the bundle of curves of scenarios #1–5. In any case, its prediction gets rather accurate, as the span number goes up towards values that are of practical interest for maximum reach. Considering the simplicity of the model and the extent of the approximations, the degree of accuracy is surprising. Remarkably, the exact same behavior as shown in Figure 7.12 was found in a broader test encompassing 12 test sets (of which Figure 7.12 is one) generated by combining the following options in all possible ways: 15 or 41 WDM channels, NZDSF or SMF, 33.6 or 37.5 or 50 GHz channel spacing, all other parameters as in Figure 7.12. Further test sets were obtained over SMF at 80 and 60 km spans. The qualitative appearance of Figure 7.12 is thoroughly maintained

at the receiver, the scenario of Figure 7.12 might vary somewhat. At the time this manuscript goes to press, this topic is being actively investigated.

in all these sets, including the feature consisting of the incoherent GN model curve essentially merging into case #5 at large span count.[21]

Subtler but perhaps crucial advantages of the incoherent accumulation approximation Equation 7.30 are the following:

- *it makes the non-linear effect of each span independent of that of any other span;*
- *it makes the overall signal degradation due to ASE and NLI additive along the link.*

The first of the above aspects means that, to evaluate each of the summation terms of Equation 7.30, one only needs link features "local" to each particular span. Specifically, one needs the WDM PSD of the signal launched into the span, the characteristics of the fiber used in the span and the span amplification features. No "non-local" information is needed regarding the path previously travelled by the either the CUT or INTs, not even their respective formats. If NLI is evaluated through Equation 7.32, all that is needed regarding the WDM signal at each span is in fact the center frequencies of each channel f_n, their respective bandwidths and symbol rates B_n and R_n, and their flat-top PSDs G_n. The latter can typically be approximated as P_n/R_n.

Regarding the second aspect, we point out that under the incoherent accumulation approximation, not just NLI, but the total "signal degradation" due to both NLI and ASE together can be computed fully locally in each span, in the form of the reciprocal of a "span-local OSNR" [38]. Such signal degradations turn out to be additive, so that the total degradation for a given channel is the sum of its per-span degradations ([38], Equations 80 and 81). Being local, the signal degradation can be minimized locally, which is equivalent to maximizing the span-local OSNR. This can be done by intervening locally on the launch power of each channel into that span.

The possibility of performing optimization for each span independently of all others has been called LOGO (for local-optimization, global-optimization) and was proposed in Refs [38, 75]. The specific optimization criteria can be varied, and will not be discussed here. We will only focus on one possible strategy, which has the merit of further drastically simplifying network management and ensuring that no lightpath disruption occurs under any circumstances, at the cost of some loss of efficiency. The goal is that of achieving the *"optical ether"* regime.

7.4.5.3 The LOGON Approach and the "Optical Ether" Any span-local OSNR maximization strategy for a given channel would require that the launch power of the channels present in that span be optimized, which in turn requires the detailed knowledge of the spectral loading of that span. A drastic simplification occurs if such optimization can simply pre-suppose that *all* channels are always present in the span. In fact, an even more drastic approximation is to assume that the whole available WDM band is always fully and seamlessly saturated.

[21] If future systems employed significantly smaller symbol rates than 28–32 GBaud, to take advantage of the NLI mitigation effect discussed in Section 7.4.4, then the accuracy of the incoherent GN model would have to be re-assessed. If needed, it is conceivable that suitable corrections could be worked out, for instance derived from the asymptotic formula Equation 7.22.

The details of this approach, called LOGON (for LOGO-Nyquist), can be found in Ref. [38], Section X. One of its advantages is that the local optimization of each span relies on a simple pre-computation which results in a single number: the optimum launch PSD level G_n, which is the same for all channels and which all channels must comply with.

Another substantial advantage of the LOGON strategy is the following. We assume that the CP has routed the channels in the DRN so that they are operational according to the performance predicted under the LOGON full-spectral-load assumption. Then, the insertion of one or more channels in an already partially populated link cannot cause any disruption nor can it require any re-routing of the channels already present in the link. This is because the NLI generated by the insertion of new channels was already factored in, to its worst (full-load) case, by the CP. The actual performance of the already present channels would of course degrade, but would be better, or at worst equal, to what the CP has already considered.

The LOGON strategy tends to achieve an operating regime which we would call "*optical ether.*" The term was used in the late 1980s and early 1990s (see for instance [76, 77]). In essence, adapted to the current DRN scenario, the concept is that of reducing the complexities of the physical layer to a care-free transparent medium, where any turn-on, turn-off or re-routing of lightpaths is allowed and is without consequences for the already lit paths.[22]

The strength of LOGON is also its main weakness: by always assuming full spectral loading, when a lightpath travels across a sparsely populated network, its degradation is substantially overestimated, possibly causing the CP to enact regeneration when it is not necessary. An analysis of the potential underestimation was shown in Ref. [38]. More sophisticated approaches than LOGON are currently being studied to attenuate this drawback.

The LOGON strategy has been recently used in experimental testbeds with good results in terms of predictivity and overall network optimization [49, 78].

In general, the use of advanced NLI modeling in the control and management of next generation flexible DRNs is a very active field of research and new strategies and results are appearing at a fast pace. There is no doubt however that the recent progress in NLI modeling, of which this chapter presents a partial overview, have already dramatically impacted this field and are bound to further significantly influence the future of next generation DRNs.

7.5 CONCLUSION

The field of non-linearity modeling has been very productive in recent years and will likely be so for the next few. Having been written in the midst of this very prolific

[22]It is interesting to quote [76] about the "optical ether" definition, which sounds close to the current DRN environment: "The networks in question perform only linear operations on optical signals, essentially operating as a dynamically controlled 'ether' within which light beams are selectively directed between prescribed source-destination pairs. Because these Linear Lightwave Networks are controllable, they are capable of being reconfigured in response to changing load conditions or component failures."

and evolutionary phase, this chapter is probably destined to age quickly and be superseded by a number of advancements. Hopefully, some of the basic results and remarks reported here will endure and be useful for some extended period of time, at least as primer background material. One of the latest developments, the EGN model, appears to have now superseded the limitations of previous models (such as the GN and XPM models) which suggests that it might take some time for it to obsolesce, though there is no guarantee of this either.

Three very prominent topics, which were only touched on or briefly introduced here, are likely to provide very interesting results in the near future: one is that of the detailed study and the accurate assessment of the actual system impact of the phase-noise component of NLI; another is that of the mitigation of NLI by choosing the optimum per-carrier symbol rate, which appears to be a non-negligible effect which the EGN model allows to study in detail; yet another is that of flexible and reconfigurable-network physical-layer-aware control and management strategies.

Many other exciting topics exist and still more will certainly develop and come to the forefront in the coming years, perhaps having to do with spatial division multiplexed systems, or with non-conventional approaches which deal with fiber non-linear effects by actually exploiting them to *increase* capacity [79].

So, our conclusive advice is for the interested readers to keep monitoring this field closely for the new disruptive results which are certain to emerge.

ACKNOWLEDGEMENTS

The authors would like to thank Dr. Marco Schiano of Telecom Italia, as well as Prof. Gabriella Bosco and Prof. Vittorio Curri of Politecnico di Torino, for reviewing this chapter and providing very valuable guidance and advice.

APPENDIX

A.1 BER FORMULAS FOR THE MOST COMMON QAM SYSTEMS

Gray coding is assumed for all formats. For PM-BPSK the exact formula is:

$$\text{BER} = 1/2 \cdot \text{erfc}\left(\sqrt{\text{OSNR}_{\text{NL}}}\right) \tag{A.1}$$

For PM-QPSK the exact formula is:

$$\text{BER} = 1/2 \cdot \text{erfc}\left(\sqrt{\text{OSNR}_{\text{NL}}/2}\right) \tag{A.2}$$

For PM-16QAM and PM-64QAM, respectively, the following formulas are approximate, but their accuracy is better than ±0.05 dB of OSNR_{NL} over the range 10^{-1}– 10^{-4}:

$$\text{BER} = 3/8 \cdot \text{erfc}\left(\sqrt{\text{OSNR}_{\text{NL}}/10}\right) \tag{A.3}$$

$$\text{BER} = 7/24 \cdot \text{erfc}\left(\sqrt{\text{OSNR}_{\text{NL}}/42}\right) \tag{A.4}$$

A.2 THE LINK FUNCTION μ

Under the assumption of lumped amplification and identical spans, the link function can be written as:

$$\mu(f_1,f_2,f) = \zeta(f_1,f_2,f) \cdot \nu(f_1,f_2,f) \tag{A.5}$$

where:

$$\zeta(f_1,f_2,f) = \gamma \frac{1 - e^{-2\alpha L_s} e^{j4\pi^2 \beta_2 (f_1 - f)(f_2 - f) L_s}}{2\alpha - j4\pi^2 \beta_2 (f_1 - f)(f_2 - f)} \tag{A.6}$$

$$\nu(f_1,f_2,f) = \frac{\sin(2\beta_2 \pi^2 (f_1 - f)(f_2 - f) N_s L_s)}{\sin(2\beta_2 \pi^2 (f_1 - f)(f_2 - f) L_s)} e^{j2\beta_2 \pi^2 (f_1 - f)(f_2 - f)(N_s - 1) L_s} \tag{A.7}$$

All the symbols were defined at the beginning of Section 7.4. The $\zeta(f_1,f_2,f)$ factor physically represents the efficiency of non-degenerate FWM occurring among three spectral components of the signal placed at frequencies $f_1, f_2, f_3 = (f_1 + f_2 - f)$, producing a beat disturbance at frequency f. The factor ν accounts for the coherent interference of the NLI field contributions produced in different spans, when they are summed up at the receiver location. Note also that only the effect of the second-order dispersion parameter β_2 is taken into account in the above formulas.

In the more general case of distributed amplification and/or non-identical spans, and including the effect of third-order dispersion β_3, the link function μ takes on a more complex form which is shown in the following:

$$\begin{aligned}
\mu(f_1,f_2,f) = {} & \sum_{n=1}^{N_s} \gamma_n \left[\prod_{k=1}^{n-1} \exp\left(\int_0^{L_{s,k}} 3g_k(z')\, dz' \right) \exp(-3\alpha_k L_{s,k}) \Gamma_k^{3/2} \right] \cdot \\
& \left[\prod_{k=n}^{N_s} \exp\left(\int_0^{L_{s,k}} g_k(z')\, dz' \right) \exp(-\alpha_k L_{s,k}) \Gamma_k^{1/2} \right] \cdot \\
& \exp\left(j4\pi^2 (f_1 - f)(f_2 - f) \sum_{k=1}^{n-1} [\beta_{2,k} L_{s,k} + \pi (f_1 + f_2) \beta_{3,k} L_{s,k} + \beta_{\text{DCU},k}] \right) \cdot \\
& \int_0^{L_{s,n}} \exp\left(\int_0^{z} 2g_n(z')\, dz' - 2\alpha_n z \right) \cdot \\
& \exp(j4\pi^2 (f_1 - f)(f_2 - f)[\beta_{2,n} + \pi \beta_{3,n} (f_1 + f_2)] z)\, dz
\end{aligned} \tag{A.8}$$

where, not previously defined:

- $g(z)$: fiber *field* gain coefficient (km^{-1}), possibly z-dependent, such that the signal *power* is amplified over a stretch of z km as $\int_0^z \exp(2g(z'))dz'$
- Γ: lumped power gain, such as due to an EDFA
- β_3: third-order dispersion (or dispersion slope), ($\text{ps}^3\cdot\text{km}^{-1}$)
- β_{DCU}: lumped accumulated dispersion (ps^2)

Equation A.8 is quite general. For the generic nth span it accounts for: span length; span fiber parameters, including third-order dispersion $\beta_{3,n}$; span amplification, including any combination of lumped and distributed amplification, the latter with any spatial profile $g_n(z)$; dispersion management, by means of a lumped dispersive element of accumulated dispersion $\beta_{\text{DCU},n}$ placed at the end of the span. Note that the possible lumped amplification element (typically an EDFA) of gain Γ_n is also assumed to be placed at the end of the span.

For more details on these link factors, see [35, 58].

A.3 THE EGN MODEL FORMULAS FOR THE X2-X4 AND M1-M3 ISLANDS

In the following, the complete set of formulas making up the EGN model correction term $G_{\text{NLI}}^{\text{corr}}(f)$ is provided, with the exception of those already shown in Section 7.4.3 for SCI and the X1 integration islands.

Note that, with reference to Figure 7.1, to obtain the full correction term $G_{\text{NLI}}^{\text{corr}}(f)$ requires adding *all* of the NLI contributions, as shown here below:

$$G_{\text{NLI}}^{\text{corr}}(f) = G_{\text{SCI}}^{\text{corr}}(f) + \sum_{i=1}^{4} G_{\text{X}i}^{\text{corr}}(f) + \sum_{i=1}^{3} G_{\text{M}i}^{\text{corr}}(f) \tag{A.9}$$

The X2-X4 and M1-M3 are as follows:

$$G_{\text{X2}}^{\text{corr}}(f) = \Phi_m P_m^2 \sum_{\substack{n=m-1\\ n\neq m}}^{m+1} P_n\, \kappa_{\text{X2}}^{n}(f) \tag{A.10}$$

$$\begin{aligned}\kappa_{\text{X2}}^{n}(f) = \frac{80}{81} R_m R_n \int_{f_n-B_n/2}^{f_n+B_n/2} df_1 \int_{f_m-B_m/2}^{f_m+B_m/2} df_2 \int_{f_m-B_m/2}^{f_m+B_m/2} df_2' \cdot \\ |s_n(f_1)|^2 s_m(f_2) s_m^*(f_2') s_m^*(f_1+f_2-f) s_m(f_1+f_2'-f) \cdot \\ \mu(f_1,f_2,f)\mu^*(f_1,f_2',f)\end{aligned} \tag{A.11}$$

$$G_{\text{X3}}^{\text{corr}}(f) = \Phi_m P_m^2 \sum_{\substack{n=m-1\\ n\neq m}}^{m+1} P_n \kappa_{\text{X3}}^{n}(f) \tag{A.12}$$

$$\kappa_{\text{X3}}^{n}(f) = \frac{16}{81} R_m R_n \int_{f_m-B_m/2}^{f_m+B_m/2} df_1 \int_{f_m-B_m/2}^{f_m+B_m/2} df_2 \int_{f_m-B_m/2}^{f_m+B_m/2} df_2' \cdot$$

$$|s_n(f_1+f_2-f)|^2 s_m(f_1)s_m(f_2)s_m^*(f_2^{'})s_m^*(f_1+f_2-f_2^{'})\cdot$$
$$\mu(f_1,f_2,f)\mu^*(f_1+f_2-f_2^{'},f_2^{'},f) \quad \text{(A.13)}$$

$$G_{\text{X4}}^{\text{corr}}(f)=\sum_{\substack{n=m-1\\n\neq m}}^{m+1} P_n^3[\Phi_n\kappa_{\text{X4}}^n(f)+\Psi_n\varsigma_{\text{X4}}^n(f)] \quad \text{(A.14)}$$

$$\kappa_{\text{X4}}^n(f)=\frac{80}{81}R_n^2\int_{f_n-B_n/2}^{f_n+B_n/2}df_1\int_{f_n-B_n/2}^{f_n+B_n/2}df_2\int_{f_n-B_n/2}^{f_n+B_n/2}df_2^{'}\cdot$$
$$|s_n(f_1)|^2 s_n(f_2)s_n^*(f_2^{'})s_n^*(f_1+f_2-f)s_n(f_1+f_2^{'}-f)\cdot$$
$$\mu(f_1,f_2,f)\mu^*(f_1,f_2^{'},f)$$
$$+\frac{16}{81}R_n^2\int_{f_n-B_n/2}^{f_n+B_n/2}df_1\int_{f_n-B_n/2}^{f_n+B_n/2}df_2\int_{f_n-B_n/2}^{f_n+B_n/2}df_2^{'}\cdot$$
$$|s_n(f_1+f_2-f)|^2 s_n(f_1)s_n(f_2)s_n^*(f_1+f_2-f_2^{'})s_n^*(f_2^{'})\cdot$$
$$\mu(f_1,f_2,f)\mu^*(f_1+f_2-f_2^{'},f_2^{'},f) \quad \text{(A.15)}$$

$$\varsigma_{\text{X4}}^n(f)=\frac{16}{81}R_n\int_{f_n-B_n/2}^{f_n+B_n/2}df_1\int_{f_n-B_n/2}^{f_n+B_n/2}df_2\int_{f_n-B_n/2}^{f_n+B_n/2}df_1^{'}\int_{f_n-B_n/2}^{f_n+B_n/2}df_2^{'}\cdot$$
$$s_n(f_1)s_n(f_2)s_n^*(f_1+f_2-f)s_n^*(f_1^{'})s_n^*(f_2^{'})s_n(f_1^{'}+f_2^{'}-f)\cdot$$
$$\mu(f_1,f_2,f)\mu^*(f_1^{'},f_2^{'},f) \quad \text{(A.16)}$$

$$G_{\text{M1}}^{\text{corr}}(f)=\sum_{\substack{n=m-1\\n\neq m}}^{m+1} P_n\sum_{l=l_{\min}}^{l_{\max}} P_l^2\Phi_l\kappa_{\text{M1}}^l(f)$$

when $n=m-1, l_{\min}=n+2, l_{\max}=N_{\text{ch}}$;

when $n=m+1, l_{\min}=1, l_{\max}=n-2$ (A.17)

$$\kappa_{\text{M1}}^l(f)=\frac{80}{81}R_nR_l\int_{f_n-B_n/2}^{f_n+B_n/2}df_1\int_{f_l-B_l/2}^{f_l+B_l/2}df_2\int_{f_l-B_l/2}^{f_l+B_l/2}df_2^{'}\cdot$$
$$|s_n(f_1)|^2 s_l(f_2)s_l^*(f_2^{'})s_l^*(f_1+f_2-f)s_l(f_1+f_2^{'}-f)\cdot$$
$$\mu(f_1,f_2,f)\mu^*(f_1,f_2^{'},f) \quad \text{(A.18)}$$

$$G_{\text{M2}}^{\text{corr}}(f)=\sum_{\substack{n=m-1\\n\neq m}}^{m+1} P_n\sum_{l=l_{\min}}^{l_{\max}} P_l^2\Phi_l\kappa_{\text{M2}}^l(f)$$

when $n=m-1, l_{\min}=1, l_{\max}=n-1$;

when $n=m+1, l_{\min}=n+1, l_{\max}=N_{\text{ch}}$ (A.19)

$$\kappa^{l}_{\rm M2}(f) = \frac{80}{81} R_n R_l \int_{f_n-B_n/2}^{f_n+B_n/2} df_1 \int_{f_l-B_l/2}^{f_l+B_l/2} df_2 \int_{f_l-B_l/2}^{f_l+B_l/2} df_2' \cdot$$
$$|s_n(f_1)|^2 s_l(f_2) s_l^*(f_2') s_l^*(f_1+f_2-f) s_l(f_1+f_2'-f) \cdot$$
$$\mu(f_1,f_2,f)\mu^*(f_1,f_2',f) \tag{A.20}$$

$$G^{\rm corr}_{\rm M3}(f) = \sum_{\substack{n=1 \\ n\neq m,m\pm1}}^{N_{\rm ch}} P_n P_l^2 \Phi_l \kappa^{l}_{\rm M3}(f)$$
$$\text{when } n \text{ is odd}, l = (n+5)/2;$$
$$\text{when } n \text{ is even}, l = n/2+2, n/2+3 \tag{A.21}$$

$$\kappa^{l}_{\rm M3}(f) = \frac{16}{81} R_n R_l \int_{f_l-B_l/2}^{f_l+B_l/2} df_1 \int_{f_l-B_l/2}^{f_l+B_l/2} df_2 \int_{f_l-B_l/2}^{f_l+B_l/2} df_2' \cdot$$
$$|s_n(f_1+f_2-f)|^2 s_l(f_1) s_l(f_2) s_l^*(f_2') s_l^*(f_1+f_2-f_2') \cdot$$
$$\mu(f_1,f_2,f)\mu^*(f_1+f_2-f_2',f_2',f) \tag{A.22}$$

A.4 OUTLINE OF GN–EGN MODEL DERIVATION

We report here a *simplified outline* of the main steps involved in the derivation of the GN–EGN model. A more general and detailed derivation along the lines reproduced here can be found in Ref. [58] for the GN model and in Ref. [35] for the EGN model. Alternative derivations can be found in Refs [12, 13] for the GN model and in Ref. [23] for the EGN model.

For simplicity we assume that all WDM channels have same symbol rate $R_{\rm ch}$ and a uniform frequency spacing Δf. We then make the assumption that the overall WDM signal can be modeled through a complex *periodic* random process. The period will be later stretched to infinity.[23] To formally represent such random process, we first write the expression of a generic periodic signal as a Fourier-series:

$$E(t) = \sum_{i=-\infty}^{\infty} q_i \; e^{j2\pi i f_0 t} \tag{A.23}$$

[23]Strictly speaking, to make sure that the overall WDM signal can be assumed to be periodic, and can therefore be written as Equation A.23, two constraints must be satisfied. Specifically, the channel frequency spacing, Δf, and the channel symbol rate, $R_{\rm ch}$, must be an exact multiple of f_0. On the other hand, these constraints are immaterial, as one could assume f_0 to be an arbitrarily small quantity. They disappear completely when, later on, the signal period is stretched to infinity. As a result, no actual loss of generality is incurred when using the signal representation Equation A.23. Note also that the method can be generalized to the case of non-uniform channel spacing and different channel symbol rates, but this generalization is not dealt with here.

We then define the coefficients q_i as:

$$q_i = \sqrt{f_0\, G_{\text{WDM}}(if_0)}\; e^{j2\pi\; \phi_{\text{WDM}}(if_0)}\; \xi_i \tag{A.24}$$

where T_0 is the period of the signal, $f_0 = 1/T_0$, and the ξ_i's are suitable complex random variables with zero-mean and unit variance $\sigma^2_{\xi_i} = 1$, which will be discussed later. The quantity $G_{\text{WDM}}(f)$ is the overall WDM signal PSD. It acts as a spectral-shaping function which imposes the WDM PSD on $E(t)$. In fact, the PSD of $E(t)$ is easily found to be:

$$G_{\text{E}}(f) = f_0 \sum_{i=-\infty}^{\infty} G_{\text{WDM}}(if_0)\; \delta(f - if_0) \tag{A.25}$$

So, apart from being made up of discrete spectral lines, due to the periodic assumption on the signal, the power spectrum of $E(t)$ follows $G_{\text{WDM}}(f)$. The phase $\phi_{\text{WDM}}(f)$ in Equation A.24 accounts for the WDM signal phase-features, ranging from the channels' data-bearing pulses $s_n(t)$ possibly being complex, to possible chromatic dispersion pre-compensation.[24]

We then consider the single-polarization NLSE, in frequency domain, for a homogenous fiber stretch, which is given by:

$$\begin{aligned}\frac{\partial}{\partial z}E(z,f) &= -j2\pi^2\beta_2 f^2 E(z,f) - \alpha E\,(z,f) - \\ &\quad - j\gamma E(z,f) * E^*(z,-f) * E(z,f)\end{aligned} \tag{A.27}$$

where "$*$" means convolution. The last term of the NLSE is the non-linear Kerr term, which we call $Q_{\text{Kerr}}(z,f)$. We first find the *linear* solution of Equation A.27, i.e., we disregard the Kerr term. One easily gets:

$$\begin{aligned}E_{\text{LIN}}(z,f) &= e^{-j2\beta_2\pi^2 f^2 z} e^{-\alpha z} E(f) \\ &= e^{-\alpha z}\sqrt{f_0} \sum_{i=-\infty}^{+\infty} e^{-j2\beta_2\pi^2 (if_0)^2 z} \sqrt{G_{\text{WDM}}(if_0)}\; e^{j2\pi\phi_{\text{WDM}}(if_0)} \xi_i \delta(f - if_0)\end{aligned} \tag{A.28}$$

[24] Under the assumption that different WDM channels do not overlap and that the transmitted data symbols on all channels are zero-mean and all independent of one another, then the notation used here relates to the notation used in Section 7.4.2 as follows:

$$\sqrt{G_{\text{WDM}}(f)}\; e^{j2\pi\phi_{\text{WDM}}(f)} = \sum_{n=1}^{N_{\text{ch}}} \sqrt{P_n R_{\text{ch}}}\; s_n(f - f_n) \tag{A.26}$$

Note that any possible phase-shift related to the nth channel is incorporated in the Fourier transform of its data-carrying pulse, $s_n(f)$, in agreement with what specified in Section 7.4.2.

where $E(f)$ is the Fourier transform of $E(t)$ from Equation A.23. We then *approximate* the Kerr term by expressing it through the *linearly propagating signal only*, that is, as:

$$Q_{\text{Kerr}}(z,f) \approx -j\gamma E_{_{\text{LIN}}}(z,f) * E^*_{_{\text{LIN}}}(z,-f) * E_{_{\text{LIN}}}(z,f) \tag{A.29}$$

Now $Q_{\text{Kerr}}(z,f)$ has a fully determined expression. The signal obtained by solving Equation A.27 using $Q_{\text{Kerr}}(z,f)$ as a "source" term, with initial condition $E(0,f) = 0$, is the "perturbation" $E_{_{\text{NLI}}}(z,f)$. The overall approximate solution is then given by:

$$E(z,f) \approx E_{_{\text{LIN}}}(z,f) + E_{_{\text{NLI}}}(z,f) \tag{A.30}$$

This solution procedure is equivalent to a fist-order regular-perturbation approach (see Section 7.2.1.3). In Equation A.30, $E_{_{\text{LIN}}}(z,f)$ is as shown in Equation A.28, whereas:

$$E_{_{\text{NLI}}}(z,f) = \sum_{i=-\infty}^{\infty} \mathcal{E}_{_{\text{NLI},i}}(z)\ \delta(f - if_0)$$

with:

$$\mathcal{E}_{_{\text{NLI},i}}(z) = -jf_0^{\frac{3}{2}}\, e^{-j2\beta_2\pi^2 i^2 f_0^2 z} e^{-\alpha z} \sum_{m,n,k \in \tilde{A}_i} \xi_m \xi_n^* \xi_k\ \zeta\ (mf_0, kf_0, nf_0)$$
$$e^{j2\pi[\phi_{\text{WDM}}(mf_0)+\phi_{\text{WDM}}(kf_0)-\phi_{\text{WDM}}(nf_0)]} \sqrt{G_{_{\text{WDM}}}(mf_0) G_{_{\text{WDM}}}(nf_0) G_{_{\text{WDM}}}(kf_0)} \tag{A.31}$$

The factor ζ is Equation A.6 with L_s replaced by z. The summation index set $\tilde{A}_i$ is:

$$\tilde{A}_i = \{(m,n,k) : m - n + k = i, m \neq n, k \neq n\}$$

Note that the index combinations for which $m = n$ or $k = n$ are removed because they can be shown to correspond to terms that produce no actual signal degradation (see [58] for the details).

The NLI noise PSD due to propagation over a *single span* (acronym "ss"), assuming that $z = L_s$ is the length of a span, can then be written as:

$$G_{_{\text{NLI},ss}}(f) = \sum_{i=-\infty}^{\infty} P_{_{\text{NLI},ss,i}} \cdot \delta(f - if_0) \tag{A.32}$$

where

$$P_{_{\text{NLI},ss},i} = \mathbf{E}\{|\ \mathcal{E}_{_{\text{NLI},ss},i}|^2\} \tag{A.33}$$

is the NLI power carried by each frequency (if_0) in Equation A.32, and $\mathcal{E}_{_{\text{NLI},ss,i}}$ is $\mathcal{E}_{_{\text{NLI},i}}(z)$ from Equation A.31 for $z = L_s$. The symbol $\mathbf{E}$ denotes the statistical expectation operator. Statistical averaging is necessary because the $\mathcal{E}_{_{\text{NLI},ss,i}}$'s are RVs, since they contain the ξ_i's, as shown in Equation A.31.

Substituting Equation A.31 into Equation A.33 yields a fairly complex formula, which shows that $P_{\text{NLI},ss,i}$ consists of many cross-products of different contributions. The key point is that each one of such cross-products has a factor of the type:

$$\mathrm{E}\{\xi_m \xi_n^* \xi_k \xi_{m'}^* \xi_{n'} \xi_{k'}^*\} \tag{A.34}$$

with the two triples of indices (m, n, k) and (m', n', k') each independently spanning the set $\tilde{A}_i$. Whether a GN model or an EGN model estimate of $P_{\text{NLI},ss,i}$ is found, it depends exclusively on the assumptions made on the statistical properties of the ξ_i's. For the GN model, the simplifying assumption that the ξ_i's are Gaussian and independent of one another is made, which causes the factor Equation A.34 to be zero for a large subset of all the possible index combinations. For the EGN model, the ξ_i's are instead characterized *exactly*, based on the formulas that rigorously relate the ξ_i's to the actual modulated signal.[25] It turns out that all the contributions to $P_{\text{NLI},ss,i}$ of the GN model are also present in the EGN model, with the same strength. However, the EGN model adds many more contributions, arising from Equation A.34 no longer being zero for many of the index combinations.

Concentrating on the GN model, the result for Equation A.33 is, quite simply:

$$P_{\text{NLI},ss,i} = 2f_0^3 e^{-2\alpha L_s} \cdot \sum_{m,n,k \in \tilde{A}_i} G_{\text{WDM}}(nf_0) G_{\text{WDM}}(mf_0) G_{\text{WDM}}(kf_0) |\zeta(kf_0, mf_0, nf_0)|^2 \tag{A.36}$$

Subsequently, the assumed time-periodicity of Equation A.23 is stretched to infinity, by letting $T_0 \to \infty$ or, equivalently, $f_0 \to 0$. The discrete-line PSD Equation A.32 then gets transformed into a continuous-frequency PSD:

$$G_{\text{NLI}}^{\text{GN}}(f) = 2 \int_{-\infty}^{\infty} \int_{-\infty}^{\infty} G_{\text{WDM}}(f_1)\, G_{\text{WDM}}(f_2)\, G_{\text{WDM}}(f_1 + f_2 - f) \cdot |\zeta(f_1, f_2, f)|^2 \, df_2 \, df_1 \tag{A.37}$$

[25] These formulas can be found for instance in the extended-appendix version of [35], available on arXiv (App. A, Equations 63 and 64), or in Ref. [40]. However, those papers use a different notation than here. According to this chapter's notation, the ξ_i's are found as:

$$\xi_i = \sqrt{f_0/(R_{\text{ch}} P_n)} \sum_{w=0}^{W-1} a_n^w e^{-j2\pi w(i - f_n/f_0)/W} \tag{A.35}$$

with the formal constraint that both the symbol rate R_{ch} and the center frequency of the nth channel f_n are exact multiples of f_0 (see Footnote A.4). The positive integer W is given by $T_0 R_{\text{ch}}$. The index n identifies one of the WDM channels and is not a free index. It is the index of the channel where the frequency (if_0) falls on. As a result, as the index i is spanned, Equation A.35 addresses, one after the other, all WDM channels.

For more details on this transformation, see [58]. In addition, to transit from a single homogenous span to an overall multi-span system with possibly non-identical spans and any loss/amplification profile, it is enough to replace the factor ζ with the link function μ, described in Section A.2. With this substitution, the single-polarization GNRF is found.

Finally, to account for dual-polarization, all results must be recalculated using the Manakov equation (see Section 7.1). Remarkably, under the assumption that the transmitted signal on one polarization is statistically independent of the signal transmitted on the other, and that the total transmitted power is evenly split between the two polarizations, accounting for dual-polarization only generates a constant multiplying factor[26] equal to 8/27. As a result, at the end of the derivation, instead of Equation A.37, we obtain Equation 7.11, that is, the dual-polarization GNRF, which is formally identical to Equation A.37, except for the multiplication by said factor 8/27 and, again, the replacement of the factor ζ with the link function μ to convert to multi-span. Incidentally, it turns out that the contribution to $G_{\rm NLI}^{\rm GN}(f)$ due to same-polarization NLI is 2/3 of the total whereas the contribution of cross-polarization NLI is 1/3. In other words, same-polarization NLI is twice as powerful as cross-polarization NLI.

For the EGN model, the steps are similar, but due to the more complex results of the averages in Equation A.34, the whole procedure is much lengthier. For a fully detailed derivation, see the arXiv version of [35].

A.5 LIST OF ACRONYMS

AGN: additive Gaussian noise
AWGN: additive white Gaussian noise
ASE: amplified spontaneous-emission noise
BER: bit error-rate
BP: backward-propagation
CD: chromatic dispersion
CIAGN: circular independent additive Gaussian noise
CP: network control-plane
CUT: channel-under-test
DAC: digital-to-analog converter
DCF: dispersion-compensating fiber
DM: dispersion-managed
DRN: dynamically reconfigurable network
DSP: digital signal processing

[26] We employ the definition of the fiber non-linearity coefficient as in Ref. [4], that is: $\gamma = n_2\, k_0$, where n_2 is the Kerr non-linearity coefficient and k_0 is the wavenumber. As a consequence, we used the Manakov equation with a factor 8/9 multiplying the Kerr non-linearity term, as shown in Ref. [4]. Other authors define $\gamma = 8/9 \cdot n_2\, k_0$ and drop the 8/9 factor from the Manakov equation [9]. According to this latter definition of γ, the factor appearing in Equation 7.11 should then be 3/4 rather than 16/27. Throughout this chapter we adopt the conventions of [4].

EDFA: erbium-doped fiber amplifier
EGN model: enhanced Gaussian-noise model
FD-XPM: the XPM model proposed in Ref. [36] as "frequency-domain analysis," from which the "FD" prefix
FEC: forward error-correcting code
FWM: four-wave mixing
GN model: Gaussian-noise model
GNRF: GN-model reference formula
INT: interfering channel in a WDM comb (as opposed to the CUT)
LC-NLPN: long-correlated non-linear phase noise
LOGO: local-optimization, global optimization
LOGON: LOGO with Nyquist-WDM
LWN: locally-white noise
MCI: multi-channel interference
ME: Manakov equation
NLC: non-linearity compensation
NLI: non-linear interference
NLSE: non-linear Schroedinger equation
NZDSF: non-zero dispersion-shifted fiber
OFDM: orthogonal frequency-division multiplexing
OSNR: optical signal-to-noise ratio
PM: polarization-multiplexed
PSCF: pure-silica-core fiber
PSD: power spectral density
PTP: point-to-point links
QAM: quadrature amplitude modulation
QPSK: quadrature phase-shift keying
SCI: self-channel interference
SE: spectral efficiency
SMF: standard single-mode fiber
SNR: signal-to-noise ratio
SOP: state-of-polarization
SpS: spectral slicing
TD: time-domain
UT: uncompensated transmission
VS: Volterra series
WDM: wavelength-division multiplexing
XCI: cross-channel interference
XPM: cross-phase-modulation
XPolM: cross-polarization modulation

REFERENCES

1. Menyuk CR. Nonlinear pulse propagation in birefringent optical fibers. *IEEE J Quantum Electron* 1987;23:174–176.
2. Menyuk CR. Pulse propagation in an elliptically birefringent Kerr medium. *IEEE J Quantum Electron* 1989;25(12):2674–2682.
3. Agrawal GP. Non-Linear Fiber Optics. 5th ed. Academic Press: New York; 2012. ISBN-13: 978-0123970237, ISBN-10: 0123970237.
4. Marcuse D, Menyuk CR, Wai PKA. Application of the Manakov-PMD equation to studies of signal propagation in optical fibers with randomly varying birefringence. *J Lightwave Technol* 1997;15(9):1735–1746.
5. Vannucci A, Serena P, Bononi A. The RP method: a new tool for the iterative solution of the nonlinear Schrodinger equation. *J Lightwave Technol* 2002;20(7):1102–1112.
6. Holzlöhner R, Grigoryan VS, Menyuk CR, Kath WL. Accurate calculation of eye diagrams and bit error rates in optical transmission systems using linearization. *J Lightwave Technol* 2002;20(3):389–400.
7. Reis J, Teixeira A. Unveiling nonlinear effects in dense coherent optical WDM systems with Volterra series. *Opt Express* 2010;18(8):8660–8670.
8. Ellis AD, Zhao J, Cotter D. Approaching the non-linear Shannon limit. *J Lightwave Technol* 2010;28(4):423–433.
9. Shieh W, Chen X. Information spectral efficiency and launch power density limits due to fiber nonlinearity for coherent optical OFDM systems. *IEEE Photonics J* 2011;3(2):158–173.
10. Kilmurray S, Fehenberger T, Bayvel P, Killey RI. Comparison of the nonlinear transmission performance of quasi-Nyquist WDM and reduced guard interval OFDM. *Opt Express* 2012;20(4):4198–4205.
11. Mecozzi A, Essiambre R-J. Nonlinear Shannon limit in pseudolinear coherent systems. *J Lightwave Technol* 2012;30(12):2011–2024.
12. Johannisson P, Karlsson M. Perturbation analysis of nonlinear propagation in a strongly dispersive optical communication system. *J Lightwave Technol* 2013;31(8):1273–1282.
13. Serena P, Bononi A. An alternative approach to the Gaussian noise model and its system implications. *J Lightwave Technol* 2013;31(22):3489–3499.
14. Secondini M, Forestieri E, Prati G. Achievable information rate in nonlinear WDM fiber-optic systems with arbitrary modulation formats and dispersion maps. *J Lightwave Technol* 2013;31(23):3839–3852.
15. Winzer PJ, Essiambre R-J. Advanced modulation formats for high-capacity optical transport networks. *J Lightwave Technol* 2006;24(12):4711–4728.
16. Essiambre R-J, Kramer G, Winzer PJ, Foschini GJ, Goebel B. Capacity limits of optical fiber networks. *J Lightwave Technol* 2011;28(4):662–701.
17. Wai PKA, Menyuk CR, Chen HH. Stability of solitons in randomly varying birefringent fibers. *Opt Lett* 1991;16(16):1231–1233.

18. Evangelides SG Jr., Mollenauer LF, Gordon JP, Bergano NS. Polarization multiplexing with solitons. *J Lightwave Technol* 1992;10(1):28–35.
19. Rossi N, Serena P, Bononi A. Polarization-dependent loss impact on coherent optical systems in presence of fiber nonlinearity. *IEEE Photonics Technol Lett* 2014;26(4):334–337.
20. Kuschnerov M, Chouayakh M, Piyawanno K, Spinnler B, Alfiad MS, Napoli A, Lankl B. On the performance of coherent systems in the presence of polarization-dependent loss for linear and maximum likelihood receivers. *IEEE Photonics Technol Lett* 2010;22(12):920–922.
21. Sun H, Vanleeuwen H, Dangui V, Nilsson A, Pan Z, Mertz P, Rahn J, Mitchell M, Wu K-T. System penalty in coherent receiver considering distributed PMD, PDL, and ASE. *IEEE Photonics Technol Lett* 2013;25(9):885–887.
22. Beygi L, Agrell E, Johannisson P, Karlsson M, Wymeersch H. A discrete-time model for uncompensated single-channel fiber-optical links. *IEEE Trans Commun* 2012;60(11):3440–3450.
23. Serena P, Bononi A. A time-domain extended Gaussian noise model. *J Lightwave Technol* 2015;33(7):1459–1472.
24. Dar R, Feder M, Mecozzi A, Shtaif M. Pulse collision picture of inter-channel nonlinear interference in fiber-optic communications. *J Lightwave Technol*, available as pre-print on IEEE Xplore.
25. Kumar S, Yang D. Second-order theory for self-phase modulation and cross-phase modulation in optical fibers. *J Lightwave Technol* 2005;23(6):2073–2080.
26. Peddanarappagari KV, Brandt-Pearce M. Volterra series transfer function of single-mode fibers. *J Lightwave Technol* 1997;15(12):2232–2241.
27. Mecozzi A, Balslev Clausen C, Shtaif M. Analysis of intrachannel nonlinear effects in highly dispersed optical pulse transmission. *IEEE Photonics Technol Lett* 2000;12(4):392–394.
28. Narimanov EE, Mitra PP. The channel capacity of a fiber optics communication system: perturbation theory. *J Lightwave Technol* 2002;20(3):530–537.
29. Forestieri E, Secondini M. Solving the non-linear Schroedinger equation. In: Optical Communication Theory and Techniques. Boston (MA): Springer-Verlag; 2005, p 3–11.
30. Secondini M, Forestieri E, Menyuk CR. A combined regular-logarithmic perturbation method for signal-noise interaction in amplified optical systems. *J Lightwave Technol* 2009;27(16):3358–3369.
31. Secondini M, Forestieri E. Analytical fiber-optic channel model in the presence of cross-phase modulations. *IEEE Photonics Technol Lett* 2012;24(22):2016–2019.
32. Poggiolini P, Carena A, Jiang Y, Bosco G, Curri V, Forghieri F. Impact of low-OSNR operation on the performance of advanced coherent optical transmission systems. Proceedings of ECOC 2014; Cannes (FR), Sept 2014. Available with corrections on www .arXiv.org, paper arXiv:1407.2223.
33. Louchet H, Hodzic A, Petermann Klaus. Analytical model for the performance evaluation of DWDM transmission systems. *IEEE Photonics Technol Lett* 2003;15(9):1219–1221.
34. Splett A, Kurzke C, Petermann K. Ultimate transmission capacity of amplified optical fiber communication systems taking into account fiber nonlinearities. Proceedings of ECOC 1993, Volume 2; Montreux (CH); Sept 1993. p 41–44.
35. Carena A, Bosco G, Curri V, Jiang Y, Poggiolini P, Forghieri F. EGN model of non-linear fiber propagation. *Opt Express* 2014;22(13):16335–16362. Extended appendices with full

formulas derivations can be found in the version available on www.arXiv.org.
36. Dar R, Feder M, Mecozzi A, Shtaif M. Properties of nonlinear noise in long, dispersion-uncompensated fiber links. *Opt Express* 2013;21(22):25685–25699.
37. Carena A, Bosco G, Curri V, Poggiolini P, Forghieri F. Impact of the transmitted signal initial dispersion transient on the accuracy of the GN-model of non-linear propagation. Proceedings of ECOC 2013; paper Th.1.D.4; London (UK); Sept 2013.
38. Poggiolini P, Bosco G, Carena A, Curri V, Jiang Y, Forghieri F. The GN model of fiber non-linear propagation and its applications. *J Lightwave Technol* 2014;32(4):694–721.
39. Poggiolini P, Carena A, Curri V, Bosco G, Forghieri F. Analytical modeling of non-Linear propagation in uncompensated optical transmission links. *IEEE Photonics Technol Lett* 2011;23(11):742–744.
40. Carena A, Curri V, Bosco G, Poggiolini P, Forghieri F. Modeling of the impact of non-linear propagation effects in uncompensated optical coherent transmission links. *J Lightwave Technol* 2012;30(10):1524–1539.
41. Pan J, Isautier P, Filer M, Tibuleac S, Ralph SE. Gaussian noise model aided in-band crosstalk analysis in ROADM-enabled DWDM networks. Proceedings of OFC 2014; paper Th1I.1; San Francisco (CA); Mar 2014.
42. Poggiolini P, Bosco G, Carena A, Curri V, Jiang Y, Forghieri F. A simple and effective closed-form GN model correction formula accounting for signal non-Gaussian distribution. *J Lightwave Technol* 2015;33(2):459–473.
43. Torrengo E, Cigliutti R, Bosco G, Carena A, Curri V, Poggiolini P, Nespola A, Zeolla D, Forghieri F. Experimental validation of an analytical model for nonlinear propagation in uncompensated optical links. *Opt Express* 2011;19(26):B790–B798.
44. Cai J-X, Zhang H, Batshon HG, Mazurczyk M, Sinkin OV, Foursa DG, Pilipetskii AN, Mohs G, Bergano NS. 200 Gb/s and dual wavelength 400 Gb/s transmission over transpacific distance at 6.0 b/s/Hz spectral efficiency. *J Lightwave Technol* 2014;32(4):832–839.
45. Cai J-X, Sinkin OV, Zhang H, Batshon HG, Mazurczyk M, Foursa DG, Pilipetskii A, Mohs G. Nonlinearity compensation benefit in high capacity ultra-long haul transmission systems. Proceedings of ECOC 2013; paper We.4.D.2; London (UK), Sept 2013.
46. Stark AJ, Hsueh Y-T, Detwiler TF, Filer MM, Tibuleac S, Ralph SE. System performance prediction with the Gaussian noise model in 100G PDM-QPSK coherent optical networks. *J Lightwave Technol* 2013;31(21):3352–3360.
47. Nespola A, Straullu S, Carena A, Bosco G, Cigliutti R, Curri V, Poggiolini P, Hirano M, Yamamoto Y, Sasaki T, Bauwelinck J, Verheyen K, Forghieri F. GN-model validation over seven fiber types in uncompensated PM-16QAM Nyquist-WDM links. *IEEE Photonics Technol Lett* 2014;26(2):206–209.
48. Cai J-X, Batshon HG, Zhang H, Mazurczyk M, Sinkin OV, Foursa DG, Pilipetskii AN. Transmission performance of coded modulation formats in a wide range of spectral efficiencies. Proceedings of OFC 2014; paper M2C.3; San Francisco (CA); Mar 2014.
49. Pastorelli R, Bosco G, Nespola A, Piciaccia S, Forghieri F. Network planning strategies for next-generation flexible optical networks. Proceedings of OFC 2014; paper M2B.1; San Francisco (CA); Mar 2014.
50. Dar R, Feder M, Mecozzi A, Shtaif M. Accumulation of nonlinear interference noise in fiber-optic systems. *Opt Express* 2014;22(12):14199–14211.
51. Dar R, Geller O, Feder M, Mecozzi A, Shtaif M, Mitigation of inter-channel nonlinear interference in WDM systems. Proceedings of ECOC 2014; Cannes (FR); Sept 2014.

52. Jiang Y, Carena A, Poggiolini P, Forghieri F. On the impact of non-linear phase-noise on the assessment of long-haul uncompensated coherent systems performance. Proceedings of ECOC 2014; Cannes (FR); Sept 2014.
53. Secondini M, Forestieri E. On XPM mitigation in WDM fiber-optic systems. *IEEE Photonics Technol Lett* 2014;26(22):2252–2255.
54. Poggiolini P. The GN model of non-linear propagation in uncompensated coherent optical systems. *J Lightwave Technol* 2012;30(24):3857–3879.
55. Tang Jau. The channel capacity of a multispan DWDM system employing dispersive nonlinear optical fibers and an ideal coherent optical receiver. *J Lightwave Technol* 2002;20(7):1095–1101.
56. Nazarathy M, Khurgin J, Weidenfeld R, Meiman Y, Cho Pak, Noe R, Shpantzer I, Karagodsky V. Phased-array cancellation of nonlinear FWM in coherent OFDM dispersive multi-span links. *Opt Express* 2008;16:15778–15810.
57. Chen X, Shieh W. Closed-form expressions for nonlinear transmission performance of densely spaced coherent optical OFDM systems. *Opt Express* 2010;18:19039–19054.
58. Poggiolini P, Bosco G, Carena A, Curri V, Jiang Y, Forghieri F. A detailed analytical derivation of the GN model of non-linear interference in coherent optical transmission systems posted on arXiv, www.arxiv.org, paper identifier 1209.0394. First posted Sept 2012.
59. Savory SJ. Approximations for the nonlinear self-channel interference of channels with rectangular spectra. *IEEE Photonics Technol Lett* 2013;25(10):961–964.
60. Johannisson P, Agrell E. Modeling of nonlinear signal distortion in fiber-optical networks posted on arXiv.org, paper arXiv:1309.4000; Sept 2013.
61. Bononi A, Beucher O, Serena P. Single- and cross-channel nonlinear interference in the Gaussian noise model with rectangular spectra. *Opt Express* 2013;21(26):32254–32268.
62. Serena P, Bononi A. On the accuracy of the Gaussian nonlinear model for dispersion-unmanaged coherent links. Proceedings of ECOC 2013; paper Th.1.D.3; London (UK); Sept 2013.
63. Poggiolini P, Jiang Y, Carena A, Forghieri F. A Simple and Accurate Closed-Form EGN Model Formula paper arXiv:1503.04132; Mar 2015.
64. Carena A, Jiang Y, Poggiolini P, Bosco G, Curri V, Forghieri F. Electronic dispersion pre-compensation in PM-QPSK systems over mixed-fiber links. Proceedings of ECOC 2014; Cannes (FR); Sept 2014.
65. Nespola A, Bertignono L, Bosco G, Carena A, Jiang Y, Bilal SM, Poggiolini P, Abrate S, Forghieri F. Experimental demonstration of fiber nonlinearity mitigation in a WDM multi-subcarrier coherent optical system. Proceedings of ECOC 2015; Valencia (ES); Sept 2015.
66. Poggiolini P, Jiang Y, Carena A, Bosco G, Forghieri F. Analytical results on system maximum reach increase through symbol rate optimization. Proceedings OFC 2015; paper Th3D.6; Los Angeles (CA); Mar 2015.
67. http://en.wikipedia.org/wiki/Closed-form&uscore;expression. Accessed 2015 Sept 23.
68. Tang Jau. A comparison study of the Shannon channel capacity of various nonlinear optical fibers. *J Lightwave Technol* 2006;24(5):2070–2075.
69. Zhuge Q, Châtelain B, Plant DV. Comparison of intra-channel nonlinearity tolerance between reduced-guard-interval CO-OFDM systems and Nyquist single carrier systems. Proceedings of OFC 2012; paper OTh1B.3; Los Angeles (CA); Mar 2012.

70. Qiu M, Zhuge Q, Xu X, Chagnon M, Morsy-Osman M, Plant DV. Subcarrier multiplexing using DACs for fiber nonlinearity mitigation in coherent optical communication systems. Proceedings of OFC 2014; paper Tu3J.2; San Francisco (CA); Mar 2014.
71. Shieh W, Tang Y. Ultrahigh-speed signal transmission over nonlinear and dispersive fiber optic channel: the multicarrier advantage. *IEEE Photonics J* 2010;2(3):276–283.
72. Du LB, Lowery AJ. Optimizing the subcarrier granularity of coherent optical communications systems. *Opt Express* 2011;19(9):8079.
73. Bononi A, Rossi N, Serena P. Performance dependence on channel baud-rate of coherent single-carrier WDM systems. Proceedings of ECOC 2013; paper Th.1.D.5; London (UK); Sept 2013.
74. Poggiolini P, Carena A, Jiang Y, Forghieri F, Pastorelli R. What is the right physical layer model for a highly dynamic reconfigurable optical network? Proceedings of Photonics in Switching; Florence (IT); Sept 2015.
75. Poggiolini P, Bosco G, Carena A, Cigliutti R, Curri V, Forghieri F, Pastorelli R, Piciaccia S. The LOGON strategy for low-complexity control plane implementation in new-generation flexible networks. Proceedings of OFC 2013; paper OW1H.3; Los Angeles (CA); Mar 2013.
76. Stern TE. Linear lightwave networks: how far can they go? Proceedings of Globecom '90; Dec 1990.
77. Sabry M, Midwinter J. Towards an optical ether. *IEE Proc Optoelectron* 1994; 141(5):327–335.
78. Pastorelli R, Piciaccia S, Galimberti G, Self E, Brunella M, Calabretta G, Forghieri F, Siracusa D, Zanardi A, Salvadori E, Bosco G, Carena A, Curri V, Poggiolini P. Optical control plane based on an analytical model of non-linear transmission effects in a self-optimized network. Proceedings of ECOC 2013; paper We.3.E.4; London (UK); Sept 2013.
79. Sorokina MA, Turitsyn SK. Regeneration limit of classical Shannon capacity. *Nat Commun* 2014;5:3861. DOI: 10.1038/ncomms4861, www.nature.com/naturecommunications.

8

DIGITAL EQUALIZATION IN COHERENT OPTICAL TRANSMISSION SYSTEMS

SEB SAVORY

Department of Engineering, University of Cambridge, Cambridge, United Kingdom

8.1 INTRODUCTION

Digital equalization within a coherent optical receiver has been critical to the wide-scale adoption of the digital coherent transceiver in core networks. Digital equalization has not only allowed optical chromatic dispersion to be removed from the line, but more critically it removed the limits imposed by the polarization-mode dispersion (PMD) to upgrading legacy systems [1–7].

Given this chapter is concerned with digital equalization, before going in to detail we segue to discuss what is meant by digital equalization in contrast to mitigation. Mitigation is defined by the Oxford English Dictionary as "the action of reducing the severity, seriousness, or painfulness of something" [8]; in contrast, equalization is concerned with "the action or process of equalizing" [9]. In this context, both chromatic dispersion and PMD, being fundamentally lossless processes, can be equalized. In contrast, the impact of filtering and polarization-dependent loss (PDL) in the line, or nonlinear impairments can only be mitigated due to the loss of information. As such our focus is on equalization of dispersion, both chromatic and polarization mode; however, the algorithms described herein can also be applied to the design of matched filters or the mitigation of PDL.

Throughout this chapter, we assume the digital filtering is partitioned into two blocks as illustrated in Figure 8.1. The first of these blocks implements a set of filters

Enabling Technologies for High Spectral-efficiency Coherent Optical Communication Networks, First Edition. Edited by Xiang Zhou and Chongjin Xie.
© 2016 John Wiley & Sons, Inc. Published 2016 by John Wiley & Sons, Inc.

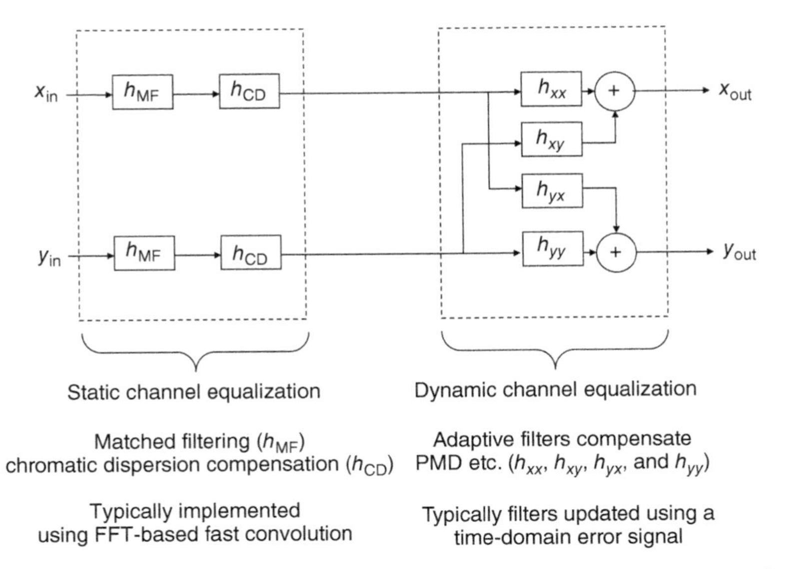

FIGURE 8.1 Functional partitioning of the equalization in a digital coherent receiver. It should be noted that the matched filter (h_{MF}) and the chromatic dispersion compensating filter (h_{CD}) can be combined and as such in the subsequent we only discuss the design of the chromatic dispersion compensating filter.

that equalize for the static channel properties such as chromatic dispersion, with the second block implementing a set of adaptive filters that compensate for the dynamic channels properties such as polarization rotations or PMD. This partitioning allows the two blocks to be implemented and updated in very different manners; for example, a large chromatic dispersion filter might be implemented via the frequency domain using an overlap and save method while a shorter adaptive equalizer could be implemented in the time domain, so as to reduce the overall complexity of the digital signal processing (DSP) and the associated power consumption.

In this chapter, we begin by detailing the necessary mathematics related to digital equalization before discussing the compensation of chromatic dispersion and then PMD. We conclude the chapter by discussing ongoing research challenges related to digital equalization.

8.2 PRIMER ON THE MATHEMATICS OF LEAST SQUARES FIR FILTERS

In this section, we discuss the underlying mathematics of finite impulse response (FIR) filters and their optimization (including differentiation with respect to a complex vector). Readers may skip this section if they are familiar with much of this information; however, it is included to introduce the notation employed and to ensure its consistency throughout this chapter.

8.2.1 Finite Impulse Response Filters

An FIR filter is a nonrecursive filter having tap weights $h[n]$ where $n \in [0, 1, \ldots, N-1]$ and the time step between taps is given by T_s such that we sample at a rate of $f_s = 1/T_s$ with corresponding angular frequency $\omega_s = 2\pi/T_s$. In the time domain, the taps may be written as

$$h(t) = \sum_{n=0}^{N-1} h[n]\delta(t - nT_s) \tag{8.1}$$

Alternatively, if we define a vector $\boldsymbol{h}^T = [h[0], h[1], \ldots, h[N-1]]$ and $\boldsymbol{\delta}_s^T = [\delta(t), \delta(t - T_s), \ldots, \delta(t - [N-1]T_s)]$ where superscript T denotes the transpose operation

$$h(t) = \boldsymbol{h}^T \boldsymbol{\delta}_s \tag{8.2}$$

Hence, the Fourier transform is given by

$$H(\omega) = \int_{-\infty}^{\infty} h(t)e^{-j\omega t}\, dt = \sum_{n=0}^{N-1} h[n]e^{-jn\omega T_s} = \boldsymbol{h}^T \boldsymbol{e}(\omega) \tag{8.3}$$

where we assume throughout this chapter the Fourier transform pair

$$X(\omega) = \int_{-\infty}^{\infty} x(t)e^{-j\omega t} dt \quad \text{and} \quad x(t) = \frac{1}{2\pi}\int_{-\infty}^{\infty} X(\omega)e^{j\omega t} d\omega \tag{8.4}$$

and we have defined $\boldsymbol{e}(\omega) = [1,\ e^{-j\omega T_s}, e^{-2j\omega T_s}, \ \ldots\ , e^{-j\omega[N-1]T_s}]^T$.

Given the above definitions it readily follows that for $m \in \mathbb{Z}$ that

$$H(\omega + m\omega_s) = \sum_{n=0}^{N-1} h[n]e^{-jn(\omega+m\omega_s)T_s} = \sum_{n=0}^{N-1} h[n]e^{-jnm2\pi}e^{-jn\omega T_s} = H(\omega) \tag{8.5}$$

since $e^{-jnm2\pi} = 1$ indicating that the spectrum is aliased to multiples of the sampling frequency as expected from sampling theory.

The process of generating an output signal $y[k]$ by filtering an input vector $\boldsymbol{x}^T = (x[k], x[k-1], \ldots, x[k-(N-1)])$ of N samples by the filter $\boldsymbol{h}$ may be written as

$$y[k] = \boldsymbol{h}^T \boldsymbol{x} = \sum_{i=0}^{N-1} h[i]x[k-i] \tag{8.6}$$

indicating that the output is obtained via the discrete time convolution of the input vector with the tap weights, which may be realized in the frequency domain as multiplication, underpinning fast techniques such as the overlap and save method [10].

8.2.2 Differentiation with Respect to a Complex Vector

If z is a complex number such that $z = x + jy$, then we can define differentiation with respect to a complex number as [11]

$$\frac{\partial}{\partial z} = \frac{1}{2}\frac{\partial}{\partial x} - \frac{j}{2}\frac{\partial}{\partial y} \tag{8.7}$$

with

$$\frac{\partial}{\partial z^*} = \left(\frac{\partial}{\partial z}\right)^* = \frac{1}{2}\frac{\partial}{\partial x} + \frac{j}{2}\frac{\partial}{\partial y} \tag{8.8}$$

which in turn gives

$$\frac{\partial z}{\partial z} = 1 \quad \frac{\partial z^*}{\partial z^*} = 1 \quad \frac{\partial z^n}{\partial z} = nz^{n-1} \tag{8.9}$$

which is in line with expectations from usual calculus; however, it also follows that

$$\frac{\partial z^*}{\partial z} = \frac{\partial z}{\partial z^*} = 0 \tag{8.10}$$

indicating that a complex variable and its conjugate may be considered as independent variables insofar as differentiation is concerned. If we extend the concept to a vector and such that

$$\frac{\partial}{\partial \mathbf{z}} = \frac{1}{2}\frac{\partial}{\partial \mathbf{x}} - \frac{j}{2}\frac{\partial}{\partial \mathbf{y}} \tag{8.11}$$

where $\frac{\partial}{\partial \mathbf{x}} = \left(\frac{\partial}{\partial x_0}, \frac{\partial}{\partial x_1}, \ldots, \frac{\partial}{\partial x_{N-1}}\right)^T$ and so on, then it follows that

$$\frac{\partial \mathbf{z}^T}{\partial \mathbf{z}} = \frac{\partial}{\partial \mathbf{z}}\mathbf{z}^T = \mathbf{I} = \frac{\partial \mathbf{z}^{*T}}{\partial \mathbf{z}^*} \quad \text{and} \quad \frac{\partial \mathbf{z}^{*T}}{\partial \mathbf{z}} = 0 = \frac{\partial \mathbf{z}^T}{\partial \mathbf{z}^*} \tag{8.12}$$

where $\mathbf{I}$ and 0 are $N \times N$ identity and null matrices, respectively.

8.2.3 Least Squares Tap Weights

If the desired frequency response of a filter is $H_d(\omega)$ over the frequency range $\omega \in (-\omega_s/2, \omega_s/2)$ then squared error ε^2 between the desired and the actual response $H(\omega)$ is given by Kidambi and Ramachandran [12]

$$\varepsilon^2 = \int_{-\frac{\omega_s}{2}}^{\frac{\omega_s}{2}} |H(\omega) - H_d(\omega)|^2 d\omega \tag{8.13}$$

Substituting our definition for $H(\omega) = \boldsymbol{h}^T\boldsymbol{e}(\omega)$, we obtain

$$\begin{aligned}\varepsilon^2 &= \int_{\frac{\omega_s}{2}}^{\frac{\omega_s}{2}} |\boldsymbol{h}^T\boldsymbol{e}(\omega) - H_d(\omega)|^2 d\omega = \int_{-\frac{\omega_s}{2}}^{\frac{\omega_s}{2}} (\boldsymbol{h}^T\boldsymbol{e}(\omega) - H_d(\omega))^*(\boldsymbol{h}^T\boldsymbol{e}(\omega) - H_d(\omega))d\omega \\ &= \int_{-\frac{\omega_s}{2}}^{\frac{\omega_s}{2}} (\boldsymbol{h}^T\boldsymbol{e}(\omega) - H_d(\omega))^*(\boldsymbol{e}^T(\omega)\boldsymbol{h} - H_d(\omega))d\omega \end{aligned} \tag{8.14}$$

since $\boldsymbol{e}^T(\omega)\boldsymbol{h} = \boldsymbol{h}^{\boldsymbol{T}}\boldsymbol{e}(\omega)$ and hence

$$\frac{d\varepsilon^2}{d\boldsymbol{h}} = 0 = \int_{-\frac{\omega_s}{2}}^{\frac{\omega_s}{2}} (\boldsymbol{h}^T\boldsymbol{e}(\omega) - H_d(\omega))^*\boldsymbol{e}(\omega)d\omega \tag{8.15}$$

Hence, rewriting we obtain

$$\int_{-\frac{\omega_s}{2}}^{\frac{\omega_s}{2}} \boldsymbol{e}^*(\omega)(\boldsymbol{e}(\omega)^T\boldsymbol{h} - H_d(\omega))d\omega = 0 \tag{8.16}$$

giving

$$\left(\int_{-\frac{\omega_s}{2}}^{\frac{\omega_s}{2}} \boldsymbol{e}^*(\omega)\,\boldsymbol{e}(\omega)^T d\omega\right)\boldsymbol{h} = \int_{-\frac{\omega_s}{2}}^{\frac{\omega_s}{2}} \boldsymbol{e}^*(\omega)H_d(\omega)d\omega \tag{8.17}$$

but given $\boldsymbol{e}(\omega) = [1,\ e^{-j\omega T_s}, e^{-2j\omega T_s},\ \dots\ , e^{-j\omega[N-1]T_s}]^T$ then $\left(\int_{-\frac{\omega_s}{2}}^{\frac{\omega_s}{2}} \boldsymbol{e}^*(\omega)\,\boldsymbol{e}(\omega)^T d\omega\right)$ $= \boldsymbol{I}$; hence, we have

$$\boldsymbol{h}_{\text{opt}} = \int_{-\frac{\omega_s}{2}}^{\frac{\omega_s}{2}} \boldsymbol{e}^*(\omega)H_d(\omega)d\omega \tag{8.18}$$

where the tap weights $\boldsymbol{h}_{\text{opt}}$ are optimal in a least squares sense.

To demonstrate the theory, let us consider a rectangular Nyquist filter with support $\omega \in (-\omega_s/4, \omega_s/4)$. While the amplitude is straightforward given that the filter is complex we must also consider the phase response. In order to simplify the phase response, we modify the basis functions to be symmetric about the origin such that $\boldsymbol{e}\ (\omega) = e^{j\omega[N-1]T_s/2} \times [1,\ e^{-j\omega T_s}, e^{-2j\omega T_s},\ \dots\ , e^{-j\omega[N-1]T_s}]^T$ ensuring that a symmetric FIR filter has zero group delay.

In this case for a rectangular Nyquist filter, the tap weights are given by

$$\boldsymbol{h}_{\text{opt}} = \int_{-\frac{\omega_S}{4}}^{\frac{\omega_S}{4}} e^{-\frac{j\omega[N-1]T_S}{2}} \times [1,\ e^{j\omega T_s}, e^{2j\omega T_s},\ \ldots\ , e^{j\omega[N-1]T_s}]^T d\omega \tag{8.19}$$

$$\begin{aligned} h_{\text{opt}}[n] &= \int_{-\frac{\omega_S}{4}}^{\frac{\omega_S}{4}} e^{j\omega nT_s - \frac{j\omega[N-1]T_S}{2}} d\omega = \frac{2\sin\left(\frac{\omega_S nT_S}{4} - \frac{\omega_s[N-1]T_S}{8}\right)}{nT_s - \frac{[N-1]T_S}{2}} \\ &= \frac{2\sin\left(\frac{\pi n}{2} - \frac{\pi[N-1]}{4}\right)}{nT_s - \frac{[N-1]T_S}{2}} = \frac{\pi}{T_s}\text{sinc}\left(\frac{n}{2} - \frac{[N-1]}{4}\right) \end{aligned} \tag{8.20}$$

where we have defined $\text{sinc}(x) = \sin(\pi x)/(\pi x)$ in accordance with the usual convention for electronic engineers.

8.2.4 Application to Stochastic Gradient Algorithms

In an adaptive equalizer, we frequently have a cost function whose gradient is stochastically estimated and used to update the tap weights. For a complex valued set of taps $\boldsymbol{h}$, the stochastic gradient algorithm is applied to the real ($\mathfrak{R}\{\boldsymbol{h}\}$) and imaginary ($\mathfrak{I}\{\boldsymbol{h}\}$) components of the taps independently and hence may be written as [11]

$$\mathfrak{R}\{\boldsymbol{h}\} := \mathfrak{R}\{\boldsymbol{h}\} - \frac{\mu}{2}\left(\frac{\partial\varepsilon^2}{\partial\mathfrak{R}\{\boldsymbol{h}\}}\right) \quad \text{and} \quad \mathfrak{I}\{\boldsymbol{h}\} := \mathfrak{I}\{\boldsymbol{h}\} - \frac{\mu}{2}\left(\frac{\partial\varepsilon^2}{\partial\mathfrak{I}\{\boldsymbol{h}\}}\right) \tag{8.21}$$

where := denotes the assignment operation, such that $x := y$ indicates that x is assigned to be the value y. These two equations may, however, be written more compactly in terms of the conjugate derivative as

$$\boldsymbol{h} := \boldsymbol{h} - \mu\frac{\partial\varepsilon^2}{\partial\boldsymbol{h}^*} \tag{8.22}$$

To illustrate the approach, we consider the least mean squares equalizer whose error term is given by $\varepsilon = d - \boldsymbol{h}^T\boldsymbol{x}$, where d is the desired output, which gives the squared error term as $\varepsilon^2 = |d - \boldsymbol{h}^T\boldsymbol{x}|^2 = (d - \boldsymbol{h}^T\boldsymbol{x})^*(d - \boldsymbol{h}^T\boldsymbol{x})$.

$$\begin{aligned} \frac{\partial\varepsilon^2}{\partial\boldsymbol{h}^*} &= \frac{\partial}{\partial\boldsymbol{h}^*}\{(d - \boldsymbol{h}^T\boldsymbol{x})^*(d - \boldsymbol{h}^T\boldsymbol{x})\} = \frac{\partial}{\partial\boldsymbol{h}^*}\{(d^* - \boldsymbol{h}^{*T}\boldsymbol{x}^*)(d - \boldsymbol{h}^T\boldsymbol{x})\} \\ &= (d^* - \boldsymbol{h}^{*T}\boldsymbol{x}^*)\frac{\partial}{\partial\boldsymbol{h}^*}\{(d - \boldsymbol{h}^T\boldsymbol{x})\} + (d - \boldsymbol{h}^T\boldsymbol{x})\frac{\partial}{\partial\boldsymbol{h}^*}\{(d^* - \boldsymbol{h}^{*T}\boldsymbol{x}^*)\} \\ &= -\varepsilon^* 0\boldsymbol{x} - \varepsilon\boldsymbol{I}\boldsymbol{x}^* \\ &= -\varepsilon\,\boldsymbol{x}^* \end{aligned} \tag{8.23}$$

where we have used the product rule for differentiation and the relationships related to differentiation with respect to a complex vector and its conjugate.

Hence, the tap weight adaption algorithm is given by

$$\boldsymbol{h} := \boldsymbol{h} + \mu\varepsilon\, \boldsymbol{x}^* \tag{8.24}$$

equally for the constant modulus algorithm (CMA) [13]

$$\varepsilon = 1 - |\boldsymbol{h}^T\boldsymbol{x}|^2 \tag{8.25}$$

Hence

$$\frac{\partial\varepsilon^2}{\partial\boldsymbol{h}^*} = \frac{\partial}{\partial\boldsymbol{h}^*}(1 - |\boldsymbol{h}^T\boldsymbol{x}|^2)^2 = -2\varepsilon\frac{\partial}{\partial\boldsymbol{h}^*}(\boldsymbol{h}^T\boldsymbol{x})^*\,(\boldsymbol{h}^T\boldsymbol{x}) = -2\varepsilon\,\boldsymbol{x}^*(\boldsymbol{h}^T\boldsymbol{x}) \tag{8.26}$$

giving the update algorithm for the CMA as

$$\boldsymbol{h} := \boldsymbol{h} - \mu\frac{\partial\varepsilon^2}{\partial\boldsymbol{h}^*} = \boldsymbol{h} + 2\mu\varepsilon\,\boldsymbol{x}^*(\boldsymbol{h}^T\boldsymbol{x}) \tag{8.27}$$

8.2.5 Application to Wiener Filter

We have already shown that if $y = \boldsymbol{h}^T\boldsymbol{x}$ and the error term is given by $\varepsilon = d - y$, then

$$\frac{\partial\varepsilon^2}{\partial\boldsymbol{h}^*} = -\varepsilon\,\boldsymbol{x}^* = -(d - \boldsymbol{h}^T\boldsymbol{x}\,)\boldsymbol{x}^* \tag{8.28}$$

While we have showed how it is possible to iteratively solve this using a stochastic gradient technique, an alternative is to solve this analytically, by setting the expected value of derivative to zero so as to give

$$E\left\{\frac{\partial\varepsilon^2}{\partial\boldsymbol{h}^*}\right\} = E\{-(d - \boldsymbol{h}^T\boldsymbol{x}\,)\boldsymbol{x}^*\} = E\{-\boldsymbol{x}^*(d - \boldsymbol{x}^T\boldsymbol{h})\} = 0 \tag{8.29}$$

where $E\{\cdot\}$ is the expectation operator. By exploiting the linearity of the expectation operator, Equation 8.29 can be simplified to give

$$E\{\boldsymbol{x}^*d\} = E\{\boldsymbol{x}^*\boldsymbol{x}^T\boldsymbol{h}\} = E\{\boldsymbol{x}^*\boldsymbol{x}^T\}\boldsymbol{h} \tag{8.30}$$

giving the Wiener filter solution of the tap weights as

$$\boldsymbol{h} = R_{xx}^{-1}\boldsymbol{P} \tag{8.31}$$

where we have defined $\boldsymbol{P} = E\{\boldsymbol{x}^*d\}$ as the cross-correlation vector between the desired signal and the distorted signal, and $R_{xx} = E\{\boldsymbol{x}^*\boldsymbol{x}^T\}$ to be the autocorrelation matrix of the distorted signal. Often, in a coherent optical communication system, the situation is further complicated by the presence of phase noise or the frequency difference between the signal and the local oscillator. Nevertheless, it may be readily applied in a simulation environment where frequency offset correction and carrier recovery is not required.

8.2.6 Other Filtering Techniques and Design Methodologies

There is a wealth of literature in the area of filter design albeit much of this is focused on the design of linear phase filters, for example, using the Parks–McClellan variant [14] of the Remez exchange algorithm [15] to minimize maximum error. In contrast, in optical communication systems nonlinear phase responses are often desired, in particular as we discuss in the subsequent section, filters with quadratic phase are required for the compensation of chromatic dispersion. As such, in this chapter, we focus on those techniques that have proved to be useful for optical communication systems, in particular the least squares criterion since it allows for closed form solutions without recourse to iterative techniques.

8.3 EQUALIZATION OF CHROMATIC DISPERSION

8.3.1 Nature of Chromatic Dispersion

Chromatic dispersion is a consequence of the frequency-dependent group delay in the optical fiber. If two wavelengths are separated by $\Delta\lambda$ nm, then the temporal spread Δt (in ps) is given by Agrawal [16]

$$\Delta t = Dz\Delta\lambda \tag{8.32}$$

where D is the dispersion coefficient of the fiber given in ps/nm/km and z is the length of the link in kilometer. Given $c = f\lambda$, it follows that $\Delta\lambda = -\Delta f \times \lambda/f$; hence, for $f \approx 193$ THz ($\lambda \approx 1553$ nm), then $\Delta\lambda/\Delta f = 8$ pm/GHz. Hence, a signal occupying 35 GHz has a spectral width of 0.28 nm. If the dispersion coefficient is 16.7 ps/nm/km then for every 1000 km of fiber the signal disperses by at least 165 symbol periods, with the minimum value obtained by assuming a rectangular spectrum (in general for a symbol rate of B_s Gbaud, then the minimum number of $T/2$ spaced taps is $0.27B_s^2$ per 1000 km of SMF with $D = 16.7$ ps/nm/km). While these are crude "back of the envelope" calculations they give insight as to the expected number of taps.

8.3.2 Modeling of Chromatic Dispersion in an Optical Fiber

In the absence of fiber nonlinearity, the effect of chromatic dispersion on the envelope $A(z, t)$ of a pulse may be modeled by the following partial differential equation, which is based on the electronic engineer's definition of phase compared with the physicist's convention [16]

$$\frac{\partial A(z,t)}{\partial z} = \frac{j\beta_2}{2}\frac{\partial^2 A(z,t)}{\partial t^2} \tag{8.33}$$

where z is the distance of propagation, t is time variable in a frame moving with the pulse $A(z, t)$, and β_2 group delay dispersion of the fiber. Taking the Fourier transform of Equation 8.33 and solving gives the frequency domain transfer function $G(z, \omega)$ given by

$$G(z, \omega) = \exp\left(-\frac{j\beta_2}{2}\omega^2 z\right) \tag{8.34}$$

The dispersion compensating filter is, therefore, given by the all-pass filter $1/G(z, \omega) = G(-z, \omega)$, which can be approximated using an FIR filter.

8.3.3 Truncated Impulse Response

Herein, we discuss a simple but intuitive means of designing the chromatic dispersion compensating FIR filter [17], providing a basis for the discussion of more complex techniques. In contrast to a frequency-domain approach not only does this give a simple closed-form solution for the tap weights but also it provides bounds on the number of taps required for a given value of dispersion. We begin by obtaining the impulse response $g(z, t)$ of the dispersive fiber by applying the inverse Fourier transform to the frequency domain transfer function $G(z, \omega)$ to give

$$g(z, t) = \frac{1}{\sqrt{2\pi j\beta_2 z}} \exp\left(\frac{j}{2\beta_2 z}t^2\right) \tag{8.35}$$

For an arbitrary input, the output can be obtained by convolving this impulse response with the input and as expected the impulse response itself satisfies Equation 8.33. By inverting the sign of the chromatic dispersion, we obtain the impulse function of the chromatic dispersion compensating filter $g_c(z, t)$, given by

$$g_c(z, t) = \frac{1}{\sqrt{-2\pi j\beta_2 z}} \exp(-j\phi(t)), \text{ where } \phi(t) = \frac{t^2}{2\beta_2 z} \tag{8.36}$$

The impulse response given by Equation 8.36 presents a number of issues for digital implementation, not only is it infinite in duration but since it passes all frequencies for a finite sampling frequency aliasing will occur. The solution to all of these problems is to truncate the impulse response to a finite duration. To determine the length of the truncation window, we note that if we sample every T_s seconds then aliasing will occur for frequencies which exceed the Nyquist frequency given by $\omega_n = \pi/T$ and that the impulse response may be considered as a rotating vector whose angular frequency is given by

$$\omega = \frac{\partial\phi(t)}{\partial t} = \frac{t}{\beta_2 z} \tag{8.37}$$

when the magnitude of this frequency exceeds the Nyquist frequency, aliasing will occur, giving the criterion that $|\omega| < \omega_n$ and hence

$$-|\beta_2|z\frac{\pi}{T_s} \leq t \leq |\beta_2|z\frac{\pi}{T_s} \tag{8.38}$$

Since the impulse response is of finite duration, this can be implemented digitally using an FIR filter. If we assume the number of taps is large, then the sampled impulse response will approximate the continuous time impulse response. Hence, if we consider a filter with N_{TI} taps then the tap weights will be given by

$$h_{\text{TI}}[n] = \frac{1}{\sqrt{\rho}} \exp\left(-\frac{j\pi}{\rho}\left(n - \frac{N_{\text{TI}} - 1}{2}\right)^2\right) \tag{8.39}$$

where

$$\rho = 2\frac{\pi\beta_2 z}{T_s^2}, \quad N_{\text{TI}} = |\rho| \quad \text{and} \quad n \in [0, 1, 2, \ldots, N_{\text{TI}} - 1]$$

where x is the integer part of x rounded toward minus infinity. These tap weights form the basis for the compensation of chromatic dispersion using an FIR filter.

8.3.4 Band-Limited Impulse Response

In the previous example, one of the problems that arose was due to the aliasing that occurred, resulting in an upper bound on the number of taps that can be employed. One obvious solution to overcome this restriction is to band-limit the signal to the Nyquist bandwidth $\omega_n = \pi/T$ such that

$$g_{c_{\text{bl}}}(z,t) = \frac{1}{2\pi}\int_{-\omega_n}^{\omega_n} \exp\left(\frac{j\beta_2}{2}\omega^2 z\right)\exp(j\omega t)d\omega = w(z,t) \times g_c(z,t) \tag{8.40}$$

where

$$w(z,t) = \frac{1}{2j}\left(\text{erfi}\left(\frac{t + z\beta_2\omega_n}{\sqrt{-2jz\beta_2}}\right) - \text{erfi}\left(\frac{t - z\beta_2\omega_n}{\sqrt{-2jz\beta_2}}\right)\right) \tag{8.41}$$

where $\text{erfi}(x)$ is the we imaginary error function given by $\text{erfi}(x) = -j\text{erf}(j\,x)$ where

$$\text{erf}(z) = \frac{2}{\sqrt{\pi}}\int_0^z e^{-t^2}\,dt \tag{8.42}$$

The fact that we can write the response as $g_{c_{\text{bl}}}(z,t) = g_c(z,t)w(z,t)$ indicates that the band-limited response may be obtained by multiplying the impulse response by a window function $w(z,t)$. As mentioned earlier, we may sample the impulse response signal in order to estimate the FIR filter coefficients. However, rather than detail this we turn our attention to the least squares formulation of the FIR filter since we see subsequently that they are equivalent.

8.3.5 Least Squares FIR Filter Design

As previously discussed, the least squares criterion may be applied to the design of a complex FIR filter, giving optimal tap weights in a least squares sense:

$$\boldsymbol{h}_{\mathrm{LS}} = \int_{-\frac{\omega_S}{2}}^{\frac{\omega_S}{2}} \boldsymbol{e}^*(\omega) H_d(z,\omega) d\omega \tag{8.43}$$

If as discussed in the previous section, we define our taps to be symmetrically defined such that $\boldsymbol{e}(\omega) = e^{j\omega[N-1]T_s/2} \times [1,\ e^{-j\omega T_s}, e^{-2j\omega T_s},\ \ldots\ , e^{-j\omega[N-1]T_s}]^T$, and so as to neglect the combination of any subsequent filtering we chose $H_d(z,\omega) = 1/G(z,\omega) = G(-z,\omega)$ then we obtain [18]

$$h_{\mathrm{LS}}[n] = \frac{1}{2j}\left(\mathrm{erfi}\left(\sqrt{\frac{j\pi}{\rho}}\left(n - \frac{N-1}{2} + \frac{\rho}{2}\right)\right) - \mathrm{erfi}\left(\sqrt{\frac{j\pi}{\rho}}\left(n - \frac{N-1}{2} - \frac{\rho}{2}\right)\right)\right) \times \frac{1}{\sqrt{\rho}} \exp\left(-\frac{j\pi}{\rho}\left(n - \frac{N-1}{2}\right)^2\right) \tag{8.44}$$

where $\rho = 2\pi\beta_2 z/T_s^2$, which we note is identical to the sampled version of the band-limited impulse response. As mentioned earlier, we may factorize this to give

$$h_{\mathrm{LS}}[n] = w[n] \times h_{TI}[n] \tag{8.45}$$

where

$$w[n] = \frac{1}{2j}\left(\mathrm{erfi}\left(\sqrt{\frac{j\pi}{\rho}}\left(n - \frac{N-1}{2} + \frac{\rho}{2}\right)\right) - \mathrm{erfi}\left(\sqrt{\frac{j\pi}{\rho}}\left(n - \frac{N-1}{2} - \frac{\rho}{2}\right)\right)\right) \tag{8.46}$$

8.3.6 Example Performance of the Chromatic Dispersion Compensating Filter

We now consider the performance of the chromatic dispersion compensating filter. In order to assess this, we consider a 35 Gbaud signal with a near-rectangular Nyquist-shaped spectrum (root raised cosine shape with $\beta = 0.01$). While the results are given for a specific distance, they can readily be scaled by calculating the maximum length of the truncated impulse response given by $N = 2\pi\beta_2 z/T_s^2$ and scaling accordingly.

As can be seen from Figure 8.2, while for PDM-QPSK operating over long-haul distances the signal-to-noise penalty incurred as a result of the design is negligible, for shorter distances or as the cardinality of the modulation format is increased the truncated impulse response has significant limitations. To overcome these limitations the least squares formulation is employed.

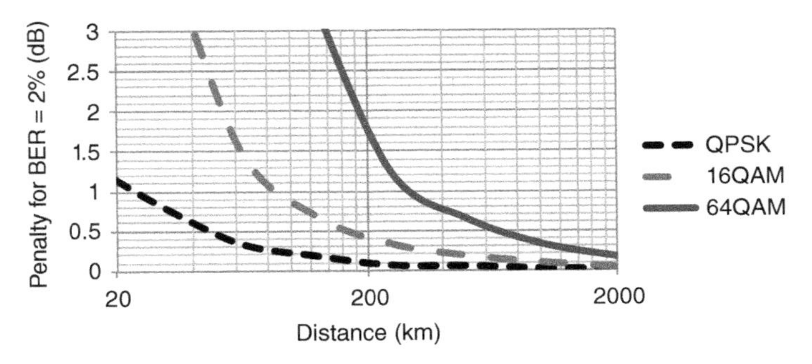

FIGURE 8.2 Performance of the truncated impulse response FIR filter design.

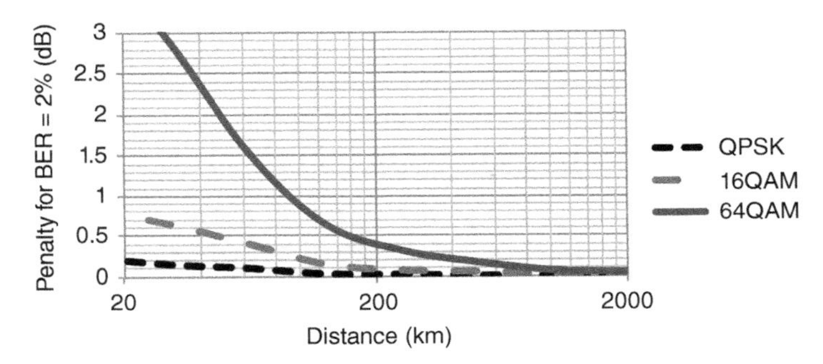

FIGURE 8.3 Performance of the least squares FIR filter design.

From Figure 8.3 we note that the least squares formulation results in a significantly reduced penalty even though in both cases the same number of taps are the same provided $N = 2\pi\beta_2 z/T_s^2$. Nevertheless, for highly spectrally efficient formats such as 64 QAM, the penalty can be significant for short distances. The penalty may, however, be mitigated by allowing the number of taps in the least squares design to increase beyond $N = 2\pi\beta_2 z/T_s^2$. To illustrate this, we again consider a 35-Gbaud signal with a near rectangular Nyquist-shaped spectrum (root raised cosine shape with $\beta = 0.01$) transmitting PDM-64QAM over a distance of 31.25 km of single-mode fiber with $D = 17$ ps/km/nm. While this gives a minimum number of taps required of 20 (or 21 if an odd number of taps are employed) as can be seen in Figure 8.4 by increasing the number of taps by 66.7% the penalty can be made negligible (<0.1 dB from the initial value of 2.9 dB with the minimum number of taps).

Before closing this section, it is illustrative to consider the shape of the window function $w[n]$. In Figure 8.5, we plot the window function $w[n]$ for a range of distances normalized to the length of the truncated impulse response (such that the truncated impulse response only extends from −0.5 to 0.5).

As can be seen in Figure 8.5, as the distance increases the shape of the window function changes such that for long distances the window function converges to the

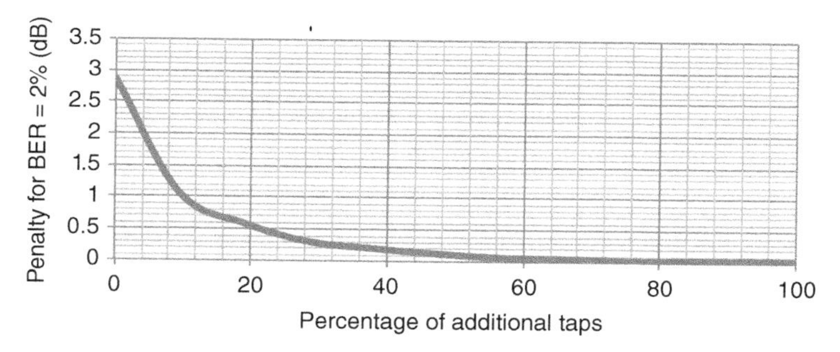

FIGURE 8.4 Reduction in penalty achieved by allowing the number of taps to increase beyond the minimum for the least squares formulation.

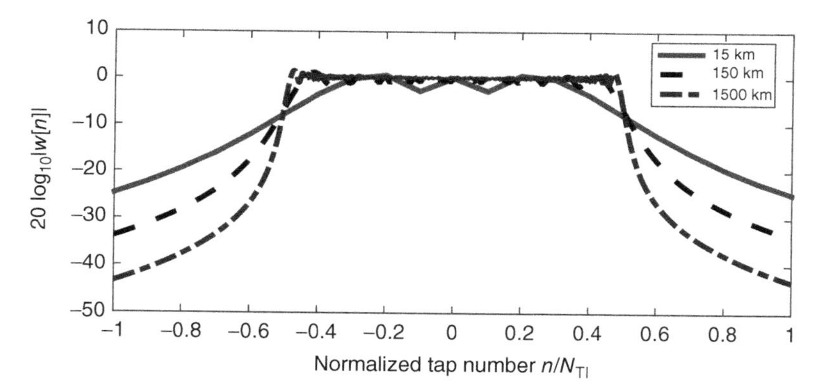

FIGURE 8.5 Shape of the window function for a range of distances with 35 Gbaud signals, with $|n/N_{\mathrm{TI}}| \leq 0.5$ corresponding to the support of the truncated impulse response.

simple window used in the truncated impulse response. This reveals why the performance of the truncated impulse response improves for longer distances, as this converges to the least squares solution.

8.4 EQUALIZATION OF POLARIZATION-MODE DISPERSION

PMD arises due to variations in the circular symmetry of the optical fiber resulting in localized birefringence [19]. While the PMD is a unitary operation, in general compensation of PMD is included within a subsystem that includes mitigation of PDL, relaxing the unitary requirement. By removing the unitary condition, polarization-independent effects such as nonideal-matched filtering can also be mitigated.

8.4.1 Modeling of PMD

PMD results in information being coupled from one polarization to another such that the information in the x and y polarization at the output $\boldsymbol{E}_{\text{out}}(\omega) = [X_{\text{out}}(\omega), Y_{\text{out}}(\omega)]^T$ is related to the input states $\boldsymbol{E}_{\text{in}}(\omega) = [X_{\text{in}}(\omega), Y_{\text{in}}(\omega)]^T$ via

$$\begin{bmatrix} X_{\text{out}}(\omega) \\ Y_{\text{out}}(\omega) \end{bmatrix} = U \begin{bmatrix} X_{\text{in}}(\omega) \\ Y_{\text{in}}(\omega) \end{bmatrix} = e^{j\phi(\omega)} \begin{bmatrix} u_1(\omega) & u_2(\omega) \\ -u_2^*(\omega) & u_1^*(\omega) \end{bmatrix} \begin{bmatrix} X_{\text{in}}(\omega) \\ Y_{\text{in}}(\omega) \end{bmatrix} \tag{8.47}$$

with $|u_1(\omega)|^2 + |u_2(\omega)|^2 = 1$. The simplest manifestation of PMD is as a differential group delay (DGD) of τ such that

$$\begin{bmatrix} u_1(\omega) & u_2(\omega) \\ -u_2^*(\omega) & u_1^*(\omega) \end{bmatrix} = \begin{bmatrix} \cos(\theta) & \sin(\theta) \\ -\sin(\theta) & \cos(\theta) \end{bmatrix} \begin{bmatrix} e^{\frac{j\omega\tau}{2}} & 0 \\ 0 & e^{-\frac{j\omega\tau}{2}} \end{bmatrix} \begin{bmatrix} \cos(\theta) & -\sin(\theta) \\ \sin(\theta) & \cos(\theta) \end{bmatrix} \tag{8.48}$$

with the DGD being obtainable from the Jones matrix via $\tau = 2\sqrt{|u_1'|^2 + |u_2'|^2}$ where $'$ denotes differentiation with respect to angular frequency ω. From this, we note that the inverse Jones matrix in the time domain is given by

$$\begin{bmatrix} \cos(\theta) & \sin(\theta) \\ -\sin(\theta) & \cos(\theta) \end{bmatrix} \begin{bmatrix} \delta\left(t - \frac{\tau}{2}\right) & 0 \\ 0 & \delta\left(t + \frac{\tau}{2}\right) \end{bmatrix} \begin{bmatrix} \cos(\theta) & -\sin(\theta) \\ \sin(\theta) & \cos(\theta) \end{bmatrix} \tag{8.49}$$

This reveals that if the equalizer is to correct for a DGD of τ then the temporal span $(N-1)T_s$ of an N tap FIR filter must exceed τ requiring $N \geq \tau/T_s + 1$. As we discuss in the subsequent sections, this is very much a lower bound on the number of taps and in practice when the convergence time of the equalizer is of concern more taps may be required. Since PMD is due to random coupling between the polarization modes, DGD has a statistical variation with a Maxwellian distribution [20] with a probability density function given by

$$f(\tau) = \frac{32}{\tau\pi^2}\left(\frac{\tau}{\tau}\right)^2 \exp\left(-\frac{4}{\pi}\left(\frac{\tau}{\tau}\right)^2\right) \tag{8.50}$$

where $\langle\tau\rangle$ is the mean DGD and the resulting outage probability given by the corresponding tail distribution $F_c(\tau)$ is given by

$$F_c(\tau) = \int_{\tau}^{\infty} f(x)dx = 1 + \frac{4}{\pi}\left(\frac{\tau}{\tau}\right)\exp\left(-\frac{4}{\pi}\left(\frac{\tau}{\tau}\right)^2\right) - \text{erf}\left(\frac{2}{\sqrt{\pi}}\left(\frac{\tau}{\tau}\right)\right) \tag{8.51}$$

By considering a minmax approximation of $\log_{10}F_c(\tau)$ in the region $1.6\langle\tau\rangle \leq \tau \leq 6\langle\tau\rangle$, we obtain the following simpler expression for the effect of DGD on the outage probability $F_c(\tau)$

$$\log_{10}F_c(\tau) = 0.2591\left(\frac{\tau}{\langle\tau\rangle}\right) - 0.5722\left(\frac{\tau}{\langle\tau\rangle}\right)^2 \tag{8.52}$$

with the maximum relative error of $\log_{10}F_c(\tau/\langle\tau\rangle)$ being less than 0.15%. By way of an example, we consider data from a field trial of 100 Gbit/s with 31.25 Gbaud PDM-QPSK with $\langle\tau\rangle = 36.6$ ps with an adaptive equalizer with 13 taps [7]. This gives $\tau/\langle\tau\rangle = 5.2$ and hence a negligible outage probability of 4×10^{-15}. While it was also noted that no penalty was observed with $\langle\tau\rangle$ up to 55 ps then the corresponding outage probability is 8×10^{-7} and as such a penalty is not expected to be observed.

8.4.2 Obtaining the Inverse Jones Matrix of the Channel

The impact of polarization-dependent effects on the propagation may be modeled by a Jones matrix. In general, this matrix is not unitary due to PDL and, furthermore, it will be frequency dependent due to PMD. The task is, therefore, to estimate the Jones matrix and obtain the inverse to compensate for the impairments incurred. In contrast to the chromatic dispersion that may be considered relatively constant, the Jones matrix may evolve in time due to effects such as rapid variations in the polarization state, and therefore the compensation scheme must be adaptive. The problem of compensating polarization rotations digitally was first considered by Betti et al. [21], and later demonstrated utilizing the formalism of multiple-input-multiple-output (MIMO) systems [22]. For inputs $\boldsymbol{x} = (x[k], x[k-1], \ldots, x[k-(N-1)])^T$ and $\boldsymbol{y} = (y[k], y[k-1], \ldots, y[k-(N-1)])^T$, the outputs $x_o[k]$ and $y_o[k]$ are given by

$$x_o[k] = \boldsymbol{h}_{xx}^T\boldsymbol{x} + \boldsymbol{h}_{xy}^T\boldsymbol{y} \quad \text{and} \quad y_o[k] = \boldsymbol{h}_{yx}^T\boldsymbol{x} + \boldsymbol{h}_{yy}^T\boldsymbol{y} \tag{8.53}$$

where $\boldsymbol{h}_{xx}$, $\boldsymbol{h}_{xy}$, $\boldsymbol{h}_{yx}$, and $\boldsymbol{h}_{yy}$ are adaptive filters each of which have length N taps. While there are a number of methods for adapting the equalizer in MIMO systems, we restrict ourselves to a specific example that exploits properties of the data, namely that for polarization-division multiplexed QPSK (PDM-QPSK) the signal for each polarization should have a constant modulus. This CMA has also been shown to be effective even when the modulus is not constant such as higher-order quadrature-amplitude modulation.

8.4.3 Constant Modulus Update Algorithm

For signals of unit amplitude the equalizer will attempt to minimize, in a mean squares sense, the magnitude of $\varepsilon_x = 1 - |x_o|^2$ and $\varepsilon_y = 1 - |y_o|^2$. Hence, to obtain the optimal tap weights a set of stochastic-gradient algorithms with convergence parameter μ are used

$$\begin{aligned}\boldsymbol{h}_{xx} := \boldsymbol{h}_{xx} - \mu\frac{\partial\varepsilon_x^2}{\partial\boldsymbol{h}_{xx}^*} &= \boldsymbol{h}_{xx} - 2\mu\varepsilon_x\frac{\partial\varepsilon_x}{\partial\boldsymbol{h}_{xx}^*} = \boldsymbol{h}_{xx} + 2\mu\varepsilon_x\frac{\partial|x_o|^2}{\partial\boldsymbol{h}_{xx}^*}\\ &= \boldsymbol{h}_{xx} + 2\mu\varepsilon_x x_o\frac{\partial}{\partial\boldsymbol{h}_{xx}^*}x_o^* = \boldsymbol{h}_{xx} + 2\mu\varepsilon_x x_o\boldsymbol{x}^*\end{aligned} \tag{8.54}$$

Similarly,

$$\boldsymbol{h}_{xy} := \boldsymbol{h}_{xy} - \mu \frac{\partial \varepsilon_x^2}{\partial \boldsymbol{h}_{xy}^*} = \boldsymbol{h}_{xy} + 2\mu\varepsilon_x x_o \frac{\partial}{\partial \boldsymbol{h}_{xy}^*} x_o^* = \boldsymbol{h}_{xy} + 2\mu\varepsilon_x x_o \boldsymbol{y}^* \tag{8.55}$$

Interchanging the variables x and y, we then obtain

$$\boldsymbol{h}_{yx} := \boldsymbol{h}_{yx} - \mu \frac{\partial \varepsilon_y^2}{\partial \boldsymbol{h}_{yx}^*} = \boldsymbol{h}_{yx} + 2\mu\varepsilon_y y_o \frac{\partial}{\partial \boldsymbol{h}_{yx}^*} y_o^* = \boldsymbol{h}_{yx} + 2\mu\varepsilon_y y_o \boldsymbol{x}^* \tag{8.56}$$

and finally we have

$$\boldsymbol{h}_{yy} := \boldsymbol{h}_{yy} - \mu \frac{\partial \varepsilon_y^2}{\partial \boldsymbol{h}_{yy}^*} = \boldsymbol{h}_{yx} + 2\mu\varepsilon_y y_o \frac{\partial}{\partial \boldsymbol{h}_{yy}^*} y_o^* = \boldsymbol{h}_{yy} + 2\mu\varepsilon_y y_o \boldsymbol{y}^* \tag{8.57}$$

where $\boldsymbol{x}^*$ and $\boldsymbol{y}^*$ denote the complex conjugate of $\boldsymbol{x}$ and $\boldsymbol{y}$, respectively. In order to initialize the algorithm, all tap weights are set to zero with the exception of the central tap of $\boldsymbol{h}_{xx}$ and $\boldsymbol{h}_{yy}$, which are set to unity. Given the equalizer is unconstrained with respect to its outputs, it is possible for the equalizer to converge on the same output, corresponding to the Jones matrix becoming singular; however, there are many well-established means overcoming this limitation [23–25].

8.4.4 Decision-Directed Equalizer Update Algorithm

Once the equalizer has converged, then the equalizer may move into a decision-directed mode, such that if $D(x)$ is the symbol closest to x then the decision-directed least mean squared (DD-LMS) algorithm minimizes $\varepsilon_x = D(x_0) - x_o$ and $\varepsilon_y = D(y_0) - y_o$ giving the update algorithm as [26]

$$\boldsymbol{h}_{xx} := \boldsymbol{h}_{xx} - \mu \frac{\partial |\varepsilon_x|^2}{\partial \boldsymbol{h}_{xx}^*} = \boldsymbol{h}_{xx} - \mu\varepsilon_x \frac{\partial \varepsilon_x^*}{\partial \boldsymbol{h}_{xx}^*} = \boldsymbol{h}_{xx} + \mu\varepsilon_x \frac{\partial x_o^*}{\partial \boldsymbol{h}_{xx}^*} = \boldsymbol{h}_{xx} + \mu\varepsilon_x \boldsymbol{x}^* \tag{8.58}$$

$$\boldsymbol{h}_{xy} := \boldsymbol{h}_{xy} - \mu \frac{\partial |\varepsilon_x|^2}{\partial \boldsymbol{h}_{xy}^*} = \boldsymbol{h}_{xy} + \mu\varepsilon_x \frac{\partial x_o^*}{\partial \boldsymbol{h}_{xy}^*} = \boldsymbol{h}_{xy} + \mu\varepsilon_x \boldsymbol{y}^* \tag{8.59}$$

$$\boldsymbol{h}_{yx} := \boldsymbol{h}_{yx} - \mu \frac{\partial |\varepsilon_y|^2}{\partial \boldsymbol{h}_{yx}^*} = \boldsymbol{h}_{yx} + \mu\varepsilon_y \frac{\partial y_o^*}{\partial \boldsymbol{h}_{yx}^*} = \boldsymbol{h}_{yx} + \mu\varepsilon_y \boldsymbol{x}^* \tag{8.60}$$

$$\boldsymbol{h}_{yy} := \boldsymbol{h}_{yy} - \mu \frac{\partial |\varepsilon_y|^2}{\partial \boldsymbol{h}_{yy}^*} = \boldsymbol{h}_{yy} + \mu\varepsilon_y \frac{\partial y_o^*}{\partial \boldsymbol{h}_{yy}^*} = \boldsymbol{h}_{yy} + \mu\varepsilon_y \boldsymbol{y}^* \tag{8.61}$$

One of the challenges of the decision-directed equalizer is the need to combine the carrier recovery with the equalization since the decisions are made on the phase-corrected signal. The resulting feedback path can, therefore, be more challenging for CMOS ASIC implementation than the blind equalizer that partitions the equalization from the carrier recovery.

8.4.5 Radially Directed Equalizer Update Algorithm

While the CMA is well suited to constant modulus formats such at QPSK many of the formats considered for future optical networks are not constant modulus such as PDM-16QAM, such that the CMA will never converge to zero error [27, 28]. Nevertheless, the CMA can be adapted to a radially directed equalizer. In this case, $\varepsilon_x = Q_r(|x_o|^2) - |x_o|^2$ and $\varepsilon_y = Q_r(|y_o|^2) - |y_o|^2$, where $Q_r(r^2)$ is a function that quantizes the radius according to the number of possible points. Once the notional radius is determined the CMA algorithm is used to update the tap weights accordingly. One of the key benefits of using a radially directed equalizer is that it is invariant to the phase of the incoming signal and hence allows the equalization and the carrier recovery to be partitioned; however, for more dense modulation formats, often it is preferable to use the CMA initially and then switch to a decision-directed equalizer.

8.4.6 Parallel Realization of the FIR Filter

One of the key benefits of the FIR is that it can readily be implemented in CMOS. Thus far, all of the implementations discussed have operated on a symbol-by-symbol basis rather than based on a lower speed CMOS bus. In this case, the DSP is similar to the serial DSP given by Equation 8.53, but now we write [29]

$$\boldsymbol{x_o}[k] = \boldsymbol{h}_{xx}^T\boldsymbol{X} + \boldsymbol{h}_{xy}^T\boldsymbol{Y} \quad \text{and} \quad \boldsymbol{y_o}[k] = \boldsymbol{h}_{yx}^T\boldsymbol{X} + \boldsymbol{h}_{yy}^T\boldsymbol{Y} \tag{8.62}$$

where $\boldsymbol{x_o}[k] = [x_o[kN_b + 1], x_o[kN_b + 2], \ldots ., x_o[kN_b + N_b]$ and $\boldsymbol{X} = [\boldsymbol{x}_1, \boldsymbol{x}_2, \ldots, \boldsymbol{x}_{N_b}]$ and so on. In this case, the error term for the CMA becomes $\varepsilon_{\boldsymbol{x}} = 1 - \boldsymbol{x_o} \circ \boldsymbol{x_o^*}$ and similarly $\varepsilon_{\boldsymbol{y}} = 1 - \boldsymbol{y_o} \circ \boldsymbol{y_o^*}$ where $\circ$ denotes the Hadamard product being the element-by-element multiplication with the resulting update algorithm for the CMA becoming

$$\boldsymbol{h}_{xx} := \boldsymbol{h}_{xx} + 2\mu(\varepsilon_x \circ \boldsymbol{x}_o)^T \boldsymbol{X}^* \tag{8.63}$$

$$\boldsymbol{h}_{xy} := \boldsymbol{h}_{xy} + 2\mu(\varepsilon_x \circ \boldsymbol{x}_o)^T \boldsymbol{Y}^* \tag{8.64}$$

$$\boldsymbol{h}_{yx} := \boldsymbol{h}_{yx} + 2\mu(\varepsilon_y \circ \boldsymbol{y}_o)^T \boldsymbol{X}^* \tag{8.65}$$

$$\boldsymbol{h}_{yy} := \boldsymbol{h}_{yy} + 2\mu(\varepsilon_y \circ \boldsymbol{y}_o)^T \boldsymbol{Y}^* \tag{8.66}$$

Once the CMA has converged, a decision-directed equalizer is employed with

$$\boldsymbol{h}_{xx} := \boldsymbol{h}_{xx} + \mu\varepsilon_x^T \boldsymbol{X}^* \tag{8.67}$$

$$\boldsymbol{h}_{xy} := \boldsymbol{h}_{xy} + \mu\varepsilon_x^T \boldsymbol{Y}^* \tag{8.68}$$

$$\boldsymbol{h}_{yx} := \boldsymbol{h}_{yx} + \mu\varepsilon_y^T \boldsymbol{X}^* \tag{8.69}$$

$$\boldsymbol{h}_{yy} := \boldsymbol{h}_{yy} + \mu\varepsilon_y^T \boldsymbol{Y}^* \tag{8.70}$$

8.4.7 Generalized 4 × 4 Equalizer for Mitigation of Frequency or Polarization-Dependent Loss and Receiver Skew

Provided the equalizer is not constrained to be unitary then all of the equalizers discussed in this section can also be used to mitigate the impact of frequency or PDL. The equalizer can, however, be generalized further to the 4 × 4 equalizer by relaxing the assumption in the 2 × 2 complex equalizer that the real and imaginary signals are sampled synchronously and are orthonormal. In this case, the equalizer structure becomes [30–32]

$$x_{o_r}[k] = \boldsymbol{h}_{x_r x_r}^T \boldsymbol{x_r} + \boldsymbol{h}_{x_r x_i}^T \boldsymbol{x_i} + \boldsymbol{h}_{x_r y_r}^T \boldsymbol{y_r} + \boldsymbol{h}_{x_r y_i}^T \boldsymbol{y_i} \tag{8.71}$$

$$x_{o_i}[k] = \boldsymbol{h}_{x_i x_r}^T \boldsymbol{x_r} + \boldsymbol{h}_{x_i x_i}^T \boldsymbol{x_i} + \boldsymbol{h}_{x_i y_r}^T \boldsymbol{y_r} + \boldsymbol{h}_{x_i y_i}^T \boldsymbol{y_i} \tag{8.72}$$

$$y_{o_r}[k] = \boldsymbol{h}_{y_r x_r}^T \boldsymbol{x_r} + \boldsymbol{h}_{y_r x_i}^T \boldsymbol{x_i} + \boldsymbol{h}_{y_r y_r}^T \boldsymbol{y_r} + \boldsymbol{h}_{y_r y_i}^T \boldsymbol{y_i} \tag{8.73}$$

$$y_{o_i}[k] = \boldsymbol{h}_{y_i x_r}^T \boldsymbol{x_r} + \boldsymbol{h}_{y_i x_i}^T \boldsymbol{x_i} + \boldsymbol{h}_{y_i y_r}^T \boldsymbol{y_r} + \boldsymbol{h}_{y_i y_i}^T \boldsymbol{y_i} \tag{8.74}$$

where $x_{o_r} = \Re\{x_o\}$, $x_{o_i} = \Im\{x_o\}$, $\boldsymbol{x_r} = \Re\{\boldsymbol{x}\}$, $\boldsymbol{x_i} = \Im\{\boldsymbol{x}\}$ and so on. For the CMA, the update algorithm is

$$\begin{aligned}\boldsymbol{h}_{x_r x_r} &:= \boldsymbol{h}_{x_r x_r} - \mu \frac{\partial \varepsilon_x^2}{\partial \boldsymbol{h}_{x_r x_r}^*} = \boldsymbol{h}_{x_r x_r} - 2\mu\varepsilon_x \frac{\partial \varepsilon_x}{\partial \boldsymbol{h}_{x_r x_r}} = \boldsymbol{h}_{x_r x_r} + 2\mu\varepsilon_x \frac{\partial |x_o|^2}{\partial \boldsymbol{h}_{x_r x_r}} \\ &= \boldsymbol{h}_{x_r x_r} + 2\mu\varepsilon_x \frac{\partial (x_{o_r}^2 + x_{o_i}^2)}{\partial \boldsymbol{h}_{x_r x_r}} = \boldsymbol{h}_{x_r x_r} + 4\mu\varepsilon_x x_{o_r} \frac{\partial x_{o_r}}{\partial \boldsymbol{h}_{x_r x_r}} \\ &= \boldsymbol{h}_{x_r x_r} + 4\mu\varepsilon_x x_{o_r} \boldsymbol{x_r}\end{aligned} \tag{8.75}$$

with the updates for the other filters being obtained in a similar manner such that if we define the set $k \in \{x_r, x_i, y_r, y_i\}$ then the 16 updates may be written compactly as

$$\boldsymbol{h}_{\boldsymbol{x_r k}} := \boldsymbol{h}_{\boldsymbol{x_r k}} + 4\mu\varepsilon_x x_{o_r} \boldsymbol{k} \tag{8.76}$$

$$\boldsymbol{h}_{\boldsymbol{x_i k}} := \boldsymbol{h}_{\boldsymbol{x_i k}} + 4\mu\varepsilon_x x_{o_i} \boldsymbol{k} \tag{8.77}$$

$$\boldsymbol{h}_{\boldsymbol{y_r k}} := \boldsymbol{h}_{\boldsymbol{y_r k}} + 4\mu\varepsilon_y y_{o_r} \boldsymbol{k} \tag{8.78}$$

$$\boldsymbol{h}_{\boldsymbol{y_i k}} := \boldsymbol{h}_{\boldsymbol{y_i k}} + 4\mu\varepsilon_y y_{o_i} \boldsymbol{k} \tag{8.79}$$

8.4.8 Example Application to Fast Blind Equalization of PMD

Consider again Nyquist-shaped 35 Gbaud PDM-QPSK signal using an FEC with a BER limit of 2×10^{-2} (dBQ = 6.25 dB). To allow for blind acquisition, the signal is recovered using a CMA, targeting a mean DGD of 10 ps (corresponding to the typical link budget for a legacy 10 Gbit/s), requiring just five taps in order to neglect the outage probability. We assume a highly parallel implementation with 256 degrees of parallelism such that the CMOS DSP operates as 273 MHz while the ADC samples

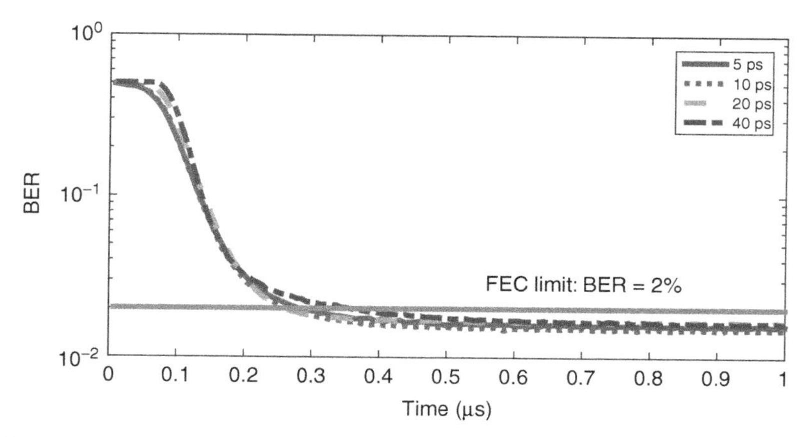

FIGURE 8.6 Convergence of a five-tap equalizer with differing levels of DGD. The BER is time resolved to the block period (7 ns) with the block averaged BER averaged over 1000 different realizations corresponding to one million bits with an SNR = 7.2 dB (being 1 dB greater than that required for BER = 2%).

at 70 GSa/s for 35 Gbaud PDM-QPSK. We include 1 dB of margin (i.e., the signal has an signal-to-noise ratio (SNR) of approximately 7.2 dB, being 1 dB better OSNR than required for a BER = 2×10^{-2}), so as to reduce the convergence time and we consider the performance for varying levels of DGD, 5, 10, 20, and 40 ps.

One of the particular challenges of improving the FEC to increase the tolerance to noise is that the SNR is significantly reduced in the equalizer. In Figure 8.6, while $\mu = 2^{-8} \approx 0.004$ to ensure that the equalizer converges with an SNR of 7.2 dB the stochastic gradient is averaged over two successive CMOS clock cycles, corresponding to a total of 512 samples. Nevertheless, even with this low SNR a convergence time of less than 400 ns is achievable for a DGD < 40 ps, being 70% of the maximum that the five tap equalizer could compensate. As the DGD increases toward the maximum theoretical value, the convergence time increases significantly highlighting the need for more research into the area of the convergence of equalization in the presence of distortions with low SNR.

8.5 CONCLUDING REMARKS AND FUTURE RESEARCH DIRECTIONS

In this chapter, we have outlined the key algorithms for equalization, both in terms of the design of fixed filters but also in the adaptive updating of filters. The key subsystems for digital equalization of chromatic and PMD have been extensively studied for PDM-QPSK with an SNR in the region of 10 dB; however, as we have discussed in this chapter transceiver technology is rapidly moving beyond this toward more dense constellations with reduced SNR.

As discussed in Section 8.3, while the truncated impulse response FIR filter design for chromatic dispersion equalization was adequate for PDM-QPSK as systems moved toward PDM-64QAM this had to be revisited requiring alternative techniques such as the least squares design to be employed. Likewise for the adaptive equalization while the CMA has been extensively utilized for PDM-QPSK systems allowing the equalization to be partitioned from the carrier recovery, an equivalent to the CMA that is robust for high levels of QAM has yet to be determined. The situation is further complicated by the move to soft-decision FEC and the associated reduced SNR. Therefore, the first area of future research to highlight is that of equalization for higher-level modulation formats with low SNR, with a key consideration being that the algorithms should be optimized for highly parallel DSP to permit realization in CMOS.

A second area of future research arises with the development of dynamic elastic optical networking in which transceivers can vary their rate but also be dynamically dropped and added, for example, to cope to the so-called "elephant flows" between data centers. This second area of research calls for both rate adaptive DSP and also research into fast acquisition algorithms so as to minimize the time taken to establish a wavelength on demand service.

A third area of research is that of reduced complexity equalization. Not only does this reduce the power consumption of the DSP but also it permits the use of DSP in applications that are more cost sensitive such as access networks. While typically the chromatic dispersion compensation filter is implemented using techniques such as the overlap and save method, similar frequency domain approaches can be utilized for the adaptive equalizers.

Ultimately, as optical fiber communication systems continue to evolve the need for research into equalization will continue to be a fertile area of research, particularly as the capabilities of the transceivers increase both in terms of maximum data rate and flexibility to adapt to network demands.

ACKNOWLEDGMENTS

The Royal Academy of Engineering and the Leverhulme Trust are thanked for financial support, and Dr Benn Thomsen, Dr David Millar, Dr Domaniç Lavery, Dr Ishaad Fatadin, Mr Milen Paskov and Dr Amir Eghbali are also thanked for the numerous stimulating discussions relating to the topics contained in this chapter.

REFERENCES

1. Savory SJ, Stewart AD, Wood S, Gavioli G, Taylor MG, Killey RI, Bayvel P. Digital equalisation of 40Gbit/s per wavelength transmission over 2480km of standard fibre without optical dispersion compensation. Proceedings of ECOC 2006; Cannes, France, paper Th2.5.5; Sep 2006.
2. Laperle C, Villeneuve B, Zhang Z, McGhan D, Sun H, O'Sullivan M. Wavelength division multiplexing (WDM) and polarization mode dispersion (PMD) performance of a coherent 40Gbit/s dual-polarization quadrature phase shift keying (DP-QPSK) transceiver. Proceedings of Optical Fiber Communications Conference 2007, paper PDP16; 2007.

3. van den Borne D, de Waardt H, Khoe G-D, Duthel T, Fludger CRS, Schulien C, Schmidt E-D. Electrical PMD compensation in 43-Gb/s POLMUX-NRZ-DQPSK enabled by coherent detection and equalization. Proceedings ECOC 2007; Berlin, Germany, invited paper 8.3.1; 2007.
4. Charlet G, Renaudier J, Salsi M, Mardoyan H, Tran P, Bigo S. Efficient mitigation of fiber impairments in an ultra-long haul transmission of 40Gbit/s polarization-multiplexed data, by digital processing in a coherent receiver. Proceedings of Optical Fiber Communications Conference 2007, paper PDP17; 2007.
5. Fludger CRS, Duthel T, van den Borne D, Schulien C, Schmidt E-D, Wuth T, de Man E, Khoe GD, de Waardt H. 10 × 111 Gbit/s, 50 GHz spaced, POLMUX-RZ-DQPSK transmission over 2375 km employing coherent equalisation. Proceedings of Optical Fiber Communications Conference 2007, paper PDP22; 2007.
6. Ip E, Kahn JM. Digital equalization of chromatic dispersion and polarization mode dispersion. *J Lightwave Technol* 2007;25:2033–2043.
7. Yamamoto S et al. PMD tolerance of 100-Gbit/s digital coherent PDM-QPSK in DSF-installed field testbed. Opto-Electronics and Communications Conference (OECC); 2011 16th, 4–8 Jul 2011. p 212, 213.
8. "mitigation, n." OED Online. Oxford University Press, March 2015. Web. 19 Mar 2015.
9. "equalization, n." OED Online. Oxford University Press, March 2015. Web. 19 Mar 2015.
10. Proakis JG, Manolakis DG. Digital Signal Processing. 3rd ed. Prentice Hall; 1996. p 430–432.
11. Haykin S. Adaptive Filter Theory. 4th ed. Prentice Hall; 2001.
12. S. S. Kidambi and R. P. Ramachandran, Complex coefficient nonrecursive digital filter design using a least-squares method. *IEEE Trans Signal Process*, vol. 44, no. 3, pp. 710–713, 1996.
13. Godard D. Self-recovering equalization and carrier tracking in two-dimensional data communication systems. *IEEE Trans Commun* 1980;28:1867–1875.
14. T.W. Parks and J.H. McClellan, Chebyshev approximation for nonrecursive digital filters with linear phase. *IEEE Trans Circuit Theory*, vol. CT-19, no. 2, pp. 189–194, 1972.
15. Remez EYa. General computational methods of Tchebycheff approximation. Kiev (Atomic Energy Commission Translation 4491); 1957. p 1–85.
16. Agrawal GP. Fiber-Optic Communication Systems. 4th ed. Wiley; 2010.
17. Savory SJ. Digital filters for coherent optical receivers. *Opt Express* 2008;16:804–817.
18. Eghbali A, Johansson H, Gustafsson O, Savory S. Optimal least-squares FIR digital filters for compensation of chromatic dispersion in digital coherent optical receivers. *J Lightwave Technol* 2014;32:1449–1456.
19. Poole CD, Wagner RE. Phenomenological approach to polarization dispersion in long single-mode fibers. *Electron Lett* 1986;22:1029–1030.
20. Foschini GJ, Poole CD. Statistical theory of polarization dispersion in single mode fibers. *J Lightwave Technol* 1991;9:1439–1456.
21. Betti S, Curti F, De Marchis G, Iannone E. A novel multilevel coherent optical system: four quadrature signaling. *J Lightwave Technol* 1991;9:514–523.
22. Han Y, Li G. Coherent optical communication using polarization multiple-input-multiple-output. *Opt Express* 2005;13:7527–7534.

23. Liu L, Tao Z, Yan W, Oda S, Hoshida T, Rasmussen JC. Initial tap setup of constant modulus algorithm for polarization de-multiplexing in optical coherent receivers. Optical Fiber Communication Conference (OFC2009),OMT2, March 22–26; 2009.
24. Chongjin X, Chandrasekhar S. Two-stage constant modulus algorithm equalizer for singularity free operation and optical performance monitoring in optical coherent receiver. Proceedings of OFC 2010; 2010.
25. Rozental V, Portela T, Souto D, Ferreira H, Mello D. Experimental analysis of singularity-avoidance techniques for CMA equalization in DP-QPSK 112-Gb/s optical systems. *Opt Express* 2011;19:18655–18664.
26. Savory SJ, Gavioli G, Killey R, Bayvel P. Electronic compensation of chromatic dispersion using a digital coherent receiver. *Opt Express* 2007;15:2120–2126.
27. Fatadin I, Ives D, Savory SJ. Blind equalization and carrier phase recovery in a 16-QAM optical coherent system. *J Lightwave Technol* 2009;27(15):3042–3049.
28. Savory SJ. Digital coherent optical receivers: algorithms and subsystems. *IEEE J Sel Top Quant* 2010;16(5):1164–1179.
29. Thomsen BC, Maher R, Millar DS, Savory SJ. Burst mode receiver for 112 Gb/s DP-QPSK with parallel DSP. *Opt Express* 2011;19(26):770–776.
30. M. S. Faruk and K. Kikuchi, Compensation for in-phase/quadrature imbalance in coherent-receiver front end for optical quadrature amplitude modulation. *IEEE Photon J*, vol. 5, no. 2, p. 7800110, 2013.
31. Paskov M, Lavery D, Savory SJ. Blind equalization of receiver in-phase/quadrature skew in the presence of Nyquist filtering. *IEEE Photon Technol Lett* 2013;25(24):2446–2449.
32. R. Rios-Muller, J. Renaudier, and G. Charlet Blind receiver skew compensation and estimation for long-haul non-dispersion managed systems using adaptive equalizer. *J Lightwave Technol*, vol.33, no.7, pp.1315–1318, 2015.

9

NONLINEAR COMPENSATION FOR DIGITAL COHERENT TRANSMISSION

GUIFANG LI

CREOL, The College of Optics and Photonics, University of Central Florida, Orlando, FL, USA

9.1 INTRODUCTION

The degree to which fiber impairments are compensated determines the transmission capacity of fiber-optic transmission systems. Dispersion-compensating fiber (DCF) is commonly used to compensate chromatic dispersion. Wavelength-division multiplexing (WDM) systems suffer from both intra- and interchannel nonlinearities such as cross-phase modulation (XPM) and four-wave mixing (FWM). These effects can be suppressed using dispersion management. Compensation of nonlinear impairments in fiber has become the next logical step in increasing the capacity of WDM systems. A few optical nonlinearity compensation schemes have been demonstrated such as lumped compensation of self-phase modulation (SPM) [1] and optical phase conjugation for the compensation of both chromatic dispersion and Kerr nonlinearity in fibers [2].

Recently, coherent detection opened new venues for long-haul optical communication systems. Among these are the possibility of using higher-order modulation formats [3, 4], the ability to pack channels more tightly using orthogonal frequency-division multiplexing and orthogonal wavelength-division multiplexing [5, 6]. Also, it was noticed early on that since coherent detection provided complete information about the electric field, including its intensity, phase, and even polarization, fiber-induced linear impairments such as dispersion could be eliminated using

Enabling Technologies for High Spectral-efficiency Coherent Optical Communication Networks, First Edition. Edited by Xiang Zhou and Chongjin Xie.
© 2016 John Wiley & Sons, Inc. Published 2016 by John Wiley & Sons, Inc.

digital signal processing [7, 8]. Dispersion, being a linear and scalar impairment, can be compensated in a single step, which is commonly referred to as lumped compensation [9]. Subsequently, it was shown that if the dispersion of the fiber was small enough, even the impairments caused by the Kerr nonlinearity, which is often data dependent, could be compensated [10]. However, in the more general case where both dispersion and nonlinearity have appreciable impact on the signal, these impairments cannot be removed in a single step.

In this chapter, we describe a universal postcompensation scheme to compensate all deterministic impairments in fiber called digital backward propagation (DBP).

9.2 DIGITAL BACKWARD PROPAGATION (DBP)

9.2.1 How DBP Works

To understand how DBP works, first let us look in detail how the electric field of the signal, $A(z, t)$, propagates in the forward direction in fiber, which is governed by the nonlinear Schrodinger equation (NLSE) [11]

$$\frac{\partial}{\partial z}A = (\hat{D} + \hat{N})A \tag{9.1}$$

with the dispersion operator

$$\hat{D} = -\frac{j}{2}\beta_2\frac{\partial^2}{\partial t^2} + \frac{1}{6}\beta_3\frac{\partial^3}{\partial t^3} - \frac{\alpha}{2} \tag{9.2}$$

and the nonlinear operator

$$\hat{N} = j\gamma_{\mathrm{NL}}|A|^2 \tag{9.3}$$

where β_j represent the jth-order dispersion, α is the absorption coefficient, γ is the nonlinear parameter, and t is the retarded time.

In general, the NLSE does not have an analytical solution. To model the nonlinear dispersive propagation, the fiber is divided into small sections in the split-step method (SSM), as shown in Figure 9.1(a). If each section is short enough, the effect of dispersion and nonlinearity can be calculated independently. In one step, the NLSE is solved ignoring the nonlinear term, and in the next step it is solved ignoring the dispersion term. The dispersion effect can be calculated in the time domain using finite-impulse response (FIR) filtering or in the frequency domain using the split-step Fourier method (SSFM) [12]. The interplay between dispersion and nonlinearity manifests after the first dispersion and nonlinearity computation pair. At the end of the fiber, the signal will be different from the transmitted signal. If the received signal is detected coherently, preserving both the amplitude and phase, it can be sent into a virtual fiber in the digital domain whose dispersion and nonlinearity are exactly opposite to those of the real transmission fiber. The dispersion and nonlinear

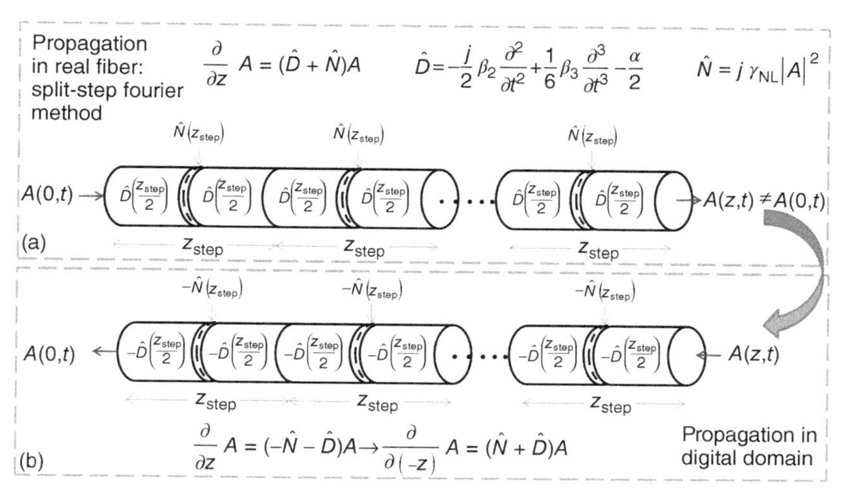

FIGURE 9.1 Schematic of the SSM for simulating forward propagation in real fiber (a) and DBP in virtual fiber (b).

effects in the real fiber are then piecewise canceled by the virtual fiber as shown in Figure 9.1(b). At the end of propagation in the virtual fiber, the signal will be the same as the transmitted signal except with a delay. Since propagation in a fiber with opposite dispersion and nonlinearity is the same as propagating backwards in the real fiber

$$\frac{\partial}{\partial z}A = (-\hat{N} - \hat{D})A \rightarrow \frac{\partial}{\partial(-z)}A = (\hat{N} + \hat{D})A \tag{9.4}$$

this digital nonlinear compensation method is called DBP. Even though physical fiber with negative nonlinearity does not exist, the beauty of digital signal processing is that the virtual fiber can assume any dispersion and nonlinearity.

9.2.2 Experimental Demonstration of DBP

The first DBP experiment was demonstrated in 2008 [13]. Three distributed-feedback lasers were used as WDM carriers. The center channel (λ_C) and two adjacent channels (λ_R and λ_L) are binary phase-shift keying (BPSK) modulated using Mach–Zehnder modulators (MZM) driven at 6 GBd by a pattern generator (PG) with a pseudorandom bit sequence (PRBS) of length $2^{23} - 1$. A symbol rate of 6 GBd was chosen in order to fit three WDM channels within the double-sided analog bandwidth of 24 GHz of the real-time oscilloscope (RTO). Larger analog bandwidth would allow higher symbol rates and more WDM channels. The WDM channels are tightly spaced at 7 GHz rather than 6 GHz, which was found to give the lowest linear crosstalk due to nonideal waveforms [14]. The adjacent channels have synchronized symbol times (using radio frequency delay) but decorrelated data content compared with the center channel (through an optical delay in the center channel path). Polarization controllers (PCs) were placed at the appropriate locations to insure that all the channels are copolarized. The launch power in each channel was equalized using optical attenuators.

To demonstrate the feasibility of digital impairment compensation for the WDM environment with low channel count limited by the bandwidth of the RTO, a relatively high launching power was used so significant nonlinear effects accumulate. A total launching power ($P_L = 6\,\text{dBm}$) into the optical fiber was set using an erbium-doped fiber amplifier (EDFA), calibrated to the insertion losses associated with the recirculating loop components. The recirculating loop consists of two NZ-DSF spools, with a combined length of 152.82 km. The fiber parameters are $\alpha = 0.2\,\text{dB/km}$, $\beta_2 = -4.9\,\text{ps}^2/\text{km}$, and $\gamma_{\text{NL}} = 1.9/\text{km/W}$. A phase diversity receiver is used to beat the transmitted optical signal with the local oscillator (LO) tuned to the center channel. The RTO was used to digitize the two signal quadratures at 40 GSamples/s. DSP was performed offline.

After employing backward propagation, proper orthogonal WDM filtering [14], phase estimation, and decision threshold Q-factor calculation for the center channel were performed. To evaluate the benefit of using SSM/FIR, other possible compensation techniques are considered; chromatic dispersion compensation (CDC) only and lumped compensation which includes CDC followed by a phase shift proportional to the signal power. The eye diagrams for the various compensation schemes after 760 km are shown in Figure 9.2. The effectiveness of DBP can be clearly seen.

9.2.3 Computational Complexity of DBP

Once the feasibility of DBP is demonstrated experimentally, the number of operations, which translates to power consumption, required for DBP becomes the critical issue for practical applications. The number of operations for a given optical link depends on the number of steps (n_s), and hence, on the step size ($h = L/n_s$). The SSFM accuracy depends fundamentally on the mutual influence of dispersion and nonlinearity within the step length. Due to the nature of the dispersion and nonlinearity operators, the step size has to be chosen such that (i) the nonlinear phase shift is small enough to preserve the accuracy of the dispersion operation and (ii) the optical power fluctuations due to dispersion effects are small enough to preserve the accuracy of the nonlinear operation.

For the discussion here [15], the total optical field of the WDM channels are written as

$$A = \sum_m A_m exp(ik_m z) \tag{9.5}$$

where A_m and k_m are the complexity amplitude and the linear propagation constant of the mth channel, respectively. One way to set the upper bound for the step size is to identify the characteristic physical lengths, which correlate the optical field fluctuations with the propagation distance. Four physical lengths are of interest here, namely, the dispersion, L_D, the nonlinear length L_{nl}, the walk-off length L_{wo}, and the FWM length L_{fwm}. The nonlinear and walk-off lengths can be defined, for a multichannel

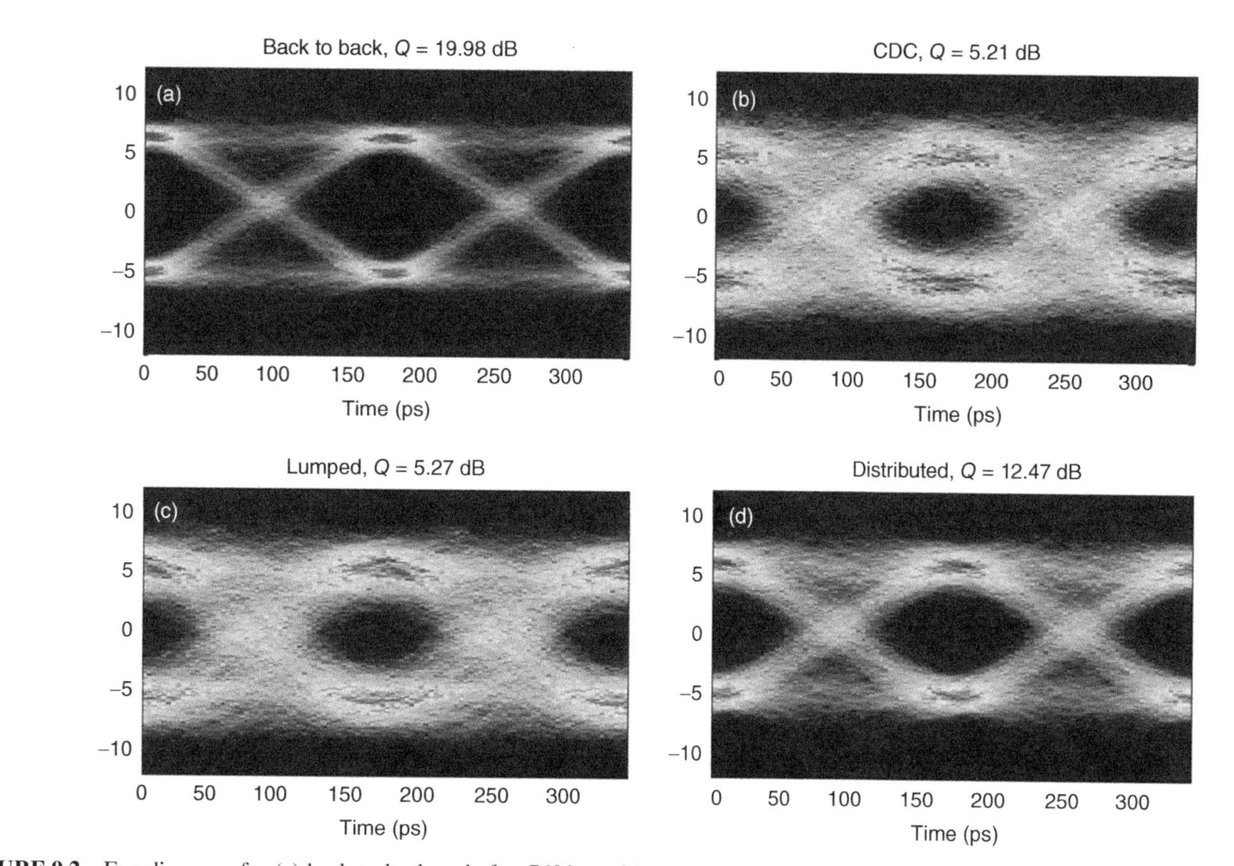

FIGURE 9.2 Eye diagrams for (a) back-to-back and after 760 km with compensation using (b) CDC, (c) lumped nonlinear compensation corresponding to one-step DBP, (d) distributed SSM/FIR DBP.

system, as follows:

$$L_{\text{nl}} = \frac{1}{\gamma P_T \frac{2N-1}{N}}, \quad L_{\text{wo}} = \frac{1}{2\pi|\beta_2|(N-1)\Delta f B} \tag{9.6}$$

where $P_T = \sum_m |A_m|^2$ is the total launched power, B is the symbol rate (effectively the inverse of the pulse width), and Δf is the channel spacing. The nonlinear length has been defined as the length after which an individual channel achieves a 1 radian phase shift due to SPM and XPM. The walk-off length is defined as the distance after which the relative delay of pulses from the edge channels is equal to the pulse width. The above-mentioned characteristic lengths are well known and widely used to qualitatively describe the optical field behavior through fiber propagation. However, when FWM is considered, the nonlinear and walk-off lengths are not enough to qualitatively identify the range where the *fastest* field fluctuations take place. The nonlinear term, including FWM, can be expressed as follows for the mth channel.

$$-i\gamma\left(2\sum_{q\in I}|A_q|^2 - |A_m|^2\right)A_m - i\gamma\left[\sum_{[rslm]\in I} A_r A_s A_l^* exp(i\delta k_{rslm} z)\right] \tag{9.7}$$

with the following conditions: $l = r + s - m$, $[m, r, s] \in I$, and $r \neq s \neq m$. The first condition neglects fast time-oscillating terms (*frequency matching*). The second condition forces the newly generated waves to lay within the WDM band. Finally, the third condition excludes SPM and XPM terms. δk_{rslm} is the phase mismatch parameter, given by

$$\delta k_{rslm} = k_r + k_s - k_l - k_m = \frac{1}{2}\beta_2 \Delta\omega^2 [r^2 + s^2 - (r + s - m)^2 - m^2] \tag{9.8}$$

In order to identify the *fastest* z-fluctuations for the mth channel, let us set $r = 1$ and $s = N$ corresponding to the indexes of the edge channels. By maximizing Equation 9.8, the expression for the maximum phase-mismatch is given by

$$\delta k_{\max} = \frac{1}{4}|\beta_2|(N-1)^2 \Delta\omega^2 \tag{9.9}$$

This expression leads to the following definition for the FWM length:

$$L_{\text{FWM}} = \frac{1}{\pi^2 |\beta_2|(N-1)^2 \Delta f^2} \tag{9.10}$$

This expression represents the length after which the argument of the *fastest* FWM term is shifted by 1 radian; hence, it can be understood as the distance after which power fluctuations due to FWM start to take place. The definition of the FWM length assumes that the FWM-induced variations on a given channel are governed by the linear (dispersive) phase mismatch. However, nonlinearity also contributes to the overall phase mismatch through SPM and XPM. This contribution is only relevant in high-power regimes and it is not expected to play a role in the analysis of fiber transmission.

9.3 REDUCING DBP COMPLEXITY FOR DISPERSION-UNMANAGED WDM TRANSMISSION

For the experimental demonstration described in Section 9.2, the dispersion length is $\cong$ 355 km, where the full bandwidth 24 GHz is taken. Considering the effects of SPM and XPM, and neglecting loss, the nonlinear length is $\cong$ 282 km, while the FWM length is $\cong$ 105 km. Typically, especially for large channel count WDM systems, the FWM length is much shorter than other physical lengths. Yet, FWM can be effectively suppressed with sufficient local dispersion in a well-designed WDM system. Hence, it is possible to reduce the DBP computational load for WDM transmission by ignoring FWM.

In a coherent detection system, a full reconstruction of the optical field can be achieved by beating the received field with a copolarized local oscillator. The reconstructed field will be used as the input for backward propagation in order to compensate the transmission impairments. Let $\widehat{E}_m$ be the envelope of optical field of the$m$th channel, where $m \in I$, $I = \{1, 2, \ldots, N\}$ and N is the total number of WDM channels. By rewriting the field expression as $A_m = \widehat{A}_m exp[im\Delta\omega t]$, where $\Delta\omega = 2\pi\Delta f$, the expression of the full optical field can be expressed as $A = \sum_m A_m$. The total-field backward propagation equation, that is, T-NLSE, is given by Taylor [9]

$$-\frac{\partial A}{\partial z} + \frac{\alpha}{2}A + \frac{i\beta_2}{2}\frac{\partial^2 A}{\partial t^2} - \frac{\beta_3}{6}\frac{\partial^3 A}{\partial t^3} - i\gamma|A|^2 A = 0 \tag{9.11}$$

which is the backward propagation equation corresponding to the forward propagation equation (Eq. 9.4). Equation 9.11 governs the backward propagation of the total field including second- and third-order dispersion, SPM, XPM, and FWM compensation.

Alternatively, the effect of FWM can be omitted in backward propagation by introducing the expression for A into Equation 9.11, expanding the $|A|^2$ term and neglecting the so-called FWM terms, that is,

$$-\frac{\partial A_m}{\partial z} + \frac{\alpha}{2}A_m + \frac{i\beta_2}{2}\frac{\partial^2 A_m}{\partial t^2} - \frac{\beta_3}{6}\frac{\partial^3 A_m}{\partial t^3} - i\gamma\left(2\sum_{q\in I}|A_q|^2 - |A_m|^2\right)A_m = 0 \tag{9.12}$$

The system of coupled equations (Eq. 9.12) describe the backward propagation of the WDM channels where dispersion, SPM, and XPM are compensated [15].

Figure 9.3 illustrates simulation results comparing selective compensation of transmission impairments for 12 channels × 25 GBd (16QAM) WDM transmission with a 50 GHz spacing. The fiber dispersion is $D = 4.4$ ps/km/nm, and the total length is 1000 km. As can be seen, the benefits of FWM compensation are not as significant as XPM compensation.

The benefits of selective compensation of fiber nonlinearity using DBP can be appreciated by defining a complexity parameter

$$F = \frac{\text{Number of equations} \times \text{Samples per symbol}}{\text{Step size}}$$

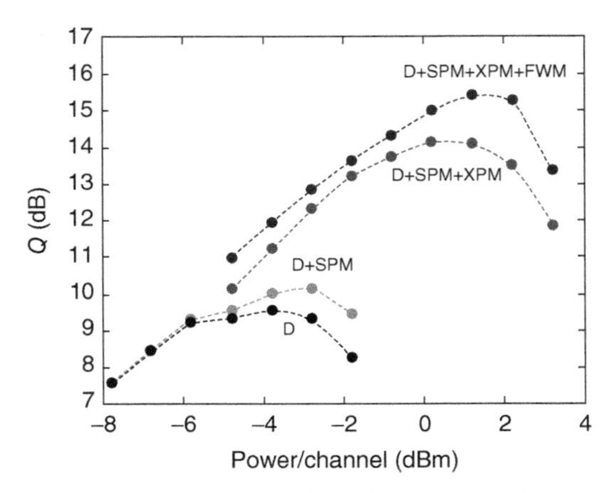

FIGURE 9.3 Q-factor as a function of per channel input power for selective compensation of fiber impairments. XPM compensation provides the largest benefit for WDM impairments.

The number of equations for CDC and total field DBP is one while the number equations for SPM and XPM compensation is equal to the number of WDM channels. Samples per symbol for total field compensation are larger by a factor equal to the channel space Δf divided by the Baud rate B. As a result, the overall computational complexity for XPM compensation, as shown in Table 9.1 is at least one order of magnitude smaller than total-field compensation that addresses FWM penalties.

In solving Equation 9.12 using the SSM, the nolinear part is computted in the time domain, having the following solution:

$$A_m(t, z + h) = A_m(t, z)\exp(i\phi_{m,\mathrm{SPM}} + i\phi_{m,\mathrm{XPM}}) \tag{9.13}$$

where

$$\phi_{m,\mathrm{SPM}}(t, z + h) = \gamma \int_{z}^{z+h} P_m(t, \hat{z})e^{\alpha z} d\hat{z} \tag{9.14}$$

$$\phi_{m,\mathrm{XPM}}(t, z + h) = 2\gamma \sum_{q \neq m} \int_{z}^{z+h} P_q(t, \hat{z})e^{\alpha \hat{z}} d\hat{z} \tag{9.15}$$

TABLE 9.1 Comparison of computational complexity F for SPM, XPM, and FWM compensation using DBP normalized to that for lump dispersion compensation ($F_D = 1$)

	ΔQ (dB)	Step-size	F/F_D
Dispersion compensation	–	1000 km	1
SPM compensation	0.6	50 km	20
XPM compensation	4.6	3.4 km	~300
FWM compensation	5.9	330 m	~6000

Step size and improvement in Q are also listed.

and $P_q = |A_q|^2$. In the conventional SSM, the above-mentioned integrals are approximated by

$$\phi_{m,\mathrm{SPM}}(t, z+h) = \gamma h_{\mathrm{eff}} P_m(t, z) \tag{9.16}$$

$$\phi_{m,\mathrm{XPM}}(t, z+h) = 2\gamma h_{\mathrm{eff}} \sum_{q \neq m} P_q(t, z) \tag{9.17}$$

where $h_{\mathrm{eff}} = [\exp(\alpha h) - 1]/\alpha$ is the effective step size. Equations 9.16 and 9.17 are valid provided variations of the optical fields due to dispersion take place in a length scale much longer than the step size. In the case of SPM, such a length scale is given by the intrachannel pulse broadening, that is, $L_D = 1/(\beta_2 B^2)$, where B is the Baud-rate. For XPM, the dispersive walk-off between channels has to be considered. This delay occurs in a length scale given by $L_{\mathrm{wo}} = 1/(2\pi\beta_2 N\Delta f B)$. Typically, N is large and $L_{\mathrm{wo}} \ll L_D$, which results in XPM being the limiting effect for the step-size. To relax the step-size requirements for XPM compensation, the effects of pulse broadening and walk-off must be separated [16]. For that, Equation 9.17 can be rewritten, by including the time delay caused by the dispersive walk-off, as follows:

$$\phi_{m,\mathrm{XPM}}(t, z+h) = 2\gamma \sum_{q \neq m} \int_z^{z+h} P_q(t - d_{mq}\hat{z}, \hat{z}) e^{\alpha \hat{z}} d\hat{z} \tag{9.18}$$

where $d_{mq} = \beta_2(\omega_m - \omega_q)$ is the walk-off parameter. By Fourier transforming Equation 9.18, the following expression is obtained for the XPM phase shift, now in the frequency domain:

$$\phi_{m,\mathrm{XPM}}(\omega, z+h) = 2\gamma \sum_{q \neq m} \int_z^{z+h} P_q(\omega, \hat{z}) e^{-id_{mq}\omega\hat{z} + \alpha\hat{z}} d\hat{z} \tag{9.19}$$

After some algebra, and by grouping the SPM and XPM contributions, the total nonlinear phase shift is given by the following relations:

$$\phi_m(t, z+h) = F^{-1}\left[\sum_{\forall q} \mathrm{F}\left(P_q(t, z)\right) W_{mq}(\omega, h)\right] \tag{9.20}$$

$$W_{mq}(\omega, h) = \begin{cases} \gamma h_{\mathrm{eff}}, & \text{for } q = m \\ 2\gamma \dfrac{e^{(\alpha + id_{mq}\omega)h} - 1}{\alpha + id_{mq}\omega}, & \text{for } q \neq m \end{cases} \tag{9.21}$$

Equation 9.19 is valid provided that the individual power spectra do not change significantly over the step-size, which is now limited by the minimum of the intrachannel dispersion length and the XPM nonlinear length. Despite the increased complexity of the above-mentioned approach (hereafter advanced-SSM), the walk-off effect factorization substantially increases the step size in WDM systems, leading to remarkable savings in computation.

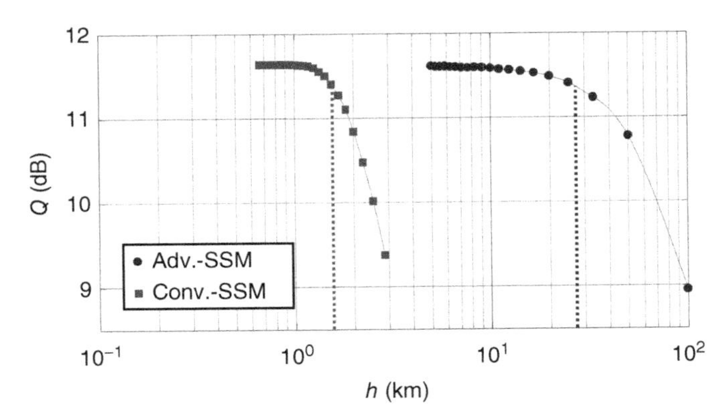

FIGURE 9.4 Step-size for the advanced and conventional implementation of the split-step method for XPM compensation.

Figure 9.4 compares the step-size requirements for the XPM compensation using the conventional and the advanced-SSMs introduced in the previous section. Results in Figure 9.4 correspond to the optimum power, whereas vertical markers indicate the optimum step-size. This value is obtained by cubic interpolation of the simulation results and by choosing the step-size value corresponding to a Q-factor penalty of 0.25 dB with respect to the plateau value. The advantage of the walk-off factorization is clear in terms of step size, which can be increased by a factor of 17 using the advanced-SSM.

It should be noted that low-pass filtering can be applied to the computation of SPM [17, 18], and XPM [19] to increase the step size and thus reduce the computational load. Intrachannel FWM can also be effectively compensated using a perturbation approach, especially for QPSK modulation since multiplication of QPSK symbols remains in the constellation and can be computed using a look-up table [20]. Despite the optimization methods presented earlier for reducing computational complexity, DBP for WDM transmission is still expected to have computational complexities at least two orders of magnitude higher than that of electronic dispersion compensation. Further reduction of DBP computational complexity will likely require clever optics to assist DSP. Such is the case of DBP for dispersion-managed transmission.

9.4 DBP FOR DISPERSION-MANAGED WDM TRANSMISSION

For dispersion-managed fiber-optic transmission systems, it can be assumed, without loss of generality, that each fiber span with a length of L is one period of the dispersion map. For long-haul fiber-optic transmission, an optimum power exists as a result of the trade-off between optical signal-to-noise ratio (OSNR) and nonlinear impairments. The total nonlinear phase shift at the optimum power level is on the order of 1 radian [21]. Therefore, for transoceanic fiber transmission systems, which consist of many (>100) amplified spans, the nonlinear effects in each span are weak. As a result, chromatic dispersion is the dominant factor that determines the evolution of the waveform within each span.

One can analyze the nonlinear behavior of the optical signal in dispersion-managed transmission systems using a perturbation approach [22]. The NLSE governing the propagation of the optical field, $A_j(z,t)$, in the jth fiber span can be expressed as

$$\frac{\partial A_j(z,t)}{\partial z} = [D + \varepsilon \cdot N(|A_j(z,t)|^2)] \cdot A_j(z,t) \tag{9.22}$$

where $0 < z < L$ is the propagation distance within each span, D is the linear operator for dispersion, fiber loss, and amplifier gain, $N(|A_j(z,t)|^2)$ is the nonlinear operator, ε (to be set to unity) is a parameter indicating that the nonlinear perturbation is small for the reasons mentioned earlier. The boundary conditions are

$$A_1(0,t) = a(0,t) \tag{9.23}$$

$$A_j(0,t) = A_{j-1}(L,t) \quad \text{for} \quad j \geq 2 \tag{9.24}$$

where $a(0,t)$ is the input signal at the beginning of the first span. The solution of Equation 9.22 can be written as

$$A_j(z,t) = A_{j,l}(z,t) + \varepsilon \cdot A_{j,\text{nl}}(z,t) \tag{9.25}$$

Substituting Equation 9.25 into Equation 9.22 and expanding the equation in power series of ε yields

$$\frac{\partial A_{j,l}(z,t)}{\partial z} - D \cdot A_{j,l}(z,t) + \varepsilon \cdot \left[\frac{\partial A_{j,\text{nl}}(z,t)}{\partial z} - D \cdot A_{j,\text{nl}}(z,t) - N(|A_{j,l}(z,t)|^2) \cdot A_{j,l}(z,t)\right] + O(\varepsilon^2) = 0 \tag{9.26}$$

Equating to zero the successive terms of the series lead to

$$\frac{\partial A_{j,l}(z,t)}{\partial z} = D \cdot A_{j,l}(z,t) \tag{9.27}$$

$$\frac{\partial A_{j,\text{nl}}(z,t)}{\partial z} = D \cdot A_{j,\text{nl}}(z,t) + N(|A_{j,l}(z,t)|^2) \cdot A_{j,l}(z,t) \tag{9.28}$$

The boundary conditions are

$$A_{1,l}(0,t) = a(0,t) \tag{9.29}$$

$$A_{j,l}(0,t) = A_{j-1}(L,t) \quad \text{for} \quad j \geq 2 \tag{9.30}$$

and

$$A_{j,\text{nl}}(0,t) = 0 \tag{9.31}$$

First, it is assumed that dispersion is completely compensated in each span. As a result, at the end of the first span

$$A_{1,l}(L,t) = a(0,t) \tag{9.32}$$

and

$$A_2(0,t) = A_1(L,t) = a(0,t) + \varepsilon \cdot A_{1,\mathrm{nl}}(L,t) \tag{9.33}$$

where $A_{1,\mathrm{nl}}(z,t)$ is the solution of Equation 9.28 with $j = 1$. In the second span,

$$A_{2,l}(z,t) = A_{1,l}(z,t) + \varepsilon \cdot \overline{A}(z,t) \tag{9.34}$$

where the first and second terms are solutions to Equation 9.27 with boundary conditions $A_{2,l}(0,t) = a(0,t)$ and $A_{2,l}(0,t) = \varepsilon \cdot A_{1,\mathrm{nl}}(L,t)$, respectively, as a result of the principle of superposition. At the end of the second span, because of complete dispersion compensation,

$$A_{2,l}(L,t) = a(0,t) + \varepsilon \cdot A_{1,\mathrm{nl}}(L,t) \tag{9.35}$$

The nonlinear distortion in the second span is governed by Equation 9.28 with $j = 2$. Since

$$|A_{2,l}|^2 = |A_{1,l} + \varepsilon \cdot \overline{A}|^2 = |A_{1,l}|^2 + O(\varepsilon) \tag{9.36}$$

the differential equation and the boundary conditions for $A_{2,\mathrm{nl}}(z,t)$ and $A_{1,\mathrm{nl}}(z,t)$ are identical, so

$$A_{2,\mathrm{nl}}(L,t) = A_{1,\mathrm{nl}}(L,t) \tag{9.37}$$

As a result, the optical field at the end of the second span is given by

$$A_2(L,t) = A_{2,l}(L,t) + \varepsilon \cdot A_{2,\mathrm{nl}}(L,t) = a(0,t) + \varepsilon \cdot 2A_{1,\mathrm{nl}}(L,t) \tag{9.38}$$

That is, the nonlinear distortion accumulated in 2 spans is approximately the same as the nonlinear distortion accumulated in 1 span with the same dispersion map and *twice* the nonlinearity. It follows that, assuming weak nonlinearity and periodic dispersion management, the optical field after K spans of propagation can be written as

$$A_N(L,t) = a(0,t) + \varepsilon \cdot KA_{1,\mathrm{nl}}(L,t) \tag{9.39}$$

which is the solution of the NLSE

$$\frac{\partial A_j(z,t)}{\partial z} = [D + \varepsilon \cdot KN(|A_j(z,t)|^2)] \cdot A_j(z,t) \tag{9.40}$$

The nonlinear term $N(|A_j(z,t)|^2)$ in Equation 9.22 is proportional to the fiber nonlinear parameter γ, so the NLSE describing optical propagation in a fiber span where the nonlinearity is K times of that in the original fiber but with the same dispersion map can be written as

$$\frac{\partial A_j(z,t)}{\partial z} = [D + \varepsilon \cdot K \cdot N(|A_j(z,t)|^2)] \cdot A_j(z,t) \tag{9.41}$$

The equivalence of 9.40 and 9.41 suggests that DBP for K spans can be folded into a single span and K times the nonlinearity. This method is distance-folded DBP [22]. Assuming that the step size for the split-step implementation of DBP is unchanged, the computational load for the folded DBP can be saved by the folding factor K.

The above-mentioned derivation is based on the assumption that waveform distortion due to fiber nonlinearity and the residual dispersion per span (RDPS) are negligible, and consequently, the nonlinear behavior of the signal repeats itself in every span. This assumption is not exactly valid since, first, fiber nonlinearity also changes the waveform, and second, dispersion is not perfectly periodic if the RDPS is nonzero or the dispersion slope is not compensated. In order for the nonlinearity compensation to be accurate, it might be necessary to divide the entire long-haul transmission system into segments of multiple dispersion-managed spans so that the accumulated nonlinear effects and residual dispersion is small in each segment. Moreover, in order to minimize the error due to residual dispersion, folded DBP should be performed with a boundary condition calculated from lumped dispersion compensation for the first half of the segment. For a fiber link with $M \times K$ spans, the distance-folded DBP is illustrated in Figure 9.5.

A better way of solving disparities in dispersion characteristics in the amplified spans is the so-called dispersion-folded DBP [23]. The dispersion map of a typical dispersion-managed fiber transmission system with RDPS is illustrated in Figure 9.6. After the dispersion-managed fiber transmission and coherent detection, conventional DBP can be performed in the backward direction of the fiber propagation. Multiple steps are required for each of the many fiber spans, resulting in a large number of steps.

Dispersion-folded DBP exploits the fact that, under the weakly nonlinear assumption, the optical waveform repeats at locations where accumulated dispersions are identical. Since the Kerr nonlinear effects are determined by the instantaneous optical field, the nonlinear behavior of the optical signal also repeats at locations of identical

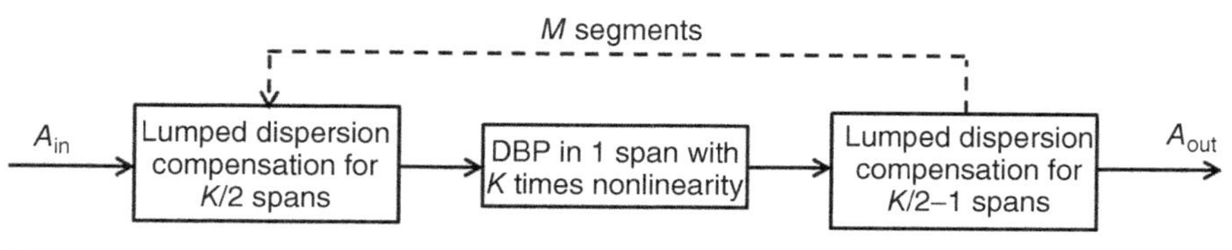

FIGURE 9.5 Distance-folded DBP for a periodically dispersion-managed fiber link with $M \times K$ spans.

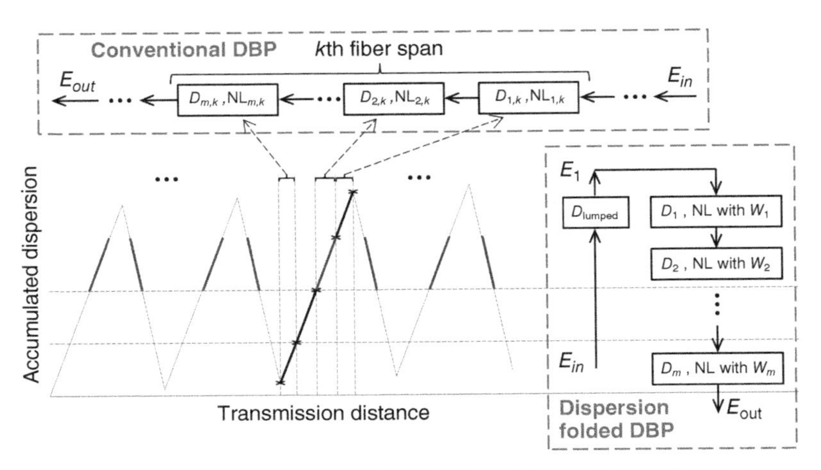

FIGURE 9.6 Conventional DBP and dispersion-folded DBP for a dispersion-managed coherent fiber link.

accumulated dispersion. Hence, it is possible to fold the DBP according to the accumulated dispersion [23].

The propagation of the optical field, $A(z, t)$, is governed by Equation 9.22. The solution of Equation 9.22 can be written as

$$A(z, t) = A_l(z, t) + \varepsilon \cdot A_{\text{nl}}(z, t) \tag{9.42}$$

Substituting Equation 9.42 into Equation 9.22, expanding the equation in power series of ε, and equating to zero the successive terms of the series yields

$$\frac{\partial A_l(z, t)}{\partial z} = D \cdot A_l(z, t) \tag{9.43}$$

$$\frac{\partial A_{\text{nl}}(z, t)}{\partial z} = D \cdot A_{\text{nl}}(z, t) + N(|A_l(z, t)|^2) \cdot A_l(z, t) \tag{9.44}$$

which describes the linear evolution and the nonlinear correction, respectively. It is noted that the nonlinear correction $A_{\text{nl}}(z, t)$ is governed by a linear partial differential equation with nonzero forcing, which depends on the linear solution only.

Therefore, as shown in Figure 9.6, the dispersion map can be divided into m divisions as indicated by the horizontal dashed lines. The fiber segments within a division have the same accumulated dispersion. Based on the principle of superposition, the total nonlinear correction is the sum of nonlinear corrections due to nonzero forcing at each fiber segment. In conventional DBP, the contribution from each fiber segment is computed separately. However, it is advantageous to calculate the total nonlinear correction as the sum of nonlinear corrections due to nonzero forcing at different accumulated dispersion divisions, each having multiple fiber segments. This is because, with the exception of different input power levels and effective lengths, the linear component $A_l(z, t)$ that generates the nonlinear correction and the total dispersion for the generated nonlinear perturbation to reach the end of the transmission are identical for the fiber segments with the same accumulated dispersion. Therefore, the nonlinear corrections due to these multiple fiber segments with the same accumulated dispersion are identical except a constant and can be calculated all at once using a weighting factor, as described later.

In dispersion-folded DBP, the fiber segments with the same accumulated dispersion (e.g., the thick gray lines) can be folded into one step. For a fiber link with positive RDPS, a lumped dispersion compensator (D_{lumped}) can be used to obtain the optical field (A_1) in the first dispersion division. Then, dispersion compensation (D) and nonlinearity compensation (NL) are performed for each of the subsequent dispersion divisions. To take into account the different power levels and effective lengths of the fiber segments, a weighting factor (W_i) is used in the nonlinearity compensator of each step. The nonlinear phase shift in the ith step of dispersion-folded DBP is given by $\varphi_i = W_i \cdot |\overline{A}_i(t)|^2$, where $\overline{A}_i(t)$ is the optical field with the power normalized to unity. The weighting factor is given by $W_i = \sum_k \gamma \int P_{i,k}(z)dz$, where $P_{i,k}(z)$ is the power level as a function of distance within the kth fiber segment in the ith dispersion division. The effect of loss for each fiber segment is taken into account in the calculation of this weighting factor.

Nonlinear compensation for multiple fiber segments with the same accumulated dispersion is performed in a single step in dispersion-folded DBP, resulting in orders-of-magnitude savings in computation. In the SSM, the linear and nonlinear effects can be decoupled when the step size is small enough. The dispersion within a fiber segment is neglected in a nonlinearity compensation operator. Meanwhile, the power level, effective length, and nonlinear coefficient of a fiber segment have been taken into account in the calculation of the nonlinear phase shift. Thus, the DBP can be folded even if the fiber link consists of multiple types of fibers. Note that calculating the weighting factors does not require real-time computation.

To demonstrate the effectiveness of the dispersion-folded DBP, single-channel 6084 km transmission of NRZ-QPSK signal at 10 Gbaud was demonstrated in [23]. The experimental setup is shown in Figure 9.7(a). At the transmitter, carrier from an external-cavity laser is modulated by a QPSK modulator using a $2^{23} - 1$ PRBS. The optical signal is launched into a recirculating loop controlled by two acousto-optic modulators (AOMs). The recirculating loop consists of two types of fibers: 82.6 km SSMF with 0.2 dB/km loss and 70 μm^2 effective area, and 11 km DCF with 0.46 dB/km loss and 20 μm^2 effective area. By optimizing the performance in the training experiments of EDC, the dispersion of the SSMF and the DCF are determined as 17.06 ps/nm/km and −123.35 ps/nm/km, respectively. The RDPS is 53 ps/nm. Two EDFAs are used to completely compensate for the loss in the loop. An optical band-pass filter (BPF) is used to suppress the EDFA noise. At the receiver, the signal is mixed with the local oscillator from another external cavity laser in a 90° hybrid. The I and Q components of the received signal are detected using two photodetectors (PDs). A RTO is used for analog-to-digital conversion and

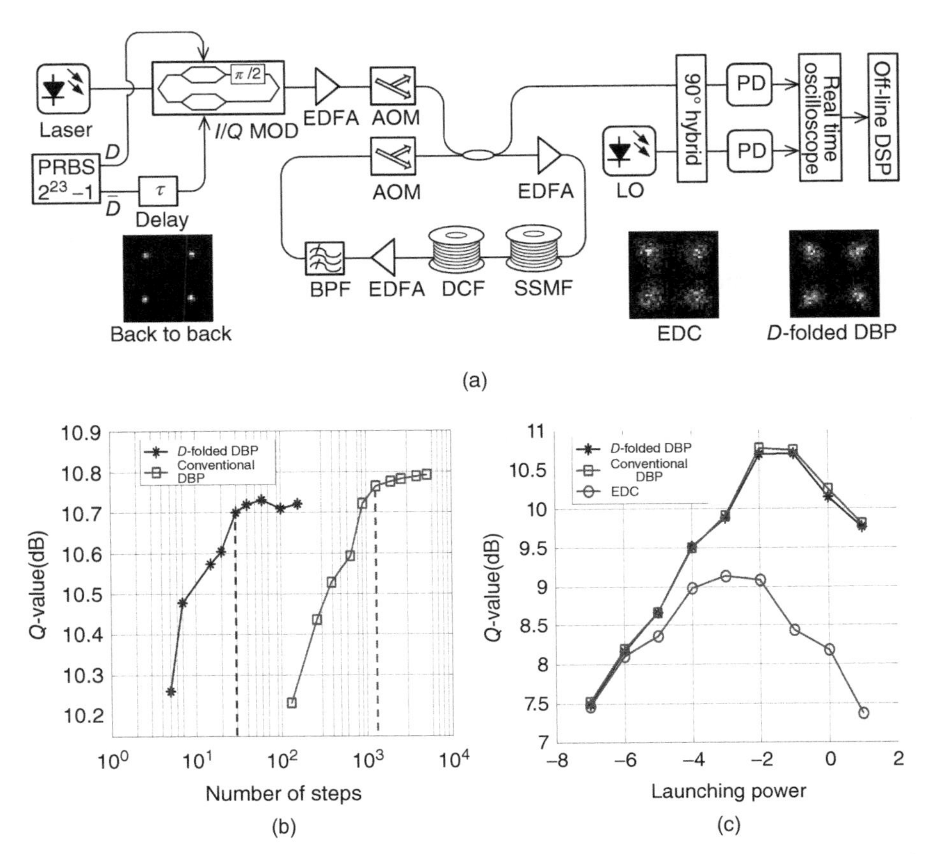

FIGURE 9.7 Experimental demonstration of dispersion-folded DBP. (a) Experimental setup. Inset: constellations after back-to-back detection, EDC and DBP at the corresponding optimum power levels. (b) Q-value as a function of the number of steps using conventional DBP (line with square) and dispersion-folded DBP (line with asterisk). (c) Q-value as a function of optical launching power after EDC (line with circle), 30-step dispersion-folded DBP (line with asterisk) and 1300-step conventional DBP (line with square).

data acquisition at 40 Gsamples/s. The DSP is performed offline with Matlab. Note that for realistic terrestrial systems, the parameters of the fiber spans may not be available with good accuracy. It is expected that the parameters for dispersion-folded DBP can also be obtained via adaptive optimization.

For long-haul transmission, the DBP step size is usually limited by dispersion [20]. In this experiment, DBP steps with equal dispersion per step were used for simplicity. The Q-value as a function of the number of steps is shown in Figure 9.7(b). The required number of steps to approach the maximum Q-value can be reduced from 1300 to 30 by using the dispersion-folded DBP. The number of multiplications per sample for DBP is reduced by a factor of 43.

Figure 9.7(c) shows the Q-value as a function of the launching power. With only EDC for the accumulated residual dispersion, the maximum Q-value is 9.1 dB. With nonlinearity compensation using dispersion-folded DBP, the maximum Q-value is increased to 10.7 dB. The performance after the 30-step dispersion-folded DBP is almost the same as that after the 1300-step conventional DBP. There is a trade-off between complexity and performance using either conventional DBP or dispersion-folded DBP. A Q-value of 10.2 dB, corresponding to a 1.1 dB improvement in comparison with EDC, can be achieved using 130-step conventional DBP or 5-step dispersion-folded DBP.

9.5 DBP FOR POLARIZATION-MULTIPLEXED TRANSMISSION

In many WDM transmission systems, not all WDM channels have the same state of polarization, such as in the case of polarization multiplexing or polarization interleaving. In those systems, the effects of polarization on the nonlinear interactions have to be taken into account. Whether the total electric field is polarized or not, its propagation in a fiber can be described by the vectorial form of the NLSE [11]:

$$\begin{aligned} \frac{\partial A_x}{\partial z} &= ib_x(z)A_x - \frac{\alpha}{2}A_x + \frac{i\beta_2}{2}\frac{\partial^2 A_x}{\partial t^2} + i\gamma\left(\left|A_x\right|^2 + \frac{2}{3}|A_y|^2\right)A_x + \frac{i\gamma}{3}A_x^* A_y^2 \\ \frac{\partial A_y}{\partial z} &= -ib_y(z)A_y - \frac{\alpha}{2}A_y + \frac{i\beta_2}{2}\frac{\partial^2 A_y}{\partial t^2} + i\gamma\left(\left|A_y\right|^2 + \frac{2}{3}|A_x|^2\right)A_y + \frac{i\gamma}{3}A_y^* A_x^2 \end{aligned} \quad (9.45)$$

where A_x and A_y are the two orthogonal polarization components of the electric field, b_x and b_y represent linear birefringence of the fiber. According to Equation 9.45, the strength of the nonlinear processes, SPM, XPM, and FWM depends not only on the relative orientations of different channels, but also on the state of polarizations. For instance, a channel polarized linearly accumulates more nonlinear phase due to SPM than another channel polarized elliptically [24, 25].

Optical transmission fibers are nominally not birefringent; however, they still exhibit the so-called residual birefringence that randomly scatters the polarization of the electric field in length scales less than 100~m [26]. This polarization scattering length is much smaller than the nonlinear interaction length, which is typically tens of kilometers. Since the polarization state of the electric field changes so fast, the resulting nonlinearity is not what is expected from a linearly or circularly polarized field, but an average over the entire Poincaré sphere. By substituting the polarization-rotating transformation

$$\begin{bmatrix} A_x' \\ A_y' \end{bmatrix} = \begin{bmatrix} \cos\theta & \sin\theta e^{i\phi} \\ -\sin\theta e^{-i\phi} & \cos\theta \end{bmatrix} \begin{bmatrix} A_x \\ A_y \end{bmatrix} \quad (9.46)$$

into Equation 9.45 and averaging over θ and ϕ, the nonlinear effects are averaged over the fast polarization changes resulting in the Manakov equation given by Agrawal

[11], Marcuse et al. [26], and Wai et al. [27]

$$\frac{\partial A_x}{\partial z} = -\frac{\alpha}{2}A_x + \frac{i\beta_2}{2}\frac{\partial^2 A_x}{\partial t^2} + \frac{8i\gamma}{9}(|A_x|^2 + |A_y|^2)A_x$$
$$\frac{\partial A_y}{\partial z} = -\frac{\alpha}{2}A_y + \frac{i\beta_2}{2}\frac{\partial^2 A_y}{\partial t^2} + \frac{8i\gamma}{9}(|A_y|^2 + |A_x|^2)A_y \tag{9.47}$$

The Manakov equation is simpler than the full vectorial NLS. Since the fast polarization rotations are averaged already, it is not necessary to follow these changes in the fiber. Moreover, since polarization changes are so fast and random, it does not matter anymore what the input polarization is. For instance, if only two copolarized channels propagate through the fiber, it does not matter whether both channels have linear polarization or circular polarization and the accumulated nonlinearity will be the same at the end of the fiber. However, the strength of the nonlinear interaction still depends on the relative orientations of the channels. If the channels have the same polarization, they interact more strongly than if they have orthogonal polarizations. The Manakov equation as given in Equation 9.47 assumes that the relative orientations of the polarizations of different channels remain the same throughout the fiber. This assumption is true as long as the bandwidth of the total field is narrow enough so that polarization-mode dispersion (PMD) can be ignored.

A consequence of the fast and random polarization rotations in the fiber is that, at the receiver, the electric field is rotated with respect to the transmitter. To demultiplex the orthogonal channels properly, this random rotation has to be corrected. As these random rotations are slow, several electronic polarization demultiplexing methods based on digital signal processing have been devised to track these rotations and correct them [28, 29]. However, most of these methods are either data-aided or Q-value directed and, therefore, they rely on high signal-to-noise ratio. Because of the linear and nonlinear impairments, the signal at the receiver end may be significantly distorted. This may make it difficult to separate the polarization-multiplexed channels using the data-aided or Q-value directed methods. However, a closer look at the Manakov equation shows that it is not necessary to know on what polarization basis the data are encoded to implement backward propagation. This can be observed easily by verifying that the Manakov equation remains the same under the unitary transformation:

$$B = \mathbf{U}A \tag{9.48}$$

where $\mathbf{U}$ is an arbitrary 2×2 unitary polarization-rotation matrix and $A = [A_x, A_y]^T$. Therefore, DBP can be applied first and the electronic polarization demultiplexing techniques can be used subsequently to demultiplex the polarization channels correctly.

9.6 FUTURE RESEARCH

Even though tremendous progress has been made to understand the techniques of optical fiber nonlinearity compensation including DBP – the focus of this chapter – there is still much to be desired before DBP becomes commercially indispensable. The two main reasons are computational load and reliability. In terms of computational load, DBP is still too intense for dispersion-unmanaged transmission systems. In terms of reliability, DBP for transmission systems with large PMD becomes ineffective. Future research must address these two main issues. It is possible to track the effect of PMD in real time and incorporate the channel state of PMD into DBP [30].

In terms of computational load, it might be worthwhile to consider dispersion-folded DBP not only for installed dispersion-managed systems but also for future transmission systems. It is possible for a dispersion-managed transmission system using dispersion-folded DBP to outperform a dispersion-unmanaged transmission system using electronic dispersion compensation, both of which are within the realm of current DSP capability. This is true especially if the low-loss of dispersion compensation modules can be further improved.

REFERENCES

1. Liu X, Wei X, Slusher RE, McKinstrie CJ. Improving transmission performance in differential phase-shift-keyed systems by use of lumped nonlinear phase-shift compensation. *Opt Lett* 2002;27:1616–1618.
2. Watanabe S, Shirasaki M. Exact compensation for both chromatic dispersion and Kerr effect in a transmission fiber using optical phase conjugation. *J Lightwave Technol* 1996;14:243–248.
3. Noe R. Phase noise-tolerant synchronous QPSK/BPSK baseband-type intradyne receiver concept with feedforward carrier recovery. *J Lightwave Technol* 2005;23: 802–808.
4. Dischler R, Buchali F. Transmission of 1.2 Tb/s Continuous Waveband PDM-OFDM-FDM Signal with Spectral Efficiency of 3.3 bit/s/Hz over 400 km of SSMF. Optical Fiber Communication Conference and National Fiber Optic Engineers Conference; Optical Society of America, San Diego, California; 2009. p PDPC2.
5. Ellis AD, Gunning FCG. Spectral density enhancement using coherent WDM. *IEEE Photonics Technol Lett* 2005;17:504–506.
6. Qi Y, Kaneda N, Xiang L, Chandrasekhar S, Shieh W, Chen YK. Real-time coherent optical OFDM receiver at 2.5-GS/s for receiving a 54-Gb/s multi-band signal. Conference on Optical Fiber Communication – incudes post deadline papers, 2009; 2009. p 1–3.
7. Takachio N, Iwashita K. Compensation of fibre chromatic dispersion in optical heterodyne detection. *Electron Lett* 1988;24:108–109.
8. Winters JH, Gitlin RD. Electrical signal processing techniques in long-haul, fiber-optic systems. IEEE International Conference on Communications, 1990. ICC '90, Including Supercomm Technical Sessions. SUPERCOMM/ICC '90. Conference Record., vol. 392; 1990. p 397–403.
9. Taylor MG. Coherent detection method using DSP for demodulation of signal and subsequent equalization of propagation impairments. *IEEE Photonics Technol Lett* 2004;16:674–676.

10. Du LBY, Lowery AJ. Fiber nonlinearity precompensation for long-haul links using direct-detection optical OFDM. *Opt Express* 2008;16:6209–6215.
11. Agrawal GP. Nonlinear Fiber Optics. Academic Press; 2007.
12. Sinkin OV, Holzlöhner R, Zweck J, Menyuk CR. Optimization of the Split-Step Fourier Method in Modeling Optical-Fiber Communications Systems. *J Lightwave Technol* 2003;21:61.
13. Goldfarb G, Taylor MG, Guifang L. Experimental demonstration of fiber impairment compensation using the split-step finite-impulse-response filtering method. *IEEE Photonics Technol Lett* 2008;20:1887–1889.
14. Goldfarb G, Li G, Taylor MG. Orthogonal wavelength-division multiplexing using coherent detection. *IEEE Photonics Technol Lett* 2007;19:2015–2017.
15. Mateo E, Zhu L, Li G. Impact of XPM and FWM on the digital implementation of impairment compensation for WDM transmission using backward propagation. *Opt Express* 2008;16:16124–16137.
16. Mateo EF, Yaman F, Li G. Efficient compensation of inter-channel nonlinear effects via digital backward propagation in WDM optical transmission. *Opt Express* 2010;18:15144–15154.
17. Lei L, Zhenning T, Liang D, Weizhen Y, Oda S, Tanimura T, Hoshida T, Rasmussen JC. Implementation efficient nonlinear equalizer based on correlated digital backpropagation. Optical Fiber Communication Conference and Exposition (OFC/NFOEC), 2011 and the National Fiber Optic Engineers Conference; 2011. p 1–3.
18. Du LB, Lowery AJ. Improved single channel backpropagation for intra-channel fiber nonlinearity compensation in long-haul optical communication systems. *Opt Express* 2010;18:17075–17088.
19. Liang Bangyuan D, Lowery AJ. Practical XPM compensation method for coherent optical OFDM systems. *IEEE Photonics Technol Lett* 2010;22:320–322.
20. Zhenning T, Liang D, Weizhen Y, Lei L, Hoshida T, Rasmussen JC. Multiplier-free intrachannel nonlinearity compensating algorithm operating at symbol rate. *J Lightwave Technol* 2011;29:2570–2576.
21. Gordon JP, Mollenauer LF. Phase noise in photonic communications systems using linear amplifiers. *Opt Lett* 1990;15:1351–1353.
22. Zhu L, Li G. Folded digital backward propagation for dispersion-managed fiber-optic transmission. *Opt Express* 2011;19:5953–5959.
23. Zhu L, Li G. Nonlinearity compensation using dispersion-folded digital backward propagation. *Opt Express* 2012;20:14362–14370.
24. Qiang L, Agrawal GP. Vector theory of cross-phase modulation: role of nonlinear polarization rotation. *IEEE J Quantum Electron* 2004;40:958–964.
25. Karlsson M, Sunnerud H. Effects of nonlinearities on PMD-induced system impairments. *J Lightwave Technol* 2006;24:4127–4137.
26. Marcuse D, Manyuk CR, Wai PKA. Application of the Manakov-PMD equation to studies of signal propagation in optical fibers with randomly varying birefringence. *J Lightwave Technol* 1997;15:1735–1746.
27. Wai PKA, Kath WL, Menyuk CR, Zhang JW. Nonlinear polarization-mode dispersion in optical fibers with randomly varying birefringence. *J Opt Soc Am B* 1997;14: 2967–2979.

28. Han Y, Li G. Coherent optical communication using polarization multiple-input-multiple-output. *Opt Express* 2005;13:7527–7534.
29. Zhang H, Tao Z, Liu L, Oda S, Hoshida T, Rasmussen JC. Polarization demultiplexing based on independent component analysis in optical coherent receivers. 34th European Conference on Optical Communication, 2008; 2008. p 1–2.
30. Yaman F, Guifang L. Nonlinear impairment compensation for polarization-division multiplexed WDM transmission using digital backward propagation. *IEEE Photonics J* 2010;2:816–832.

10

TIMING SYNCHRONIZATION IN COHERENT OPTICAL TRANSMISSION SYSTEMS

HAN SUN AND KUANG-TSAN WU
Infinera, Ottawa, Canada

10.1 INTRODUCTION

A fundamental building block of modern coherent optical transport system is the timing recovery or timing synchronization circuit. Recovering the transmitted clock from the received signal is a first step in recovering the data. Only when the receive-side VCO (voltage-controlled oscillator) is phase-locked to the transmit-side VCO, the other DSP functions such as equalization and carrier recovery can commence. A typical receiver acquisition sequence will start with locking the receive VCO, followed by blind equalization (such as the constant modulus algorithm, CMA), and then finally carrier phase recovery using equalized data. After system acquisition, the timing synchronization circuit must operate continuously and robustly as long as a valid signal is present at the input to the receiver. Any slip of the recovered phase or momentary loss of lock will result in catastrophic data failure and reacquisition of the equalizer and other control loops that depend on synchronous data. In a connected network, it triggers reframe of multiple nodes downstream, and enormous amount of data is lost. For this reason, timing synchronization is often designed to be the most robust control loop in the receiver.

Timing synchronization is a topic that has been studied extensively in almost all areas of digital transmission. There is an abundance of literature on topics related to

Enabling Technologies for High Spectral-efficiency Coherent Optical Communication Networks, First Edition. Edited by Xiang Zhou and Chongjin Xie.
© 2016 John Wiley & Sons, Inc. Published 2016 by John Wiley & Sons, Inc.

phase detectors and phase lock loops stemming from early developments in voice-band digital modems [1–12]. Yet this fundamental building block is studied again and again in every corner of digital transmission, and in every generation of new technology and design. The reason is that there can be many variables that impact the design of timing synchronization, such as the type of signal and the type of distortion that the signal experiences as it goes through the communication channel; the amount of noise that is present; the sources of jitter and jitter tracking requirements; the environment in which it is implemented; the spectral occupancy of the signal; and the modulation format. These factors are all important in that they affect not only the design of timing recovery, but also the resulting robustness and jitter performance.

Fiber-optic transmission systems have evolved tremendously since the late 1990s. Modulation formats since then have evolved from intensity modulation and direct detection (IMDD) to QPSK format with differential detection [13], and then to polarization-multiplexed QPSK (Pol-Mux QPSK) and polarization-diverse coherent detection [14–19]. The first commercialization of Pol-Mux QPSK and coherent detection was at 40 Gbit/s data rate and was reported in 2008 [20]. Compared with IMDD systems, the modern coherent receiver receives signals of a much different quality. For instance, the signal received on each polarization is a linear combination of two independently modulated signals. Within each received polarization, the interference from the other polarization is just as strong as the signal itself. In addition, the signal is also dispersed by large amount of chromatic dispersion (CD). Because of the paradigm change in the method of modulation and detection, the timing synchronization block is impacted in a fundamental way. Timing synchronization methods developed in IMDD systems cannot be directly applied to modern coherent systems.

The goal of this chapter is to describe some of the methods and techniques used to recover timing in modern coherent systems. To do so, each section of this chapter is devoted to a particular aspect of timing recovery. The first section describes the overall system environment. The next section is devoted to jitter penalty and jitter sources in coherent optical systems, with particular emphasis on jitter generated from FM noise of the laser and digital dispersion compensation. The next section describes different types of phase detectors and their jitter performances. As described elsewhere in this book, propagation in the single-mode fiber produces two important distortion effects that require significant amount of effort in equalization, namely chromatic dispersion (CD) and polarization-mode dispersion (PMD). These distortions are also harmful to phase detectors. Their effects and their remedy are discussed in the last sections of this chapter.

Timing recovery is usually implemented as feedback second-order phase-locked loops (PLLs). The principles of PLLs are not covered in this chapter. The principle is the same for all feedback control loops. The receive VCO is assumed to have a tuning port for feedback tuning of the phase. The second-order loop filter is a standard component that does not depend on input signal properties. The reader is referred to prior literature on the topic of second-order PLLs [1, 2, 7, 8].

10.2 OVERALL SYSTEM ENVIRONMENT

The overall system diagram is shown in Figure 10.1. On the left side, the Tx digital signal processor (DSP) filters the input data and up-samples the data to a sampling rate determined by the D/A (digital-to-analog) converter. Four D/A converters drive a pair of *I*/*Q* modulators that modulates the in-phase and, quadrature components of both TE and TM polarizations of the light. Note that the D/A converters at the Tx are clocked using a free running oscillator (Tx VCO). The VCO is typically very high frequency (e.g., 16 GHz) in order to meet tight jitter requirements dictated by the high sampling rate of the D/A and A/D converters (e.g., 64 GSamples/s [21, 22], and 90 GSamples/s [23]). Since the digital circuits implemented in the Tx DSP can only operate at 100s of MHz (e.g., 500 MHz), the VCO output is divided down by an integer ratio N (e.g., 16G/32 = 500 M) to be used to clock all the digital circuits. Essentially, the RF signal output of the D/A converter and hence the optical signal is modulated at a symbol rate (a.k.a. baud rate) that is synchronous with the frequency of the Tx VCO. The symbol rate is exactly defined by the frequency of the Tx VCO scaled by a constant integer factor. At 32G symbol rate and 16 GHz VCO, that integer factor is 2. Therefore, any jitter or frequency variation of the Tx VCO is directly modulated onto the light going out of the transmitter as baseband delay. Instantaneously, if the Tx VCO phase is advanced or retarded by 1 ps, then the modulated envelop carried by the 192 THz optical carrier is also advanced or retarded by 1 ps. The characteristics of this time-varying phase wander of the VCO will be discussed in more detail in the next section. Once down converted to baseband by a coherent receiver, the advancement or retardation of the VCO manifests as a positive or negative delay on the baseband electrical signal. If the delay is significantly large compared with the symbol period, then significant distortion results at the decision device and system performance will degrade. The function of the clock recovery in the Rx side is to track out this phase variation of the Tx VCO by high-speed tuning of the Rx VCO. However, the remaining jitter still has system impact, and this is also discussed in detail in the next section.

On the Rx side, a local oscillator (LO) beats with the input signal to produce in-phase and quadrature components of the TE, and TM polarizations of the received light. These four components are sampled by four A/D (analog-to-digital) converters. They feed the sampled data to an Rx DSP circuit that will filter the input signal to perform functions such as equalization and carrier recovery. Somewhere in the signal processing chain, a phase detector uses the sampled data and produces a digital output that is proportional to the delay of the input signal. The delay or timing error feeds a loop filter that typically implements a second-order PLL. The loop filter output drives the tuning port of the Rx VCO. Positive and negative changes to the input of the Rx VCO directly translate to advancement or retardation of the sampling phase of the A/D converters and that affects the delay of the digital signal input to the Rx DSP. This forms a feedback loop.

In previous generations of IMDD transmission systems where the signal arrives in the receiver with minimal distortion, the phase detector and clock recovery are

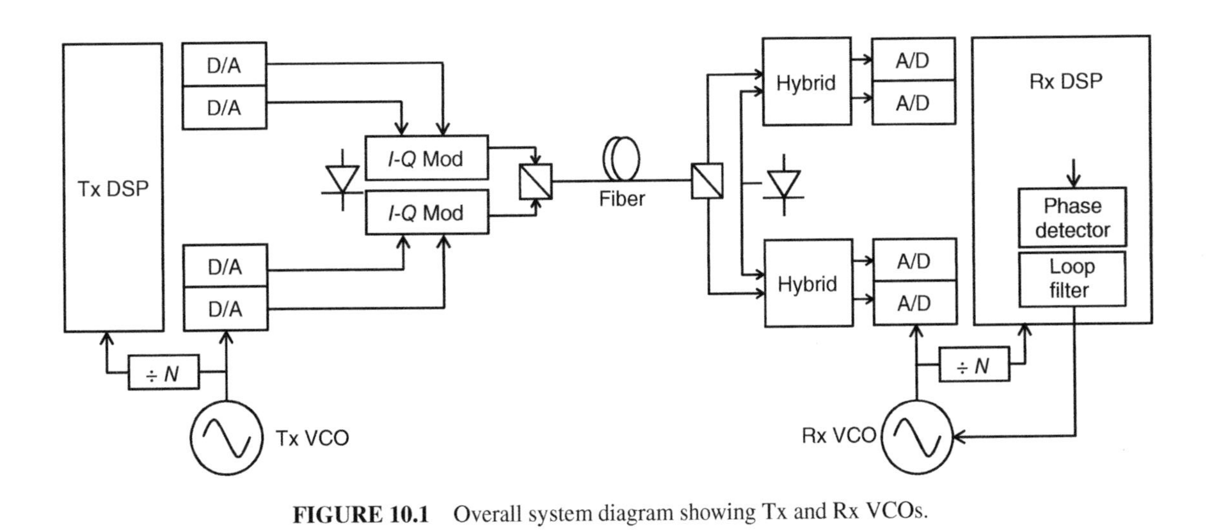

FIGURE 10.1 Overall system diagram showing Tx and Rx VCOs.

implemented as analog components. In contrast, modern coherent receiver uses A/D converters and digital signal processing, phase detectors and loop filter components are all digital implementations.

10.3 JITTER PENALTY AND JITTER SOURCES IN A COHERENT SYSTEM

10.3.1 VCO Jitter

Like in most digital transmission systems, the first and foremost source of jitter is the VCO themselves. The clocking architecture shown in Figures 10.1 and 10.2 is representative of state-of-the-art transmission system at 100G rate. The Rx VCO component in Figure 10.1 is actually a PLL shown in Figure 10.2. The internal VCO is designed to naturally oscillate at 16 GHz; however, temperature and CMOS process variations cause the VCO to power up with very large frequency error. This frequency error can be as large as ±10%, making clock recovery very difficult. The solution is to provide an external crystal oscillator as a stable reference. In Figure 10.2, the crystal reference is at 2 GHz shown on the bottom left as an input signal. The 16G VCO is always phase-locked to the 2 GHz reference using a standard phase detector and loop filters. A similar PLL is also used in the transmitter to generate a stable 16 GHz clock. Since each crystal reference in the Tx and Rx is very stable within tens of PPM over life, locking the Rx VCO to the Tx VCO requires a pull-in range of only tens

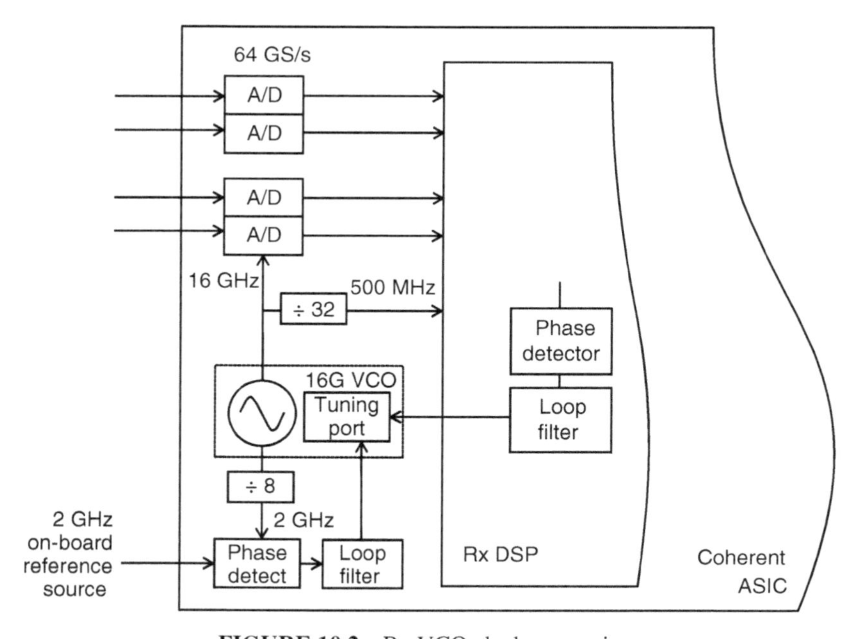

FIGURE 10.2 Rx VCO clock generation.

of PPM. This greatly simplifies the task of clock phase synchronization. Typically, a tuning port is provided, so that the VCO can be tuned away from its lock point by the control of the phase detector and loop filter components inside the Rx DSP. The VCO, PLL, and A/D converters are analog components that are built inside a coherent receiver ASIC.

The jitter of the Rx VCO is derived from the phase noise spectrum of the Rx VCO. The spectral purity of the Rx VCO can be defined as the phase noise spectrum of the 16 GHz output clock measured in a spectrum analyzer, while the VCO is phase-locked to the 2 GHz reference, and with no tuning from the DSP. The 16G output clock directly drives the sampling aperture of the A/D converter and, therefore, directly impacts system performance. Let the VCO center frequency be f_0, and let the phase noise of the VCO be modeled as a single sinusoidal phase modulation of amplitude α in radians "$\phi(t)$." The voltage of the VCO can be expressed as $V(t)$.

$$\begin{aligned}\phi(t) &= \alpha \cdot \sin(2 \cdot \pi \cdot f_m \cdot t)\\ V(t) &= \sin(2 \cdot \pi \cdot f_0 \cdot t + \phi(t))\\ &= \sin(2 \cdot \pi \cdot f_0 \cdot t) \cdot \cos(\phi(t)) + \cos(2 \cdot \pi \cdot f_0 \cdot t) \cdot \sin(\phi(t))\end{aligned} \tag{10.1}$$

If the phase modulation $\phi(t)$ is small, then the $V(t)$ can be simplified ($\cos(\phi(t)) = 1$, $\sin(\phi(t)) = \phi(t)$), and using trigonometric identities, $V(t)$ can be written as the following three terms that correspond to three spectral lines.

$$\begin{aligned}V(t) &\approx \sin(2 \cdot \pi \cdot f_0 \cdot t) + \alpha \cdot \cos(2 \cdot \pi \cdot f_0 \cdot t) \cdot \sin(2 \cdot \pi \cdot f_m \cdot t)\\ &\approx \sin(2 \cdot \pi \cdot f_0 \cdot t) + \frac{\alpha}{2} \cdot \sin(2 \cdot \pi \cdot (f_m - f_0) \cdot t) + \frac{\alpha}{2} \cdot \sin(2 \cdot \pi \cdot (f_m + f_0) \cdot t)\end{aligned} \tag{10.2}$$

Spectrally, the phase modulation generates two tones at $\pm f_m$ from the center frequency f_0. When integrated in a spectrum analyzer, the two side tones sum into a single tone at f_m away from center frequency f_0. The ratio of the modulation at f_m compared with the carrier at f_0 is defined as the carrier to modulation ratio expressed in dBC, which is commonly known as the single-sided phase noise spectrum. This quantity is shown in Equation 10.3 by summing the powers of the two modulation tones divided by the power of the carrier.

$$\text{dBC} = 10 \cdot \log_{10}\left(\frac{0.5 \cdot \alpha^2/4 + 0.5 \cdot \alpha^2/4}{0.5}\right) = 10 \cdot \log_{10}\left(\frac{\alpha^2}{2}\right) \tag{10.3}$$

The phase modulation in units of radians RMS can be expressed as a function of dBC.

$$\text{Jitter [radians RMS]} = \alpha/\sqrt{2} = \sqrt{10^{\text{dBC}/10}} \tag{10.4}$$

The absolute jitter in picosecond RMS can be defined in terms of α, and by scaling of the center frequency f_0.

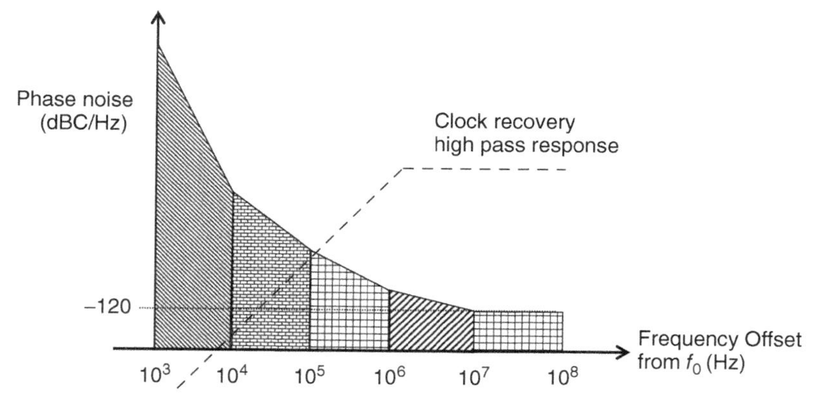

FIGURE 10.3 Rx VCO phase noise spectrum.

$$\text{Jitter [ps RMS]} = \frac{\sqrt{10^{\text{dBC}/10}}}{2 \cdot \pi \cdot f_0} \times 10^{12} \tag{10.5}$$

An illustrative phase noise spectrum of a VCO is shown in Figure 10.3 in units of dBC/Hz versus frequency offset from center frequency f_0. To calculate the total jitter between 10 and 100 MHz, the above-mentioned equation is used with the phase noise density at −120 dBC/Hz integrated over 90 MHz of frequency using center frequency f_0 of 16 GHz. This gives 0.0944 ps RMS of jitter in that part of the spectrum.

Overlaid on Figure 10.3 is a dotted line that represents the high-pass filtering function of a clock recovery loop operating at about 1 MHz loop bandwidth (BW). The clock recovery loop essentially operates as a high-pass filter, rejecting phase noise at frequencies slower than the loop BW. The VCO phase noise spectrum integrated over the first-order high-pass function represents the untracked jitter, and this directly impacts system performance.

In order to minimize this jitter source, the clock recovery loop bandwidth needs to be as high as possible. Aside from the implementation difficulties associated with a very high loop BW, the power of the phase detector jitter (or self-noise) increases linearly with loop BW, and therefore limiting the loop BW to a range of a few megahertz. Phase detectors and the jitter that they produce are discussed in a later section. The overall minimum jitter and optimum loop BW is achieved when the untracked VCO jitter balances out the detector jitter. One should note that in practice, the operating point may not be at the optimum due to other reasons such as round-trip loop latency that may limit the loop BW on the high side, and specific phase transient tolerance may limit the loop BW on the low side.

10.3.2 Detector Jitter Definitions and Method of Numerical Evaluation

The performance of a phase detector is directly defined by the jitter that it produces. The source of this noise is inherent in the data itself. The modulated data itself is random, with phase error imparted on the modulated signal. The detector must estimate the phase error in the presence of modulated data, which produces estimation noise. Theoretical analysis and derivation of detector noise in some detectors can be done resulting in analytical solutions, but in more complicated detectors, or with signals that have distortion, analytical methods become very difficult. Often one resorts to simulation to numerically evaluate the amount of noise the detector produces. This section is devoted to the methodology used in comparing two different types of phase detectors.

The detector output at a given delay error "τ_{err}" can be written as shown in Equation 10.6, where the symbol time interval is denoted as "τ_{UI}", detector strength is denoted as "K_d", and detector noise "$n[i]$" can be modeled as additive white Gaussian with zero mean and variance σ^2, and "i" denotes time index. The time between successive evaluations of the phase detector output is the inverse of phase detector rate "R_{PD}." Here, we assume a detector having sinusoidal output with period equal to one symbol time interval. For 32 Gbaud symbol rate, τ_{UI} is 31.3 ps. In some other detectors, the detector period is half of τ_{UI}, in which case it needs to be factored into calculation.

$$P[i]|_{\tau_{err}} = K_d \cdot \sin(2 \cdot \pi \cdot \tau_{err}/\tau_{UI}) + n[i] \tag{10.6}$$

The expected value of P is plotted on the vertical axis in Figure 10.4. What is relevant to the detector performance is the ratio of the power of the detector noise $\langle n^2 \rangle = \sigma^2$ compared with the detector strength K_d. Furthermore, to calculate how much detector jitter will be present with closed-loop BW (BW_{Loop}) of 1 MHz, one must also factor in the rate at which the random variable $P[i]$ is generated (R_{PD}).

The amount of detector jitter present when the PLL is in closed loop can be approximated as follows [3]:

$$\text{Jitter [ps RMS]} = \frac{\sigma}{K''_d} \cdot \sqrt{\frac{BW_{LOOP}}{R_{PD}}} \tag{10.7}$$

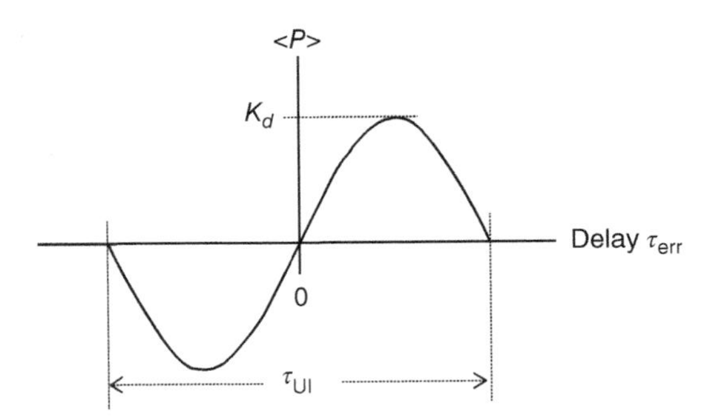

FIGURE 10.4 Typical detector sensitivity curve.

K_d'' is defined as the slope of the sine function in Figure 10.4 at $\tau_{\mathrm{err}} = 0$, which is the lock point of the PLL. Taking derivative of the sine function, K_d'' can be defined as a function of K_d and τ_{UI} [ps] in Equation 10.8.

$$K_d'' = \frac{2 \cdot \pi}{\tau_{\mathrm{UI}}} \cdot K_d \tag{10.8}$$

The equation shows that as loop BW is reduced by a factor of 2, the detector jitter standard deviation is reduced by $\sqrt{2}$. When comparing different detectors, the jitter value must be normalized to a common closed-loop BW by taking into account the rate of the phase detector. For example, frequency-domain phase detector output is generated every FFT clock cycle, which is in hundreds of megahertz, whereas a time-domain phase detector output can be generated every symbol, and so R_{PD} are very different between the two. Therefore, it is important to normalize the jitter to a closed-loop BW of 1 MHz as an example.

10.3.3 Laser FM Noise- and Dispersion-Induced Jitter

In a coherent receiver where a large amount of chromatic dispersion (CD) is compensated digitally, there is an additional timing jitter source that is unique to the coherent receiver and did not exist in traditional direct-detection systems. It is related to the convolution of FM noise of the local laser (LO) and the amount of CD compensated in the digital receiver. Consider the system diagram in Figure 10.5, where the Rx (LO) laser is offset from the Tx laser by an amount of frequency ω_0. Dispersion in the fiber is perfectly compensated by the digital dispersion filter in the receiver. In a process that is discussed later, this produces a measurable delay on the signal at the output of the dispersion compensation filter. Note that if ω_0 is absent in the model, then the digital compensation is perfect. However, laser always has instantaneous frequency variations characterized by a linewidth parameter as well as low-frequency FM noise. "ω_0" is in general a stochastic quantity unknown at the receiver. The presence of Rx laser phase noise convolved with digital dispersion compensation produces a well-documented noise source [24]. The effect results in bit error rate degradation in a coherent transmission system. Here we consider another aspect of the same problem, reflected as a new stochastic jitter source for coherent systems.

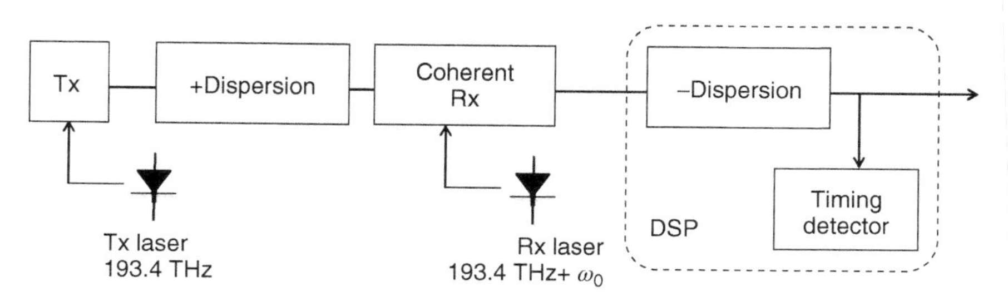

FIGURE 10.5 System diagram illustrating Rx laser and dispersion compensation.

Consider the frequency-domain representation of the received digital signal after A/D converter as written in the following equation. The transmitted signal is represented as $H(\omega)$. The received signal contains a static LO frequency offset (ω_0). The fiber dispersion is a group delay distortion and is modeled as multiplication in frequency domain by a parabolic phase with coefficient "κ."

$$H(\omega - \omega_0) \cdot e^{j \cdot \kappa \cdot (\omega - \omega_0)^2} \tag{10.9}$$

The coefficient "κ" is related to the fiber dispersion coefficient D (ps/nm/km), fiber distance L (km), wavelength of the light λ (nm), and speed of light c (m/s) according to the following:

$$\kappa = \frac{D \cdot L \cdot \lambda^2}{4 \cdot \pi \cdot c \cdot 10^{21}} \tag{10.10}$$

The carrier recovery function in the receiver can extract instantaneous frequency that is imparted on the signal, but it cannot separate the Rx laser contribution from the Tx laser contribution. According to Figure 10.5, Rx laser frequency variations are the source of the problem. Therefore, practically, the only choice for the receiver is to compensate for dispersion assuming a frequency offset of zero. The following equation describes the signal after dispersion compensation.

$$\underbrace{H(\omega - \omega_0) \cdot e^{j \cdot \kappa \cdot (\omega - \omega_0)^2}}_{\text{Received Signal}} \cdot \underbrace{e^{-j \cdot \kappa \cdot (\omega)^2}}_{\text{Dispersion compensation}} = H(\omega - \omega_0) \cdot \underbrace{e^{-j \cdot (2 \cdot \kappa \cdot \omega_0) \cdot \omega}}_{\text{Induced delay or jitter}} \cdot e^{j \cdot \kappa \cdot \omega_0^2} \tag{10.11}$$

The group delay distortion is removed, but the resultant signal is frequency shifted by ω_0 (to be compensated by carrier recovery), a phase offset term $\kappa \cdot \omega_0{}^2$ (also compensated by carrier recovery), and a delay term remaining that is a linear function of both the amount of frequency offset and the dispersion ($\tau = 2 \cdot \kappa \cdot \omega_0$). As the frequency of the LO (centered at 193.4 THz) is slowly varying, this delay is also time varying. At 50,000 ps/nm of compensation, the conversion from FM noise to timing jitter has efficiency of ~0.4 ps/MHz. If the frequency of the LO varies up to ±50 MHz, that translates to ±25 ps of jitter. If un-tracked, this jitter is detrimental to a 32 Gbaud system.

One should note that this is a low-frequency effect. If the LO FM modulation rate is fast compared to the time-span of the CD compensation filter, then the phase fluctuation of the input data is convolved into noise seen at the output of the CD filter, and this causes a well-documented system penalty [24]. If the FM modulation rate is static compared to the CD filter, then timing jitter is produced. Once tracked out by the clock recovery circuit, this causes no system penalty. For a 32 Gbaud transmission rate, the time-span of a CD compensation filter may be on the order of 10 ns, and thus for FM modulation rates significantly slower than 100 MHz, timing jitter can be produced. While jitter less than 1 MHz can be efficiently tracked out by the clock recovery loop, FM rates between 1 and 100 MHz is a sensitive region, where the LO laser FM noise must be carefully controlled.

The conversion of FM noise to timing jitter has been studied in literature [25–27]. Recasting their results [25, 27], a jitter conversion efficiency ("JCE") can be defined as the amount of jitter induced in units of picoseconds per MHz of Rx laser frequency excursion (ps/MHz). The JCE is a function of FM modulation rate "f," and it can be written as shown in the following equation. A parameter $\Delta\tau$ is introduced that is inversely related to the cutoff frequency f_{CUTOFF}. The frequency-dependent JCE can be written as a function of the DC conversion efficiency JCE_0, and parameter $\Delta\tau$. The parameter $\Delta\tau$ is a function of the baud rate of the data under modulation and dispersion in units of ps/nm. The DC conversion efficiency (JCE_0) is only a function of the dispersion.

$$
\begin{aligned}
\text{JCE}(f) &= \frac{\text{JCE}_0}{\sqrt{2}\cdot\pi\cdot\Delta\tau}\,\frac{\sqrt{1-\cos(2\cdot\pi\cdot f\cdot\Delta\tau)}}{f} \quad (\text{ps/MHz}) \\
\Delta\tau &= \frac{F_{\text{baud}}}{125\times 10^9} D\cdot L\cdot 10^{-12} \quad (\text{sec}) \\
f_{\text{CUTOFF}} &= \frac{1}{\Delta\tau} \quad (\text{Hz}) \\
\text{JCE}_0 &= 2\cdot\kappa\cdot 2\cdot\pi\cdot 10^{12}\cdot 10^{6} \quad (\text{ps/MHz})
\end{aligned}
\tag{10.12}
$$

The JCE function is plotted in Figure 10.6 against Rx laser FM modulation rate for symbol rate of 32 GHz, and dispersion of 50,000 ps/nm. At frequency components below the cutoff frequency f_{CUTOFF} (~80 MHz), the jitter conversion is efficient at ~0.4 ps/MHz. At higher frequencies, the phase jitter interacts with the dispersion compensation filter in the digital receiver and is converted into an additive noise process. Overlaid on top in Figure 10.6 are simulated values (squares) using 32 Gbaud

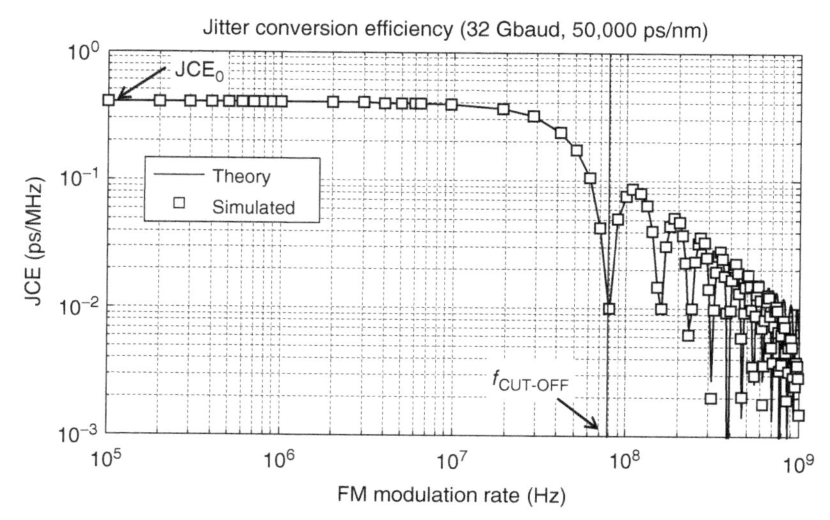

FIGURE 10.6 Frequency-dependent jitter conversion efficiency (JCE) at 50,000 ps/nm and 32G symbol rate.

binary toggle sequence as the modulated data, going through 50,000 ps/nm of dispersion and with sinusoidal FM modulation on the Rx laser.

10.3.4 Coherent System Tolerance to Untracked Jitter

For coherent systems using fractionally spaced equalizer and clock recovery, the VCO jitter is tracked out mainly by the clock recovery loop since the clock loop generally has much wider loop bandwidth compared to the equalizer. As a result, the residual untracked jitter is quite fast. For instance, if a 1 MHz clock loop BW is implemented, then untracked jitter is faster than or equal to 1 MHz. The equalizer tracking speed is quite slow compared to this rate, and this can be due to a variety of reasons: loop latency, number of taps, ASIC resource etc. The effect of the equalizer is negligible in determining system jitter tolerance, and it is ignored in this analysis. Instead we focus on performance impact due to residual untracked jitter from the clock recovery loop.

The tolerance of a system to untracked jitter strongly depends on the symbol rate. In general, 1 ps of jitter affects BER at 32G symbol rate same as 2 ps of jitter at 16G symbol rate. In usual practice, the system BER performance is evaluated against jitter normalized to the symbol period. The symbol period is denoted as "UI" or unit-interval. A system's tolerance to untracked jitter is then a strong function of the following three parameters:

1. system pulse shape
2. FEC coding gain
3. modulation format.

Figure 10.7 shows three eye diagrams. The one on the left is generated using raised-cosine filter with roll-off factor $\alpha = 0.01$. Spectrally, this pulse shape generates a signal with near brick-wall power spectral density with excess bandwidth only 1% more than the minimum Nyquist bandwidth. The power spectral density is plotted beneath the eye diagram. It is ideal for dense frequency multiplexing of multiple channels in order to achieve high spectral efficiency. The eye diagram in the middle figure is generated using roll-off factor $\alpha = 1.0$. Spectrally this pulse shape occupies twice the minimum Nyquist bandwidth, and therefore it is spectrally less-efficient. The eye diagram on the right figure illustrates a non-return-to-zero (NRZ) pulse shape with practical rise/fall time. Spectrally, the signal has strong component at 1.5× the symbol rate (around the middle of the first side lobe). Plotted above these figures are three Gaussian probability density functions of the same variance, representing jitter in each system. It is easy to see that NRZ pulse shape is most tolerant to jitter, while the most spectrally efficient system is the least tolerant. This is because as the sampling phase is moved away from the optimum point, the sampled signal does not reduce in amplitude in the NRZ pulse shape due to a very wide flat top. In contrast, for the Nyquist pulse shape, the sampled signal quickly reduces in amplitude, and therefore also signal-to-noise ratio (SNR). For Nyquist systems that deliver the highest fiber capacity, jitter and its system impacts remains a problem that requires careful design and component specification.

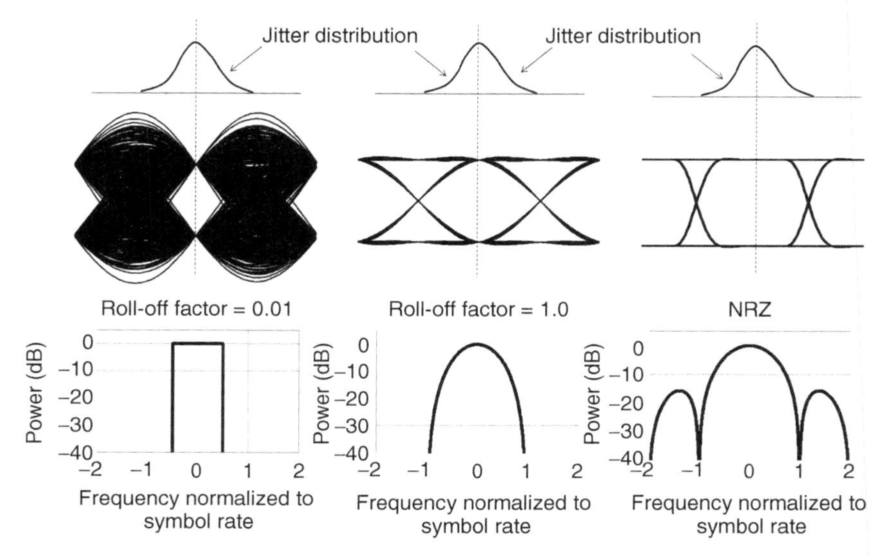

FIGURE 10.7 Three examples of pulse shapes used in optical transmission, each having different jitter tolerance and spectral occupancy.

Historically, in un-amplified transmission systems that pre-date 1995, forward-error-correction (FEC) technology had not made its way into optical transmission. This was largely due to difficulties in implementing complex decoding algorithms at high data rates using CMOS technology at the time. In un-coded transmission systems, jitter is one of the dominant sources of BER degradation. Since the advent of EDFAs and the wide spread adoption of first, second and third generation of FEC [28, 29], system degradation due to jitter is an increasingly small portion of the overall degradation budget. The dominant source of BER degradation is the noise accumulated in the repeatedly amplified fiber transmission line itself, going over tens and hundreds of spans of fiber all over the world. Jitter tolerance of a system depends on the FEC employed in the system.

In modern coherent systems, the use of higher-order modulation formats allows higher spectral efficiency and fiber capacity. However it comes at the cost of reduced Euclidean distance between neighboring symbols, and therefore reduces the system's tolerance to noise of all types, including jitter. A 64QAM modulation is a lot more sensitive to jitter than QPSK modulation.

Figure 10.8(a) and (b) exemplifies system impact of all three of these effects. QPSK, 16QAM and 64QAM modulations are analyzed using simulation and brute-force error counting. Two pulse shapes are considered: roll-off factor $\alpha = 0.01$ and $\alpha = 1.0$. Two different FEC schemes are considered: a 7% overhead hard decision FEC with BER threshold of 0.001, and a 20% overhead soft decision FEC with BER threshold of 0.015. In Figure 10.8(a), QPSK and 16QAM formats are shown where signal-to-noise ratio (SNR, measured in symbol-rate bandwidth) penalty at the FEC threshold is plotted against jitter in units of UI RMS. Pulse shape with roll-off factor

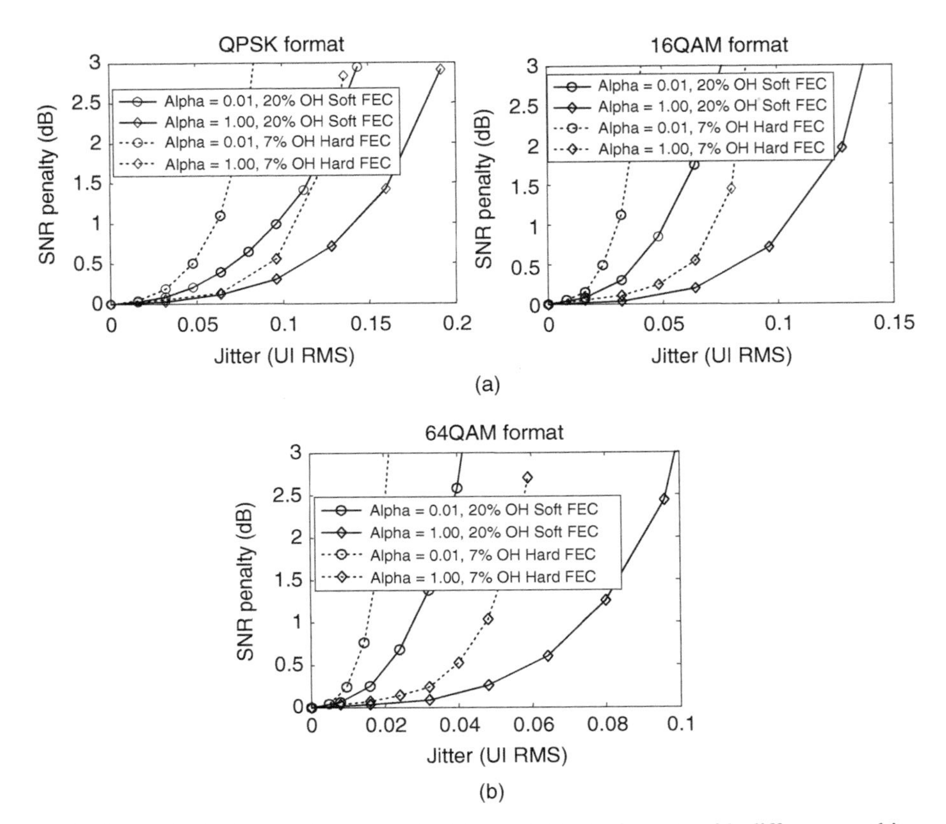

FIGURE 10.8 (a) Jitter tolerance of QPSK and 16QAM formats with different combinations of pulse shapes and FEC schemes. (b) Jitter tolerance of 64QAM formats with different combinations of pulse shapes and FEC schemes.

of 1.0 and with 20% soft FEC tolerates the most amount of jitter (see black diamond curve). Roll-off factor of 0.01 with 7% hard FEC tolerates the least amount of jitter (see dotted circles). The same trend is true for 16QAM and 64QAM formats (Figure 10.8(b)). QPSK is the most tolerant format and 64QAM is the most sensitive of the three formats.

10.4 DIGITAL PHASE DETECTORS

With the use of A/D converters and extensive signal processing for compensation of linear distortions, the phase detector is most naturally implemented in the digital domain inside the Rx DSP as shown in Figure 10.1. This section focuses only on digital phase detectors using digitally sampled signals. Furthermore, the A/D converters are practically always over-sampling. The sampling rate is always higher than the symbol rate. Although theoretically it doesn't have to be, this feature is necessary in order to implement perfect compensation of dispersion and PMD using FIR filter lengths that are within practical bounds. Therefore, we have at our disposal signals that are over-sampled in the DSP and equalized symbols at the output of the DSP available at the symbol rate. This section reviews two types of phase detectors based on these two types of signals that are readily available inside the DSP.

The first class of phase detector is based on a direct measure of the clock phase using signal components above and below the Nyquist frequency ($f_b/2$, where f_b is the symbol rate). Examples of this type of phase detector include the conventional analog squaring phase detector [3], Gardner's phase detector [11], Godard's pass-band timing detector [12], and a frequency-domain phase detector [30]. In contrast, decision directed algorithms like Mueller–Muller [9], minimum mean-squared error [10] and probabilistic algorithms like maximum-likelihood [4] do not use frequency components above Nyquist, and are referred to as the second class of phase detectors.

In the description of the first class of phase detectors, we first introduce a frequency-domain phase detector and show how it is able to detect the clock phase. Then, we demonstrate that the analog squaring phase detector, Gardner, Godard, and the frequency-domain phase detectors are in essence equivalent to each other, in that they all use frequency components above and below the Nyquist frequency to measure directly the clock phase from the input signal.

10.4.1 Frequency-Domain Phase Detector

Let the transmitted data symbols ($a[i]$) be zero mean, complex, and random. In the case of QPSK, $a[i]$ is uniformly distributed on an alphabet of four phases (0°, 90°, 180°, 270°). After modulation and transmission, the signal is sampled by a set of A/D converters that sample at two samples per symbol. We assume transmission of a periodic sequence $a[i]$ in order to define signals in the frequency domain. The principles derived can be directly applied to aperiodic signals for a system that carries live traffic. A/D converters can sample slower than two samples per symbol, but it must sample at a rate faster than one sample per symbol. The sampled data $x[n]$ has the length N, and its frequency-domain representation $X[k]$ can be written as the following equation. Sampling rate of two samples per symbol is assumed throughout our derivation, and the length of $a[i]$ must be $N/2$. $A[k]$ is the N-point FFT of time sequence $a'[m]$, where $a'[m]$ is 2× up-sampled from $a[i]$ by zero insertion. Zero insertion can be defined as the following: $a'[0, 2, 4, \ldots] = a[0, 1, 2, \ldots]$, and $a'[1, 3, 5, \ldots] = 0$. The channel response is defined as $H[k]$ contains clock phase error τ (normalized to symbol period $1/f_b$), and the system symbol rate is f_b. $H[k]$ can be modeled by a simple amplitude weighting function $H'[k]$ with clock phase error. This process of mathematically describing the received signal can be found in standard communication texts [6, 31].

$$
\begin{aligned}
X[k] &= A[k] \cdot H[k] \\
H[k] &= H'[k] \cdot e^{-j \cdot 4 \cdot \pi \cdot k \cdot \tau / N}
\end{aligned}
\tag{10.13}
$$

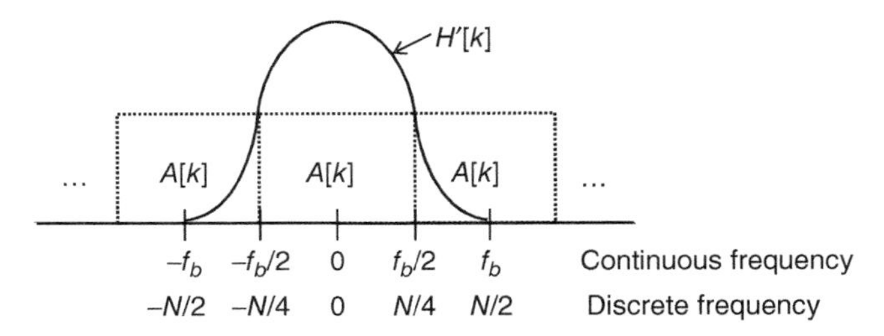

FIGURE 10.9 Illustration of received signal spectrum showing correlated frequency components.

Because of the up-sampling, $A[k]$ repeats in spectrum according to sampling theorem [31]. The frequency components of $A[k]$ between $f_b/2$ and f_b ($N/4$ to $N/2$) are the same as those components between $-f_b/2$ and 0 ($-N/4$ to 0). The data spectrum is periodic with period f_b (or $N/2$). This important consequence of sampling theorem is pictorially illustrated in Figure 10.9.

Because the frequency components of the sampled data $A[k]$ are correlated across f_b, there is opportunity to estimate the phase difference in the received signal $X[k]$ across f_b (or across $N/2$ samples in frequency). Consider the frequency component at k_o, estimating the clock phase from that single-frequency component can be defined as:

$$\tau_{\text{est}} = \left\langle \text{Im}\left\{X\left[k_o\right] \cdot X^*[k_o - N/2]\right\}\right\rangle \tag{10.14}$$

where $\langle\cdot\rangle$ denotes time averaging, $\text{Im}\{\cdot\}$ is the imaginary component, * denote complex conjugation. Substitute $X[k]$ as $A[k]$ multiplied with $H[k]$:

$$\tau_{\text{est}} = \left\langle \text{Im}\left\{A\left[k_o\right] \cdot A^*[k_o - N/2] \cdot H[k_o] \cdot H^*[k_o - N/2]\right\}\right\rangle \tag{10.15}$$

Realizing that $A[k_o]$ is equal to $A[k_o - N/2]$ from sampling theorem, and substitute $H[k]$ as an amplitude weighting function $H'[k]$ and clock phase error τ (normalized to symbol period):

$$\tau_{\text{est}} = \left\langle \left|A\left[k_o\right]\right|^2 \cdot H'[k_o] \cdot H'[k_o - N/2] \cdot \text{Im}\left\{e^{-j\cdot 4\cdot\pi\cdot k_o\cdot\tau/N} \cdot e^{+j\cdot 4\cdot\pi\cdot(k_o - N/2)\cdot\tau/N}\right\}\right\rangle$$

$$\tau_{\text{est}} = -\left\langle \left|A\left[k_o\right]\right|^2\right\rangle \cdot H'[k_o] \cdot H'[k_o - N/2] \cdot \sin(2\cdot\pi\cdot\tau)$$

$$\tau_{\text{est}} \propto \sin(2\cdot\pi\cdot\tau) \tag{10.16}$$

The derivation shows that Equation 10.14 produces a phase estimate that is linearly proportional to sine of the clock phase error. The proportionality constant is related to the magnitude of the channel response $H'[k]$ at frequencies k_o and k_o–$N/2$. It is also important that the transmitted random data $A[k]$ at frequency component k_o should have non-zero power. This is usually guaranteed since the modulated data $a[i]$ is assumed random and has white power spectral density.

In order to reduce the detector noise, all frequency components can be used. Here the expectation is dropped, and the averaging is done through summation over all $N/2$ frequency-domain samples.

$$\tau_{\text{est}} = \sum_{k=0}^{N/2-1} \text{Im}\{X[k] \cdot X^*[k - N/2]\} \tag{10.17}$$

This detector relies on using correlated spectral components around $\pm f_b/2$ to yield a phase estimate. The following figure shows the placement of this phase detector in the Rx DSP. The phase detector can use frequency-domain samples after an N-point FFT block. FFT and I-FFT engines are typically used to efficiently compensate for dispersion using overlap-and-add (OLA) or overlap-and-save (OLS) algorithm [32], and therefore we have readily available frequency-domain samples. The size of the FFT ("N") is related to the dispersion compensation range. Figure 10.10 shows the possibility to use frequency data from both polarizations $X[k]$ and $Y[k]$ and sum the results together to reduce detector jitter. In a later section, we shall show that it is not as simple as directly adding phase detector outputs on both polarizations in this way. With fiber impairment known as Polarization Mode Dispersion (PMD), data from two polarizations must be combined differently.

10.4.2 Equivalence to the Squaring Phase Detector

The squaring phase detector is used extensively in recovering symbol rate timing wave for signals of NRZ type spectral shaping. In this section, we show that this commonly used phase detector has working principles that are equivalent to the frequency-domain phase detector described in the previous section.

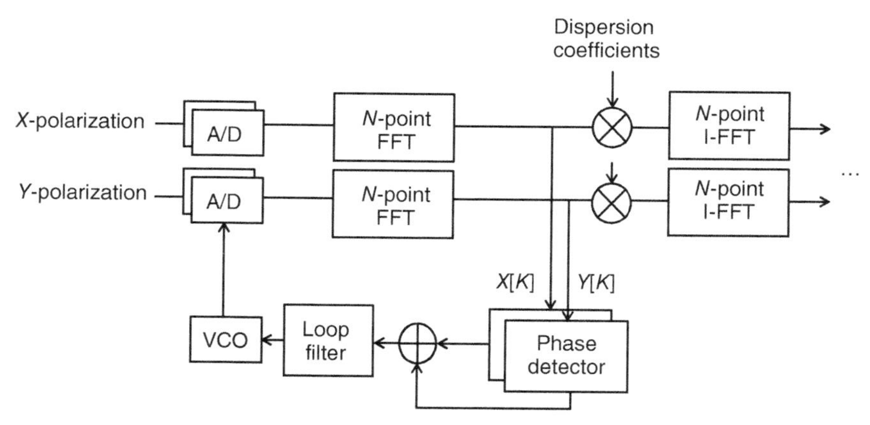

FIGURE 10.10 Placement of the frequency-domain phase detector inside Rx DSP.

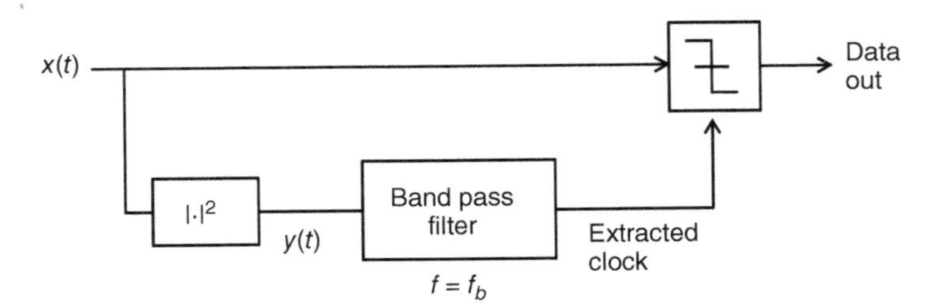

FIGURE 10.11 The squaring phase detector extracting timing wave for decoding data.

Figure 10.11 illustrates the conventional squaring method [3] that generates the timing wave using an input signal $x(t)$. The timing wave or the extracted clock is used by the decision device to recover data from the input signal. As shown in the diagram, the signal is first squared and is then filtered by a narrow-band band-pass-filter (BPF) centered at the symbol rate frequency f_b. The output is a symbol rate timing wave that recovers the timing of the input signal. In the figure, the squaring device is replaced with absolute squared device in anticipation of complex signal $x(t)$.

To simplify analysis, let's consider a periodic signal $x(t)$. If $X(f)$ is the frequency-domain representation of $x(t)$, then the frequency-domain representation of $x(t) \times x*(t)$ can be written in terms of $X(f)$ using the Fourier multiplication and conjugation property (⊗ denote convolution).

$$Y(f) = X(f) \otimes X^*(-f)$$
$$Y(f) = \int_{-\infty}^{\infty} X(\lambda) \cdot X^*(\lambda - f) \cdot d\lambda \tag{10.18}$$

The band-pass filter is a filter with real-valued input signals. The extraction of the f_b component by the BPF is equivalent to the evaluation of $\text{Re}\{Y(f = f_b)\}$. Assuming a very narrow-band filtering, and by using the convolution integral, the output of the BPF can be written as:

$$\text{Re}\left\{Y\left(f_b\right)\right\} = \int_{-\infty}^{\infty} \text{Re}\left\{X\left(\lambda\right) \cdot X^*(\lambda - f_b)\right\} \cdot d\lambda \tag{10.19}$$

Comparing Equation 10.19 to Equation 10.17, we see that they are essentially the same. Equation 10.19 is the continuous-time version of the discrete-time formulation of Equation 10.17. One difference is that Equation 10.19 extracts cosine of the clock phase error ($\text{Re}\{\cdot\}$), while Equation 10.17 extracts the sine of the clock phase error ($\text{Im}\{\cdot\}$). This difference leads to a lock point difference of 90° or ¼ of the symbol period. This can be adjusted by delaying the output timing wave.

Therefore, similar to the frequency-domain phase detector, the commonly used squaring phase detector must also use correlated frequency components above and below the Nyquist frequency.

10.4.3 Equivalence to Godard's Maximum Sampled Power Criterion

Let us hypothetically consider a receiver that uses A/D converter sampling at one sample per symbol. Such a receiver will be sensitive to clock phase error. For optimum performance with such a converter, the sampling phase should be chosen to maximize sampled signal power [12], which then maximizes sampled signal to noise ratio. A clock phase detector derived using this criterion will also work for a receiver system with sampling rate of two samples per symbol, in which one of the two samples will contain the highest sampled signal power. This criterion was adopted by Godard in his search for an optimum phase detector.

To describe the maximum sampled power criterion, let the transmitted data symbols "$a[i]$" be zero mean, complex, and random. After modulation and transmission, the signal is sampled by a set of A/D converters at a sampling rate of two samples per symbol. Assume the digitally sampled data "$x[n]$" is periodic with period N, and its frequency-domain description is $X[k]$ $(k = 0, 1, \ldots, N-1)$. The even samples of $x[n]$ are T-spaced sampled data $(x_e[n], n = 0, 1, \ldots, N/2-1)$, and its frequency-domain description $(X_e[k], k = 0, 1, \ldots, N/2-1)$ can be written in terms of $X[k]$, where sampling theorem is invoked.

$$X_e[k] = X[k] + X[k + N/2] \tag{10.20}$$

The power of the T-spaced sampled signal $x_e[n]$ can be calculated in the frequency domain by Parseval's theorem (* denote complex conjugation):

$$\sum_{n=0}^{N/2-1} x_e[n] \cdot x_e^*[n] = \sum_{k=0}^{N/2-1} (X[k] + X[k+N/2]) \cdot (X^*[k] + X^*[k+N/2]) \tag{10.21}$$

As was discussed earlier, the received signal spectrum $X[k]$ contains the original zero-inserted, random data sequence spectrum $A[k]$ multiplied with the channel response $H[k]$. $H[k]$ consists of a real-valued amplitude weighting function $H'[k]$ and clock phase error τ. Similar to that mentioned previously, τ is the channel delay (or the clock phase error) normalized to the symbol interval $(1/f_b)$. The goal is to maximize the sampled power of the signal $x_e[n]$, with respect to τ. Therefore, in the product expansion of the above-mentioned equation, the two terms that are not a function of τ will be neglected $(X[k] \times X^*[k]$ and $X[k+N/2] \cdot X^*[k+N/2])$, and the two cross terms will be retained in the analysis that follows:

$$\sum_{k=0}^{N/2-1} x_e[n] \cdot x_e^*[n] = \sum_{k=0}^{N/2-1} X[k] \cdot X^*[k+N/2] + X^*[k] \cdot X[k+N/2]$$

$$\sum_{k=0}^{N/2-1} x_e[n] \cdot x_e^*[n] = \sum_{k=0}^{N/2-1} \mathrm{Re}\{X[k] \cdot X^*[k+N/2]\} \tag{10.22}$$

Substitute $X[k] = A[k] \times H'[\text{k}] \cdot e^{-j \cdot 4 \cdot \pi \cdot k \cdot \tau / N}$ ($H'[k]$ is assumed real value):

$$\sum_{n=0}^{N/2-1} x_e[n] \cdot x_e^*[n] = \sum_{k=0}^{N/2} \text{Re}\left\{A[k] \cdot A^*[k+N/2] \cdot H'[k] \cdot H'[k+N/2] \cdot e^{-j \cdot 4 \cdot \pi \cdot k \cdot \tau / N} \cdot e^{j \cdot 4 \cdot \pi \cdot (k+N/2) \cdot \tau / N}\right\}$$

$$\sum_{n=0}^{N/2-1} x_e[n] \cdot x_e^*[n] = \sum_{k=0}^{N/2} \text{Re}\left\{A[k] \cdot A^*[k+N/2] \cdot H'[k] \cdot H'[k+N/2] \cdot e^{j \cdot 2 \cdot \pi \cdot \tau}\right\} \tag{10.23}$$

Recalling that $A[k]$ is periodic with period f_b (or $N/2$), and $A[k] = A[k+N/2]$. The equation is simplified to

$$\sum_{n=0}^{N/2-1} x_e[n] \cdot x_e^*[n] = \cos(2\pi \cdot \tau) \cdot \sum_{k=0}^{N/2-1} A[k] \cdot A^*[k] \cdot H'[k] \cdot H'[k+N/2] \tag{10.24}$$

The equation shows that maximizing the sampled signal power (the even samples) is the same as zeroing the clock phase error τ. At $\tau = 0$, $\cos(2 \cdot \pi \cdot \tau)$ is maximized. The four terms in the summation are properties of the input random data sequence and channel magnitude response. Furthermore, minimizing τ is the same as driving $\sin(2 \cdot \pi \cdot \tau)$ to zero. Combining the results of Equations 10.22 and 10.24

$$\sum_{k=0}^{N/2-1} \text{Re}\left\{X[k] \cdot X^*[k+N/2]\right\} = \cos(2\pi \cdot \tau) \cdot \sum_{k=0}^{N/2-1} A[k] \cdot A^*[k] \cdot H'[k] \cdot H'[k+N/2] \tag{10.25}$$

It is then easy to see that the following equation is also true, where the real component of the beating of $X[k]$ and $X * [k+N/2]$ extracts the cosine of the phase error and the imaginary component extracts the sine of the phase error:

$$\sum_{k=0}^{N/2-1} \text{Im}\left\{X[k] \cdot X^*[k+N/2]\right\} = \sin(2\pi \cdot \tau) \cdot \sum_{k=0}^{N/2-1} A[k] \cdot A^*[k] \cdot H'[k] \cdot H'[k+N/2] \tag{10.26}$$

Equations 10.17 and 10.26 are essentially the same. This proves that the lock point of the frequency-domain phase detector of Equation 10.17 is optimum in the sense that the sampled data contain the highest sampled signal power and maximize the sampled signal-to-noise ratio. Godard's phase detector [12] also has the same working principle as the frequency-domain phase detector.

10.4.4 Equivalence to Gardner's Phase Detector

The Gardner's phase detector is also a very commonly used phase detector [11]. In what follows, the digital Gardner's phase detector in the time domain is shown. "$x[n]$"

$(n = 0, 1, 2, \ldots, N-1)$ is the complex discrete time data sampled at two samples per symbol. "$x_R[n]$" and "$x_I[n]$" are the real and imaginary components, respectively.

$$\tau_{\text{err}} = \sum_{n=0}^{N/2-1} (x_R[2n-1] - x_R[2n+1]) \cdot x_R[2n] + (x_I[2n-1] - x_I[2n+1]) \cdot x_I[2n] \tag{10.27}$$

The equation can be written in terms of complex quantity $x[n]$ and its conjugate:

$$\tau_{\text{err}} = \text{Re}\left\{ \sum_{n=0}^{N/2-1} (x[2n-1] - x[2n+1]) \cdot x^*[2n] \right\} \tag{10.28}$$

Similar to earlier derivations, we first define periodic sequences for the transmitted and received data; therefore, we can represent signals easily in the frequency domain. The frequency-domain representation of the received samples $x[n]$ is defined as $X[k]$. $x[n]$ and $X[k]$ are periodic with period equal to N. "$x[2n]$" is a 2× decimated version of $x[n]$. It represents the even samples of the received data. It can be represented in the frequency domain using $X[k]$ as shown in Equation 10.29. Sampling theorem is invoked, and the double arrow indicates transformation from time domain to frequency domain.

$$\begin{gathered} x[2n] \Longleftrightarrow X[k] + X[k + N/2] \\ k = 0, 1, \ldots, N/2 - 1 \end{gathered} \tag{10.29}$$

The differencing function in Equation 10.28 $(x[2n-1] - x[2n+1])$ can be thought of as two steps. First, the signal $x[n]$ is filtered by a differencing filter, and then down-sampled by a factor of two. The filtering function is a differencing filter with impulse response $\begin{bmatrix} 1 & 0 & -1 \end{bmatrix}$. It has magnitude and phase response as shown in Figure 10.12.

We note that in order to derive and show the principle of the phase detection process, the magnitude response of this filter can be ignored. The phase response of j in the upper-side band and $-j$ in the lower-side band is important. The magnitude response will affect detector sensitivity. For the sake of simplicity, we approximate this differencing filter with $H[k]$, where $H[k]$ has a flat magnitude response and only the phase response is retained.

$$H[k] = \begin{cases} +j & \text{where } k = 0 \rightarrow N/2 - 1 \\ -j & \text{where } k = N/2 \rightarrow N - 1 \end{cases} \tag{10.30}$$

Using Equation 10.30 and sampling theorem illustrated in Equation 10.29, the differencing function in Equation 10.28 can be written in the frequency domain through the following derivation:

$$\begin{aligned} x[n-1] - x[n+1] &\Longleftrightarrow X[k] \cdot H[k] \\ x[2n-1] - x[2n+1] &\Longleftrightarrow X[k] \cdot H[k] + X[k + N/2] \cdot H[k + N/2] \end{aligned} \tag{10.31}$$

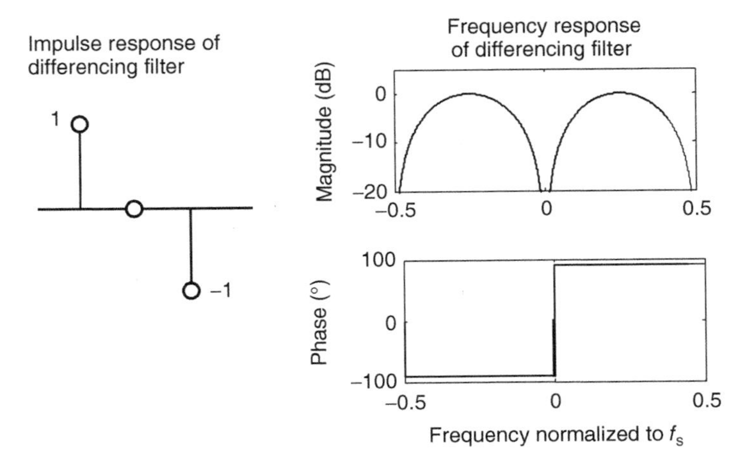

FIGURE 10.12 The differencing filter in Gardner's phase detector.

We note another discrete time Fourier property (assume $A[k]$ is the FFT of $a[n]$, and $B[k]$ the FFT of $b[n]$):

$$\sum_N a[n] \cdot b^*[n] \Longleftrightarrow \sum_N A[k] \cdot B^*[k] \tag{10.32}$$

Using Equations 10.29, 10.31, and 10.32, we derive the frequency-domain equivalent function of Equation 10.28:

$$\mathrm{Re}\left\{ \sum_N (x[2n-1] - x[2n+1]) \cdot x^*[2n] \right\}$$
$$\Longleftrightarrow \mathrm{Re}\left\{ \sum_k (X[k] \cdot H[k] + X[k+N/2] \cdot H[k+N/2]) \cdot (X^*[k] + X^*[k+N/2]) \right\} \tag{10.33}$$

Using $H[k]$ definition in Equation 10.30, the frequency-domain phase detector can be simplified:

$$\mathrm{Re}\left\{ \sum_k j \cdot (X[k] - X[k+N/2]) \cdot (X^*[k] + X^*[k+N/2]) \right\} \tag{10.34}$$

Multiplying out the product terms, and removing the quadrature components, we show that this equation can be equivalently written as

$$-2 \cdot \sum_k \mathrm{Im}\left\{ X[k] \cdot X^*[k+N/2] \right\} \tag{10.35}$$

This shows that the Gardner's phase detector under the simplification of the differencing filter $H[k]$ (represented by Eq. 10.30) is equivalent to the frequency-domain phase detector introduced earlier in Equation 10.17, and that it also uses correlated frequency components that are outside of the Nyquist bandwidth.

One caveat is that the digital implementation of Gardner's phase detector is most easily implemented with A/D sampling rate of two samples per symbol. For systems that sample less than two samples per symbol, the phase detector implementation is not obvious in the time domain, whereas the frequency-domain phase detector can be easily adapted to fractional sampling rates of less than two samples per symbol.

10.4.5 Second Class of Phase Detectors

The second class of phase detectors uses equalized symbols at the output of the equalizer and carrier recovery. These signals are also readily available inside the Rx DSP. The equalized symbols are represented at one sample per symbol, and therefore no information is available above the Nyquist frequency. The principle in which they operate is quite different from the first class of phase detectors. Some examples include Mueller–Muller detector [9], minimum mean-squared error [10], and maximum-likelihood detector [4]. Because they derive phase error from equalized symbols, the jitter performance of these detectors is not a function of the distortion in the channel. The distortions such as dispersion and PMD are assumed compensated. Figure 10.13 shows the placement of this type of phase detector in the Rx DSP. As the diagram shows, this placement adds significant latency compared with the frequency-domain phase detector, and that can impose a lower operating loop bandwidth.

Figure 10.14(a) shows an implementation of the Mueller–Muller phase detector [9]. Input sequence x_k is real valued and is available at one sample per symbol. For complex-valued signals such as QPSK, one can use the circuit for the real and imaginary parts separately and combine the results. The decision device makes decision on x_k to form a_k. "a_k" represents the transmitted bits (or symbols), except that it may contain decision errors. Decision errors will degrade the average detector sensitivity,

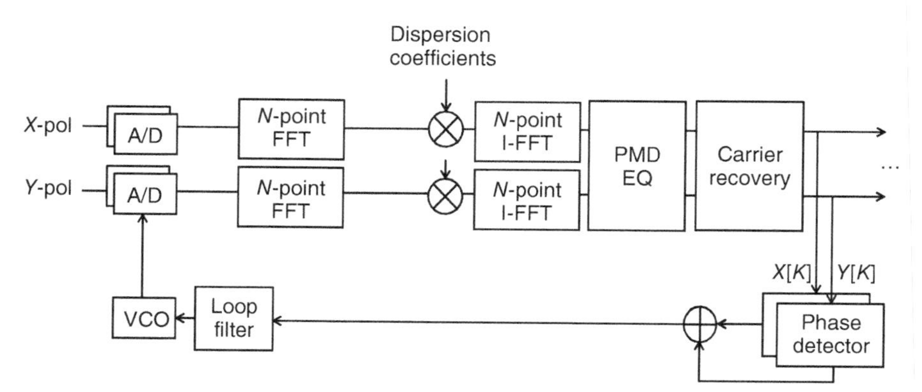

FIGURE 10.13 Detecting clock phase after equalization using a second class of phase detectors.

but at meaningful bit error rates of 10^{-3} to 10^{-2}, the sensitivity degradation is small. The two multipliers correlate x_k with a_{k-1}, and x_{k-1} with a_k. The time-averaged correlation of x_k with a_{k-1} is the same as time-averaged correlation of x_{k+1} with a_k. In the absence of decision errors, the correlators calculate the net system impulse response $h[n]$ for $n = 1$ and $n = -1$. The following equation shows this relationship:

$$\begin{aligned} \langle a_k \cdot x_{k-1} \rangle &\approx h[-1] \\ \langle a_{k-1} \cdot x_k \rangle = \left\langle a_k \cdot x_{k+1} \right\rangle &\approx h[1] \end{aligned} \tag{10.36}$$

Since the phase detector is placed after equalization, the net impulse response without any clock phase error can be assumed white (i.e., fully equalized by the adaptive equalizer), and can be represented with three taps: $h[n] = [0, 1, 0]$. Furthermore, if the equalizer is slowly responding, then fast clock phase jitter will not cause any adjustment of the equalizer coefficients, and the clock phase error can be measured after equalizer. Assuming a static equalizer, the clock phase error is seen as a delay ("τ") on the net impulse response, and will create odd symmetry in $h[n]$ [9]. Figure 10.14(b) illustrates the odd symmetry behavior of $h[n]$ under positive delay "τ" and negative delay "τ". By measuring the difference between $h[1]$ and $h[-1]$ ($z_k \approx h[1] - h[-1]$), a phase detector can be formed. This is the working principle of the Mueller–Muller phase detector.

10.4.6 Jitter Performance of the Phase Detectors

In this section, the jitter performances of the phase detectors described in the previous sections are compared together. The modulation format assumed is 32 Gbaud QPSK with raised cosine spectral shaping using roll-off factor $\alpha = 1.0$ and $\alpha = 0.01$. The different phase detectors may operate in different rates. The Mueller–Muller and Gardner phase detectors produce one estimate of phase for every input symbol, whereas the frequency-domain phase detector produces one estimate for every FFT cycle. We compare them by normalizing the detector jitter assuming a common

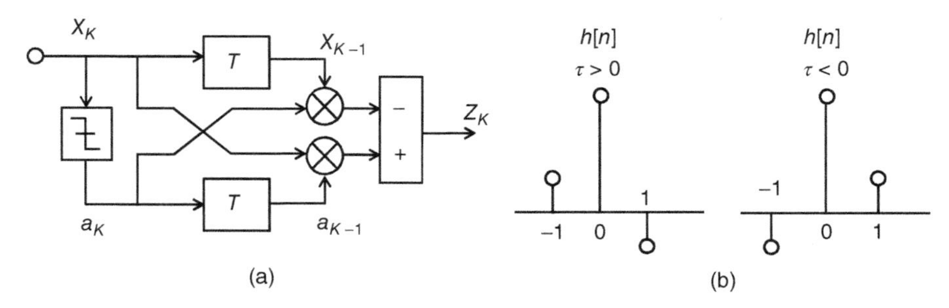

FIGURE 10.14 Mueller–Muller phase detector.

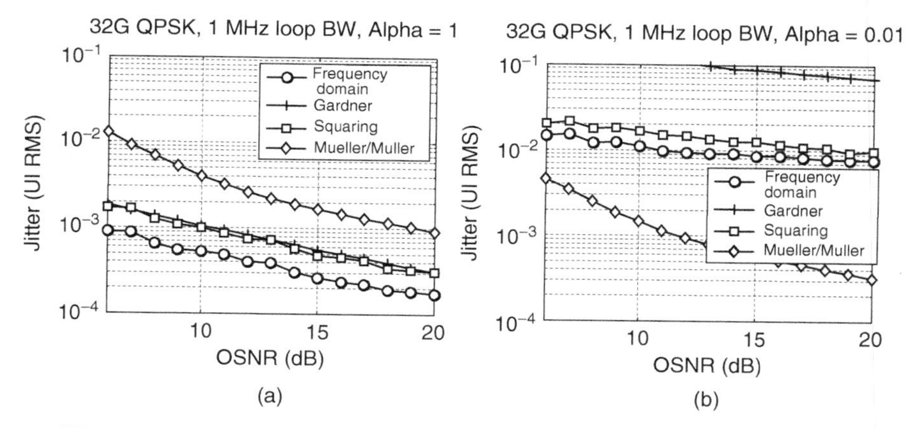

FIGURE 10.15 Jitter performance of phase detectors ((a) $\alpha = 1.0$, (b) $\alpha = 0.01$).

closed-loop BW of 1 MHz. A previous section has details on the methodology of evaluating the detector jitter. Signals from both polarizations are averaged to reduce jitter by 3 dB. The received signal is assumed to have no residual distortions such as CD and PMD.

The jitter performance of the different phase detectors versus OSNR (optical signal-to-noise ratio measured in 0.1 nm) is shown in Figure 10.15. In Figure 10.15(a), the signal pulse shape is raised cosine with roll-off factor $\alpha = 1.0$. The signal occupies twice the Nyquist frequency, and there is plenty of signal content above the Nyquist frequency. The frequency-domain phase detector performs the best. Gardner and squaring phase detectors perform similarly well. The Mueller–Muller detector performs the worst amongst the four due to decision errors. However, over the entire OSNR range, it is able to achieve jitter <0.01 UI RMS, and this is sufficiently good for QPSK modulation. For higher-order modulation Mueller–Muller detector is also sufficiently good since the operating OSNR is also higher.

In Figure 10.15(b), the pulse shape is changed to one that is most spectrally efficient, using $\alpha = 0.01$. The Mueller–Muller detector continues to perform well with near-Nyquist-shaped signals. Its jitter improves compared with α factor of 1.0 because of the stronger $h[1]$ and $h[-1]$ component in the impulse response. However, the frequency-domain phase, squaring, and Gardner phase detectors have all suffered greatly. Frequency-domain and squaring detectors are marginal, and Gardner phase detector is not functioning. The reason is that all three detectors rely on frequency components of the signal above and below the Nyquist frequency. As the pulse shape reduces from $\alpha = 1.0$ to $\alpha = 0.01$, the power of the signal components above the Nyquist frequency is reduced dramatically. At $\alpha = 0.0$, which is Nyquist transmission, the first class of phase detector will fail, and an alternative phase detector is needed. This is the subject of the next section.

10.4.7 Phase Detectors for Nyquist Signals

Nyquist signals are spectrally shaped to fit as closely as practically possible within the minimum Nyquist bandwidth. For 32G symbol rate modulation, the spectral occupancy would ideally be between −16 and 16 GHz, centered on the optical carrier. This allows dense packing of WDM channels, and thereby achieving higher spectral efficiency. Practically, the transmitters in these systems are constructed using DSP and D/A converter driving into a set of *I/Q* modulators operating in the linear regime. The DSP is capable of using a large number of tap coefficients to filter the signal to be spectrally as close to Nyquist as possible. One type of digital filter that achieves this goal is the class of raised-cosine filters defined by roll-off factor α [5, 31]. Although a filter that achieves $\alpha = 0$ is not possible to implement, DSP computation power available at the time of this writing allows long enough filters to implement α factor near 0.01. As the previous section shows, detecting clock phase using signals of this type is particularly difficult due to lack of signal content above the Nyquist frequency. While the frequency-domain phase detector has a lot of advantages at $\alpha = 1$, it struggles at $\alpha = 0.01$, and does not function at $\alpha = 0.001$.

Another type of phase detector concept has emerged in order to resolve this shortcoming. The key is first realizing that the instantaneous power envelop of the input Nyquist signal can be used as an input to a conventional phase detector such as the Gardner phase detector. In [33], it is shown that by using the familiar Gardner phase detector, and replacing the input complex signal $x_I[n] + j \cdot x_Q[n]$ by the instantaneous power $P[n] = x_I{}^2[n] + x_Q{}^2[n]$, the phase detector sensitivity can be recovered. Because the input to the Gardner phase detector is a power signal, this phase detector is referred to as the fourth-power time-domain phase detector. The phase detector equation assumes input signal $(x_I[n] + j \cdot x_Q[n])$ sampled at two samples per symbol, and is shown in the following equation:

$$\begin{aligned} P[n] &= x_I^2[n] + x_Q^2[n] \\ \tau_{\mathrm{err}}[n] &= (P[2n-1] - P[2n+1]) \cdot P[2n] \end{aligned} \tag{10.37}$$

A frequency-domain implementation of this method is highly desirable. The first signal available inside the DSP after dispersion compensation is in the frequency domain as shown in Figures 10.10 and 10.16. The placement of frequency-domain phase detector directly after dispersion compensation coefficients achieves the lowest feedback latency. Furthermore, A/D converters may or may not be sampled at the convenient two samples per symbol, and with fractional sampling, Equation 10.37 may be difficult to implement without compromising performance. A frequency-domain implementation of the fourth-power method is possible since the A/D converters capture the signal in its entirety, and therefore mathematical manipulation of the frequency-domain samples can be equivalent to the time-domain description of Equation 10.37.

A frequency-domain implementation can be derived by first calculating the frequency-domain representation of $P[n]$ by convolving frequency-domain data samples at the output of the dispersion compensation coefficients as shown in

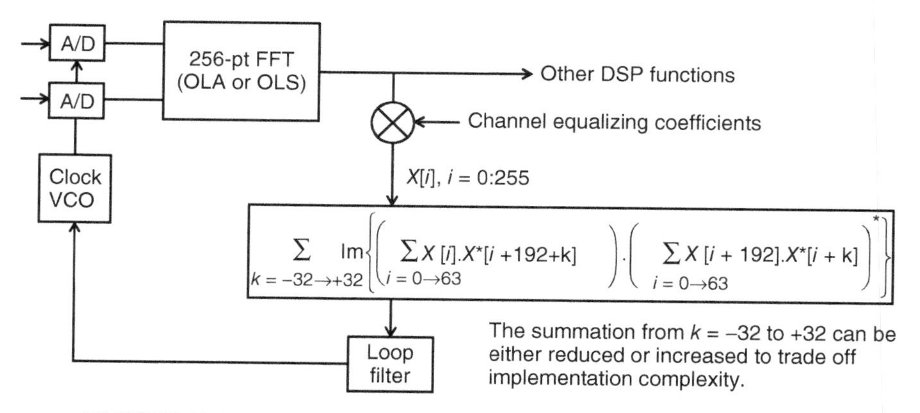

FIGURE 10.16 Fourth-power phase detector in the frequency domain [34].

Figure 10.16 [34]. The spectral components of $P[n]$ need to be calculated at frequency bins $f_b/2$ (upper-side band) and $-f_b/2$ (lower-side band), and other pairs of frequency bins separated by f_b. This is consistent with the principle of the frequency-domain phase detector described in Equation 10.17.

The received X-pol signal is first digitized by two ADCs sampled at two samples per symbol, followed by (as an example) a 256-point FFT. We note that the FFT size can be larger or smaller than 256. It is assumed that distortions impacted on the signal such as chromatic dispersion are largely compensated by the frequency-domain coefficient multiplication. The frequency-domain equalized signal $X[i]$ ($i = 0, 1, \ldots, 255$) is supplied to a phase detector that operates on the frequency-domain samples. As shown in Figure 10.16, its output is further filtered by a loop filter which then drives the VCO tuning port, affecting the timing of the sampled signal. The phase detector algorithm is shown below. "*" denotes complex conjugation, "Im{·}" denotes imaginary part, and "T" means one symbol period.

$$\mathrm{PD} = \sum_{k=-32\to+32} \mathrm{Im}\left\{\left(\sum_{i=0\to63} \underbrace{X[i]\cdot X^*[i+192+k]}_{\text{freq separation} = \frac{1}{2T}-\frac{k}{128T}}\right)\cdot\left(\sum_{i=0\to63} \underbrace{X[i+192]\cdot X^*[i+k]}_{\text{freq separation} = \frac{-1}{2T}-\frac{k}{128T}}\right)^*\right\} \tag{10.38}$$

As indicated in this equation, there is an outer summation and two inner summations inside the Im{·} function. The two inner summations essentially calculate the frequency bins of the power envelope of the input signal. The method outlined in [33] (see Eq. 3 in [33]) also uses the power envelope of the input signal, but in the time domain. The frequency separation between two terms within the first inner summation is $1/(2T) - k/(128T)$, whereas it is $-1/(2\,T) - k/(128T)$ in the second inner summation, and the difference between them is always $1/T$. If there is delay in the signal, the multiplication of the two summation terms (with the second term conjugated) averages to a term proportional to $\exp(j2\pi\tau)$, where τ is the delay normalized

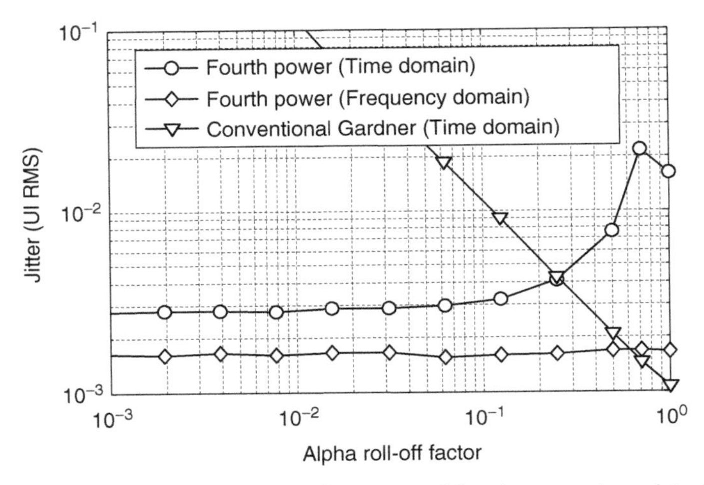

FIGURE 10.17 Comparison of jitter performances of fourth-power phase detector and the conventional Gardner phase detector.

to T. The $\text{Im}\{\cdot\}$ function then extracts the $\sin(2\pi\tau)$ portion. The outer summation from $k = -32$ to 32 is to combine all possible terms that average to $\exp(j2\pi\tau)$, and that reduces jitter. The detector output can be approximated as $K_d \cdot \sin(2\pi\tau)$, where "τ" is the timing error (normalized to T) and K_d is the detector sensitivity.

The performance of the phase detector is evaluated numerically assuming 32G symbol rate QPSK, X-pol, and Y-pol phase detectors are averaged, noise loaded at OSNR = 10 dB with results summarized in Figure 10.17. Note on the legend:

- The RMS jitter is calculated based on a 1-MHz 3-dB loop bandwidth and is normalized to UI RMS (unit interval = symbol interval).
- Fourth power (time domain) refers to the method of Equation 10.37.
- Conventional Gardner (time domain) refers to the method of Equation 10.28.
- Fourth power (frequency domain) refers to the method of Equation 10.38.

The following can be observed:

- The jitter increases as α decreases for the conventional Gardner (time domain). Jitter-induced system penalties are too high for α factors significantly below 0.1.
- The jitter increases as α increases above 0.1 for the fourth-power time-domain method of Equation 10.37.
- The fourth-power frequency-domain approach has better performance than the conventional Gardner for $\alpha < 0.6$. Its jitter performance is insensitive for all α factors.

The fourth-power frequency-domain method of Equation 10.38 has the best jitter performance, but it is also more complex to implement than the fourth-power time-domain method. Both approaches are good in solving the problem of phase detection for Nyquist-type signals. The increase of jitter at $\alpha > 0.1$ for the fourth-power time-domain method can be improved by prefiltering the signal before phase detection. Channel distortions such as dispersion and PMD can degrade both of these phase detectors, and they are active research topics.

10.5 THE CHROMATIC DISPERSION PROBLEM

With the use of DSP to compensate for chromatic dispersion in the fiber, the signal input to the receiver can be dispersed by large amounts of dispersion. The entire fiber plant can be built without dispersion compensation (or dispersion management), and the receive DSP can be made to equalize 50,000 ps/nm [18, 20], more than enough for compensating terrestrial long-haul fiber plants of the largest dispersion coefficient. New fibers with larger CD coefficient (large core area and lower loss) are being planned in the near future with total CD as large as 250,000 ps/nm for transpacific links. The discussion provided later applies generally to the effect of dispersion on phase detectors. Dispersion affects each signal component differently. Modulated signal components at $+f_b/2$ and $-f_b/2$ propagate at slightly different speeds due to dispersion, and when they arrive at the receiver, they can be time-skewed by 1 symbol period or more, and this can severely affect clock phase sensitivity.

Chromatic dispersion can be modeled as an all-pass filter $H[k]$ on the E-field with parabolic phase response (or linear group delay). Consistent with previous assumptions, "f_b" is baud rate, N is the number of samples in the FFT, "k" is the frequency-domain index for frequency bins, and sampling rate of two samples per symbol is assumed. Note that the exponential coefficient "κ" in $H[k]$ is related to the amount of dispersion, and is not to be confused with the frequency bin index "k."

$$H[k] = e^{j\cdot\kappa\cdot(2\cdot\pi\cdot k/N\cdot 2\cdot f_b)^2}$$

$$\kappa = \frac{\mathrm{D}\cdot L\cdot\lambda^2[\mathrm{nm}^2]\cdot 10^{-21}}{4\cdot\pi\cdot c} \tag{10.39}$$

In order to understand how dispersion reduces clock phase sensitivity, we first define a function that measures detector strength. Based on the frequency-domain phase detector derived earlier (see Equation 10.17), the following equation produces output proportional to cosine of the phase error (τ normalized to symbol period). At the lock point, τ is zeroed, and this function is maximized. The maximized value is an indication of the phase detector strength. In the PLL theory, detector strength is K_d and is an important parameter in determining the loop BW.

$$K_d\cos(2\cdot\pi\cdot\tau) \quad \propto \quad \sum_{k=0}^{N/2-1} \mathrm{Re}\{X[k]\cdot X^*[k-N/2]\} \tag{10.40}$$

Similar to the derivation earlier for Equation 10.17, we substitute input signal $X[k] = A[k]\cdot H[k]\cdot H'[k]$ into the equation for detector strength. $A[k]$ is the random data,

$H'[k]$ is an amplitude weighting function, and $H[k]$ is the parabolic phase response in Equation 10.39. Similar to previous derivation with sampling theorem, $A[k]$ is periodic with period $N/2$ and is the same as $A[k - N/2]$.

$$k_d = \sum_{k=0}^{N/2-1} \mathrm{Re}\{X[k] \cdot X^*[k - N/2]\}$$

$$k_d = \sum_{k=0}^{N/2-1} \mathrm{Re}\{A[k] \cdot A^*[k - N/2] \cdot e^{j\cdot\kappa\cdot(2\cdot\pi\cdot k/N\cdot 2\cdot f_b)^2}$$

$$\cdot\, e^{-j\cdot\kappa\cdot(2\cdot\pi\cdot(k-N/2)/N\cdot 2\cdot f_b)^2} \cdot H'[k] \cdot H'[k - N/2]\}$$

$$k_d = \sum_{k=0}^{N/2-1} |A[k]|^2 \cdot H'[k] \cdot H'[k - N/2] \cdot \cos\left[\kappa \cdot \left(2 \cdot \pi/N \cdot 2 \cdot f_b\right)^2 \cdot (k \cdot N - N^2/4)\right] \tag{10.41}$$

Redefine frequency bin index k in the summation, by centering the summation at the half-baud-rate frequency ($N/4$ or $f_b/2$), the $N^2/4$ term can be dropped:

$$k_d = \sum_{k=-N/4}^{N/4} |A\,[k]|^2 \cdot H'[k] \cdot H'[k - N/2] \cdot \cos\left[\kappa \cdot \left(2 \cdot \pi/N \cdot 2 \cdot f_b\right)^2 \cdot k \cdot N\right] \tag{10.42}$$

This equation shows that the term inside the cosine function governs the sensitivity of the phase detector. The dispersion coefficient "κ" induces a phase ramp between frequency components in the summation. This is expected, because dispersion is a linear group delay distortion, and therefore each frequency component will have different delays. If the dispersion is large enough such that the terms in the cosine function between frequency bins $-N/4 < k < N/4$ approaches 2π, then the summation approaches zero. Thus, for large dispersion, the clock phase sensitivity can be NULL.

On the contrary, if the summation range is halved, then it would take a larger κ (or dispersion) to induce a full 2π inside the cosine function, and therefore reducing the range of summation can increase dispersion tolerance of the phase detector. Figure 10.18 shows normalized detector sensitivity K_d versus dispersion generated using Equation 10.42. A 1024-point FFT is assumed ($N = 1024$). The data spectrum $A[k]$ is assumed to be white. The amplitude weighting function $H'[k]$ is assumed to be a full raised cosine function with α factor of 1. Figure 10.18(a) shows that the phase detector loses sensitivity at 1000 ps/nm of dispersion at 16G symbol rate. At 32G symbol rate, the dispersion tolerance is reduced to one-quarter at 250 ps/nm. This result is generated using all 1024 frequency bins. Figure 10.18(b) shows the benefit of using only 32 frequency bins instead of 1024 for a 32 Gbaud signal. Sixteen bins in the upper-side band centered at $+f_b/2$ and 16 bins in the lower-side band centered at $-f_b/2$ can be used to extend the dispersion tolerance dramatically.

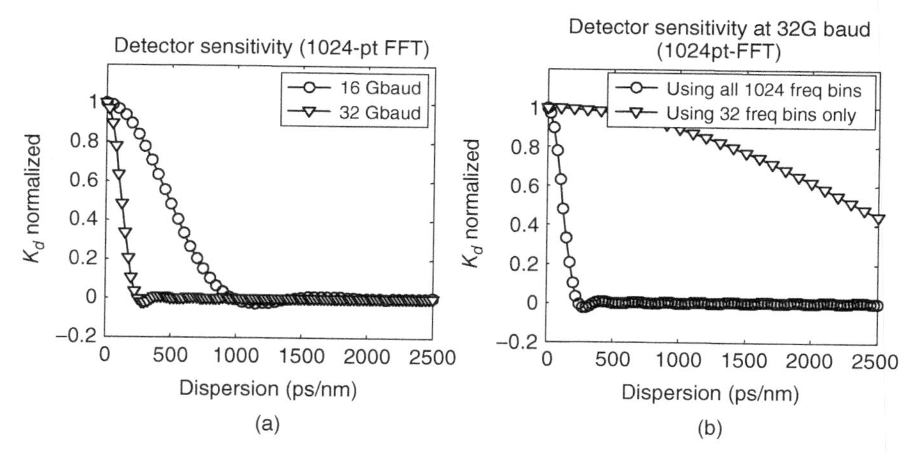

FIGURE 10.18 (a) K_d versus dispersion at 16G and 32G symbol rates using all 1024 FFT bins. (b) K_d versus dispersion at 32G symbol rate using 32 or 1024 FFT bins.

However, reducing the number of frequency components increases the detector noise in that a smaller fraction of the input data is used for averaging. One must account for the performance impact of the increased jitter in this approach. All-in-all, this is an effective way of trading off dispersion tolerance and jitter performance. A similar method using time-domain prefiltering can be found in [35].

An alternative way to compensate for the effect of dispersion in the phase detector of Equation 10.17 is first recognizing that the phase of each of the frequency component "k" in the summation is related to the delay of that frequency component. The phase (or delay) is directly related to the dispersion in the signal. One can compensate for the effect of dispersion by introducing a frequency-dependent phase rotation applying to each of the summation terms as shown in Equation 10.43. Each frequency bin has frequency separation of Δf(Hz). "$\Delta\theta$" defines a phase difference between adjacent frequency bins. "$\Delta\theta$" is related to the amount of dispersion ($D \cdot L$) in the signal as shown in the following:

$$\tau_{\text{est}} = \sum_{k=0}^{N/2-1} \text{Im}\left\{X[k] \cdot X^{*}[k-N/2] \cdot e^{j\Delta\theta \cdot k \cdot \Delta f}\right\}$$

$$\Delta\theta\ [\text{rad/Hz}] = \frac{D[\text{ps/nm/km}] \cdot L[\text{km}] \cdot 10^{-24} \cdot \lambda^2[nm] \cdot 1000 \cdot 2 \cdot \pi \cdot f_{\text{baud}}[\text{Hz}]}{c} \tag{10.43}$$

Figure 10.19 shows a simulation of system jitter (normalized to 1 MHz closed loop BW), sourced from the phase detector alone, using a 32 Gbaud PM-QPSK system with DSP designed to tolerate 50,000 ps/nm. FFT size "N" is chosen to be 1024. The black circles show jitter result applying Equation 10.17 using 128 frequency bins centered at $\pm f_b/2$. The detector can only tolerate 400 ps/nm of dispersion. By using

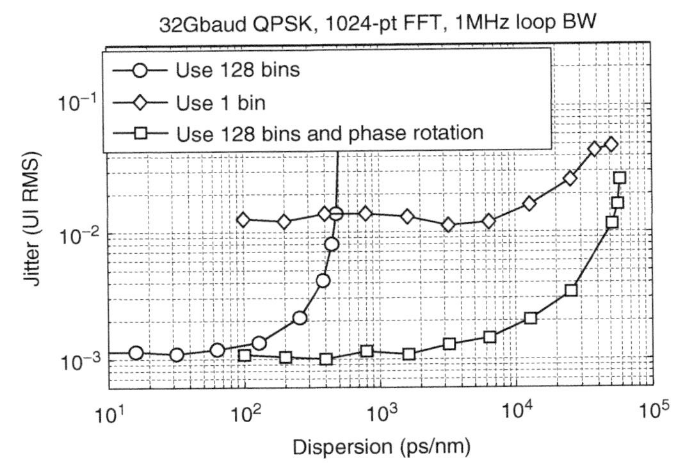

FIGURE 10.19 Dispersion tolerance of phase detectors using 128 frequency bins (line with circle), 1 frequency bin (line with diamond), and 128 bins and phase rotation (line with square) as described in Equation 10.43.

only one frequency bin (black diamond), the dispersion range is increased dramatically, but the jitter also suffers. The method implemented in Equation 10.43 (black squares) extends the detector tolerance out to 50,000 ps/nm.

10.6 THE POLARIZATION-MODE DISPERSION PROBLEM

Modern coherent systems utilize a polarization-diverse receiver and extensive signal processing to compensate for a distortion known as PMD in the fiber. By doing so, one can also increase fiber capacity by polarization multiplexing two independent modulated data streams, one on each polarization of the fiber. In the receiver, each pair of A/D converters sampling the real and imaginary components of the X-polarization contains data from both X- and Y-polarizations of the transmitted field. Since there is no effort in aligning and maintaining the alignment of polarization axis of the fiber and the receiver, the received signal on each polarization is always a linear combination of the transmitted fields. Equation 10.44 illustrates this. The transmitted fields are represented in frequency domain as $T_X[\omega]$ and $T_Y[\omega]$, the received fields as "$X[\omega]$" and "$Y[\omega]$," and the fiber channel can be modeled as a 2×2 frequency-dependent complex matrix H [14, 18].

$$\begin{bmatrix} X[\omega] \\ Y[\omega] \end{bmatrix} = \begin{bmatrix} H_{XX}[\omega] & H_{YX}[\omega] \\ H_{XY}[\omega] & H_{YY}[\omega] \end{bmatrix} \cdot \begin{bmatrix} T_X[\omega] \\ T_Y[\omega] \end{bmatrix} \tag{10.44}$$

The effect of PMD is quite complex. In general, PMD induces different polarization rotations on each frequency component of the input signal [36]. The distortion it creates is dominantly a first-order effect known as differential group delay

(DGD) [37]. As shown later, DGD can severely impact phase detector sensitivity. The second-order PMD induces differential dispersion and depolarization. It can also impact the phase detector, although the first-order PMD effect is dominant. DGD can be modeled as a concatenation of three matrices shown in Equation 10.45. The first and last matrices are polarization rotations in Jones matrix representation with parameters $\theta_1, \phi_1, \theta_2$, and ϕ_2. The diagonal matrix in the middle induces a time delay $(+\tau/2)$ on X-polarization and a matching time advancement $(-\tau/2)$ on Y-polarization. The DGD in this model is equal to τ.

$$\begin{bmatrix} H_{XX}[\omega] & H_{YX}[\omega] \\ H_{XY}[\omega] & H_{YY}[\omega] \end{bmatrix} = \begin{bmatrix} \cos(\theta_2)\, e^{-j\phi_2} & -\sin(\theta_2) \\ \sin(\theta_2) & \cos(\theta_2) e^{-j\phi_2} \end{bmatrix} \cdot \begin{bmatrix} e^{-j\omega\tau/2} & 0 \\ 0 & e^{+j\omega\tau/2} \end{bmatrix} \cdot \begin{bmatrix} \cos(\theta_1)\, e^{-j\phi_1} & -\sin(\theta_1) \\ \sin(\theta_1) & \cos(\theta_1) e^{-j\phi_1} \end{bmatrix} \tag{10.45}$$

The fiber's polarization and PMD states evolve over time due to thermal and mechanical disturbances, and therefore the amount of DGD (τ) and the polarization rotations $(\theta_1, \phi_1, \theta_2, \phi_2)$ all vary as a function of time. Consider the scenario where θ_1 and ϕ_1 are zero and θ_2 and ϕ_2 are also zero, and the DGD is equal to half symbol period $(T/2)$. The delay on X and Y polarizations are $+T/4$ and $-T/4$.

$$\begin{bmatrix} X \\ Y \end{bmatrix} = \begin{bmatrix} e^{-j\omega T/4} & 0 \\ 0 & e^{+j\omega T/4} \end{bmatrix} \cdot \begin{bmatrix} T_X \\ T_Y \end{bmatrix} \tag{10.46}$$

In the receiver, if the phase detector outputs on X and Y polarizations are summed as illustrated in Figure 10.10, then the pair of timing waves of opposite sign are summed ($+\pi/2$ from X and $-\pi/2$ from Y). The phase detector sensitivity is NULLed in this case. Figure 10.20 illustrates the two timing waves of opposite phase on each polarization. Since the polarization states vary slowly, this scenario may persist for milliseconds or tens of milliseconds. This is a very long time from the point of view of clock recovery loop that has loop time constant on the order of microseconds.

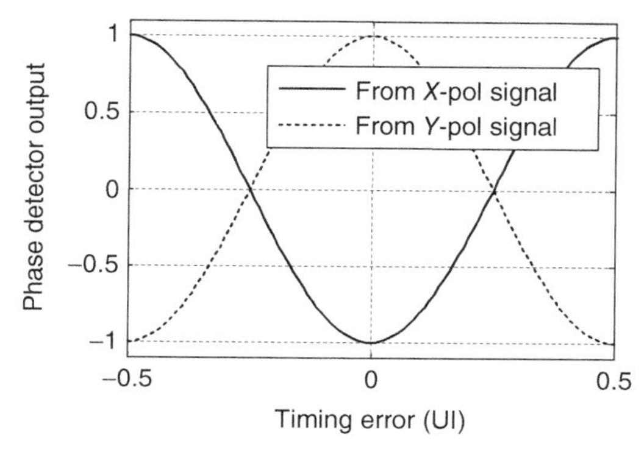

FIGURE 10.20 Phase detector output from X-pol and Y-pol signal sum to a NULL condition.

When the polarization state of the fiber evolves to a state near this condition, the clock recovery loop will lose lock. The PMD tolerance of the receiver in this case is limited by the stability in the clock recovery loop. In order to avoid DGD of half a symbol period, the mean PMD needs to be three times lower than the instantaneous DGD. At 32G symbol rate, the mean PMD is at most 5.2 ps.

One may wonder that if only the received X-polarization signal is used (ignoring the Y-polarization signal) then this particular NULL condition can be avoided. But consider a different polarization state where θ_1, ϕ_1, and ϕ_2 are still zero, but θ_2 is at 45°. The received signal on X polarization contains a linear combination of both the transmitted signals T_X and T_Y as illustrated in Equation 10.47. Applying phase detector on the composite signal is equivalent to detecting clock phase on $T_X \cdot e^{-j\omega T/4}$ and $T_Y \cdot e^{+j\omega T/4}$ individually. The result is also the NULL condition as shown in Figure 10.20.

$$\begin{bmatrix} X \\ Y \end{bmatrix} = \frac{1}{\sqrt{2}} \begin{bmatrix} 1 & 1 \\ -1 & 1 \end{bmatrix} \cdot \begin{bmatrix} e^{-j\omega T/4} & 0 \\ 0 & e^{+j\omega T/4} \end{bmatrix} \cdot \begin{bmatrix} T_X \\ T_Y \end{bmatrix}$$

$$X(\omega) = \frac{1}{\sqrt{2}} \left(T_X \cdot e^{-j\omega T/4} + T_Y \cdot e^{+j\omega T/4} \right) \tag{10.47}$$

This NULL condition always exists and is a consequence of multiplexing two independent signals on two polarizations of the fiber and the effect of the first-order PMD delaying one signal with respect to the other. Note that this NULL condition does not depend on θ_1 and ϕ_1, because the original transmitted fields T_X and T_Y are assumed to be time aligned with respect to each other. This problem of half symbol period DGD is also reported in [38, 39].

To solve this problem, the phase detector of Equation 10.17 is modified to include a special linear combination of received signals from both polarizations [30]. When adding them, positive-frequency components of Y-pol signal ($Y[k]$) are rotated by a scalar phase ϕ_U, and negative-frequency components of Y-pol signal ($Y(k + N/2)$) are rotated by a scalar phase ϕ_L. The final phase detector equation is shown in the following. Similar to earlier assumptions and derivations, the received digital signals $X(k)$ and $Y(k)$ are samples in the frequency domain after an N-point FFT. "k" indexes the discrete frequency bins. The notations used are adopted from Equation 10.17.

$$\tau_{\text{err}} = \sum_{k=0}^{N/2-1} \text{Im} \left\{ \left[X(k) + Y(k) \cdot e^{j\phi_U} \right] \cdot \left[X(k + N/2) + Y(k + N/2) \cdot e^{j\phi_L} \right]^* \right\} \tag{10.48}$$

It can be shown that for all values of rotation θ_2 and ϕ_2, there exists a combination of ϕ_U and ϕ_L that restores the detector sensitivity. Figure 10.21 shows the detector sensitivity K_d as a function of ϕ_U and ϕ_L at $\theta_2 = 30°$ and $\phi_2 = 40°$, and DGD of half symbol period. The troughs are the regions of degradation in K_d and should be avoided.

ϕ_U and ϕ_L needs to be adapted in real-time in order to track the polarization state of the fiber and continuously restore the detector sensitivity. To do so, the detector

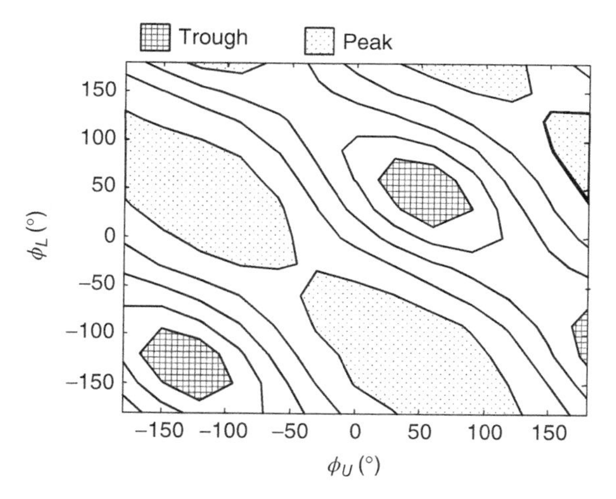

FIGURE 10.21 Phase detector sensitivity contour as a function of ϕ_U and ϕ_L at DGD = $T/2$.

strength can simply be defined as the "real" component of the complex function in 10.48. This definition of detector strength is what is used for Equation 10.40 and stems from derivations in Equation 10.17.

$$K_d = \sum_{k=0}^{N/2-1} \mathrm{Re}\left\{ \left[X(k) + Y(k) \cdot e^{j\phi_U}\right] \cdot \left[X(k+N/2) + Y(k+N/2) \cdot e^{j\phi_L}\right]^* \right\} \tag{10.49}$$

By taking derivative of the K_d function with respect to variables ϕ_U and ϕ_L independently yields the following update equations for ϕ_U and ϕ_L. A short notation X_U is used to represent the X-pol signal in the positive frequencies $X(k)$, and X_L is used to represent the X-pol signal in the negative frequencies $X(k + N/2)$, and similarly for Y_U & Y_L. A digital feedback integrator can be used to continuously update ϕ_U and ϕ_L, using the gradient functions shown in the following equation.

$$\begin{aligned}
\phi_U &= \phi_U + \mu \cdot \frac{dK_d}{d\phi_U} \\
\phi_L &= \phi_L + \mu \cdot \frac{dK_d}{d\phi_L} \\
\frac{dK_d}{d\phi_U} &= \sum_k -\mathrm{Re}\left\{Y_U Y_L^*\right\} \sin\left(\phi_U - \phi_L\right) - \mathrm{Im}\left\{Y_U Y_L^*\right\} \cos\left(\phi_U - \phi_L\right) \\
&\quad - \mathrm{Re}\left\{Y_U X_L^*\right\} \sin\left(\phi_U\right) - \mathrm{Im}\left\{Y_U X_L^*\right\} \cos\left(\phi_U\right) \\
\frac{dK_d}{d\phi_L} &= \sum_k \mathrm{Re}\left\{Y_U Y_L^*\right\} \sin\left(\phi_U - \phi_L\right) + \mathrm{Im}\left\{Y_U Y_L^*\right\} \cos\left(\phi_U - \phi_L\right) \\
&\quad - \mathrm{Re}\left\{X_U Y_L^*\right\} \sin(\phi_L) + \mathrm{Im}\left\{X_U Y_L^*\right\} \cos\left(\phi_L\right)
\end{aligned} \tag{10.50}$$

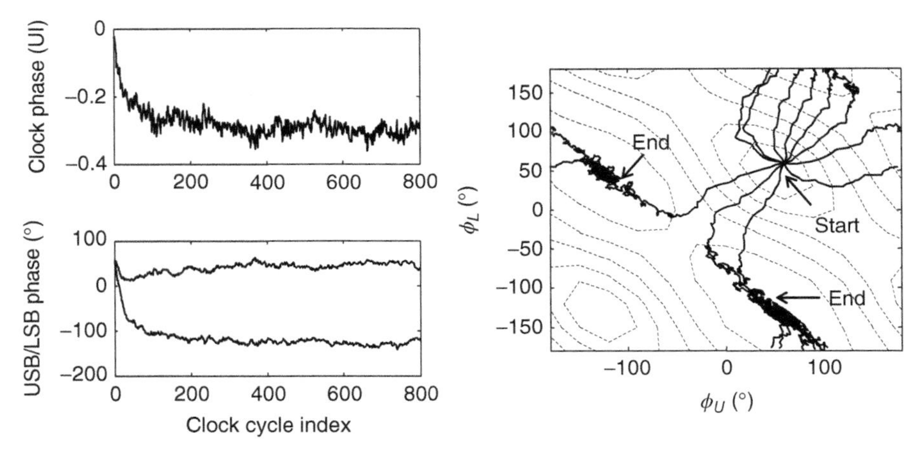

FIGURE 10.22 (a) Convergence curves of clock phase, ϕ_U and ϕ_L. (b) Convergence of ϕ_U and ϕ_L plotted on top of the detector sensitivity surface [30]. From Ref. [30].

Figure 10.22 shows a simulation of all three controls in close loop: clock phase recovery, ϕ_U and ϕ_L updates. Channel model is a *T*/2 DGD followed by rotation $\theta_2 = 30°$, and $\phi_2 = 40°$. Figure 10.22(a) shows convergence of all three parameters. Figure 10.22(b) shows the K_d sensitivity contour as a 2D function of ϕ_U and ϕ_L. The contour surface is the same as that shown in Figure 10.21. Drawn on top of the contour are multiple convergence traces. Each convergence trace is a separate simulation where the parameters ϕ_U and ϕ_L are initialized at the surface minimum. Each trace shows the convergence from a different initial clock phase (−0.5 to 0.5 UI in steps of 0.1 UI). Each trace converges all three loop parameters simultaneously. In all cases, the sensitivity peak is found starting from the minimum sensitivity. The loops should maintain lock as polarization angle is a time-varying function.

10.7 TIMING SYNCHRONIZATION FOR COHERENT OPTICAL OFDM

Coherent optical OFDM has recently emerged as an exciting and promising approach in fiber-optic transmission. Its high spectral efficiency and robustness to channel distortion effects has sparked significant amount of research in this area and produced a number of demonstrations of real-time decoding of coherently demodulated OFDM systems [40–45]. In OFDM transmission, the subcarrier symbol rate is made small by the use of a large number of subcarriers. The data in the subcarriers are multiplexed together at the subcarrier symbol rate by the use of an IDFT (inverse discrete Fourier transform) function at the transmitter, and they are demultiplexed using DFT at the receiver. Owing to the low symbol rate, an OFDM signal is inherently very tolerant to distortion effects such as uncompensated residual dispersion and PMD. The intersymbol-interference (ISI) caused by dispersion and PMD are compensated by the use of a cyclic prefix [41], where an end portion of the OFDM word is copied identically to the beginning of the OFDM word, forming a slightly longer OFDM word. Uncompensated channel ISI does not degrade system performance as long as the time spread of the ISI remains within the guard interval defined by the cyclic prefix. Effects of timing error on an OFDM signal is treated the same way as other channel distortions. Time delay (or jitter) as well as dispersion and PMD have their respective allocations in the guard interval. An OFDM system with guard interval designed to accommodate dispersion can also tolerate a significant amount of channel delay. As an example, a 128 Gbit/s PM-QPSK system using 256 subcarriers and guard interval of 250 ps will tolerate 1000 ps/nm of dispersion or alternatively $\pm$125 ps of timing error with negligible system penalty. In this sense, an OFDM system can be much more tolerant to timing error compared with a single carrier system. However, because the use of guard interval reduces spectral efficiency and capacity, the guard interval needs to be minimized as much as possible by minimizing the timing error. It is to be noted that for large CD, the coherent modem using OFDM still requires a CD compensation block using FFT and IFFT, which is different from the modulation and demodulation functions of OFDM.

Owing to the use of cyclic prefix and DFT windowing, the problem of timing synchronization for OFDM signal is essentially determining the location of the DFT window such that a DFT can be performed in the receiver matching the opposite function in the transmitter. The error in determining the location of the DFT window is a timing error in which the guard interval has to absorb. OFDM signals are also sensitive to residual carrier frequency offset in the received signal before the DFT function. DFT window synchronization and carrier frequency estimation are active research topics in practical implementations of OFDM transmission. In one particular example, a frame is defined that consists of multiples OFDM symbols and a header containing two copies of identical patterns [45, 46]. In the receiver, the known identical patterns can be used to determine the alignment of the DFT window and as well as carrier frequency offset.

10.8 FUTURE RESEARCH

The problem of timing recovery is intimately related to the adopted method of modulation and detection in a transmission system. With the ever-increasing pressure to deliver higher capacity in the fiber and higher single-channel data rates, new techniques continue to surface. Nyquist transmission poses its challenges in detecting the timing phase, and the solutions in literature are still slightly incomplete in dealing with PMD effects. Recent introduction of methods in which multiple signals are frequency-multiplexed at a fraction of the Nyquist rate [47, 48] poses further challenges for timing recovery. Modulation techniques such as optical coherent OFDM with guard interval [40–45] or without guard interval [49] each presents their own challenges.

Driven by forces of economy and market demands on data rate, fiber-optic transmission systems are expected to evolve continuously. Research activity in timing recovery needs to happen in concert with developments in modulation techniques and overall system evolution. The problem of timing recovery has been a fertile ground for research and is expected to be so in the future.

REFERENCES

1. Gardner FM. Phase Lock Techniques. 3rd ed. John Wiley and Sons; 2005.
2. Blanchard A. Phase-Locked Loops. John Wiley & Sons; 1976.
3. L. E. Franks, Statistical properties of timing jitter in a PAM timing recovery scheme. *IEEE Trans Commun*, COM-22, 7, 1974.
4. Franks LE. Carrier and bit synchronization in data communication – a tutorial review. *IEEE Trans Commun* 1980;COM-28(8).
5. Lucky RW, Salz J, Weldon EJ Jr. Principles of Data Communication. New York: McGraw-Hill; 1968.
6. Proakis J. Digital Communications. 2nd ed. McGraw-Hill; 1983.
7. Gupta SC. Phase-locked loops. *Proc IEEE* 1975;63(2).
8. Lindsey WC. A survey of digital phase-locked loops. *Proc IEEE* 1981;69(4).
9. Mueller KH, Muller M. Timing recovery in digital synchronous data receivers. *IEEE Trans Commun* 1976;COM-24(5).
10. Sari H, Desperben L, Moridi S. Minimum mean-squared error timing recovery schemes for digital equalizers. *IEEE Trans Commun* 1986;COM-34(7).
11. Gardner FM. A BPSK/QPSK timing-error detector for sampled receivers. *IEEE Trans Commun* 1986;COM-34(5).
12. Godard D. Passband timing recovery in an all-digital modem receiver. *IEEE Trans Commun* 1978;COM-26(5).
13. Griffin RA , et al. 10Gb/s optical differential quadrature phase shift key (DQPSK) transmission using GaAs/AlGaAs integration. OFC 2002 Post deadline, FD6-1; 2002.
14. Taylor MG. Coherent detection method using DSP for demodulation of signal and subsequent equalization of propagation impairments. *IEEE Photon Technol Lett* 2004;16(2):674–676.
15. Noe R. PLL-free synchronous QPSK polarization multiplex/diversity receiver concept with digital I&Q baseband processing. *IEEE Photon Technol Lett* 2005;17(4):887–889.
16. Han Y, Li G. Coherent optical communication using polarization multiple-input -multiple-output. *Opt Express* 2005;13(19):7527–7534.
17. Tsukamoto S, Ly-Gagnon D-S, Katoh K, Kikuchi K. Coherent demodulation of 40-Gbit/s polarization-multiplexed QPSK signals with 16-GHz spacing after 200-km transmission. Proceedings of OFC; 2005. PDP29.
18. Savory SJ, Stewart AD, Wood S, Gavioli G, Taylor MG, Killey RI, Bayvel P. Digital equalisation of 40 Gbit/s per wavelength transmission over 2480 km of standard fibre without optical dispersion compensation. Proceedings of ECOC 2006; Cannes, France, paper Th2.5.5; 2006.
19. Pfau T, Hoffmann S, Peveling R, Bhandare S, Ibrahim SK, Adamczyk O, Porrmann M, Noé R, Achiam Y. First real-time data recovery for synchronous QPSK transmission with standard DFB lasers. *IEEE Photon Technol Lett* 2006;18(9):1907–1909.
20. Sun H, Wu K-T, Roberts K. Real-time measurements of a 40 Gb/s coherent system. *Opt Express* 2008;16(2):873–879.

21. Murmann B. A/D converter circuit and architecture design for high-speed data communication. Custom Integrated Circuits Conference (CICC); 2013. p 1–78.
22. Fujitsu Semiconductor Europe, LUKE-ES 55–65 GSa/s 8 bit ADC, Mar. 2012. Available at http://www.fujitsu.com/downloads/MICRO/fme/documentation/c63.pdf. Accessed 2013 Sep.
23. Kull L, et al. A 90GS/s 8b 667mW 64x interleaved SAR ADC in 32 nm digital SOI CMOS. ISSCC 2014, Session 22.1; 2014.
24. Xie C. Local oscillator phase noise induced penalties in optical coherent detection systems using electronic chromatic dispersion compensation. OFC 2009, OMT4; 2009.
25. Griffin RA. Laser FM noise impact on DCF-free transmission utilizing electronic dispersion compensation. ECOC 2005, Tu.4.2.5; 2005.
26. Oda S. Interplay between local oscillator phase noise and electrical chromatic dispersion compensation in digital coherent transmission system. ECOC 2010, Mo.1.C.2; 2010.
27. Sun H, Wu K-T. Clock recovery and jitter sources in coherent transmission systems. OFC 2012, OTh4C.1; 2012.
28. Xie C. FEC for high speed optical transmission. Communications and Photonics Conference and Exhibition; ACP, Asia; 2011.
29. Mizuochi T. Next generation FEC for optical communication. OFC 2008; 2008. p 1–33.
30. Sun H, Wu K-T. A novel dispersion and PMD tolerant clock phase detector for coherent transmission systems. OFC 2011, OMJ4; 2011.
31. Haykin S. Communication Systems. 3rd ed. John Wiley and Sons; 1994.
32. Geyer JC, Fludger CRS. Efficient frequency domain chromatic dispersion compensation in a coherent Pol-Mux QPSK-receiver. OFC 2010, OWV5; 2010.
33. Yan M, et al. Digital clock recovery algorithm for Nyquist signal. OFC 2013, OTu2I.7; 2013.
34. Wu K-T, Sun H. Frequency-domain clock phase detector for Nyquist WDM systems. OFC 2014, Th3E.2; 2014.
35. M. Kuschnerov, et al., DSP for coherent single-carrier receivers," *J Lightwave Technol*, 27, 16, 2009.
36. Gordon JP, Kogelnik H. PMD fundamentals: Polarization mode dispersion in optical fibers. *Proc Natl Acad Sci U S A* 2000;97(9):4541–4550.
37. Poole CD, Wagner RE. Phenomenological approach to polarization dispersion in long single-mode fibers. *Electron Lett* 1986;22(19):1029–1030.
38. Zibar D. Analysis and dimensioning of fully digital clock recovery for 112Gb/s coherent polmux QPSK systems. ECOC 2009, 7.3.4; 2009.
39. Hauske FN. Impact of optical channel distortions to digital timing recovery in digital coherent transmission systems. International Conference on Transparent Optical Networks (ICTON) 2010, We.D1.4; 2010.
40. Shieh W, Athaudage C. Coherent optical orthogonal frequency division multiplexing. *Electron Lett* 2006;42:587–589.
41. Shieh W, Bao H, Tang Y. Coherent optical OFDM: theory and design. *Opt Express Lett* 2008;16(2).
42. Lowery AJ, Du L, Armstrong J. Orthogonal frequency division multiplexing for adaptive dispersion compensation in long haul WDM systems. OFC; Anaheim, CA, paper PDP; 2006. p 39.

43. Dischler R, Buchali F. Transmission of 1.2 Tb/s continuous waveband PDM - OFDM – FDM signal with spectral efficiency of 3.3 bit/s/Hz over 400 km of SSMF. OFC; San Diego, USA, paper PDP C2; 2009.
44. Jansen SL, Morita I, Takeda N, Tanaka H. 20-Gb/s OFDM transmission over 4,160-km SSMF enabled by RF-Pilot tone phase noise compensation. OFC; Anaheim, CA, USA, paper PDP15; 2007.
45. Shieh W, Yi X, Ma Y, Yang Q. Coherent optical OFDM: has its time come? *J Opt Netw* 2008;7:234–255.
46. Schmidl TM, Cox DC. Robust Frequency and Timing Synchronization for OFDM. *IEEE Trans Commun* 1997;45:1613–1621.
47. Cai Y, et al. High spectral efficiency long-haul transmission with pre-filtering and maximum a posteriori probability detection. ECOC 2010, September 2010; Torino, Italy, We.7.C.4; 2010.
48. Li L, et al. 20x224Gbps (56Gbaud) PDM-QPSK transmission in 50GHz grid over 3040km G.652 fiber and EDFA only link using soft output faster than Nyquist technology. OFC 2014; San Francisco, W3J.2; 2014.
49. Sano A et al. No-guard-interval coherent optical OFDM for 100-Gb/s long-haul WDM transmission. *J Lightwave Technol* 2009;27(16).

11

CARRIER RECOVERY IN COHERENT OPTICAL COMMUNICATION SYSTEMS

XIANG ZHOU
Platform advanced technology, Google Inc., Mountain View, CA, USA

11.1 INTRODUCTION

Carrier recovery is another important building block for modern coherent optical communication systems. Since both amplitude and phase of an optical carrier are used for carrying the data, phase and frequency synchronization between the signal source and the local oscillator (LO) is required in order to correctly demodulate a coherently modulated optical signal. Early coherent optical communication experiments (1980s to early 1990s) employed optical phase-locked loops (PLLs) for phase synchronization; however, this type of optical method was too complex for practical implementation. With the recent advancement of high-speed electrical processing techniques, digital signal processing (DSP)-based digital carrier recovery methods have recently been introduced, from lower-speed wireless systems to high-speed coherent optical systems [1–3], where phase and frequency deviations are estimated and removed in the digital domain.

Digital carrier recovery in a high-speed coherent optical system is more challenging than a lower-speed wireless system, especially when higher-order modulation formats are employed. Unlike the wireless system in which the frequency and phase offset changes are relatively similar and slow, the characteristics of frequency and phase offsets in a high-speed optical system are very different: the frequency change is relatively slow (in the milliseconds for a high-quality laser) but the range can be

Enabling Technologies for High Spectral-efficiency Coherent Optical Communication Networks,
First Edition. Edited by Xiang Zhou and Chongjin Xie.
© 2016 John Wiley & Sons, Inc. Published 2016 by John Wiley & Sons, Inc.

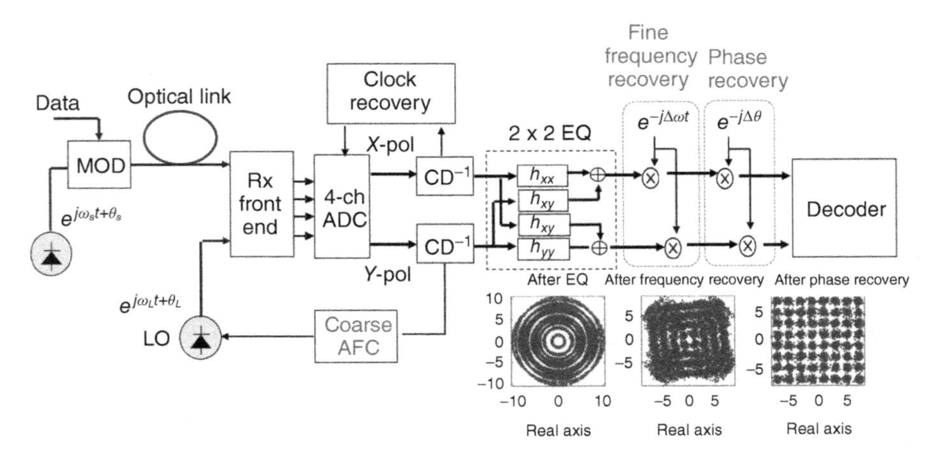

FIGURE 11.1 Digital carrier recovery for a single-carrier modulated high-speed coherent optical system (square 64QAM used as an example). ADC: analog-to-digital converter; MOD: modulator; Rx: receiver; CD: chromatic dispersion; EQ: equalizer.

large (up to 5 GHz for a free-running signal source and LO [4]), while the phase noise varies rapidly in comparison to the wireless system (in the nanosecond range). Furthermore, due to limited complementary metal–oxide–semiconductor (CMOS) speed, a high-speed optical system usually demands high-degree of parallelism which makes feedback-based digital carrier phase recovery algorithms (widely used in wireless communication systems) much less effective for high-speed optical systems. Also due to the CMOS capability constraint, more hardware-efficient algorithms are required for high-speed optical systems.

The basic concept of digital carrier recovery for a high-speed coherent optical transmission system is illustrated in Figure 11.1, in which square 64QAM is used as an example. Due to the differing characteristics of carrier frequency and phase offset, carrier frequency recovery should be performed before phase recovery. Furthermore, since the frequency of a free running laser can drift up to ± 2.5 GHz end-of-the-life [4], without controlling the LO frequency up to 5 GHz extra electrical bandwidth is needed in order to preserve all the needed signal components for the following processing. To avoid such an increased receiver bandwidth requirement, a feedback-based coarse auto-frequency control (AFC) circuit is usually introduced before the nominal frequency recovery. Using a coarse AFC to control the LO could reduce the frequency offset from several gigahertz to the tens of megahertz range. Coarse AFC should be performed before timing recovery because a large frequency offset may cause a problem for timing synchronization [5]. The nominal (fine) frequency and phase recovery are realized after AFC, timing recovery, and signal equalization. The fine frequency recovery unit estimates the residual frequency offset ($\Delta\omega$) down to megahertz range (the required frequency-estimate accuracy depends on the used modulation format), and then removes this offset from the signal. The phase recovery unit estimates the combined phase noise ($\Delta\theta$) from the LO and the signal source and then removes it from the signal. The phase-recovered signal is then sent to the decision and decoding unit.

11.2 OPTIMAL CARRIER RECOVERY

11.2.1 MAP-Based Frequency and Phase Estimator

The optical signal electric field envelope arriving at the receiver is the sum of the modulated transmitter laser and additive noise. Assuming that the transmit pulse shape and receiver impulse response are chosen so there is no intersymbol interference (ISI) and the signal polarization is aligned with the LO, then the received electrical signal after coherent detection can be arrived by

$$y_k = x_k e^{-j\Delta\omega_k t_k + j\theta_k} + n_k \tag{11.1}$$

where y_k and x_k denote the received and transmitted signal at the kth time instant, while $\Delta\omega_k$ and θ_k represent the frequency and phase offset, respectively. n_k is the additive white Gaussian noise (AWGN), a circular Gaussian noise with zero mean and variance of $N_0/2$ per dimension. The conditional probability density function (pdf) of the received signal y_k is given by

$$P(y_k|x_k) = \frac{1}{\pi\sqrt{N_0}} \exp\left(-\frac{\left|y_k - x_k e^{-j\Delta\omega_k t_k - j\theta_k}\right|^2}{N_0}\right) \tag{11.2}$$

The best possible estimate of the carrier frequency and phase that can be made, given the observed received values y_k, is the maximum *a posteriori* (MAP) estimate. Since x_k, $\Delta\widehat{\omega}_k$, and θ_k are statistically independent, the MAP estimate is the sequence of values $\widehat{x}_k$, $\Delta\widehat{\omega}_k$, and $\widehat{\theta}_k$ that maximize the probability function

$$(\widehat{x}_k, \Delta\widehat{\omega}_k, \widehat{\theta}_k) = \max_{x_k, \Delta\omega_k, \theta_k} \left\{ \prod \frac{1}{\pi N_0} \exp\left(-\frac{\left|y_k - x_k e^{-j\Delta\omega_k t_k - j\theta_k}\right|^2}{N_0}\right)\right\} P(x_k)P(\Delta\omega_k)P(\theta_k) \tag{11.3}$$

Since the pdf $P(x_k)$ is known *a priori* and θ_k a Wiener process, taking the logarithm of Equation 11.3 yields

$$(\widehat{x}_k, \Delta\widehat{\omega}_k, \widehat{\theta}_k) = \max_{x_k, \Delta\omega_k, \theta_k} \left\{ \sum \exp\left(-\frac{\left|y_k - x_k e^{-j\Delta\omega_k t_k - j\theta_k}\right|^2}{N_0}\right) - \frac{(\theta_k - \theta_{k-1})^2}{2\pi\tau\Delta\nu} + c\ln(P(\Delta\omega_k))\right\} \tag{11.4}$$

where c is a constant τ denotes the time interval between two consecutive samples, and $\Delta\nu$ denotes the laser linewidth. Equation 11.4 is a joint estimation of the carrier frequency, phase, and data. In general, this joint maximization problem does not yield a closed-form solution, and computationally intense numerical methods have to be used to solve for $\widehat{x}_k$, $\Delta\widehat{\omega}$, and $\widehat{\theta}_k$, which are nearly impossible to be implemented in any practical high-speed communication system. However, the maximization in Equation 11.4 can be performed offline, and the results can be used as a baseline to evaluate the effectiveness of different carrier-recovery algorithms.

For actual real-time implementation, it is necessary to estimate the carrier frequency and phase independently from data recovery, in order to reduce the computational load. Much progress has been made toward developing hardware-efficient suboptimal carrier recovery algorithms, which is described in detail in Sections 11.3 and 11.4.

11.2.2 Cramér–Rao Lower Bound

Before diving into the detail of hardware-efficient carrier recovery algorithms, here we give a brief introduction to the *Cramér–Rao lower bound (CRLB)* [6]. *CRLB* provides us with a more fundamental way to evaluate the effectiveness of various carrier recovery algorithms. In estimation theory and statistics, the CRLB expresses a lower bound on the variance of estimators of a deterministic parameter. The bound is also known as the Cramér–Rao inequality or the information inequality.

In its simplest form, the bound states that the variance of any unbiased estimator is at least as high as the inverse of the Fisher information [7]. The Fisher information is a way of measuring the amount of information that an observable random variable X carries about an unknown parameter θ, upon which the probability of X depends. The probability function for X, which is also the likelihood function for θ, is a function $f(X|\theta)$; it is the probability density of the random variable X conditional on the value of θ. Fisher information is given by [7]

$$I(\theta) = E\left[\left(\frac{\partial}{\partial\theta}\log f\,(X|\theta)\right)^2\right] \tag{11.5}$$

where E denotes expected value (over X) and log denotes natural logarithm.

An unbiased estimator that achieves this lower bound is said to be (fully) efficient. Such a solution achieves the lowest possible mean-squared error among all unbiased methods, and is therefore the minimum variance unbiased (MVU) estimator. Suppose θ is an unknown deterministic parameter that is to be estimated from measurements X, and is distributed according to some pdf $f(X|\theta)$. The variance of any unbiased estimator $\widehat{\theta}$ of θ is then bounded by the reciprocal of the Fisher information $I(\theta)$, that is the CRLB,

$$\mathrm{Var}(\widehat{\theta}) \geq \frac{1}{I(\theta)} = \mathrm{CRLB}(\widehat{\theta}) \tag{11.6}$$

The efficiency of an unbiased estimator $\widehat{\theta}$ measures how close this estimator's variance comes to this lower bound, and estimator efficiency is defined as

$$e(\widehat{\theta}) = \frac{\mathrm{CRLB}(\widehat{\theta})}{\mathrm{Var}(\widehat{\theta})} \tag{11.7}$$

For a general QAM-modulated carrier, there is no closed-form expression for CRLB [8]. But closed-form expressions do exist for an unmodulated carrier [9]. The frequency CRLB for an unmodulated subcarrier (assuming known carrier phase or joint frequency and phase estimation with unknown carrier phase) is given by Rife and Boorstyn [9]

$$\mathrm{CRLB}(\widehat{\varpi}) = \frac{6}{N(N^2-1)\frac{E_s}{N_0}} \tag{11.8}$$

The phase CRLB for an unmodulated subcarrier (assuming known frequency) is given by Rife and Boorstyn [9]

$$\mathrm{CRLB}(\widehat{\theta}) = \frac{1}{2N\frac{E_s}{N_0}} \tag{11.9}$$

In Equations 11.8 and 11.9, N denotes the number of samples used for carrier estimation, while E_s/N_0 denotes the classic signal-to-noise ratio (SNR, E_s representing the average signal energy over a sample period). As compared with the MAP-based method shown in Equation 11.4, Equations 11.8 and 11.9 allow a simpler benchmark calculation method to evaluate the effectiveness of various carrier recover algorithms.

11.3 HARDWARE-EFFICIENT PHASE RECOVERY ALGORITHMS

As mentioned in Section 11.2.1, to reduce the computational load, it is necessary to estimate the carrier frequency and phase independently from the data recovery. Before proceeding to discuss various frequency offset estimation schemes, this section focuses on several hardware-efficient phase estimation algorithms by assuming that the frequency offset has been largely compensated-for before phase recovery.

11.3.1 Decision-Directed Phase-Locked Loop (PLL)

When very narrow linewidth lasers are used for a coherent optical communication system, a decision-directed (DD) PLL may be used to track the carrier phase in a feedback manner without employing any training sequence (i.e., fully blind recovery). Especially, a second-order DD-PLL [10] allows us to track not only the phase change, but also to some degree the frequency change.

Figure 11.2 shows a typical decision-directed second-order PLL for single-carrier-modulated systems. Here, the feedback phase error ϕ_{error} is calculated as follows:

$$\phi_{\mathrm{error}}(k) = \frac{\mathrm{Im}\{\widehat{a}_k^* \cdot y_k e^{-j\Delta\widehat{\phi}(k)}\}}{|\widehat{a}_k^* \cdot y_k|} \tag{11.10}$$

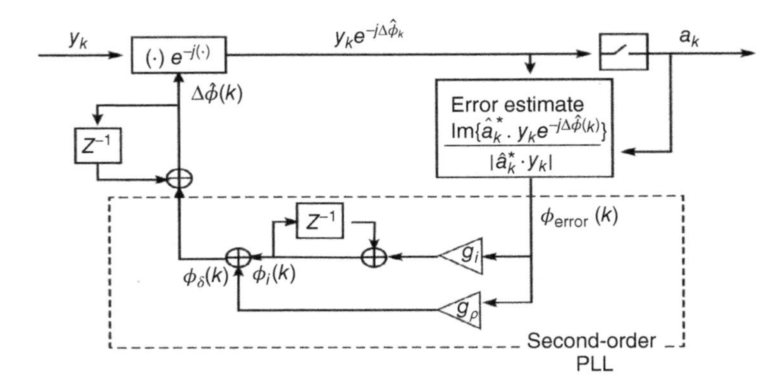

FIGURE 11.2 Decision-directed second-order PLL.

where k is the time index, y_k is the kth received sample (assuming one sample per symbol, after equalization is performed), $\widehat{a}_k^*$ is the conjugate of the kth decided symbol, and $\Delta\widehat{\phi}(k)$ is the kth estimated phase offset. Here, $y_k e^{-j\Delta\widehat{\phi}(k)}$ is the received sample with phase correction based on an estimated phase offset. By multiplying it with $\widehat{a}_k^*$, any phase information encoded in the sample is removed, and the symbol rotated to the x-axis (real-axis). Any deviation of this result from the x-axis is, therefore, representative of the error in phase estimation. For small angles, $\phi \approx \sin(\phi)$, so to simplify calculations, the angle is estimated by measuring the imaginary component, and normalized for magnitude scaling. This normalization is not present in some literature; however, it has recently been found that the use of amplitude normalization could improve PLL performance in low signal-to-noise ratio (SNR) region [11]. The remaining PLL equations are as follows:

$$\phi_i(k) = \phi_i(k-1) + g_i\phi_{\text{error}}(k) \tag{11.11}$$

$$\phi_\delta(k+1) = g_p\phi_{\text{error}}(k) + \phi_i(k) \tag{11.12}$$

$$\Delta\widehat{\phi}(k+1) = \Delta\widehat{\phi}(k) + \phi_\delta(k+1) \tag{11.13}$$

The performance of a second-order DD-PLL in terms of both phase noise and frequency offset tolerance under ideal conditions (the feedback delay is assumed to be a single-symbol clock cycle) has been numerically studied in [10]. It is shown that a second-order DD-PLL could recover both carrier frequency and phase for a 14 Gbaud 16QAM system when the lasers-LO beat linewidth is smaller than 1 MHz and the signal-LO frequency offset is smaller than 140 MHz.

For a realistic high-speed optical system, however, a high-degree of parallelism has to be utilized, since the optimal CMOS clock rate is only about 500 MHz [12]. Furthermore, there are operations within the feedback loop that cannot be completed in a single clock cycle, so pipeline architectures usually have to be employed. The use of parallel and pipeline architectures (assume with the conventional time-interleaving architecture) greatly increases the feedback delay, which will result in much reduced frequency and phase noise tolerance (as compared with the ideal case with feedback delay = 1) [13, 14]. Even with the use of external cavity-based narrow linewidth lasers (linewidth ~100 kHz), the classic DD-PLL may still be useful only for certain lower-order modulation formats such as BPSK or QPSK, which have a relatively large phase error tolerance as compared with higher-order QAMs.

Recently, several variants of DD-PLL-based phase recovery schemes have been proposed by employing different types of loop filters and/or phase detection schemes [15–18]. Some of these modified DD-PLLs may have the potential to achieve slightly better phase estimation performance at the cost of increased implementation complexity. But all of these feedback-based algorithms suffer from the same problem of poor linewidth tolerance due to the extended feedback delay. Although the extended feedback delay inherent in the use of symbol-by-symbol time-interleaving could be avoided by using a block-by-block time-interleaved architecture (i.e., the so-called superscalar parallelization) [19–21]. For this architecture, however, not only does extra overhead have to be introduced within each data block to start the PLL operation on a block-by-block basis, substantial memory/buffer units are also needed to realize block-by-block parallelization. This not only complicates the circuit design, it also introduces considerable latencies.

11.3.2 *M*th-Power-Based Feedforward Algorithms

11.3.2.1 Principle One of the key problems for carrier phase recovery is how to remove the effect of data modulation. For DD-PLL, the effect of data modulation is removed by making a decision-directed (decision feedback) phase estimate. For equal-phase encoded communication systems, the effect of data modulation can also be removed by raising the *M*-ary PSK (phase-shift keying) signal to the *M*th power [22]. To illustrate this principle, a quadrature phase-shift keying (QPSK) signal is used, in which the signal can be represented as

$$y_k = A\ \exp\{j[\theta_{d,k} + \theta_{c,k}]\} \tag{11.14}$$

in which the optical carrier phase $\theta_{c,k}$ is the phase of the transmitter laser referenced to the LO, and the data phase takes on four values, $\theta_{d,k} = 0, \pm\pi/2, \pi$. In Equation 11.14, k denotes the kth symbol. When the received signal is raised to the fourth power as is shown in Figure 11.3, one obtains

$$A^4\ \exp\{j[4(\theta_{d,k} + \theta_{c,k})]\} = A^4\ \exp\{j[4\theta_{c,k}]\} \tag{11.15}$$

because $\exp\{j[4\theta_{d,k}]\} = 1$, that is, the fourth-power operation strips off the data phase. The carrier phase can then be computed and subtracted from the phase of the received signal to recover the data phase as shown in Figure 11.3. Such a feedforward phase estimation scheme lends itself well to real-time digital implementation [24].

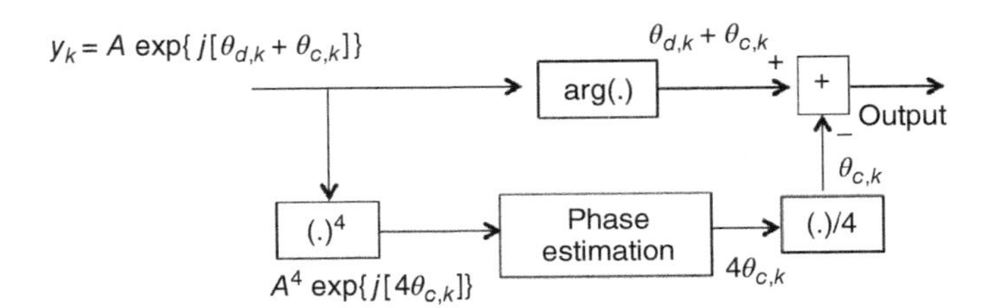

FIGURE 11.3 Schematic of the fourth-power-based feedforward phase estimation algorithm for QPSK. From Ref. [23]. Reproduced with permission of OSA.

Figure 11.3, however, is an idealization in which no additive noise is present in the received signal.

In realistic systems, the received signal will contain noise dominated by either ASE–LO beat noise (where ASE is amplified spontaneous emission) or shot noise of the LO. The impact of additive noise on *M*th power algorithm can be analyzed using a small-signal approximation-based method [24] as briefly described later.

With additive noise included, the received QPSK signal can be written as

$$y_k = A \ \exp\{j[\theta_{d,k} + \theta_{c,k}]\} + n_k \tag{11.16}$$

where the amplitude of the received signal has been normalized, and n_k is the additive noise, which is assumed to be a complex zero-mean Gaussian distribution random variable characterized with a variance σ_n^2. The impact of raising the received QPSK signal to the fourth power yields

$$y_k^4 = \exp\{j4\theta_{c,k}\} + 4\exp\{j[3\theta_{d,k} + 3\theta_{c,k}]\}n_k + O(n_k^2) \tag{11.17}$$

and, in the small-angle approximation,

$$\arg(y_k^4) = 4\theta_{c,k} + \delta(\theta_{c,k})n_k + O(n_k^2) \tag{11.18}$$

where $\delta(\theta_{c,k})$ is a small quantity. It is apparent that the phase estimate is no longer accurate in the presence of additive noise. The phase estimation error will be of the order of $\delta(\theta_{c,k})n_k/4$, which is inversely proportional to the SNR.

A typical method used to reduce the effect of additive noise on phase estimation error is to average the estimated phase over a sequence of symbols by filtering the per-symbol phase estimate through an equal-tap-weight transversal filter [25]

$$\theta_{c,\text{est}} = \frac{1}{4} \arg\left(\sum_{k=1}^{N_b} y_k^4\right) \tag{11.19}$$

This implementation is referred to as block-window filtering as sequences of data are processed in blocks. Assuming the carrier phase is constant over the sequence of symbols, the variance of the phase estimation error due to additive noise is then reduced by a factor equal to the symbol sequence length N_b. The filtering process itself introduces an error in phase estimation, as the carrier phase is actually not constant owing to the finite beat linewidth of the transmitter and LO lasers. This error thus increases with N_b. A trade-off between these two effects dictates that the tap number for the additive noise filter must be optimized. In the case of block-window filtering, the carrier-phase estimate is the same for the entire sequence of symbols. In [26], it was shown through a series of approximations that the phase estimation error for QPSK using a block window can be modeled as a zero-mean Gaussian random variable with a variance depending on the beat linewidth, electrical SNR, and block size. The equal-tap-weight filter can also be implemented by using gliding window filtering [27].

Since the phase estimate is forced in value to the range $-\pi/4 \leq \theta_{c,\mathrm{est}}(k) \leq \pi/4$, there is a fourfold phase ambiguity. Fundamentally, the observed fourfold phase ambiguity originates from the rotational symmetry of the QPSK constellation as is discussed in Section 3.2.3. Thus, the phase estimator cannot discern if the recovered phase is accidently rotated by $\pi/2$. This problem is widely referred to as the "cyclic slip problem" in the literature. If left without addressing, catastrophic error propagation may occur because a single cyclic slip may cause errors for all following symbols, until another reverse-direction cyclic slip occurs as reversion.

11.3.2.2 Differential Bit Coding Quadrant-based differential coding/decoding can resolve the error propagation problem caused by cyclic phase slips [2, 28]. Using QPSK as an example, this technique can be described by the following formulae [2]:

$$
\begin{aligned}
q_{o,k} &= (q_{r,k} - q_{r,k-1} + q_{j,k}) \bmod 4 \\
q_{o,k}, q_{r,k}, q_{j,k} &\in \{0, 1, 2, 3\}
\end{aligned}
\tag{11.20}
$$

where $q_{o,k}$ denotes the differentially decoded quadrant number (see Figure 11.4), $q_{r,k}$ is the received quadrant number and $q_{j,k}$ is the quadrant jump number, which can be

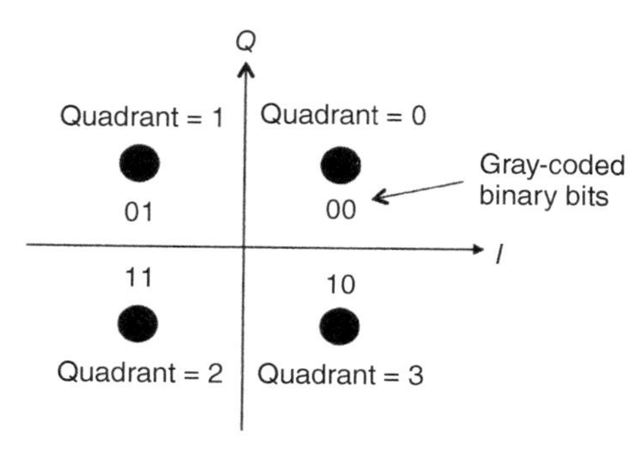

FIGURE 11.4 Schematic illustration of quadrant number and Gray coding.

detected by using the following criteria:

$$\left|\theta_{c,\text{est}}(k) - \theta_{c,\text{est}}(k-1) - \frac{\pi}{2}q_{j,k}\right| < \frac{\pi}{4} \tag{11.21}$$

In Equation 11.21, k denotes the kth blocks for the block-by-block phase estimation. Quadrant-based differential encoding at the transmitter side is given by

$$q_{c,k} = (q_{d,k} + q_{c,k-1}) \bmod 4 \tag{11.22}$$

where $q_{c,k}$ denotes the differentially encoded quadrant number while $q_{d,k}$ denotes the original quadrant number (before differential encoding). For a more generalized M-PSK, the $\pi/4$ in Equation 11.21 should be replaced by π/M, and mod 4 operation should be replaced by mod M.

11.3.2.3 Wiener Filtering Because the laser phase noise is a Wiener process and the additive noise is Gaussian, the Wiener filter that is applied to the unwrapped phase estimate provides the best phase estimation according to estimation theory [29]. The detailed derivation can be found in [30]. The Wiener filter can make an estimate $\theta_{c,\text{est}}(k)$ based on all samples up to and including y_k, in which case the filter is referred to as the zero-lag Wiener filter. Alternatively, the Wiener filter can make an estimate $\theta_{c,\text{est}}(k)$ based on all samples up to and including y_{k+D}, where D is a positive integer, in which case the filter is referred to as the finite-lag Wiener filter. The finite-lag Wiener filter has been shown to perform better because it estimates the phase both forward and backward in time. The z transfer function of the zero-lag and finite-lag Wiener filters are given by Taylor [30]

$$H_{\text{ZL}}(z) = \frac{1-\alpha}{1-\alpha z^{-1}} \tag{11.23}$$

$$H_{\text{FL}}(z) = \frac{(1-\alpha)\alpha^D + (1-\alpha)^2 \sum_{k=1}^{D} \alpha^{D-k} z^{-k}}{1-\alpha z^{-1}} \tag{11.24}$$

from which the coefficients of the finite impulse response (FIR) filters can be obtained accordingly. The parameter α depends on the variance of phase noise and additive noise and is given by

$$\alpha = \frac{M^2\sigma_w^2 + 2\sigma_q^2 - M\sigma_w\sqrt{M^2\sigma_w^2 + 4\sigma_q^2}}{2\sigma_q^2} \tag{11.25}$$

for an M-ary PSK signal. In Equation 11.24, σ_q^2 is the variance of the noise term $\delta(\theta_{c,k})n_k$ in Equation 11.18 and $\sigma_w^2 = 2\pi\Delta\nu T$, where $\Delta\nu$ is the beat linewidth of the transmitter and LO lasers, and T is the symbol period.

While the power law combined with finite-lag Wiener filters produces an optimal phase estimate, its real-time implementation in a parallel architecture digital processor may require the use of complicated look ahead-based computation algorithms [30] since there are places where feedback of the immediately preceding result is needed for utilizing finite-lag Wiener filter. As an alternative suboptimal solution, a gliding window filter might be more compatible with real-time implementation.

11.3.2.4 Discussion To reduce the implementation complexity, the Mth-power operation used for an M-ary PSK signal, which requires M complex multipliers, can be replaced by a much simpler angle-based Mod operation as [31]

$$\theta_{c,\mathrm{est}}(k) = \arg\{r_k\} \bmod \frac{\pi}{M} \tag{11.26}$$

because Equation 11.26 shares the same property as

$$\theta_{c,\mathrm{est}}(k) = \frac{1}{M}\left[\arg\left\{r_k^M\right\} \bmod 2\pi\right] \tag{11.27}$$

The Mth-power-based algorithm allows the use of a feedforward configuration (see Figure 11.3), so its phase noise tolerance performance will not be limited by the use of parallel and pipeline processing. Although this algorithm is originally proposed for M-PSK systems, it can be extended to QAM-modulated systems by using constellation partition-based methods [32–35]. The basic idea is to classify QAM symbols and select only those symbols having equal phase spacing for phase estimation. For a high-order QAM, however, only a small portion of QAM symbols can be used for phase estimation. This inevitably results in reduced phase noise tolerance.

11.3.3 Blind Phase Search (BPS) Feedforward Algorithms

11.3.3.1 Principle For arbitrary QAM modulation formats, robust phase recovery can be achieved by using a minimum distance-based blind phase search (BPS) algorithm. The BPS algorithm was first introduced for coherent optical transmission systems in [14]. For this algorithm, the carrier phase is scanned over a limited phase range ($[0, \pi/2]$ for a square QAM) at fixed or variable phase increments, and the decisions made following each trial phase are approximated as the correct/reference signal for mean square distance error (MSDE) calculation. The optimal phase is the one that gives the minimum MSDE.

The principle of this algorithm is illustrated in Figure 11.5. For convenience, we denote the digitized signal (one sample per symbol) entering into the carrier phase recovery module as y_k. To recover the carrier phase in a pure feedforward approach, BPS requires y_k to be rotated by multiple test carrier phase angles ϕ_m. If the constellation is rotationally symmetric by γ, the trial phase angle can be selected by

$$\phi_m = \frac{m-1}{B} \cdot \gamma, \quad m \in \{1, 2, \ldots, B\} \tag{11.28}$$

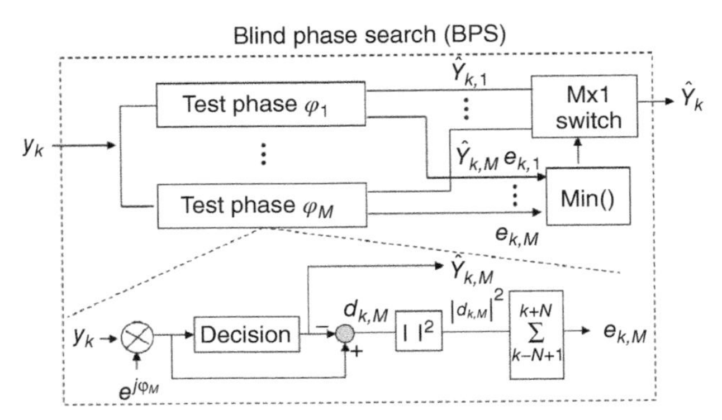

FIGURE 11.5 Schematic illustration of minimum-distance-based BPS algorithm. From Ref. [36].

where M denotes the total number of selected trial phase angles. For square QAM constellations, $\gamma = \pi/2$ holds. Without rotational symmetry $\gamma = 2\pi$ must be used. Then, all rotated symbols are fed into a decision circuit and the squared distance $|d_{k,m}|^2$ to the closest constellation point is calculated. In order to remove distortions from additive noise, the distances of $2N$ consecutive test symbols rotated by the same carrier phase angle ϕ_m are summed

$$e_{k,m} = \sum_{n=-N+1}^{N} |d_{k-n,m}|^2 \tag{11.29}$$

and the "optimum" phase angle is determined by searching for the minimum sum of the distance values. As the decoding was already executed in each phase test unit, the decoded output symbol $\widehat{Y}_k$ can be selected from the $\widehat{Y}_{k,m}$ by a switch controlled by the index $m_{k,\min}$ of the minimum distance sum.

11.3.3.2 Generalized Differential Bit Coding Due to the fourfold ambiguity of the recovered phase in the square QAM constellation, the receiver cannot uniquely assign the $\log_2(M)$ bits to the recovered symbol. This problem can be resolved by applying a generalized differential coding and decoding technique [13, 28]. For square M-QAM constellations, the differential encoding and decoding process is the same as for QPSK since M-QAM also exhibits fourfold phase ambiguity. Thus, it is sufficient to differentially Gray-encode the two bits that determine the quadrant of the complex plane. The only required modification of the decoding process compared with Equations 11.20 and 11.21 is that quadrant jumps are detected according to the following formula [14]:

$$q_{j,k} = \begin{cases} 1, \text{ if } m_{k,\min} - m_{k-1,\min} > B/2 \\ 3, \text{ if } m_{k,\min} - m_{k-1,\min} < -B/2 \\ 0, \text{ otherwise} \end{cases} \tag{11.30}$$

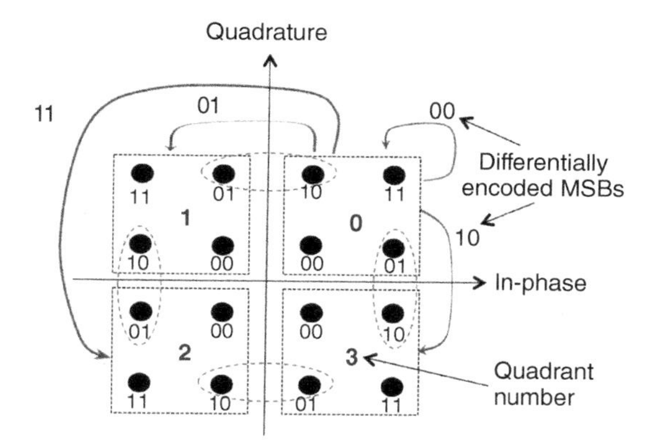

FIGURE 11.6 A 16QAM bit-to-symbol assignment: the encircled symbol pairs have a Hamming distance larger than 1 due to differential encoding of the quadrant.

All other bits that determine the symbol within the quadrant of the complex plane are Gray-encoded without any differential encoding or decoding. Figure 11.6 exemplifies the bit-to-symbol assignment including differential encoding/decoding for square 16QAM.

For arbitrary QAM constellations with k-fold phase ambiguity ($k = 2\pi/\gamma$), $[\log_2^k]$ bits should be differentially encoded/decoded. If $[u]$ is the smallest integer larger than or equal to u, then the differential decoding formulae given in Equations 11.20 and 11.21 should be modified as

$$\begin{aligned} q_{o,k} &= (q_{r,k} - q_{r,k-1} + q_{j,k}) \bmod k \\ & q_{o,k}, q_{r,k}, q_{j,k} \in \{0, \ldots, k-1\} \end{aligned} \tag{11.31}$$

Although the differential-bit-encoded QAM resolves the phase ambiguity and the associated error propagation problem, this capability comes with an intrinsic bit error rate (BER) penalty. For example, for QPSK, an isolated single symbol error will result in two continuous symbol errors after differential decoding. But recent progress in coded modulation techniques reveal that this BER penalty can be minimized by using an iterative differential decoding technique as described in Chapter 3.

11.3.3.3 Performance Discussions As a feedforward phase estimation algorithm, the BPS allows all constellation points to be used for phase estimation with arbitrary QAMs. As a result, for higher-order QAMs, this method can achieve much better linewidth tolerance than constellation-partition-based Mth-power algorithms [14, 32]. With this algorithm, the carrier phase estimator efficiency, which is defined as the ratio between the CRLB and the mean squared error of the phase estimator output, can reach 80% for square 16QAM by using 6-bit phase quantization ($B = 64$) [14], which is close to an optimal phase estimator. The linewidth tolerance achieved

TABLE 11.1 Achievable linewidth tolerance using the BPS algorithm with differing square QAM constellations

Constellations	Max tolerable $\Delta f \cdot T_s$ for 1 dB penalty at BER = 10^{-3}	Max tolerable Δf for a 32 Gbaud system (MHz)
4QAM	4.1×10^{-4}	13.12
16QAM	1.4×10^{-4}	4.48
64QAM	4.0×10^{-5}	1.28
256QAM	8.0×10^{-6}	0.256

by using the BPS algorithm with different square QAM constellations is summarized in Table 11.1, where Δf denotes the signal-LO beat linewidth.

11.3.3.4 Hardware Efficiency Discussions The BPS algorithm requires many vector rotation operations. The rotation of a symbol in the complex plane normally requires a complex multiplication, consisting of four real-valued multiplications with subsequent summation. This would lead to a large number of multiplications having to be executed, in order to achieve a sufficient resolution for the carrier phase values φ_m. The hardware effort would, therefore, become prohibitive. Applying the CORDIC (coordinate rotation digital computer) algorithm [37] greatly reduces the hardware effort needed for calculating the B-rotated test symbols since this algorithm computes vector rotations by simply using summation and shift operations. The hardware efficiency can be further improved by using a lookup table-based mean square distance calculation method [14]. By using the CORDIC and the lookup table-based distance calculation method, the required overall hardware effort for the BPS is roughly B times higher than QPSK phase recovery using the Mth-power-based algorithm [14] – however, the required B increases with the modulation order. For example, in order to achieve a close-to-optimal performance, the required number of trial phases needs to be greater than 16 for 16QAM, and greater than 64 for 64QAM [14]. So the implementation complexity can still be high for high-order QAMs.

11.3.4 Multistage Carrier Phase Recovery Algorithms

To address the performance and implementation complexity challenges facing higher-order coherent QAM systems, several multistage phase recovery algorithms have recently been proposed [38–45]. The core idea of these algorithms is to use a hardware-efficient but less-accurate phase estimator at the first stage to perform coarse-phase estimation, and then refine the estimated phase with a more accurate fine-phase estimator. To further improve this performance, more than one stage of fine-phase estimator may be applied. The coarse-phase estimation stage could be a BPS estimator with coarse trial-phase resolution [38, 41, 42], it could also be a decision-directed PLL [39, 40] or a constellation partition-based Mth-power algorithm [43–45]. If training symbols are allowed, coarse phase can also be estimated from sparsely and periodically inserted training symbols [46]. Fine-phase estimation can be realized by using the BPS estimator with finer-phase resolution over a narrower phase-varying range, the constellation-assisted maximum likelihood (ML) phase estimator [38], or some constellation transformation-based algorithms [43], in which the regular QAM constellation after coarse-phase recovery is first transformed into an M-ary PSK-like constellation and then the Mth power algorithm is applied to the transformed constellations for a more accurate phase estimation. In the following, we describe in more detail three hardware-efficient multistage phase recovery algorithms that have been demonstrated for high-SE 100 Gb/s and beyond transmission experiments. These are (i) the multistage hybrid BPS/ML phase recovery algorithm, (ii) the hybrid PLL/ML algorithm, and (iii) the training-assisted two-stage BPS/ML algorithm.

11.3.4.1 Multistage Hybrid BPS and ML Algorithm The multistage hybrid BPS/ML algorithm was first proposed in [38]. The principle of this algorithm is illustrated in Figure 11.7. For this method, the BPS algorithm with coarse trial-phase resolution is used in the first stage to find a rough location of the optimal phase angle. The decoded/decided signal $\widehat{Y}_n^{(1)}$ based on this rough phase estimation (along with the original signal y_k) are then fed into the second stage where an ML phase estimate is employed to find a more accurate phase estimate φ_k^{ML} by Proakis [47]

$$H_k = \sum_{n=k-N+1}^{k+N} y_n [Y_n^{(1)}]^* \tag{11.32}$$

$$\phi_k^{\mathrm{ML}} = \tan^{-1}\{\mathrm{Im}[H_k]/\mathrm{Re}[H_k]\} \tag{11.33}$$

where $\widehat{Y}_n^{(1)}$ serves as the (approximate) reference signal and N denotes the block length used for phase averaging. The decoded signal $\widehat{Y}_n^{(2)}$ based on this ML phase estimate

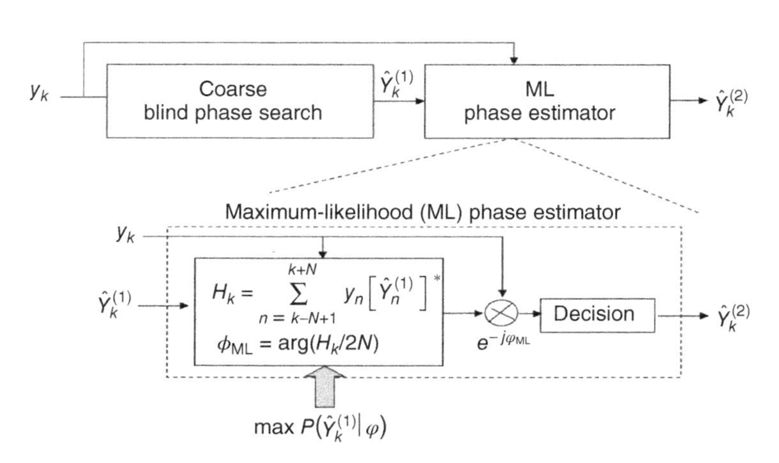

FIGURE 11.7 Two-stage hybrid BPS/ML algorithm.

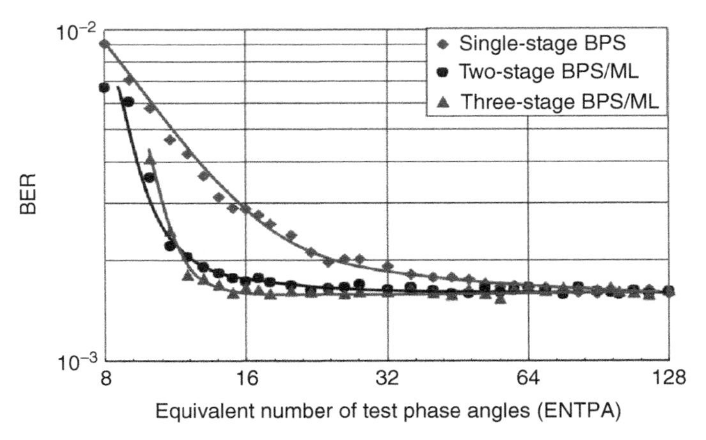

FIGURE 11.8 Required ENTPA for a square 64QAM system using three different phase recovery scenarios (simulation results). From Ref. [49]. Reproduced with permission of OSA.

along with the original signal y_n may be passed into another ML phase estimation stage to further refine the phase estimation.

The effectiveness of this method has been demonstrated through both simulation [38] and experiments [48]. Figure 11.8 shows the simulated results for a 38 Gbaud square PDM-64QAM system, where the required equivalent number of test phase angles (ENTPA) is used to measure the relative implementation complexity for three different phase recovery algorithms – the single-stage BPS, and the two- and three-stage hybrid BPS/ML. For this study, the laser phase noise for both the signal source and the LO is assumed to be 100 kHz. The received OSNR in 0.1 nm noise bandwidth and for a single polarization is 28 dB. Thus, it can be seen that, in order to achieve a performance that is close to optimum, the single-stage BPS method needs to test approximately 64 different phase angles, while the three-stage hybrid BPS/ML algorithm only needs to equivalently test 18 different phase angles (14 test phase angles used in the first coarse BPS stage plus two cascaded ML phase estimation stages). This results in a reduction of computational effort by more than a factor of 3.

The simulated BER performance versus OSNR is given in Figure 11.9, in which we compare two different phase recovery methods: the single-stage BPS using 64 test phase angles and the three-stage hybrid BPS/ML using an equivalent 18 test phase angles. Two laser linewidths, 100 kHz and 1 MHz, are investigated here, with corresponding phase block lengths of 28 and 16, respectively. The results with 0 kHz laser linewidth using ideal phase recovery are also displayed as a reference. For the 100 kHz laser linewidth, the multistage method (using equivalent 18 test phase angles) can achieve almost identical performance to the single-stage BPS method using three times more test phase angles for a wide range of OSNR values. The impact of ASE noise on the performance of the ML phase estimate introduced in the new multistage method is quite small even for OSNR down to 23 dB (corresponding

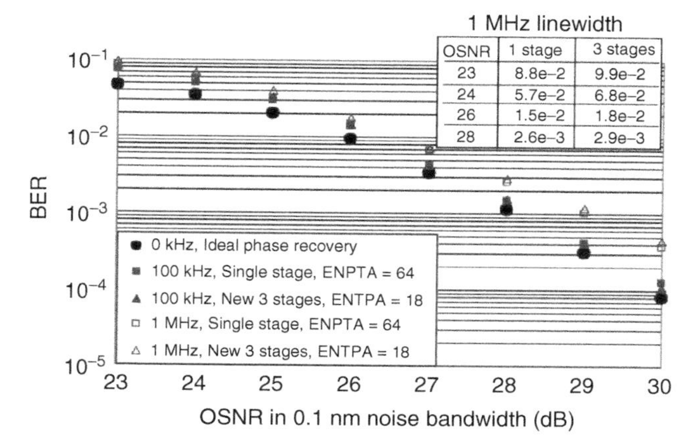

FIGURE 11.9 BER versus OSNR using different phase recovery scenarios.

to a BER 7.6×10^{-2}). For a 1 MHz laser linewidth, however, the multistage method exhibits a slightly worse performance compared with the single-stage BPS. This may indicate that the ML phase estimate is more sensitive to the residual phase error than the BPS method because a larger linewidth implies a faster-changing symbol phase, resulting in larger residual phase errors after block-by-block-based phase recovery.

11.3.4.2 Multistage Hybrid PLL/ML Algorithm To further reduce the implementation complexity, [39] presents a detailed study of a different multistage strategy, in whicha first-order decision-directed PLL is used as a coarse phase estimator and the ML phase estimator is used for fine phase recovery. This design is illustrated in Figure 11.10. It is shown that such a multistage configuration can reduce the implementation complexity by more than one order of magnitude as compared with

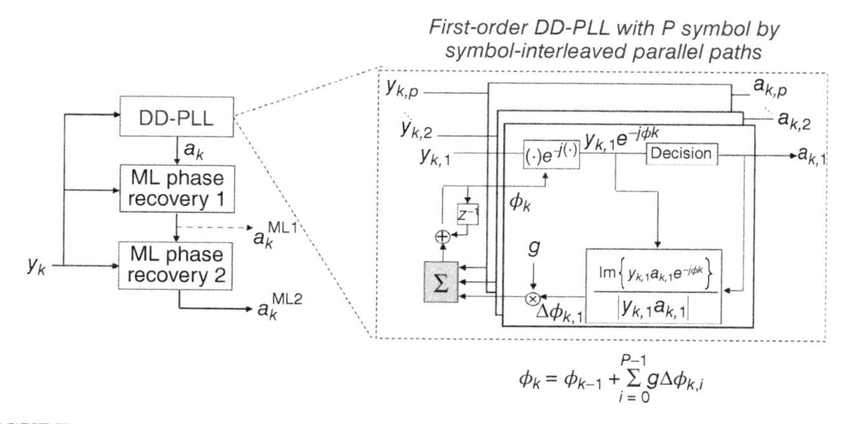

FIGURE 11.10 Multi-stage hybrid PLL/ML phase estimator. *P*: symbol-by-symbol interleaved parallel paths. DD-PLL: decision-directed phase locked loop.

the single-stage BPS for a 64QAM system, but at the cost of reduced laser phase noise tolerance when the degree of parallelism is high. But as compared with the traditional single-stage DD-PLL, such a multistage algorithm can improve the linewidth tolerance by more than two orders of magnitude.

The effectiveness of this algorithm has been tested in a 9.4 Gbaud 64QAM (single polarization) back-to-back experiment as is shown in Figure 11.11(a) and (b). Figure 11.11(a) shows the impact of parallel processing on the proposed algorithms for a constant 23 dB OSNR. One can observe that one PLL followed by two stages of MLs can achieve the same BER performance as the single-stage BPS for the symbol-by-symbol interleaved parallel path P up to 20. The BER performance versus OSNR level for $P = 16$ is given in Figure 11.11(b). This multistage algorithm can achieve a performance similar to the BPS for a wide range of OSNR levels with BER ranging from 2×10^{-2} to close to 10^{-4}.

Figure 11.12 shows the simulated linewidth tolerance performance for using different phase recovery algorithms for a 38 Gbaud square 64QAM system (laser linewidth = 100 kHz, OSNR = 25 dB and PLL pipe line delay $D = 5$). One can see that the linewidth tolerance of the multistage algorithm is more than two orders better than PLL-only method.

11.3.4.3 Training-Assisted Two-Stage Phase Recovery Algorithm As mentioned earlier, if training symbols are allowed, the coarse phase can be estimated from the sparsely and periodically inserted training symbols. Figure 11.13 shows the functional block illustration for a training-assisted two-stage phase recovery algorithm proposed and demonstrated in [46]. For this method, training symbols (known at the receiver) are periodically inserted into the data stream to assist in the phase recovery. These training symbols may also be used for other purposes, such as frame synchronization. To reduce overhead, training symbols are only sparsely inserted at the transmitter. At the receiver, the received data are processed block by block, where each block consists of at least two training symbols. For each block of data,

- First, the average phase over this block is estimated by using the inserted training symbols through an ML phase estimator [47] and
- second, each block is divided into multiple groups, and then the phase of each group is refined by using a BPS-based phase estimator over a small phase-varying range that is centered at the average phase estimated through the training symbols.

One significant advantage for this training-assisted algorithm is its robustness against cyclic phase slips, and thus may remove the need for differential encoding/decoding. Since the baseline phase is recovered from the training symbols, there is inherently no phase ambiguity problem for this algorithm. There still exists chances that the recovered phase may deviate from the true phase by as much as pi/2 [$\pi/2$] due to the impact of large linear or nonlinear noise, but this large phase error or phase jump will only impact a single data block, because the baseline phase of different blocks

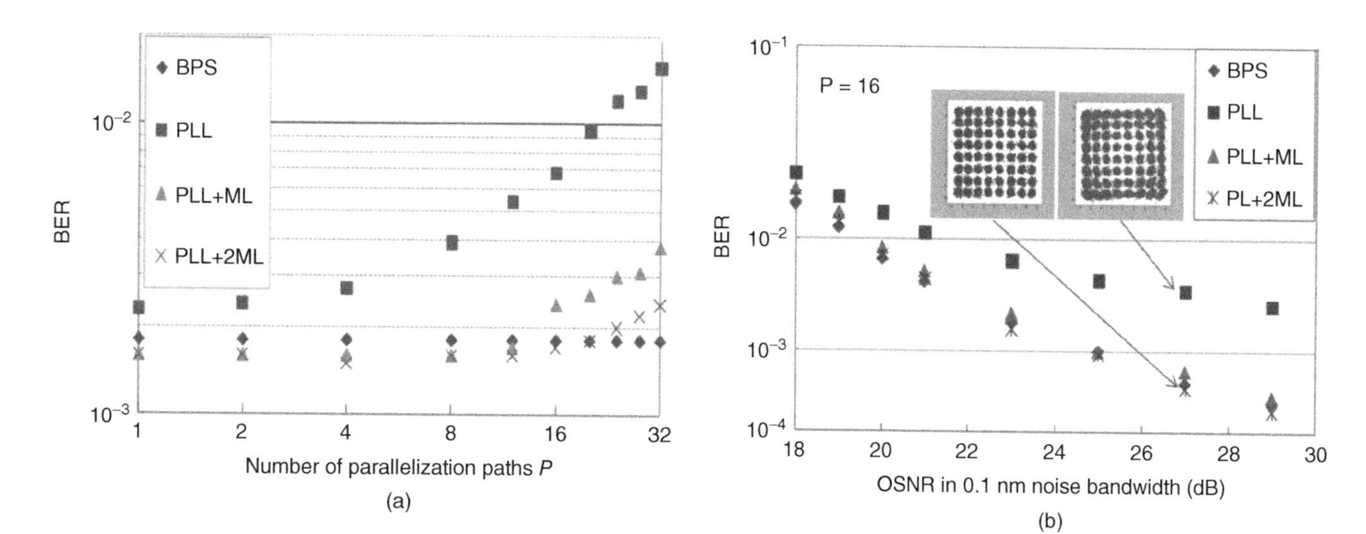

FIGURE 11.11 Experimental results for a 9.4 Gbaud 64QAM system, where (a) shows the BER performance for different degrees of parallelization with 23 dB OSNR and (b) shows the BER performance at different OSNR levels. PLL delay $D = 5$. From Ref. [49]. Reproduced with permission of OSA.

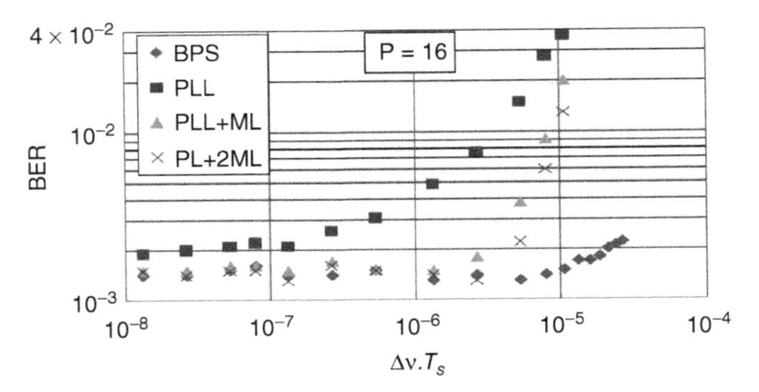

FIGURE 11.12 Simulated linewidth tolerance performance for using different phase recovery scenarios for a 38 Gbaud square 64QAM system (laser linewidth = 100 kHz and PLL pipe line delay $D = 5$).

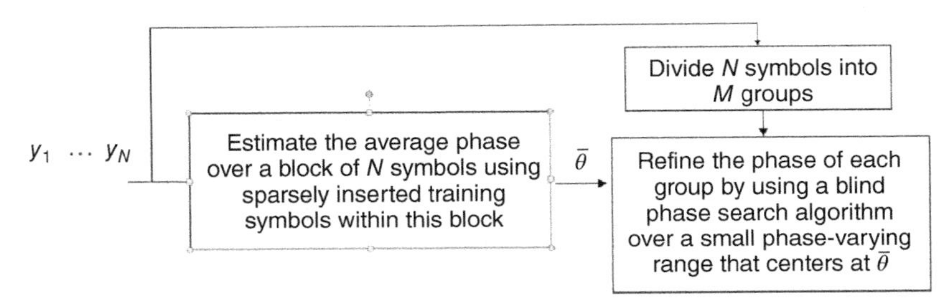

FIGURE 11.13 Functional block illustration for a training-assisted two-stage phase recovery algorithm.

are estimated independently. This effectively prevents errors from propagating from one block to another.

The validity of this new method has been verified by a 400 Gb/s experiment using a time-domain hybrid 32–64QAM [46]. Its robustness against cyclic phase slips is shown in Figure 11.14, with the recovered carrier phases using two different algorithms: the training–assisted two-stage algorithm and the conventional BPS are displayed for a back-to-back measurement with OSNR = 24.2 dB (corresponding to a bit error ratio 2×10^{-2} for using this training-assisted algorithm). One can see that there was no phase jump with the training-assisted two-stage algorithm, whereas the phase-jump problem was severe (due to low OSNR, nonideal equalization, and signal constellation in this experiment) when using the conventional single-stage BPS algorithm, which mandates use of differential coding.

This new two-stage algorithm can achieve comparable or even better (in the low OSNR region) phase noise tolerance than the single-stage BPS method with much

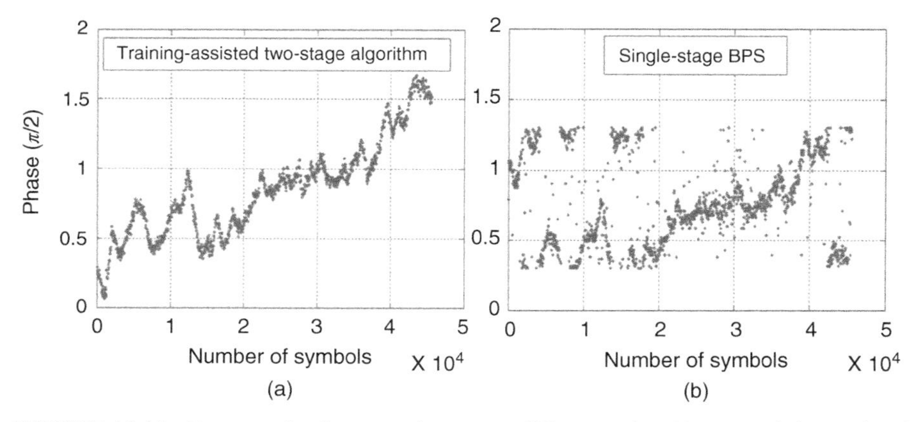

FIGURE 11.14 Recovered phases using two different algorithms: training-assisted two-stage algorithm (a) and single-stage BPS (b). From Ref. [46].

TABLE 11.2 A comparison for several recently demonstrated carrier phase recovery algorithms

	Algorithm	Hardware complexity	Linewidth tolerance	Applicable modulation	Training overhead	Differential coding
Single-stage	DD-PLL	Very low	Low	Any QAMs	No	Yes
	*M*th-power	Low	High	MPSK	No	Yes
	BPS	High	High	Any QAMs	No	Yes
Multi-stage	BPS/ML, BPS/BPS, *M*th power/ML, *M*th power/BPS etc.	Moderate	High	Any QAMs	No	Yes
	PLL/ML, PLL/BPS	Low	Moderate	Any QAMs	No	Yes
	Training-assisted ML/BPS	Moderate	High	Any QAMs	≤2%	No

lower implementation complexity. The required (approximate) 2% training overhead in [46] can be further reduced by exploring joint phase recovery over two orthogonal polarization states for current polarization-multiplexed transmission systems or joint phase recovery over multiple spatial channels for future space division multiplexed systems [50]. Joint phase recovery over multiple spatial channels can also be explored to improve the linewidth tolerance.

As a summary, Table 11.2 gives a brief comparison of several phase recovery algorithms discussed in this section in terms of the achievable hardware efficiency, linewidth tolerance, and several other metrics.

11.4 HARDWARE-EFFICIENT FREQUENCY RECOVERY ALGORITHMS

11.4.1 Coarse Auto-Frequency Control (ACF)

As described in Section 11.1, before the nominal (fine) frequency and phase recovery, a feedback-based coarse AFC is usually required in order to lock the LO frequency close to the incoming signal source frequency (typically within tens or a few hundreds of megahertz range). The key component for the coarse AFC is the frequency error detector (FED). Since nonzero frequency offset between the LO and the signal source will cause signal spectrum asymmetry, two types of FEDs based on this spectrum asymmetry have been developed. The first one is a balanced quadricorrelator (BQ) [51], and the other one is based on differential power measurement (DPM) [52].

11.4.1.1 BQ-Based FDE Figure 11.15 shows a typical balanced quadricorrelator (BQ). Assume that the signal entering into the BQ can be expressed by

$$y_k = (a_k + jb_k)e^{j2\pi\Delta f t_k} \tag{11.34}$$

where Δf denotes the frequency offset, a_k and b_k denote the real and imaginary components of the transmitted (complex) signal. For conceptual simplicity, here we ignore the additive noise as well as carrier phase noise. A straightforward analysis of the balanced quadricorrelator of Figure 11.15 yields the following for its output signal:

$$u_k = [a_k a_{k-1} + b_k b_{k-1}]\sin(2\pi\Delta f T_s) + [b_k a_{k-1} - a_k b_{k-1}]\cos(2\pi\Delta f T_s) \tag{11.35}$$

where T_s is the sampling period. The first term of Equation 11.35 is the desired frequency error signal as long as the sample period is shorter than signal correlation period such that $E\{a_k a_{k-1} + b_k b_{k-1}\}$ is nonzero. The second term is not desired because it only produces pattern jitter if random data are transmitted in an MQAM

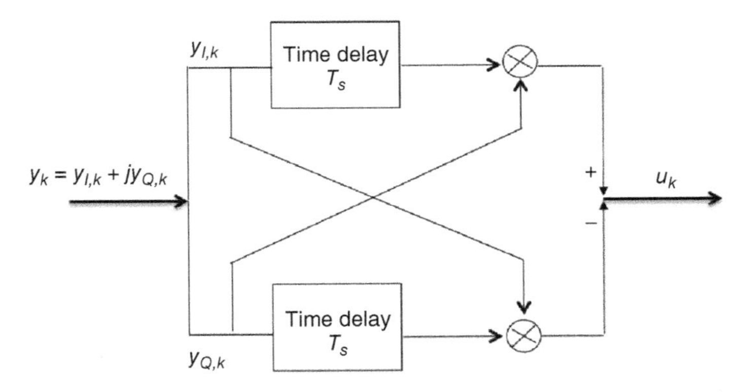

FIGURE 11.15 Balanced quadricorrelator. T_s denotes the sampling period.

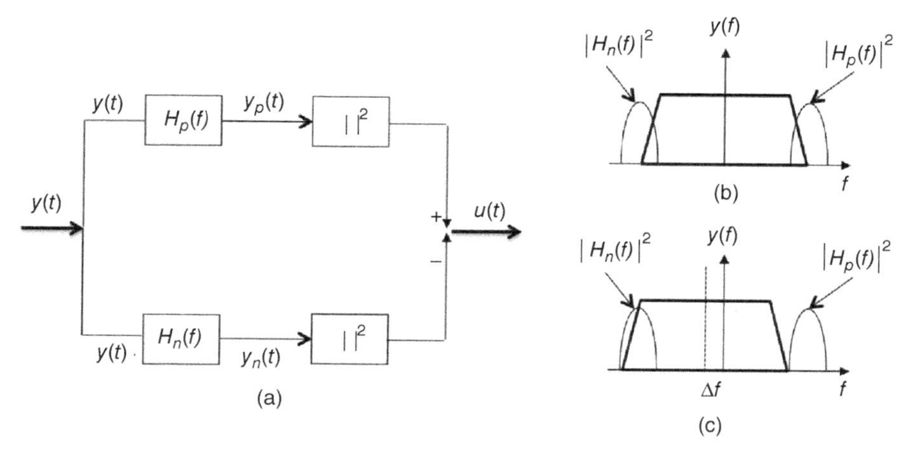

FIGURE 11.16 Frequency error detector with differential power measurement (DPM-FED). (a) Block diagram. (b) Power spectrum of incoming signal: no frequency offset. (c) Power spectrum of incoming signal: frequency offset Δf.

or MPSK transmission system in which $M > 2$. But for real-value modulated signals, such as PAM or BPSK where $b_k = 0$, the second term will be zero. Even with MQAM or MPSK where the second term is nonzero, the produced pattern jitter can still be suppressed by using a low-bandwidth loop filter as was demonstrated in [53], where ± 10 GHz frequency offset error detection was demonstrated by using BQ-based FED for a 43 Gb/s QPSK coherent receiver, with a loop filter bandwidth of 625 kHz.

11.4.1.2 DPM-Based FED Figure 11.6 shows the principle of DPM-based FED. The incoming signal $y(t)$ is fed into two bandpass filters $H_p(f)$ and $H_n(f)$. The output signal $u(t)$ of the DPM-FED is then the difference between the instantaneous powers coming out of the two bandpass filters. In order to provide a useful error signal that can be fed into a feedback network, the signal $u(t)$ must have the following properties:

1. If the incoming signal $y(t)$ is centered at zero frequency, it is necessary that the mean value $E\{u(t)\}$ be zero.
2. If there is a frequency offset Δf then $E\{u(t)\}$ must be a measure of this frequency offset.

For example, the first condition is fulfilled if the power spectrum of the incoming signal is an even function of the frequency and if the bandpass filters observe the relationship $|H_p(f)| = |H_n(-f)|$, as shown in Figure 11.16(b). In order to observe the second condition, a reasonable possibility would be to have the passbands of the bandpass filters in the range around the slopes of the incoming signal (if centered at zero frequency). In the case of a frequency offset *of* Δf, the differential power is then a measure of this frequency offset (Figure 11.16(c)).

However, a pure DPM-FED only yields on average an output signal $u(t)$, which is related to the frequency offset. The desired error signal is distributed by a pattern-dependent term. This produces a pattern-dependent frequency jitter in the AFC loop. A low-bandwidth loop filter can suppress such a pattern-dependent jitter; it can also be suppressed by properly designing the transfer functions of $H_p(f)$ and $H_n(f)$. When $\Delta f = 0$, the pattern-dependent jitter can be completely suppressed by using optimal design of $H_p(f)$ and $H_n(f)$ [54], realizing jitter-free stable operation.

For real-value modulated signal, in order to realize jitter-free operation, $H_p(f)$ and $H_n(f)$ should satisfy the following conditions [54]:

$$H_p(f) = H_n^*(-f) \tag{11.36}$$

For a complex-value modulated signal such as QAM, however, the optimal $H_p(f)$ and $H_n(f)$ depends on channel transfer function (including both transmitter and receiver filters). Let $G(f)$ denote the channel transfer function, then $H_p(f)$ and $H_n(f)$ should satisfy the following criteria to ensure jitter-free operation:

$$G(f - f_s)H_p(f - f_s) = G(f - f_s)H_n(f - f_s) \tag{11.37}$$

where $f_s = 1/2T$ and T denotes the symbol period.

For high-speed long-haul coherent optical systems that employ frequency domain-based equalization for chromatic dispersion (CD) compensation, DPM-based FDE can be easily implemented in the frequency domain by employing hardware-efficient fast Fourier transform (FFT) as was demonstrated in [55]. Since only coarse AFC control is required, there is no need for very large FFT size, implying the FFT used for CD compensation can be reused for FED.

11.4.2 *M*th-Power-Based Fine FO Estimation Algorithms

As described in Section 13.3.2, the *M*th-power algorithm can be used to erase the data modulation for *M*-ary PSK signals. After erasing the data modulation, the frequency offset (FO) can be estimated using either a time-domain-based differential phase method [56] or some FFT-based algorithms [34].

11.4.2.1 Time-Domain Differential Phase-Based Algorithm The principle of the time-domain differential phase-based FO estimation method is illustrated in Figure 11.17, in which QPSK with the fourth-power algorithm is used as an example. For this method, the FO is extracted from the average phase increment between two consecutive data-erased symbols. In order to get a reliable FO estimate at the low OSNR region, a long averaging window of thousands of symbols is typically required even for QPSK. Many more symbols are required for higher-order modulation formats.

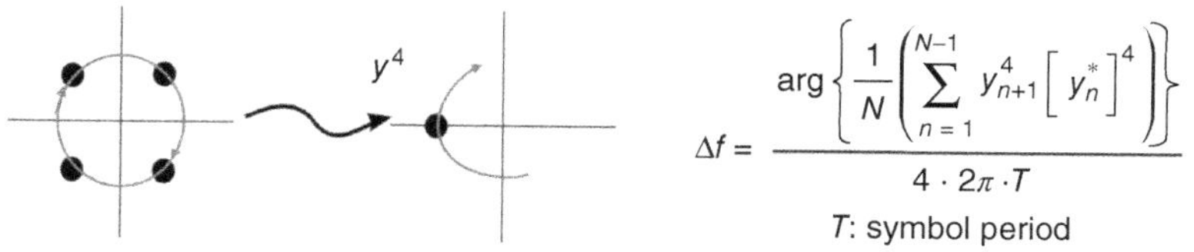

y^4

$$\Delta f = \frac{\arg\left\{\frac{1}{N}\left(\sum_{n=1}^{N-1} y_{n+1}^{4}\left[y_n^{*}\right]^{4}\right)\right\}}{4\cdot 2\pi \cdot T}$$

T: symbol period

FIGURE 11.17 Fourth-power FO estimation using time-domain-based differential phase method.

11.4.2.2 FFT-Based FO Estimation Algorithm The FFT-based method can also be used to extract the FO from the data-erased signals, because the phase angle of a data-erased signal will exhibit an FFT peak at M times the FO. FFT-based methods can achieve better FO estimation accuracy than the time-domain-based method (by using the same number of symbols) but the implementation complexity is much higher, especially for higher-order QAMs, where tens of thousands of symbols may have to be used for a reliable and accurate FO estimate [34, 57]. Furthermore, a single FFT operation can only determine the magnitude of the FO. In order to get the sign of the FO, an additional FFT may have to be used [57].

Some efforts have been made to simplify the FFT-based FO estimator for high-order QAMs [58]. For a square QAM, it is shown that, by using only the outmost four constellation points combined with the use of linear interpolation and down sampling-based methods, the implementation complexity can be greatly reduced. Figure 11.18 illustrates the schematics of this method. First, the received one sample per symbol signal is preprocessed with constellation classification, previous neighbor interpolation, and down-sampling. The *M*th-power algorithm is then used to erase the data modulation. Finally, the frequency magnitude and sign are detected by using two concurrent FFTs with a modified FFT architecture as described in detail in [58].

As an example, Figure 11.19 shows the classification applied to a square 64QAM, in which a ring-based classification method is employed. Symbols are classified as Class I if their magnitude is closest to a Class I ring, and Class II if otherwise. The Class I points are identified by four rings, which exactly intersect the transmitted symbols that lie on a perfect diagonal; these symbols can be derotated using the *M*th power algorithm.

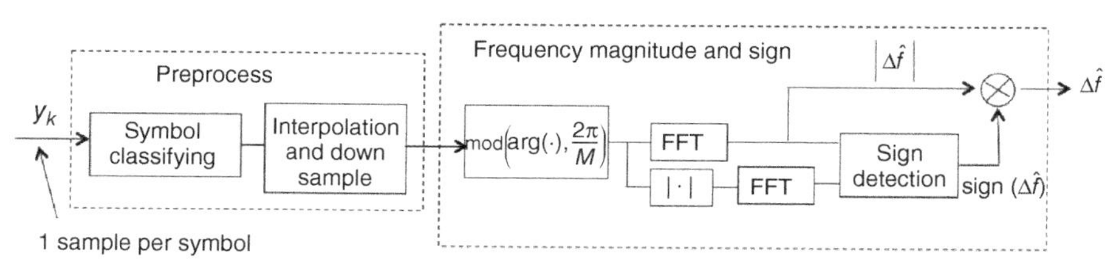

FIGURE 11.18 FFT-based FO estimation method for high-order QAM.

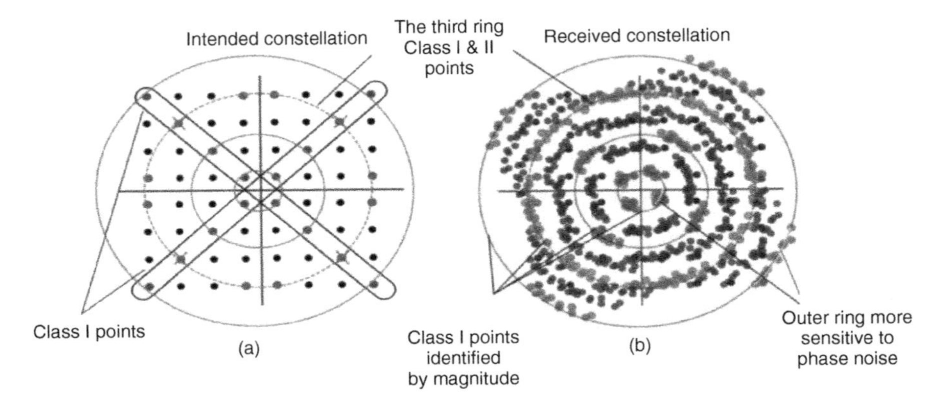

FIGURE 11.19 Illustration of classification method for a 64QAM system: (a) ideal signal and (b) corrupted with phase and additive noise.

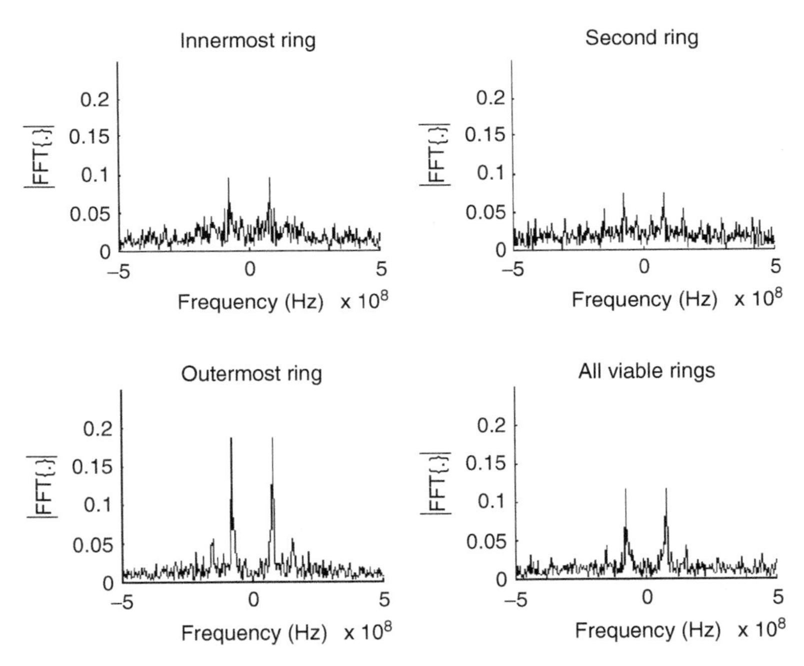

FIGURE 11.20 FFT of rotated symbol angles for each viable ring.

Figure 11.20 shows the FFT of rotated symbol angles for each viable ring. Although all show a peak at four times the frequency offset, the outermost ring shows the best peak-to-noise ratio, and is most robust to high noise.

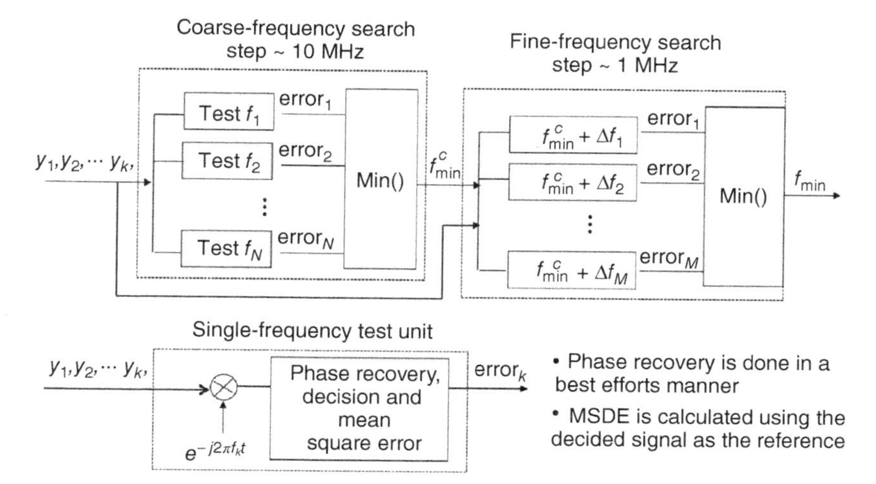

FIGURE 11.21 Operation principle for the proposed blind frequency search algorithm

11.4.3 Blind Frequency Search (BFS)-Based Fine FO Estimation Algorithm

BFS algorithm was first proposed and demonstrated in [59] as a universal carrier FO estimation method, where minimum mean square distance error (MSDE, in terms of phase or Euclidean distance) is used as the frequency-selection criteria. For this method, the frequency offset is first scanned at a coarse step size (~10 MHz) and then at a fine step size of (~1 MHz), and the optimal frequency offset is the one that gives the minimum MSDE (see Figure 11.21). For each trial frequency, the carrier phase is first recovered (with best efforts) by using the BPS algorithm, and decisions made following this phase estimation are then approximated as the reference/correct signals for MSDE calculation.

Figure 11.22 shows the simulated results for a 38Gbaud 64QAM system with laser linewidth = 100 kHz and OSNR = 25 dB. Figure 11.22(a) shows how the normalized MSDE varies with the frequency deviation (the difference between the trial FO and the actual FO) and the number of symbols used for FO estimation, while Figure 11.22(b) shows the frequency error distributions by using several different data block lengths. Here, the frequency error is defined as the difference between the estimated FO and the actual FO. BFS can reliably estimate FO to be within 20 MHz accuracy by using only 32 symbols. Increasing the number of symbols to 128 can improve the FO estimation accuracy to within 3 MHz, which is good enough even for 64QAM. As compared with the previous *M*th-power-based algorithms, BFS requires much less number of symbols for a reliable FO estimate. Thus, this algorithm can achieve very fast carrier frequency recovery if it is implemented with a parallel processing architecture as is shown in Figure 11.21. For a typical coherent receiver where the carrier frequency varies much more slowly than the symbol rate, BFS can also be

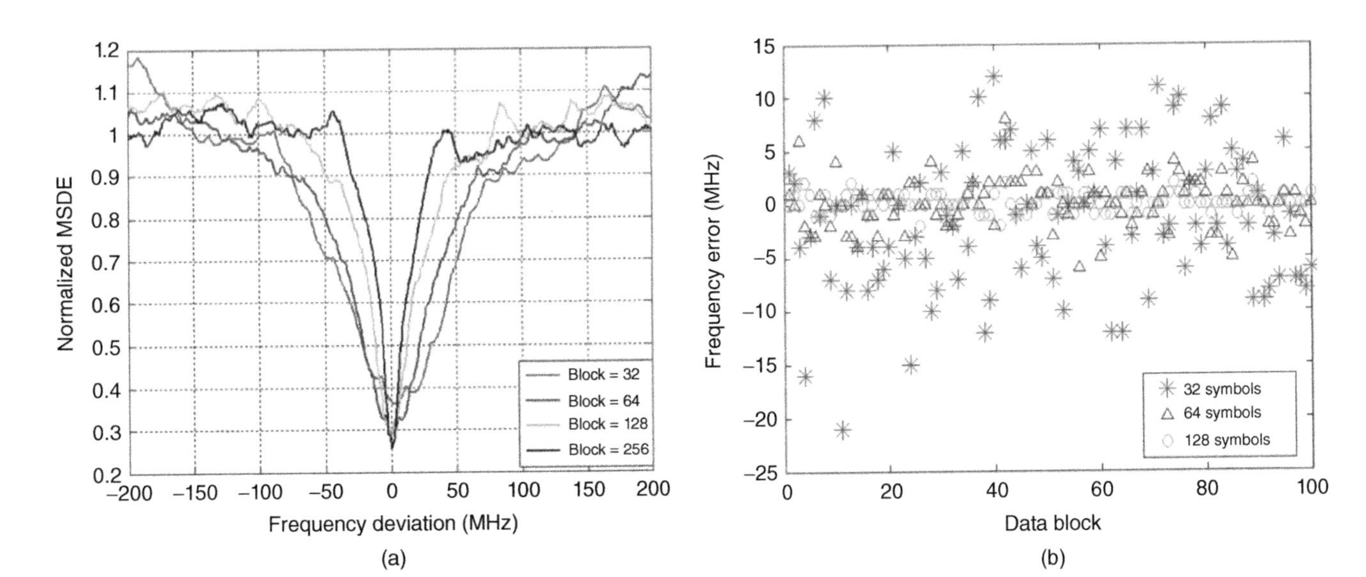

FIGURE 11.22 Simulated results for a 38-Gbaud 64QAM system with laser linewidth = 100 kHz and OSNR = 25 dB by using the BFS method. (a) Impact of frequency deviation and block lengths on MSDE and (b) frequency error distribution by using different block lengths for FO estimation.

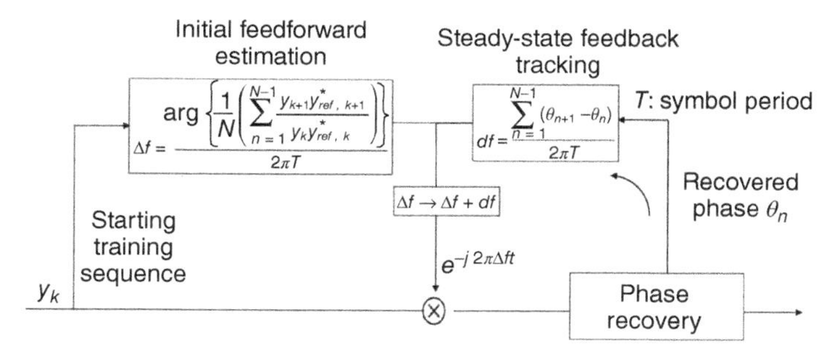

FIGURE 11.23 Schematic illustration of the proposed training-initiated frequency-recovery algorithm. T: symbol period, $\Delta\omega$: estimated carrier frequency offset.

implemented with a sequence or partial sequence processing architecture (such as the CPU in a computer) to reduce the implementation complexity. The robustness of this algorithm has been verified by multiple 100 and 400 Gb/s transmission experiments employing high-order QAMs [59, 60].

Just recently it has been shown that, in addition to MSDE, phase entropy can also be used as a reliable frequency-selection criterion for the BFS algorithm [61]. But using the phase entropy as the frequency selection criterion requires more samples for each FO estimation than the MSDE-based solution.

11.4.4 Training-Initiated Fine FO Estimation Algorithm

The training-initiated fine FO estimation algorithm was first proposed and demonstrated in [46]. This algorithm also works for arbitrary QAMs. For this method, the initial FO is estimated by using a starting training sequence in a feedforward manner and then the FO variation is tracked through a feedback configuration using the recovered carrier phases from the following phase recovery stage as is shown in Figure 11.23. Note that, unlike the fast-changing phase noise that cannot tolerate extended feedback delay in high-speed optical systems, carrier frequency typically varies much more slowly and thus it could be tracked by using a feedback-based architecture. The advantages of this method are as follows: (i) it is applicable to arbitrary QAM, (ii) its implementation complexity is very low because it requires significantly fewer complex multiplications than algorithms described previously, and (iii) the tolerable frequency offset can be very large (up to the symbol rate). As compared with the frequency recovery method based on the second-order PLL, where both the frequency and phase recovery rely on a feedback mechanism, here only the FO tracking uses a feedback configuration while the phase recovery is achieved in a feedforward manner.

In summary, a coarse AFC is typically required before the nominal fine frequency recovery. For coarse AFC, the key component is the FDE, which can be either a balanced quadricorrelator or an FFT-based differential power monitor (DPM). For fine frequency recovery, the Mth-power-based time-domain differential phase algorithm is a hardware-efficient blind-FO estimation method for M-ary PSK signals operating at high OSNR with relatively small phase noise. For more general QAM-modulated signals, however, the training-initiated feedback-based method presented in Section 11.4.4 can achieve much reliable performance with even lower implementation complexity. The BFS method presented in Section 11.4.3 has the potential to achieve much faster FO estimation at the expense of higher implementation complexity.

11.5 EQUALIZER-PHASE NOISE INTERACTION AND ITS MITIGATION

The carrier recovery algorithms described so far are mostly optimized based on the additive Gaussian noise assumption. However, in a realistic long-haul coherent transmission system without using inline optical dispersion compensation, a long-memory equalizer usually has to be employed before the carrier recovery to compensate for the accumulated fiber chromatic dispersion (CD). For future few-mode-fiber-based space-division multiplexing (SDM) systems, long-memory multi-input multi-output (MIMO) equalizers are needed for intermodal dispersion compensation. As described later, the interaction between the long-memory equalizers and the laser phase noise will not only enhance the phase noise, but will also cause additional amplitude distortions [62–64]. Especially, it is found that impairments due to equalizer and phase noise interaction (EPNI) increase with the signal baud rate and can be a significant problem for future 400 Gb/s and beyond systems operating at very high symbol rate. So designing a carrier recovery scheme capable of mitigating impairments due to EPNI is becoming more and more important.

For the case with additive Gaussian noise, phase noises from the signal source and the LO have essentially similar system impact, which can be minimized by using a fast single-tap phase rotation equalizer (i.e., the nominal phase recovery circuit), as is shown in Figure 11.24(a). For a typical coherent system using a long-memory receiver-side equalizer for CD compensation, however, the impacts of signal and LO phase noises become quite different, because the signal source experiences both positive CD from the fiber and negative CD from the digital equalizer, but the LO only sees negative CD from the digital equalizer. Due to the fact that CD can convert phase noise into amplitude noise (and may also enhance phase noise), the impact of LO phase noise becomes more severe than the signal source phase noise [62–64], and the additional amplitude distortion caused by a long-memory equalizer cannot be mitigated by using conventional phase recovery algorithms. To address this problem, a hardware-based laser phase noise compensation method has recently been proposed [66]. However, this method is very complex and costly, because it requires an additional coherent receiver to measure the laser phase noise.

In [65], a DSP-based solution is proposed to address this challenge. The basic idea is based on the following observations: if the laser linewidth is small and signal symbol rate is high, the amplitude and phase distortions caused by EPNI will be highly correlated over quite a few symbols (tens to hundreds of symbols depending on the laser linewidth and the symbol rate), and moreover, such distortions can be modeled

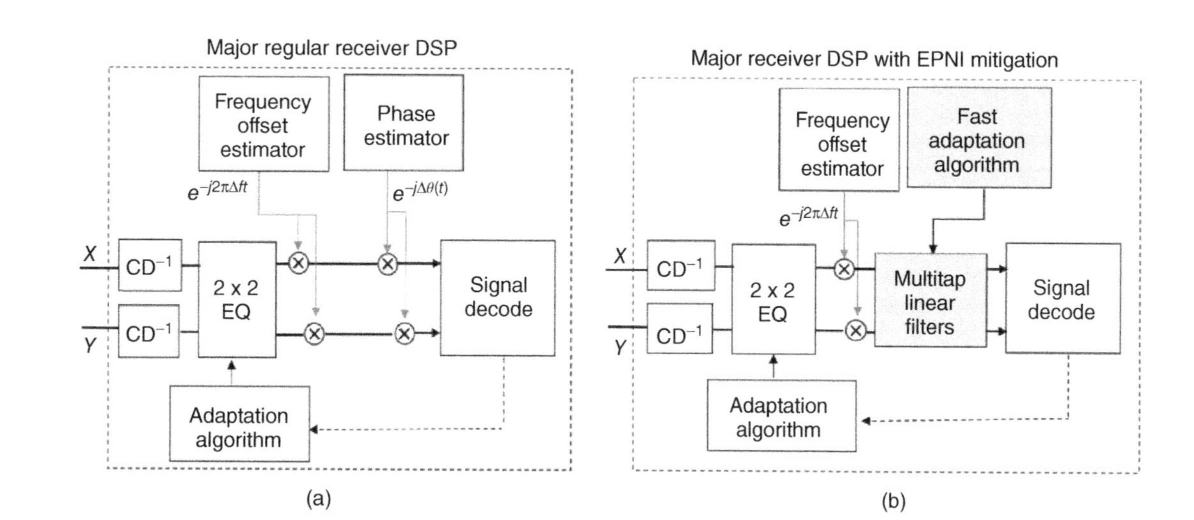

FIGURE 11.24 Major receiver functional blocks for the conventional coherent receiver (a) and the proposed improved coherent receiver (b). From Ref. [65].

as the result of a time-varying multitap linear filtering effect in which – although the filter coefficients vary over time – over every limited time block (consisting of tens to hundreds of symbols), the filter coefficients can be well approximated as unvarying constants. Thus, by replacing the commonly used fast-tracking single-tap phase rotation equalizer with a fast-tracking multitap linear equalizer, as is shown in Figure 11.24(b), both the amplitude and phase distortion caused by EPNI can be mitigated.

Because laser phase noise typically varies 2–4 orders of magnitude faster than the state of polarization change, the adaptation rate for the proposed EPNI mitigation equalizer should be much faster than the regular 2×2 polarization-compensating equalizer. Very fast adaptation rate can be realized by using a block-by-block feed-forward adaptation algorithm as is illustrated in Figure 11.25. For this algorithm, the frequency-recovered signal is first divided into blocks (with a few overlap symbols included between blocks), and the regular phase recovery method is applied to each block. After that, one can make an initial decision from the phase-recovered signal. The decided-upon signal is then approximated as the correct data for EPNI filter coefficients estimation through the classic least square (LS) algorithm. To reduce the impact of imperfect decision accuracy, multiple iterations may be applied to each data block for filter coefficients update. Note that the initial decision can be made based on performing pure phase recovery over the current data block. But the initial decision may also be made by applying the recovered phase of the prior data block to the current data block or by directly applying the EPNI filter coefficients acquired

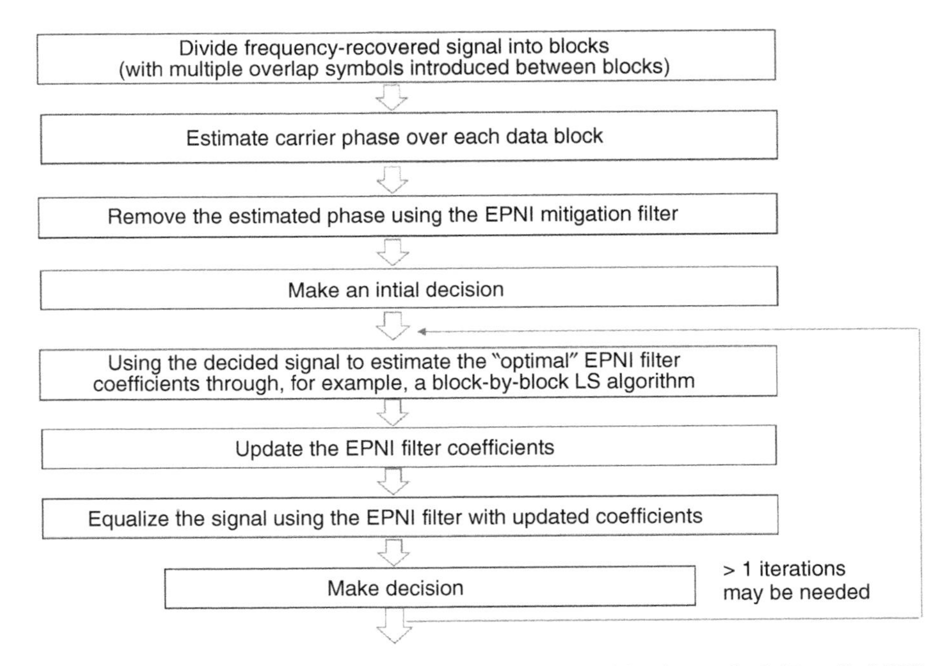

FIGURE 11.25 Illustration of the newly proposed EPNI mitigation method. From Ref. [65].

from the prior data block to the current data block (the starting phase or EPNI filter coefficients can be obtained using a starting training sequence). Because the block length cannot be too large due to the need for rapid adaptation and the time-varying nature of EPNI, the accumulated amplifier noise may degrade the performance of the proposed EPNI equalizer. This problem may be alleviated by joint optimization of the proposed EPNI filter over both polarizations, because the phase noises in *X*- and *Y*-polarizations are usually correlated (if they are from the same source).

This new impairment mitigation method has been numerically verified in a 7-channel 50 GHz-spaced 49 Gbaud PDM-16QAM system. The transmission link consists of 20 total spans with each span composed of 100 km of large area fiber (dispersion coefficient and fiber loss are assumed to be 21 ps/nm/km and 0.18 dB/km, respectively) and EDFA-only amplification (noise figure is assumed to be 5 dB). No inline optical dispersion compensation is used for this simulation. For simplicity, polarization-mode dispersion (PMD) and polarization-dependent loss are not considered. For the laser sources, we assume that the signal source and the LO have identical linewidth.

Figure 11.26 shows the BER performance of the middle channel (channel 4) versus the laser linewidth at the optimal signal launch power of 3 dBm/channel. The diamond-shaped symbols give the results using a conventional coherent receiver, where the phase is tracked by using the previously described training-assisted two-stage phase recovery algorithm. (The phase is first estimated by using three training symbols that are uniformly distributed over every 128 symbols, and then the 128 symbols are divided into four groups and the phase over each group is refined by using the BPS algorithm over a small phase-varying region.) The square-shaped symbols illustrate the result using the proposed EPNI mitigation method, in which the NPNI filter length is chosen to be five T-spaced taps. For this study, the block length is chosen to be 95 symbols (including 5 overlap symbols), and two iterations are applied for each data block, in which the initial decision for each data block is made based on the same phase recovery algorithm used for the conventional coherent receiver (i.e., for the diamond-shaped symbols).

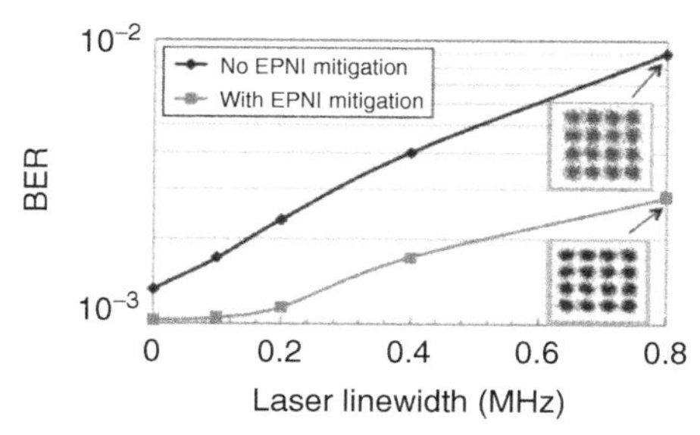

FIGURE 11.26 Simulated BER performance versus laser linewidth for a 7-channel 49-Gbaud PDM-16QAM WDM system (at 3 dBm launch power and 2000 km transmission of an ultra-large area fiber) with and without using the proposed EPNI mitigation technology. From Ref. [65].

TABLE 11.3 Simulated BER versus number of iterations (laser linewidth = 0.8 MHz) when using the proposed EPNI mitigation algorithm

Iteration	0	1	2	3	4
BER	8.86e−3	3.56e−3	2.81e−3	2.67e−3	2.65e−3

Source: From Ref. [65].

The effectiveness of this new algorithm is evident from Figure 11.26. For a laser linewidth 0.8 MHz (a typical linewidth for widely used DFB lasers), the new method improves the Q performance by 1.35 dB while using only two iterations. As we further increase the number of iterations, the performance gain is small, as can be seen from Table 11.3. It is interesting to note that, even without laser phase noise, the introduction of a five-tap EPNI equalizer improves the performance by 0.22 dB, indicating that the proposed method also helps in mitigating fiber nonlinear effects. To confirm this, we also simulated the results by switching off the fiber nonlinearity and found that use of the EPNI mitigation equalizer does not improve the performance. In fact, when the fiber nonlinearity is switched off, the use of EPNI mitigation equalizer slightly degrades the performance (BER = 2e−5 versus BER = 1.8e−5 when using the normal phase recovery algorithm).

A similar result is observed in a back-to-back simulation as is shown in Figure 11.27. One can see that (i) when using the conventional one-tap filter-based phase recovery, increasing the laser linewidth from 0 to 0.8 MHz only results in a small performance degradation while there is no EPNI effect; and (ii) the use of the EPNI mitigation filter does not improve the performance when there are no EPNI effects. The small performance degradation caused by the EPNI mitigation filter when there are no EPNI effects probably occurs because the block length of 95 symbols is not long enough for optimal estimates of the five complex EPNI filter coefficients (Note that the normal phase recovery only needs to estimate one real filter coefficient, the phase angle.)

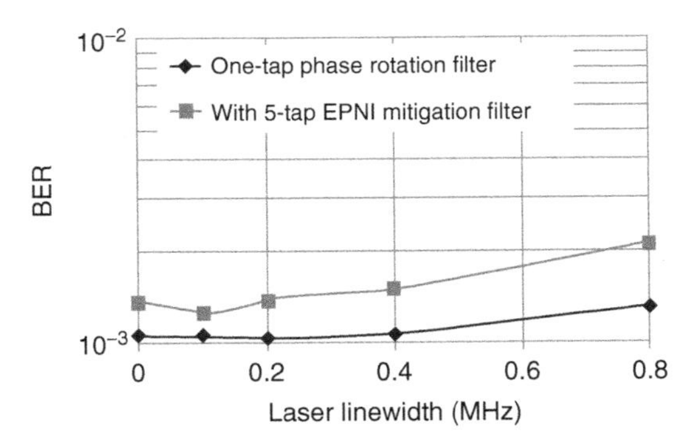

FIGURE 11.27 Back-to-back performance comparison (OSNR = 23 dB) when using the conventional one-tap filter-based phase recovery method and the proposed multitap filter-based impairment mitigation method. From Ref. [65].

11.6 CARRIER RECOVERY IN COHERENT OFDM SYSTEMS

Unlike single-carrier modulated systems in which the carrier phase noise only causes the constellation rotation of QAM symbols (assuming linear systems with negligible channel memory), carrier phase noise not only changes the common phase of OFDM subcarriers but also causes interference between subcarriers. Phase noise-induced common phase change of OFDM subcarriers can be estimated by using similar phase estimation methods described in Section 11.3, which were originally developed for single carrier modulated systems. For example, in single-carrier modulated systems, the *M*th-power-based algorithms, the BPS algorithm as well as the ML-based algorithm are applied to the symbols in the time domain. In coherent OFDM systems, these algorithms can be applied to subcarriers in the frequency domain within each OFDM frame as is illustrated in Figure 11.28, in which the ML algorithm is used as an example.

OFDM systems are very sensitive to carrier frequency offset, and the required frequency offset estimation accuracy must be much greater than the subcarrier spacing – so training or pilot tone-based frequency estimation methods [67, 68] are typically used for fine frequency offset estimation. The blind AFC algorithms described in Section 11.4.3 are also applicable for OFDM systems.

Phase-noise-induced intercarrier interference (ICI) is a significant problem for coherent OFDM systems. Effective mitigation of such a problem essentially requires us to estimate carrier phase in the time domain on a sample-by-sample basis, which is a more difficult problem than a single-carrier modulated system since, in the time domain, an OFDM signal behaves like a "Gaussian" noise. A common method used to address this problem is to employ ultra-low phase noise lasers such that the carrier phase remains constant (approximately) over the whole OFDM frame, requiring only the subcarrier common phase to be estimated. Alternatively, several recent studies have shown that an RF-pilot tone-based method is effective in mitigating phase noise-induced ICI [69, 70].

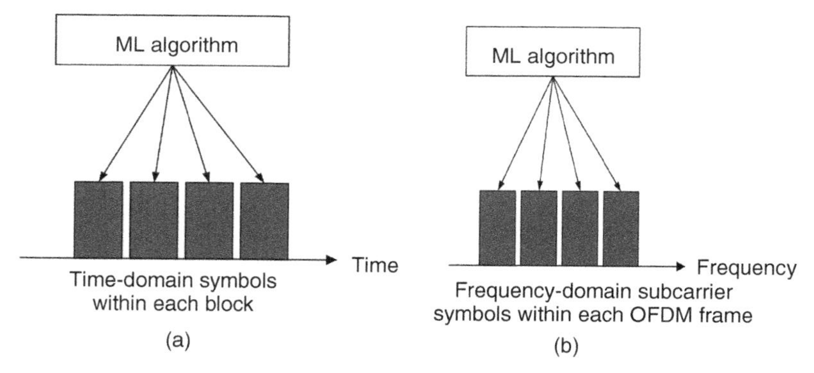

FIGURE 11.28 Schematic illustration of ML phase estimation algorithm applied for a single-carrier system (a) and an OFDM system (b).

11.7 CONCLUSIONS AND FUTURE RESEARCH DIRECTIONS

In this chapter, we first introduce the concept of carrier recovery and the challenges faced. We show that, due to differing carrier frequency and phase noise characteristics, an independent and slower carrier frequency recovery unit is typically required before the much faster carrier phase recovery unit.

Two important theoretical concepts regarding optimal frequency and phase estimation are introduced in Section 11.2: MAP estimator and the Cramér–Rao lower bound. Although the MAP estimator is too complex to be implemented in an actual system, it can be implemented with offline process to establish a baseline to estimate the efficiency of some suboptimal estimators. The Cramér–Rao lower bound provides us with another and more fundamental way to evaluate the efficiency of any estimator.

Section 11.3 is devoted to several hardware-efficient phase estimation algorithms recently demonstrated for high-speed coherent optical systems. These include the decision-directed PLL, the Mth-power-based algorithms, the BPS algorithm as well as several multistage hybrid phase estimation algorithms. Among these algorithms, the DD-PLL is the most hardware-efficient, but its linewidth tolerance is fundamentally limited by the inherent feedback delay. The Mth-power algorithm works well for M-ary PSKs but is much less efficient for higher-order QAMs. The BPS algorithm can achieve close-to-optimal phase noise tolerance for arbitrary QAM, but implementation complexity increases with modulation order. The recently proposed multistage algorithms such as the hybrid BPS/ML and the training-assisted two-stage ML/BPS can achieve a linewidth tolerance performance similar to the single-stage BPS but with significantly reduced implementation complexity. Furthermore, the training-assisted two-stage ML/BPS algorithm is very robust against the detrimental cyclic slips, and may remove the need for differential encoding/decoding.

Section 11.4 is devoted to frequency recovery algorithms. We first introduce two auto-AFC techniques developed for coarse-locking of the LO frequency to the vicinity of the transmitter source laser. Then, we describe three types of fine FO estimation algorithms.

- The first employs the Mth-power algorithm to remove data modulation, and the frequency offset is estimated by using either a time-domain-based differential phase method or FFT-based methods. The time-domain method only works for M-PSK while FFT-based methods can be effective for higher-order QAMs, but the implementation complexity is very high.
- The second is a constellation-assisted blind frequency search method that works for arbitrary QAM. The significant advantage of this method lies in the fact that it can achieve very fast frequency recovery, but the down side is its high implementation complexity.
- The third is a training-initiated feedback method, in which the acquisition training sequence is utilized to do an initial FO estimation, and then FO variations are tracked by using the recovered phases from the following phase recovery stage in a feedback configuration.

Among these fine FO estimation algorithms, the training-initialized feedback method is probably the best choice for the common continuous-mode coherent receiver, mostly due to its low implementation complexity and large FO tolerance.

In Section 11.5, a new impairment caused by interaction between the long-memory equalizer and the laser phase noise is introduced. This new impairment can be a significant problem for future 400 Gb/s and beyond systems operating at very high symbol rate. A new multitap equalizer-based carrier recovery method capable of mitigating this problem is then presented.

A brief introduction to carrier recovery in coherent OFDM systems is given in Section 11.6.

So far most of the developed carrier recovery algorithms are optimized based on the additive circular Gaussian noise assumption (i.e., laser phase noise is only corrupted by additive circular Gaussian noise) and algorithms have assumed ideal equalization, but in real optical communication systems there exist not only Gaussian noise, but also nonlinear phase and amplitude noise, as well as interaction between the equalizer and the phase noise. How to optimize the phase recovery algorithm for these realistic nonlinear optical communication systems with imperfect equalization are still a significant challenge and an interesting topic for future research. In addition, better algorithms to fight phase noise-induced ICI in coherent OFDM systems are also highly desirable. More work on these can be expected in coming years.

REFERENCES

1. Taylor MG. Coherent detection method using DSP for demodulation of signal and subsequent equalization of propagation impairments. *IEEE Photon Technol Lett* 2004;16(2).
2. Noe R. PLL-free synchronous QPSK polarization multiplex/diversity receiver concept with digital I&Q baseband processing. *IEEE Photon Technol Lett* 2005;17(4):887–889.
3. Tsukamoto S, Ly-Gagnon D-S, Katoh K, Kikuchi K. Coherent demodulation of 40-Gbit/s polarization-multiplexed QPSK signals with 16-GHz spacing after 200-km transmission. Proceedings of OFC 2005, paper PDP-29; Mar 2005.
4. OIF-ITLA-MSA-01.1, Integrable Tunable Laser Assembly MSA, Optical internetworking forum (OIF), Nov 25, 2005.
5. Zhang S, Xu L, Yu J, Huang M –F, Kam PY, Yu C, Wang T. Dual-stage cascaded frequency offset estimation for digital coherent receivers. *IEEE Photon Technol Lett* 2010;22(6):401–403.
6. Cramér H. Mathematical Methods of Statistics. Princeton University Press; 1946.
7. Rao CR. Information and the accuracy attainable in the estimation of statistical parameters. *Bull Calcutta Math Soc* 1945;37:81–89.
8. Rice F, Cowley B, Moran B, Rice M. Cramér–Rao lower bounds for QAM phase and frequency estimation. *IEEE Trans Commun* 2001;49(9):1582–1591.
9. Rife DC, Boorstyn RR. Single-tone parameter estimation from discrete-time observations. *IEEE Trans Inf Theory* 1974;IT-20(5):591–598.
10. Fatadin I, Ives D, Savory SJ. Compensation of frequency offset for differentially encoded 16- and 64-QAM in the presence of laser phase noise. *IEEE Photon Technol Lett* 2010;22(3):176–178.

11. Zhou X, Sun Y. Complexity Reduced Feed Forward Carrier Recovery Methods for M-QAM Modulation Formats. US patent 8,908,809, 2014.
12. Bower P, Dedic I. High-speed converters and DSP for 100G and beyond. *Opt Fiber Technol* 2011;17:464–471.
13. Ip E, Kahn M. Feedforward carrier recovery for coherent optical communications. *J Lightwave Technol* 2007;25(9):2675–2692.
14. Pfau T, Hoffmann S, Noe R. Hardware-efficient digital coherent concepts with feedforward carrier recovery for M–QAM constellations. *J Lightwave Technol* 2009;27(8): 989–999.
15. Tarighat A, Hsu RCJ, Sayed AH, Jalali B. Digital adaptive Phase noise reduction in coherent optical links. *J Lightwave Technol* 2006;24(3):1269–1276.
16. Tselniker I, Sigron N, Nazarathy M. Joint phase noise and frequency offset estimation and mitigation for optically coherent QAM based on adaptive multi-symbol delay detection. *Opt Express* 2012;20(10):10944–10962.
17. Mori1 Y, Zhang C, Kikuchi K. Novel FIR-filter configuration tolerant to fast phase fluctuations in digital coherent receivers for higher-order QAM signals. OFC/NFOEC 2012, paper OTh4C.4; Mar 2012.
18. Zhang S, Kam PY, Yu C, Chen J. Decision-aided carrier phase estimation for coherent optical communications. *J Lightwave Technol* 2010;28(11):1597–1607.
19. Piyawanno K, Kuschnerov M, Spinnler B, Lankl B. Low complexity carrier recovery for coherent QAM using superscalar parallelization. ECOC 2010, paper We.7.A.3; Sep 2010.
20. Tolmachev A, Tselniker I, Meltsin M, Sigron I, Nazarathy M. Efficient multiplier-free FPGA demonstration of polar-domain multi-symbol-delay-detector (MSDD) for high performance phase recovery of 16-QAM. OFC 2013, paper OM2C.8; Mar 2013.
21. Zhuge Q, Mousa-Pasandi ME, Xu X, Chatelain B, Pan Z, Osman M, Plant DV. Linewidth-tolerant low complexity pilot-aided carrier phase recovery for M-QAM using superscalar parallelization. OFC 2012, paper OTu2G.2; Mar 2012.
22. Viterbi A, Viterbi A. Nonlinear estimation of PSK-modulated carrier phase with application to burst digital transmission. *IEEE Trans Inf Theory* 1983;29(4):543–551.
23. Li G. Recent advances in coherent optical communication. *Adv Opt Photon* 2009;1(2): 279–307.
24. Noé R. PLL-free synchronous QPSK polarization multiplex/diversity receiver concept with digital I & Q baseband processing. *IEEE Photon Technol Lett* 2005;17(4):887–889.
25. Ly-Gagnon DS, Tsukarnoto S, Katoh K, Kikuchi K. Coherent detection of optical quadrature phase-shift keying signals with carrier phase estimation. *J Lightwave Technol* 2006;24:12–21.
26. Goldfarb G, Li G. BER estimation of QPSK homodyne detection with carrier phase estimation using digital signal processing. *Opt Express* 2006;14:8043–8053.
27. Noe R. Phase noise-tolerant synchronous QPSK/BPSK baseband-type intradyne receiver concept with feedforward carrier recovery. *J Lightwave Technol* 2005;23:802–808.
28. Weber W. Differential encoding for multiple amplitude and phase shift keying systems. *IEEE Trans Commun* 1978;26(3):385–391.
29. Proakis JG, Manolakis DG. Digital Signal Processing: Principles, Algorithms, and Applications. Prentice-Hall, Inc 1996.
30. Taylor MG. Phase estimation methods for optical coherent detection using digital signal processing. *J Lightwave Technol* 2009;27(7):901–912.

31. Hoffmann S, Peveling R, Pfau T, Adamczyk O, Eickhoff R, Noé R. Multiplier-free real-time phase tracking for coherent QPSK receivers. *IEEE Photonics Lett* 2009; 21(3):137–139.
32. Seimetz M. Laser linewidth limitations for optical systems with high-order modulation employing feed forward digital carrier phase estimation. Proceedings of OFC, paper OTuM2; Feb 24–28, 2008.
33. Louchet H, Kuzmin K, Richter A. Improved DSP algorithmes for coherent 16-QAM transmission. Proceedings of ECOC'08, paper Tu.1.E.6 ; Sep 2008.
34. Cao Y, Yu S, Chen Y, Gao Y, Gu W, Ji Y. Modified frequency and phase estimation for M-QAM optical coherent detection. Proceedings of ECOC 2010, paper We.7.A.1 ; Sep 2010.
35. Fatadin I, Ives D, Savory SJ. Laser linewidth tolerance for 16-QAM coherent optical systems using QPSK partitioning. *IEEE Photon Technol Lett* 2010;22(9):631–633.
36. Zhou X, Zhong K, Gao Y, Lu C, Lau A. P. T., and Long K, Modulation-format-independent blind phase search algorithm for coherent optical square M-QAM systems. *OSA*, Vol. 22, Issue 20, pp. 24044-24054 (2014).
37. Volder J. The CORDIC trigonometric computing technique. *IRE Trans Electron Comp* 1959;EC-8(3):330–334.
38. Zhou X. An improved feed-forward carrier recovery algorithm for coherent receiver with M-QAM modulation format. *IEEE Photon Technol Lett* 2010;22(14):1051–1053.
39. Zhou X, Sun Y. Low-complexity, blind phase recovery for coherent receivers using QAM modulation. Proceedings of OFC'11, paper OMJ3, Mar 2011.
40. Pfau T, Noáe R. Phase-noise-tolerant two-stage carrier recovery concept for higher order QAM formats. *IEEE J Sel Top Quant Electron* 2010;16(5):1210–1216.
41. Zhuge Q, Chen C, Plant DV. Low computation complexity two-stage feedforward carrier recovery algorithm for M-QAM. Proceedings of OFC 2011, paper OMJ5 ; Mar 2011.
42. Li J, Li L, Tao Z, Hoshida T, Rasmussen J. Laser-linewidth tolerant feed-forward carrier phase estimator with reduced complexity for QAM. *J Lightwave Technol* 2011;29(16):2358–2364.
43. Bilal SM, Fludger C, Curri V, Bosco G. Multi-stage CPE algorithms for phase noise mitigation in 64-QAM optical systems. *J Lightwave Technol* 2014;32(17):2973–2980.
44. Zhong K, Ke JH, Gao Y, Cartledge JC. Linewidth-tolerant and low-complexity two-stage carrier phase estimation based on modified QPSK partitioning for dual-polarization 16-QAM systems. *J Lightwave Technol* 2013;31(1):50–57.
45. Gao Y, Lau APT, Yan S, Lu C. Low-complexity and phase noise tolerant carrier phase estimation for dual-polarization 16-QAM systems. *Opt Express* 2011;19(22):21717–21729.
46. Zhou X, Nelson LE, Magill P, Isaac R, Zhu B, Peckham DW, Borel P, Carlson K. High spectral efficiency 400 Gb/s transmission using PDM time-domain hybrid 32-64QAM and training-assisted carrier recovery. *J Lightwave Technol* 2013;31(7):999–1005.
47. Proakis JG. Digital Communications. 4th ed. Prentice Hall; 2000. p 348Chapter 6.
48. Zhou X, Yu J. Two-stage feed-forward carrier phase recovery algorithm for high-order coherent modulation formats. ECOC 2010; Torino, Italy, paper We.7.A.6; Sep 2010.
49. Zhou X, Hardware efficient carrier recovery algorithms for single-carrier QAM systems. OSA; 2011.
50. Feuer MD, Nelson LE, Zhou X, Woodward SL, Isaac R, Zhu B, Taunay TF, Fishteyn M, Fini JF, Yan MF. Joint digital signal processing receivers for spatial superchannels. *IEEE Photon Technol Lett* 2012;24(21):1957–1960.

51. Messerschmitt DG. Frequency detectors for PLL acquisition in timing and carrier recovery. *IEEE Trans Commun* 1979;27:1288–1295.
52. Natali FD. AFC tracking algorithms. *IEEE Trans Commun* 1984;COM-32:935–947.
53. Tao Z, et al. Simple, robust, and wide-range frequency offset monitor for automatic frequency control in digital coherent receivers. Proceedings of ECOC 2007, paper 03.5.4; 2007.
54. Alberty T, Hespelt V. A new pattern jitter free frequency error detector. *IEEE Trans Commun* 1989;37(2):159–163.
55. Piyawanno K, Kuschnerov M, Spinnler B, Lankl B. Fast and accurate automatic frequency control for coherent receivers. Proceedings of ECOC 2009, paper 7.3.1; Sep 2009.
56. Leven A, Kaneda N, Koc U-V, Chen Y-K. Frequency Estimation in Intradyne Reception. *IEEE Photon Technol Lett* 2007;19(6):306–308.
57. Selmi M, Jaouen Y, Ciblat P. Accurate digital frequency offset estimator for coherent PolMux QAM transmission systems. Proceedings of ECOC'09, paper P. 3.08; Sep 2009.
58. Sun Y, Zhou X. Blind Carrier Frequency Recovery Methods for Coherent Receivers Using QAM Modulation Formats. US patent US20,120,294,630, 2012.
59. Zhou X, Yu J, Huang M-F, Shao Y, Wang T, Nelson LE, Magill PD, Birk M, Borel PI, Peckham DW, Lingle R. 64-Tb/s, 8 b/s/Hz, PDM-36QAM transmission over 320 km using both pre- and post-transmission digital signal processing. *J Lightwave Technol* 2011;29(4):571–577.
60. Zhou X, Nelson LE, Magill P, Isaac R, Zhu B, Peckham DW, Borel P, Carlson K. PDM-Nyquist-32QAM for 450-Gb/s per-channel WDM transmission on the 50 GHz ITU-T grid WDM transmission on the 50 GHz ITU-T grid. *J Lightwave Technol* 2012;30(4):553–559.
61. Dris S, Lazarou I, Bakopoulos P, Avramopoulos H. Phase entropy-based frequency offset estimation for coherent optical QAM systems. Proceedings of OFC'12, paper OTu2G.4; Mar 2012.
62. Zhou X, Yu J. Multi-level, multi-dimensional coding for high-speed and high spectral-efficiency optical transmission. *J Lightwave Technol* 2009;27(16):3641–3653.
63. Shieh W, Ho KP. Equalization-enhanced phase noise for coherent-detection systems using electronic digital signal processing. *Opt Express* 2008;16(20):15718–15727.
64. Xie C. WDM coherent PDM-QPSK systems with and without inline optical dispersion compensation. *Opt Express* 2009;17(6):4815–4823.
65. Zhou X, Nelson L. Advanced DSP for 400 Gb/s and beyond optical networks. *J Lightwave Technol* 2014;32(16):2716–2725.
66. Colavolpe G, Foggi T, Forestieri E, Secondini M. Impact of phase noise and compensation techniques in coherent optical systems. *J Lightwave Technol* 2011;29(18):2790–2799.
67. Schmidl TM, Cox DC. Robust frequency and timing synchronization for OFDM. *IEEE Trans Commun* 1997;45(12):1613–1621.
68. Zhang Z, Jiang W, Zhou H, Liu Y, Gao J. High accuracy frequency offset correction with adjustable acquisition range in OFDM systems. *IEEE Trans Wireless Commun* 2005;4(1):228–237.
69. Jansen SL, Morita I, Schenk TCW, Takeda N, Tanaka H. Coherent optical 25.8 Gb/s OFDM transmission over 4160-km SSMF. *J Lightwave Technol* 2008;26(1):6–15.
70. Peng WR, Tsuritani T, Morita I. Simple carrier recovery approach for RF-pilot-assisted PDM-CO-OFDM systems. *J Lightwave Technol* 2013;31(15):2555–2564.

12

REAL-TIME IMPLEMENTATION OF HIGH-SPEED DIGITAL COHERENT TRANSCEIVERS

TIMO PFAU
DSP and Optics, Acacia Communications Inc., Maynard, MA, USA

The real-time implementation of digital coherent receivers for 100 Gbit/s data rates and beyond pushes the limits of all the technologies involved. Digital-to-analog converters (DACs) and analog-to-digital converters (ADCs) with ultra-high sampling rates need to be co-integrated with a large-scale digital signal processing (DSP) engine, and highly sophisticated algorithms have to be mapped into digital logic with the best compromise between performance and power consumption. This chapter describes the main constraints put on the DSP algorithms by the hardware structure, and gives a brief overview on technologies and challenges for prototype and commercial real-time implementations of coherent receivers.

12.1 ALGORITHM CONSTRAINTS

Coherent optical transmission systems at 100 Gb/s use baud rates of 28 Gbaud or beyond, with the receiver ADCs usually using two times oversampling, that is 56 Gs/s and beyond, to convert the received signals from the analog to the digital domain. Standard cell logic in modern complementary metal–oxide–semiconductor (CMOS) technology cannot operate at this high sampling rate of the ADCs. The preferred clock speed for standard logic is usually in the range of a few hundred megahertz, which means that the digital samples from the ADCs cannot be processed one by one in a serial implementation, but need to be demultiplexed and processed in parallel. But even then, in order to achieve the clocking speed of several hundred megahertz, the

Enabling Technologies for High Spectral-efficiency Coherent Optical Communication Networks, First Edition. Edited by Xiang Zhou and Chongjin Xie.
© 2016 John Wiley & Sons, Inc. Published 2016 by John Wiley & Sons, Inc.

logic is only capable of processing a small fraction of the overall receiver algorithm until intermediate results need to be pipelined or buffered in flip flops or memory. This processing architecture, which is depicted in Figure 12.1, needs to be considered already during the design phase of the processing algorithms, as there might otherwise be no possibility to bring a certain algorithm into a real-time processing engine. Amongst others the algorithms need to be implementable within the power constraints of the DSP engine, they must be parallelizable, and significant feedback latencies need to be assumed. The following sections analyze these constraints in more detail.

12.1.1 Power Constraint and Hardware Optimization

The way algorithms are mathematically described in scientific publications or implemented in software simulation tools can differ significantly from the way these algorithms are most efficiently implemented in hardware.

In this section, the Viterbi & Viterbi (V&V) carrier recovery algorithm serves as an example [1]. Figure 12.2 shows a block diagram of a textbook description of the algorithm. The incoming signal is raised to the fourth power and filtered in a sliding window averaging filter. The phase angle from the filter output is negated and divided by 4, which provides the actual carrier phase estimate. This estimate is converted to a complex phasor and used to derotate the received symbol.

This description makes it easy to understand the functionality of the algorithm, but a straightforward hardware implementation would be highly inefficient. The fourth power calculation requires six multiplications, and the derotation of the received symbol requires three multiplications. As multiplications are the most costly functions in hardware, avoiding them generally reduces the hardware effort. Figure 12.3 shows a block diagram of the same algorithm that potentially reduces the hardware effort significantly.

At first glance, the processing in Figure 12.3 appears more complicated than the one in Figure 12.2, as it contains several conversion blocks from Cartesian to polar coordinates and vice versa. However, these conversions can be implemented very efficiently using the CORDIC (coordinate rotation digital computer) algorithm, which only uses simple shift and add operations [2]. In addition, the first conversion most likely already happens before the carrier recovery, as the constant modulus algorithm (CMA) update of the channel equalizer requires knowledge of the signal amplitude, and the intermediate frequency compensation is more efficiently implemented in polar coordinates as well. Also note that the calculation of $|X_k|^4$ is plotted with dotted lines. This calculation can be omitted, which has been shown to actually improve the performance of the V&V algorithm [3]. Hence, the number of required multipliers can be reduced to zero.

Optimizing all parts of the DSP chain is critical, because coherent receivers have stringent constraints on the power they can burn. This is most apparent in the development of coherent pluggable modules, which have maximum power consumptions specified as part of the form factor (e.g., 24 W for compact form-factor pluggable (CFP) [4]). But even coherent receiver line cards have this power restriction, which

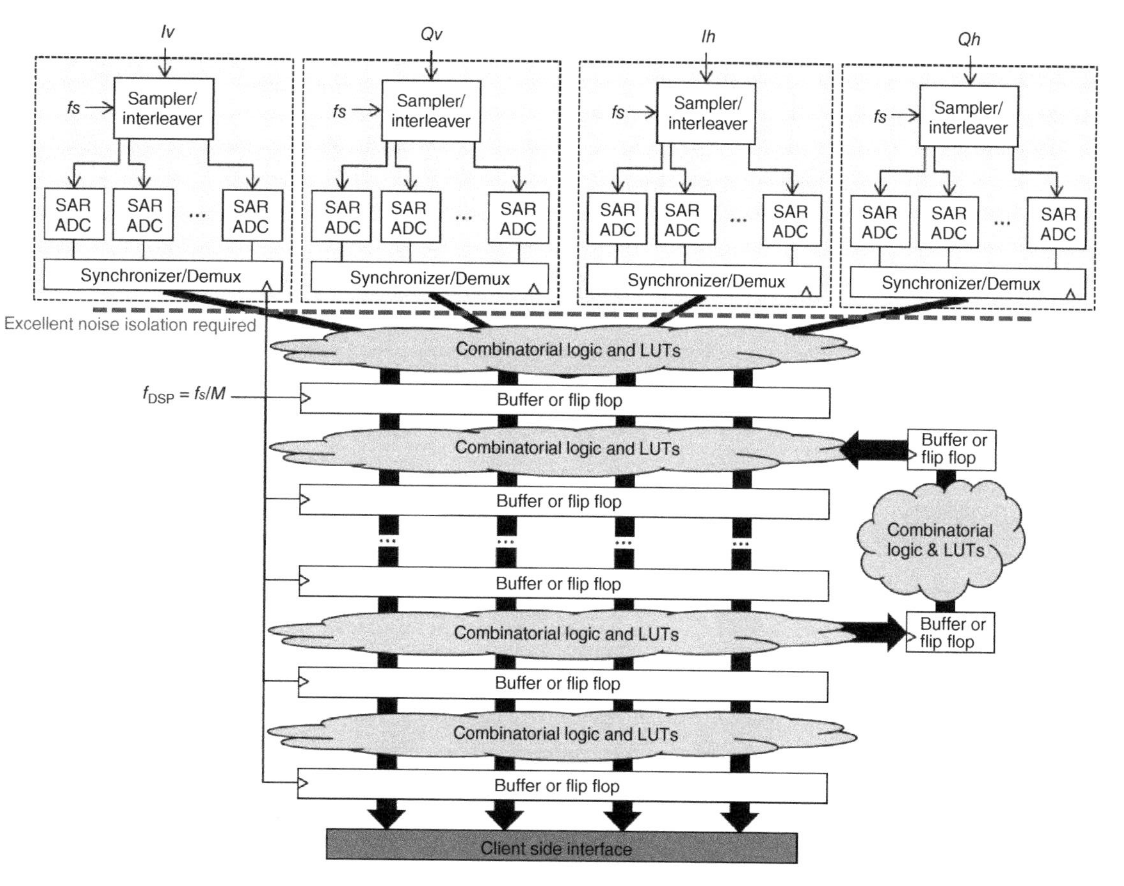

FIGURE 12.1 Hardware structure of a real-time coherent receiver DSP.

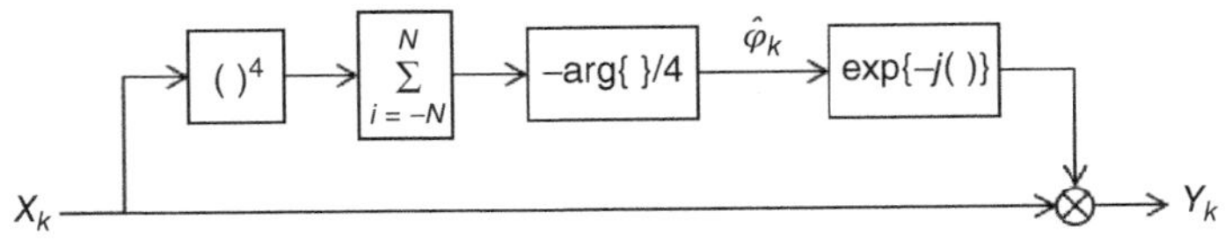

FIGURE 12.2 Viterbi & Viterbi (V&V) carrier recovery algorithm.

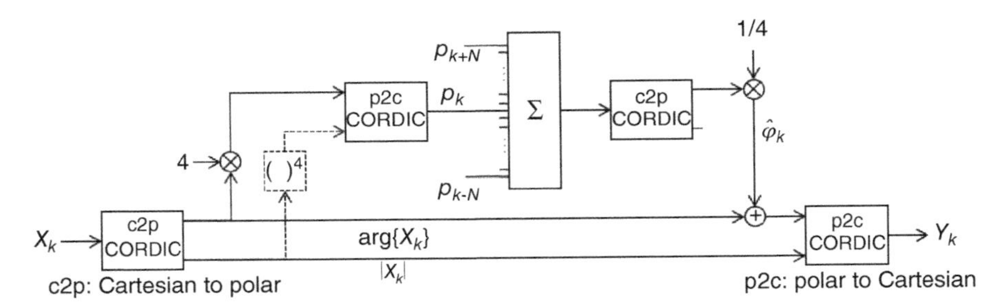

FIGURE 12.3 Hardware optimized V&V carrier recovery algorithm.

is a direct consequence of the cooling abilities on the line card. So the more the algorithms are optimized in terms of power consumption, the more features can be packed into the DSP.

12.1.2 Parallel Processing Constraint

Algorithms that support multi-Gb/s operations must allow parallel processing as was shown in Figure 12.1 [5]. As explained in the introduction, the DSP logic cannot operate directly at the sampling clock frequency of the ADC, but requires demultiplexing to process the data in parallel modules. This approach is even already used inside the ADCs, which use an architecture of a high-speed sample and hold buffer fanning out into several (e.g., 64) interleaved successive approximation (SAR) ADCs [6]. Hence, each SAR-ADC only needs a conversion rate of $<$1 GHz.

For the DSP processing, this low clock frequency allows automated generation of the layout, which is indispensable due to the complexity of the system. Algorithms for real-time applications must, therefore, allow parallel processing with a large number of demultiplexed channels. This translates into the requirement that (intermediate) results within one module cannot depend on results calculated at the same time in other parallel modules.

A good example to explain the feasibility of parallel processing is the comparison of two filter structures: Finite (FIR) and infinite (IIR) impulse response filters. Figure 12.4 depicts their structures in both serial and parallel systems. It can be seen that it is easily possible to parallelize an FIR filter. Although the output signal depends on information provided by several parallel modules, it does not depend on the results of the same calculations performed in these modules. In contrast, it is a big challenge

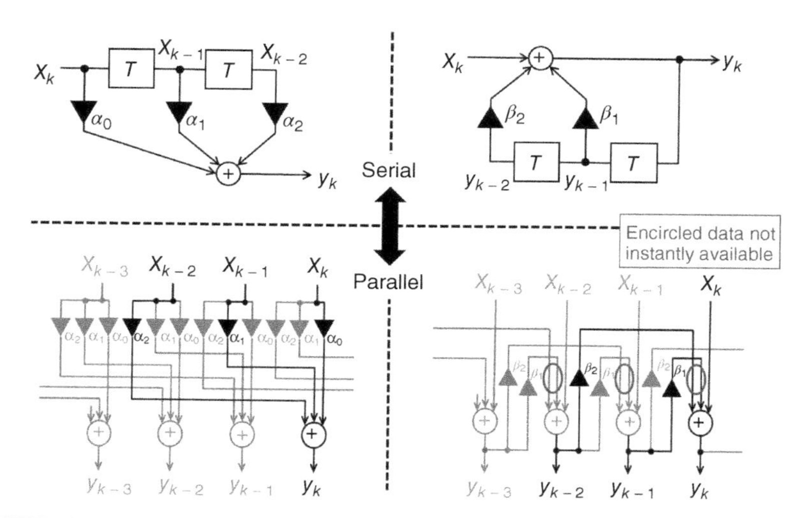

FIGURE 12.4 Finite impulse response (FIR) and infinite impulse response (IIR) filters in serial and parallel implementation.

to realize the parallel structure shown for the IIR filter, because the result depends on results calculated at the same time in other parallel modules. Similar to a carry bit in a digital adder, information has to traverse the entire parallel processing bus until the output becomes stable. A very low clock frequency or a low number of parallel channels would be needed to allow all calculations to be executed within one clock cycle. Neither of these requirements is fulfilled in coherent digital receivers for optical transmission system.

The only possible way to use algorithms that do not lend themselves for parallelization is to use parallel-to-serial conversion as shown in Figure 12.5 [7]. In this technique, the incoming data are written into a large memory column by column, and then read out for processing row by row. This allows frames of data to be processed in a truly serial fashion. However, this approach comes with large drawbacks, because a large and complicated memory structure is required, and discontinuities have to

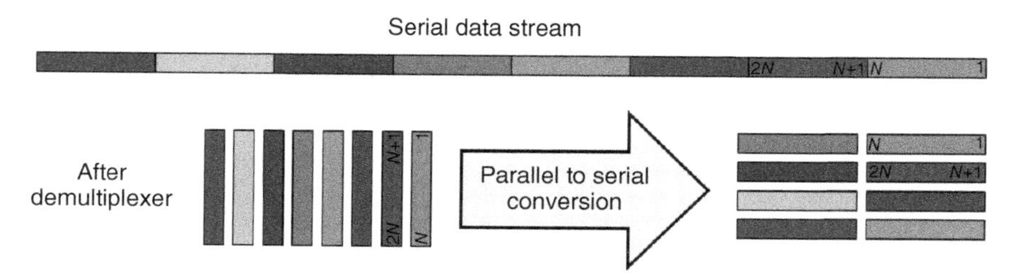

FIGURE 12.5 Concept of parallel to serial conversion to enable serial processing algorithms in a parallel processing structure.

be resolved at the start of each serial-processing block. Solving this problem requires either to have an overlap between the processing blocks, which further increases complexity, or to use a frame-based data structure with training symbols at the beginning of each processing block. Due to these drawbacks, this approach is generally avoided, and algorithms are chosen that map efficiently into a parallel processing structure.

12.1.3 Feedback Latency Constraint

In simulations or offline data processing, often feedback loops with one symbol loop latency are used. While this implementation achieves the highest performance, it does not represent a system that can be realized in an ASIC or FPGA. Hence, these results can only be considered as an upper bound for the performance that can be achieved.

Due to the massive parallelization and pipelining required in coherent real-time DSP processing, loop latencies can easily reach hundreds or thousands of symbols. Unfortunately, very few publications that analyze DSP algorithms actually study the effects of these large loop latencies. One of the reasons is that it is very difficult to estimate these latencies without detailed knowledge of the actual technology used for implementing the real-time processing. Table 12.1 provides a rough overview for the required clock cycles to implement the most common digital building blocks. The assumed clock frequencies are 500 MHz for ASIC and 200 MHz for FPGA implementations. Of course in reality, these numbers depend on a wide range of parameters (e.g., CMOS process node, resolution, digital implementation). But this table can be used by researchers and engineers who are not directly involved in hardware development to get a feel for the expected latency of their algorithms.

It can be seen that the processing latency inside an FPGAs is roughly three times larger than inside an ASIC – even despite the fact that FPGAs are also clocked at lower frequencies. This is due to the overhead in FPGAs caused by mapping the algorithms into a general purpose processing structure. This difference has to be kept in mind when FPGAs are used for the prototyping of coherent receivers.

Let us apply the latency numbers to an actual problem system designers face when selecting algorithms for a coherent receiver development. In several publications, it has been shown that the least mean square (LMS) equalizer update algorithm has faster convergence and better performance than the CMA algorithm [8, 9]. However, these studies in general do not consider the different loop latencies of these

TABLE 12.1 Real-time signal processing latencies for basic DSP functions

	ASIC	FPGA	Comment
Complex multiplier	2	5	Non-integer numbers are rounded to the next larger integer
Real multiplier	1	3	
N-point FFT	$2 \times \log_2\{N\}$	$5 \times \log_2\{N\}$	
k-bit CORDIC	$k/3$	k	
N-point summation	$\log_2\{N\}/4$	$\log_2\{N\}$	
LUT	1	1	

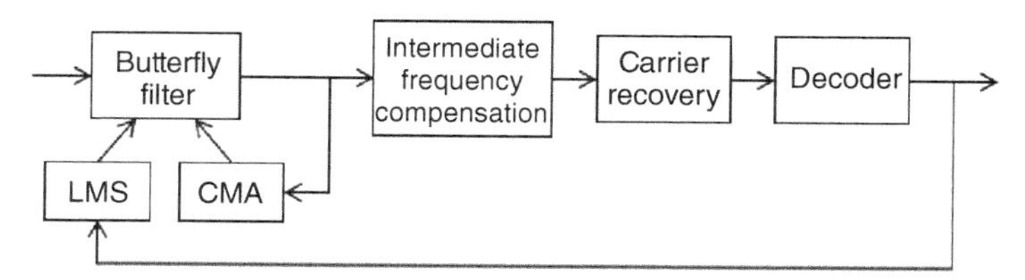

FIGURE 12.6 DSP blocks for CMA and LMS equalizer updates.

two algorithms. So how large is actually the difference in the loop latency between a CMA-based equalizer coefficient update and an LMS-based update. Therefore, we require the latencies of the following blocks: the butterfly filter, the frequency offset compensation, the carrier recovery, the decision circuit or decoder, and the equalizer coefficient update block (Figure 12.6). For simplicity 8-bit resolutions are assumed in all processing steps.

The butterfly filter consists of complex multipliers followed by a summation of the multiplier outputs. Let us assume a 16-tap filter. The latency through the equalizer, therefore, is

$$L_{\mathrm{EQ}} = 2 + \log_2\{16\} = 6$$

For the frequency offset compensation, the signal is converted from Cartesian to polar coordinates. This simplifies the subsequent processing stages. If we assume a feedback loop, then compensating the IF is a simple subtraction, that is, the latency becomes

$$L_{\mathrm{IF}} = 8/3 + 1 = 4$$

The carrier recovery uses the feedforward V&V algorithm depicted in Figure 3. Calculating the fourth power of the input symbols in polar coordinates is a simple left shift operation, which comes for free. The next part is the filter, which consists of two CORDIC blocks and one summation block. Finally, the estimated phase is subtracted from the received symbol, which is then converted back into Cartesian coordinates before it is fed into the decoder. If we assume a 50-tap filter length in the carrier recovery, the latency becomes

$$L_{\mathrm{CR}} = 8/3 + \log_2\{50\}/4 + 8/3 + 1 + 8/3 = 12$$

The decoder basically consists of comparators, so its latency can be assumed to be $L_D = 1$.

The two equalizer updates have almost identical processing. The only difference is the calculation of the error signal. While the LMS algorithm requires the distance vector between the decoded and the received sample (subtraction), the CMA algorithm weighs the input vector with the amplitude difference of the equalizer output from unity (CORDIC + subtraction + multiplication) [6]. Then, both algorithms multiply the error signal with the equalizer output, apply a control gain (can be implemented as right shift, which comes for free), and add it to the previous channel estimate. Hence, their latencies are

$$L_{\text{CMA}} = (8/3 + 1 + 2) + 2 + 1 = 9$$

$$L_{\text{LMS}} = 1 + 2 + 1 = 4$$

So in total, the latencies for the two different equalizer update implementations become

$$L_{\text{CMA,total}} = L_{\text{EQ}} + L_{\text{CMA}} = 15$$

$$L_{\text{LMS,total}} = L_{\text{EQ}} + L_{\text{IF}} + L_{\text{CR}} + L_D + L_{\text{LMS}} = 27$$

The LMS equalizer update has a >60% higher loop latency than the CMA update. If we assume a demultiplexing factor of 64, the loop latency for the CMA update is 960 symbols, for the LMS update it is 1728 symbols. And replacing some of the processing with higher performance algorithms (e.g., feedforward frequency estimation, or using BPS instead of V&V) can easily push the latency for the LMS update to >2000 samples. This latency difference has a large impact on the convergence and tracking capabilities of the algorithms, and the choice which equalizer to use has to be reviewed under this constraint. This is a crucial example of why feedback latency needs to be considered as early in the development of a next generation coherent receiver as possible.

12.2 HARDWARE IMPLEMENTATION OF DIGITAL COHERENT RECEIVERS

The different requirements for research or product development also cause different strategies in the real-time architectures for the digital transmitter and receiver. In research, the main objective is to have a highly flexible setup, which allows implementing a variety of different processing algorithms for different modulation formats. In addition, the design cycle needs to be fast, as the research has to be conducted before the start of the product development. Therefore, research prototypes are implemented using field-programmable gate arrays (FPGA). These devices are made up of up to a million so-called slices (Figure 12.7), which contain look-up tables, multiplexers, dedicated adder logic, and registers. The slices are connected through wide busses and switching nodes. By changing the content of the LUTs and the wiring of the interconnects between the slices, virtually any logic function can be implemented inside the FPGA fabric. Modern FPGAs offer a huge amount of processing capacity with hundreds of thousands of slices, thousands of dedicated multiplier blocks, clock managers, and a large number of gigabit transceivers, supporting aggregated IO bandwidths of >1 Tb/s.

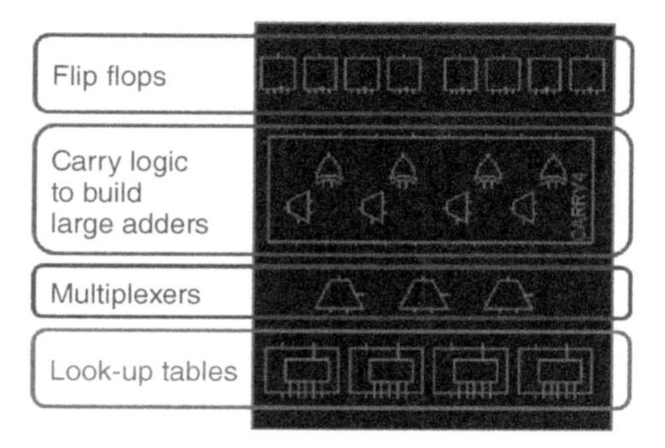

FIGURE 12.7 Example for an FPGA slice containing look-up tables, multiplexers, dedicated adder logic, and registers.

But as impressive as these numbers may be, comparing them with the requirements of a 100G digital coherent receiver reveals that these devices still imply large restrictions on possible prototype implementations. One major limitation is that no high-speed ADCs are available integrated into FPGAs. Hence, external high-speed ADCs need to be used, which require enormous interface bandwidths. A coherent receiver requires four ADCs running at ~56 Gs/s. If the resolution of the ADCs is 8 bit, then the required interface bandwidth is ~1.8 Tb/s. Hence, FPGA prototypes often use ADCs with a lower sampling rate and/or lower resolution compared with a product receiver.

Due to the internal architecture of FPGAs with slices and large reconfigurable interconnect networks, often more than 50% of the processing delay is caused by routing the signals inside the device. This increases the number of pipeline stages required in the design and reduces the optimal clock speed for the FPGA (see Table 12.1). The latter causes the design to grow in size, as a higher degree of parallelism is required to achieve the desired throughput.

In addition, due to the massive parallel processing required inside the FPGA (bus widths of 256 or larger are possible), only a few thousand slices and a few multipliers are available per parallel lane. Hence, only a fraction of the overall processing can be implemented in a single FPGA, and the algorithms need to be partitioned into multiple FPGAs. This further increases the interconnect restrictions mentioned earlier.

Figure 12.8 shows an example of an FPGA board used for real-time prototyping of a coherent receiver. It hosts four 6-bit 32-Gs/s ADCs [10], an FPGA with an IO bandwidth of 96 × 13.1-Gb/s, and VCSEL-based parallel optical interconnects for daisy-chaining of multiple boards. Four of these boards were needed to build a real-time 100G coherent OFDM receiver, which was successfully employed in a first field trial for real-time CO-OFDM [11].

Inside commercial transceiver modules with much tighter space, power, and cost restrictions than research prototypes, ASIC implementations are the only possible solution. In an ASIC, all functions are implemented exactly as they are needed, which makes the design significantly faster and more power efficient than an FPGA implementation. In addition, it is possible to monolithically integrate high-speed ADCs and

FIGURE 12.8 Real-time FPGA based coherent receiver prototype board.

DACs with huge amounts of DSP logic into a single chip. Figure 12.9 shows the layout of the very first commercial coherent ASIC taking advantage of this integration [12]. The two parts of the chip are very distinct: the ADCs on the left, and the very dense DSP section on the right. However, this tight integration leads to new design challenges.

One major challenge is the proper isolation between the ADC cores and the standard DSP logic. In order to achieve a high effective number of bits for the ADCs, the added noise and distortions that the ADCs inject into the analog input signal have to be minimal. This is not an easy task, because a massive DSP engine sits right next to it, which produces a lot of switching noise from toggling CMOS logic and can cause current spikes in the order of 100 A. This isolation requires very careful design in order to ensure that the ADCs do not degrade in performance when integrated with the DSP.

Figure 12.10 shows an example for a commercial coherent transceiver linecard. In future, different ASIC designs will offer interesting differentiation options for various systems. While one chip can only include the bare essentials of algorithms required for a coherent receiver to achieve low power consumption and enable a small form-factor-pluggable module, a different chip can be designed with the most sophisticated algorithms such as digital backpropagation or advanced soft-decision FEC for high-performance transoceanic submarine systems. With the industry moving toward higher-order modulation formats, super-channel transmission systems, and maybe even spatial division multiplexing in multimode fibers, and at the same time coherent transmission also becoming attractive for metro and shorter reach links, the diversification of coherent DSP chips optimized for different transmission scenarios has only just begun.

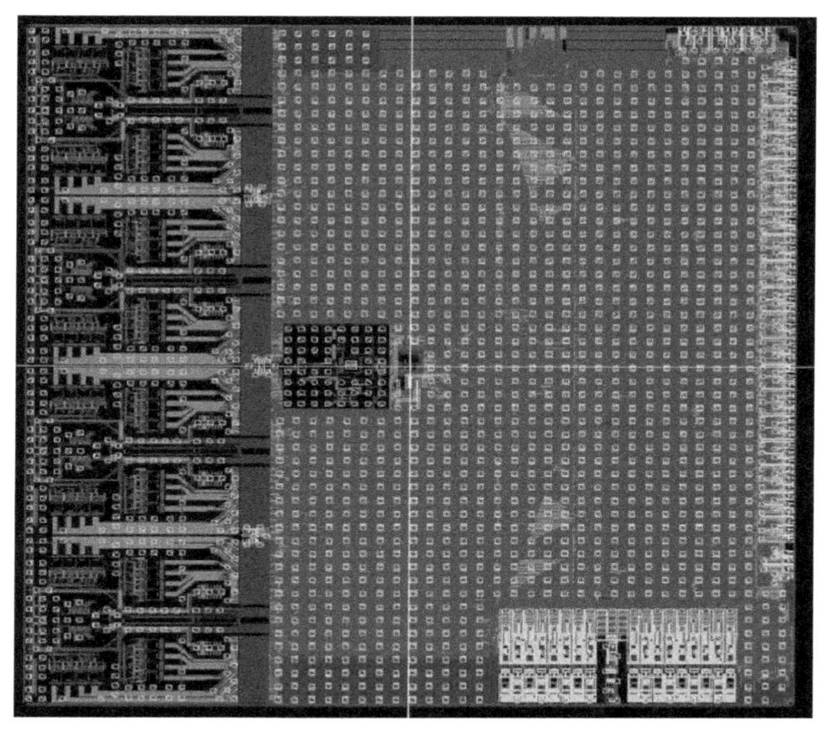

FIGURE 12.9 Layout of the first coherent receiver CMOS ASIC with four 20 Gs/s ADCs for 40 Gb/s PDM-QPSK transmission. Reproduced from Ref. [12]. With permission of Ciena Corporation.

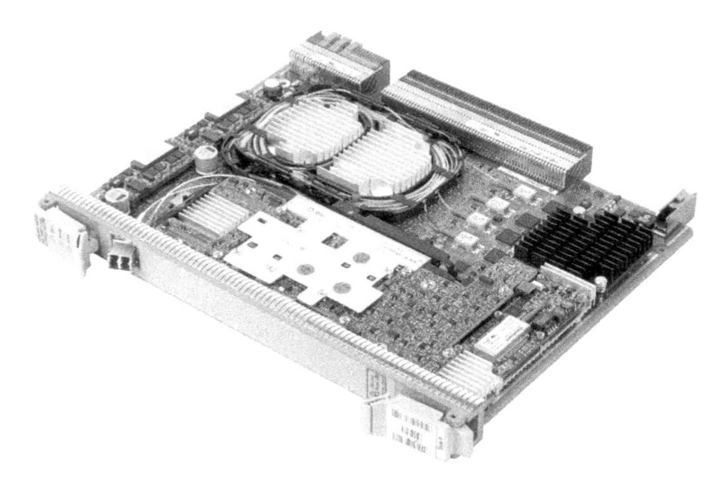

FIGURE 12.10 Commercial coherent transceiver linecard.

REFERENCES

1. Viterbi AJ, Viterbi AM. Nonlinear estimation of PSK-modulated carrier phase with application to burst digital transmission. *IEEE Trans Inf Theory* 1983;29(4):543–551.
2. Volder J. The CORDIC trigonometric computing technique. *IRE Trans Electron Comput* 1959;EC-8:330–334.
3. Pfau T. Real-Time Coherent Optical Receivers: Theoretical Description and Real-Time Implementation of Digital Signal Processing Algorithms for Coherent Optical Receivers. Suedwestdeutscher Verlag fuer Hochschulschriften; 2009.
4. CFP MSA Hardware Specification, Revision 1.4; 7 June 2010.
5. Pfau T, Peveling R, Herath V, et al. Towards real-time implementation of coherent optical communications. Proceedings of OFC/NFOEC'09, OThJ4; San Diego, CA, USA; Mar 22–26, 2009.
6. Kull L, Toifli T, Schrnatz M, et al. A 90GS/s 8b 667 mW 64× interleaved SAR ADC in 32 nm digital SOI CMOS. Proceedings of ISSCC'14, 22.1; San Francisco, CA, USA; Feb 9–13, 2014.
7. Pfau T. Carrier recovery and real-time DSP implementation for coherent receivers. Proceedings of OFC'14, W4K.1 (tutorial); San Francisco, CA, USA; Mar 9–13, 2014.
8. Kuschnerov M, Chouayakh M, Piyawanno K, et al. Data-aided versus blind single-carrier coherent receivers. *IEEE Photon J* 2010;2(3):387–403.
9. Mantzoukis N, Petrou CS, Vgenis A, et al. Performance comparison of electronic PMD equalizers for coherent PDM QPSK systems. *J Lightwave Technol* 2011;29(11):1721–1728.
10. Ellermeyer T, Müllrich J, Rupeter J, et al. DA and AD converters for 25 GS/s and above. Proceedings of IEEE/LEOS Summer Topical Meetings, TuC3.1; Acapulco, Mexico; Jul 21–23, 2008.
11. Kaneda N, Pfau T, Zhang H, et al. Field demonstration of 100-Gb/s real-time coherent optical OFDM detection. Proceedings of ECOC'14, Th.2.5.3; Cannes, France; Sep 21–25, 2014.
12. Sun H, Wu KT, Roberts K. Real-time measurements of a 40 Gb/s coherent system. *Opt Express* 2008;16(2):873–879.

13

PHOTONIC INTEGRATION

PO DONG AND SETHUMADHAVAN CHANDRASEKHAR
Bell Laboratories, Nokia, Holmdel, NJ, USA

13.1 INTRODUCTION

Advanced modulation formats, together with wavelength-division multiplexing (WDM) and polarization-division multiplexing (PDM), have become key enablers to increase the spectral efficiency and the overall capacity per fiber [1, 2]. Coherent optical transmission has been employed in long-haul communications with channel data rates at 100 G and beyond and with transmission distance typically beyond 2000 km. PDM quadrature phase-shift keying (QPSK) is the format of choice in the current deployed 100 G networks. Next-generation 400 G networks may utilize PDM 16-ary quadrature-amplitude modulation (16QAM). Coherent transmission technology is also well-suited for metro and short-reach optical networks [3], in which small transceiver footprint along with reduced power consumption is strongly required.

A coherent transponder mainly consists of three parts, namely optical components, modulator drivers, and transimpedance amplifiers (TIAs), and application-specific integrated circuits (ASICs), as shown in Figure 13.1 for a dual-polarization QPSK or 16QAM transceiver. The optical front-end consists of a number of high-performance optical components, such as narrow-linewidth lasers, high-speed modulators, high-dynamic-range photodetectors (PDs), polarization beam splitters and combiners (PBS/Cs), polarization rotators (PRs), and 90° optical hybrids. The ASICs include electronic processing elements such as analog-to-digital converters (ADCs) and digital-to-analog converters (DACs) to capture and generate optical signals, respectively, digital signal processing (DSP) core to process and condition the

Enabling Technologies for High Spectral-efficiency Coherent Optical Communication Networks, First Edition. Edited by Xiang Zhou and Chongjin Xie.
© 2016 John Wiley & Sons, Inc. Published 2016 by John Wiley & Sons, Inc.

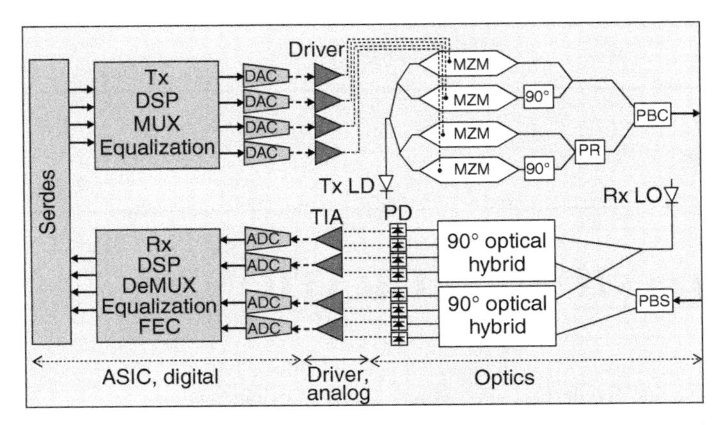

FIGURE 13.1 Schematic diagram of dual-polarization coherent optical transceiver. MZM: Mach–Zehnder modulator, PBS/C: polarization beam splitter/combiner, PR: polarization rotator, LD: laser diode, LO: local oscillator, TIA: transimpedance amplifier, DAC: digital-to-analog converter, ADC: analog-to-digital converter, DSP: digital signal processing, ASIC: application-specific integrated circuit.

signals, and multiplexers and demultiplexers to aggregate and disaggregate the information payload.

In current optical transceivers, discrete optical components with optimized performance are used for meeting the very challenging transmission requirements. Multisource-agreement (MSA) modules based on optical internetworking forum (OIF) recommendations have emerged, with the transceiver module size of 5 × 7 square inches, power consumption of less than 80 W, and the modulation format of PDM-QPSK. Nevertheless, the size and power consumption of this MSA transceivers for long-haul applications are not suitable for metro networks where port densities on both line side and client side are important. Smaller-form-factor 100 G coherent transceivers are under development by various subsystem vendors, equipment manufacturers, and module suppliers. There are a few options of form factors under consideration, which all require high level of photonic integration. One choice is a nonpluggable 4 × 5-square inch MSA module with less than 40 W of power. Other choices include the compact form-factor pluggable (CFP), CFP2, and even smaller CFP4, as shown in Figure 13.2. For CFP/CFP2, the size and power are limited to 82 × 145 mm^2/41.5 × 107 mm^2 and 32 W/12 W, respectively. Implementing a 100-G coherent transceiver in such small modules is very challenging since considerably less space is available and low heat dissipation is required. Therefore, the CFP implementation requires a high degree of optical integration, low-power driver/TIAs, and power-optimized ASICs.

The power consumption for a coherent transceiver mainly includes the power consumed by the ASICs (DACs, ADCs, DSPs), lasers, modulator drivers, and TIAs. In order to pack all the components in a smaller form factor, the power consumption need to be reduced. The power consumption of the ASICs is a significant part of the

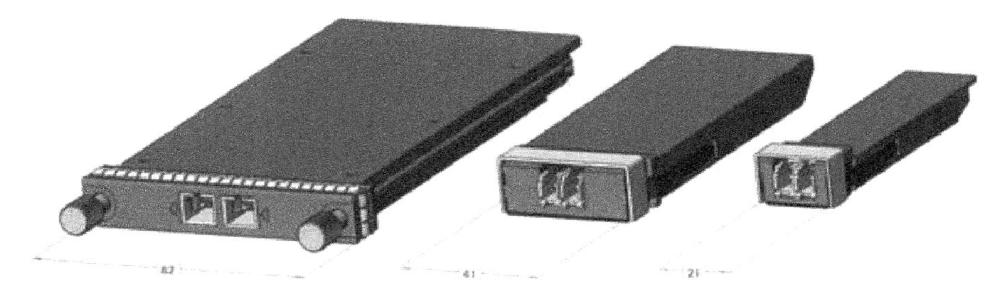

FIGURE 13.2 Possible CFP modules for pluggable coherent optical transceivers. http://www.cfp-msa.org.

overall power consumption. This power can be reduced if implemented with smaller transistor feature size. Today, 40-nm complementary metal–oxide–semiconductor (CMOS) represents the state-of-the-art for mixed-signal ASICs. However, more advanced CMOS processes such as 28-nm will be available soon and may become the preferred feature size if low-power ASICs are required. In addition, compared with long-haul application, the metro link reach may be less than a few hundred kilometers. The shorter transmission reach could require less link dispersion equalization, namely fewer filter taps and less processing are needed, leading to lower power consumption. The rest of the power consumption related to optical components become more critical once the ASIC power drops. Energy-efficient lasers and low-voltage electro-optic modulators would then be preferable.

Photonic integration appears to be a must for CFPx modules. A typical length of a $LiNbO_3$ modulator is about 100 mm which makes it difficult to fit into even the largest CFP modules. Alternative modulators based on indium phosphide (InP) and silicon photonic integrated circuits (PICs) are emerging as the most promising modulator candidates. To save space, it is also becoming critical to employ optical packaging without the use or with only limited use of micro-optic components, such as lenses, power splitters, polarizers, PBS/Cs, and PRs. In summary, photonic integration is required to achieve compact coherent optical transceivers for next-generation pluggable modules to be used in metro and even shorter-distance coherent optical systems. In Section 13.2, we review three photonic integration technologies that may fulfill the stringent requirements for this application. In Section 13.3, we review the progress in integrated transmitters, while in Section 13.4 we review the progress in integrated receivers. We finally make concluding remarks in Section 13.5.

13.2 OVERVIEW OF PHOTONIC INTEGRATION TECHNOLOGIES

There are mainly three promising optical technologies currently in development aiming for compact and low-power coherent transponders. Planar lightwave circuit (PLC) technology is used to realize passive optical components which may consist of power splitters/combiners and PBSs/PBCs. By packaging high-speed modulators/detectors with PLCs, "hybrid" coherent transmitters/receivers can be realized. InP and silicon monolithic PICs are the second and the third technologies, respectively. Most of the demonstrated coherent transmitters and receivers do not have monolithic integration of the laser. Integration of the laser with the modulators or detectors may have limited advantages. This integration significantly increases the integration complexity for InP and silicon PICs. Moreover, a laser generally requires cooling, which consumes significant power. In order to save power, it would be better not to cool other devices such as modulators and detectors. Furthermore, a single laser can be shared in the transmitter and receiver to reduce the cost and power, which makes the laser integration less attractive.

For PLC hybrid technology, silica or polymer PLCs have been co-packaged with $LiNbO_3$ modulators for transmitters [4–6], or have been assembled with III–V photodetectors for receivers [7–16]. The overall sizes for optical front-end and the packaging cost could be concerns.

Higher-degree photonic integration can be achieved by III–V (mainly InP) and silicon PICs. The PICs could significantly reduce the device sizes and simplify the packaging procedures and hence the cost. Both InP and silicon PICs are very promising, each with their own merits and limitations. InP PICs can provide integrated lasers, but it is challenging to integrate lasers, modulators, photodetectors and polarization elements all on a single chip. Moreover, its yield and cost could be concerns. Silicon PICs take the advantage of CMOS foundries with large wafer size, high yield, and low cost. Silicon waveguides also have the flexibility to implement polarization combiners and rotators, which offers advantages in polarization diversity circuits. The emerging hybrid wafer-scale integration technology to bond InP on silicon [17] offers another possibility to support integrated amplifiers and lasers.

InP-based PICs have been realized to integrate high-speed modulators or photodetectors, wavelength multiplexing filters, and even integrated lasers on the same chip for coherent applications [18–41]. However, polarization diversity is typically achieved by off-chip micro-optic polarization elements.

Silicon PICs also show very promising progress for coherent transceivers. Compact silicon microring modulators [42, 43] as well as silicon Mach–Zehnder modulators (MZMs) [44–51] have been demonstrated for in-phase/quadrature (*I*/*Q*) modulation. Dual-polarization *I*/*Q* modulators have been implemented on a single chip [45, 50]. Silicon–organic *I*/*Q* modulators with low-power consumption at 40 Gbaud have been reported [52]. In addition, monolithic silicon PICs were also demonstrated for polarization-diversity coherent receivers [46, 53–55] and a single-chip transceiver was reported in [56]. These demonstrations confirm the potential of silicon photonics for high-level integration capability for dual-polarization coherent transceivers.

Table 13.1 illustrates the comparisons of the three technologies discussed earlier. As each technology was evolving very rapidly and dramatically during the last few years, and is expected to continue to do so in the near future, this comparison serves as a general guideline rather than an accurate prediction. If one does not consider monolithic integration of the laser, the integration level is highest for silicon PICs as polarization elements can be monolithically integrated on-chip, while it is lowest for PLC hybrids, as both high-speed optical devices and polarization elements need to

TABLE 13.1 High-level comparison of different photonic integration technologies for coherent optical transceivers. As each technology evolves, this table may only serve as a guideline

	PLC hybrid	InP PIC	Si PIC
Performance	High	Medium	Low
Integration level	Low	Medium	High
Cost	High	Medium	Low
Size	Large	Medium	Small

be assembled together. Considering the integration level point of view, the packaging cost for silicon PICs maybe the lowest, with compact footprints as well. However, the optical performance of PLC hybrids may be the best as PLCs provide both low insertion loss and high-performance passive components, together with individually optimized modulators and PDs made from other materials.

13.3 TRANSMITTERS

In this section, we review *I*/*Q* modulator technologies. This section is organized as follows. Section 13.3.1 describes the transmitter circuits for the realization of dual polarization *I*/*Q* modulators. Section 13.3.2 examines the critical component in an *I*/*Q* modulator, namely the MZM. Sections 13.3.3–13.3.5 summarize previously reported *I*/*Q* modulators based on PLC hybrids, InP PICs, and silicon PICs, respectively.

13.3.1 Dual-Polarization Transmitter Circuits

The most widely used modulator for full E-field encoding is the nested MZM, also called an *I*/*Q* modulator or a vector modulator. It consists of two MZMs in parallel, one for encoding the in-phase part of light, and the other for encoding the quadrate-phase part of light. A power combiner combines the outputs of two MZMs with a $\pi/2$ phase difference. Since a single-mode fiber can support two orthogonal polarization states, a dual-polarization *I*/*Q* modulator meets the requirement for maximizing information encoding. The building blocks for such a dual-polarization *I*/*Q* modulator are a polarization-maintaining power splitter to split the continuous-wave (CW) laser power into equal powers to be launched into each *I*/*Q* modulator. At the output, one needs to combine one signal with the other signal in the orthogonal polarization using a PR followed by a PBC. Figure 13.3 shows a circuit diagram for a dual-polarization *I*/*Q* modulator. A multilevel QAM signal can be synthesized by using parallel *I*/*Q* modulators with binary electrical driving signals, or by a single *I*/*Q* modulator with multilevel electrical driving signals.

Four RF driving signals and at least six tunable phase-control elements are required. The four RF electrical driving signals are usually termed as IX, QX, IY, and QY, where *I*/*Q* represents in-phase/quadrature components, and *X*/*Y* refers to the

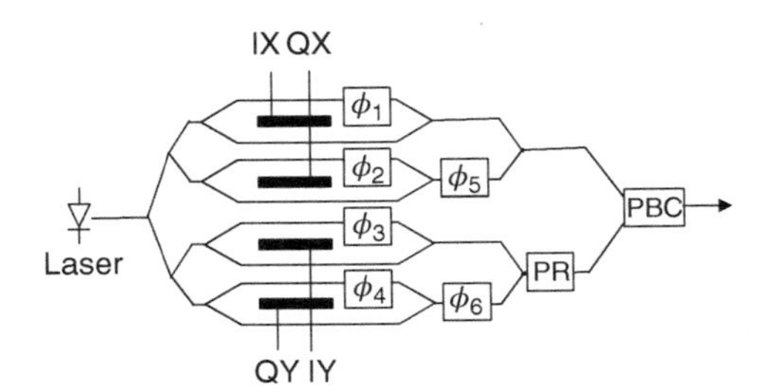

FIGURE 13.3 Optical circuit for a dual-polarization *I/Q* modulator. PR: polarization rotator, PBC: polarization beam combiner.

two polarization states. For QPSK generation, the RF driving signals typically have two levels, while for 16QAM generation, the RF driving signals typically have four levels. The MZMs inside each *I/Q* modulator are usually operated under push–pull condition, which makes them more tolerant to the variations of drive voltage and at the same time achieve zero-chirp operation. Each MZM (also sometimes referred to as "daughter" MZM) needs to be biased at its minimum transmission point. Four-phase elements (ϕ_1–ϕ_4) in the daughter MZMs are, therefore, present in order to tune the phase difference between the two arms of MZMs to be at π. In commercial *I/Q* modulators, active locking to π-phase differences are required by monitoring the light from MZMs. For $\pi/2$-phase difference between *I* and *Q* branches, two additional phase elements (ϕ_5 and ϕ_6) after the daughter MZMs are needed.

Different technologies have different monolithic integration levels. For PLC hybrids, the passive power coupler and phase elements, sometimes with variable optical attenuators (VOAs), are on PLCs, while MZMs, PRs, and PBCs are off chip. For InP PICs, PR and PBC are typically external, and the rest can be monolithically integrated. For silicon PICs, all components (except the laser) can be on a single chip.

13.3.2 High-Speed Modulators

Modulators are one of the most crucial and challenging optical devices in coherent transceivers. The performance of an MZM can be characterized by three most important parameters, that is, the 3-dB electro-optic bandwidth, the half-wave voltage swing V_π, and the optical insertion loss. For coherent transceivers, it is highly desirable to have both higher bandwidth and lower V_π. This is driven by the need for low power consumption, high data rate operation, as well as difficulties in developing high-voltage broad-band linear electronic amplifiers. However, reducing V_π usually results in reduced bandwidth and increased insertion loss because of the need of longer phase shifter in the MZM.

There are two common types of MZMs operating in push-pull configuration, that is, single drive and dual drive. For a single-drive push–pull MZM, the two arms of MZM are electrically connected and a single RF signal drives the two arms simultaneously. With a proper electrode arrangement, the push–pull operation can be obtained, which is important for zero-chirp QPSK and 16QAM modulation formats. For dual-drive push–pull, the two electrodes that drive two arms are independent. If an RF signal and its complementary are applied on the two electrodes simultaneously, push–pull operation is achieved. Typically, V_π is defined as a voltage swing for a phase change of π between two arms of the MZMs, where the voltage is applied on one arm. Therefore, for single-drive InP and silicon MZMs, a full V_π is needed to achieve a phase change of π between the two arms, while for dual-drive, only $V_\pi/2$ is required on each arm. Depending on the availability of modulator drivers and driver output, either single drive or dual drive may be preferable. For the device design, Single-drive InP and silicon MZMs are better for broad bandwidth, but may have higher insertion loss. The bandwidth benefit comes from the fact that the two arms connect to the RF transmission line are in series, resulting in the loaded capacitance being reduced to half. However, in order to use same drive voltage amplitude, the phase shifter in single-drive MZMs needs to be twice as long as the dual-drive MZMs, which can make the insertion loss higher, and also degrades the bandwidth.

The InP material system (which also includes ternary materials such as InGaAs, and quaternary materials such as InGaAsP and InGaAsAl) can provide both electro-absorption (EAM) and electro-refraction modulations. Although InP EAMs have been employed to achieve QPSK and 16QAM, I/Q modulators based on electro-refractive or phase modulation are more popular. The high-speed modulation in InP waveguides can be achieved by applying a reverse voltage across a p–i–n or n–i–n junction, as shown in Figure 13.4. There are two main electro-optic effects involved. The first one is the Pockels effect, which is similar to that in $LiNbO_3$. The Pockels effect occurs in crystals that lack inversion symmetry. The refractive index change is proportional to the strength of the electrical field and the sign of the index change depends on the orientation of the electrical field with respect to the crystal axis. This effect is an ultra-fast electro-optic effect; however, it is typically a weak effect, and may require centimeter-long device to achieve the desired π-phase shift for a low V_π. The second effect is a shifting of the band edge with an applied voltage, leading to a change in the absorption. It is called the Franz–Keldysh effect in bulk material and the quantum-confined Stark effect (QCSE) in multiple quantum wells (MQWs). The applied electrical field shifts the band edge toward longer wavelengths, which causes changes to both absorption and refractive index for a

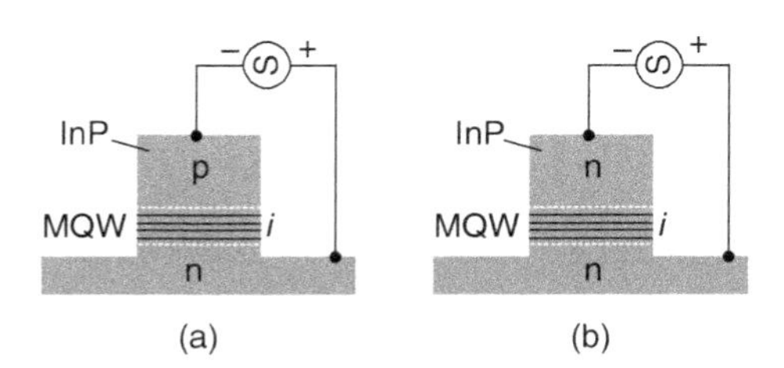

FIGURE 13.4 Typical waveguide cross-sections in InP for high-speed modulators.

signal whose wavelength is located a few tens to about one hundred nanometers larger than the band edge wavelength. Depending on the signal wavelength, EAMs can be realized if the shifted band edge reaches or is close to the signal wavelength, and therefore the absorption effect results in a modulation of light. If the signal wavelength is far away from the band edge, MZMs can be realized by exploiting the phase change that occurs with the applied voltage.

In recent years, silicon photonic MZMs have been undergoing significant developments. Silicon itself does not have strong electro-optic effects such as the Kerr effect and the Franz–Keldysh effect as the signal wavelength at 1550 or 1330 nm is too far away from the silicon band edge. Furthermore, silicon does not exhibit the Pockels effect due to inversion symmetry. Researchers have reported electro-optic modulators by employing other hybrid materials deposited on silicon, such as germanium, polymer, and III–V semiconductors. Nevertheless, having an all-silicon modulator would potentially yield a very reliable structure. High-speed modulation in silicon itself can be realized by free-carrier-induced refractive index change. The carrier-density modulation in a silicon waveguide can be obtained with carrier injection in a forward-biased p–i–n diode structure, carrier accumulation in an MOS capacitor structure, or carrier depletion in a reverse-biased p–n diode structure (see Figure 13.5). Carrier injection is the most efficient modulation mechanism, but it is difficult to achieve high speed since it suffers from high junction capacitance and slow free-carrier recombination (Figure 13.5a). Complicated driving signals with pre-emphasis can be used to speed up the free-carrier injection/recovery process and thus increase the speed of the modulator. Carrier accumulation in MOS capacitors (Figure 13.5c) has better modulation efficiency than carrier-depletion in a reverse-biased p–n junction (Figure 13.5b), but the speed of MOS-type modulators may be limited by the high capacitances from the thin oxide layer. Carrier depletion has the worst modulation efficiency [57–63], yet comes with the best high-speed performance, as the junction capacitance can be reduced by optimizing doping profiles.

Table 13.2 summarizes the performance of $LiNbO_3$, InP, and silicon MZMs reported in the literature. For both $LiNbO_3$ [64, 65] and InP modulators [26, 66], a V_π of 1–2.5 V can be achieved with a bandwidth about 30–40 GHz, while for silicon MZMs, the best reported V_π is 3.0–4.0 V with a bandwidth 10–20 GHz for reverse-biased p–n junction modulators. In addition, silicon MZMs tend to have higher insertion loss as p–n junctions in the waveguide center result in significant free-carrier-induced propagation loss. For MOS-type silicon MZMs [47], the V_π is

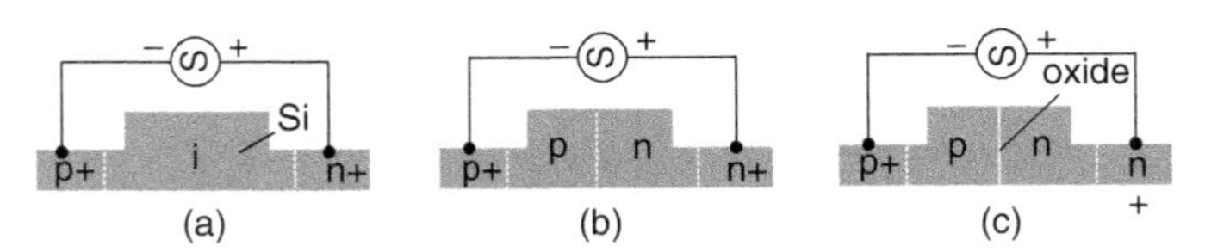

FIGURE 13.5 Typical waveguide cross-sections in all-silicon high-speed modulators. Here, only lateral junctions are shown, while vertical junctions are possible.

TABLE 13.2 Performance comparisons of MZMs with different technologies

	$LiNbO_3$ MZM		InP MZM		Si MZM[a]	
	Single drive [64]	Dual drive [65]	Single drive [26]	Dual drive [66]	Single drive [61]	Dual drive [63]
$V_\pi(V)$	~2.5	~1.0	1.5	2.2	~3.1	~4.0
Speed (GHz)	~28	~28	35	40	~15	~20
Insertion loss (dB)[b]	~4	NA	~6	NA	~9	~5.2

[a]Silicon MZMs based on depletion-mode p–n junction type.
[b]On-chip insertion loss, not including fiber coupling loss.

comparable to the p–n junction modulator but the insertion loss may be even higher. In silicon–organic modulators [49, 52], where polymers are deposited on silicon slot waveguides or photonic crystals, the V_π can be comparable with $LiNbO_3$ and InP modulators, but the insertion loss is higher and also the reliability is a concern. It is to be noted that although the bandwidth-V_π performance of InP modulators are comparable to those of $LiNbO_3$, the fiber coupling to $LiNbO_3$ MZMs is much easier due to better mode matching. In addition, InP modulators typically have phase shifters with lengths of a few millimeters, similar to that of silicon MZMs, while for $LiNbO_3$, the length can be a few centimeters.

13.3.3 PLC Hybrid *I/Q* Modulator

It was recognized that several functional modulators could be realized using multiple *I/Q* modulators by appropriately combining these modulators with passive PLCs both at the input and at the output [4]. As seen in Figure 13.6, devices such as a QPSK modulator, a DP-QPSK, a 64-QAM modulator, and a dual-carrier orthogonal frequency-division multiplexing (OFDM) transmitter have been demonstrated using the versatile hybrid integration platform. The PLC part incorporated passive elements such as splitters, combiners, VOAs, phase shifters, and polarization beam rotators/combiners, whereas the $LiNbO_3$ part incorporated several MZMs that could be combined to form multiple *I/Q* modulators. Nevertheless, with the widespread adoption of DACs at the transmitter to synthesize multilevel modulation formats using a single dual-polarization *I/Q* modulator, these functional modulators have been largely used for research demonstrations and have not made it into commercial products.

13.3.4 InP Monolithic *I/Q* Modulator

In Table 13.3, we list some examples of monolithically integrated InP *I/Q* modulators as well as transmitters that include a laser. The examples clearly show that InP modulators have made excellent progress in the last few years, with the latest InP modulators with a V_π of 1.5 V and bandwidth >40 GHz [26].

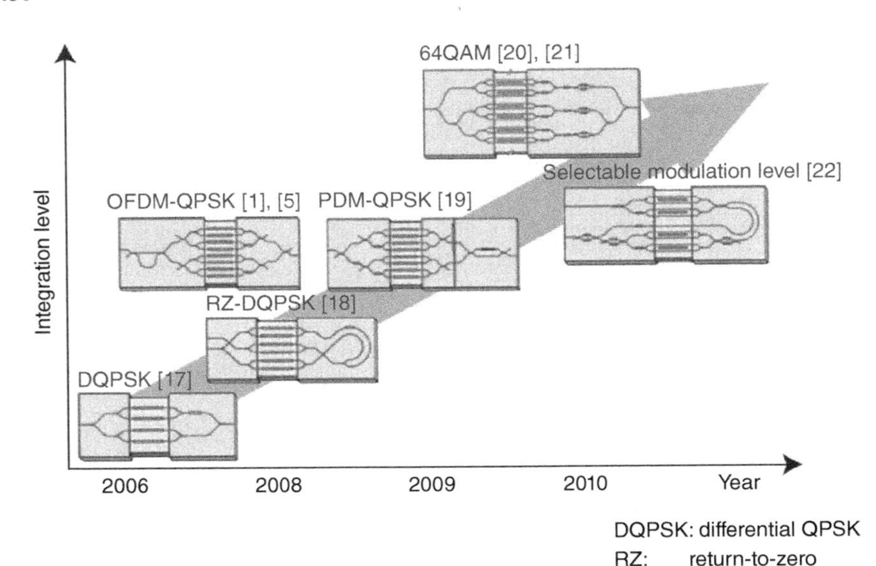

FIGURE 13.6 Functional modulators using hybrid integration of PLC with $LiNbO_3$ modulators. From Ref. [4].

TABLE 13.3 Some reported *I/Q* modulators on InP PICs

Reference	Devices	Performance
Clarke et al. [18]	DS-DBR Laser and pulse carver with Vector modulator (2 chips)	40-Gb/s (RZ) DQPSK
Shibata et al. [21]	DFB Laser + MZM and Vector modulator (2 chips)	28-Gbaud QPSK 112-Gb/s PDM-QPSK
Corzine et al. [23]	DFB and 2X Vector modulators 10 channels	14.25-Gbaud QPSK 57-Gb/s PDM-QPSK
Heck et al. [28]	InP-based polarization multiplexer 2X Vector modulators with	28-Gbaud PDM-QPSK
Prosyk et al. [19]	1X Vector modulator	40-GHz bandwidth. No data encoding demonstrated

There may exist some challenges to implementing and operating InP-based *I/Q* modulator. Unlike $LiNbO_3$ modulators, which purely rely on the electro-optic (EO) interaction via the Pockels effect, InP MZMs exhibit a more complex behavior due to the interplay of both the electro-optic effect as well as the electro-absorption effect, as discussed in Section 13.3.2. The phase modulation of an InP MZM via voltage modulation exhibits some nonlinearity, with increasing modulation efficiency at deeper bias. In addition, there is the voltage-dependent optical absorption associated with the shift and broadening of the exciton peak QCSE. The two effects make operation of

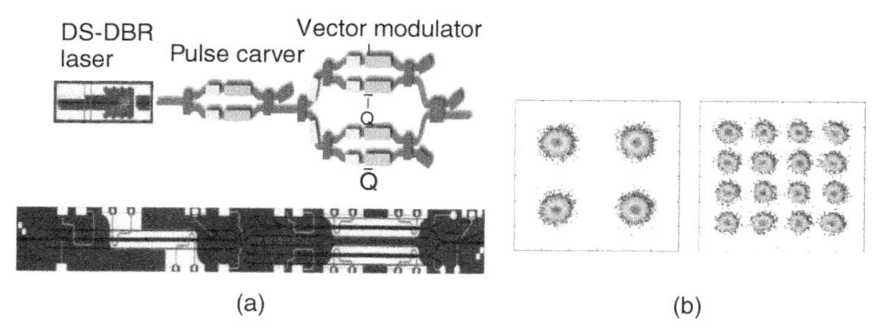

FIGURE 13.7 (a) All-InP coherent optical transmitter consisting of a laser chip coupled to a pulse carver I/Q modulator. (b) Constellations for 32-Gbaud QPSK and 16QAM.From Refs [20, 27]. Reproduced with permission of OSA.

the InP MZM in a "chirp-free" manner challenging and need calibration and look-up tables across the wavelength range. In addition, the epitaxial layer structure for modulation is quite different from that needed for a tunable laser. Nevertheless, there has been progress in fully integrated transmitter-modulator assemblies, as shown in Table 13.3.

As an example, Chandrasekhar et al. presented a two-chip integrated transmitter, which was directly driven by CMOS DACs (differential output of 1 V) to generate 32-Gbaud QPSK and 16QAM formats without the use of modulator drivers [20, 27]. The schematic of the integrated device is shown in Figure 13.7. The laser was a digital supermode (DS) distributed Bragg reflector (DBR) laser, and it was optically coupled with micro-optics to a second InP chip that consisted of a pulse carver followed by a single polarization vector modulator. The laser was tunable over the entire C-band and the analog bandwidth of the MZMs that comprised the vector modulator had a bandwidth of about 12 GHz, limited by the package. Nevertheless, using DSP at the transmitter to equalize the bandwidth roll-off, operation up to 32 Gbaud was achieved and transmission over 8000 km for the 100-Gb/s PDM-QPSK format and 900 km for the 200-Gb/s PDM-16QAM format was successfully demonstrated.

13.3.5 Silicon Monolithic *I*/*Q* Modulator

In 2012, an *I*/*Q* modulator based on nested single-drive push–pull silicon MZMs was reported in [44], where a 50-Gb/s QPSK signal was generated with only 2.7-dB optical signal-to-noise ratio (OSNR) penalty from the theoretical limit at a bit error ratio (BER) of 10^{-3}. Compared with commercial $LiNbO_3$ *I*/*Q* modulators, there is only ~1 dB OSNR penalty. This is the first successful demonstration of advanced modulation formats using silicon MZMs. By further integrating two *I*/*Q* modulators and an on-chip polarization rotator and PBC, a monolithic single-chip dual-polarization coherent modulator was implemented to generate a 112-Gb/s PDM-QPSK [45] and a 224-Gb/s PDM-16QAM signal [46], as shown in Figure 13.8(a) and (b). In [45], it was shown that the integration of on-chip polarization elements introduces an additional 0.9-dB penalty due to polarization-dependent loss (PDL). The PR is

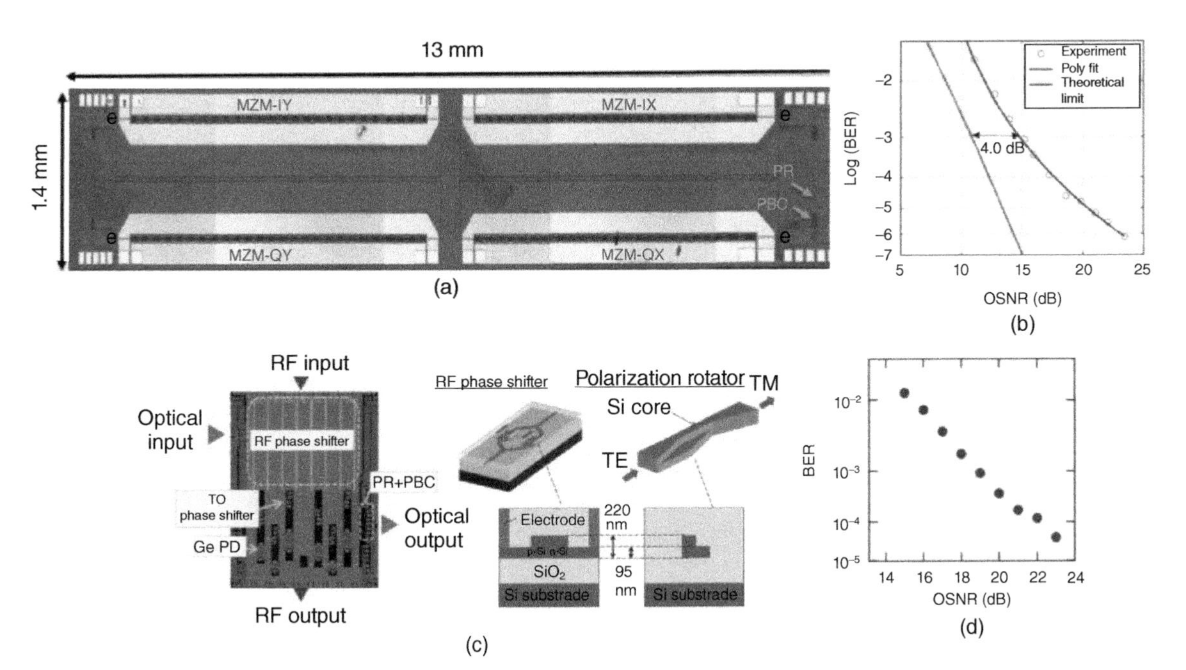

FIGURE 13.8 Reported monolithic dual-polarization silicon-PIC *I/Q* modulator in [45] and [50]. (a) Photograph of the first dual-pol *I/Q* modulator on silicon PIC in [45]. (b) BER versus OSNR for 112-Gb/s PDM-QPSK in [45]. (c) Photograph of a dual-pol *I/Q* modulator and its phase shifter and polarization rotator in [50]. (d) BER versus OSNR for 128-Gb/s PDM-QPSK in [50]. MZM: Mach–Zehnder modulator, PBC: polarization beam combiner, PR: polarization rotator, PD: photodiode. From Refs. [45] and [50]. Reproduced with permission of OSA.

realized by a SiN-assisted taper structure. This PIC is the first monolithic single-chip dual-polarization *I/Q* modulator, with highest photonic integration in this particular application. Dual-polarization *I/Q* modulators were further demonstrated in [50], where a novel polarization rotator based on silicon-only structure was employed, as shown in Figure 13.8(c). In [50], 128-Gb/s PDM-QPSK was generated, with BER performance shown in Figure 13.8(d).

Single-polarization silicon *I/Q* modulators were also reported using MOS-capacitor type silicon modulators with performance comparable to $LiNbO_3$-based *I/Q* modulators [47]. Here, a compact *I/Q* modulator with a size of less than 1 mm × 25 μm per MZM was used, and the MZM was driven by a low-power CMOS driver with less than 200 mW per channel. The 56-Gb/s QPSK signals were generated, externally polarization multiplexed to 112-Gb/s, and transmitted through 2427-km standard single mode fiber. The performance showed negligible OSNR penalties were associated with the silicon modulator compared to $LiNbO_3$ modulators for back-to-back operation, and a small penalty was found after transmission due to chromatic dispersion.

Single-polarization *I/Q* modulators were also realized using silicon–organic modulators where an electro-optic polymer is spin-coated on silicon slot waveguides [49]. High modulation efficiency could be realized through both the high EO coefficient of the polymer as well as the high E-field concentration in the slot. Up to 40-Gbaud QAM signals were generated in [52].

13.4 RECEIVERS

In this section, we review coherent optical receiver technologies. This section is organized as follows. Section 13.4.1 describes the principle of coherent detection and optical circuits for polarization diversity receivers. Sections 13.4.2–13.4.4 summarize previously reported coherent receivers based on PLC hybrids, InP PICs, and silicon PICs, respectively. Section 13.4.5 presents coherent receivers based on 120° optical hybrids.

13.4.1 Polarization Diversity Receiver Circuits

The fundamental principle of coherent detection is to measure the resulting electrical field following the beating of a signal and a CW local oscillator (LO) laser. Generation of the beat signal requires mixing the signal and LO in an optical mixer. For example, mixing the signal and LO using a 2 × 2 coupler will result in a beat signal. Balanced detection can be employed at the two outputs to suppress direct current (DC) components. However, only single-quadrature information of the signal can be measured with such 2 × 2 coupler. In order to detect the full electric field in the complex plane, one typically uses an optical 90° hybrid, which sends the mixed portions of the beat signal to four outputs. The four outputs represent a beat signal that is 0°, 90°, 180°, and 270° shifted in phase between the signal and LO. If the signal and LO are launched into a 90° optical hybrid, the four output fields become

$$E_i = \frac{1}{2}\left(E_S + e^{j(i-1)\frac{\pi}{2}}E_L\right) \tag{13.1}$$

where E_L and E_s are the electrical fields of the LO and signal, respectively; $i = 1$–4 for four outputs. Output photocurrents from balanced photodetections are then given as

$$\begin{cases} I_I = I_1 - I_3 = |E_S||E_L|\cos(\phi) \\ I_Q = I_2 - I_4 = |E_S||E_L|\sin(\phi) \end{cases} \tag{13.2}$$

Here, ϕ represents the phase difference between the signal and LO. The above-mentioned photocurrents produce the *I/Q* components of the signal without DC components, fulfilling the requirements to measure both the magnitude and phase of the signal.

For a dual-polarization signal such as PDM-QPSK, a polarization-diversity receiver is required. Figure 13.9 shows three possible optical circuits for polarization diversity coherent receiver. In the first circuit shown in Figure 13.9(a), the input polarization of the LO is 45°. Two PBSs are used to split the signal and LO into transverse-electric (TE) and transverse-magnetic (TM) components, respectively. The TE (or TM) components of signal and LO are mixed with a 90° optical hybrid designed for TE (or TM). In this circuit, a PR is not needed, but the optical hybrids and the PDs need to be designed for two polarizations, which is not trivial for InP and silicon PICs. In the second circuit shown in Figure 13.9(b), a PBS is used for the signal port only to split the signal into TE and TM components. For the TM component, a PR rotates the TM to TE polarization. In the LO side, only TE is coupled in and then a power splitter is used. Both the two 90° hybrids and the following PDs are designed for TE polarization only. The advantages of this circuit include (i) the optical hybrids and PDs are designed only for one polarization and (ii) the input LO is polarized only to TE, which matches the typical output polarization from a semiconductor laser. In this scheme, an additional component of

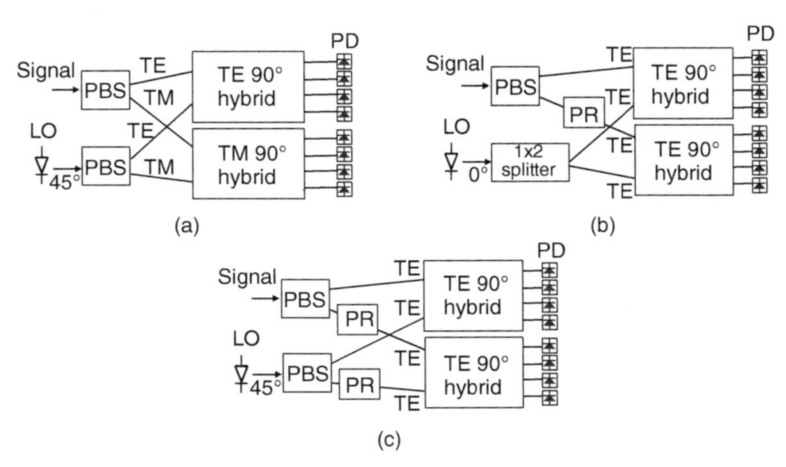

FIGURE 13.9 Various optical circuits for polarization diversity coherent receiver. LO: local oscillator, PBS: polarization beam splitter, PR: polarization rotator, PD: photodiode.

PR is required, which may further cause unwanted PDL. A more symmetric circuit shown in Figure 13.9(c) uses two PBSs and two PRs for both the signal and LO ports. Furthermore, the input polarization of LO can be fine-tuned to compensate the PDL.

13.4.2 PLC Hybrid Receivers

Silica PLCs offer high-performance passive devices such as optical power splitters/combiners, 90° hybrids, and arrayed waveguide gratings (AWGs), but not high-speed PDs. To implement coherent receivers, PD arrays can be bonded or packaged with PLCs. Such PLC-based hybrid coherent receivers have achieved excellent device and system performance in a compact footprint [7–16]. To illustrate the versatility of the platform to incorporate various optical elements, we cite the work by Kurata et al. [12]. In this paper, the authors heterogeneously integrated eight high-speed PDs on a silica-based PLC platform with a PBS, 90° optical hybrids, and a VOA. The use of a 2.5% index contrast waveguide reduced the receiver PLC size to 11×11 mm^2. The schematic of the chip layout is shown in Figure 13.10(a), and the optical circuit employed is the one shown in Figure 13.9(b).

The VOA driven by the thermo-optic effect is composed of a Mach–Zehnder interferometer (MZI) and thin-film heaters. The PBS is also composed of an MZI and two polyimide quarter waveplates. The waveplates were tilted at 0° and 90°, respectively, and inserted in the two arms of the MZI. A half-wavelength waveplate is inserted in

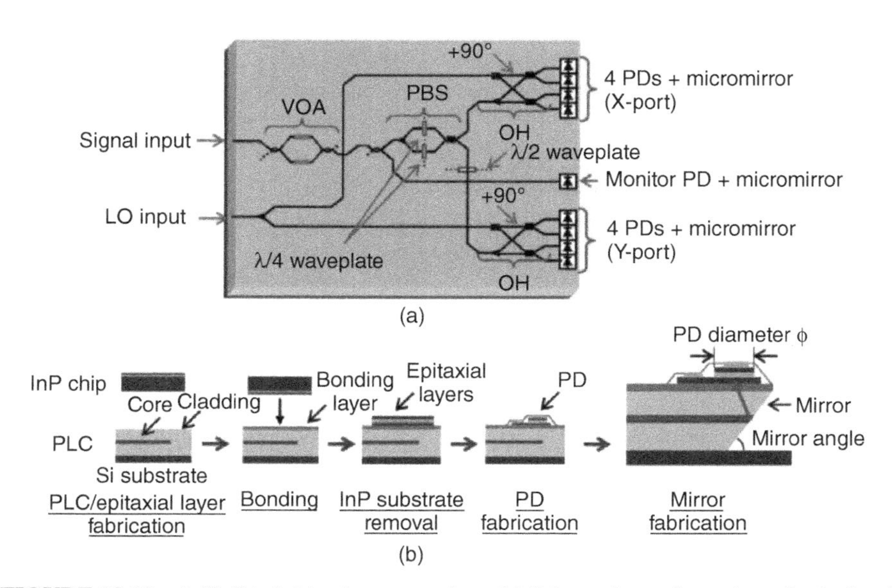

FIGURE 13.10 A PLC hybrid coherent receiver. (a) Schematic configuration of a dual-pol receiver. (b) Fabrication process of PDs and micromirror to realize heterogeneous integration. From Ref. [12].

the pathway of one polarization after the signal passes through the PBS to realize the polarization rotation. In the 90° hybrids, a 2 × 2 multimode interference (MMI) coupler was designed for a 90° phase shifter instead of a delay line, because an MMI is more tolerant to fabrication errors such as refractive index and waveguide errors. The interference signals from the hybrids are reflected into the PDs by the micromirror fabricated at the chip edge. The fabrication sequences for such heterogeneous integration are shown schematically in Figure 13.10(b). The PLC is first fabricated and is bonded to an InP chip. Next, the InP substrate is removed and only the top epitaxial layers are left for the following conventional PD fabrication. Finally, the micromirror is fabricated by dry tilt etching and aluminum deposition.

The integrated coherent receiver exhibited a total insertion loss of about 11 dB, a polarization extinction ratio (PER) of more than 20 dB, optical hybrid phase error of ± 2°, and photodetector external responsivities of about 0.015 A/W. These values are quite comparable to the state-of-the-art coherent receivers. The integrated receivers were subsequently packaged with TIA arrays into a very compact module with a volume of only 1.3 cm^3 [16].

13.4.3 InP Monolithic Receivers

The second approach to achieve a small form factor coherent receiver is to use InP PIC to implement the 90° hybrid and the balanced PDs [18–41]. In some cases, the LO laser has also been integrated. The challenges are to realize monolithic PBSs and PRs and also to achieve uniform performance across the full C-band wavelength window for such devices to become useful in commercial applications. Some examples of monolithically integrated all-InP coherent receivers that have been demonstrated are listed in Table 13.4. Micro-optic PRs and PBSs are typically employed by packaging methods, except in [33]. In [33], a four-channel dual-polarization coherent receiver was realized on a monolithic InP PIC by the use of a novel interleave-chirped AWG that acts simultaneously as a wavelength demultiplexer, 90° hybrids, and also polarization splitter. In [32], a 10-channel receiver chip monolithically integrated with 10 LO lasers and also an InP AWG was reported. The receiver chip can detect 100-Gb/s PDM-QPSK signals but external PBSs and PRs were used.

13.4.4 Silicon Monolithic Receivers

Monolithic dual-polarization coherent receivers have been demonstrated based on silicon PICs. In [53], Doerr et al. demonstrated a grating-assisted coherent receiver PIC, where two-dimensional (2D) gratings are used for fiber coupling (shown in Figure 13.11a). A 2D grating coupler is a photonic crystal that couples a wave traveling normal to a substrate to a wave guided parallel to the substrate. Light incident on the grating from the vertical direction is phase-matched to waveguide-guiding modes in a certain direction at the designed polarization state. With these gratings, the TE and TM components of signal (and LO) can be coupled and separated into different silicon waveguides. Once coupled in, all of the light on the chip is TE polarized. Therefore, the 2D gratings can also realize polarization splitting, polarization

TABLE 13.4 Some previously reported coherent receivers based on InP PICs

Reference	Devices	Features
Farwell et al. [41]	Micro ICR single-pol with 90° hybrid, balanced PDs	$12 \times 25 \times 5$ mm^3, 0.15 A/W responsivity, >32 GHz bandwidth for PD
Takeuchi et al. [30]	ICR dual-pol, 90° hybrid, external PBS, balanced PDs, external TIA	$13.6 \times 25.2 \times 5.5$ mm^3, 1W, 0.006 A/W, >22 GHz bandwidth, 128 Gb/s
Beling et al. [35]	Micro-optics+InP 90° hybrid+balanced PDs + external TIA	$27 \times 40 \times 6$ mm^3, 0.03 A/W, PER > 19 dB, 128 Gb/s reception
Yagi et al. [37]	Single polarization optical hybrid with differential outputs	12-16 dB insertion loss, 22-24 GHz bandwidth
Lu et al. [39]	Single polarization optical hybrid + single ended PDa + tunable laser	13 dBm laser output, 35 GHz bandwidth for PD
Doerr et al. [33]	4-Channel coherent receivers: optical hybrid+balanced PDs	10.7 Gbaud PDM-QPSK 4-ch WDM reception
Estrella et al. [36]	LO laser integrated with dual-pol coherent receiver (external PBS)	40 nm tuning and 10 Gbaud PDM-QPSK operation
Nagarajan et al. [32]	10-Channel dual-pol coherent receiver with PDs, micro-optic PBS	25 Gbaud PDM-QPSK reception

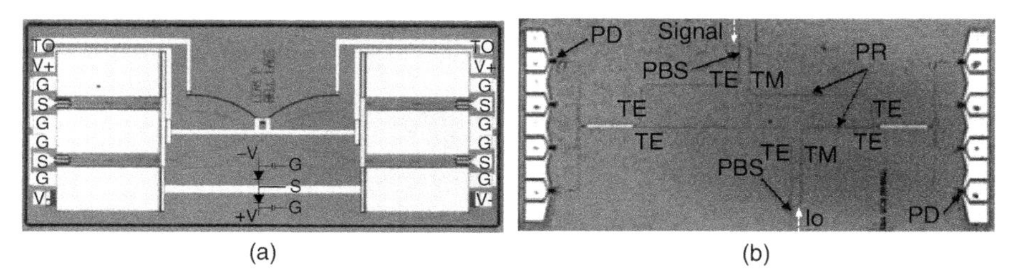

FIGURE 13.11 Reported monolithic dual-polarization silicon-PIC coherent receiver in [53] and [46]. LO: local oscillator, PBS: polarization beam splitter, PR: polarization rotator, PD: photodiode.

rotation, and 50/50 power splitting. The coupled signal and LO pass through two 90° hybrids based on 2×2 MMIs and eight germanium photodetectors. The optical circuit for polarization-diversity coherent detection is essentially the same as in Figure 13.9(b). Using this PIC packaged with in-house TIAs, a 112-Gb/s PDM-QPSK signal was successfully detected with BER performance comparable to a commercial coherent receiver.

While the 2D gratings are elegant optical elements for a polarization-diversity receiver, the coupling loss is typically high and also the 1-dB bandwidth is limited to ~40 nm. To solve these problems, edge coupling can be used, but on-chip PBSs and PRs are required. In [46], Dong et al. implemented a silicon coherent receiver by integrating on-chip polarization rotators and splitters, with the optical circuit shown in Figure 13.9(c). The optical signal enters the PIC from one facet with two polarizations, as shown in Figure 13.11(b). The optical LO is coupled into the silicon PIC from the other facet. Once coupled in, the signal and LO are divided into TE and TM polarizations by two PBSs. The TE-polarized light proceeds to the 4×4 MMI-based 90° hybrids, whose four outputs are detected by four germanium photodetectors on the left side of the PIC. The TM components from the output of the PBSs are converted to TE polarization by two polarization rotators. The converted TE light enters the right-side 4×4 MMI-based 90° hybrid, which is identical to that for the TE mode. Using this PIC co-packaged with TIAs, a 112-Gb/s PDM-QPSK signal and a 224-Gb/s PDM-16QAM signal were successfully detected [46].

The monolithic silicon coherent receivers have been used in the transmission experiment where both the transmitter and receiver front-ends are silicon PICs fabricated from the same wafers [54]. A 2560-km fiber transmission of a 112-Gb/s PDM-QPSK signal has been obtained at a BER of 3.8×10^{-3} (the typical threshold of a 7%-overhead hard decision forward error correction), as shown in Figure 13.12. This experiment validates the readiness of silicon PICs in optical coherent links.

In [56], Doerr et al. demonstrated a monolithic silicon PIC that contains all the optical front-end for a 100-Gb/s coherent transceiver except the laser (Figure 13.13). The silicon PIC was co-packaged with linear drivers and TIAs in a very compact gold-box module with a size of 27×35.5 mm^2 and power consumption <4.5 W. This PIC further verifies the high integration level of silicon photonic for this particular application, mainly by taking advantage of polarization elements on-chip.

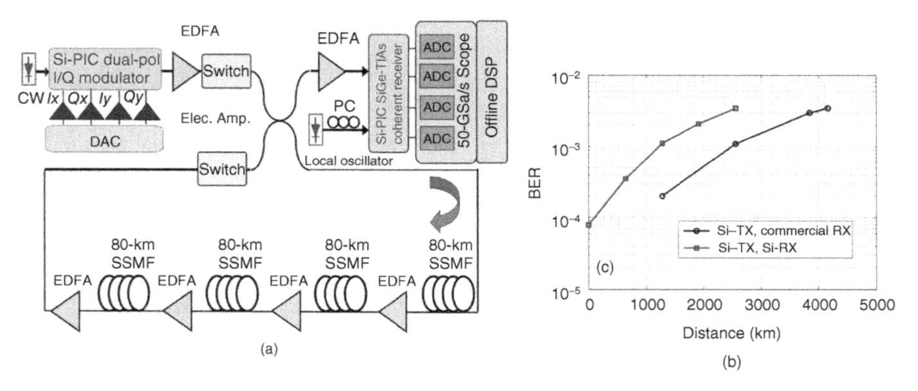

FIGURE 13.12 Transmission of 112-Gb/s PDM-QPSK signals generated and detected by monolithic Si PICs reported in [54]. (a) Experimental setup. (b) Measured bit error ratio (BER) as a function of transmission distance.

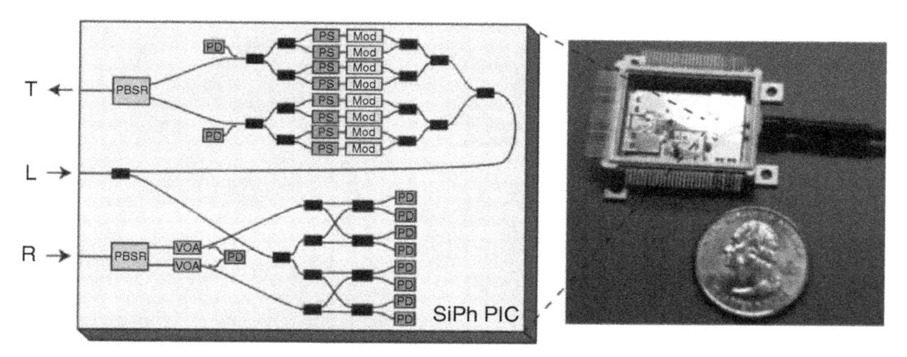

FIGURE 13.13 Single-chip coherent PIC reported in [56]. PS: phase shifter, PD: photodiode, VOA: variable optical attenuator, PBSR: polarization beam splitter and rotator.

13.4.5 Coherent Receiver with 120° Optical Hybrids

Instead of 90° hybrids, coherent detection can also be achieved by mixing a signal with an LO in 120° hybrids with three single-ended detectors. In general, balanced detection after 90° hybrids is superior in suppressing various noise sources. This is preferred for long-haul transmission driven by high-performance requirements. Single-ended detection, however, may be lower in cost because of simpler TIAs and less RF connections between optical and electrical interfaces. Compared with MMI-based 90° hybrids, 120° hybrids may have a broader optical bandwidth, larger fabrication tolerance, and the ability to mitigate hardware-induced imperfections from ADCs and digital signal processors [67]. If a signal and an LO are launched into a symmetric 3 × 3 MMI with 1:1:1 power splitting, its three output photocurrents are given by Xie et al. [68]

$$\begin{pmatrix} I_1 \\ I_2 \\ I_3 \end{pmatrix} = \frac{1}{3}\begin{pmatrix} |E_L|^2 + |E_S|^2 \\ |E_L|^2 + |E_S|^2 \\ |E_L|^2 + |E_S|^2 \end{pmatrix} + \frac{2}{3}\begin{pmatrix} |E_L|\,|E_S|cos(\phi + 2/3\pi) \\ |E_L||E_S|cos(\phi) \\ |E_L||E_S|cos(\phi - 2/3\pi) \end{pmatrix} \tag{13.3}$$

where E_L and E_s are the electrical fields of the LO and signal, respectively, and ϕ represents the phase difference between them. The first term in Equation 13.3 is the direct-detection term and the second term is the beat term, with a 120° difference among its three components. A proper rearrangement of the photocurrents can yield the currents for I/Q components with the suppression of the direct-detection term:

$$\begin{cases} I_I = I_2 - 0.5I_1 - 0.5I_3 = |E_L|\,|E_S|cos(\phi) \\ I_Q = \sqrt{3}/2(I_3 - I_1) = |E_L||E_S|sin(\phi) \end{cases} \tag{13.4}$$

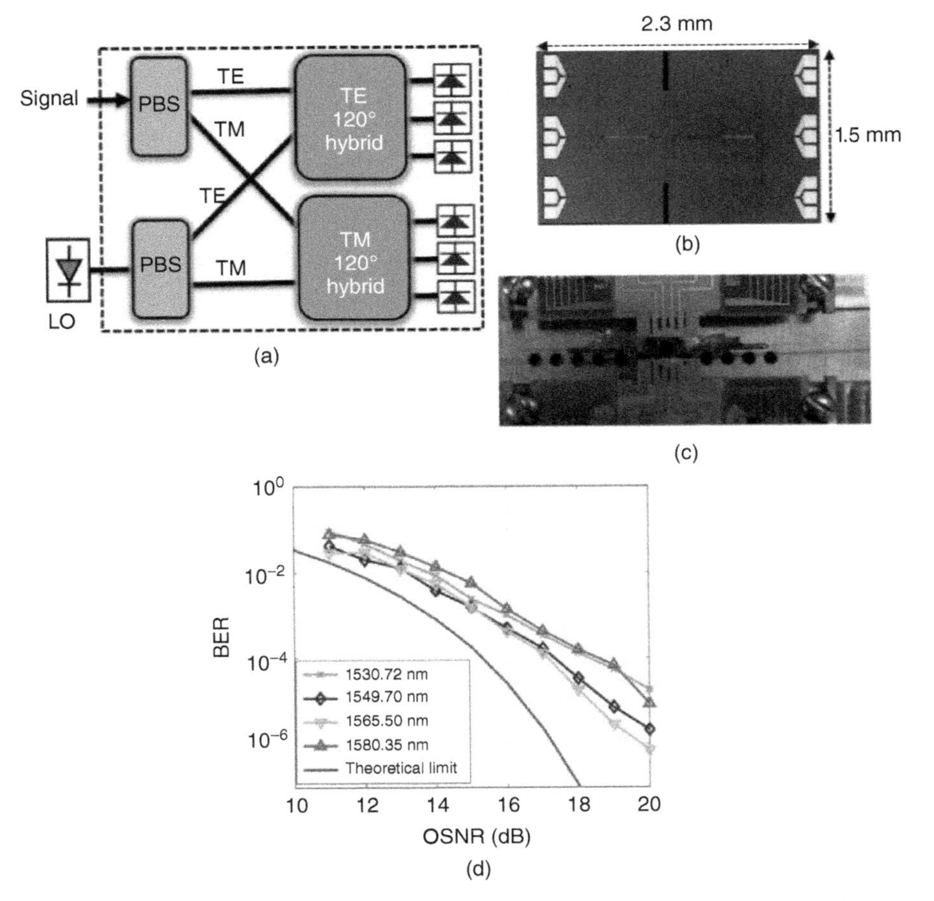

FIGURE 13.14 Dual-polarization coherent receiver based on 120° hybrids. (a) Optical circuit diagram. (b) and (c) Photo of a fully fabricated PIC and the packaged PIC with fibers and circuit boards. (d) BER as a function of OSNR for four wavelengths (differential decoding). PBS: polarization beam splitter, LO: local oscillator. From Ref. [55].

A monolithic polarization diversity coherent receiver by employing such 120° optical hybrids on a silicon PIC was reported in [55]. This PIC monolithically integrates silicon inverse tapers for fiber coupling, silicon polarization splitters, germanium high-speed photodetectors, and 120° optical hybrids based on 3×3 MMIs, as shown in Figure 13.14. The chip size is only $2.3 \times 1.5\,\mathrm{mm}^2$. The 112-Gb/s PDM-QPSK signals were detected in the wavelength range of 1530–1580 nm. The BERs as a function of OSNR, presented in Figure 13.14(d), demonstrate comparable performance to commercial receivers. The broadband operation indicates that this type of receiver may find applications in low-cost transceivers.

13.5 CONCLUSIONS

This chapter reviewed recent progress on photonic integration for coherent optical transceivers. In order to achieve compact and low-power pluggable modules for metro or short-distance applications, photonic integration is perceived as critical and necessary in order to address the need to have several optical elements in a small footprint. In order to reduce optical packaging cost, it is also highly preferable for high-level monolithic photonic integration. Three main integration technologies have been reviewed in this chapter, that is, PLC hybrids, InP, and silicon PICs. While PLC hybrids have the least monolithic integration of optical components, the performance tends to be the best. InP PICs have the advantages of monolithic integration of lasers with MZMs and PDs. Silicon PICs excel in monolithic integration of polarization elements such as PBS/Cs and PRs. While all the three technologies have demonstrated promising progresses in recent years, the cost, power, and size of the final modules may decide which technology will likely dominate. It is also possible that all these three technologies will be used in the future.

ACKNOWLEDGMENTS

The authors thank Young-Kai Chen and Peter Winzer at Bell Labs for useful comments and discussions as well as paper revisions.

REFERENCES

1. Winzer PJ, Essiambre R. Advanced optical modulation formats. *Proc IEEE* 2006;94(5):952–985.
2. Kikuchi K. Digital coherent optical communication systems: fundamentals and future prospect. *IEICE Electron Express* 2011;8(20):1642–1662.
3. Berger M. 100G challenges and solutions. Optical Fiber Communication Conference/National Fiber Optic Engineers Conference 2010; Market Watch Session; 2010.
4. Mino S, Miyazaki H, Goh T, Yamada T. Multilevel optical modulator utilizing PLC-$LiNbO_3$ hybrid-integration technology. *NTT Technical Review* 2011;9(3):1–7.
5. Yamazaki H, Yamada T, Suzuki K, Goh T, Kaneko A, Sano A, Yamada E, Miyamoto Y. Integrated 100-Gb/s PDM-QPSK modulator using a hybrid assembly technique with silica-based PLCs and $LiNbO_3$ phase modulators. European Conference on Optical Communication (ECOC 2008), paper Mo.3.C.1; 2008.
6. Sano A, Kobayashi T, Ishihara K, Masuda H, Yamamoto S, Mori K, Yamazaki E, Yoshida E, Miyamoto Y, Yamada T, Yamazaki H. 240-Gb/s polarization-multiplexed 64-QAM modulation and blind detection using PLC-LN hybrid integrated modulator and digital coherent receiver. European Conference on Optical Communication (ECOC 2009), paper PD2.2; 2009.

7. Wang J, Kroh M, Theurer A, Zawadzki C, Schmidt D, Ludwig R, Lauermann M, Zhang Z, Beling A, Matiss A, Schubert C, Steffan A, Keil N, Grote N. Dual-quadrature coherent receiver for 100G Ethernet applications based on polymer planar lightwave circuit. *Opt Express* 2011;19(26):B166–B172.
8. Inoue T, Nara K. Ultrasmall PBS-integrated coherent mixer using 1.8%-delta silica-based planar lightwave circuit. European Conference on Optical Communication (ECOC 2010), paper Mo.2.F.4; 2010.
9. Nasu Y, Mizuno T, Kasahara R, Saida T. Temperature insensitive and ultra wideband silica-based dual polarization optical hybrid for coherent receiver with highly symmetrical interferometer design. *Opt Express* 2011;19(26):B112–B118.
10. Kurata Y, Nasu Y, Tamura M, Yokoyama H, Muramoto Y. Heterogeneous integration of high-speed InP PDs on silica-based planar lightwave circuit platform. European Conference on Optical Communication (ECOC 2011), paper Th.12.5; 2011.
11. Tsunashima S, Nakajima F, Nasu Y, Kasahara R, Nakanishi Y, Saida T, Yamada T, Sano K, Hashimoto T, Fukuyama H, Nosaka H, Murata K. Silica-based, compact and variable-optical-attenuator integrated coherent receiver with stable optoelectronic coupling system. *Opt Express* 2012;20:27174–27179.
12. Kurata Y, Nasu Y, Tamura M, Kasahara R, Aozasa S, Mizuno T, Yokoyama H, Tsunashima S, Muramoto Y. Silica-based PLC with heterogeneously-integrated PDs for one-chip DP-QPSK receiver. *Opt Express* 2012;20:B264–B269.
13. Ohyama T, Ogawa I, Tanobe H, Kasahara R, Tsunashima S, Yoshimatsu T, Fukuyama H, Itoh T, Sakamaki Y, Muramoto Y, Kawakami H, Ishikawa M, Mino S, Murata K. All-in-one 112-Gb/s DP-QPSK optical receiver front-end module using hybrid integration of silica-based planar lightwave circuit and photodiode arrays. *IEEE Photon Technol Lett* 2012;24(8):646–648.
14. Painchaud Y, Pelletier M, Poulin M, Pelletier F, Latrasse C, Robidoux G, Savard S, Gagné J-F, Trudel V, Picard M-J, Poulin P, Sirois P, D'Amours F, Asselin D, Paquet S, Paquet C, Cyr M, Guy M, Morsy-Osman M, Zhuge Q, Xu X, Chagnon M, Plant DV. Ultra-compact coherent receiver based on hybrid integration on silicon. Optical Fiber Communication Conference/National Fiber Optic Engineers Conference (OFC 2013), paper OM2J.2; 2013.
15. Y. Hashizume, M. Itoh, T. Hashimoto and M. Itoh, Compact and low-loss silica-based dual polarisation optical hybrid with PBS based on large geometrical birefringence. *Elec Lett*, 49(20), 1288–1289, 2013.
16. Itoh T, Nakajima F, Ohno T, Yamanaka S, Soma S, Saida T, Nosaka H, Murata K. Ultra-compact coherent receiver with serial interface for pluggable transceiver. *Opt Express* 2014;22(19):22583–22589.
17. Fang AW, Park H, Cohen O, Jones R, Paniccia MJ, Bowers JE. Electrically pumped hybrid AlGaInAs-silicon evanescent laser. *Opt Express* 2006;14:9203–9210.
18. Clarke CF, Griffin RA, Goodall TC. Highly integrated DQPSK modules for 40 Gb/s transmission. Optical Fiber Communication Conference/National Fiber Optic Engineers Conference (OFC 2009), paper NWD3; 2009.
19. Prosyk K, Brast T, Gruner M, Hamacher M, Hoffmann D, Millett R, Velthaus K. Tunable InP-based optical IQ modulator for 160 Gb/s. European Conference on Optical Communication (ECOC 2011), paper Th.13.A.5; 2011.
20. Chandrasekhar S, Liu X, Winzer P, Simsarian JE, Griffin RA. Small-form-factor all-InP integrated laser vector modulator enables the generation and transmission of 256-Gb/s

PDM-16QAM modulation format. Optical Fiber Communication Conference/National Fiber Optic Engineers Conference (OFC 2013), paper PDP5B.6; 2013.

21. Shibata Y, Yamada E, Yasui T, Ohki A, Watanabe K, Ishii H, Iga R, Oohashi H. Demonstration of 112-Gbit/s DP-QPSK modulation using InP n-p-i-n Mach–Zehnder modulators. European Conference and Exhibition on Optical Communication (ECOC 2010), paper Mo.2.F.2; 2010.
22. Evans P, Fisher M, Malendevich R, James A, Studenkov P, Goldfarb G, Vallaitis T, Kato M, Samra P, Corzine S, Strzelecka E, Salvatore R, Sedgwick F, Kuntz M, Lal V, Lambert D, Dentai A, Pavinski D, Behnia B, Bostak J, Dominic V, Nilsson A, Taylor B, Rahn J, Sanders S, Sun H, Wu K, Pleumeekers J, Muthiah R, Missey M, Schneider R, Stewart J, Reffle M, Butrie T, Nagarajan R, Joyner C, Ziari M, Kish F, Welch D. Multi-channel coherent PM-QPSK InP transmitter photonic integrated circuit (PIC) operating at 112 Gb/s per wavelength. Optical Fiber Communication Conference/National Fiber Optic Engineers Conference (OFC 2011), paper PDPC7; 2011.
23. Corzine S, et al. Large-scale monolithic integration of PM-QPSK modulation architecture in 500 Gb/s transmitters. Presented at the Integrated Photonics Research, Silicon, and Nanophotonics, paper ITuC1; 2011.
24. Yamada E, Kanazawa S, Ohki A, Watanabe K, Nasu Y, Kikuchi N, Shibata Y, Iga R, Ishii H. 112-Gb/s InP DP-QPSK modulator integrated with a silica-PLC polarization multiplexing circuit. Optical Fiber Communication Conference/National Fiber Optic Engineers Conference (OFC 2012), PDP5A.9; 2012.
25. Korn D, Schindler PC, Stamatiadis C, O'Keefe MF, Stampoulidis L, Schmogrow RM, Zakynthinos P, Palmer R, Cameron N, Zhou Y, Walker RG, Kehayas E, Tomkos I, Zimmermann L, Petermann K, Freude W, Koos C, Leuthold J. First monolithic GaAs IQ electro-optic modulator, demonstrated at 150 Gbit/s with 64-QAM. Optical Fiber Communication Conference/National Fiber Optic Engineers Conference (OFC 2013), paper PDP5C.4; 2013.
26. http://www.teraxion.com/en/iqm?gclid=CM3lhq_W_cICFXBp7Aod-koA8g. Accessed 2015 Oct 9.
27. Chandrasekhar S, Liu X, Winzer P, Simsarian JE, Griffin RA. Compact all-InP laser-vector-modulator for generation and transmission of 100-Gb/s PDM-QPSK and 200-Gb/s PDM-16 QAM. *IEEE J Lightw Technol* 2014;32(4):736–742.
28. Heck SC, Jones SK, Griffin RA, Whitbread N, Bromley PA, Harris G, Smith D, Langley LN, Goodall T. Miniaturized InP dual I&Q mach Zehnder modulator with full monitoring functionality for CFP2. European Conference on Optical Communication (ECOC 2014), paper Tu.4.4.2; 2014.
29. Rouvalis E, Metzger C, Charpentier A, Ayling T, Schmid S, Gruner M, Hoffmann D, Hamacher M, Fiol G, Schell M. A low insertion loss and low $V\pi$ InP IQ modulator for advanced modulation formats. 2014 European Conference on Optical Communication (ECOC 2014), paper Tu.4.4.1; 2014.
30. Takeuchi H, Kasaya K, Kondo Y, Yasaka H, Oe K, Imamura Y. Monolithic integrated coherent receiver on InP substrate. *IEEE Photon Technol Lett* 1989;1(11):398–400.
31. Kunkel R, Bach H-G, Hoffmann D, Weinert CM, Molina-Fernandez I, Halir R. First monolithic InP-based 90°-hybrid OEIC comprising balanced detectors for 100GE coherent frontends. Proceedings of IPRM 2009, paper TuB2.2; 2009.
32. Nagarajan R, Kato M, Pleumeekers J, Lambert D, Lal V, Dentai A, Kuntz M, Rahn J, Tsai H, Malendevich R, Goldfarb G, Tang J, Zhang J, Butrie T, Raburn M, Little B, Nilsson A,

Reffle M, Kish F, Welch D. 10 channel, 45.6Gb/s per channel, polarization multiplexed DQPSK InP receiver photonic integrated circuit. Optical Fiber Communication Conference/National Fiber Optic Engineers Conference (OFC 2010), paper PDPB2; 2010.

33. Doerr CR, Zhang L, Winzer PJ. Monolithic InP multi-wavelength coherent receiver. Optical Fiber Communication Conference/National Fiber Optic Engineers Conference (OFC 2010), PDPB1; 2010.
34. Nagarajan R, Lambert D, Kato M, Lal V, Goldfarb G, Rahn J, Kuntz M, Pleumeekers J, Dentai A, Tsai H, Malendevich R, Missey M, Wu K, Sun H, McNicol J, Tang J, Zhang J, Butrie T, Nilsson A, Reffle M, Kish F, Welch D. 10 channel, 100Gbit/s per channel, dual polarization, coherent QPSK, monolithic InP receiver photonic integrated circuit. Optical Fiber Communication Conference/National Fiber Optic Engineers Conference (OFC 2011), paper OML7; 2011.
35. Beling A, Ebel N, Matiss A, Unterbörsch G, Nölle M, Fischer JK, Hilt J, Molle L, Schubert C, Verluise F, Fulop L. Fully-integrated polarization-diversity coherent receiver module for 100G DP-QPSK. Optical Fiber Communication Conference/National Fiber Optic Engineers Conference (OFC 2011), paper OML5; 2011.
36. Estrella SB, Johansson LA, Masanovic ML, Thomas JA, Barton JS. Widely tunable compact monolithically integrated photonic coherent receiver. *IEEE Photonics Technology Letters* 2012;24(5):365–367.
37. Yagi H, Inoue N, Onishi Y, Masuyama R, Katsuyama T, Kikuchi T, Yoneda Y, Shoji H. High-efficient InP-based balanced photodiodes integrated with 90° hybrid MMI for compact 100 Gb/s coherent receiver. Optical Fiber Communication Conference/National Fiber Optic Engineers Conference (OFC 2013), paper OW3J.5; 2013.
38. Schell M, Bach H, Janiak K, Keil N, Moehrle M, Runge P, Zhang Z. Coherent receiver PICs. Optical Fiber Communication Conference/National Fiber Optic Engineers Conference (OFC 2013), paper OW3J.6; 2013.
39. Lu M, Park H-C, Sivananthan A, Parker JS, Bloch E, Johansson LA, Rodwell MJW, Coldren LA. Monolithic integration of a high-speed widely tunable optical coherent receiver. *Photonics Technology Letters, IEEE* 2013;25(11):1077–1080.
40. Takechi M, Tateiwa Y, Ogita S. Compact 100G coherent receiver using InP-based 90° hybrid integrated with photodiodes. ECOC 2013, poster P.2.8; 2013.
41. Farwell S, Aivaliotis P, Qian Y, Bromley P, Griggs R, Hoe JNY, Smith C, Jones S. InP coherent receiver chip with high performance and manufacturability for CFP2 modules. Optical Fiber Communication Conference/National Fiber Optic Engineers Conference (OFC 2014), paper W1I.6; 2014.
42. Dong P, Xie C, Chen L, Fontaine NK, Chen Y-K. Experimental demonstration of microring quadrature phase-shift keying modulators. *Opt Lett* 2012;37:1178–1180.
43. Dong P, Xie C, Buhl LL, Chen Y. Silicon microring modulators for advanced modulation formats. Optical Fiber Communication Conference/National Fiber Optic Engineers Conference (OFC 2013), paper OW4J.2; 2013.
44. Dong P, Chen L, Xie C, Buhl LL, Chen Y-K. 50-Gb/s silicon quadrature phase-shift keying modulator. *Opt Express* 2012;20:21181–21186.
45. Dong P, Xie C, Chen L, Buhl LL, Chen Y. 112-Gb/s monolithic PDM-QPSK modulator in silicon. *Opt Express* 2012;20:B624–B629.

46. Dong P, Liu X, Sethumadhavan C, Buhl LL, Aroca R, Baeyens Y, Chen Y. 224-Gb/s PDM-16-QAM modulator and receiver based on silicon photonic integrated circuits. Optical Fiber Communication Conference/National Fiber Optic Engineers Conference (OFC 2013), paper PDP5C.6; 2013.
47. Milivojevic B, Raabe C, Shastri A, Webster M, Metz P, Sunder S, Chattin B, Wiese S, Dama B, Shastri K. 112Gb/s DP-QPSK transmission over 2427 km SSMF using small-size silicon photonic IQ modulator and low-power CMOS driver. Optical Fiber Communication Conference/National Fiber Optic Engineers Conference (OFC 2013), paper OTh1D.1; 2013.
48. Goi K, Kusaka H, Oka A, Terada Y, Ogawa K., Liow T-Y, Tu X, Lo G-Q, Kwong D-L. DQPSK/QPSK modulation at 40–60 Gb/s using low-loss nested silicon Mach–Zehnder modulator. Optical Fiber Communication Conference/National Fiber Optic Engineers Conference (OFC 2013), paper OW4J.4; 2013.
49. Korn D, Palmer R, Yu H, Schindler PC, Alloatti L, Baier M, Schmogrow R, Bogaerts W, Selvaraja SK, Lepage G, Pantouvaki M, Wouters JMD, Verheyen P, Van Campenhout J, Chen B, Baets R, Absil P, Dinu R, Koos C, Freude W, Leuthold J. Silicon-organic hybrid (SOH) IQ modulator using the linear electro-optic effect for transmitting 16QAM at 112 Gbit/s. *Opt Express* 2013;21:13219–13227.
50. Goi K, Kusaka H, Oka A, Ogawa K, Liow T, Tu X, Lo PG, Kwong DL. 128-Gb/s DP-QPSK using low-loss monolithic silicon IQ modulator integrated with partial-rib polarization rotator. Optical Fiber Communication Conference (OFC 2014), paper W1I.2; 2014.
51. Dong P, Xie C, Buhl LL, Chen Y-K, Sinsky JH, Raybon G. Silicon in-phase/quadrature modulator with on-chip optical equalizer. European Conference on Optical Communication (ECOC 2014), paper We.1.4.5; 2014.
52. Lauermann M, Schindler PC, Wolf S, Palmer R, Koeber S, Korn D, Alloatti L, Wahlbrink T, Bolten J, Waldow M, Koenigsmann M, Kohler M, Malsam D, Elder DL, Johnston PV, Phillips-Sylvain N, Sullivan PA, Dalton LR, Leuthold J, Freude W, Koos C. 40 GBd 16QAM modulation at 160 Gbit/s. European Conference on Optical Communication (ECOC 2014), paper We.3.1.3; 2014.
53. Doerr CR, Buhl LL, Baeyens Y, Aroca R, Chandrasekhar S, Liu X, Chen L, Chen Y-K. Packaged monolithic silicon 112-Gb/s coherent receiver. *IEEE Photonics Technology Letters* 2011;23(12):762–764.
54. Dong P, Liu X, Sethumadhavan C, Buhl LL, Aroca R, Baeyens Y, Chen Y. Monolithic silicon photonic circuits enable 112-Gb/s PDM-QPSK transmission over 2560-km SSMF. European Conference on Optical Communication (ECOC 2013), paper We.2.B.1; 2013.
55. Dong P, Xie C, Buhl LL. Monolithic polarization diversity coherent receiver based on 120-degree optical hybrids on silicon. *Opt Express* 2014;22:2119–2125.
56. Doerr CR, Chen L, Vermeulen D, Nielsen T, Azemati S, Stulz S, McBrien G, Xu X, Mikkelsen B, Givehchi M, Rasmussen C, Park SY. Single-chip silicon photonics 100-Gb/s coherent transceiver. Optical Fiber Communication Conference (OFC 2014), paper Th5C.1; 2014.
57. Thomson DJ, Gardes FY, Hu Y, Mashanovich G, Fournier M, Grosse P, Fedeli J-M, Reed GT. High contrast 40Gbit/s optical modulation in silicon. *Opt Express* 2011;19:11507–11516.

58. Gardes FY, Thomson DJ, Emerson NG, Reed GT. 40 Gb/s silicon photonics modulator for TE and TM polarizations. *Opt Express* 2011;19:11804–11814.
59. Xu H, Xiao X, Li X, Hu Y, Li Z, Chu T, Yu Y, Yu J. High speed silicon Mach–Zehnder modulator based on interleaved PN junctions. *Opt Express* 2012;20:15093–15099.
60. Baehr-Jones T, Ding R, Liu Y, Ayazi A, Pinguet T, Harris NC, Streshinsky M, Lee P, Zhang Y, Lim AE-J, Liow T-Y, Teo SH-G, Lo G-Q, Hochberg M. Ultralow drive voltage silicon traveling-wave modulator. *Opt Express* 2012;20:12014–12020.
61. Dong P, Chen L, Chen Y-K. High-speed low-voltage single-drive push-pull silicon Mach-Zehnder modulators. *Opt Express* 2012;20:6163–6169.
62. Chen L, Doerr C, Dong P, Chen Y-K. Monolithic silicon chip with 10 modulator channels at 25 Gbps and 100-GHz spacing. European Conference and Exposition on Optical Communications (ECOC 2011), paper Th.13.A.1; 2011.
63. Tu X, Chang K-F, Liow T-Y, Song J, Luo X, Jia L, Fang Q, Yu M, Lo G-Q, Dong P, Chen Y-K. Silicon optical modulator with shield coplanar waveguide electrodes. *Opt Express* 2014;22:23724–23731.
64. http://www.fujitsu.com/jp/group/foc/en/. Accessed 2015 Oct 9.
65. Sugiyama M, Doi M, Taniguchi S, Nakazawa T, Onaka H. Driver-less 40 Gb/s $LiNbO_3$ modulator with sub-1 V drive voltage. Optical Fiber Communications Conference (OFC 2002), paper FB6; 2002.
66. Tsuzuki K, Ishibashi T, Ito T, Oku S, Shibata Y, Iga R, Kondo Y, Tohmori Y. 40 Gbit/s n-i-n InP Mach–Zehnder modulator with a π voltage of 2.2 V. *Electron Lett* 2003;39(20):1464–1465.
67. Reyes-Iglesias PJ, Molina-Fernández I, Moscoso-Mártir A, Ortega-Moñux A. High-performance monolithically integrated 120° downconverter with relaxed hardware constraints. *Opt Express* 2012;20:5725–5741.
68. Xie C, Winzer PJ, Raybon G, Gnauck AH, Zhu B, Geisler T, Edvold B. Colorless coherent receiver using 3 × 3 coupler hybrids and single-ended detection. *Opt Express* 2012;20:1164–1171.

14

OPTICAL PERFORMANCE MONITORING FOR FIBER-OPTIC COMMUNICATION NETWORKS

FAISAL NADEEM KHAN[1], ZHENHUA DONG[2], CHAO LU[3], AND ALAN PAK TAO LAU[2]

[1]*School of Electrical and Electronic Engineering, Engineering Campus, Universiti Sains Malaysia, Penang, Malaysia*
[2]*Photonics Research Center, Department of Electrical Engineering, The Hong Kong Polytechnic University, Hung Hom, Kowloon, Hong Kong*
[3]*Photonics Research Center, Department of Electronic and Information Engineering, The Hong Kong Polytechnic University, Hung Hom, Kowloon, Hong Kong*

14.1 INTRODUCTION

Fiber-optic communication has seen tremendous growth over the last decade fueled mainly by the incessant and relentless demand for high capacity. This insatiable demand is spurred by the Internet traffic growth both in terms of number of users and the bandwidth consumed by each user. This trend of multifold increase in data traffic every year is expected to continue in the foreseeable future. In order to comply with the enormous bandwidth requirements posed by the growth in data traffic, fiber-optic communication networks have evolved drastically. For example, dense wavelength-division multiplexing (DWDM) technology, which increases the number of carrier wavelengths and hence data rates, has been introduced [1]. Advanced optical modulation formats offering high spectral efficiencies have been successfully employed in conjunction with digital coherent receivers [2]. The transmission distances between regenerative repeaters have been increased significantly with the

Enabling Technologies for High Spectral-efficiency Coherent Optical Communication Networks,
First Edition. Edited by Xiang Zhou and Chongjin Xie.
© 2016 John Wiley & Sons, Inc. Published 2016 by John Wiley & Sons, Inc.

introduction of optical amplifiers, efficient dispersion compensation techniques, and forward-error-correction (FEC) coding. Furthermore, complex network architectures utilizing reconfigurable optical add-drop multiplexers (ROADMs) have been incorporated in order to promote dynamicity, flexibility, and better utilization of available transmission capacity. All these developments have paved the way for a multifold increase in data rates currently supported by the modern fiber-optic networks [3].

The performance of optical networks operating at ultra-high data rates and over long transmission distances strongly depends on the extent of impairments introduced into the signals by the optical fiber as well as other network elements. Conventionally, the deleterious channel effects are handled by either introducing some safety margins while designing optical networks or by making attempts to compensate some of these impairments manually. Unfortunately, due to stochastic nature of some of these impairments, it is difficult if not impossible to compensate them entirely. The advent of ROADMs makes the situation overwhelmingly more complicated since it allows the signal path to change dynamically, thereby making the impairments in reconfigurable fiber-optic networks become path dependent, dynamic and hence, random in nature [4].

As the optical networks grow larger, faster and more dynamic in nature, the control and management of these networks is rapidly becoming an arduous task. Unlike wireless networks where all the necessary networking issues such as link setup, optimization, and testing are performed automatically, such tasks are currently handled manually in optical networks requiring substantial human resources and time. This is due to the fact that the existing fiber-optic networks are not capable of acquiring real-time information about the physical state of the network and the health of the signals propagating through the network. The price paid for this lacking of vital information is that the network designers are forced to keep considerable safety margins, in order to provide reasonable level of reliability, resulting in wastage of precious network resources. The designers may also be compelled to use more aggressive component specifications. In order to reduce the operating costs, ensure optimum resources utilization and guarantee adequate operation and management of dynamic optical networks, it is essential to have the capability of continuous monitoring of network performance parameters [5].

14.1.1 Optical Performance Monitoring and Their Roles in Optical Networks

The capacity of optical communication systems is increasing unremittingly and the architectures of optical networks are continuously becoming more complex, transparent, and dynamic in nature. These high-capacity fiber-optic networks are vulnerable to several transmission impairments, which can alter over time due to dynamic nature of these networks [6]. Since each fiber carries an enormous amount of data traffic, even a brief disruption of services may result in disastrous consequences. Therefore, it is imperative to incorporate effective monitoring mechanisms across the whole fiber-optic network, which could provide precise and real-time information about the health of each individual DWDM channel. Optical performance monitoring (OPM) is an enabling technology and a potential mechanism for the control, management, and maintenance of existing and future high-speed reconfigurable optical networks. Quintessentially, OPM is a set of measurements performed on an optical signal at the intermediate network nodes or inside the receiver itself so as to estimate the performance of a transmission network. Performance monitoring has been a part of fiber-optic transmission systems from the very beginning, for example, bit/block-error-rate (BER) monitoring and other quality-of-service (QoS) measurements. However, such monitoring has fundamentally been conducted in the electronic domain, that is, after optical-to-electronic (O/E) conversion of the signal being monitored. For the efficient operation of dynamic DWDM networks, it is crucial to monitor the key performance parameters directly in the optical domain. That is why, OPM is expected to assess the quality of a data channel by estimating its optical characteristics without having *a priori* knowledge of the transmitted sequence of bits as shown in Figure 14.1.

It is envisaged that OPM will be indispensable for the efficient operation and management of existing as well as future complex and dynamic optical networks. The incorporation of OPM in fiber-optic networks can enhance the network reliability and minimize the network down time. OPM can also facilitate the efficient utilization of available networks resources through significant reduction in systems' safety margins currently used to ensure error-free operation. Other important benefits of OPM include (i) reduced network operation and maintenance costs, (ii) effective control of network elements, that is, fault detection, localization, and troubleshooting of failures, and (iii) link setup and optimization. OPM can also empower the carriers and network operators to certify service-level agreements (SLA) and to guarantee certain QoS provision to their clients [5].

14.1.2 Network Functionalities Enabled by OPM

The installation of monitoring mechanisms across the whole optical network, which could continuously monitor the health of network as well as the data signals, may

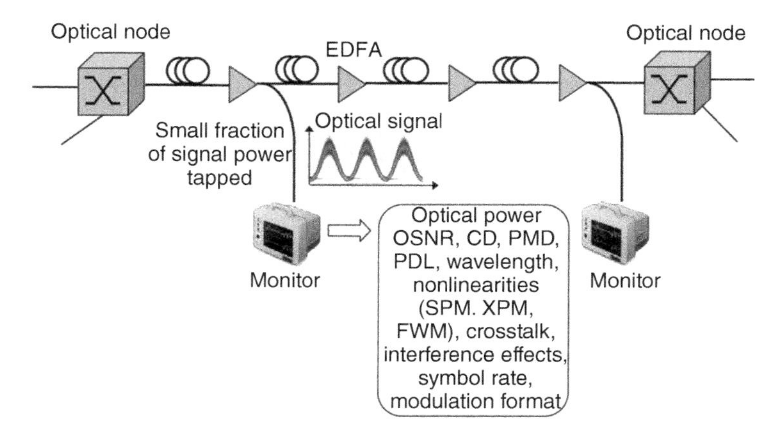

FIGURE 14.1 A fiber-optic transmission network equipped with OPM capability.

enable several advanced network functionalities. Some of the areas where OPM can play a significant role are discussed in this section.

14.1.2.1 ***Adaptive Impairments Compensation*** The transmission impairments in dynamic optical network are inherently time-variant in nature. In fixed point-to-point links, these changes occur due to certain physical effects while in dynamic networks, the impairments vary because the paths traversed by the optical signals change continuously due to network reconfiguration. This implies that the compensation techniques used in dynamic optical networks must be adaptive in nature. The realization of adaptive compensation techniques requires precise information about the extent of impairments introduced by the link. OPM can effectively assess the degradations caused by an optical link. This information available through OPM can then be utilized to provide feedback signals for the adaptive compensation of these impairments as shown in Figure 14.2.

14.1.2.2 ***Reliable Network Operation*** OPM can provide continuous and real-time information about the physical condition of an optical network and is thus capable of identifying the faults' locations as well as their causes. Furthermore, it can facilitate the acquisition of information about the extent of individual impairments contributed by the network as well as their distribution. This allows the network providers to know when the data signals are beginning to deteriorate. Therefore, preventive measures can be taken at right time to fix the problem before it starts to cause serious degradation to system performance, thus enabling reliable network operation [7] as shown in Figure 14.3.

14.1.2.3 ***Efficient Resources Allocation*** OPM can facilitate efficient utilization of available network resources. For example, if the link quality is too good in a dynamic optical network, this information can be conveyed to the transmitter, which may effectively reduce the transmitted signal power and thus decrease optical signal-to-noise

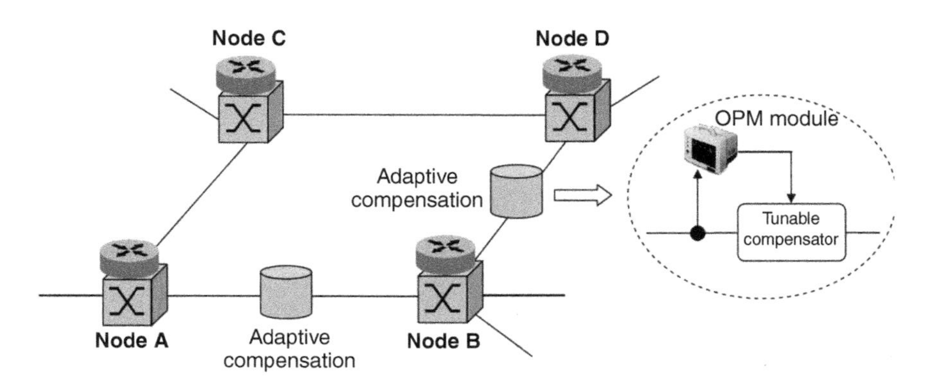

FIGURE 14.2 Adaptive impairments compensation enabled by OPM in dynamic optical networks.

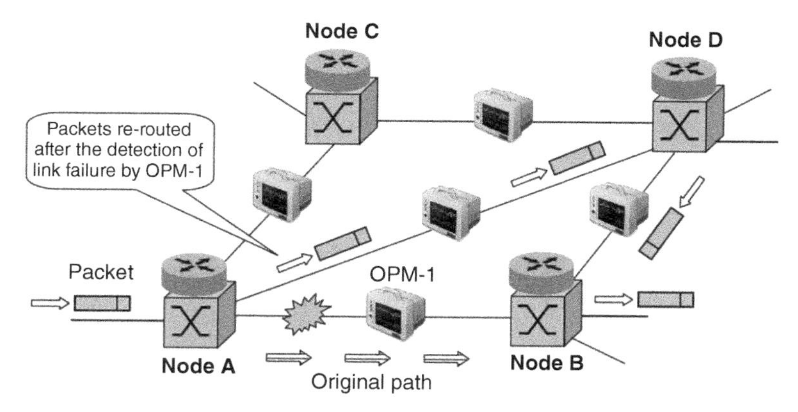

FIGURE 14.3 Reliable network operation and automatic fault management enabled through the incorporation of OPM in optical networks.

ratio (OSNR), while still meeting the desired BER requirements. The reduction in transmitted optical power may enhance the signal's robustness against the undesirable nonlinear distortions. On the contrary, the transmitted power of each channel may be increased (and hence, increase OSNR) accordingly, when the BER rises above the required specifications. Alternatively, the data rate can be increased by using higher modulation formats if the OSNR of the link is monitored to be good.

14.1.2.4 Impairment-Aware Routing The data-routing algorithms used in existing static optical networks either route traffic on shortest paths (i.e., fewest hops between source and destination) or on paths that satisfy certain minimum QoS constraints (e.g., delay, packet loss, and data rate) [8, 9]. However, such routing algorithms will perform far from optimum in dynamic optical networks since they do not take the variable physical state of the network into account. Therefore, in order to have better routing capabilities, the routing tables must be updated by taking optical layer parameters (e.g., fiber length, signal distortion, amplifier noise and transients) into consideration as shown in Figure 14.4 [4, 10–12]. The valuable information available through OPM can be provided to the network controllers, which may assign different weights to all possible paths considering various parameters. The routing decisions can subsequently be made based on these calculations.

14.1.3 Network Parameters Requiring OPM

Despite being a terrific transmission medium, an optical fiber introduces several signal impairments, which may degrade the performance of an optical network. Apart from optical fiber, other network components may also contribute several degradations as shown in Figure 14.5. The effects of all these impairments in downgrading the quality of the transmission link are strongly dependent on the data rates of the signals being transmitted. Impairments in optical networks can generally be categorized as catastrophic and noncatastrophic in nature. Catastrophic impairments result

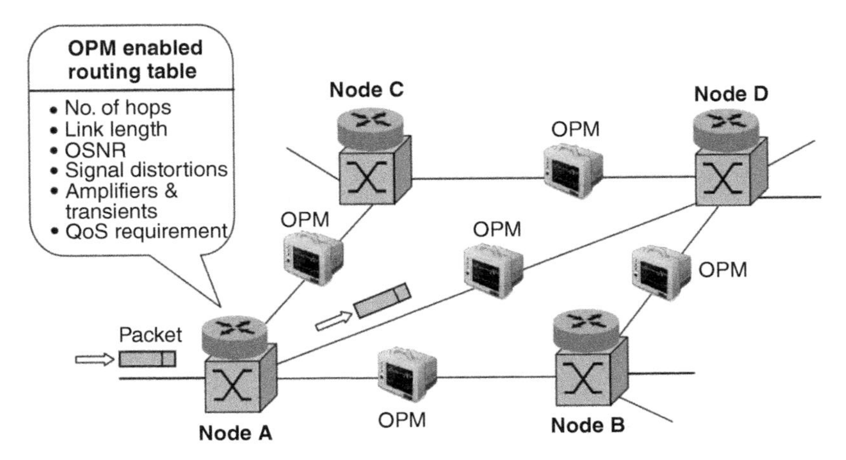

FIGURE 14.4 Impairment-aware routing in dynamic optical networks enabled by OPM.

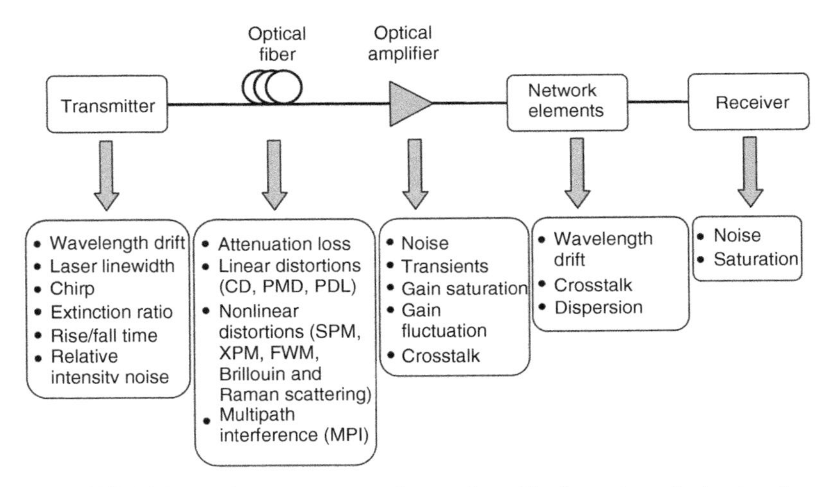

FIGURE 14.5 Major parameters to be monitored in dynamic optical networks.

in loss of optical power and include fiber breaks, network components failures, inappropriately installed network equipment, and so on. Noncatastrophic impairments do not necessarily decrease the optical signal power but they may severely distort the signal. Such distortions may be linear or nonlinear in nature and must be minimized or properly compensated so as to guarantee the desired network performance [4].

14.1.3.1 Optical Power The most fundamental parameter to be monitored in an optical network is the optical power. The optical power may decrease substantially due to fiber attenuation (caused by absorption and scattering) and losses encountered at the fiber connectors, splices, and couplers [1]. In wavelength-division multiplexed

(WDM) systems, the information about each channel's power is required so as to dynamically equalize the power in each of these channels through a feedback mechanism, thus ensuring a stable system performance [13]. This can be accomplished either by demultiplexing all optical channels and monitoring their powers using several photodetectors or alternatively in a cost-effective manner by making use of tunable optical filters in conjunction with a single photodetector [4].

14.1.3.2 OSNR Optical amplifiers such as erbium-doped fiber amplifiers (EDFAs) are normally employed in optical networks to compensate for the transmission losses over long distances. However, besides providing optical gain, EDFAs also add unwanted amplified spontaneous emission (ASE) noise into the optical signal. Furthermore, the cascading of EDFAs results in accumulation of ASE noise [14]. ASE noise is typically quantified by OSNR and is one of the most important parameters to be monitored in optical networks since the BER is directly related to the signal OSNR [15]. Furthermore, it also plays a pivotal role in fault diagnosis and as a measure of general health of links in an optical network.

14.1.3.3 Chromatic Dispersion (CD) CD belongs to the category of dispersive impairments and is one of the most crucial parameters to be monitored in dynamic optical networks. It arises due to the frequency-dependent nature of refractive index in an optical fiber and is one of the major limitations in high-speed, long-distance fiber-optic transmission systems [15]. In reconfigurable optical networks, changes in path lengths for a given channel (due to switching) may result in variable amounts of accumulated CD. Therefore, fixed CD compensation techniques are less effective in such scenarios [16]. For robust high-speed systems, it is essential to compensate CD adaptively within tight tolerances, making CD monitoring imperative in such systems.

14.1.3.4 Polarization-Mode Dispersion (PMD) Polarization-mode dispersion (PMD) is another dispersive impairment that needs to be effectively monitored since it is a major limitation in fiber-optic networks operating at data rates in excess of 40 Gbps [15]. PMD is caused by the ellipticity of fiber geometry. Since the same spectral component of an optical signal splits into two orthogonal states-of-polarization (SOP) inside a fiber, these two spectral copies have slightly different propagation speeds due to fiber birefringence. Hence, they reach the receiver at slightly different times resulting in pulse broadening. PMD effects are stochastic, time-variant, and temperature and data rate dependent [17].

14.1.3.5 Fiber Nonlinearity Fiber nonlinearity is also a critical parameter to be monitored. The nonlinear effects arise in an optical fiber due to the power dependence of refractive index and result in interference and crosstalk between different WDM channels when the transmitted power surpasses a certain limit. Fiber nonlinearities deteriorate the performance of high-speed WDM systems employing advanced modulation formats [15, 18]. Therefore, fiber nonlinearities need to be well managed and for that objective, monitoring is mandatory.

14.1.3.6 Quality-Factor (Q-Factor) Q-factor is an important parameter to assess the overall performance of an optical network. It is used to analyze the performance of transmission systems for which direct measurement of BER is impractical. Q-factor is an indicator of the quality of an optical signal due to strong correlation between Q-factor and BER [15]. It can be used to evaluate the effects of channel degradations such as ASE noise, CD, PMD, fiber nonlinearities as well as the impairments introduced by the transmitter and receiver, thus allowing effective BER estimation. Factors including non-Gaussian nature of noise, crosstalk and signal distortions may culminate in an inaccurate BER estimation using Q-factor [19].

14.1.3.7 BER The BER is an ultimate measure of the quality of an optical link. It is statistically defined as the time-averaged fraction of erroneous bits contained in a given bit stream [15]. BER monitoring has traditionally been used as a preferable tool to characterize the overall performance of a system [19]. However, BER is merely a number and does not provide any insight into the individual contributions of different impairments toward the degradation of system performance. In addition, BER monitoring requires expensive equipments (such as clock and data recovery systems).

14.1.3.8 Wavelength Shift The ever-increasing demand for larger link capacity has resulted in a significant reduction in channel spacing in DWDM systems. As the interchannel spacing decreases, the requirements for wavelength control of optical components become extremely stringent. This demands the incorporation of sophisticated monitoring mechanisms in optical networks that can detect or compare the wavelengths because a relative offset between the centre frequencies of DWDM channels, and the optical filters may cause significant power loss and may also result in crosstalk between the channels [4].

Apart from the aforementioned parameters, other factors such as optical amplifier gain and distortions, crosstalk and interference effects, SOP and polarization-dependent effects, pulse shape and timing jitter are also useful to be monitored in optical networks [4, 14, 19, 20].

14.1.4 Desirable Features of OPM Techniques

The features that a given OPM technique is expected to demonstrate are determined by several factors such as the nature of the optical network in which the OPM module is anticipated to be deployed, types and extents of impairments prevalent in the network, data rates, implementation cost, and the degree of intelligence sought to be incorporated with the inclusion of OPM modules. Some common features expected from the OPM techniques are discussed in the following sections [4, 5, 20].

14.1.4.1 Accuracy, Sensitivity and Dynamic Range The precise compensation of impairments in dynamic optical networks depends on the degree of accuracy of the monitoring technique being used. Therefore, the monitoring techniques are expected to meet the desired accuracy requirements. Apart from accuracy, the techniques are also anticipated to demonstrate good sensitivities in the whole monitoring range. In addition, the techniques are expected to exhibit broad monitoring ranges in order to enable appropriate compensation of impairments occurring in wide dynamic ranges. The accuracy, sensitivity, and dynamic range of a monitoring technique may rely on a number of factors such as the methodology being employed in the OPM module and the amount of signal power tapped for monitoring purposes.

14.1.4.2 Multichannel Operation Since fiber-optic networks encompass multiple data channels by incorporating WDM, the OPM techniques used must be capable of monitoring several data channels. This can be achieved by either using a parallel bank of monitoring devices or a tunable optical filter to select a particular channel for sequential monitoring. The parallel operation requires more number of devices and thus involves higher hardware costs. On the contrary, the sequential operation may introduce measurement latency, especially in systems with large number of data channels.

14.1.4.3 Multi-Impairment Monitoring As discussed in Section 14.1.3, several impairments may coexist in an optical network. If different techniques are employed for the monitoring of individual impairments then it will increase the monitoring costs immensely. Therefore, the monitoring techniques must be capable of monitoring multiple network impairments simultaneously as well as independently.

14.1.4.4 Data Rate and Modulation Format Transparency The future optical networks are envisioned to contain data traffic with different modulation formats and data rates in individual channels. Therefore, the developed OPM techniques must be transparent to data rates and modulation formats, thus avoiding the need for the modification of monitoring modules.

14.1.4.5 Cost-Effectiveness Since OPM may be needed at multiple locations in an optical network, a general requirement is that its cost must be relatively lower than that of conventional testing equipment. The cost of OPM module may depend on the complexity of the technique being employed for monitoring. A reduction in cost can be achieved by using techniques capable of monitoring multiple network impairments independently for several data channels and for various data rates and modulation formats.

14.1.4.6 Operation at Low Input Power For monitoring purposes, part of the signal power needs to be tapped from the optical link. The OPM module must be capable of performing its operation by exploiting only a small fraction of the signal power while still meeting the accuracy and sensitivity requirements. A general rule of thumb is that the power used for monitoring must not exceed a small percentage of the total signal power.

14.1.4.7 Fast Response Time In static optical networks, the response time of the OPM technique can be of the same order of magnitude as the network restoration time of 50 ms [20]. However, in case of dynamic optical networks, the response time must be much smaller than the network reconfiguration interval. A general rule of thumb is that the monitoring time must be in the range of a few milliseconds.

14.1.4.8 Passiveness The OPM technique must not have an adverse effect on the normal operation of an optical network. This requires the OPM technique to not modify the network components while it performs the monitoring task. Also, it should not insert additional monitoring signals into the network, which may interfere with the data signal resulting in the degradation of data signal quality.

14.2 OPM TECHNIQUES FOR DIRECT DETECTION SYSTEMS

The receivers in direct detection systems employ simple photodetectors to detect the intensity of the optical signal (e.g., in case of on-off-keying (OOK) modulation scheme) or use delay interferometers (DI) in conjunction with photodetectors to transform the differential phase information into amplitude information in the electrical domain (e.g., in case of differential binary/quaternary phase-shift keying (DBPSK/DQPSK) modulation schemes). Due to square-law detection nature of the direct detection receivers, only limited information can be retrieved from the optical signals, whereas the receivers in coherent detection systems perform down-conversion by using a combination of local oscillator (LO) lasers, optical hybrids, and photodetectors in conjunction with digital signal processing (DSP) modules so as to retrieve the information contained in the amplitude, phase, frequency, and polarization of the carrier. In this case, it is possible to retrieve all the information from an optical signal in the electrical domain [20]. Coherent detection techniques allow the use of advanced higher-order modulation formats such as m-PSK and m-QAM along with polarization-division multiplexing (PDM), thereby enabling higher spectral efficiencies. Due to the absence of high-speed and economical sampling and DSP devices in the past, most of the currently deployed fiber-optic communication networks employ direct detection receivers. Irrespective of whether the optical networks are using direct detection or coherent detection techniques, OPM is equally important for the reliable and efficient operation of these networks. However, the requirements, nature of parameters requiring monitoring, and the scope and network functionalities enabled by OPM are generally different for these two types of networks.

14.2.1 OPM Requirements for Direct Detection Optical Networks

The crucial parameters to be monitored in networks employing direct detection include residual CD, total power (i.e., signal plus noise power), OSNR, PMD, polarization-dependent loss (PDL), and fiber nonlinearities (i.e., self-phase modulation (SPM), cross-phase modulation (XPM), four-wave mixing (FWM), and scattering). The OPM techniques developed for existing direct detection systems are preferred to utilize simple direct detection. This is due to the fact that the complexity and cost of a coherent receiver is relatively higher and thus it may not be ideal for use in monitoring units deployed at the intermediate network nodes where cost is a major limitation. However, the adverse effect of using direct detection in these monitoring devices is that only limited information about the optical signal can be retrieved in the electrical domain, thereby making the accurate estimation of various network parameters quite challenging.

Direct detection fiber-optic transmission systems are typically dispersion-compensated systems. However, due to network reconfigurability enabled by ROADMs as well as due to temperature and other physical effects, the CD of the link varies dynamically. Therefore, despite the use of dispersion compensation techniques in these networks, there is always some residual CD present. Furthermore, most of the currently deployed optical fibers have reasonably high PMD coefficient values, that is, of the order of $0.5\,\text{ps/km}^{1/2}$ [20]. Hence, the optical signals are also affected by polarization-related distortions. Finally, the optical signals in direct detection systems are also subjected to degradations caused by the ASE noise of EDFAs as well as fiber nonlinearities. Since all the above-mentioned deleterious channel effects may coexist, the OPM techniques developed for direct detection optical networks must be capable of monitoring these parameters independently. For example, the monitoring of OSNR should not be affected by the presence of CD and PMD. However, as mentioned earlier, the availability of limited information about the optical signal in the OPM devices (due to the use of simple direct detection) makes simultaneous and independent monitoring of multiple performance parameters overwhelmingly complicated in direct detection optical networks.

14.2.2 Overview of OPM Techniques for Existing Direct Detection Systems

Over the past few years, a plethora of techniques for monitoring optical signal quality parameters in direct detection fiber-optic communication networks have been proposed. The general classification of these techniques is depicted in Figure 14.6.

The existing monitoring techniques can be classified as either digital or analog in nature. Digital techniques exploit the digital information content of the signal waveform in the electrical domain. Digital OPM methods, for example, BER monitoring, provide information about the overall degradation of the system caused by the network impairments but are unable to isolate their individual contributions. Analog monitoring techniques make use of the specific characteristics of the analog signal waveform to extract information about the channel impairments [5]. These techniques can be further subdivided into time-domain, frequency-domain, and polarization-domain techniques depending upon whether the monitoring information is extracted from the signal waveform, signal spectrum or the signal polarization, respectively.

Time-domain monitoring techniques can be categorized into synchronous and asynchronous sampling-based techniques depending upon whether the sampling rate is synchronized with the symbol rate. Synchronous sampling techniques require clock recovery, which is a relatively complex operation especially in networks supporting multiple data rates. Eye-diagram is a typical synchronous sampling-based technique, which qualitatively reflects the effects of all impairments on the signal quality. However, it is unable to quantify the effects of individual impairments [4]. Similarly, Q-factor monitoring is another popular synchronous

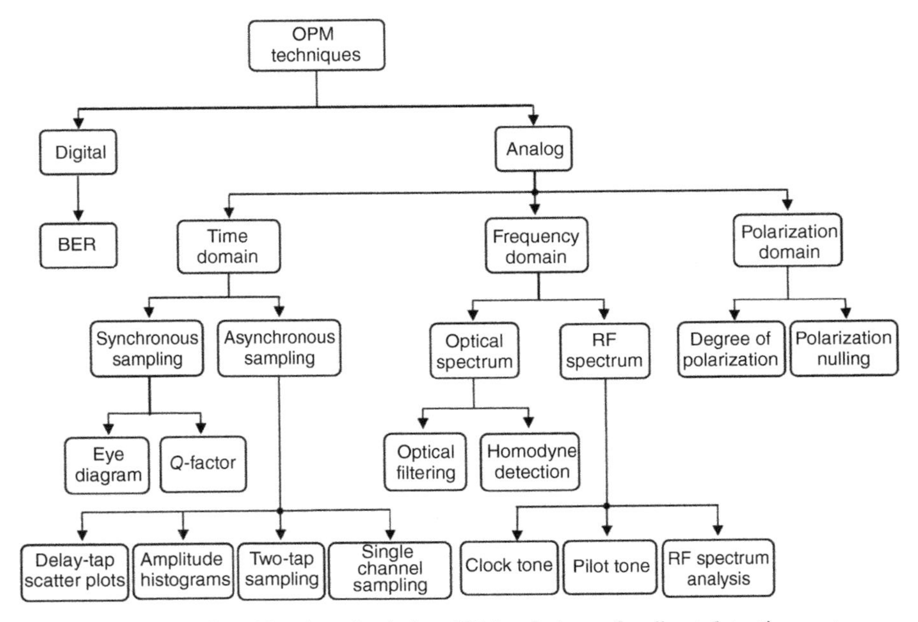

FIGURE 14.6 Classification of existing OPM techniques for direct detection systems.

sampling-based technique, which is commonly used in practice due to its strong correlation with BER [5]. Asynchronous sampling-based techniques, for example, asynchronous amplitude histograms (AAHs) [21–28], asynchronous delay-tap plots (ADTPs) [29–37], asynchronous two-tap plots (ATTPs) [38, 39], and asynchronous single-channel sampling (ASCS) [40], are considered attractive due to the fact that they do not require clock information and they are also capable of monitoring multiple impairments simultaneously, thus being cost-effective.

Frequency-domain monitoring techniques can be subdivided into optical and radio frequency (RF) spectrum-based techniques. The optical spectrum analysis techniques can make use of an optical filter, which is tuned over the channel bandwidth and the optical power is recorded [4]. The spectral resolution in this case is determined by the filter's bandwidth. Alternatively, optical spectrum can be analyzed by performing homodyne detection, where a tunable LO laser signal is mixed with the monitored signal and the interference signal is then analyzed for spectral analysis [41]. The LO laser is swept across the channel bandwidth. The spectral resolution in this case is determined by the linewidth of the LO laser and is several orders of magnitude higher than that of tunable optical filter. The optical spectrum-based techniques are capable of monitoring out-of-band OSNR, total optical power, and wavelength drift but they cannot monitor CD and PMD. These techniques can be used to monitor multiple WDM channels. Since the optical filter or the LO laser needs to be tuned for scanning the whole WDM spectrum (which may require some time), such techniques can introduce measurement latency.

Radio frequency spectrum-based monitoring techniques can provide better estimation of signal quality as compared with optical spectrum-based techniques because they analyze the spectrum of the signal that is encoded on the optical carrier. RF spectrum-based techniques can either make use of clock tones present inherently in the spectrum of various modulation formats or insert pilot tones of different frequencies in each channel at the transmitter. The clock tones-based monitoring techniques can measure CD and PMD and are data rate and modulation format dependent [42–45]. On the contrary, pilot tones-based schemes can measure various parameters such as wavelength, OSNR, CD, and PMD and are data rate as well as modulation format independent [46–48]. However, the adverse effect of pilot tones-based techniques is that the tones interfere with the data signal resulting in the deterioration of BER. Apart from monitoring the specific tones (i.e., clock and pilot tones) in the RF spectrum, changes in the spectral distribution of overall RF spectrum, due to various network impairments, may also be exploited for the monitoring of these impairments [49].

Polarization-domain monitoring techniques exploit the polarization properties of the optical signal. The alterations in the polarization characteristics due to various channel degradations can be utilized for the effective monitoring of these impairments. These techniques can monitor signal and noise powers (and hence, OSNR), for example, through polarization nulling [50–53], as well as PMD of the fiber link, for example, by measuring the degree-of-polarization (DOP) of the received signal [45, 54, 55]. These techniques have the advantage of being transparent to data rates and modulation formats. However, they cannot be applied to polarization-multiplexed signals, which severely limits their use in coherent transmission systems.

14.2.3 Electronic DSP-Based Multi-Impairment Monitoring Techniques for Direct Detection Systems

OPM techniques utilizing electronic DSP have gained substantial attention in recent years. DSP-based monitoring techniques exploit the statistical properties of the data signals after O/E conversion for the estimation of critical signal quality parameters. The reason behind the popularity of DSP-based OPM techniques is that they can facilitate cost-effective monitoring of multiple signal quality parameters simultaneously for several data rates and modulation formats and without necessitating modifications of monitoring hardware. Furthermore, the monitoring of a new parameter as well as monitoring for a different signal type can simply be enabled by downloading the relevant algorithm to the DSP-based monitor, thereby facilitating flexibility and cost-effectiveness. DSP-based monitoring techniques typically perform asynchronous sampling of electrical signal amplitude and then generate one-dimensional (1D) or two-dimensional (2D) histograms of the signal samples. The statistical features of these histograms are then exploited using statistical signal processing, artificial intelligence, and digital image processing techniques for multi-impairment monitoring [20]. Some prominent DSP-based

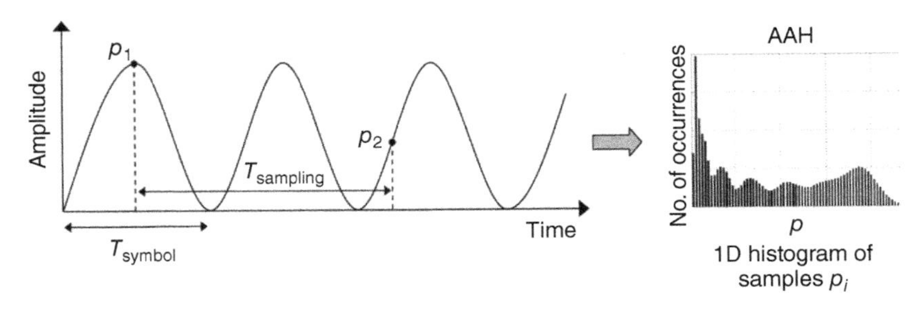

FIGURE 14.7 Conceptual diagram of asynchronous amplitude sampling with an AAH on the right. T_{sampling} is the sampling period and T_{symbol} is the symbol period.

monitoring techniques for direct detection systems include AAH, ADTP, ATTP, and ASCS techniques.

Electronic DSP-based techniques using AAHs are attractive due to their remarkable simplicity and flexibility. The concept of AAH is shown in Figure 14.7. To obtain an AAH, the electrical signal amplitude is randomly sampled (without any clock information) at a rate much lower than the symbol rate. It is important to note that the sampling period T_{sampling} has no relation with the symbol period T_{symbol}. After the acquisition of numerous samples p_i, a histogram is formed by first dividing the dynamic range of sample values into different uniformly spaced levels called histogram bins. Next, the samples are sorted out depending upon their values. The values are then mapped onto the histogram bins and the number of samples falling into each of these bins is counted. Plotting the bin count against the bin value generates an AAH [20]. The number of samples used for the synthesis of AAH must be sufficient enough to obtain the complete statistics of the signal.

The shape of an AAH reflects the signal properties. Since the signal is distorted by several optical impairments, the statistical features of an AAH also vary accordingly. The variations in AAH's statistical properties can thus be tracked to evaluate the levels of various impairments degrading the optical signal. The impairments-sensitive features of AAHs have been successfully exploited using statistical signal processing and machine learning techniques for the monitoring of multiple signal quality parameters in fiber-optic networks [21–28].

The advantages of AAH-based monitoring techniques are that they are cost-effective, have less implementation complexity, and are modulation format independent. Since they do not require clock information, they are also transparent to data rates. However, their drawback is that the effects of various impairments are often intermixed, thus prohibiting independent monitoring. The monitoring accuracy of AAH-based techniques depends on the number of samples acquired for monitoring purpose. Hence, there is a trade-off between accuracy and monitoring speed.

ADTS-based monitoring techniques are quite interesting since they offer the potential for data rate and modulation format-independent multi-impairment monitoring [29–37]. Similar to AAH, ADTS-based techniques also exploit the statistical properties of the asynchronously sampled signal. However, in contrast to AAH,

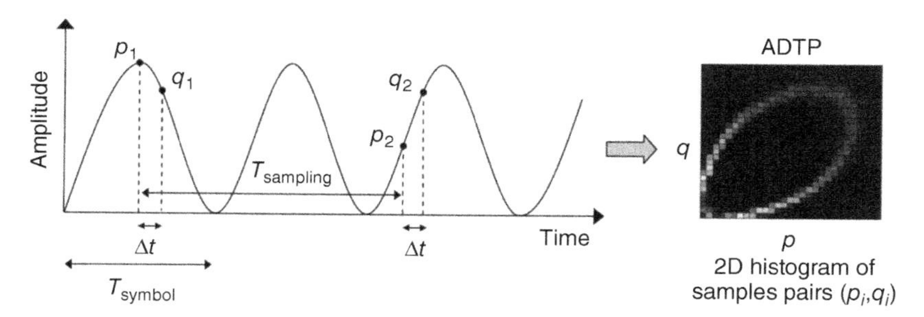

FIGURE 14.8 Conceptual diagram of ADTS technique with an ADTP on the right. Δt is the time delay within each sample pair and T_{sampling} is the sampling period.

which is a 1D histogram of signal amplitudes, ADTS produces a 2D histogram of closely located sample pairs [29]. The concept of ADTS is illustrated in Figure 14.8. The electrical signal amplitude after direct detection is asynchronously sampled in pairs (p_i, q_i) with a known constant time delay Δt between them called tap-delay. The sampling period T_{sampling} between the pairs (p_i, q_i) is not related to the symbol period T_{symbol} and can be many orders of magnitude longer. Binning the sample pairs (p_i, q_i) into a 2D histogram generates an ADTP or scatter plot as shown in Figure 14.8. An ADTP provides the information richness of an eye-diagram without requiring clock information for its generation. The shape and features of an ADTP depend on the modulation format, bit rate, tap-delay Δt as well as various signal distortions such as ASE noise, CD, PMD, and crosstalk. The unique signatures of various impairments reflected in ADTPs can be exploited for OPM purpose. In addition, the ADTPs can even distinguish between the sign of accumulated link CD and Li et al. detailed how this property can be effectively used for signed CD monitoring [30]. However, it is important to note that an ADTP is essentially a graphical representation allowing qualitative estimation of signal properties. In order to extract the quantitative information from an ADTP, a further analysis is required. Several approaches such as statistical signal processing [31–34], image processing [35], artificial intelligence, and machine learning [36, 37] have been proposed for the treatment of ADTPs for the purpose of multi-impairment monitoring in direct detection systems.

The advantage of ADTS-based OPM techniques is that they are capable of monitoring multiple impairments simultaneously and independently. Similar to AAH-based techniques, they are also transparent to data rates and modulation formats. However, the implementation complexity of ADTS-based techniques is higher than that of AAH. This is due to the fact that the tap-delay value is a function of symbol period and hence, it needs to be adjusted precisely depending upon the data rate of the signal being monitored. The monitoring accuracy of ADTS-based techniques depends on the number of sample pairs acquired for monitoring purpose as well as on the degree of correlation between the transponders used for the acquisition of calibration curves or for the training of artificial intelligence-based classifiers, and the transponders employed during the actual monitoring process.

In addition to AAH- and ADTS-based techniques, several other electronic DSP-based schemes have been proposed for OPM in direct detection optical networks. These include ASCS, asynchronous two-tap sampling (ATTS), and empirical moments based monitoring techniques. In ASCS-based technique [40], a 2D scatter plot equivalent to ADTP is produced. However, in contrast to ADTS technique, which requires two-channel sampling for the acquisition of delay-tap sample pairs, this technique needs only single-channel sampling, whereby the original and shifted versions of the acquired samples are utilized as sample-pairs for the generation of 2D scatter plot. Consequently, the implementation complexity and cost of this technique is expected to be lower as compared with ADTS-based techniques. In [39], an ATTS-based technique is proposed for CD monitoring in direct detection systems. This technique estimates the CD-induced relative group delay between the two vestigial sideband (VSB) signals by computing the differences in the sampled amplitude levels of two VSB signals, which are sampled simultaneously but asynchronously. The technique is shown to monitor CD for various modulation formats and data rates without requiring hardware modifications. Unlike ADTS-based techniques, no precise adjustment of tap-delay is needed in this case. In [6], the use of empirical moments of asynchronously sampled signal amplitudes in conjunction with artificial neural networks (ANNs) is proposed for multi-impairment (i.e., OSNR, signed CD, and PMD) monitoring in direct detection systems. This technique requires simple hardware for the acquisition of signal samples as compared with ADTP- and ATTS-based techniques. Furthermore, no hardware modifications are needed for the monitoring of several different data rates and modulation formats.

14.2.4 Bit Rate and Modulation Format Identification Techniques for Direct Detection Systems

The existing OPM techniques for direct detection fiber-optic networks assume either prior information about the signal's bit rate and modulation format or the attainment of this knowledge from the upper layer protocols. However, practically it is not feasible to introduce additional cross-layer communication for OPM purposes at the intermediate network nodes because these nodes can only handle limited complexity. Therefore, it is crucial to have the capability of joint bit rate and modulation format identification (BR-MFI) at the intermediate network nodes since the OPM techniques used at these nodes may be bit rate/modulation format dependent. The critical information about the signal's bit rate and modulation format can enable the OPM devices deployed at the intermediate network nodes to select a monitoring technique most suitable for the identified signal type [35, 56, 57].

Recently, a few techniques for modulation format identification (MFI) as well as joint BR-MFI in direct detection receivers have been proposed. Khan et al. [56] demonstrated a simple and cost-effective MFI technique utilizing an ANN trained with the features extracted from the AAHs. This technique is shown to successfully identify six different widely used modulation formats in the presence of various channel impairments. However, since an AAH lacks the timing/slope information necessary for distinguishing between different bit rates of the signals, this technique is limited to just MFI. A modification of this technique is presented in [57], whereby joint BR-MFI is demonstrated by using an ANN trained with the features extracted from ADTPs. Since ADTPs contain information about the slopes of the signal pulses (which in turn depend on the bit rate) as well as unique signatures of different modulation formats, this technique is capable of simultaneous identification of bit rates and modulation formats of the signals. Tan et al. [35] proposed a simple technique for joint BR-MFI as well as multi-impairment monitoring by using principal component analysis (PCA)-based pattern recognition on ADTPs. This technique can successfully identify the bit rate and modulation format of the signal from a known set of bit rates and modulation formats. In addition, it can enable simultaneous and independent monitoring of OSNR, CD, and DGD without necessitating information about the signal type during the online monitoring process.

14.2.5 Commercially Available OPM Devices for Direct Detection Systems

OPM is regarded as a key enabling technology for high-speed self-managed optical networks. However, currently, the commercially available OPM devices are nothing more than a simplified version of optical spectrum analyzers [58]. These devices are capable of monitoring a few parameters such as wavelengths of WDM channels, total optical power, and out-of-band OSNR, and are normally referred to as optical channel monitors (OCM). These devices typically employ a tunable band-pass filter or a diffraction grating combined with a single detector to monitor the above-mentioned parameters [4]. The information about the power level of each WDM channel is typically used to balance the optical amplifier gain across the spectrum. Similarly, the wavelength measurement provides information about whether the optical signal is properly placed in its channel.

The limited information provided by the current OPM devices can only facilitate the detection of sudden faults and thus enable system alarms and error warnings for lost or out-of-specification optical channels. These devices are far from achieving the real objectives of OPM, that is, to enable error root cause analysis and to provide early failure identification so that the operators can initiate fast error cancellation. Therefore, to fulfill the ultimate potential of OPM, the commercial OPM devices need to undergo significant improvement in order to be able to constantly monitor the signal dynamics, observe system functionality, detect performance changes, and provide feedback information to the network elements for the optimization of operational performance of the optical networks.

14.2.6 Applications of OPM in Deployed Fiber-Optic Networks

Recently, there have been a few attempts to realize the OPM-enabled functionalities in practical fiber-optic networks. Morgan et al. [59] reported the in-service measurement of OSNR, CD, and PMD on a live 140 km, 10 Gbps WDM optical link between Bromsgrove and Shrewsbury in Western England without interrupting the actual data traffic. An ADTS-based technique is used for the simultaneous monitoring of above-mentioned parameters in a field trial. The real-time information provided by the OPM module is then used to effectively diagnose the underlying system issues. For example, it is revealed that the root cause of the higher than expected pre-FEC BER and reduced system margin of the monitored link is the high amount of residual CD resulting from the use of an incorrect dispersion compensation module. Thus, an in-service diagnosis of underlying network problems, as demonstrated in this work, can facilitate the early identification of root causes and permit the network operators to resolve the issues.

14.3 OPM FOR COHERENT DETECTION SYSTEMS

The explosive growth in data traffic and demand for higher bandwidth has led to the evolution of employing advanced optical modulation formats with coherent receiver and DSP in fiber-optic communication systems. Since it is possible to retrieve all the information from an optical signal in the electrical domain, coherent detection techniques allow the use of advanced higher-order modulation formats such as m-PSK and m-QAM along with PDM [60], thereby enabling higher spectral efficiencies. On the contrary, since higher-order modulation signal with high baud rate are more sensitive to the channel impairments and noise [61], the link margin may be reduced.

The requirements and parameters to be monitored for coherent systems are quite different from noncoherent systems. With the help of DSP, linear channel impairments such as CD and PMD can be fully compensated [60]. Meanwhile, it also enables a promising and comprehensive built-in OPM at the receiver for free [62, 63]. Assume that the optical channel is linear with the channel transfer matrix $\mathrm{H}(f)$, the equalization filter $\mathrm{W}(f)$ obtained through the zero-forcing (ZF) solution or minimum mean square estimation (MMSE) solution is the inverse impulse response of the channel that can be expressed as [63]

$$\mathrm{W}(f) = \mathrm{H}^{-1}(f) = D^{-1}(f) \prod_{i=N,-1}^{1} \mathrm{U}_i^{-1}(f)\mathrm{E}_i^{-1} \tag{14.1}$$

where $\mathrm{H}(f)$ is composed of the transfer function $D(f)$ and concatenated elements E_i and $\mathrm{U}_i(f)$ accounting for PDL and higher-order PMD.

After some algebraic manipulations, the following equations can be obtained:

$$\arg(\widehat{H}_{\mathrm{CD}}^{-1}) = \arg(\sqrt{\det(\mathrm{W}(f))}) = -f^2\varphi \tag{14.2}$$

$$\widehat{H}_{\mathrm{PDL}}(f) = \left|\sqrt{\det(\mathrm{W}(f))}\right| = |H_{\mathrm{AF}}(f)|^{-1}\prod_{i=1}^{N}(k_i)^{-1/2} \tag{14.3}$$

$$\mathrm{W}_{\mathrm{UE}}(f) = \frac{\mathrm{W}(f)}{\sqrt{\det(\mathrm{W}(f))}} = \prod_i \begin{pmatrix} u_i^* & -v_i \\ v_i^* & u_i \end{pmatrix} \begin{pmatrix} k_i^{1/2} & 0 \\ 0 & k_i^{-1/2} \end{pmatrix} \tag{14.4}$$

which are responsible for residual CD, PDL, and PMD. $\mathrm{W}_{\mathrm{UE}}(f)$ is the normalized $\mathrm{W}(f)$ by the square root of its determinant. u_i and v_i form the PMD matrices and k_i is the attenuation factor accounting for PDL.

Since those linear impairments can be estimated and compensated, they no longer limit system performance. On the contrary, as ASE noise cannot be compensated, the system performance is largely determined by the OSNR of received signals. Accurate, reliable and low-cost in-band OSNR monitoring is still highly desired for coherent systems.

Although OSNR monitoring is not as easy as reading off filter taps, ASE-noise-induced distortions can be separated from all the other linear transmission impairments in a digital coherent receiver and reliable OSNR can still be estimated with further processing of the received signals. There are two types of approaches to OSNR monitoring: non-data-aided estimation [61, 64–69] and data-aided estimation [70–72] that is based on the property of training sequences (TS). Data-aided estimation gives high estimation accuracy and faster estimation speed with the trade-off of reduced bandwidth efficiency compared with non-data-aided estimation [71].

14.3.1 Non-Data-Aided OSNR Monitoring for Digital Coherent Receivers

One simple approach for non-data-aided OSNR monitoring utilizes the statistical moments of the equalized signal [64–67], the signal used to estimate the OSNR is taken from just after the adaptive equalization, before the carrier phase recovery stage (as shown in Figure 14.9(a)). The adaptive equalization can either utilize "blind" non-data-aided channel acquisition by gradient algorithms such as constant-modulus algorithm (CMA) or data-aided channel estimation based on a periodically transmitted training sequence.

After adaptive equalization, the linear distortions such as CD and PMD ideally can be fully compensated, and thus the variations of the equalized signal envelope are mainly caused by the ASE noise. The envelope of the output signal from the adaptive filter (as shown in Figure 14.9(b)) can be approximated as [65]

$$y_n \approx \sqrt{C} a_n e^{j\theta_n} + \sqrt{N} w_n \tag{14.5}$$

where a_n is the m-PSK or m-QAM symbol amplitude, C is the signal-power scale factor, N is the noise power scale factor, w_n is the ASE noise, θ_n is the phase noise stemming from phase fluctuations of a transmitter laser and a local oscillator, and n is the number of samples.

In a practical system, the second- and fourth-order moments from a received data block of L symbols can be calculated as

$$\mu_2 \approx \frac{1}{L} \sum_{n=0}^{L-1} |y_n|^2 \tag{14.6}$$

$$\mu_4 \approx \frac{1}{L} \sum_{n=0}^{L-1} |y_n|^4 \tag{14.7}$$

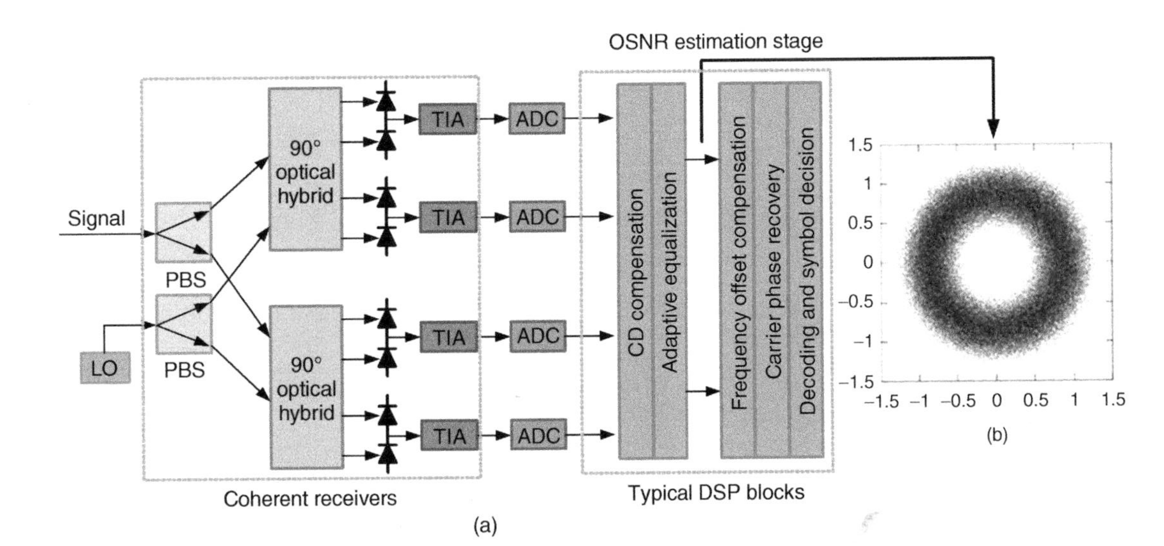

FIGURE 14.9 Schematic of the coherent optical receiver and typical DSP blocks for data recovery (a). Adaptive equalizer outputs are used for OSNR monitoring (b). LO: local oscillator, PBS: polarization-beam splitter, TIA: transimpedance amplifier, ADC: analog-to-digital converter.

respectively. And the carrier-to-noise ratio (CNR) for QPSK is expressed as

$$\mathrm{CNR}_{\mathrm{QPSK}} = \frac{\sqrt{2\mu_2^2 - \mu_4}}{\mu_2 - \sqrt{2\mu_2^2 - \mu_4}} \tag{14.8}$$

and CNR for QPSK is expressed as

$$\mathrm{CNR}_{16-\mathrm{QAM}} = \frac{\sqrt{2\mu_2^2 - \mu_4}}{\mu_2\sqrt{0.68} - \sqrt{2\mu_2^2 - \mu_4}} \tag{14.9}$$

When the launched power is so low that fiber nonlinear effects can be neglected, the OSNR value in dB can be estimated from the CNR value as

$$\mathrm{OSNR}_{\mathrm{dB}} = 10\log_{10}(\mathrm{CNR}) + 10\log_{10}\left(\frac{R_s}{B_r}\right) \tag{14.10}$$

where R_s is the symbol rate and R_s/B_r is a scaling factor adjusting the measured noise bandwidth to the reference bandwidth B_r. The bandwidth B_r is usually set to 12.5 GHz, which is equivalent to the 0.1-nm OSA resolution bandwidth. As shown in Equations 14.8 and 14.9, measuring second- and fourth-order moments does not include any effect of the phase noise and thus the proposed scheme operates phase insensitively.

Another approach is to utilize the error vector magnitude (EVM) of fully equalized signals as an OSNR estimator [61, 68, 69]. In this approach, the signal is taken after decoding and symbol decision stage where both the linear impairments such as CD and PMD and carrier-phase are recovered. The distributions of fully equalized QPSK signal are shown in Figure 14.10.

In this case, the kth received symbol in one particular polarization can be represented as

$$r_k = s_k + n_k \tag{14.11}$$

where s_k is the transmitted QPSK symbol and n_k models the collective ASE noise generated by inline optical amplifiers, which is a band-limited complex circularly

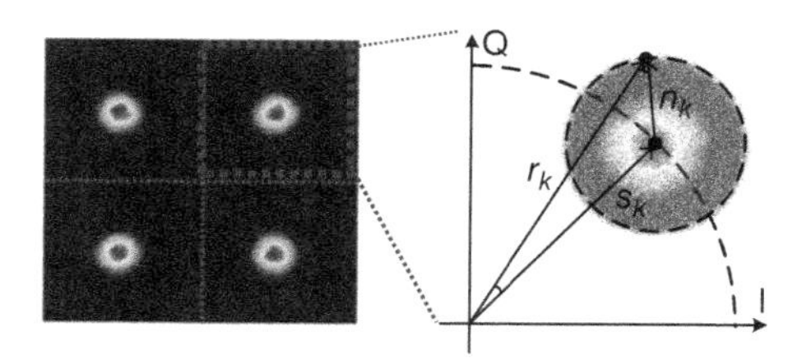

FIGURE 14.10 Graphical illustration of distributions of fully equalized QPSK signal.

symmetric zero-mean Gaussian random process with covariance matrix $\sigma^2 I$. The OSNR can be estimated through an EVM-based approach [2]

$$\text{OSNR}_{\text{Estimated}} = \frac{P_{\text{in}}}{P_{\text{ASE}}} = \frac{\mathbf{E}(|\hat{s}_k|^2)}{\mathbf{E}(|n_k|^2)} \tag{14.12}$$

where P_{in} is the signal power, P_{ASE} accounts for the ASE noise power, and $\hat{s}_k$ is the symbol after decoding and symbol decision stage and $\mathbf{E}(\cdot)$ denotes expectation. However, since the signals used for estimation are taken after carrier-phase recovery, the accuracy of the estimation may be affected by the frequency offset and phase noise.

Non-data-aided OSNR estimation is not modulation format independent [61, 68, 69] and suffers from a relatively longer acquisition time, preventing the receiver from fast switching protection.

14.3.2 Data-Aided (Pilot Symbols Based) OSNR Monitoring for Digital Coherent Receivers

Data-aided OSNR estimation has the advantage of being independent of modulation format as it only utilizes the properties of the equalized TSs, and the modulation format can be altered arbitrarily in between the fixed training patterns. Furthermore, channel equalization based on the periodically transmitted TSs allows instantaneous filter acquisition and thus enables faster OPM [70].

In [71], a method is proposed to employ a moving average filter based on the estimated channel transfer function to reserve the noise information in the received training sequences, and then OSNR can then be estimated through the SNR from the training sequences. Figure 14.11 shows a constellation plot of the equalized training sequences for QPSK system before and after channel filtering using Golay sequences with OSNR = 22 dB. In case of no channel filtering, small variations in the equalized training sequences are due to misalignment between the overlap-cut equalizer and the training blocks, while it can be seen that, with estimation filtering, the noise information is reserved and can be used for SNR estimation.

After equalization of the training sequences using the estimated channel information with filtering, the added noise can then be subtracted from the signal and the SNR is measured as

$$\text{SNR} = \sum_{k=1}^{N} \frac{s[k]^2}{w[k]^2} \tag{14.13}$$

where $s[k]$ is the original sequence and $w[k]$ is the added noise sample. Finally, the OSNR is estimated by first measuring the electrical RF noise SNR_{RF} at a reference OSNR point. This is done by measuring the reference point in a back-to-back configuration and without any added ASE and then the OSNR is calculated as

$$\text{OSNR}_{\text{dB}} = 10\log_{10}\left(\frac{B_{\text{ref}}}{R_s}\right) - 10\log_{10}(\text{SNR}^{-1} - \text{SNR}_{\text{RF}}^{-1}) \tag{14.14}$$

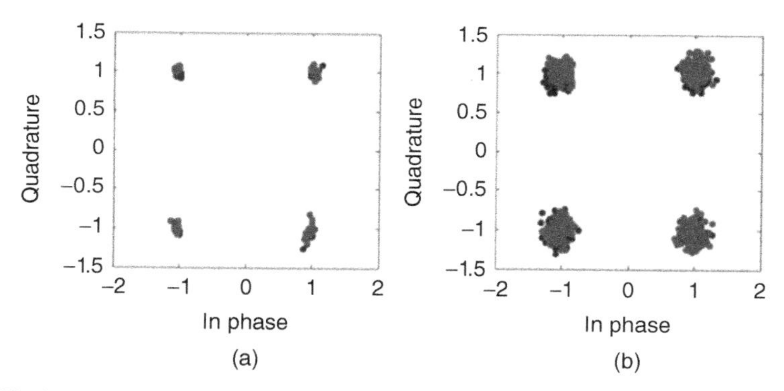

FIGURE 14.11 Constellation of equalized TS (QPSK) with (a) no filtering and (b) channel filtering. From Ref. [71], with permission of IEEE.

where B_{ref} and R_s are the reference bandwidth and baud rate, respectively, and SNR_{RF} is the measured system SNR without any added ASE noise.

14.3.3 OPM at the Intermediate Network Nodes Using Low-Cost Structures

As discussed in the previous section, coherent receiver with DSP enables a comprehensive built-in OPM at the receiver end for free. However, OPM devices are supposed to be deployed ubiquitously across the network including intermediate nodes and it is simply too costly and impractical to use full digital coherent receivers with symbol-rate bandwidth for that purpose. A low-cost monitoring solution utilizing reduced-complexity and low speed hardware for distributed monitoring of optical network is in demand.

In [72], an in-band OSNR estimation technique with data-aided DSP utilizing low-bandwidth coherent receivers and low sampling rate is proposed and demonstrated. It is known that Golay sequences are a pair of complimentary sequences $S_1[k]$ and $S_2[k]$ that satisfy the power spectrum property:

$$G[k] = |S_1[k]|^2 + |S_2[k]|^2 = L \tag{14.15}$$

where $S_1[k]$ and $S_2[k]$ are the discrete Fourier transform of the original S_1 and S_2 sequence, respectively, and L is a constant related to the length of each sequence. Neglecting the effect of linear impairments, considering Gaussian distributed noise with zero mean and after some mathematical manipulation and simplification, it is shown that the variance of Golay power spectrum of the received TSs $G[k]$ is proportional to the expected value of noise power spectral density and thus is related to the system SNR as shown in Figure 14.12.

Since the proposed technique utilizes spectral property in frequency domain, SNR can be estimated by using a low sampling speed and low-bandwidth receiver. Experimental verification was demonstrated to monitoring the OSNR of 10 Gbaud PDM-QPSK and PDM-16QAM signals utilizing a low-bandwidth receiver with

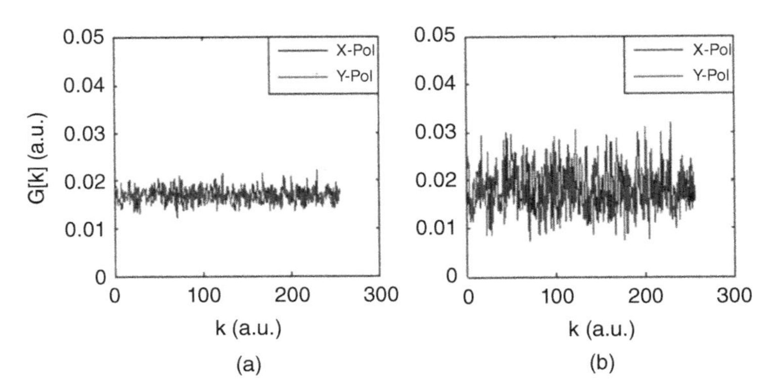

FIGURE 14.12 Frequency spectrum $G[k]$ of Golay pair on both polarizations with (a) OSNR =22 dB and (b) OSNR=12 dB. From Ref. [72], with permission of IEEE.

800-MHz filter working at 2.5 GHz sampling rate. A wide OSNR monitoring range with accuracy within 1 dB for up to 1000-km transmission is achieved.

In [73–75], another three cost-effective OPM techniques that can be deployed at network intermediate nodes for coherent systems are proposed. In [73], an ANN is used in combination with RF spectrum of a directly detected PDM-return-to-zero (RZ)-QPSK signal for OSNR monitoring in the presence of large CD. This is motivated by the fact that new coherent transmissions links will not have inline dispersion compensation and hence OSNR monitoring in the presence of a wide range of unknown CD become a new and unprecedented challenge in OPM research. The input to the ANN is part of the RF spectrum power which can be measured in practice by simple power meters and only direct detection is used as shown in Figure 14.13. Therefore, this proposed technique is low-cost and may serve as a key step toward practical realization of ubiquitous OSNR monitors for coherent links.

In [74], a delay-line interferometer (DLI)-based OSNR monitor is proposed. Since the signal is coherent and experiences constructive and destructive interference in the DLI whereas in-band noise is noncoherent and insensitive to constructive and destructive interference, the power splitting ratio of signal and noise between the constructive and the destructive ports of the DLI are different. Therefore, the OSNR can be estimated from the power ratios of the two ports. In [75], the authors propose an OSNR monitoring approach based on uncorrelated signal generated by optical bandpass filtering and balanced subtraction after photodetection. The power ratio of the uncorrelated signal and original signal can be a measure of OSNR. The method is experimentally verified in 100-Gb/s PDM-QPSK and 50-Gb/s QPSK systems in the presence of CD and PMD effects.

14.3.4 OSNR Monitoring in the Presence of Fiber Nonlinearity

Most of the currently deployed long-haul optical communication systems operate in the weakly nonlinear regime, which is a trade-off between mitigating the effect

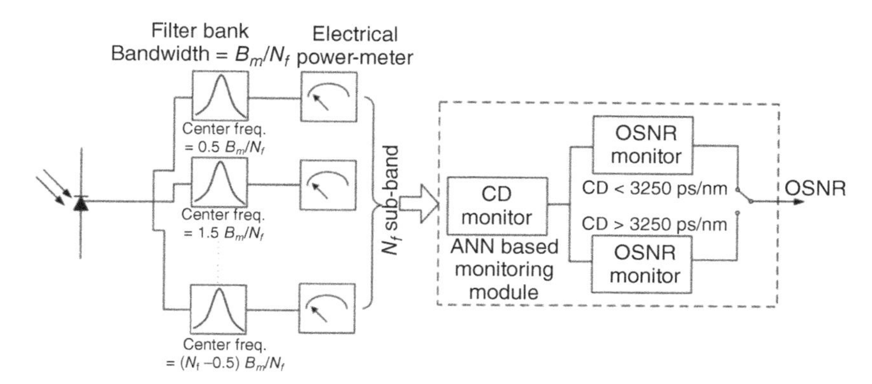

FIGURE 14.13 OSNR monitoring setup for 112 Gbps PDM-RZ-QPSK systems with large inline CD using a bank of electrical filters and power meters covering an overall bandwidth of several GHz. An ANN-based CD monitor is used to determine whether the CD exceeds 3250 ps/nm followed by two separate OSNR monitors covering the low and high CD ranges. (from T. S. R. Shen, Q. Sui, and A. P. T. Lau, OSNR Monitoring for PDM-QPSK Systems With Large Inline Chromatic Dispersion Using Artificial Neural Network Technique, IEEE Photon. Technol. Lett. 24(17), 1564 -1567 (2012), with permission of IEEE).

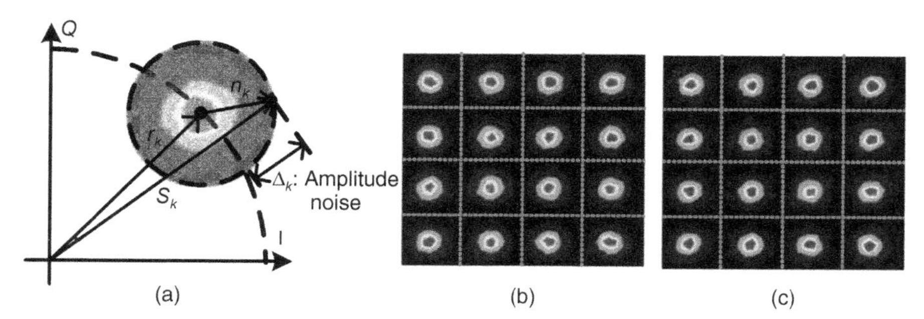

FIGURE 14.14 (a) Graphical illustration of received signal and amplitude noise; Received 16QAM distributions with (b) −4 dBm signal launched power and 18 dB OSNR (c) 4 dBm signal launched power and 26 dB OSNR over an 800-km link. As evident from the figures, amplifier noise and fiber nonlinearity effects will induce similar distortions to the received signal distributions and, therefore, it is not easy to distinguish between them for accurate OSNR monitoring. From Ref. [69].

of ASE noise and fiber nonlinearities. Nonlinear distortions are typically treated as noise and are indistinguishable from amplifier noise (as shown in Figure 14.14) by the standard DSP platform since fiber nonlinearity compensation algorithms such as digital backpropagation [76] is too complex to be realized at present. Therefore, those above-mentioned OSNR estimation techniques based on the SNR of equalized signal or TSs will considerably underestimate the OSNR for long-haul transmission systems.

In [69], the fiber nonlinearity-induced amplitude noise correlation among neighboring symbols is characterized as a quantitative measure of nonlinear distortions which is shown to depend only on signal-launched power but not OSNR and hence fiber nonlinear distortions can be isolated from ASE noise. In this case, nonlinearity-insensitive OSNR monitoring is achieved by incorporating/calibrating such amplitude noise correlations into an EVM-based OSNR estimator.

Since nonlinear distortions can be modeled as complex circularly symmetric additive Gaussian noise with zero-mean for long-haul coherent transmission links without in-line dispersion compensation. Equation 14.11 can be rewritten as

$$r_k = s_k + n_k' = s_k + n_k + v_k \tag{14.16}$$

where $n_k' = n_k + v_k$ consists of ASE noise n_k and nonlinearity-induced distortions v_k.

With the EVM methodology, v_k become addition distortions and thus if we naively use the EVM method by simply measuring the "size" of the "clouds" in the received signal distributions, the OSNR estimates in Equation 14.12 can be rewritten as

$$\mathrm{OSNR}_{\mathrm{Estimated}} = \frac{\mathbf{E}(|\widehat{s}_k|^2)}{\mathbf{E}(|n_k'|^2)} = \frac{\mathbf{E}(|\widehat{s}_k|^2)}{\mathbf{E}(|n_k|^2) + \underbrace{\mathbf{E}(|v_k|^2) + \mathbf{E}(n_k v_k^*) + \mathbf{E}(n_k{}^* v_k)}_{P_{\mathrm{NL}}}}$$

$$= \frac{P_{\mathrm{in}}}{P_{\mathrm{ASE}} + P_{\mathrm{NL}}} \tag{14.17}$$

which can significantly underestimate the true OSNR.

The interaction of fiber nonlinearity, CD and ASE noise will produce distortions such as IFWM that are shown to be correlated across neighboring symbols even after appropriate linear impairment compensation [77]. Denoting Δ_k as the amplitude noise of the kth received symbol, let the autocorrelation function (ACF) of amplitude noise across neighboring symbols be

$$R_\Delta(m) = \mathbf{E}[\Delta_k \Delta_{k+m}] \tag{14.18}$$

It is shown in [77] that the amplitude noise is correlated across neighboring symbols and $|R_\Delta(1)|$ can be used by multiplied by a calibration factor ξ as a measure/estimate of the amount of nonlinear distortions P_{NL} in the received signal r_k. The calibration factor ξ only depends on the transmission distance L. Incorporate the term $|R_\Delta(1)| \times \xi$ in the OSNR estimator in Equation 14.17 and thus a nonlinearity-insensitive OSNR estimation can be obtained by

$$\mathrm{OSNR}_{\mathrm{Estimated}} = \frac{\mathbf{E}(|\widehat{s}_k|^2)}{\mathbf{E}(|n_k'|^2) - |R_\Delta(1)| \times \xi} \tag{14.19}$$

Typical OSNR monitoring results before and after calibration are shown in Figure 14.15. When $|R_\Delta(1)| \times \xi$ is not incorporated, the OSNR is significantly

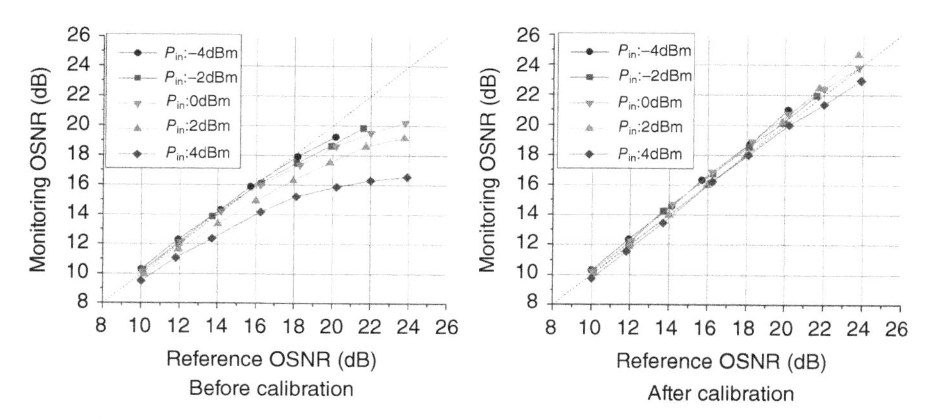

FIGURE 14.15 Monitoring OSNR versus reference OSNR experimentally obtained from a 112 Gb/s PDM-QPSK system for various signal launched powers and OSNR values after 800 km transmission.

underestimated as the nonlinear distortions are treated as ASE noise in the OSNR estimates and the monitoring error generally increases with input power due to enhanced nonlinearity effects. With the calibration based on $|R_{\Delta}(1)| \times \xi$, the OSNR monitoring error is largely reduced.

14.4 INTEGRATING OPM FUNCTIONALITIES IN NETWORKING

One of the major application and motivation for OPM is to obtain real-time detailed conditions of the network to realize impairment-aware routing and improve overall network efficiency. In this regard, it is vital that information from OPM devices be integrated in network management. To this end, Lai et al. [78] demonstrated the use of OSNR monitoring for efficient network routing and enabling packet protection for critical data flows. In their work, OSNR information obtained thorough a DLI-based OPM module is used to send feedback signals to the higher layers for effective packet rerouting and protection. A low OSNR value detected by the OPM device indicates degraded signal quality and vice versa. Depending upon the measured OSNR and the packet-encoded priority, the optical packets are either discarded and rerouted on the alternate path, or forwarded to the final destination port. This approach helps to reduce the penalties incurred due to re-transmission of critical, high-priority data packets. The DLI approach is also extended to WDM system setups with different signal modulation formats [74, 79].

14.5 CONCLUSIONS AND OUTLOOK

OPM continues to be an integral part of optical network operation and imperative for their evolution toward higher speed and improved reliability. As we move toward digital coherent transmissions and beyond, more tools are at our disposal for OPM and their underlying principles inherently merge with other well-researched disciplines such as channel estimation in traditional copper-wire/wireless communications. In addition, the emerging dominance of data centers/cloud computing driven traffic has fueled the need for OPM to not only manage network faults, but also provide information on real-time network conditions for implement impairment-aware routing and software-defined networking (SDN). OPM and related optical network functionalities are expected to play an increasing role in shaping next generation optical networks.

ACKNOWLEDGMENTS

The authors acknowledge the support of the Universiti Sains Malaysia under Research University grant (1001/PELECT/814203).

REFERENCES

1. Agrawal GP. Fiber-Optic Communication Systems. 3rd ed. New York: John Wiley and Sons; 2002.
2. Essiambre RJ, Winzer PJ. Advanced optical modulation formats. In: Kaminow IP, Li T, Willner AE, editors. Optical Fiber Telecommunications V-B. New York: Academic Press; 2008. (Chapter 2).
3. Ramaswami R, Sivarajan KN. Optical Networks: A Practical Perspective. 2nd ed. San Francisco: Morgan Kaufmann; 2002.
4. Willner AE, Pan Z, Yu C. Optical performance monitoring. In: Kaminow IP, Li T, Willner AE, editors. Optical Fiber Telecommunications V-B. New York: Academic Press; 2008. (Chapter 7).
5. Kilper DC, Bach R, Blumenthal DJ, Einstein D, Landolsi T, Ostar L, Preiss M, Willner AE. Optical performance monitoring. *IEEE/OSA Journal of Lightwave Technology* 2004;22(1):294–304.
6. Khan FN, Shen TSR, Zhou Y, Lau APT, Lu C. Optical performance monitoring using artificial neural networks trained with empirical moments of asynchronously sampled signal amplitudes. *IEEE Photonics Technology Letters* 2012;24(12):982–984.
7. Willner AE. The optical network of the future: can optical performance monitoring enable automated, intelligent and robust systems? Optics and Photonics News, Mar 2006.
8. Huang Y, Heritage JP, Mukherjee B. Connection provisioning with transmission impairment consideration in optical WDM networks with high-speed channels. *IEEE/OSA Journal of Lightwave Technology* 2005;23(3):982–993.
9. Martinez R, Pinart C, Cugini F, Andriolli N, Valcarenghi L, Castoldi P, Wosinska L, Comellas J, Junyent G. Challenges and requirements for introducing impairment-awareness into the management and control planes of ASON/GMPLS WDM networks. *IEEE Communications Magazine* 2006;44:76–85.
10. Strand J, Chiu A, Tkach R. Issues for routing in the optical layer. *IEEE Communications Magazine* 2001;39(2):81–87.

11. Bernstein G, Rajagopalan B, Saha D. Optical Network Control. Boston: Addison-Wesley; 2004.
12. Bouillet E, Ellinas G, Labourdette J-F, Ramamurthy R. Path Routing in Mesh Optical Networks. Chichester, West Sussex: John Wiley and Sons; 2007.
13. http://www.optoplex.com/Optical_Channel_Monitor.htm. Accessed 2015 Oct 5.
14. Becker PC, Olsson NA, Simpson JR. Erbium-Doped Fiber Amplifiers. Boston: Academic Press; 1999.
15. Chomcyz B. Planning Fiber Optic Networks. New York: McGraw Hill; 2009.
16. Willner AE, Hoanca B. Fixed and tunable management of fiber chromatic dispersion. In: Kaminow IP, Li T, editors. Optical Fiber Telecommunications IV B. New York: Academic Press; 2002. (Chapter 14).
17. Kogelnik H, Jopson RM, Nelson LE. Polarization-mode dispersion. In: Kaminow IP, Li T, editors. Optical Fiber Telecommunications IV B. New York: Academic Press; 2002. (Chapter 15).
18. Forgheiri F, Tkach RW, Chraplyvy AR. Fiber nonlinearities and their impact on transmission systems. In: Kaminow IP, Koch TL, editors. Optical Fiber Telecommunications III A. New York: Academic Press; 1997. (Chapter 8).
19. Hui R, Sullivan MO'. Fiber Optic Measurement Techniques. New York: Academic Press; 2009.
20. Chan CCK. Optical Performance Monitoring. New York: Academic Press; 2010.
21. Hanik N, Gladisch A, Caspar C, Strebel B. Application of amplitude histograms to monitor performance of optical channels. *Electronics Letters* 1999;35(5): 403–404.
22. Shake I, Takara H, Uchiyama K, Yamabayashi Y. Quality monitoring of optical signals influenced by chromatic dispersion in a transmission fiber using averaged Q-factor evaluation. *IEEE Photonics Technology Letters* 2001;13(4):385–387.
23. Shake I, Takara H, Kawanishi S. Technology for flexibly monitoring optical signal quality in transparent optical communications. *Journal of Optical Networking* 2007;6(11):1229–1235.
24. Li Z, Lu C, Wang Y, Li G. In-service signal quality monitoring and multi-impairment discrimination based on asynchronous amplitude histogram evaluation for NRZ-DPSK systems. *IEEE Photonics Technology Letters* 2005;17(9):1998–2000.
25. Li Z, Li G. Chromatic dispersion and polarization-mode dispersion monitoring for RZDPSK signals based on asynchronous amplitude-histogram evaluation. *IEEE/OSA Journal of Lightwave Technology* 2006;24(7):2859–2866.
26. Kozicki B, Takuya O, Hidehiko T. Optical performance monitoring of phase-modulated signals using asynchronous amplitude histogram analysis. *IEEE/OSA Journal of Lightwave Technology* 2008;26(10):1353–1361.
27. Luís RS, Teixeira A, Monteiro P. Optical signal-to-noise ratio estimation using reference asynchronous histograms. *IEEE/OSA Journal of Lightwave Technology* 2009;27(6):731–743.
28. Shen TSR, Meng K, Lau APT, Dong ZY. Optical performance monitoring using artificial neural network trained with asynchronous amplitude histograms. *IEEE Photonics Technology Letters* 2010;22(22):1665–1667.

29. Dods SD, Anderson TB. Optical performance monitoring technique using delay tap asynchronous waveform sampling. Proceedings of Optical Fiber Communication Conference; Anaheim, CA, paper OThP5; Mar 2006.
30. Li Z, Jian Z, Cheng L, Yang Y, Lu C, Lau APT, Yu C, Tam HY, Wai PKA. Signed CD Monitoring of 100Gbit/s CS-RZ-DQPSK Signal by Evaluating the Asymmetry Ratio of Delay-tap Sampling. *Opt Express* 2010;18(3):3149–3157.
31. Khan FN, Lau APT, Li Z, Lu C, Wai PKA. Statistical analysis of optical signal-to-noise ratio monitoring using delay-tap sampling. *IEEE Photonics Technology Letters* 2010;22(3):149–151.
32. Khan FN, Lau APT, Li Z, Lu C, Wai PKA. OSNR monitoring for RZ-DQPSK systems using half-symbol delay-tap sampling technique. *IEEE Photonics Technology Letters* 2010;22(11):823–825.
33. Kozicki B, Maruta A, Kitayama K. Transparent performance monitoring of RZ-DQPSK systems employing delay-tap sampling. *Journal of Optical Networking* 2007;6(11):1257–1269.
34. Choi HY, Takushima Y, Chung YC. Optical performance monitoring technique using asynchronous amplitude and phase histograms. *Opt Express* 2009; 17(26):23953–23958.
35. Tan MC, Khan FN, Al-Arashi WH, Zhou Y, Lau APT. Simultaneous optical performance monitoring and modulation format/bit-rate identification using principal component analysis. *IEEE/OSA Journal of Optical Communications and Networking* 2014;6(5):441–448.
36. Wu X, Jargon JA, Wang C-M, Paraschis L, Willner AE. Experimental comparison of performance monitoring using neural networks trained with parameters derived from delay-tap plots and eye diagrams. Proceedings of Optical Fiber Communication Conference; San Diego, CA, paper JThA17; Mar 2010.
37. Anderson TB, Kowalczyk A, Clarke K, Dods SD, Hewitt D, Li JC. Multi-impairment monitoring for optical networks. *IEEE/OSA Journal of Lightwave Technology* 2009;27(16):3729–3736.
38. Jong KC, Tsao HW, Lee SL. Q-factor monitoring of optical signal-to-noise ratio degradation in optical DPSK transmission. *Electronics Letters* 2008;44(12):761–763.
39. Khan FN, Lau APT, Lu C, Wai PKA. Chromatic dispersion monitoring for multiple modulation formats and data rates using sideband optical filtering and asynchronous amplitude sampling technique. *Opt Express* 2011;19(2):1007–1015.
40. Yu Y, Zhang B, Yu C. Optical signal to noise ratio monitoring using single channel sampling technique. *Opt Express* 2014;22(6):6874–6880.
41. Amrani A, Junyent G, Prat J, Comellas J, Ramdani I, Sales V, Roldán J, Rafel A. Performance monitor for all-optical networks based on homodyne spectroscopy. *IEEE Photonics Technology Letters* 2000;12(11):1564–1566.
42. Yu Q, Pan Z, Yan L-S, Willner AE. Chromatic dispersion monitoring technique using sideband optical filtering and clock phase-shift detection. *IEEE/OSA Journal of Lightwave Technology* 2002;20(12):2267–2271.
43. Pan Z, Xie Y, Havstad SA, Yu Q, Willner AE, Grubsky V, Starodubov DS, Feinberg J. Real-time group-velocity dispersion monitoring and automated compensation without modifications of the transmitter. *Optics Communications* 2004;230(1–3):145–149.
44. Lize′ YK, Christen L, Yang J-Y, Saghari P, Nuccio S, Willner AE, Kashyap R. Independent and simultaneous monitoring of chromatic and polarization-mode dispersion in OOK and DPSK transmission. *IEEE Photonics Technology Letters* 2007;19(1):3–5.

45. Wang Y, Hu S, Yan L, Yang J-Y, Willner AE. Chromatic dispersion and polarization mode dispersion monitoring for multi-level intensity and phase modulation systems. *Opt Express* 2007;15(21):14038–14043.
46. Rossi G, Dimmick TE, Blumenthal DJ. Optical performance monitoring in reconfigurable WDM optical networks using subcarrier multiplexing. *IEEE/OSA Journal of Lightwave Technology* 2000;18(12):1639–1648.
47. Petersen MN, Pan Z, Lee S, Havstad SA, Willner AE. Online chromatic dispersion monitoring and compensation using a single inband subcarrier tone. *IEEE Photonics Technology Letters* 2002;14(4):570–572.
48. Liu A, Pendock GJ, Tucker RS. Improved chromatic dispersion monitoring using single RF monitoring tone. *Opt Express* 2006;14(11):4611–4616.
49. Zhao J, Lau APT, Qureshi KK, Li Z, Lu C, Tam HY. Chromatic dispersion monitoring for DPSK systems using RF power spectrum. *IEEE/OSA Journal of Lightwave Technology* 2009;27(24):5704–5709.
50. Jung DK, Kim CH, Chung YC. OSNR monitoring technique using polarization-nulling method. Proceedings of Optical Fiber Communication Conference; Baltimore, MD; Mar 2000.
51. Cheung M-H, Chen L-K, Chan C-K. PMD-insensitive OSNR monitoring based on polarization-nulling with off-center narrow-band filtering. *IEEE Photonics Technology Letters* 2004;16(11):2562–2564.
52. Lee JH, Choi HY, Shin SK, Chung YC. A review of the polarization-nulling technique for monitoring optical-signal-to-noise ratio in dynamic WDM networks. *IEEE/OSA Journal of Lightwave Technology* 2006;24(11):4162–4171.
53. Sui Q, Lau APT, Lu C. OSNR Monitoring in the presence of first-order PMD using polarization diversity and DSP. *IEEE/OSA Journal of Lightwave Technology* 2010;28(15):2105–2114.
54. Kikuchi N. Analysis of signal degree of polarization degradation used as control signal for optical polarization mode dispersion compensation. *IEEE/OSA Journal of Lightwave Technology* 2010;19(4):480–486.
55. Yang J-Y, Chitgarha MR, Zhang L, Wang J, Paraschis L, Willner AE. Optical monitoring of PMD accumulation on a Pol-MUX phase-modulated signal using degree-of-polarization measurements. *Opt Lett* 2011;36(16):3215–3217.
56. Khan FN, Zhou Y, Lau APT, Lu C. Modulation format identification in heterogeneous fiber-optic networks using artificial neural networks. *Optical Society of America (OSA) Optics Express* 2012;20(11):12422–12431.
57. Khan FN, Zhou Y, Sui Q, Lau APT. Non-data-aided joint bit-rate and modulation format identification for next-generation heterogeneous optical networks. *Optical Fiber Technology* 2014;20(2):68–74.
58. Chung YC. A review of optical performance monitoring techniques. International Conference on Photonics in Switching; Sapporo; Aug 2008.
59. Morgan T, Zhou YR, Lord A, Anderson T. Non-intrusive simultaneous measurement of OSNR, CD and PMD on live WDM system. Proceedings of Optical Fiber Communication Conference; Los Angeles, CA, paper NTu2E.4; Mar 2012.
60. Ip E, Lau APT, Barros DJF, Kahn JM. Coherent detection in optical fiber systems. *Opt Express* 2008;16(2):753–791.

61. Schmogrow R, Nebendahl B, Winter M, Josten A, Hillerkuss D, Koenig S, Meyer J, Dreschmann M, Huebner M, Koos C, Becker J, Freude W, Leuthold J. Error vector magnitude as a performance measure for advanced modulation formats. *IEEE Photon Technol Lett* 2012;24(1):61–63.
62. Woodward SL, Nelson LE, Feuer MD, Zhou X, Magill PD, Foo S, Hanson D, Sun H, Moyer M, O'Sullivan M. Characterization of real-time PMD and chromatic dispersion monitoring in a high-PMD 46-Gb/s transmission system. *IEEE Photon Technol Lett* 2008;20(24):2048–2050.
63. Hauske FN, Kuschnerov M, Spinnler B, Lankl B. Optical performance monitoring in digital coherent receivers. *J Lightwave Technol* 2009;27(16):3623–3631.
64. Ives DJ, Thomsen BC, Maher R, Savory S. Estimating OSNR of equalised QPSK signals. Proceedings of European Conference and Exhibition on Optical Communication (ECOC), paper Tu.6.A.6; 2011.
65. Faruk MS, Kikuchi K. Monitoring of optical signal-to-noise ratio using statistical moments of adaptive-equalizer output in coherent optical receivers. Opto-Electronics and Communications Conference; 2011. p 233–234.
66. Faruk MS, Mori Y, Kikuchi K. In-band estimation of optical signal-to-noise ratio from equalized signals in digital coherent receivers. *Photonics Journal, IEEE* 2014;6(1):1–9.
67. Zhu C, Tran AV, Chen S, Du L, Do C, Anderson T, Lowery A, Skafidas E. Statistical moments-based OSNR monitoring for coherent optical systems. *Opt Exp* 2012;20(16):17711–17721.
68. Nebendahl B, Schmogrow R, Dennis T, Josten A, Hillerkuss D, Koenig S, Meyer J, Dreschmann M, Winter M, Huebener M, Freude W, Koos C, Leuthold J. Quality metrics in optical modulation analysis: EVM and its relation to Q-factor, OSNR, and BER. Asia Communications and Photonics Conference(ACP); Guangzhou China, paper AF3G.2; 2012.
69. Dong Z, Lau APT, Lu C. OSNR monitoring for QPSK and 16-QAM systems in presence of fiber nonlinearities for digital coherent receivers. *Opt Exp* 2012;20(17): 19520–19534.
70. Pittalà F, Hauske FN, Ye Y, Gonzalez NG, Monroy IT. Joint PDL and in-band OSNR monitoring supported by data-aided channel estimation. Proceedings of OFC'12 ; Los Angeles, paper OW4G; Mar 2012.
71. Do C, Tran AV, Zhu C, Hewitt D, Skafidas E. Data-aided OSNR estimation for QPSK and 16-QAM coherent optical system. *Photonics Journal, IEEE* 2013;5(5):6601609–6601618.
72. Do C, Zhu C, Tran AV. Data-aided OSNR estimation using low-bandwidth coherent receivers. *IEEE Photon Technol Lett* 2014;26(13):1291–1294.
73. Shen TSR, Sui Q, Lau APT. OSNR monitoring for PM-QPSK systems with large inline chromatic dispersion using artificial neural network technique. *IEEE Photon Technol Lett* 2014;26(13):1291–1294.
74. Chitgarha MR, Khaleghi S, Daab W, Ziyadi M, Mohajerin-Ariaei A, Rogawski D, Tur M, Touch JD, Vusirikala V, Zhao W, Willner AE. Demonstration of WDM OSNR performance monitoring and operating guidelines for Pol-Muxed 200-Gbit/s 16-QAM and 100-Gbit/s QPSK data channels. OFC, OTh3B.6; 2013.
75. Yu Y, Yu C. Dispersion insensitive optical signal to noise ratio monitoring of PDM signal by using uncorrelated signal power. *Opt Exp* 2014;22(11):12823–12828.

76. Ip E. Nonlinear compensation using backpropagation for polarization-multiplexed transmission. *J Lightwave Technol* 2010;28(6):939–951.
77. Lau APT, Rabbani S, Kahn JM. On the statistics of intra-channel four-wave mixing in phase-modulated optical communication systems. *J Lightwave Technol* 2008;26(14):2128–2135.
78. Lai CP, Yang J-Y, Garg A, Wang MS, Chitgarha M, Willner AE, Bergman K. Experimental demonstration of packet-rate 10-Gb/s OOK OSNR monitoring for QoS-aware cross-layer packet protection. *Opt Express* 2011;19(16):14871–14882.
79. A. S. Ahsan, M. S. Wang, M. R. Chitgarha, D. C. Kilper, A. E. Willner and K. Bergman, Autonomous OSNR Monitoring and Cross-Layer Control in a Mixed Bit-Rate and Modulation Format System Using Pilot Tones, *Proceedings of Advanced Photonics for Communications*, San Diego, CA, Jul. 2014, paper NT4C.3.

15

RATE-ADAPTABLE OPTICAL TRANSMISSION AND ELASTIC OPTICAL NETWORKS

PATRICIA LAYEC, ANNALISA MOREA, YVAN POINTURIER, AND JEAN-CHRISTOPHE ANTONA

Bell Laboratories, Nokia, Nozay, France

15.1 INTRODUCTION

Fiber optical systems are now present in multiple segments of the network, as depicted in Figure 15.1. The segments are derived from different requirements and functionalities associated with each segment. This leads to a plurality of technology selections; for instance, the core segment relies on the coherent transponder technology while the access segment with passive optical networks (PONs) uses low-cost transponders with noncoherent detection.

Though different technologies are used, each segment faces the need of high-capacity networks in order to support the growth of traffic as well as the change in customer usage with the new era of cloud networking and connected devices. Due to the limited available bandwidth in an optical fiber, a higher data rate translates into a need for higher spectral efficiency. This can drastically reduce the reach of optical signals and requires the use of more optoelectronic (OEO) regeneration resources. As a result, traditional optical networks with a fixed transmission rate do not scale well and are, therefore, not economically viable. To meet these challenges, future optical networks have generated interest in scalable, reconfigurable, and sustainable solutions, known as elastic optical networks (EONs). This elastic concept can be applicable to the various segments and some example of applications are shown later in this chapter.

Enabling Technologies for High Spectral-efficiency Coherent Optical Communication Networks, First Edition. Edited by Xiang Zhou and Chongjin Xie.
© 2016 John Wiley & Sons, Inc. Published 2016 by John Wiley & Sons, Inc.

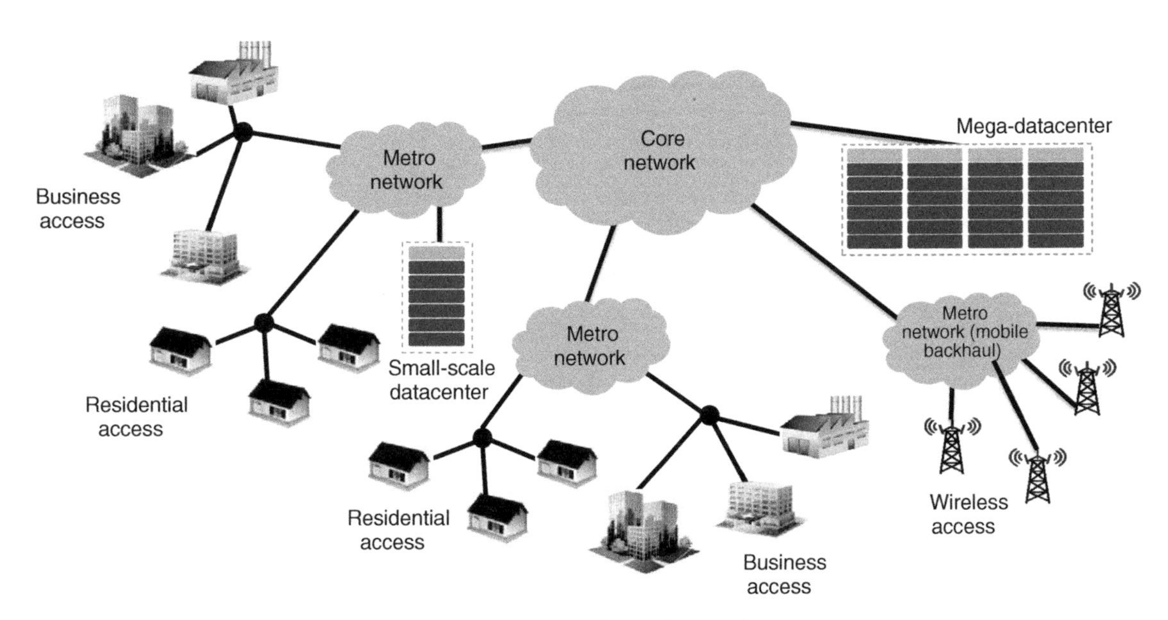

FIGURE 15.1 Overview of optical network segments.

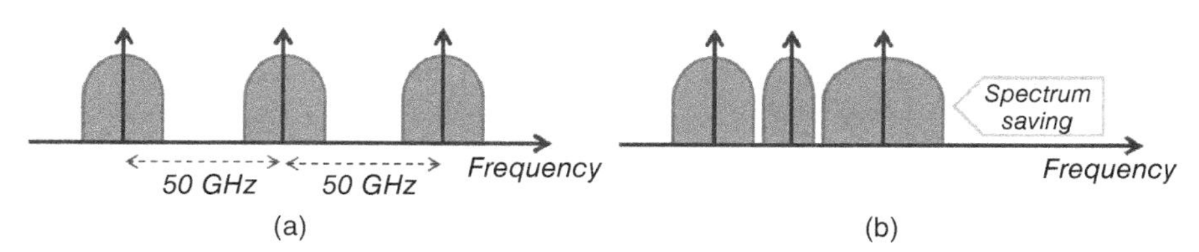

FIGURE 15.2 (a) Traditional fixed networks with a 50 GHz frequency grid; (b) elastic networks with flex-grid using variable channel spacing and symbol rate.

Focusing now on core networks, unless specified otherwise, traditional fixed networks divide the spectrum bandwidth into a set of parallel channels of fixed channel spacing by steps of 50 or 100 GHz. In addition, the modulation format and symbol rate (or equivalently the spectrum occupancy) are fixed for a selected generation and are also generally associated with a data rate such as legacy 10 Gb/s or current 100 Gb/s. In the latter case, a 100 Gb/s transmission classically uses a polarization division multiplexed (PDM) QPSK modulation with a baud rate of 28 Gbaud and a hard-decision forward error correction (FEC) with 7% overhead. In contrast, EONs are capable of tuning one or several transmission parameters to adjust their data rate and their reliability to the connection characteristics such as lightpath distance, transmission impairments, capacity demand. As shown in Figure 15.2, one of the most attractive handles for elastic networking is the possibility to use a flexible frequency spacing between channels, also called flex-grid, together with a tunable symbol rate. This allows to optimize the spectrum usage and to gain in spectral efficiency. Other options of key interest are the ability to adjust modulation format and/or FEC overhead to (dynamically) match the data rate to the transmission quality.

Needless to say, elasticity is empowered by software-defined devices such as transponders where the advent of coherent detection with digital signal processing, high-speed digital-to-analog converter (DAC) and analog-to-digital converter (ADC) elements have drastically enhanced the programmable flexibility. In an EON, the software-defined devices are under the supervision of a management tool with a local controller to adapt the different elements of the network in a harmonized fashion after, for instance, the selection of flex-grid and modulation format. The local controller may also be run by a control plane for more automated processes.

Such elasticity concept is very common and successful in other telecommunication industries, such as wireless and copper-based fixed access networks, but is a newcomer in optical transport. To summarize, elasticity allows both a more dynamic resource management and operation as close as possible to physical limits; the benefits are one or a combination of the following: (i) increased capacity, (ii) cost reduction, (iii) reduced power consumption, and (iv) enhanced scalability. These benefits are assessed in a number of scenarios later in this chapter.

15.1.1 History of Elastic Optical Networks

Optical networks have a fourfold heterogeneity in terms of (i) connection length, from a few hundred to several thousands of kilometers, (ii) deployed infrastructures, with the coexistence of terminal or transmission equipments from different generations (different data-rates, generations and/or types of fiber, amplifier type or technology), (iii) capacity demand, from tenths to tens of Gb/s, and (iv) connection duration, from hour-long (or shorter) to quasi-permanent connection. This has generated interest in rate-adaptive networks.

The connection length and capacity demand heterogeneities have first been addressed by mixed-line rate networks, that is, low bit rate transponders are provisioned for the longest lightpath while higher bit rate transponders are used for shorter distances [1]. However, managing multiple types of devices (one for each data rate) is cumbersome for the operators. Committed transponders are provisioned for a given traffic and network conditions; hence, mixed-line rate networks suffer from a lack of flexibility and scalability when traffic and connections evolve.

To ensure a competitive solution, EONs have been introduced in 2008 [2] where one transponder device is able to deliver multiple bit rates per connection needs. This is also known as a universal transponder. Initially, the transponder was made tunable by delivering "just-enough" spectrum to each connection demand with the ability to increase or decrease the bit rate dynamically [2]. This concept was based on OFDM technology and the network architecture called SLICE. Next, the ability to select between different modulation formats [3, 4] has been shown to be well coupled with a distance adaptation according to the physical impairments. Indeed, higher-order modulations achieve high data rate but require a good signal-to-noise ratio (SNR); hence, 16QAM and even more 64QAM are best suited for short distances. In addition, the FEC overhead may also be dynamically adjusted [5] to match the channel quality and to avoid the over-provisioning of margins. The larger the FEC overhead, the better the transmission reliability.

However, when an elastic transponder configured for short-reach high-capacity, it is not operating at its maximum data rate and it may look like part of the available capacity is wasted. To address this issue, the concept of multiflow transponders (or also known as sliceable transponders) has been introduced in 2011 [6]. The aim is to transmit from one source node to one or multiple destination nodes by having a (dynamic) sharing of bandwidth. It is worth noting that different modulations, FEC overheads, and baud rates may be selected for each of the flows. This multiflow concept is especially relevant for very high bit rate transponders such as 400G, 1 T to benefit from cost-savings [6, 7]. Indeed, the network roll-out of an operator will most likely use these devices at a low bit rate with multiple destinations during the first years of deployment before upgrading them to a higher bit rate, hence with a lower number of destinations, when the traffic is growing.

In the remainder of this chapter, we discuss the changes in technology option or design to support the elastic concept in an optical network. Next, we introduce in detail some practical aspects of the elastic transponder and aggregation devices and show a first green application of elasticity. Leveraging on dimensioning resource allocation tools and techno-economic studies, we dedicate the next section to the foreseen short- to medium-term opportunities in core networks before opening on longer term opportunities with elastic burst optical networks.

15.2 KEY BUILDING BLOCKS

As previously mentioned, EONs would bring some benefits in terms of cost, capacity, and energy consumption, but making them real requires evolutions of hardware, software as well as control plane, as illustrated by Figure 15.3.

The hardware elements contain three major blocks – depicted from the optical layer to the IP router layer: (i) optical cross-connects (OXCs) with the optional capability to handle flexible channel spacing and spectrum allocation so as to support flex-grid scenarios, (ii) transponders for which the choice of an appropriate modulation, symbol rate, FEC overhead is critical to achieve a good efficiency, (iii) elastic aggregation interfaces that have the ability to deliver variable bit rate by switching off lanes and/or by sending a variable bit rate per lane; this is in complement to existing grooming and aggregation features using digital capabilities of the optical transport network (OTN) [8]. Indeed, the OTN comprises an optical layer and a digital layer: the optical channel payload unit (OPU), the optical channel digital unit (ODU) that is a method for encapsulating data and the optical channel transport unit (OTU) corresponding to the resulting line rate. OTN may encompass flexible functionalities such as switching, multiplexing, and inverse multiplexing. For details, the reader should refer to [8], as it is out of the scope of this chapter. The OTN layers are depicted in Figure 15.4.

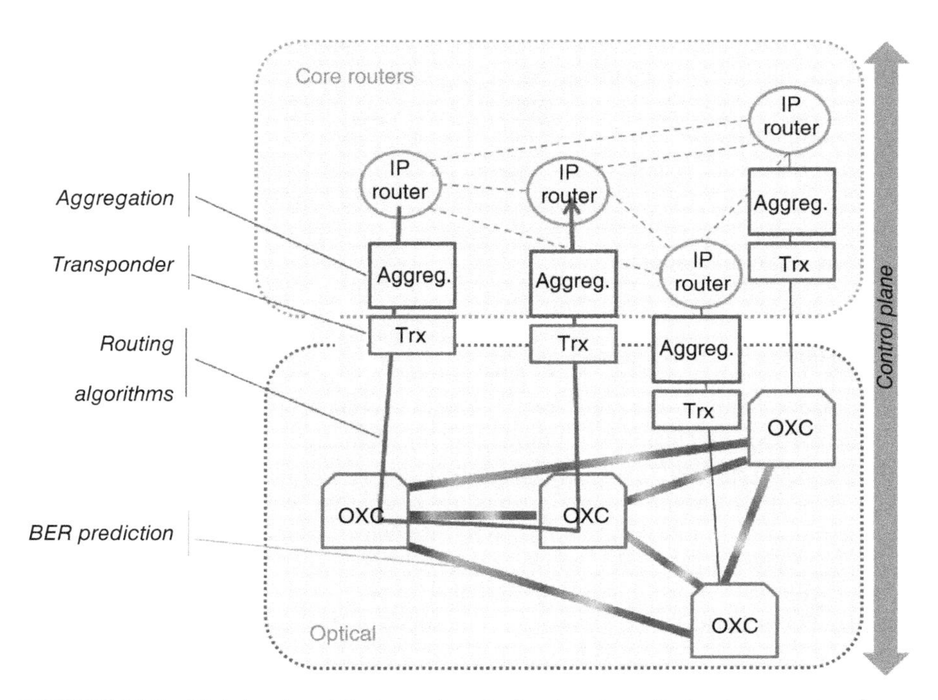

FIGURE 15.3 High-level architecture of optical networks with hardware and software challenges.

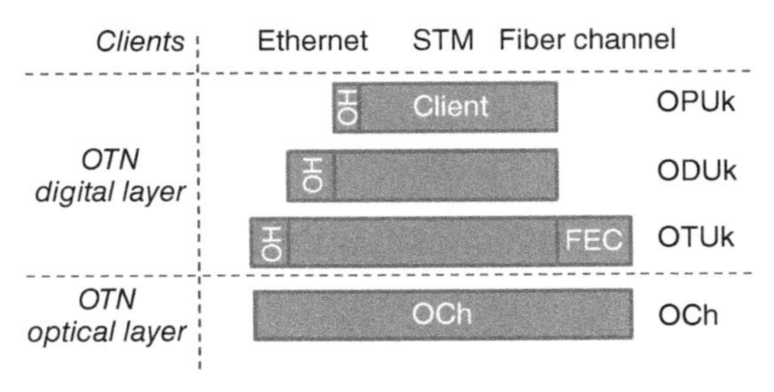

FIGURE 15.4 OTN hierarchy. The payload in the OTUk for $k = \{1,2,3, \text{ or } 4\}$ is approximately 2.5, 10, 40, and 100 Gb/s respectively.

The software modules also need to evolve as the hardware elements become more flexible. In particular, during the planning phase it is of high importance to understand the impairments along the candidate lightpaths in order to be able to select the right transmission parameters (e.g., data rate, bandwidth, path characteristics) that allow the digital client to be carried with a satisfactory bit error ratio (BER). This allows the optimization of the elastic system design on a point-to-point lightpath. At the network level, the development of novel impairment-aware algorithms for routing that include rate selection and spectrum allocation is essential to estimate the number of optoelectronic interfaces to be provisioned and optimize the cost of the whole network deployment. In addition, online optimization can be seen as an extension of planning tool with the new capability of handling dynamic management of the network connections.

In this context of dynamic behavior (e.g., accommodating new connection requests or traffic variations), the control plane would also need some extensions.

Specific implementation issues and challenges of the building blocks are further described in the following sections.

15.2.1 Optical Cross-Connect

OXCs, made essentially of optical amplifiers and interconnected wavelength-selective switches (WSS), are capable of multiplexing and demultiplexing signals in both the wavelength and space domain. The incoming signals from direction d at wavelength w can be switched to direction d' potentially at another wavelength w'. OXCs are important in a mesh network to improve its efficiency and capacity through a better ability for grooming, an improved network reliability and scalability.

Deployed WSS are switching devices that filter the spectrum based on a 50 or 100 GHz ITU grid (hence creating penalties if the signal is larger than allowed) while the new WSS generation (already commercially available) supports a dynamic bandwidth allocation. The spectrum can now be switched by steps of about 12.5 GHz. The most predominant technology to support this fine bandwidth granularity involves the liquid crystal on silicon (LCoS) technology [9].

The accommodation of flex-grid technology (see [10] for standard flexible grid definition) with the new WSS generation allows the spectrum slots to be concatenated so as to create much larger spectrum chunks with no filtering within it. It thus offers a good spectral efficiency if the channels are compactly packed. In addition, EONs relying on the flex-grid handle may need a mechanism, also known as spectrum defragmentation, which is capable of removing unused small-frequency blocks by moving the central frequency of established connections. This permits to generate a spectrum block that is sufficiently large to accommodate a new demand.

15.2.2 Elastic Transponder

Transponders are capable of sending the optical signals on the fiber media channel as well as performing the signal processing both at the emission and reception. When a transponder changes one or several of its transmission parameters, the robustness to channel impairments also changes. Therefore, we present hereafter a few examples of trade-offs between data rate and optical reach.

15.2.2.1 Modulation Formats It is well-known that the distance to be covered is highly dependent upon the selected modulation. This is due to the reduction of the minimal distance between two points of the constellation, which reduces the resilience to channel impairments. For instance, going from a PDM-QPSK up to a PDM-16QAM transmission doubles the data rate at the cost of an optical reach divided by a factor of 5. To alleviate this steep trade-off, the flexible transponder may support additional modulation formats and in particular more complex formats such as recently investigated four-dimensional coded modulations [11] making use of set partitioning (SP). Set partitioning comes from Ungerboeck [12] where the idea is to partition the constellation points into smaller subsets that fulfill an increase of the minimal Euclidean distance of the original constellation as well as a decrease of the resulting data-rate. This type of coded modulation exhibits the advantage of reusing some DSP algorithms designed for classical 8QAM, 16QAM, and other formats: for instance, 8-SP-16QAM utilizes half the symbols of the 16QAM constellation, thus reducing the data-rate by 25%; it is rather equivalent to 8QAM in terms of performance and data-rate while requiring the same DSP than 16QAM. Alternatively, time-domain hybrid-QAM [13] is capable to show a very fine granularity of spectral efficiency due to its own construction. Indeed, the principle is to split the frame into multiple time slots, each slot is filled from the set of x-QAM modulations so as to allocate a variable portion of low- and high-order QAM (e.g., 2 out of 3 slots filled with 32QAM and 1 slot filled with 64QAM results in a spectral efficiency of 5.33 bit/symbol).

Figure 15.5 presents a set of candidate modulation formats with a fine incremental data rate. It can be seen that the optical reach progressively decreases as the data rate grows; a transponder with this set of modulations is therefore suitable for a wide variety of field conditions.

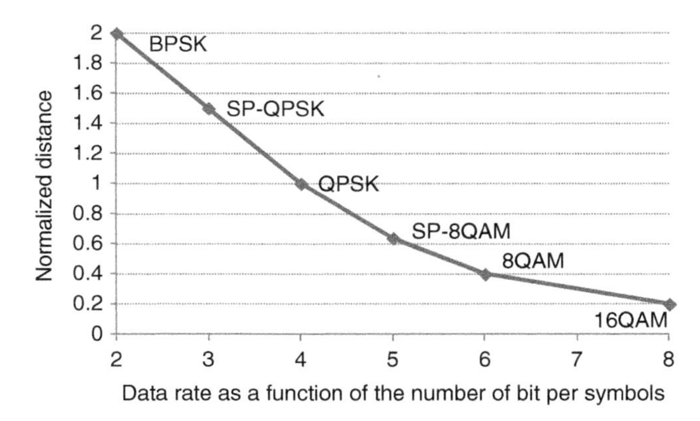

FIGURE 15.5 Example of trade-off between data rate and distance for standard modulations with dual polarization.

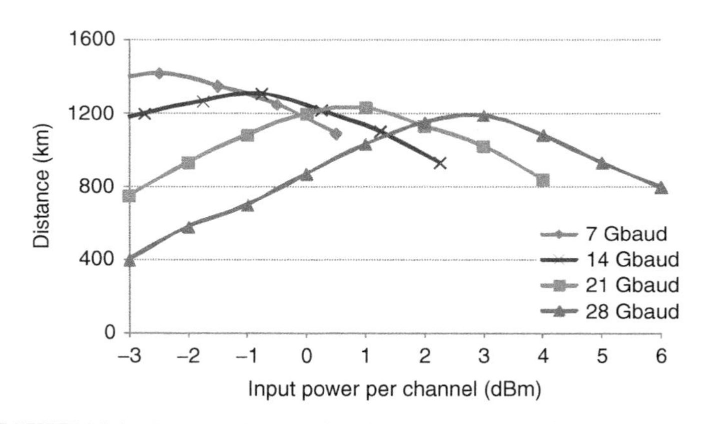

FIGURE 15.6 Reach estimation for different symbol rate in a 50 GHz grid.

15.2.2.2 Symbol Rate The variation in symbol rate (equivalently called baud rate) changes the spectrum occupancy. Therefore, in nondispersion-managed system, a PDM-QPSK transmission system over a fixed channel spacing of 50GHz grid is only weakly changing the optical reach [14]. This is illustrated in Figure 15.6, where the reach is estimated at optimal power based on split-step Fourier method numerical simulation. It can be seen that a lower symbol rate is more robust to the tolerance of amplified spontaneous emission (ASE) noise while a higher symbol rate shows less nonlinear degradation, which interestingly almost cancels out the variation of sensitivity to ASE noise.

15.2.2.3 Forward Error Correction (FEC) The FEC inserts parity bits to the initial information bits in order to improve the reliability of the transmission but at the cost of a lower effective throughput. This is also called an FEC coding gain that translates into an increased SNR in order to achieve the same target BER performance.

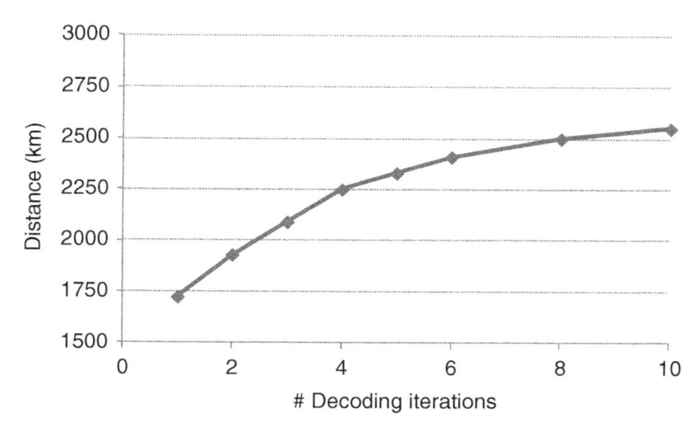

FIGURE 15.7 Impact of the number of decoding iterations on the optical reach.

Playing on the coding gain, either by varying the number of decoding iterations in a soft-decision FEC [15] or directly by adapting the overhead of the FEC [16], results in the ability to trade-off capacity versus distance and power-consumption.

As evidenced in Figure 15.7, the optical reach depends on the number of LDPC decoding iterations [15]. This was computed for an LDPC (23,125, 20,000) concatenated with a Reed–Solomon of 7% overhead (the hard FEC defined in the OTN standard). The optical reach was obtained for PDM-QPSK transmission at 100 Gb/s with 16 dB OSNR sensitivity (within 0.1 nm) in back to back, 100 km-long spans with 22 dB loss, separated by optical amplifiers with 6 dB noise figure.

15.2.3 Elastic Aggregation

Aggregation interconnects several ports of an IP router to a WDM fiber pair. Today's metro/core optical networks are optimized for (relatively) static conditions and in the event of client traffic changes (e.g., bandwidth, new demand) unforeseen in the planning process, transport WDM signals may operate well below their maximum capacity in some portions of the network, while demonstrating insufficient capacity in others. With the introduction of elastic WDM transmission with flexible data rates, the transport side can become more responsive to data rate requirements from the client side, thus bringing additional opportunities.

Two major setups of the IP core network on top of the optical WDM transport can be distinguished:

- IP over WDM – a two-layer network interconnecting directly the IP layer with core routers upon the WDM layer. The flexible switching/grooming capabilities are in the IP layer while the WDM signals are just carried within an OTUk ($k = 1, 2, 3$, or 4) [8]. At an intermediate node, the transit traffic that needs sub-wavelength flexibility is processed at the IP layer.

- IP over OTN over WDM – a three-layer network relying on an additional intermediate OTN layer. The aim is to reduce the number of expensive router ports and to offload the transit traffic from the IP routers.

  The OTN layer possesses digital capabilities such as ODU multiplexing, which can be seen as a first level of adaptation to the client data rate. The goal of ODU multiplexing is to aggregate lower-order ODUs into a higher-order ODU. In addition, more flexibility has been introduced in ODU by varying the size of the aggregate packet flow mapped into a resizable ODUflex (GFP) [8, 17], where the goal was to fill the gap of the ODU hierarchy. However, the OTU part is still not yet specified to support flexibility. Thus, ODUflex has to be carried within an OTUk ($k = 1 \ldots 4$), which means that the WDM transport bit rate is still not adaptable.

In both scenarios, the incoming traffic from the client interface is sent in a parallel fashion with a lanes transporting b Gb/s so as to reach a total bit rate of $a \times b$ Gb/s. As an illustrative example, the standard 100GbE [18] signal defines a transmission of 4 lanes of 25 Gb/s each. Therefore, from a fixed input bit rate the elastic aggregation is able to deliver a variable output bit rate. To this end, two candidate options are available, which can possibly be used in combination.

- The number of active lanes: lanes are able to switch between on and off states. This operation should be done without losing any data.
- The bit rate per lane: lanes are able to vary the bit rate at a given granularity (e.g., 1 Gb/s). To this end, an electric module should process the input flows to generate flexible OTU frames.
- The combination of the two previous modes – this offers the most flexible and promising approach, but also results in the highest complexity.

15.2.4 Performance Prediction

Performance predictions are of utmost importance in elastic network for routing and management systems in order to optimize the overall network functionality. For nondispersion managed systems [19–21], have presented a system model, validated both numerically and experimentally for coherent optical links, in which nonlinear Kerr effects and linear Amplified Spontaneous Emission (ASE) noise are both well-approximated by an additive white Gaussian noise (AWGN) model. The combination of this Gaussian noise modeling of signal distortions associated to coherent detection with possibly matched transmitter and receiver filters make the system impact of such distortions being quite accurately captured by the total noise variance, or more precisely the signal to noise ratio.

15.2.4.1 Generic Multi-Impairments BER Prediction Models As a result, some simple and accurate expressions to estimate the BER before error correction [22] (for

BER below 10^{-2}) can be derived whatever the modulation format:

$$\mathrm{BER} \approx \frac{x}{2}\mathrm{erfc}\left(\frac{d_{\min}}{2\sqrt{\overline{P}}}\sqrt{p\,\mathrm{SNR}}\right) \tag{15.1}$$

where x is a scaling factor representing the modulation constellation, the bit mapping (such as Gray coding) and the average number of bit errors per symbol error. $d_{\min}$ is the minimum Euclidian distance between the symbols of the noiseless constellation. Eventually, SNR is the electrical signal to noise ratio, equal to the ratio of average channel power per polarization $\overline{P}/p$ (with p modulated polarizations) over the total electrical noise variance per received polarization. SNR can be related to optical signal to noise ratio (OSNR) expressed within a reference bandwidth B_{ref} (typically 12.5 GHz) by:

$$\mathrm{SNR} = \frac{1}{p}\frac{B_{\mathrm{ref}}}{B_{\mathrm{elec}}}\mathrm{OSNR}_{B_{\mathrm{ref}}} \tag{15.2}$$

where B_{elec} is the receiver electrical noise-equivalent bandwidth (typically half the symbol-rate if Nyquist matched filters are used), such that (15.1) can be written in a generic way, also applicable for legacy intensity-modulation direct detection (IMDD) systems [23]:

$$\mathrm{BER} \approx \frac{x}{2}\mathrm{erfc}\left(\sqrt{\eta.\mathrm{OSNR}_{B_{\mathrm{ref}}}}\right) \approx \frac{x}{2}\mathrm{erfc}\left(\frac{Q'}{\sqrt{2}}\sqrt{\frac{OSNR_{B_{\mathrm{ref}}}B_{\mathrm{ref}}}{B_{\mathrm{elec}}}}\right)$$

$$\text{with } Q' = \frac{d_{\min}}{\sqrt{2\overline{P}}},\ Q'_{\mathrm{PDM}} = \frac{Q'_{\text{single polarization}}}{\sqrt{2}},\ \text{and } \eta = \frac{Q'^2 B_{\mathrm{ref}}}{2B_{\mathrm{elec}}} \tag{15.3}$$

Q' can be seen as a geometrical eye aperture, signature of the modulation format. For instance $Q' = (\sqrt{P_1} - \sqrt{P_0})/\sqrt{2\overline{P}}$ for legacy IMDD formats, while for PDM-QPSK with coherent detection at baud rate $R = 28$ Gbaud and Nyquist matched filters $Q' = 1/\sqrt{2}$, and $B_{\mathrm{elec}} = R/2$, leading to $\eta = 0.22$ in line with experiments from [21].

Equation 15.3 can be generalized to the case of a nonideal transceiver and of signal impairment over a typical transmission link, through an extension of the sources of Gaussian noise into play leading to the total equivalent OSNR: transmitter imperfections can be captured by an $\mathrm{SNR_{TRx}}$ term, optical amplifiers Amplified Spontaneous Emission noise can be related to an $\mathrm{OSNR_{ASE}}$ term, the interplay of nonlinear Kerr effect and Group-Velocity Dispersion can be related to an $\mathrm{SNR_{NL}}$ term, in-band and out-of-band crosstalks due to the nonideal rejection of other signals throughout the optical network or distortions stemming from signal filtering can be related to an additional $\mathrm{SNR_X}$ term, such that Equation 15.3 becomes:

$$\mathrm{BER} \approx \frac{x}{2}\mathrm{erfc}\left(\sqrt{\frac{\eta}{\frac{1}{\mathrm{OSNR_{ASE}}} + \frac{1}{\mathrm{SNR_{TRX}}} + \frac{1}{\mathrm{SNR_{NL}}} + \frac{1}{\mathrm{SNR}_X}}}\right) \tag{15.4}$$

In [21], for 28 Gbaud PDM-QPSK, an $\mathrm{SNR_{TRX}}$ equal to 23.5 dB within 0.1 nm associated to an effective $\eta = 0.2$ yielded very good fit with experimental data.

In the context of an EON, η is about to change from one format / baud-rate to another, as well as the SNR terms.

15.2.4.2 Generic Nonlinear Models for Elastic Optical Networks Beside the amplifier noise, which is straightforward to model, the nonlinear noise is usually the second dominant effect that limits system reach. Its modeling is considerably simplified by the use of high modulation rates, complex 2D-4D formats and the absence of inline dispersion management [19, 21, 24]. The resulting signal distortions can be considered as an additive Gaussian noise, which variance normalized to received signal power is proportional to the square of fiber input power per channel (thus leading to simple scaling rules and tools to predict penalties or to set powers). Hence perturbative propagation theories of the nonlinear Schrödinger equation and even simplified closed-form expressions of the nonlinear noise variance now appear quite predictive conversely to past 10 Gb/s IMDD systems.

Most models accounting for Kerr effect stem from perturbative approaches and share a few assumptions: each span is considered as a source of additive Gaussian noise, with contributions stemming from intra-channel and inter-channel nonlinear effects. This enables to compute an end to end total noise variance (or power spectral density), or to separate contributions from the different spans or channels, depending on the needs.

First Example: Computing Nonlinear Noise in a Cumulative Way The span to span accumulation of nonlinear noise can be written as in [21]: considering that total noise n is the sum of the contributions n_k of each span k, the total noise variance becomes after N spans:

$$\frac{\overline{P}}{\mathrm{SNR_{NL}}} = \mathrm{var}(n) = \mathrm{var}\left(\sum_{k=1}^{N} n_k\right) = \sum_{k=1}^{N} \mathrm{var}(n_k) + 2\sum_{k=1}^{N}\sum_{k'=1}^{k-1} \Re e(cov(n_k, n_{k'})) \quad (15.5)$$

Fortunately, it has been shown both numerically and experimentally in [21, 25, 26] that the covariance terms can be neglected in WDM dispersion unmanaged systems, such that the total nonlinear noise variance can be accurately modeled by a sum of single-span variances. Without loss of generality, (consistent with [21, 24, 26–28]), this single-span variance can be written (at least for inter-channel nonlinearities) as:

$$\mathrm{var}(n_k) \propto \overline{P}_{\mathrm{Rx}} * {P_{\mathrm{in},k}}^2 * {\gamma_k}^2 f(D_{\mathrm{in},k}, \text{format, fiber type}) \quad (15.6)$$

with $P_{\mathrm{in},k}$ the span input power per channel, γ_k the span nonlinear coefficient, and f a function of the modulation format, fiber type (local dispersion, attenuation, length) and $D_{\mathrm{in},k}$ the cumulated dispersion at input of span k. It is also possible to decouple this equation into summed contributions from different channels in a pump-probe approach, considering that the noise variance contributions stemming a neighbouring

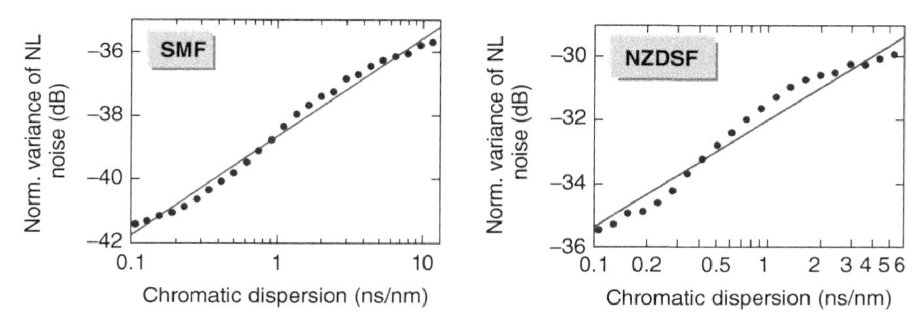

FIGURE 15.8 Nonlinear noise variance generated on a single span as a function of the chromatic dispersion at span input. Simulations of nine 50 GHz spaced 100 Gbit/s PDM-QPSK channels.

channel spaced by Δf from the impaired channel is proportional to $1/\Delta f$ with good accuracy [28, 29].

A typical evolution of f with D_{in} is depicted in Figure 15.8, after numerical simulation of the propagation of nine 50 GHz-spaced 112 Gb/s NRZ-PDM-QPSK channels over 100 km-long fiber. f is an increasing function of the absolute value of input cumulated dispersion, with a rather linear evolution (in dB × dB scale) for dispersion values up to 5–10 ns/nm and a saturation for higher dispersion values, leading to a plateau roughly 6 dB higher than for zero input dispersion. Such an evolution can be correlated to the increase of signal peak-to-average power ratio with input cumulated dispersion. This generic approach has been shown accurate for single-fiber type as well as mixed-fiber type terrestrial and submarine systems such as in [21, 25] for various modulation formats (BPSK, QPSK, 16QAM), provided an experimental determination of the function f. Besides, it allows explaining the impact of fiber order in a mixed-fiber type system [26], and the quasi-linear dependence of f (with positive slope ε dB/dB) in logarithmic scale for low cumulated dispersions allows capturing why supra-linear evolution of variance with number of spans N (proportional to $N^{1+\varepsilon}$) for terrestrial systems [21], while for very high input cumulated dispersions, the saturation of f leads to an effective $\varepsilon = 0$ and a linear increase of total noise variance with distance, also in agreement with experiments. Eventually, the simplicity of this generic model makes it very attractive for path allocation in elastic networking and the cumulative nature of the model renders it particularly well-adapted to a distributed control plane.

Second Example: End-to-End Computation of Nonlinear Noise End to end models from [28, 30] are interesting since they do not rely a priori on an in-depth experimental calibration of this function f, even though a calibrated scaling factor is eventually recommended (one per format for [28]). They allow seamlessly changing fiber dispersion, spectral efficiency, modulation format, amplification type (localized, distributed) and enable to build quite accurate BER estimation tools or system reach estimation tools, essentially for homogeneous long reach systems. The derivation of the function *f* is straightforward, yet in [28] or the GN model from [30], the nonlinear noise variance is independent of span input cumulated dispersion, or even modulation format in [30], thus such models fail to capture accurately the supra linear evolution of variance with distance or the impact of fiber order observed in [21, 26] for terrestrial distances. The main origin of such limitations for the GN model lies in the simplified modeling of the input signal as a stationary Gaussian random process. A more advanced input signal modeling considering third and fourth moments of the equivalent random process (depending on modulation scheme) is considered in [27, 31], leading to the so-called extended Gaussian noise model (EGN). The first validations suggest that this modeling successfully overcomes the abovementioned limitations, at least as far as inter-channel nonlinearities are concerned, however at the expense of computing time. A closed-form expression of the variance is proposed in [30], accounting for the modulation format, not for fiber input cumulated dispersion.

In the context of EONs with a wide variety of possibly coexisting modulation formats, symbol rates, channel spacing, the use of such models, with the necessary calibrations, becomes paramount. More details on those models can be found in Chapter 7.

15.2.5 Resource Allocation Tools

Resource allocation tools are used by network providers and operators for computing the number of resources necessary to transport a given traffic by ensuring their quality of service and minimizing as much as possible the whole network cost.

Depending on their application, resource allocation tools are classified into two categories: off-line and on-line. Off-line tools address the network planning phases [32, 33], occurring initially to deploy the resources necessary to transport the forecast traffic ("greenfield") and later on during the network life to cope with planned changes of the network state, such as network upgrades (set-up/tear-down of optical connections) or maintenance operations. On-line tools apply to unpredictable dynamic provisioning of optoelectronic resources during the network life, either to cope with changes in the network state (e.g., upon network failures) or to set-up on-demand services [34, 35]; on-line tools are implemented in control planes. The main differences between these two categories of tools lie in the computational time constraints and on the distance of the algorithm output from the optimal solution; complex route and resource allocation problems are decomposed in several subparts at the expense of the global solution accuracy. Off-line tools have to ensure the transport of the forecast traffic at a minimal cost, this operation can require millisecond to minutes per connection. Off-line algorithms can be solved either with mathematical optimization methods, such as integer linear programming (ILP) and mixed ILP (MILP), which advantage is the optimality of the provided solution for the solved (sub)problem, resulting in high computation times; or with heuristic approaches, which ensure low computation time but return suboptimal solutions; faster computation times typically stem from simplified solutions that are further to the optimal solution [36, 37]. Meta-heuristic solutions strike a balance between computation time and solution optimality [32, 38]. (M)ILP and meta-heuristic solutions are not suitable for the implementation of on-line algorithms because of their time constraints, and suboptimal heuristics are preferred.

Thanks to the advances in the physical layer technologies triggered by the introduction of flex-grid WSS and elastic transponders, it will be possible to select the rate of an optical connection according to the network state by choosing the best combination among: modulation format, coding rate, and spectrum width. This problem is known as routing, data-rate and wavelength assignment (RDWA) if fixed grid networks are considered [39], and as routing, modulation level and spectrum allocation (RMLSA) in flex-grid scenarios [32]. Consequently, in the RMLSA case, traditional resource allocation tools, which are based on routing and wavelength assignment (RWA) for fixed grids and for a limited set of data-rate devices with given physical properties, have to evolve into routing and spectrum assignment (RSA) routines accounting for the variable frequency slot occupancy of a channel and its physical properties depending on the chosen channel configuration [40, 41]; taking into account the physical constraints at the optical layer during the routing phase is known as Impairment-Aware RWA/RSA [42, 43]. RMLSA accounts for the physical impairments acting on the path before choosing transmission channel parameters, such as modulation format, coding rate and symbol rate [40].

The complexity of resource allocation tools does not limit to the introduction of a flex grid and to the choice of the channel configuration, but also depends on additional dimensions to be considered while computing the number of resources and the total network chose, such as traffic grooming (i.e., how to aggregate multiple low-rate flows to larger traffic units by means of electronic processing) [44], network resiliency, blocking probability, and energy efficiency.

All such additional dimensions increase the number of network configurations to be explored, increasing the complexity of the resource allocation tools. In [45] the complexity of RSA algorithms is investigated when traffic grooming and/or regeneration and/or variable modulation formats and baud rates are taken into account, and also under which network and traffic conditions the resource utilization savings are worth the additional complexity.

To simplify the complexity of these tools, the resource computation procedure is decomposed in several subproblems at the expense of the solution optimality; in [32, 40, 43, 46] the RMLSA is decomposed in RML and SA problems; in [40] the proposed sequential heuristic combined with an appropriate ordering gives solutions close to the ILP ones in low running times.

Table 15.1 shows a summary on the main approaches of resource assignment tools for EONs based on fixed- and flex-grids; symbol "+" ("&") means that two successive phases are performed separately (jointly, resp.). Each resource assignment approach can be solved with mathematical formulations such as (M)ILP, meta-heuristics or heuristics as a function of their application (off-line or on-line resolution). We remember that independently on the adopted solution method, decomposing a problem in several subproblems provides suboptimal solutions with respect to a problem formulation considering all phases jointly.

In dynamic EONs, connections having different bandwidth occupation are continuously set-up and released, such that an optimal resource allocation cannot be

TABLE 15.1 Main approaches for resource assignment for fixed- and flex-grid elastic networks, their application scopes and solution methods

	Mono-layer	Multi-layer	Off-line methods	On-line methods
Fixed-grid	R + D + WA R&D&WA R&D + WA R + IA&D + WA R& IA&D&WA R& IA&D + WA	R + G + D + WA R + G&D + WA R&G&D + WA R&G&D&WA R + G+ IA&D + WA R + G& IA&D + WA R&G& IA&D + WA R&G& IA&D&WA	(M)ILP Meta-heuristic Heuristic	Heuristic Meta-heuristic
Flex-grid	R+ IA&ML + SA R& IA&ML&SA R& IA&ML + SA R+ IA&ML + SA R& IA&ML&SA R& IA&ML + SA	R + G + ML + WA R + G&ML + WA R&G&ML + WA R&G&ML&WA R + G+ IA&ML + WA R + G& IA&ML + WA R&G& IA&ML + WA R&G& IA&ML&WA	(M)ILP Meta-heuristic Heuristic	Heuristic

R: routing; D: data-rate assignment; WA: wavelength assignment; G: grooming; ML: modulation level assignment; SA: spectrum assignment; IA: impairment aware; "+": sequential solution; "&" joint solution.

performed for each change of the network state. For dynamic scenarios, a main concern is the spectrum fragmentation, which occurs when empty spectrum slots are not contiguous within a network kink or over adjacent links as a result of several set-up and tear-down of dynamic connections. Such unused slots cannot be used by future connections and their presence worsens both the performance (increase of blocking ratio) and the resource utilization (spectrum and spectrum converters) efficiency of the network. The spectrum fragmentation is similar to the wavelength fragmentation in legacy RWA problems, but now the availability of not only a single slot, but that of a group of contiguous slots has to be guaranteed along the path.

Figure 15.9 illustrates an example of demand blocking due to spectrum fragmentation: on the considered link a certain number of connections, with different bandwidth requirements, are established over a given network link (Figure 15.9(a)). Later in the network life, one of these connections is released, freeing up the corresponding spectrum (Figure 15.9(b)). Then, a new connection requiring six slots has to be set up, this connection is blocked even though six slots are free on the link, as those slots are not contiguous.

In the literature many works aim to solve this issue, known as spectrum defragmentation, while setting-up a new request like in [47, 48] or by re-configuring existing paths periodically or according to a fragmentation threshold [49]. In [47]

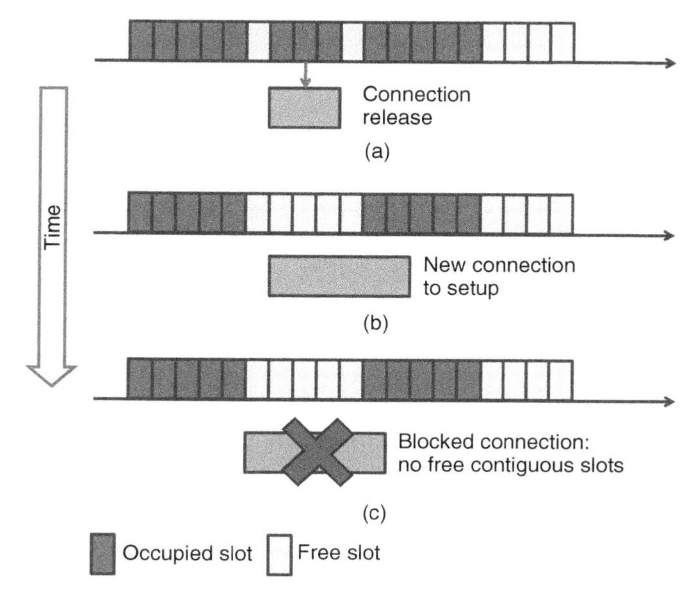

FIGURE 15.9 Example of spectrum fragmentation on a given link for a dynamic scenario.

the maximum common larger segment (MCLS) tries to maximize the number of consecutive unused slots on the frequency axis. In [48] the entropy state associated to a link, path and network is introduced; this entropy describes the fragmentation state of each one of these entities by assessing the percentage of empty slots before and after a path set-up; in this manner the spectrum allocation minimizing the entropy of the whole network is selected. Concerning dynamic spectrum defragmentation, a set of existing connections has to be reconfigured by a rerouting and/or spectral reallocation performed sequentially or at the same time. This operation is generally time consuming and is not suitable because service disruption can arise during the reconfiguration phase. Some defragmentation methods have been proposed and demonstrated in research laboratories. Such methods mainly rely on the "make-before-break" paradigm [50], whereby an additional path is set-up before switching the signal that has to be reallocated, or on push-pull defragmentation [50, 51] (also known as wavelength sweeping), whereby the spectrum of a channel is shifted on free contiguous slots belonging to the same path with no service interruption. Similarly, hop-tuning techniques [52–54] appear as an extension of push-pull by allowing channel spectrum shifts to noncontiguous slots. A qualitative comparison of the diverse defragmentation techniques is provided in [41]. Another manner to cope with the fragmentation issue is to add spectrum converters along the path; this solution seems to be the more straightforward, but involves expensive optoelectronic devices making elastic networking less cost-attractive.

15.2.6 Control Plane for Flexible Optical Networks

The affirmation of cloud computing services [55] is creating more and more dynamic traffic requests, where on-demand bandwidth at Internet speeds have to be set-up. Thanks to elastic/flexible optical solutions, it becomes possible to adapt the optical resources to these dynamic bandwidth services. The dynamic set-up of on-demand service is no longer compatible with manual interventions, hence automatic processes and resource orchestration become mandatory across multiple network layers (from IP to optical) and between diverse vendors and operators.

This orchestration and its automatic reconfigurations need advanced control framework in order to guarantee the update of the network state information and to be able to route connections whenever a reconfiguration is demanded. Presently, the implementation of such framework relies on two main approaches: the first extends the existing generalized multiprotocol label switching (GMPLS) and path computation element (PCE) based protocols, while the second opts for the newly proposed software-defined network (SDN)/OpenFlow (OF) architecture.

15.2.6.1 GMPLS/PCE-Based Control Plane To support elasticity and to guarantee the establishment and the maintenance of connections control plane enhancements consists in the extensions of GMPLS routing and signaling protocols [56], that is, open shortest path first (OSPF) and resource reservation protocol traffic engineering (RSVP-TE). The former protocol, also known as routing protocol, disseminates the network state information (e.g., frequency slot utilization) required by RDWA/RMLSA on-line algorithms, whereas the latter enables establishment/tear-down of resources from a source node to a destination node and provide the acknowledgement about resource reservation.

GMPLS abstracts the network in logical representations of each of its layers (one layer per used switching capability, e.g. one for IP and another for WDM).

The path computation can be performed either in a centralized manner, by the means of the PCE, or in a distributed way. Although the use of PCE is computationally more powerful than the distributed solution, it reduces the scalability of the network for larger network sized; to cope with this problem, several PCEs can be placed in the network, providing a logically centralized system with a specified set of protocols, called PCE protocol (PCEP).

Currently in standard bodies extensions of the control plane protocols to support elasticity are discussed [57]. More specifically, OSPF protocol now needs to disseminate not only the number of free wavelengths but information about free spectral slots along with the value of the slot width [57]. In a flex-grid, the reservation of a channel through RSVP-TE requires to specify the nominal central frequency of the channel and its spectral slot width [57–59]. In [39] there are examples of protocol extensions required in flex-grid networks where resource allocation and path establishment are performed in both centralized and distributed manners, whether the routing and spectrum allocation are performed jointly or separately.

Beside standards, many methods have been proposed in the scientific literature to aggregate information about the spectral occupancy. For example in [60] a

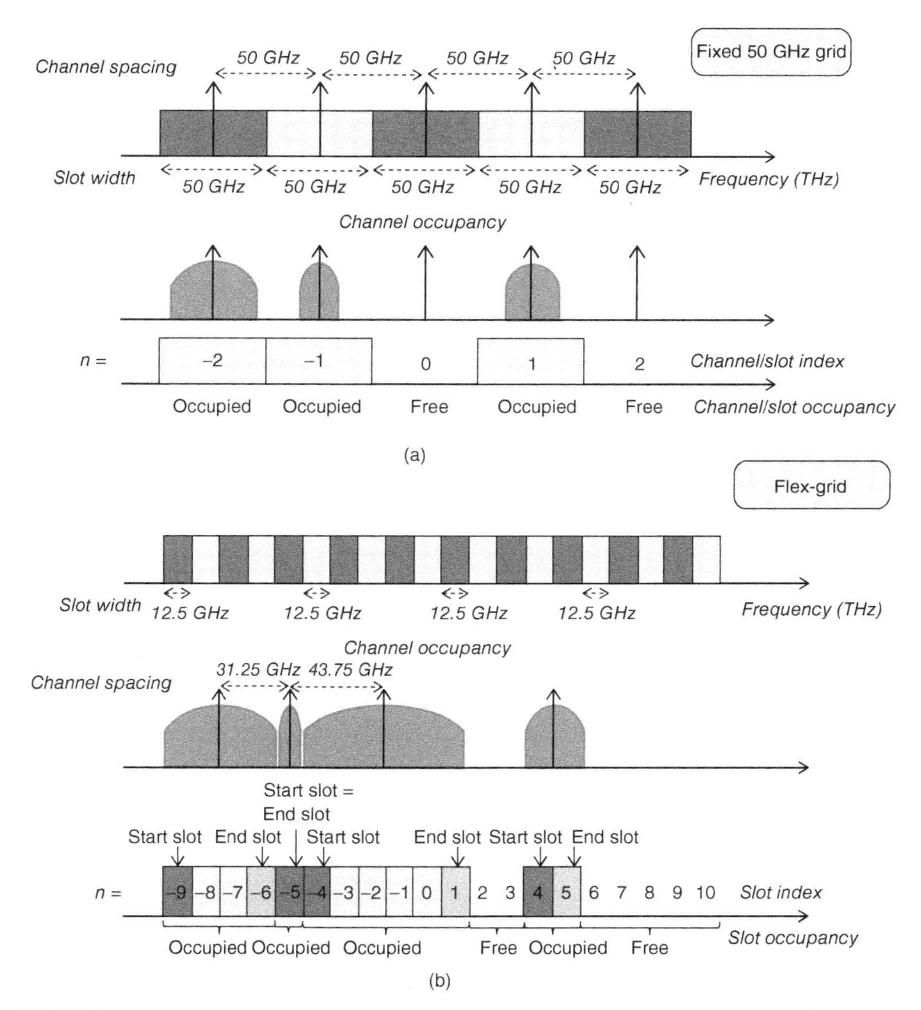

FIGURE 15.10 Representation of current fixed-grid (a), where each channel is equally spaced and occupies a unique 50 GHz-slot of the grid, and of the proposed flex-grid (b), where the grid is made of 12.5 GHz slots and a channel occupies a variable number of slots depending on its spectral occupation.

redefinition of the RSVP-TE protocol is proposed to specify the total spectrum occupancy of a channel by specifying its extremity frequencies, as indicated in Figure 15.10.

Figure 15.10(a) provides an example of the current 50 GHz fixed-grid network where each channel occupies a slot of the grid and is equally spaced; the spectral occupation of a channel is indicated by the number of the occupied slots. In a flex-grid network, the slot width is no longer equal to the channel spacing or to the spectrum occupied by a channel. Indeed, a channel can occupy several slots and the channel spacing depends on the width of two contiguous channels as shown in Figure 15.10(b). As a result it becomes mandatory to specify the total number of occupied slots and where they are placed; this is solved by indicating the start and end slots associated to each channel.

Moreover, a channel can be realized with a single carrier or with multiple carriers, hence among the forwarded information also the number of subcarriers, their symbol-rate and their modulation format have to be specified.

15.2.6.2 SDN Architecture SDN is based on the concept of decoupling data and control planes. It relies on the abstraction of underlying network infrastructure that can be used by applications and network services as virtual entity [61]. This abstraction enables the coexistence of various network slices relying on diverse transport technologies, domains and control protocols. SDN is based on a centralized entity, named SDN controller, which implements the configuration of the underlying network devices. To this purpose, the best suited protocol for the SDN architecture is OpenFlow (OF).

OpenFlow is a vendor and technology agnostic protocol that relies on traffic flow tables, enabling software/user-defined flow-based routing, control and management in the SDN controller, outside the data path. Indeed any data-plane entity is connected to external software controllers managing the network operating system (netOS) [62], which sends the operation messages to be executed by the selected data-plane.

The SDN network paradigm facilitates the packet-optical integration (also known as packet-optical integration convergence, PAC.C [63]) as packet-switches and routers can operate jointly with optical transport elements, which are circuit-based. The packet-optical integration will facilitate the provision of services fitted to specific application requirements (e.g., on demand bandwidth services) at the optical layer. Current SDN cannot fully manage the optical layer, based on circuit and wavelength-based architectures, meanwhile protocol extensions are largely proposed. Today to control heterogeneous technology domains, OpenFlow protocol has to be able to interact directly with existing GMPSL/PCE control plane. This is possible by reusing the existing GMPLS encodings and the link state protocol by introducing additional circuit flow table for circuit provisioning [64] in SDN for optical transport layer. It results in a distributed GMPLS control plane under the centralized management performed by the SDN Controller, as proposed in [65, 66] and exploited in various Metro/Core Border nodes [67].

For EONs, both GMPLS/PCE and SDN control plane approaches are investigated, extended and compared [68], but a complete comparison study is not yet available because problems like network scalability and inter-operability are not taken into account.

To conclude, SDN combined with EONs are a promising solution enabling the network virtualization [69], for delivering user-adapted services; elastic devices can be seen as virtual and programmable devices (e.g., sliceable transponders [6]) and SDN controller allows the orchestration of a heterogeneous network.

15.3 PRACTICAL CONSIDERATIONS FOR ELASTIC WDM TRANSMISSION

We outline the key required technology concepts with an emphasis on proposed solutions built upon existing the standard coherent (Nyquist) WDM transmission techniques. However, the concepts are also largely applicable to an orthogonal frequency division multiplexing (OFDM) transmission.

15.3.1 Flexible Transponder Architecture

Numerous elastic transponder designs have been proposed, notably based on single-carrier technologies [4, 70] or OFDM [3, 71]. We chose to focus on implementations associated with 100 Gb/s PDM-QPSK coherent detection since this technology can be made adaptive in various ways with little additional complexity.

Before depicting the various adaptations on the hardware design, Figure 15.11 briefly recalls the architecture of a 100 Gb/s PDM-QPSK transceiver.

The 100 Gb/s transport frame can carry a single 100 Gb/s client signal (e.g., 100GbE or OTU4) or ten 10 Gb/s client signals (e.g., 10GbE or OTU2). This transport frame includes FEC encoding to mitigate physical impairments arising along the lightpath. The encoded bit stream is mapped onto four electrical 28 Gb/s lanes going to the optical module, including the 100 Gb/s payload plus additional framing and FEC overhead. Each signal, represented by I1, Q1, I2, and Q2 on the

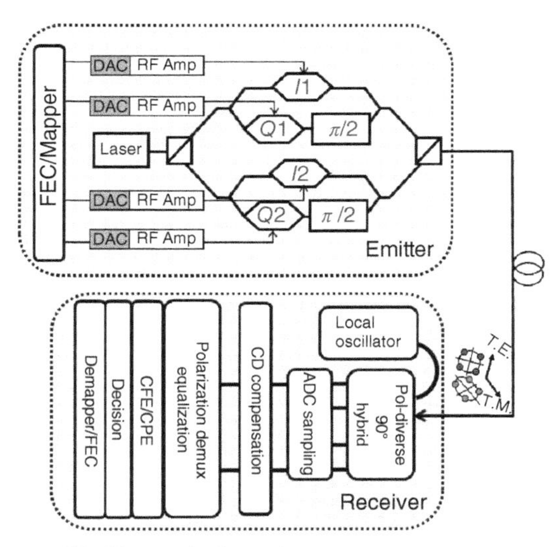

FIGURE 15.11 Architecture of the 100 Gb/s transponder that will be made elastic.

block diagram, is then independently modulated to binary phase shift keying (BPSK) modulation thanks to Mach–Zender modulators. The resulting four BPSK signals are then combined through phase shifters and polarization beam combiners so as to produce a 28 Gbaud PDM-QPSK signal.

After transmission, the signal is detected by means of a coherent receiver: it is first combined with a local oscillator within a polarization diversity 90° hybrid mixer which outputs allow the linear sampling of the in-phase and quadrature components of the optical signal along two arbitrary orthogonal polarizations. The entire optical field (amplitude, phase, and polarization states) can thus be reconstructed. A clock frequency is then extracted from the digital signal before going to the DSP stage. The DSP allows a wide range of techniques to compensate for signal distortions such as chromatic dispersion and polarization mode dispersion (PMD), and also enables the recovery of an estimate of the frequency and phase carrier. Then, a symbol-to-bit decision stage is performed before decoding the FEC.

Due to the introduction of elasticity, the signal processing of one or multiple blocks needs to be adjusted depending upon the level of flexibility one wants to achieve. Elasticity is indeed enabled either by modulation format, symbol rate, channel spacing, FEC adaptation, or a combination of these options.

- *Modulation format adaptation*. The emitter side can straightforwardly support various modulations by adjusting the sequences feeding the four modulators (for in-phase and in-quadrature of both polarizations). For instance, dual polarization BPSK is generated by I1 = Q1 and I2 = Q2 while SP-QPSK is generated by Q2 = I1 ⊕ Q1 ⊕ I2 where ⊕ denotes the XOR operator. At the receiver, much of the DSP processing can be reused between different modulations and in particular in the chromatic dispersion block which accounts for a very large part of the total DSP. Higher-order modulations such as 8QAM, 16QAM require DAC modules prior to the Mach–Zehnder modulators in order to generate I, Q multi-level signals inherent to dual polarization modulation larger than 4 bit/symbols.
- *Symbol rate adaptation*. The generation of variable-bandwidth signals requires a tunable clock or alternatively it can be emulated with a fixed clock in combination with symbol repetition at the transmitter and decimation at the receiver. The former has the advantage of offering a large flexibility thanks to the tunable clock while the latter does not require a new phase, bit or frame synchronization after a symbol rate change but is less flexible since repetition/decimation uses an integer scaling factor.
- *Channel spacing adaptation*. The challenge for the transceiver is rather limited to the need for a tunable laser (potentially fully tunable if over the entire C or L band) and local oscillator. However, all filtering elements such as optical filters and WSS must be compatible with the nonstandard ITU grid. If such optical filters are not sharp enough or if the chosen channel spacing is very tight to reach high spectral efficiency, additional constraints can be put on the transceiver. In the former case, the emitter with the addition of a pre-compensation block can indeed help to compensate for physical impairments, while the receiver can

enhance the equalization. On the other hand, the latter case can rely on Nyquist WDM pulse shaping.

- *FEC adaptation.* By varying the overhead and hence the reliability and coding gain, the bit rate can be adjusted to match the transmission impairments and distance. Both the coding and decoding blocks should be adapted to reflect the overhead change. The other DSP blocks can be reused as such.

15.3.2 Example of a Real-Time Energy-Proportional Prototype

This section considers the realization of an energy-efficient optical point-to-point transmission with elastic-enabled hardware elements as a proof-of-concept example. A muxponder combines the functions of aggregation and transponder. We present a real-time muxponder prototype that is able to follow the traffic fluctuations in order to reduce its energy-consumption during low traffic periods [72].

The experimental setup consists of a fully equipped demonstration built upon an original elastic device capable of aggregating and disaggregating partially-filled 10GbE Ethernet clients, and of a real-time coherent elastic transponder with symbol-rate adaptation, as depicted in Figure 15.12.

The elastic aggregation unit has ten classical (fixed) 10GbE interfaces with the client side (left of the module) filled by random traffic but where we can control the ratio of useful frames (i.e., representative of real traffic) to dummy frames (added to reach the 10 Gb/s Ethernet nominal data rate). At the output of the aggregation unit (right of the module), the interface is elastic with a variable bit rate per output lane so as to deliver the proper bit rate to the transponder unit. We use a voltage controlled oscillator to generate a centralized clock and distribute it all across the aggregation module and the transponder. The transponder is a typical PDM-QPSK with 7% FEC overhead (using Reed–Solomon) [73], which is fed with 4 lanes corresponding to in-phase and in-quadrature of the two polarizations. The transponder adapts its data rate by tuning the symbol rate from 1 to 7.5 GHz due to hardware limitations. This translates in a maximum bit rate of 30 Gb/s with PDM-QPSK. The detailed experimental setup and performance results can be found in [72].

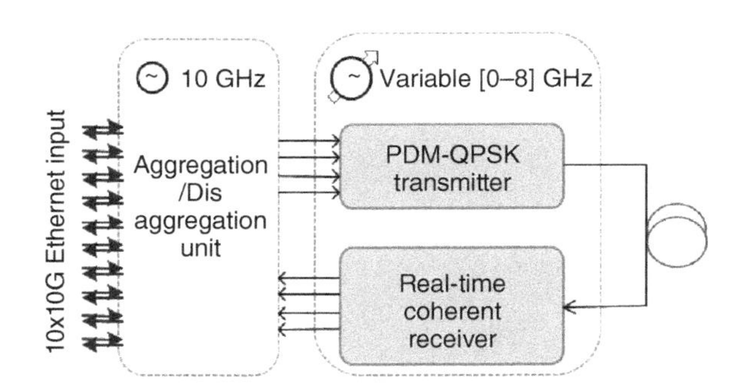

FIGURE 15.12 High-level muxponder prototype architecture.

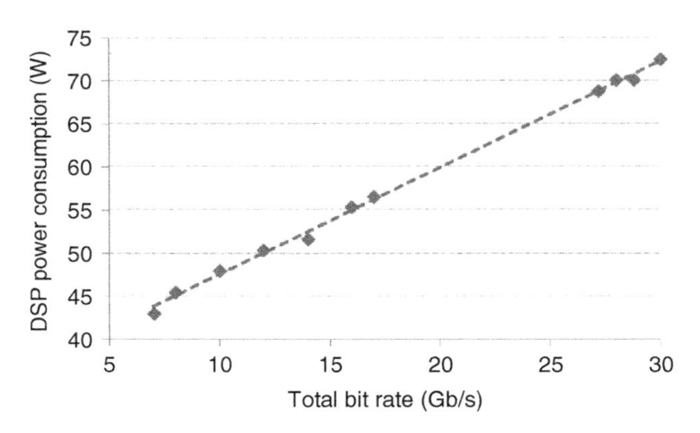

FIGURE 15.13 Impact of the symbol rate on the power consumption.

We measure the energy consumption of the real-time DSP, which includes the sampling of data in field programmable gate arrays (FPGAs) where polarization demultiplexing is performed with 9-tap finite impulse response filters, arranged in a butterfly structure and updated by the constant modulus algorithm. Carrier frequency and phase estimation are performed using the Viterbi algorithm. The measurement was performed by direct inspection of the voltage and current supplied to the receiver board and is shown in Figure 15.13. A linear relationship is demonstrated between power consumption and actual transported traffic. This is typical of the power consumption of logical gates versus clock rate, when gate voltage is held constant for all clock rates. Overall, we found a 41% reduction in power consumption between the highest investigated bit rate (30 Gb/s) and the lowest (7 Gb/s).

Even though, qualitatively speaking, these results are extremely promising, quantitatively they are specific to our hardware implementation, and to the specific technological choices for our receiver. For example, our implementation is based on FPGA, whereas higher bit rate transponders are typically based on application specific integrated circuits. The research in the area of energy proportional optical network elements is still relatively young, and several improvements to our proposal can be imagined, for example the use of dynamic voltage and frequency scaling.

15.4 OPPORTUNITIES FOR ELASTIC TECHNOLOGIES IN CORE NETWORKS

EON results in more cost- and energy-efficient solutions because of an *ad-hoc* utilization of resources. The opportunities offered by elastic networks are multiple and depend on the improvements brought with both hardware and software solutions, as well as at the control plane. Usually, such modifications are introduced gradually by network operators following the trade-off between the near-term cost impact of the new adopted solution and the expected benefits for future network evolutions. In the following we quantify the various opportunities enabled by the adoption of EON by investigating network-level dimensioning *via* resource allocation tools as described in Section 15.2.5.

15.4.1 More Cost-Efficient Networks

EONs allow significant cost savings, depending on the network scenarios. In the short term, rate-adaptive devices will likely be deployed in legacy fixed-grid networks. In this context, the capability of selecting an appropriate modulation format to the connection capacity and the distance it has to cover will increase the network capacity by approximately 30% [74].

Today, to fit the capacity of a connection to the capacity required by a demand, multiple types of devices are deployed, each carrying a specific rate and with a specific design [75]. The use of a single type of rate-tunable technology to handle all types of connections simplifies the design of the network (e.g., the same dispersion map can be used for all links) and allows the sharing of resources in dynamic networking scenarios. Surely, elastic devices are able to configure their data-rate to meet planned (e.g., upgrades) or unexpected (e.g., failures) evolution of the network conditions: when a new connection has to be established, any available elastic optoelectronic interface can be used (on unplanned failure recovery or dynamic demand set-up) or reconfigured (on upgrade or pre-computed failure recovery). In restorable networks (with online rerouting of the connection and new light-path establishment after planned and/or unplanned failure, as opposed to protected networks where protection light-paths are always lit-on) it has been demonstrated that the use of elastic interfaces limit the over-provisioning of the number of spare resources, as these resources are no longer associated to a specific rate and can be shared between any failed connections, regardless of their rate. The number of spare resources in an elastic network reduces by 30–70% with respect to mixed-data rate scenarios (where a specific device per each data rate exists) [76], yielding up to 37% in cost reduction [77]-[75]. In case of a network upgrade, the use of rate adaptive resources ensures a higher resilience of the deployed resources to traffic evolution and a postponement of the deployment of regenerators; thanks to this capability, when a diverse network upgrades are considered, the cumulative cost of the network over the total period is up to 18% less expensive for elastic scenarios than for mixed-line rate and 34% lesser expensive than single rate scenarios [78].

Though flexible equipment may cost more than conventional WDM equipment [77–79], possible fast price erosion with time due to large-scale production can be observed. In [7] the maximum additional cost of a sliceable transponder with respect to a fixed one is estimated to lie between 20% and 60% of the cost of a fixed transponder to render a flexible network more cost-attractive than a fixed one.

Economic interests are also provided by elastic networks due to flex-grid [79]. Cost savings are compared in terms of equipment but also in terms of saved spectrum [79–81]. Savings are also demonstrated for resilient networks, both in case of pre-planned and on-line restoration [81, 82].

15.4.2 More Energy Efficient Network

Elastic networks are also attractive for their eco-sustainability, which has been important due to environmental awareness and the pressure to reduce the operational expenditure of network operators [83–85]; due to the introduction of high-capacity devices that are more and more power greedy; and due to the increase of the cost of energy.

The capability of elastic interfaces to tune the data rate to the carried capacity minimizes the energy per bit as just enough power is used. Data rate adaptation can be provided by two different methods: (i) by adjusting the modulation format, hence increasing the optical reach of systems if less complex modulation format are used and skipping useless regenerators; and (ii) by adapting the symbol rate of the connection, thereby reducing the energy consumption if lower symbol rates are transmitted, because the energy consumption decreases proportionally to the frequency clock of electronic devices (like DSP, line cards and framer/deframer) as considered in [15] and shown in Section 15.3.2 from the realization in [72]. Figure 15.14 depicts an example of the aforementioned two rate adaptation allowing energy savings.

This dynamic capability of changing the connection data-rate can be used for greening the network in static operation and for adapting the energy consumption of each single connection to its daily and weekly traffic variations. As an example [86], in a European-sized network and with unprotected traffic, following daily traffic variations brings up to 20% of energy savings when using modulation format adaptation, and up to 24% when symbol rate adaptation is implemented; gains reach 32% when both are simultaneously implemented. Energy savings have also been demonstrated by setting to a low power mode the spare devices planned to be used only for failure recovery [87].

15.4.3 Filtering Issues and Superchannel Solution

Most advantages associated to flex-grid exist under the assumption that filters with finer spectrum granularity have sharp profiles and do not affect the transmission performance of the optical channels. Many of the previously cited studies have shown the interest of adopting narrower channel spacing when comparing the overall extra-capacity of flex-grid networks with respect to fixed-grid ones (33% when 37.5 GHz grid systems replaces legacy 50 GHz-grid systems).

With commercially available flex-grid filters, OSNR penalties induced by tight optical filtering when individual channel steering is performed are not negligible [88]. In [89] the total network throughput in a transparent network featuring 37.5 GHz channel spacing is compared with the whole throughput in a 50 GHz grid network. It is shown that the penalties due to tight filtering reduce the optical reach of transparent signals proportionally to the number of traversed nodes. It follows that, under the assumption of a constant cost per transported bit, the expected 33% gain brought by narrowing the channel spacing from 50 down to 37.5 GHz strongly depends on network size: it can actually be halved for nation-wide networks or vanish for larger ones [89].

To cope with the filtering-induced penalties and improve the spectral efficiency while keeping as low as possible the cost per bit some transmission paradigms have

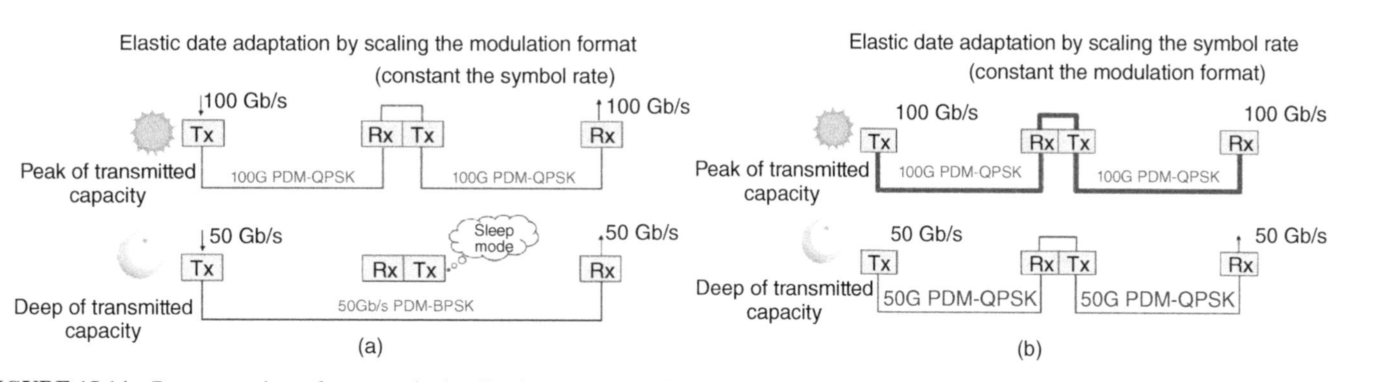

FIGURE 15.14 Representation of two methods allowing energy savings thanks to the reconfiguration of elastic transponders: by means of modulation rate adaptation (a) and symbol rate adaptation (b).

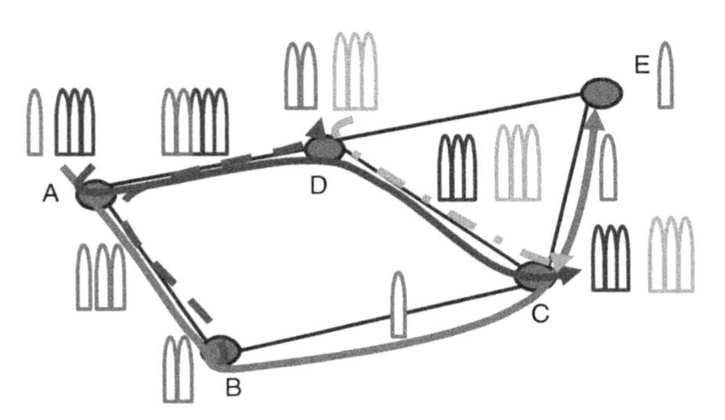

FIGURE 15.15 Example of the presence of superchannels in the network. The connection is composed of one or more subcarriers that are inserted, transported and dropped together.

to change. A solution to this problem consists in adding to the channel bandwidth a guard band between narrower channels (or conversely a half-guard band at each channel extremity) so as to mitigate the penalties due to both filtering functions and the cross-talk from partially blocked additional channels [90], at the expense of the overall spectral efficiency [91].

To improve the spectral efficiency and cope with filtering issues at the same time, superchannel routing may be adopted. A superchannel consists of a group of adjacent optical subcarriers, as shown in Figure 15.15, that propagate together along the same optical path: they are inserted, transported, and extracted together [92, 93]. All subcarriers belonging to the same superchannel have the same modulation format and symbol rate, which are adapted to the physical degradations occurring along the path [92, 93]. The number of subcarriers depends on the capacity of the connection and on the spectral resources available along the selected path.

The flexibility and reconfiguration of high-capacity superchannels could also make them prime candidates for inter-data center transmissions, which require a tremendous amount of bandwidth [94].

15.5 LONG TERM OPPORTUNITIES

15.5.1 Burst Mode Elasticity

We have considered so far elastic networking solutions aimed at circuit-switched core networks, where enhanced flexibility is most needed and where the high cost of the coherent technology compounded by the equipment cost overhead incurred by elasticity are compatible with the overall network cost. In the segments that are closer to the end-user – access networks and metro networks – and inside datacenter networks (see Figure 15.1), the tremendous bandwidth made available by optical fibers is not (yet) fully leveraged and optimizing the spectrum using flex-grid technologies is not of interest today and in the near term. However, as capacity requirement is growing (Bell Labs predicts a 560% in traffic increase in the metro segment alone

[95]), coherent technology and elasticity could eventually be needed to optimize equipment utilization, and thus be deployed in access, metro and datacenter networks. This section shows how elasticity can increase the capacity made available to end-users.

In core networks, due to the aforementioned fundamental trade-off between reach (in terms of distance or number of nodes) and data rate, either several transponders are installed on the network nodes, or OEO regeneration is used, so that any node can communicate with any other node. In access, metro and datacenter networks, the cost of the terminating equipment such as the transponders is crucial and the cost of deploying several transponders per node or of deploying regenerators can be prohibitive. Thus the same transponder should send different data flows to different destinations, or receive different data flows from different sources through time-sharing of the channel capacity. A transponder may have to be reconfigured many times every second or even millisecond so that communication between any source and any destination is possible at virtually any time. Such fast reconfiguration capability is denoted by "burst-mode," where a burst typically consisting of a stream of data going from a source to destination for a limited duration of a few tens of nanoseconds to a few seconds depending on the implementation. Microsecond-scale bursts are typically considered [96–100]. Burst-mode coherent reception is challenging because not only clock but also polarization must be recovered for each burst, and DSP algorithms must be reset at each burst and must converge within a duration that is much lower than the burst duration, in order to minimize the overhead due to burst-mode reception. Additional details about coherent burst mode reception can be found in [101–104].

Coherent transmission enables rate-adaptation and permits a single transponder to communicate with both close nodes (at a high data rate) and further nodes (at a lower data rate). In the segments considered here, elasticity is typically implemented by varying the modulation format for each burst (potentially every microsecond) based on the distance or number of nodes between source and destination. Observe nonetheless that, when energy efficiency is primordial, baud rate elasticity may be used as explained in the previous sections, as a complement to modulation format elasticity.

The concept of flex-rate burst-mode networking is illustrated in Figure 15.16, which depicts how one node (A) communicates with all other nodes. Node A sends data to D via nodes B and C, and to node G via node F. Node B is sufficiently close to node A such that A can send high rate, PDM-16QAM modulated bursts to B; node C is slightly further and a lower modulation format (e.g., PDM-8QAM) is used while an even lower modulation format (e.g., PDM-QPSK) is needed by node A to send bursts to node D. Unlike with circuit switching, with optical burst switching node A can use the same transponder using time-sharing access to send data to nodes B, C and D without requiring nodes B and C to perform optoelectronic conversions.

In the following sections we review two examples of flex-rate burst-mode networks and show their potential benefits: a coherent PON, and a ring-based metro or datacenter network.

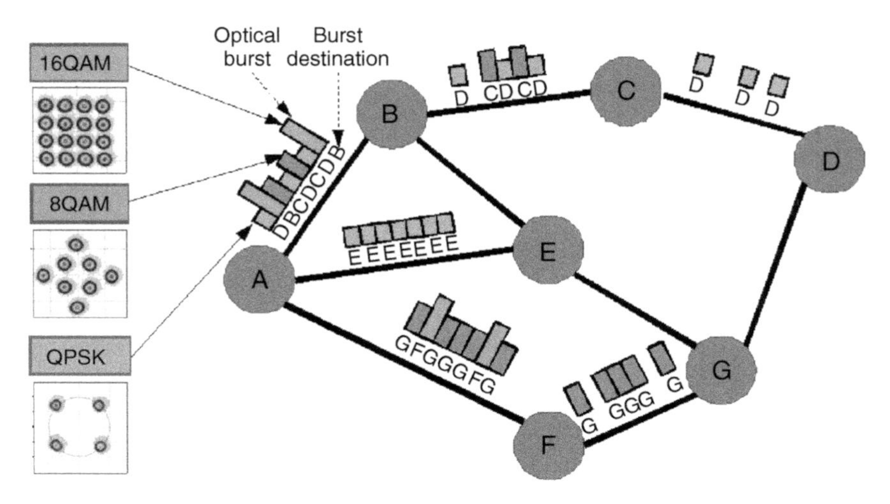

FIGURE 15.16 Burst-mode elastic network.

15.5.2 Elastic Passive Optical Networks

PONs can be viewed as a star-shaped optical burst switching network where the node located in the operator's central office, the Optical Line Terminal (OLT), and N nodes located at the customers' premises, the Optical Network Units (ONU), communicate via a 1:N optical splitter. Today's PONs use noncoherent modulation formats at data rates up to 10 Gb/s both downstream and upstream [105]. Although coherent modulation is currently prohibitively expensive for a deployment in the highly cost-sensitive access segment, the combination of technological advances in coherent transmission equipment and of the demand growth could lead to the introduction of the coherent technology in PONs in the medium to long term.

In a PON, the distance between the operator's central office and the end users varies widely: the data rate in the whole PON is driven by the data rate achievable by the ONU that is most distant from the OLT; closer ONUs have transmission margins that are not exploited. In a flex-rate coherent PON (flex-PON), terminals adjust their rate depending on the distance to be covered in order to exploit the aforementioned margins, as shown in Figure 15.17. To ensure fairness across all users, that is that all ONUs experience, on average, the same service (in terms of sending or receiving data rate) irrespective of their physical distance to the OLT, a generalized proportional-fair scheduler that allocates more sending time to the more distant users experiencing more impairments can be used as explained in [102]. Assuming 256 ONUs per OLT with a maximum OLT-ONU distance of 20 km, a standard, fixed-rate coherent PON would operate at 75 Gb/s (14 Gbaud signals with a PDM-8QAM modulation and a standard 12% hard decision FEC) corresponding to an effective throughput of 292 Mb/s per user, while flex-PON with possible dual polarization modulations up to 16QAM would deliver on average 318 Mb/s to each user, a 9% premium over fixed-rate PON. When the OLT-ONU distances are more widely spread, flex-PON

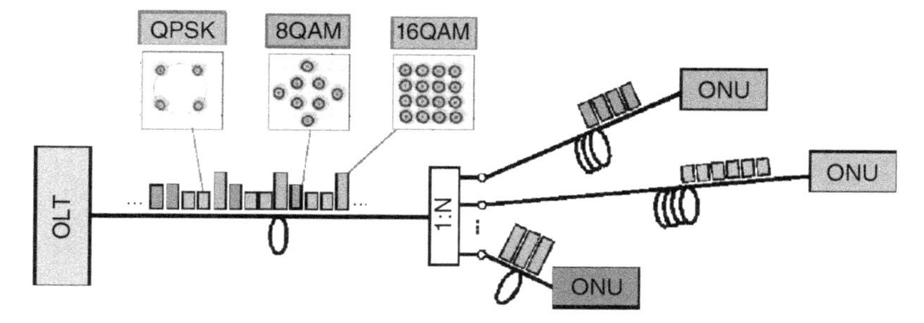

FIGURE 15.17 Flexible passive access optical network (flex-PON).

becomes significantly more efficient than a fixed-rate PON; for instance, if the most distant user is now 50 km away from the OLT, the average gain jumps to 120% (220 Mb/s per user instead of 100 Mb/s). Since those gains are per-wavelength, they do not change if several wavelengths are used in the PON.

15.5.3 Metro and Datacenter Networks

An even more forward-looking application to burst-mode elasticity is the intra-datacenter interconnection network, that is a network within a datacenter. WDM optical technologies such as optical burst switching may find a promising application in the datacenter segment to improve scalability [106], for instance in terms of energy consumption or cabling woes. Although this section focuses on datacenters, the same technology could be used for large scale metro networks.

Datacenters are currently built with electronic switches and many point-to-point copper or fiber interconnections: a typical datacenter consists of servers set into racks; within each rack the servers are connected to a Top of Rack (ToR) Ethernet switch, connected in turn to a hierarchy of electronic switches following for instance the Folded Clos topology shown in Figure 15.18.

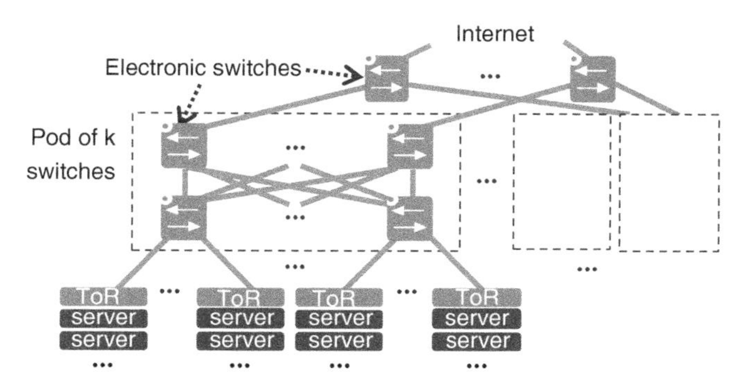

FIGURE 15.18 Folded Clos-based datacenter network.

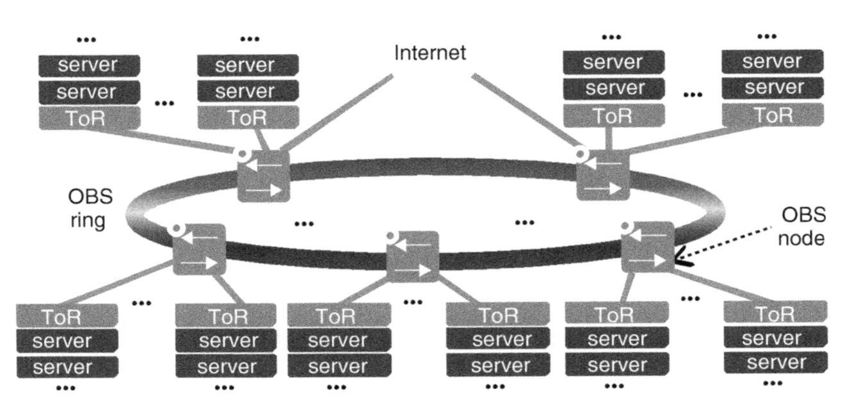

FIGURE 15.19 Optical burst switching ring-based datacenter network.

Consider the following alternative topology. ToRs are connected to large optical burst switching nodes, arranged into a ring network, as shown in Figure 15.19. In datacenters, propagation distances are so small that they cause few physical layer impairments, however, as signals traverse nodes potentially many optical filters (which are located within the nodes) need to be traversed, resulting in signal distortion due to filter concatenation. Experimental measurements for an implementation of an optical burst switching node shows that PDM-QPSK signals could cross 120 nodes but PDM-16QAM signals are limited to a reach of only 20 nodes [107]. If for instance a ring of 88 nodes is deployed then the transponder capacity is limited to PDM-QPSK signals with an effective throughput of 100 Gb/s (after FEC) per wavelength, in order to ensure that each node is reachable from any other node. With elastic-rate transponders capable of rates from PDM-QPSK (100 Gb/s) to PDM-32QAM (250 Gb/s), each node can adjust its sending rate on a per-burst basis to the number of nodes to be crossed to reach a destination, resulting in an average operating rate of 156 Gb/s, a 56% premium over fixed-rate transponders. This of course comes at the cost of deploying more complex transponders that are PDM-32QAM capable. The same applies to metro networks, where ToRs are replaced with PON OLTs or DSLAMs.

To better grasp the benefits of burst-mode coherent transmission and elasticity within a datacenter, consider the following example; we refer the interested reader to [107] for the full details. With the standard 10 Gb/s, single-wavelength per fiber technology, a datacenter with 140,000 servers each equipped with a gigabit Ethernet network adapter and a 10:1 oversubscription ratio (i.e., some links are under-provisioned by a factor 10) would require around 30,000 linecards (and cables) each operating at 10 Gb/s to interconnect the large switches above the ToR level. The same datacenter would require only 88 linecards (and cables) each operating at up to 250 Gb/s if it was built using a coherent optical burst switching ring. Although no real datacenter will ever be built using a single ring due to protection issues, this small example gives a good sense of what could be achieved using the elastic burst-mode coherent technology.

15.6 CONCLUSIONS

This chapter reported on the recently introduced concept of EONs. Elasticity encompasses various means for providing flexibility, such as FEC or modulation format adaptation, variable symbol rate, and/or variable channel spacing. The latter handle corresponds to a flex-grid network, which offers significant improvements in spectrum usage but requires new hardware deployment (in particular optical filtering) and a more sophisticated control plane.

We reviewed the evolution of optical hardware and software elements for future EONs. Coherent detection with digital signal processing, high-speed DAC and ADC enable an easy adoption of first levels of flexibility towards future improvements with dynamic EONs. New ground in exploiting the opportunities of elasticity allows not just an increase of the network capacity but also a better management of resources in a cost and eco-sustainability way.

Burst-mode solutions may eventually come in a next step as elastic optical technology matures. This disruptive approach in today's circuit-switched based optical communication still leaves substantial work for further developments.

ACKNOWLEDGMENTS

The authors would like to thank Dominique Verchère for useful comments and discussions as well as the following European projects: CELTIC EO-NET, CELTIC+ SASER SAVENET, and FP7 IDEALIST which partly support this work.

REFERENCES

1. Nag A, Tornatore M, Mukherjee B. Optical network design with mixed line rates and multiple modulation formats. *J Lightw Technol* 2010;28(4):466–475.
2. Jinno M, et al. Demonstration of novel spectrum-efficient elastic optical path network with per-channel variable capacity of 40Gb/s to over 400Gb/s. Proceeding of ECOC'08 Brussels, Th.3.F.6; Sep 2008.
3. Jinno M et al. Spectrum-efficient and scalable elastic optical path network: architecture, benefits, and enabling technologies. *IEEE Commun Mag* 2009;47(11):66–73.
4. O. Rival et al. Elastic optical networks with 25-100G format-versatile WDM transmission systems. OECC'10; 2010.
5. Gho G-H et al. Rate-adaptive coding for optical fiber transmission systems. *J Lightwave Technol* 2011;29(2), p. 222–233.
6. Jinno M et al. Multiflow optical transponder for efficient multilayer optical networking. *IEEE Com Mag* 2012;50(5):56–65.
7. Lopèz V et al. Target cost for sliceable bandwidth variable transponders in a real core network. In: Future Network and Mobile Summit (FutureNetworkSummit). IEEE; 2013. p 1–9.

8. International Telecommunication Union, Telecommunication Standardization Sector. Interfaces for the optical transport network. ITU-T Rec. G.709/Y.1331, Edition 4.2; Feb 2012.
9. Poole S. Flexible ROADM architectures for future optical networks. Workshop OSUE, in Proceedings of OFC 2010; San Diego; 21–25 Mar 2010.
10. International Telecommunication Union, Telecommunication Standardization Sector. Spectral grids for WDM applications: DWDM frequency grid. ITU-T Rec. G.694.1, Edition 2; Feb 2012.
11. Renaudier J, Voicila A, Bertran-Pardo O, Rival O, Karlsson M, Charlet G, Bigo S. Comparison of set-partitioned two-polarization 16QAM formats with PDM-QPSK and PDM-8QAM for optical transmission systems with error-correction coding. Proceedings of 38th European Conference on Optical Communication, paper We.1.C.5; 2012.
12. Ungerboeck G. Channel coding with multilevel/phase signals. *IEEE Trans Inform Theory* 1982;IT-28:55–67.
13. X. Zhou, et al. Rate-adaptable optics for next generation long-haul transport networks. IEEE Commun. Mag., 51, 3, 41-49, 2013.
14. Morea A, Rival O, Brochier N, Le Rouzic E. Datarate adaptation for night-time energy savings in core networks. *J Lightwave Technol* 2013;31(5):779–785.
15. Dorize C, Rival O, Costantini C Power scaling of LDPC decoder stage in long haul networks. Proceedings of Photonics in Switching Conference (PS '12); 2012.
16. Gho G-H, Kahn JM. Rate-adaptive modulation and coding for optical fiber transmission systems. *J Lightwave Technol* 2012;30(12):1818–1828.
17. International Telecommunication Union, Telecommunication Standardization Sector. Hitless Adjustment of ODUflex(GFP). ITU-T Rec. G.7044/Y.1347: Amendment 1, Edition 1.1; Feb 2012.
18. Institute of Electrical and Electronics Engineers. Amendment 4: media access control parameters, physical layers and management parameters for 40 Gb/s and 100 Gb/s operation. IEEE 802.3ba-2010; 2010.
19. Poggiolini P. The GN model of non-linear propagation in uncompensated coherent optical systems. *J Lightwave Technol* 2012;30:3857–3859.
20. Grellier E, Bononi A. Quality parameter for coherent transmissions with Gaussian-distributed nonlinear noise. *Opt Express* 2011;19:12781–12788.
21. Vacondio F, Rival O, Simonneau C, Grellier E, Bononi A, Lorcy L, Antona J-C, Bigo S. On nonlinear distortions of highly dispersive optical coherent systems. *Opt Express* 2012;20:1022–1032.
22. Agrell E et al. Power-efficient modulation format in coherent transmission systems. *J Lightwave Technol* 2009;27(22):5115–5126.
23. Penninckx D, et al. Optically preamplified systems: defining a new eye aperture. Proceedings of OFC, WM36; San Jose; 1998.
24. Poggiolini P et al. The GN-model of fiber non-linear propagation and its applications. *J Lightwave Technology* 2014;32:694–721.
25. Lavigne B et al. System design tool for high bit rate terrestrial transmission systems with coherent detection. *Bell Labs Tech J* 2013;18:251–266.
26. Seve E, Ramantanis P, Antona J-C, Grellier E, Rival O, Vacondio F, Bigo S. Semi-analytical model for the performance estimation of 100Gb/s PDM-QPSK optical

transmission systems without inline dispersion compensation and mixed fiber types. Proceedings of European Conference on Optical Communications, ECOC 2013; London, paper Th1.D.2; 2013.

27. Carena A et al. EGN model of non-linear fiber propagation. *Opt Express, OSA* 2014;22:16335–16362.
28. DeMuth B et al. Semi-empirical system scaling rules for DWDM system design. *Opt Express, OSA* 2012;20:2992–3004.
29. Rival O, Mheidly K. Accumulation rate of inter and intra-channel nonlinear distortions in uncompensated 100G PDM-QPSK systems. Optical Fiber Communication Conference, OSA Technical Digest (Optical Society of America, 2012), paper JW2A.52; 2012.
30. Poggiolini P, et al. Simple and effective closed-form GN-model correction formula accounting for signal non-Gaussian distribution. ArXiv e-prints, arXiv1402.3528P; 2014.
31. Serena P, Bononi A, Rossi N. The impact of the modulation dependent nonlinear interference missed by the Gaussian noise model. Proceedings of ECOC; 2014.
32. Christodoulopoulos K, Tomkos I, Varvarigos E. Elastic bandwidth allocation in flexible OFDM-based optical networks. *J Lightwave Technol* 2011;29(9):1354–1366.
33. Klinkowski M, Walkowiak K. Routing and spectrum assignment in spectrum sliced elastic optical path network,'. *IEEE Commun Lett* Aug. 2011;15(8):884–886.
34. Wan X, Wang L, Hua N, Zhang H, Zheng X.. Dynamic routing and spectrum assignment in flexible optical path networks. Proceedings of OFC/NFOEC, paper JWA55, Mar 2011.
35. Rofoee BR, Zervas GS, Yan Y, Simeonidou D. Online repurposing and dimensioning of a programmable fixed-grid and flex-grid optical network. ECOC, paper P.5.2; Sep 2013.
36. K. Zhu and B. Mukherjee, Traffic grooming in an optical WDM mesh network. *IEEE J Sel Area Commun*, 20, No. 1, pp. 122–133, 2002.
37. Dacomo A, De Patre S, Maier G, Pattavina A, Martinelli M. Design of static resilient WDM mesh networks with multiple heuristic criteria. Infocom, paper 1793–1802, vol.3; 2002.
38. Zhang G, Leenheer MD, Mukherjee B. Optical traffic grooming in OFDM-based elastic optical networks. *J Opt Commun Netw* 2012;4(11):B17–B25.
39. Layec P, Morea A, Vacondio F, Rival O, Antona JC. Elastic optical networks: the global evolution to software configurable optical networks. *Bell Lab Tech J* Dec. 2013;18(3):133–152.
40. Christodoulopoulos K, Tomkos I, Varvarigos E. Routing and spectrum allocation in OFDM-based optical networks with elastic bandwidth allocation. Proceedings of IEEE GlobCom; Dec 2010.
41. Christodoulopoulos K, Tomkos I, Varvarigos E. Spectrally/data-rate flexible optical network planning. Proceedings of ECOC, paper We.8.D.3; Sep 2010.
42. Geisler DJ, Proietti R, Yin Y, Scott RP, Cai X, Fontaine NK, Paraschis L, Gerstel O, Yoo SJB. Experimental demonstration of flexible bandwidth networking with real-time impairment awareness. *Opt Express* 2011;19(26):B736–B745.
43. Christodoulopoulos K, Manousakis K, Varvarigos EA, Angelou M, Tomkos I. A multi-cost approach to online impairment-aware RWA. Proceedings of ICC; Jul 2009.
44. Modiano E, Lin PJ. Traffic grooming in WDM networks. *IEEE Commun Mag* 2001;39(7):124–129.
45. Tornatore M, Rottondi C, Rizzelli G, Morea A. Complexity and flexible grid networks. Proceedings of OFC/NFOEC, paper W3A.1; Mar 2014.

46. Zhang G, Leenheer MD, Morea A, Mukherjee B.A survey on OFDM-based elastic core optical networking, *IEEE Commun Surv Tut*, 15, 1, pp. 65–87, First Quarter; 2013.
47. Sone Y, Hirano A, Kadohata A, Jinno M, Ishida O. Routing and spectrum assignment algorithm maximizes spectrum utilization in optical networks. Proceedings of ECOC, paper Mo.1.K.3; Sep 2011.
48. Wang X, Zhang Q, Kim I, Palacharla P, Seliya M. Utilization entropy for assessing resource fragmentation in optical networks. Proceedings of OFC/NFOEC 2012, paper OTh1A.2; Mar 2012.
49. Yu X, et al. Spectrum compactness based defragmentation in flexible bandwidth optical networks. Proceedings of OFC/NFOEC, paper JTh2A.35; 2012.
50. Takagi T Hasegawa H, Sato K-i, Sone Y, Hirano A, Jinno M. Disruption minimized spectrum defragmentation in elastic optical path networks that adopt distance adaptive modulation. Proceedings of ECOC, paper Mo.2.K.3; Sep 2011.
51. Cugini F et al. Push-pull defragmentation without traffic disruption in flexible grid optical networks. *J Lightwave Technol* 2013;31(1):125–133.
52. Proietti R, Yu R, Wen K, Yin Y, Yoo S. Quasi-hitless defragmentation technique in elastic optical networks by a coherent RX LO with fast TX wavelength tracking. Proceedings of Photonic Switching; Sep 2012.
53. Wang X, Kim I, Zhang Q, Palacharla P, Sekiya M. A hitless defragmentation method for self-optimizing flexible grid optical networks. Proceedings of ECOC, paper P5.04; Sep 2012.
54. Wang R, Mukherjee B. Provisioning in elastic optical networks with non-disruptive defragmentation. *J LightwTechnol* 2013;31(15):2491–2500.
55. Azodolmolky S, Wieder P, Yahyapour R. Cloud computing networking: challenges and opportunities for innovations. *IEEE Commun Mag* Jul. 2013;51(7):54–62.
56. E. Mannie et al., Generalized multi-protocol label switching (GMPLS) architecture, IETF RFC 3945; Oct 2004.
57. O. Gonzales de Dios, R. Casellas, F. Zhang, X. Fu, D. Ceccarelli, and I. Hussain, Framework and requirements for GMPLS based control of flexi-grid DWDM Networks. IETF, internet draft; Feb 2013.
58. Meloni G, Paolucci F, Sambo N, Cugini F, Secondini M, Gerardi L, Potì L, Castoldi P. PCE architecture for flexible WSON enabling dynamic rerouting with modulation format adaptation. Proceedings of ECOC, paper Tu.5.K.3; Sep 2011.
59. Li Y, Fei Z, Casellas R. Flexible grid label format in wavelength switched optical network. IETF draft; Jan 2012.
60. Jinno M et al. Elastic and adaptive optical networks: Possible adoption scenarios and future standardization aspects. *IEEE Commun Mag* 2011;49(10):164–172.
61. Channegowda M, Nejabati R, Simeonidou D. Software-defined optical networks technology and infrastructure: enabling software-defined optical network operations. *J Opt Commun Netw* 2013;5(10):A274–A282.
62. Channegowda M, Kostecki P, Efstathiou N, Azodolmolky S, Nejabati R, Kaczmarek P, Autenrieth A, Elbers J-P, Simeonidou D. Experimental evaluation of extended openflow deployment for high-performance optical networks. Proceedings of ECOC, paper Tu.1.D.2; Sep 2012.
63. Das S, Yiakoumis Y, Parulkar G, McKeown N, Singh P, Getachew D, Desai PD.

Application-aware aggregation and traffic engineering in a converged packet-circuit network. in Proceedings of OFC/NFOEC, paper NThD.3; Mar 2011.

64. Shirazipour M, John W, Kempf J, Green H, Tatipamula M. Realizing packet-optical integration with SDN and OpenFlow 1.1 extensions. Proceedings of ICC; Jul 2012. p 6633–6637.
65. Liu L, Muñoz R, Casellas R, Tsuritani T, Martínez R, Morita I. OpenSlice: an OpenFlow-based control plane for spectrum sliced elastic optical path networks. *Opt Express* 2013;21(4):4194–4204.
66. Crabbe E, Minei I, Sivabalan S, Varga R. PCEP extensions for PCE-initiated LSP setup in a stateful PCE model. IETF draft; Jun 2014.
67. Penna MC, Jamhour E, Miguel MLF. A clustered SDN architecture for large scale WSON. Proceedings of of AINA; May 2014. p 374–381.
68. Liu L, Tsuritani T, Morita I, Casellas R, Martínez R, Muñoz R. Control plane techniques for elastic optical networks: GMPLS/PCE vs OpenFlow. Proceedings of Globecom Workshops; Dec 2012.
69. ITU-T Recommendation Y.301,1. Framework of network virtualization for future networks; Jan. 2012.
70. Schmogrow R, Hillerkuss D, Dreschmann M, Huebner M, Winter M, Meyer J, Nebendahl B, Koos C, Becker J, Freude W, Leuthold J. Real-time software-defined multiformat transmitter generating 64QAM at 28 GBd. *IEEE Photon Technol Lett* 2010;22(21):1601–1603.
71. Klekamp A, Rival O, Morea A, Dischler R, Buchali F. Transparent WDM network with bitrate tunable optical OFDM transponders. Proceedings of Optical Fiber Communication Conference (OFC'10); San Diego, CA, paper NTuB5; 2010.
72. Vacondio F, El Falou A, Voicila A, Le Bouëtté C, Tanguy J-M, Simonneau C, Dutisseuil E, Pamart J-L, Schoch L, Rival O. Real-time elastic coherent muxponder enabling energy proportional optical transport. Proceedings of Optical Fiber Communication Conference (OFC '13); Anaheim, CA, paper JTh2A; 2013.
73. Dutisseuil E, Tanguy J-M, Voicila A, Laube R, Bore F, Takeugming H, de Dinechin F, Cerou F, Charlet G. 34 Gb/s PDM-QPSK coherent receiver using SiGe ADCs and a single FPGA for digital signal processing. Proceedings of Optical Fiber Communication Conference (OFC'12); Los Angeles, CA, paper OM3H.7; 2012.
74. O. Rival, G. Villares, and A. Morea, Impact of inter-channel nonlinearities on the planning of 25–100 Gb/s elastic optical networks. *J Lightwave Techn*, 29, Issue 9, pp. 1326–1334, 2011.
75. Morea A, Rival O. Efficiency gain from elastic optical networks. Proceedings of ACP; Nov 2011.
76. Morea A, Rival O. Advantages of elasticity versus fixed data-rate schemes for restorable optical networks. Proceedings of ECOC, paper Th.10.F.5; Sep 2010.
77. Rival O, Morea A. Cost-efficiency of mixed 10-40-100Gb/s networks and elastic optical networks. Proceedings of OFC/NFOEC, paper OTuI4; Mar 2011.
78. Rival O, Morea A, Brochier N, Drid H, Le Rouzic E. Upgrading optical networks with elastic ransponders. Proceedings of ECOC, paper P5.12; Sep 2012.
79. Christodoulopoulos K, Angelou M, Klonidis D, Zakynthinos P, Varvarigos M, Tomkos I. Value analysis methodology for flexible optical networks. Proceedings of of ECOC, paper We.10.P1; Sep 2011.

80. Patel A, Ji P, Jue J, Wang T. Survivable transparent flexible optical WDM (FWDM) networks. Proceedings of OFC/NFOEC, paper O.Tu.I.2; Mar 2011.
81. Liu M, Tornatore M, Mukherjee B. Survivable traffic grooming in elastic optical networks-shared protection. *J Lightwave Technol* 2013;31(6):903–909.
82. Ruan L, Xiao N. Survivable multipath routing and spectrum allocation in OFDM-based flexible optical networks. *IEEE/OSA J Opt Commun Netw* 2013;5(3):172–182.
83. Gosselin S, Saliou F, Bourgart F, Le Rouzic E, Le Masson S, Gati A. Energy consumption of ICT infrastructures: an operator's viewpoint. ECOC Symposium on "Energy Consumption of the Internet"; Sept 2012.
84. Palkopoulou E, Angelou M, Klonidis D, Christodoulopoulos K, Klekamp A, Buchali F, Varvarigos E, Tomkos I. Quantifying spectrum, cost, and energy efficiency in fixed-grid and flex-grid networks. *J Opt Commun Netw* 2012;4(11):B42–B51.
85. Lopez Vizcaino J, Ye Y, López V, Jiménez F, Duque R, Krummrich P. On the energy efficiency of survivable optical transport networks with flexible-grid. Proceedings of ECOC, paper P5.05; Sep 2012.
86. Morea A, Rival O, Brochier N, Le Rouzic E. Data-rate adaptation for night-time energy savings in core networks. *J Lightwave Techn* 2013;31(5):779–785.
87. Perello J, Morea A, Spadaro S, Pagès A, Ricciardi S, Gunkel M, Junyent G. Power consumption reduction through elastic data rate adaptation in survivable multi-layer optical networks. *J Photon Netw Commun* 2014;28(3):276–286.
88. Ghazisaeidi A, Tran P, Brindel P, Bertran-Pardo O, Renaudier J, Charlet G, and Bigo S. Impact of tight optical filtering on the performance of 28 Gbaud Nyquist-WDM PDM-8QAM over 37.5 GHz Grid. Proceedings of OFC, paper OTu3B.6; Mar 2013.
89. Morea A, Renaudier J, Ghazisaeidi A, Zami T, Bertran-Pardo O. Impact of reducing channel spacing from 50GHz to 37.5GHz in fully transparent meshed networks. Proceedings of OFC/NFOEC, paper OTh1E.4; Mar 2014.
90. Kozicki B, Takara H, Tsukishima Y, Yoshimatsu T, Yonenaga K, Jinno M. Experimental demonstration of spectrum-sliced elastic optical path network (SLICE). *Opt Express* 2010;18(21):22105–22118.
91. Morea A, Rival O, Fen Chong A. Impact of transparent network constraints on capacity gain of elastic channel spacing. Proceedings of OFC/NFOEC, paper J.W.A062; Mar 2011.
92. Bosco G, Curri V, Carena A, Poggiolini P, Forghieri F. On the performance of Nyquist-WDM Terabit superchannels based on PM-BPSK, PM-QPSK, PM-8QAM or PM-16QAM Subcarriers. *J Lightwave Technol* 2011;29(1), p. 53–61.
93. Zami T. Comparison of elastic implementations for WDM networks. Proceedings of ECOC'2011, paper We.10.P1.103; Sep 2011.
94. Liou C. Next-generation inter-data center networking. ECOC Special Symposia2; Sep 2013.
95. Alcatel-Lucent's Bell Labs forecasts a 560 percent increase in data traffic on metro networks by 2017, driving a major shift in network design, press release, 3 December 2013, http://www.alcatel-lucent.com/press/2013/002957.
96. Widjaja I, Saniee I, Giles R, Mitra D. Light core and intelligent edge for a flexible, thin-layered, and cost-effective optical transport network. *IEEE Commun Mag* May 2003;41(5):S30–S36.

97. Stavdas A et al. A novel scheme for performing statistical multiplexing in the optical layer. *OSA J Opt Netw* 2005;4(5):237–47.

98. M.C. Yuang, I-F. Chao and B.C. Lo, HOPSMAN: an experimental optical packet-switched metro WDM ring network with high-performance medium access control. *IEEE/OSA J Opt Commun Netw*, 2, 2, 91–101, Feb. 2010.

99. Chiaroni D, Buforn Santamaria G, Simonneau C, Etienne S, Antona J-C, Bigo S, Simsarian J. Packet OADMs for the next generation of ring networks. *Bell Lab Tech J* Winter 2010;14(4):265–283.

100. Cao S, Ma T, Shi X, Luo X, Shen S, Xiong Q. A novel optical burst ring network with optical-layer aggregation and flexible bandwidth provisioning. Proceedings of the IEEE/OSA Optical Fiber Communication Conference (OFC), paper O.Th.R.5; Mar 2012.

101. Maher R, Millar DS, Savory SJ, Thomsen BC. Widely tunable burst mode digital coherent receiver with fast reconfiguration time for 112 Gb/s DP-QPSK WDM networks. *IEEE/OSA J Lightwave Technol* 2012;30(24):3924–3930.

102. Vacondio F, Bertran-Pardo O, Pointurier Y, Fickers J, Ghazisaeidi A, de Valicourt G, Antona J-C, Chanclou P, Bigo S. Flexible TDMA access optical networks enabled by burst-mode software defined coherent transponders. Proceedings of the European Conference on Optical Communication (ECOC); London, UK, paper We.1.F.2; Sept 2013.

103. Gripp J, Simsarian JE, Corteselli S, Pfau T. Wavelength-tunable burst-mode receiver with correlation-based polarization separation. Proceedings of ECOC 2013, paper Th.2.A.3; Sept 2013.

104. Li M, Deng N, Xue Q, Gong G, Feng Z, Cao S. A 100-Gb/s Real-time Burst-mode Coherent PDM-DQPSK Receiver. Proceedings of the European Conference on Optical Communication (ECOC); , London, UK, paper PDP.2.D.4; Sept 2013.

105. ITU-T, recommendation G987. 10-Gigabit-capable passive optical network (XG-PON) systems: Definitions, abbreviations and acronyms; Jun. 2012, https://www.itu.int/rec/T-REC-G.987/en.

106. Deng N, Xue Q, Li M, Gong G, Qiao C. An Optical Multi-ring Burst Network for a Data Center. Proceedings of the IEEE/OSA Optical Fiber Communication Conference (OFC), paper O.Th.1.A.5; Mar 2013.

107. Mestre MA, de Valicourt G, Jennevé P, Mardoyan H, Bigo S, Pointurier Y. Optical Slot Switching-Based Datacenters With Elastic Burst-Mode Coherent Transponders. Proceedings of the European Conference on Optical Communication (ECOC), paper Th.2.2.3, Cannes, France; Sept 2014.

16

SPACE-DIVISION MULTIPLEXING AND MIMO PROCESSING

ROLAND RYF AND NICOLAS K. FONTAINE
Bell Laboratories, Nokia, Holmdel, NJ, USA

16.1 SPACE-DIVISION MULTIPLEXING IN OPTICAL FIBERS

The capacity of fiber-optic communication systems has been increasing continuously by two to three orders of magnitude over the last two decades. Coherent optical communication has provided the most recent capacity improvements, reaching a performance close to the theoretical limit imposed on standard single-mode fibers (SMFs) by Shannon's formula in combination with nonlinear effects [1]. In order to substantially increase the capacity of optical fibers, only one option, involving the introduction of multiple parallel optical paths, is left. The general technique is referred to a space-division multiplexing (SDM) and is the obvious choice for any interconnection technology if the capacity of a single serial channel reaches a technological barrier. Space-division multiplexed system can be implemented by using multiple parallel SMFs; in this form, however, no significant reduction of cost-per-bit can be expected and, therefore, around 2009 a new research effort in fiber optics communication started with the aim of identifying cost-efficient and scalable high-capacity SDM fiber systems [2–4] Two fiber types of particular interest emerged: The multicore fibers (MCFs), where multiple fiber guiding cores are introduced in a common cladding area and multimode fibers (MMFs), where the fiber modes are exploited to transmit multiple parallel channels. Both fiber types are well known but the design was now revisited and optimized for high-capacity transmission of parallel channels.

All SDM transmission systems can essentially be divided into two categories depending on the way the multiple parallel channels are processed at the receiver. In

Enabling Technologies for High Spectral-efficiency Coherent Optical Communication Networks, First Edition. Edited by Xiang Zhou and Chongjin Xie.
© 2016 John Wiley & Sons, Inc. Published 2016 by John Wiley & Sons, Inc.

uncoupled systems, each path is processed individually and any crosstalk from other channels will appear as impermanent limiting either the reach or the capacity of the individual channels. The second approach is based on joint processing of the signals from the multiple parallel paths. The transmission system can then be described as a multiple-input multiple-output (MIMO) channel and well-known digital signal processing (DSP) techniques developed for wireless communication can be adopted. In combination with coherent detection, it is then possible to practically undo any crosstalk between the channels as long as none of the channels is substantially attenuated in respect to the others. The latter condition, which is typically not fulfilled in wireless systems, is often valid for low-loss optical components and multimode optical fibers.

Chapter 16 consists of two main parts: SDM transmission and components. In the first part starting with Section 16.2 we describe various optical fibers that can support SDM transmission. Section 16.3 reviews SDM transmission techniques that can be utilized in case the crosstalk between the spatial channels is small, for example, in MCFs. Section 16.5 introduces the general principles for SDM transmission over optical fibers in the presence of coupling between the spatial channels, describes the basic properties of the fiber-optic multimode channel, such as differential group delay (DGD) and mode-dependent loss (MDL), and describes strategies to reduce the impact of the DGD. Section 16.6 discusses experimental results for MIMO transmission and MIMO DSP techniques specific to SDM systems. The second part of the chapter describes methods for SDM component characterization and components required for SDM transmission systems such as mode couplers, SDM wavelength-selective switches (WSSs), and SDM optical amplifiers.

16.2 OPTICAL FIBERS FOR SDM TRANSMISSION

In SDM transmission systems, multiple parallel optical paths are introduced to provide an increased link capacity. There are numerous ways to introduce multiple optical paths, and some representative approaches are graphically represented (drawn in scale) in Figure 16.1 for a spatial channel with a spatial multiplicity of 7.

Figure 16.1(a) shows the state of the art approach consisting of conventional optical cable containing multiple standard SMFs. Optical cables are widely available and can contain any number of fibers ranging from 1 to well over 1000.

The second solution (see Figure 16.1(b)) consists of using fiber ribbons. This allows for an increase in fiber density of optical cables, and can reduce the time required to splice the optical cable, as fusion splicer for fiber ribbons are available and can splice a entire ribbon during a single splice operation.

A more recent approach that is not yet commercially available consists in the use of so-called multielement fibers [5], which are multiple fibers that after the draw are brought together and covered with a single common coating (see Figure 16.1(c)). Composite fibers can be packed much more densely together than individual SMFs, and basically retain the same optical properties. The individual fibers are accessed by stripping the coating and are then handled similarly to regular SMFs.

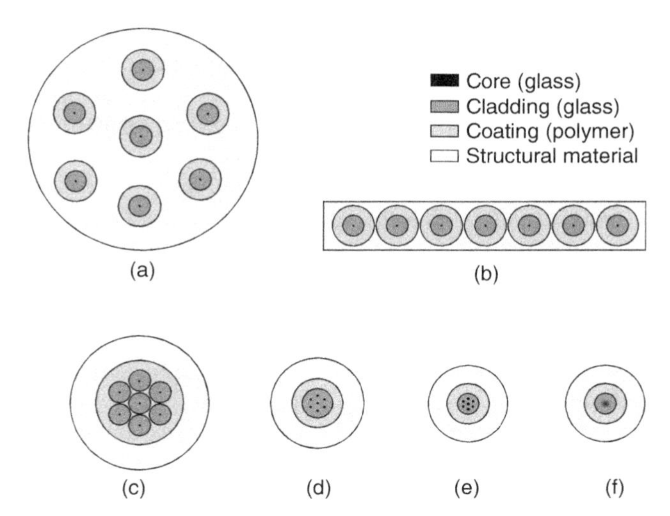

FIGURE 16.1 Fibers supporting SDM. (a) Cable composed of SMFs. (b) Fiber ribbons composed of SMFs. (c) Composite fibers manufactured with a common coating. (d) Multicore fibers composed of multiple single-mode cores embedded in a common cladding. (e) Coupled-core multicore fibers, also referred to as microstructured fibers. (f) Multimode fiber composed of a single core, which supports multiple spatial modes.

Further integration can be reached by using MCFs [6, 7], where multiple fiber cores share a common cladding and are, therefore, embedded physically in a single fiber (Figure 16.1(d)). Splicing MCFs can be achieved by using a fusion splicer that is similar in functionality to fusion splicers designed for splicing polarization maintaining fibers, and therefore allows for splicing multiple spatial channels similarly as with ribbon splicer. Connectors for MCFs have also been demonstrated [8] confirming that technologically MCFs are a viable solution for parallel interconnects. The placement of multiple core in a single cladding, however, also poses some new challenges. To keep the crosstalk negligibly small, the cores have to be placed at a distance >40 μm between each other, but the maximum practical cladding diameter is limited to 250 μm as the fiber becomes fragile and the bending radius tolerance becomes unpractical. The realistic number of cores is currently limited to around 19 [9], but new index profiles and various type of trances such as air holes, are currently under investigation to increase the maximum number of cores.

The MCF design can be dramatically simplified if crosstalk between cores is allowed. The fibers are then referred to as coupled-core multicore fibers (CC-MCFs) or microstructured fibers (Figure 16.1(e)) and the cores can be placed much closer together. The fiber will then behave similarly to a conventional MMF, and the fiber modes are defined by the so-called "super-modes" that are the modes of the composite fiber refractive index profile including all cores. CC-MCFs currently under study have a core-to-core spacing from 22 to 38 μm (see [10] and [11], respectively), but closer spaced cores have been proposed and analyzed in numerical studies. In order to transmit parallel channels over CC-MCFs MIMO DSP techniques are required, and CC-MCFs will typically exhibit a strong mixing between the super-modes, which is advantageous to reduce the impact of core-to-core variation of the fiber optical properties. The implication of strong coupling on transmission is discussed in Section 16.5.3.

Finally, the fibers that can achieve the highest density of spatial channels are MMFs (Figure 16.1(f)). MMFs have a core diameter or a core refractive index such as the resulting waveguide can support multiple spatial modes; this is in contrast to SMFs that only support a single spatial mode, however, with two polarizations. The possibility to transmit multiple signals by making use of fiber modes has been proposed by Berdague and Facq [12]. The concept was revitalized in 2000 by Stuart [13] by proposing to apply MIMO DSP techniques to optical transmission over MMFs in order to increase either capacity or reach, followed by work form numerous other researchers [14–18]. Finally, in 2011, the idea was applied to the so-called few-mode fibers (FMFs) [19–21] that are MMFs with only a small number of spatial modes (typically three or six spatial modes), and numerous long-distance and high-capacity demonstration have followed [22–24]. In contrast to the MCF approach where the challenge involves keeping the coupling between the core small, the challenge in MMF transmission is to minimize the transmission delay difference between the signals traveling over different modes. This delay difference has to be minimized to contain the complexity of the MIMO DSP that is required to undo the coupling between modes and is addressed in detail in Section 16.6.1. MMFs are not just advantageous because of their high spatial modal density, but offers several other significant advantages:

- Three to over 45 times the capacity of SMFs is possible while maintaining a standard cladding diameter of 125 μm.
- FMFs are compatible with conventional fusion splices and also compatible with existing cabling and connector technologies [25].
- Most optical components such as splitters, isolators, circulators, wavelength combiners, and wavelength switches [26] can be modified to support MMFs with a simple design modification.
- The strong-mode overlap in FMFs allows for efficient pump sharing in FMF-based optical amplifiers [27].

These are all important advantages that are partially offset, however, by the need of MIMO DSP to successfully recover the transmitted data. Note, however, that the complexity of MIMO DSP is addressable with current state-of-the-art ASIC technology (see also Section 16.6.1).

The list of approaches to implement SDM fibers shown in Figure 16.1 should give a good representative overview of the most promising technologies; additionally, it is noteworthy that many of the presented approaches can be combined. Of particular interest are for example MCFs with few-mode cores as demonstrated in Refs [28–31], where up to 36 spatial channels have been demonstrated for a combination of uncoupled and CC-MCFs as reported in Ref. [10].

16.3 OPTICAL TRANSMISSION IN SDM FIBERS WITH LOW CROSSTALK

To overcome the capacity bottleneck represented by the SMFs, multiple spatial paths have to be introduced. In order for the resulting SDM system to be cost effective in terms of cost of bit/s, further integration steps are essential. Integration can happen in any part of the optical transmission system, but will be most relevant in the part that contributes significantly to cost of the system, such as transceivers, reconfigurable optical add/drop multiplexers (ROADMs), optical amplifiers, and the optical fibers. An often encountered side effect of integration is the increase of crosstalk between the spatial channels, which will typically grow significantly as function of the length of the transmission system. Managing the system crosstalk is, therefore, essential for SDM systems operated without digital crosstalk mitigation. In addition, the acceptable crosstalk level also depends on the format of the transmitted signal [32, 33]. In Table 16.1, we report the maximum acceptable crosstalk level for a given added system penalty at a bit-error rate (BER) of 10^{-3} [32]. Quadrature phase-shift keying (QPSK) is the most tolerant format and requires the crosstalk to be < -17 dB for a added system penalty of 1 dB. The requirements are more severe for higher-order modulation formats such as quadrature-amplitude modulation (QAM) format with 16 symbol (16QAM) where a minimum crosstalk of -23 dB is required.

The effect of accumulated crosstalk is particularly important in MCFs, as the spatial channels share a common cladding. The theory describing coupling between parallel waveguides is referred in the literature as "coupled-mode theory" [34, 35]; however, early experimental results clearly indicated that coupling in MCFs was not following the coupled-mode theory, which for the case of two cores predicts a sinusoidal transfer of the signals between cores as function of the distance. The reason for the observed discrepancy is that multiple effects, such as fiber imperfection, and macro- and microbending, generally cause an almost random phase relation between the light propagating in different cores. A new theory, the "coupled-power theory" was proposed by Fini et al. [36] and Hayashi et al. [7] and applied to the design MCFs with minimum crosstalk also in practical bending scenarios. Many different approaches to design MCFs, such as the use of trenches, air holes, or even the use of cores with diverse phase velocity (referred to as heterogeneous cores), have been proposed by numerous groups and fibers with up to 19 cores [9] have been successfully demonstrated. The main outcome from the coupled-power theory is that

TABLE 16.1 Maximum acceptable crosstalk level for a given added system penalty observed at a bit-error rate of 10^{-3} as a function of the modulation format (according to [32] for a 21.4-Gbaud signal)

Acceptable system penalty (dB)	Maximum crosstalk (dB)			
	QPSK	16QAM	64QAM	256QAM
1	−17	−23	−29	−35
3	−12	−18.5	−25	−31

TABLE 16.2 Summary of relevant SDM transmission in multicore fibers

Fiber type	Nr spatial channels	Spectral efficiency (bit/s/Hz)	Distance (km)	Spectral efficiency distance (bit/s/Hz km)	Capacity (Tbit/s)	References
MCF	7	28	7326	205128	140.7	[37]
MCF[a]	12	73.6	1500	110374	688	[38]
MCF	7	14.4	6160	88704	28.8	[39]
MCF	7	15	2688	40320	9	[6]
MCF	12	91.4	55	4753	1010	[40]
MCF[b]	14	109	3	327	1050	[31]
MCF	19	30.5	10.1	307	305	[9]
MCF	7	11.2	16.8	188.2	109	[41]

[a]Core interleaved bidirectional transmission.
[b]Twelve single-mode cores and two few-mode cores with three spatial modes.

the crosstalk accumulates linearly as function of the fiber length, which allowed the design of MCFs for long-distance transmission. Let us for example consider a system with 5000 km transmission based on QPSK signals. The resulting crosstalk requirements for a single 100 km span is then a maximum crosstalk < -34 dB, which can be easily achieved with seven core fibers [37].

Numerous high-performance digital optical transmission experiments have been performed over MCFs by various groups. Some of the most significant results are summarized in Table 16.2. The maximum demonstrated capacity was over 1 Pbit/s in two cases, which are also the experiments with record high spectral efficiency, above 100 bit/s/Hz in one case. In addition, there have also been significant long-distance results reaching distances up to 7326 km and record spectral-efficiency distance products above 200,000 bit/s/Hz km.

The MCFs research activities have been covering various aspects of MCFs: New OTDR measurement techniques to characterize the MCF crosstalk [42], study of splices for MCFs, connectors [8] and even interoperability studies between MCFs from various fiber manufacturers [43], indicating that the technology is ready for real applications.

16.3.1 Digital Signal Processing Techniques for SDM Fibers with Low Crosstalk

DSP techniques developed for SMFs can be directly applied to MCFs by treating every core independently and practically scale the capacity of the link up to the number of cores. In MCFs, it comes naturally to consider all signals in the individual cores as a single SDM channel, and the term "spatial super channel" has been coined [44] to enforce the idea. Spatial super channels offer significant advantages: for example, signals at a single wavelength can be transmitted over a spatial super channel by using a single wavelength. This can be used to significantly reduce the cost of the transceivers and to decrease the DSP complexity at the receiver. In fact, all signals in the spatial channels that originate from a single laser will also exhibit similar phase noise, and it was proposed by Feuer et al. [45] to use a single joint frequency offset compensation circuit for all spatial channels. Similar work was also reported in Ref. [46], and Sakaguchi et al. [47] even proposed a device that can synchronize the phase noise between spatial channels.

The presence of multiple spatial channels can also be exploited to implement a self-homodyne transmission, where one spatial channel is used to transmit a copy of the unmodulated transmit laser as pilot tone to be used as local oscillator at the receiver [48, 49]. The idea was originally proposed in SMFs where, for example, a polarization orthogonal to the signal can be used to transmit a pilot tone. The concept becomes more attractive in MCFs because by sacrificing a single core for the pilot tone transmission, only a modest reduction in the capacity has to be accepted. Self-homodyne transmission in MCFs has multiple benefits, and optimized phase recovery algorithms have been studied and proposed in Ref. [50]. Self-homodyne detection makes the system much more insensitive to laser phase noise, allowing the use of low-cost distributed feedback laser (DFB) instead of the traditionally employed external cavity laser (ECL), which is a significant cost advantage [51].

Finally, the presence of multiple spatial modes can also be exploited to increase the sensitivity of the transmitted signal by sending multiple copies of the signal over different cores and digitally coherently superposing the multiple copies after receiving. The scheme was proposed and investigated in detail by Liu et al. [52]. Furthermore, Liu et al. discovered that the sensitivity improvement could be further enhanced by sending multiple scramble copies of the signal. The received signals are then unscrambled and added coherently in the digital domain. The scheme works because the scrambled signals will be impacted differently by the nonlinear fiber effects, and the reconstructed channels have less distortions than in the case where the signals are unscrambled and basically present highly correlated nonlinear distortions. Using the technique, signal quality factor improvement of almost 5 dB by using three scrambled transmitted copies over a distance of 2688 km has been demonstrated.

16.4 MIMO-BASED OPTICAL TRANSMISSION IN SDM FIBERS

The main driver for SDM system cost reduction is integration of multiple spatial channels along all possible components of the transmission system [2]. In many occasions, integration will also result in added crosstalk between the SDM channels. Crosstalk in low-loss fibers and optical components does not fundamentally degrade the information transmitted, but just mixes the transmitted signals. For moderate transmission powers, the system can be approximated by a linear transmission function and the transmitted signals can then be recovered by applying the corresponding inverse linear transformation that undoes the mixing that occurs during transmission. The principle is shown in Figure 16.2: Multiple signals $s_1 \ldots s_M$ are generated from a common laser and coupled into an SDM fiber by using an SDM multiplexer. The transmission system looks similar to a conventional single-mode system, except that all components such as fibers, optical amplifiers, or WSSs are

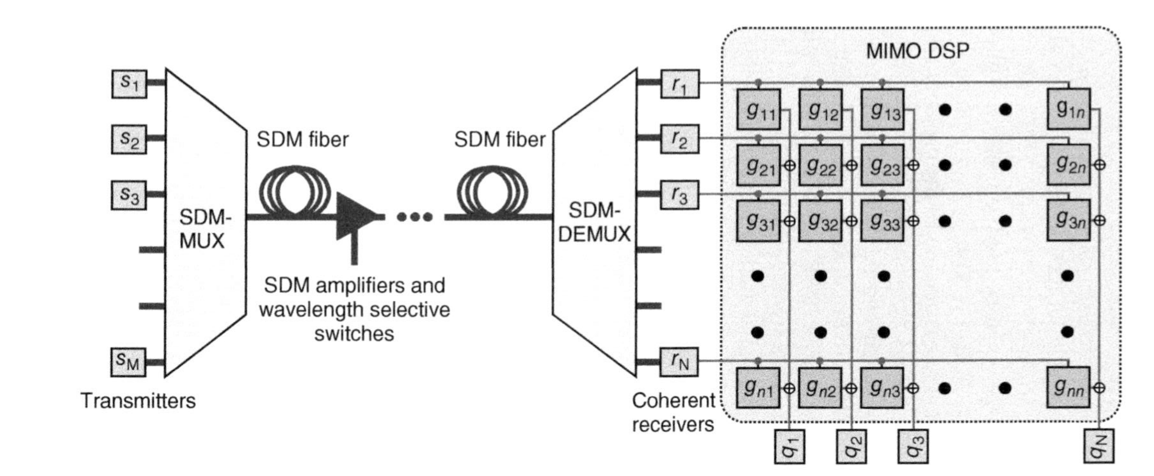

FIGURE 16.2 Basic concept of MIMO-based transmission over SDM fibers: The multiple channels are coupled into the SDM fiber using spatial multiplexers (SDM-MUX) and the transmitted signals are recovered using MIMO digital signal processing consisting of an $N \times N$ array of equalizers.

replaced with their SDM counterparts. At the receiver, an SDM demultiplexer separates the signals by spatial channels, and the signals $r_1 \ldots r_N$ are received by coherent receivers and fed into the MIMO DSP block that is capable of undoing all linear impairments of the transmission system and giving the reconstructed signals $q_1 \ldots q_N$ as output. The MIMO DSP found in SDM systems can be considered an extension of the 2×2 MIMO DSP encountered in polarization multiplexed system as described, for example, in Chapter 6, and is addressed in more detail in Section 16.6.1.

If nonlinear effects are neglected, the SDM transmission system can be described by the linear $N \times M$ channel matrix $h_{j,k}$, where each element of the matrix contains the complex amplitude impulse response for a particular pair of M input signals and N output signals, which can be explicitly written as [53]

$$r_j(l) = \sum_{m=1}^{L_h} h_{j,k}(m) \cdot s_k(l + m - m_o) \tag{16.1}$$

where $s_k(l)$ is the signal transmitted on channel k at discrete time index l, $r_j(l)$ is the signal received on channel j at discrete time index l, and m_0 is a constant that accounts for the overall propagation delay occurring in the MIMO channel. For practical purposes, it is convenient to rewrite Equation 16.1 in a form that can be evaluated using conventional matrix multiplication rules. We define the matrix $\mathbf{H}$ describing the SDM system and the vectors $\mathbf{s}(l)$ and $\mathbf{r}(l)$ describing the input and output signals, respectively, as

$$\mathbf{H} = \begin{bmatrix} h_{1,1}(1) & h_{1,1}(2) & \ldots & h_{1,1}(L_h) & h_{1,2}(1) & h_{1,2}(2) & \ldots & h_{1,2}(L_h) & \ldots & \ldots & h_{1,M}(1) & h_{1,M}(2) & \ldots & h_{1,M}(L_h) \\ h_{2,1}(1) & h_{2,1}(2) & \ldots & h_{2,1}(L_h) & h_{2,2}(1) & h_{2,2}(2) & \ldots & h_{2,2}(L_h) & \ldots & \ldots & h_{2,M}(1) & h_{2,M}(2) & \ldots & h_{2,M}(L_h) \\ \vdots & \vdots & \ddots & \vdots & \vdots & \vdots & \ddots & \vdots & \ddots & \ddots & \vdots & \vdots & \ddots & \vdots \\ h_{N,1}(1) & h_{N,1}(2) & \ldots & h_{N,1}(L_h) & h_{N,2}(1) & h_{N,2}(2) & \ldots & h_{N,2}(L_h) & \ldots & \ldots & h_{N,M}(1) & h_{N,M}(2) & \ldots & h_{N,M}(L_h) \end{bmatrix} \tag{16.2}$$

$$\mathbf{s}(l) = \begin{bmatrix} s_1(l - m_0 + 1) \\ s_1(l - m_0 + 2) \\ \vdots \\ s_1(l - m_0 + L_h) \\ s_2(l - m_0 + 1) \\ s_2(l - m_0 + 2) \\ \vdots \\ s_2(l - m_0 + L_h) \\ \vdots \\ \vdots \\ s_M(l - m_0 + 1) \\ s_M(l - m_0 + 2) \\ \vdots \\ s_M(l - m_0 + L_h) \end{bmatrix}, \text{ and} \tag{16.3}$$

$$\mathbf{r}(l) = \begin{bmatrix} r_1(l) \\ r_2(l) \\ \vdots \\ r_N(l) \end{bmatrix} \tag{16.4}$$

and Equation 16.1 can then be rewritten in matrix form as

$$\mathbf{r}(l) = \mathbf{H} \cdot \mathbf{s}(l) \tag{16.5}$$

Equation 16.5 calculates the output vector $\mathbf{r}(l)$ for only one particular time index l, and in practice it is desirable to expand the equation to include values for multiple time indices. This can be achieved by expanding the vectors $\mathbf{s}(l)$ and $\mathbf{r}(l)$ into the corresponding matrices $\mathbf{S}(l)$ and $\mathbf{R}(l)$ defined as

$$\mathbf{S}(l) = \begin{bmatrix} s_1(l-m_0+1) & s_1(l-m_0+2) & \dots & s_1(l-m_0+L_p) \\ s_1(l-m_0+2) & s_1(l-m_0+3) & \dots & s_1(l-m_0+L_p+1) \\ \vdots & \vdots & \ddots & \vdots \\ s_1(l-m_0+L_h+1) & s_1(l-m_0+L_h+2) & \dots & s_1(l-m_0+L_h+L_p) \\ s_2(l-m_0+1) & s_2(l-m_0+2) & \dots & s_2(l-m_0+L_p) \\ s_2(l-m_0+2) & s_2(l-m_0+3) & \dots & s_2(l-m_0+L_p+1) \\ \vdots & \vdots & \ddots & \vdots \\ s_2(l-m_0+L_h+1) & s_2(l-m_0+L_h+2) & \dots & s_2(l-m_0+L_h+L_p) \\ \vdots & \vdots & \ddots & \vdots \\ \vdots & \vdots & \ddots & \vdots \\ s_M(l-m_0+1) & s_M(l-m_0+2) & \dots & s_M(l-m_0+L_p) \\ s_M(l-m_0+2) & s_M(l-m_0+3) & \dots & s_M(l-m_0+L_p+1) \\ \vdots & \vdots & \ddots & \vdots \\ s_M(l-m_0+L_h+1) & s_M(l-m_0+L_h+2) & \dots & s_M(l-m_0+L_h+L_p) \end{bmatrix}, \text{ and} \tag{16.6}$$

$$\mathbf{R}(l) = \begin{bmatrix} r_1(l) & r_1(l+1) & \dots & r_1(l+L_p-1) \\ r_2(l) & r_2(l+1) & \dots & r_2(l+L_p-1) \\ \vdots & \vdots & \ddots & \vdots \\ r_N(l) & r_N(l+1) & \dots & r_N(l+L_p-1) \end{bmatrix} \tag{16.7}$$

Finally, the SDM system can be described as multiplication between two matrices in the following simple form:

$$\mathbf{R} = \mathbf{H} \cdot \mathbf{S} \tag{16.8}$$

In a practical SDM transmission experiment, the transmitted signal $\mathbf{S}(l)$ is formed by the chosen test pattern, for example, a De Bruijn bit sequence (DBBS), and the received signal $\mathbf{R}(l)$ captured by a multichannel real-time digital oscilloscope, and the system channel matrix **H** can then be estimated by multiplying Equation16.8 from the right side with the pseudoinverse of the matrix **S**. We obtain

$$\hat{\mathbf{H}} = \mathbf{R} \cdot \mathbf{S}^* \cdot (\mathbf{S} \cdot \mathbf{S}^*)^{-1} \tag{16.9}$$

where * denotes the conjugate transpose operation and $\hat{\mathbf{H}}$ is the SDM channel estimation according to the least square criterion.

The estimated matrix $\hat{\mathbf{H}}$ contains all combination of input–output impulse responses and is an important tool to study the transmission performance limitations in SDM systems. It contains information such as the maximum delay spread and the maximum capacity estimate [54], and it can also be used to analyze the temporal evolution of the SDM system [54, 55].

In addition, the channel matrix can be further analyzed by performing a singular value decomposition (SVD), where the channel matrix is decomposed according to

$$\hat{\mathbf{H}} = \mathbf{U} \cdot \mathbf{\Lambda} \mathbf{V}^* \tag{16.10}$$

where **U** and **V** are unitary matrices and the rectangular diagonal matrix $\mathbf{\Lambda}$ contains the real and positive singular values $\lambda_1 \ldots \lambda_N$. The singular values squared λ_i^2 represent the intensity transfer functions for the spatial paths transmitted through the SDM system. The ratio between the smallest and the largest singular values squared is defined as MDL according to

$$\mathrm{MDL}_{\mathrm{AVG}} = 10 * \log 10 \left(\frac{\max(\lambda_i^2)}{\min(\lambda_i^2)} \right) \tag{16.11}$$

and is an important indicator for the system performance that can be expected for the SDM channel [56, 57].

The linear MIMO channel matrix as defined in Equation 16.1 implicitly includes a convolution between the input signal and the output signal pairs. When transforming the channel matrix from the time domain into the frequency domain, the convolution will be replaced with a simple multiplication and the SDM system can then be described as

$$\tilde{\mathbf{r}}(\omega) = \tilde{\mathbf{H}}(\omega) \cdot \tilde{\mathbf{s}}(\omega) \tag{16.12}$$

where the transmitted signal vectors $\tilde{\mathbf{s}}(\omega)$ has M elements and the received signal vector $\tilde{\mathbf{r}}(\omega)$ has N elements corresponding to the input and output channels, respectively. The frequency channel matrix is related to the time-domain channel matrix by the Fourier transform along the time index m of $h_{j,k}(m)$ for each combination of input and output channels j, k. The channel matrix represented in the frequency domain also offers the advantage that the total channel matrix of multiple cascaded SDM subsystems can be obtained by the matrix multiplication of the subsequent subsystem matrices (see Section 16.7.1).

The concept of MDL is often also applied to the channel matrix in the frequency domain. The singular values are then evaluated for each individual frequency components and the resulting MDL becomes frequency dependent

$$\mathrm{MDL}(\omega) = 10 * \log 10\left(\frac{\max(\tilde{\lambda}_i^2(\omega))}{\min(\tilde{\lambda}_i^2)(\omega)}\right) \tag{16.13}$$

which is particularly relevant for characterization of components as described in Section 16.7.1. The MDL of the channel matrix can exhibit strong frequency dependence, and fading dips are often observed in long-distance MIMO transmission experiments that can dramatically impact the performance of orthogonal frequency-division multiplexed (OFDM) MIMO system as reported in Refs [58] and [59].

16.5 IMPULSE RESPONSE IN SDM FIBERS WITH MODE COUPLING

In order to better understand mode coupling in a SDM fiber, it is instructive to study the spatial modes supported by the fiber. Modes in optical fibers are solutions of the Maxwell equation in a dielectric material that are invariant along the propagation direction [60]. The solutions of the mode equation can be calculated analytically for a few special cases, such as for fibers with a step-index profile where the solutions are Bessel functions, or for graded-index fibers where the solutions are Laguerre–Gaussian functions. The latter are reported as examples in Figure 16.3. In general, the modes can be calculated numerically using mode solvers that are based on finite-element methods. The solution of the mode equation provides not only the spatial amplitude and phase, but also the propagation constant β_n of the corresponding mode n. The propagation constant is wavelength dependent and depends both on the material dispersion and the refractive index profile of the waveguide. In general, the propagation constant can be approximated by the Taylor expansion

$$\beta_n(\omega) = \beta_{0,n} + \beta_{1,n}(\omega - \omega_0) + \frac{1}{2}\beta_{2,n}(\omega - \omega_0)^2 + \frac{1}{6}\beta_{3,n}(\omega - \omega_0)^3, \tag{16.14}$$

where ω is the angular frequency of the light. The phase velocity that describes the speed at which a constant phase front travels along the waveguide is given by

$$v_{ph,n} = \frac{\omega_0}{\beta_{0,n}}. \tag{16.15}$$

The second term of Equation 16.14 is related to the group velocity, which describes the propagation speed of light pulses in the fiber and is defined as

$$v_{gr,n} = \frac{1}{\beta_{1,n}}. \tag{16.16}$$

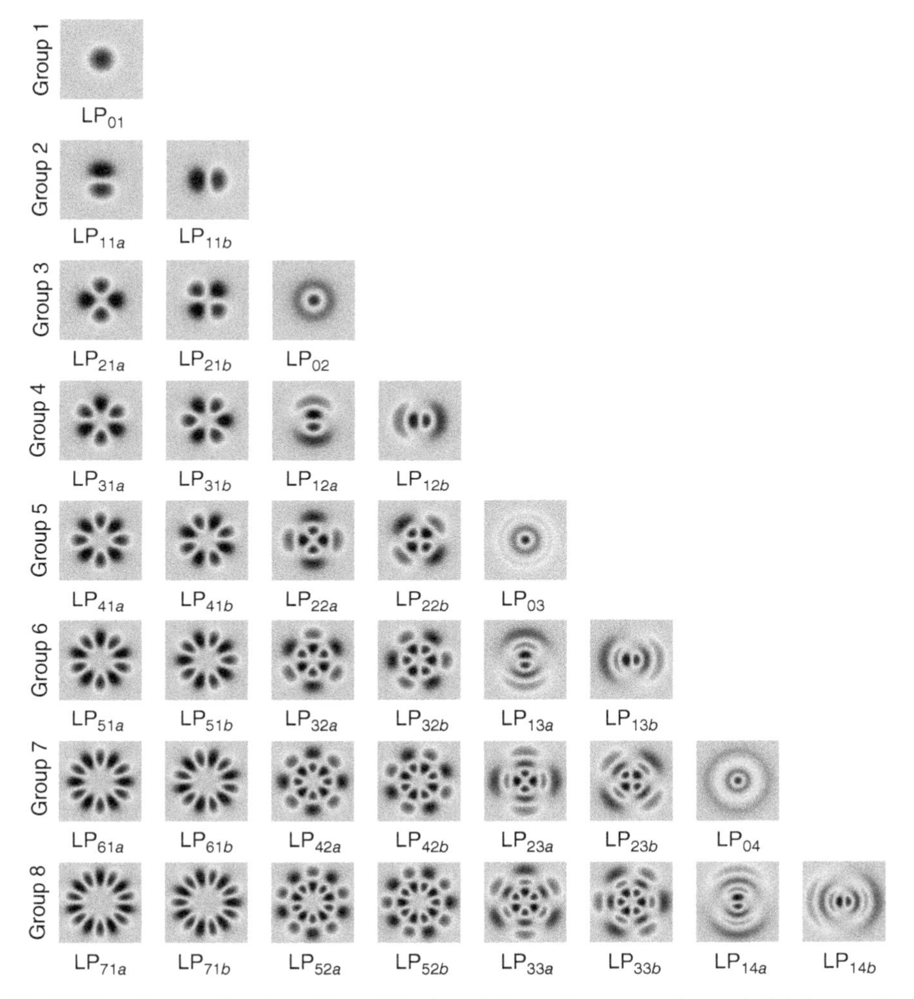

FIGURE 16.3 Amplitude profiles of the first eight mode groups of a graded-index profile multimode fibers. Modes that are part of the same group are degenerate.

Finally, the third term of Equation 16.14 is related to the chromatic dispersion coefficient, which is defined as

$$cd_n = \frac{2\pi c}{\lambda^2}\beta_{2,n}, \tag{16.17}$$

where c is the speed of light in vacuum, and $\lambda = 2\pi c/\omega$ is the wavelength of the light.

Modes are independent solutions of the mode equation and are orthogonal to each other, and form a full orthogonal set describing all the possible waves that can propagate through the waveguide. Perfect waveguides do not show couplings between the modes; however, in real fibers, small variations in refractive index, and fiber geometry or bends in the fibers can cause modes to couple. The relative phase velocity between two modes is an important parameter in determining the coupling between the modes. Modes that have almost the same phase velocity and also degenerate modes that have exactly the same phase velocity can be easily coupled by small perturbation, whereas generally less coupling is observed between modes with larger difference in phase velocity. Alternatively, the effect of the difference in phase velocity can be transformed into the so-called beat length $L_{n,m}$ defined as

$$L_{n,m} = \frac{2\pi}{|\beta_{0,n} - \beta_{0,m}|} \tag{16.18}$$

which is indicative of the length scale over which the coupling occurs. Because mode coupling is caused by multiple distributed fiber imperfections, it is difficult to identify the exact cause of the coupling. Numerous models have been proposed to describe mode coupling in SDM fibers under various assumptions such as slight offset of the cores in different fiber sections [61, 62] or the introduction of a generalized Stokes space and relative coupling matrix as presented in Ref. [63].

The group velocity difference is also a very important parameter, as it determines the maximum time delay between the received signals and, therefore, directly determines the number of equalizer taps necessary to completely undo the effect of mode coupling. The DGD between two modes of a fiber can be calculated by

$$\mathrm{DGD}_{n,m} = L_F|\beta_{1,n} - \beta_{1,m}| \tag{16.19}$$

where L_F is the fiber length. MMF designs that minimize the DGD across a wide wavelength range are, therefore, essential. The conventional step-index fiber modified to support multiple mode has typically more than 1 ns/km of DGD, which makes MIMO transmission impractical to more than 50 km. Various fiber designs to minimize the DGD of MMFs have been proposed; the most common are depressed cladding (DC)-based designs [64], ring-core fibers (RC) [65, 66], and graded-index (GI) profile fibers [67]. MMFs with graded-index profiles are also popular in short-distance interconnects and can theoretically have very small DGD values. They also offer other interesting properties, such as having a very similar difference in phase velocity between subsequent mode groups, and therefore optimally utilize the index contrast provided by the doping profile [68]. The first eight mode groups of a graded-index fiber are shown in Figure 16.3. Each subsequent mode group contains one additional spatial mode, and all modes within a mode group are nominally degenerate. Graded-index MMFs are the most often used fibers in MIMO-based SDM experimental demonstrations. Optimized MMFs have typical DGD values around 30–100 ps/km for fiber with six or less spatial modes. Consistent fabrication of MMFs with smaller DGD values is difficult as the relative group velocity error has to be controlled to an accuracy of better than 1 ppm.

In practice, when measuring the total intensity impulse response of an MMF, three different scenarios, represented graphically in Figure 16.4 for an hypothetical MMF with two spatial modes, can occur. Note that all three scenarios can be encountered in

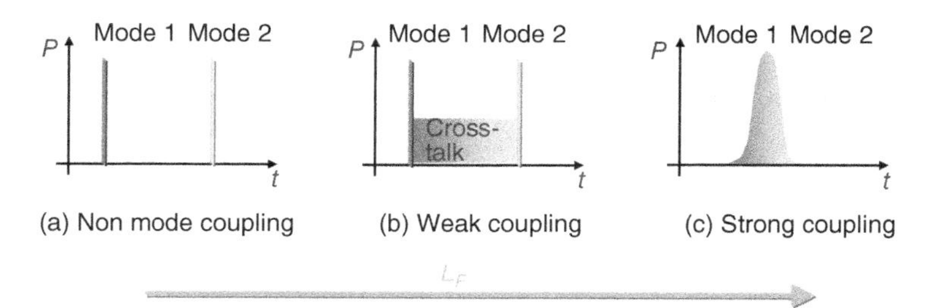

FIGURE 16.4 Possible total intensity impulse response regimes for a hypothetical fibers with two spatial modes. (a) Uncoupled modes. (b) Moderately coupled modes. (c) Strongly coupled modes (see text for description).

the same MMF just by varying the transmission length, which is graphically indicated by an arrow in Figure 16.4. All three scenarios have been experimentally investigated and are discussed later in detail.

16.5.1 Multimode Fibers without Mode Coupling

In the first scenario, mode coupling is negligible, and the impulse response consists of two delayed peaks separated by DGD between the two modes. This situation can be observed in short sections of MMFs. It is then possible to transmit independent signals over each mode without the need of MIMO DSP. Early SDM demonstrations used this scheme to transmit over distances up to 10 km [69] using direct modulation and detection. A more advanced scheme employing LP_{01} and LP_{11} modes used 2×2 MIMO DSP to transmit over the LP_{01} mode and 4×4 MIMO DSP to transmit over the degenerate LP_{11} modes of an MMF with three spatial modes, reaching a transmission distance of up to 40 km [70] based on a QPSK transmission signal. The scheme was also applied to a fiber with five spatial modes reaching a distance of 40 km [71]. The scheme requires mode couplers with high modal selectivity, but ultimately the crosstalk build-up in the fibers themselves are limiting the transmission distance to <100 km.

16.5.2 Multimode Fibers with Weak Coupling

In the second scenario, a weak crosstalk is present and produces an almost random uniform crosstalk plateau between the peaks corresponding to the undisturbed modes. The crosstalk plateau is generated by the light coupling between modes, which happens all along the fiber propagation, and a particular time position on the plateau can then be mapped to a location in the fiber where the coupling occurs. For homogeneous fibers, the plateau will have a constant envelop, and defects, such as splices present in the fiber span, will generate localized peaks. In this scenario, the accumulated crosstalk of the whole plateau is large enough to require MIMO DSP to compensate for it, and the distance between the peaks is still determined by the DGD between the fiber modes and, therefore, will determine the minimum depth of the MIMO equalizer memory. For long-distance communication (1000 km or more), the DGD of an optimized MMF is still too large, and additional strategies are required to mitigate the effects of DGD. The strategy involves cascading the MMF with a second optical element that has the inverse DGD effect: The compensating element can be an MMF with corresponding negative DGD values or a DGD compensator according to Figure 16.18. The DGD compensation is presented graphically in Figure 16.5, where the group delay as function of the distance is shown in a DGD-compensated fiber span, showing the modes separating in time during the first section of fiber and then overlapping at the end of the fiber span after traversing a section of the fiber with the inverse DGD.

Understanding the benefits of DGD compensation in the presence of modal crosstalk is somewhat counterintuitive. In fact, the mode coupling between non-degenerate modes, which can be as large as 8% for a 100 km GI-MMF span, can happen at any location along the compensated span and, therefore, produce copies of the signal delayed by amounts up to the DGD of the uncompensated fiber. The inherent advantage of DGD compensation, which is to limit the spread of the system impulse response, is therefore not apparent anymore. Nevertheless, the following simulations and experimental results confirm DGD compensation to be effective, but only in multispan systems, and under the condition that the majority of the optical power stays in the mode it was launched into. If the last condition is not fulfilled, strong coupling will reduce the impact of DGD as described in Section 16.5.3. DGD compensation can be performed at a span level or also within a span. Deployed fibers typically consist of multiple cables that are spliced every kilometer, and it is, therefore, conceivable to perform DGD compensation by splicing cables with DGD values with alternating sign. The effect of DGD compensation is demonstrated in Figure 16.6, where the sum of all intensity responses is simulated for link lengths

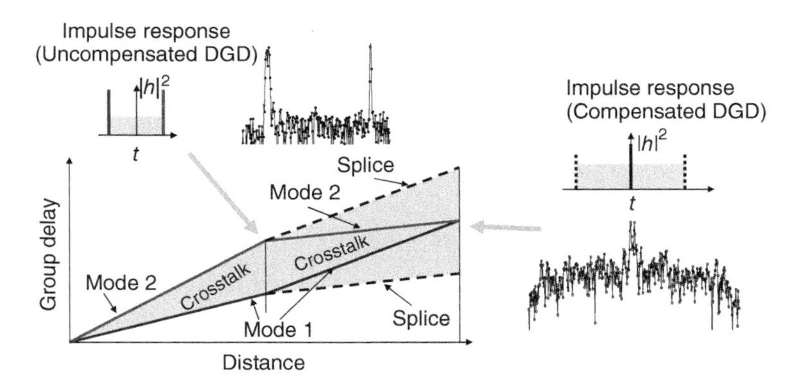

FIGURE 16.5 Principle of DGD compensation: signal pulses in two different modes travel at two different speeds. At the middle of the span the DGD is inverted and the pulse arrive at the end of the fiber at the same time. The effect of crosstalk and mode coupling in splices is also indicated. Furthermore, the impulse response in the middle and at the end of the span is indicated by arrows.

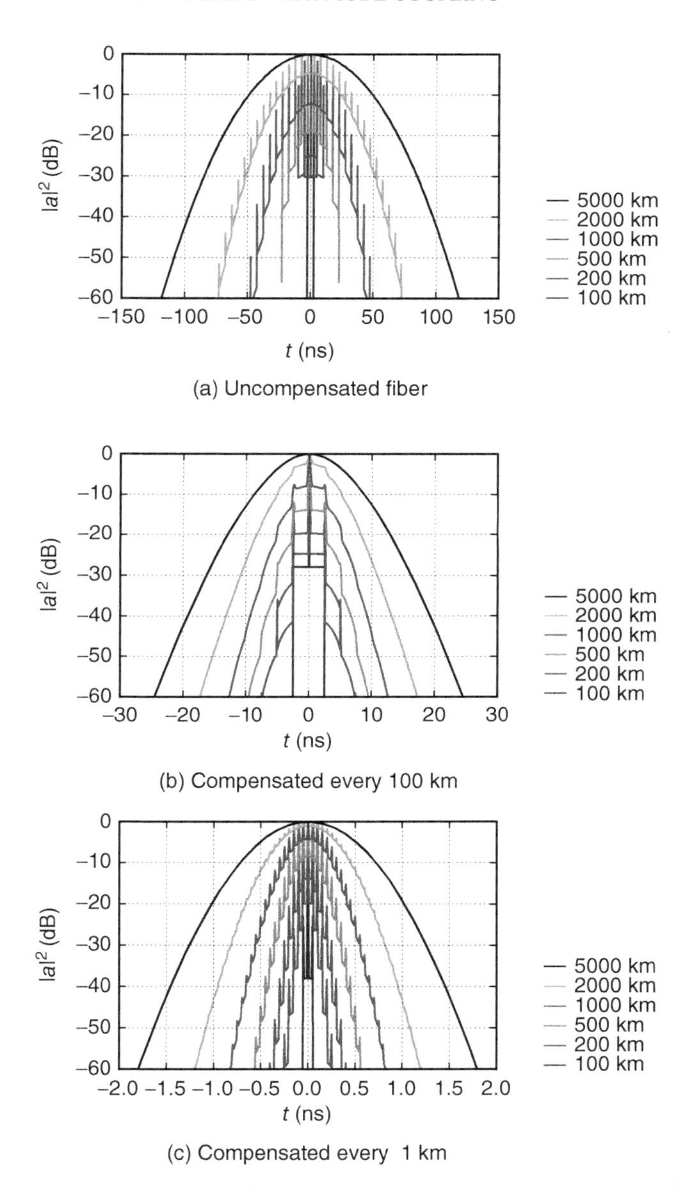

FIGURE 16.6 Impulse response simulations for a multispan system after 100, 200, 500, 1000, 2000, and 5000 km distance for (a) DGD uncompensated spans, (b) DGD compensated spans, and (c) DGD compensated every kilometer.

of 100, 200, 500, 1000, 2000, and 5000 km, based on a 100-km spans, and reported for (a) DGD uncompensated spans, (b) DGD compensated spans, and (c) spans compensated every kilometer. The simulations are based on a simple model [72], which uses incoherent convolutions to determine the effects of cascaded spans and additionally mode couplers with strong coupling are assumed after each span. Other simulation parameters are as follows: 8% distributed mode coupling between nondegenerate modes per span, 1% mode coupling at splice locations, fibers with 50 ps/km DGD, and a residual not compensated DGD of 100 ps after each span. The results demonstrate the efficacy of DGD compensation, as can be observed by comparing the different time scale for (a), (b), and (c) in Figure 16.6. When DGD compensation is applied at every span, the spread of the impulse response is reduced by a factor of 5. Furthermore, when DGD compensation is performed every kilometer, the impulse response spread is reduced by an additional factor of 50. DGD compensation was also confirmed experimentally, performing recirculating loop experiments in FMF with three and six spatial modes. The total intensity responses are reported for different propagation distances in Figure 16.7(a) and (b), for 30-km FMF with three spatial modes and 59-km FMF with six spatial modes, respectively. Both fibers were DGD compensated and show a strong central peak after traversing the fiber the first time. In the subsequent loop traversals, the width of the impulse response stays contained within the DGD compensation windows as long as the central peak is dominating. At longer distances, the impulse response changes to a bell-shaped curve and will transit into the strong coupling regime as described in Section 16.5.3.

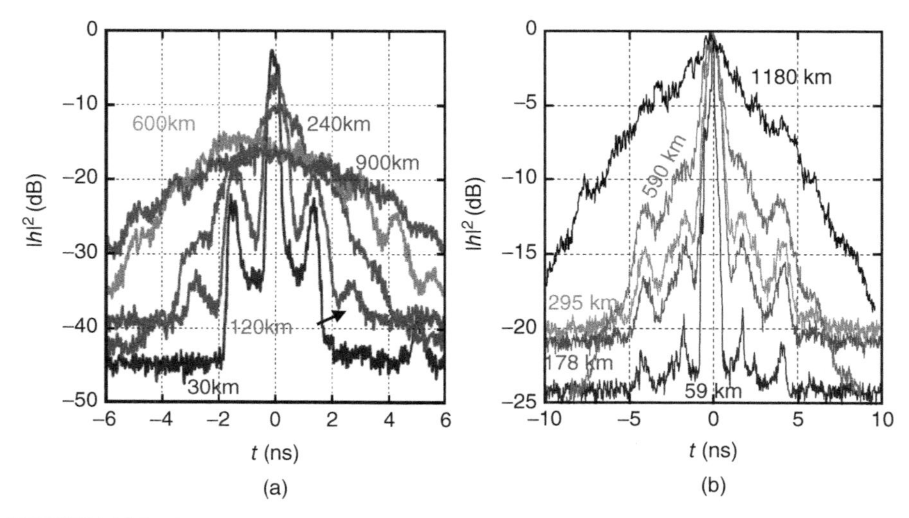

FIGURE 16.7 Experimental impulse response plotted for different transmission distances for (a) 30-km span DGD compensated few-mode fiber with three spatial modes [73] and (b) 59-km span DGD compensated few-mode fiber with six spatial modes [74].

16.5.3 Multimode Fibers with Strong Mode Coupling

The third scenario that can be encountered in MMFs is the so-called strong coupling regime, where the coupling between the modes is so strong to continuously mix the signals between the modes. The strong coupling regime is beneficial for reducing the effect of DGD, because before any significant amount of delay between modes can build up, the signals are mixed all over again and the resulting width of the impulse response will stay narrower than the value of the DGD. Furthermore, strong coupling will cause the system impulse response width to grow as a square root of the propagation distance, instead of a linear growth as observed in weakly coupled FMFs without DGD compensation. Strong coupling offers similar advantages in case MDL is present in the transmission system. The benefits of the strong coupling regime in terms of tolerance to MDL and reduction of the impulse response spread have been theoretically investigated in detail by Kahn et al. [75, 76]. In conventional MMFs, strong coupling occurs at transmission distances starting typically above 500 km as can be observed in Figure 16.7, where the impulse response starts to become bell shaped. Intentionally introducing mode coupling into MMFs by perturbing the fibers along propagation has been proposed, but an effective method that does not introduce excess loss has not been demonstrated yet. A fiber where strong coupling between super-modes was observed is the CC-MCF where MIMO-based transmission distances up to 4200 km have been demonstrated [72]. The impulse response and the width of the impulse response as function of the transmission distance are shown in Figure 16.8(a) and (b), respectively. The experiment shows that the FWHM of the impulse response growth form 230 ps after 60 km, to only 2.8 ns after a distance of

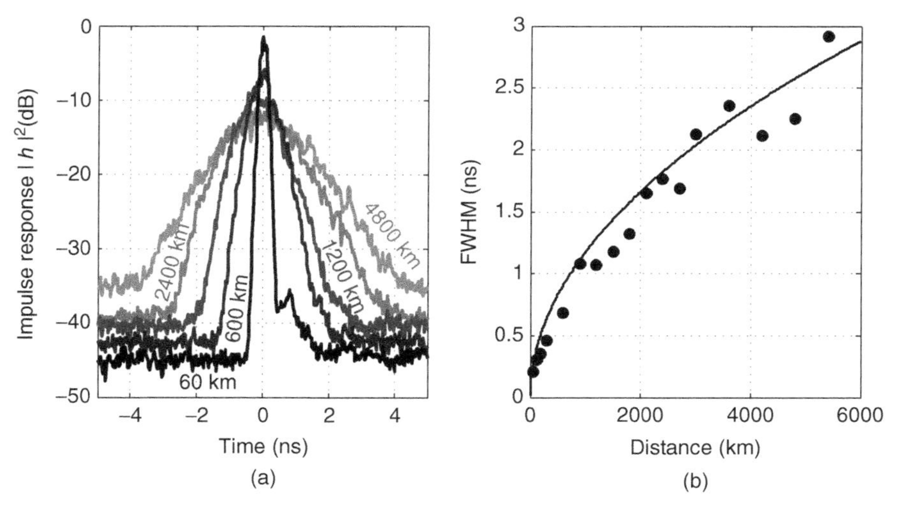

FIGURE 16.8 Experimental impulse response for a three-core coupled-core fiber. (a) The total intensity response is plotted for various distances for a 60 km long span. (b) The full-width-half-max (FWHM) of the impulse response is plotted as function of the distance and fitted with a square-root function.

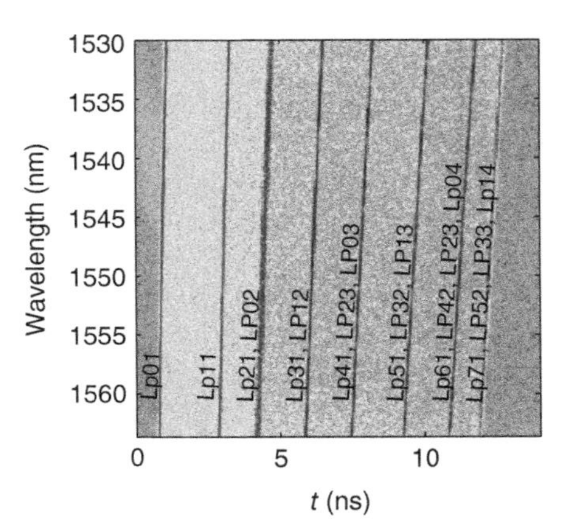

FIGURE 16.9 Spectrogram of an 8 km long graded-index multimode fibers. The lines are labeled according to the corresponding mode groups.

6000 km, which is 100 times larger than the initial distance, therefore, clearly outperforming DGD compensated FMF in terms of width of the impulse response (see Figure 16.7). This has a direct impact on the memory depth required for the MIMO equalizer or the reach of the system [77].

16.5.4 Multimode Fibers: Scaling to Large Number of Modes

Most reported MIMO SDM long-distance transmission experiments are reported for FMFs supporting only three or six spatial modes. A scalable approach for MIMO SDM has recently been proposed and demonstrated [78], where only a subset of the mode groups of a conventional GI MMF with 36 spatial modes (see Figure 16.3) is used for transmission and the undesired higher-order modes are suppressed by spatial modes filters. The approach was demonstrated experimentally by using three spatial modes (first two mode groups) [79] and for six spatial modes (first three mode groups) [78], for distances up to 305 and 17 km, respectively. The approach requires the coupling between the mode groups to stay moderate ($< 20\%$) over the entire fiber span. Mode coupling between groups was investigated using swept laser interferometry (described in detail in Section 16.7.2), and the results reported in Figure 16.9 indicates that the modal groups are clearly separated and only moderate modal crosstalk is observed between the groups identified as sharp lines in Figure 16.9.

16.6 MIMO-BASED SDM TRANSMISSION RESULTS

MIMO-based SDM transmission experiments have been performed by multiple research groups in FMFs with three and six spatial modes and conventional MMFs by using a subset of the mode groups supported by the fiber. In addition, MIMO-based transmission experiments were also performed in CC-MCFs. Some representative

TABLE 16.3 Summary of relevant MIMO-based transmission results in SDM fibers

Fiber type	Nr spatial channels	Spectral efficiency (bit/s/Hz)	Distance (km)	Spectral efficiency distance (bit/s/Hz km)	Capacity (Tbit/s)	References
CC-MCF	6	18	1705	30690	18	[80]
CC-MCF	3	4	4200	16800	1.0	[72]
FMF	6	16	708	11328	6.1	[74]
FMF-MCF[a]	3 × 12	247.9	40	9916	5.1	[29]
FMF	3	7.6	1000	7600	13.3	[27]
FMF	6	32	176	5632	24.6	[24]
MMF	3	9	310	2745	18	[81]
FMF	3	3	900	2700	9.6	[73]
FMF	6	10	74	740	41.6	[82]
MMF	6	7	17	119	23	[81]

[a]The fiber consisted of multiple few-mode cores.

results are summarized in Table 16.3. The longest transmission distances and highest spectral-efficiency-distance products were demonstrated in CC-MCFs, clearly confirming the advantages of the strongly coupled regime. The maximum capacity demonstrated in MIMO SDM transmission are still below the largest value reported for SMFs, but the largest demonstrated spectral efficiency per core of 32 bit/s/Hz is well above the nonlinear Shannon limit of the SMF, indicating that capacities well above 100 Tbit/s should be feasible in FMF-based SDM systems.

16.6.1 Digital Signal Processing for MIMO Transmission

MIMO DSP is a mature technology and has been studied in detail for wireless applications. Also, 2 × 2 MIMO DSP is widely spread in commercial coherent communication equipment to perform polarization multiplexed transmission. In polarization multiplexing, the delay between the transmitted polarizations is caused by polarization-mode dispersion (PMD), which is typically much smaller than the DGD encountered in DGD-optimized MMFs. For polarization multiplexing, around 10 equalizer taps are usually sufficient, whereas for SDM transmission up to 1000 taps have been utilized in some experiments. For such large number of taps, the complexity of the time-domain equalizer [83] scales very unfavorably, and it is essential that frequency-domain block equalizers are used [84–88]. An analysis comparing the complexity of six uncoupled polarization multiplexed SMFs channels versus a 12 × 12 MIMO equalizer shows that the complexity of the MIMO equalizer is increased by a factor of 4 if an unconstrained frequency-domain equalizer [89] is used. Because in current coherent DSP systems, the MIMO DSP only takes around 10% of the overall complexity, which is typically dominated by the complexity of the chromatic dispersion compensation and the forward error correction, the overall DSP complexity for a 12 × 12 MIMO system will only increase by a modest 30%, which is within reach of current ASIC technology.

The MIMO frequency domain equalizer is typically optimized according to the least mean square (LMS) algorithm in data-aided mode to achieve good initial convergence of the equalizer weights. After convergence, the equalizer is adapted according to the constant-modulus algorithm (CMA) or the multimodulus algorithm (MMA) depending on the format of the transmitted signal. Several optimized algorithms have been proposed to improve the convergence of the equalizer, for example, the use of adaptive feedback weight by using the normalized LMS algorithm [90] or the recursive least square (RLS) algorithm [87].

In addition, blind equalization algorithms, that allow to reconstruct the transmitted channels without a known test pattern or the use of a cyclic prefix, have also been studied and applied to experimental data in Refs [91] and [92].

The number of taps to optimize can also be reduced by decreasing the baud rate of the transmitted signal, the concept was demonstrated in Ref. [29] and does not significantly reduce the total overall complexity of the MIMO DSP, as more wavelength channels have to be processed, assuming that the same overall spectral efficiency is maintained. Nevertheless, the reduction of the number of taps can help improving the convergence of the equalizer.

OFDM has also been proposed to reduce the complexity of the MIMO DSP [82, 93], but has not been experimentally proven for long-distance experiments where local fading dips may cause severe penalties to some of the individual OFDM channels [59].

16.7 OPTICAL COMPONENTS FOR SDM TRANSMISSION

For SDM to be adopted, it must provide a significant cost advantage over duplicating SMF systems. These cost savings in SDM components and devices are achieved through component sharing and reuse rather than by the novel fibers themselves. In fact, such shared SDM components can be used as drop in replacements for N SMF components well before new SDM fibers are installed. For example, an N-core amplifier could replace N separate SMF amplifiers while using far fewer components: one fiber to amplify N signals, one isolator to isolate N signals, and one pump combiner [9, 94]. Other examples of components that provide N times as much capacity as a single SMF component include WSSs that simultaneously switch all N modes [26, 95–98], and FMF and MCF cladding-pumped amplifiers [99, 100] that provide gain to N modes with a single pump.

Figure 16.10 shows an SDM system supporting N modes that includes many components as seen in SMF systems, including amplifiers and wavelength routing and/or switching elements. Some components, such as free-space devices, can often be adapted from SMF systems with minimal changes. Some examples include isolators, filters for MCFs [101], and WSSs [26, 97]. SDM also requires some new components that are not present in SMF systems, but are required specifically for SDM. These include the spatial-multiplexers [102] necessary to couple into SDM fibers, optical spatial-channel equalizers [103], and also the $N \times N$ MIMO processing unit, which can be made all optical [104] or with electronic digital

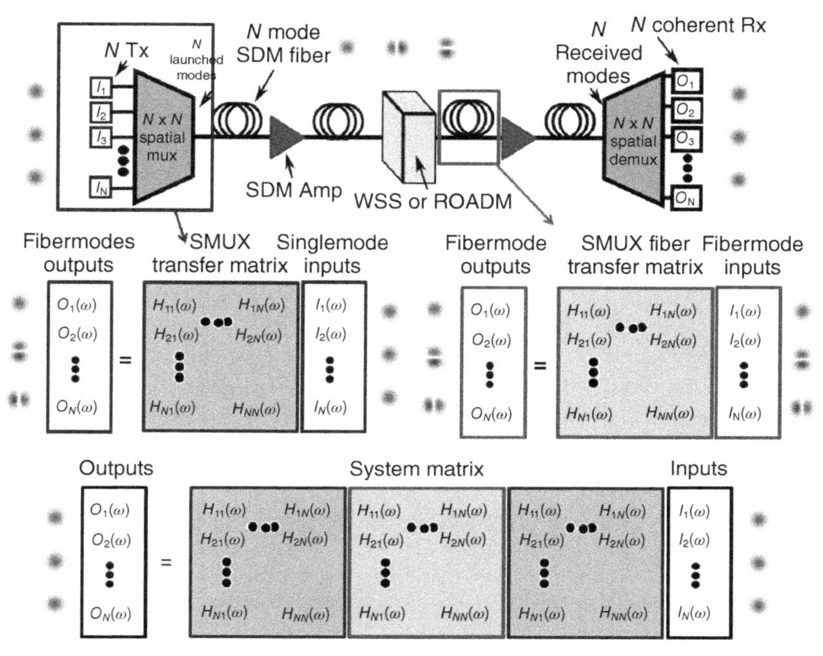

FIGURE 16.10 Equivalent matrix representation of SDM components and the entire SDM system.

processing [84]. These components represent an additional complexity over SMF systems and their use should be minimized. Fortunately, many of them, such as the spatial-multiplexers or the MIMO processing unit, are only required at the input and/or the output of the fiber link.

The new challenges for SDM components are handling multiple spatial modes, in particular minimizing mode dependencies (i.e., attenuation, gain, and delay in propagation time between modes). Furthermore, a fundamental question arises: Should the modes be treated as individual channels that can be routed, or should they stay bundled from transmitter to receiver? The most significant cost savings over parallel SMF systems are only achievable when spatial channels are treated as an inseparable end-to-end unit. The main motivation is that joint processing and switching of all modes will significantly simplify components and, therefore, boost cost savings. In fact, switching N modes/spatial channels as a unit reduces the number of switching components by a factor of N as described in more detail in Section 16.7.5. In addition, crosstalk is unavoidable in integrated SDM long-haul systems, and if the spatial channels are bundled, the crosstalk can be undone by MIMO DSP if desired. Finally, if channels are routed together and MIMO DSP is used, coupling or strong mixing between signals becomes acceptable and the requirements on SDM components can be further relaxed, which allows for additional cost saving. This applies for SDM components such as spatial multiplexers that excite orthogonal combinations of modes, WSSs that mix modes, and amplifiers with cores that couple while traversing the amplifier.

The following sections are dedicated to various aspects regarding SDM components: Section 16.7.1 describes how SDM components can be described as matrices and show how to derive the important system parameters from the matrices. In Section 16.7.2, we describe characterization techniques used to measure the transfer matrix of components and fibers. In particular, we focus on using swept-wavelength interferometry (SWI) [105–107] to rapidly and accurately characterize the transfer matrix of SDM fibers and systems. Section 16.7.3 describes the different types of spatial-multiplexers that can be used to couple into MMFs and components. Section 16.7.6 discusses the switching capacity and interconnectivity of SDM switching elements and how jointly switching spatial channels can greatly increase the switching capacity (throughput) of SDM switches if the spatial channels are routed jointly. Section 16.7.7 describes various types of amplifiers for MCF and MMF and their practical challenges.

16.7.1 Characterization of SDM Systems and Components

Compared with SMF components, SDM components and fibers present new effects such as spatial-coupling/mode-mixing, mode-dependent loss (MDL), modal dispersion, and spatial mode-profiles other than simple Gaussian beams. Many powerful measurement techniques for characterizing the modal content of fibers, including S^2 imaging [108], C^2 imaging [109], low-coherence imaging [110], interferometry [111], and selective launching and receiving using spatial-light modulators [16], have been proposed. S^2 imaging (Spatial and Spectral) obtains phase and amplitude images of the modes by launching light into all modes and measuring the spatially resolved output at every wavelength. The fiber acts as a common path interferometer. All fiber modes have a unique group velocity and have a distinct phase velocity (see also Section 16.5), and thus their relative phases change with wavelength. Each pixel records a complex interference pattern versus wavelength of all the modes at that spatial location. Using the fundamental mode as a reference, and assuming low mode-mixing, the modes and group delays can be extracted. C^2 imaging is similar to S^2 except that the reference path is sent through a separate fiber. C^2 has better dynamic range, less ambiguity in the interpretation of the results, but requires stabilization of the reference path since the reference and signal travel different paths.

While these techniques provide important information about fibers, from a transmission perspective, the three most important linear parameters that affect capacity and reach are the MDL, insertion loss (IL), and the DGD. The parameters can be calculated from the systems transfer matrix, $\boldsymbol{H}(\omega)$, where each element describes the phase and amplitude coupling from an input spatial channel to an output spatial channel. The parameters are mathematically defined and do not necessarily correspond to the individual fiber modes. In general, transmission performance improves with the minimization of (i) the MDL, which in the presence of noise reduces capacity [56], (ii) the IL, which limits the length of the span or causes more noise to build up, and (iii) the DGD between the spatial paths, which has to be kept shorter than the memory of the MIMO equalizer. These three parameters can be determined by the measurement of the linear transfer matrix between the N input and N output modes of the system.

We need techniques that can rapidly characterize the entire system transfer matrix, as the matrix can change over time due to, for example, mechanical vibration or thermal fluctuations in the fiber [55]. As described in Section 16.4, this information can be estimated through data transmission measurements and subsequent pseudoinversion of the equalization matrix. However, transmission experiments are equipment-intensive requiring arrays of high-speed coherent receivers and digitizers and only cover a limited spectral range. Section 16.7.2 describes a low-cost and fast technique based on a modified swept-wavelength interferometer that can acquire the transfer matrix across the entire C-band and L-band.

Figure 16.10 shows the transfer matrices of different elements in the SDM system. A transfer matrix describes the phase and amplitude coupling of each input mode to each output mode at every wavelength. The input and output modes could be different, and maybe even change with wavelength. For instance, spatial multiplexers describe the coupling between N spatially separated Gaussian like beams guided by multiple SMFs to the N spatial modes of a single multimode core. Fibers and fiber pig-tailed components have identical input and output mode basis sets. Also, splices between dissimilar fibers, for example, can have a different set of input and output modes. Even though the modes can be different through the SDM system, the inputs and outputs are typically Gaussian like modes of SMFs.

To understand how an individual SDM component impacts the system, one must realize that the entire system is the concatenation of all individual components, $\mathbf{H}(\omega) = \mathbf{H_1}(\omega)\mathbf{H_2}(\omega) \ldots \mathbf{H_N}(\omega)$ (see Figure 16.10). The concatenation has an import consequence for system design—only the transfer matrix from the N SMF inputs to the N SMF outputs matter. If any element in the system has mixing (i.e., is not a diagonal matrix), then the entire system appears mixed even if the majority of components have isolated spatial channels. This means that the MIMO is advantageously employed for a majority of systems, especially long-haul links.

Manufacturing of SDM components will require unambiguous and measurable specifications such as IL, MDL, and DGD similarly as found today in single-mode components. As not all of these parameters are accessible by direct measurement, a technique that fully characterizes the components such as SWI is essential for SDM components.

16.7.2 Swept Wavelength Interferometry for Fibers with Multiple Spatial Paths

SWI is well established for rapid characterization of SMF optical components in both amplitude and phase over a broad wavelength range (100 nm) with picometer spectral resolution, high sensitivity, and large dynamic range [105, 106]. In SMF systems, which have two polarizations, SWI measures the Jones matrices, which enables the calculation of PMD and polarization-dependent loss (PDL). A simple modification using time delays and splitter boxes enables the measurement of SDM components [107, 112].

The simplest form of SWI characterizes the amplitude and phase transfer function, $H(\omega)$, between an input and output across all wavelengths in a single sweep of a laser. Once $H(\omega)$ is obtained, a Fourier transform of $H(\omega)$ produces the impulse response, $h(t)$. Figure 16.11(a) shows this single-polarization SWI that comprises a frequency swept laser, an interferometer with a reference fiber with delay of τ_R and a fiber or device under test (FUT/DUT) in the different arms, a balanced receiver, and a digitizer. A single-sweep of the laser produces a beat signal on the photodetector that encodes both the phase and amplitude transmission, $H(\omega)$, in an interferogram.

SMF fibers and components are already multispatial channel by having two-input and two-output polarizations. Their transmissions are represented by a 2×2 Jones matrix with four elements: $H_{11}(\omega)$ launch x polarization, receive x polarization $H_{12}(\omega)$ launch x polarization, receive y polarization $H_{21}(\omega)$ launch y polarization, receive x polarization, and $H_{22}(\omega)$ launch y polarization receive y polarization. Figure 16.11(b) shows a modified SWI configuration with polarization diversified detection to receive both polarizations, and a polarization multiplexer with a fiber delay of τ to transmit both polarizations. The delay separates the impulse responses for the two launched polarizations. Time selection windows centered at $h(t)$ and $h(t-\tau)$ can extract the two polarizations from the Fourier transform of the interferogram.

Using additional delay fibers, we further extend the SWI for N inputs and outputs. Time delays at the input enable launching additional modes. Time delays at the receiver enable receiving many additional modes with a single balanced photodetector. Figure 16.11(c) and (d) shows splitter boxes with fiber delays in front and after the SDM device for both transmissive or reflective components. The SDM device or component needs to be interfaced to these splitter boxes with a spatial-multiplexer at the input and a second spatial-multiplexer at the output. Several types of spatial-multiplexers are described in Section 16.7.3. Each impulse corresponding to a unique spatial path between an input and output, $h_{ij}(t)$, can be extracted from $h(t)$ if the fiber delays are longer than the individual impulse responses. The next section provides an example measurement of a 4.7-km FMF to show how the time delays separate individual impulse responses.

16.7.2.1 Transfer Matrix, DGD, and MDL Measurements of a 4.7 km Few-Mode Fiber We used a swept-wavelength interferometer to measure a 4.7-km FMF supporting three spatial modes. The FMF is interfaced to the swept-wavelength interferometer using the transmissive splitter box (see Figure 16.11) and two phase mask SMUXs (Figure 16.13). Including the two polarizations of each spatial mode, this system has 6 input and 6 outputs or 36 possible spatial paths through the system. Figure 16.12(a) shows the impulse response measured on each polarization's balanced receiver. The 36 spatial paths (18 per receiver) are separated from each other by the polarization multiplexer delay and splitter box delays. The label nearby each impulse denotes its position inside the transfer matrix and each $H_{ij}(\omega)$ can be reconstructed by a Fourier transform of each extracted $h_{ij}(t)$.

This FMF has two mode groups: the LP_{01} group contains two polarization modes, and the LP_{11} group contains four spatial and polarization modes (LP_{11a} and LP_{11b}

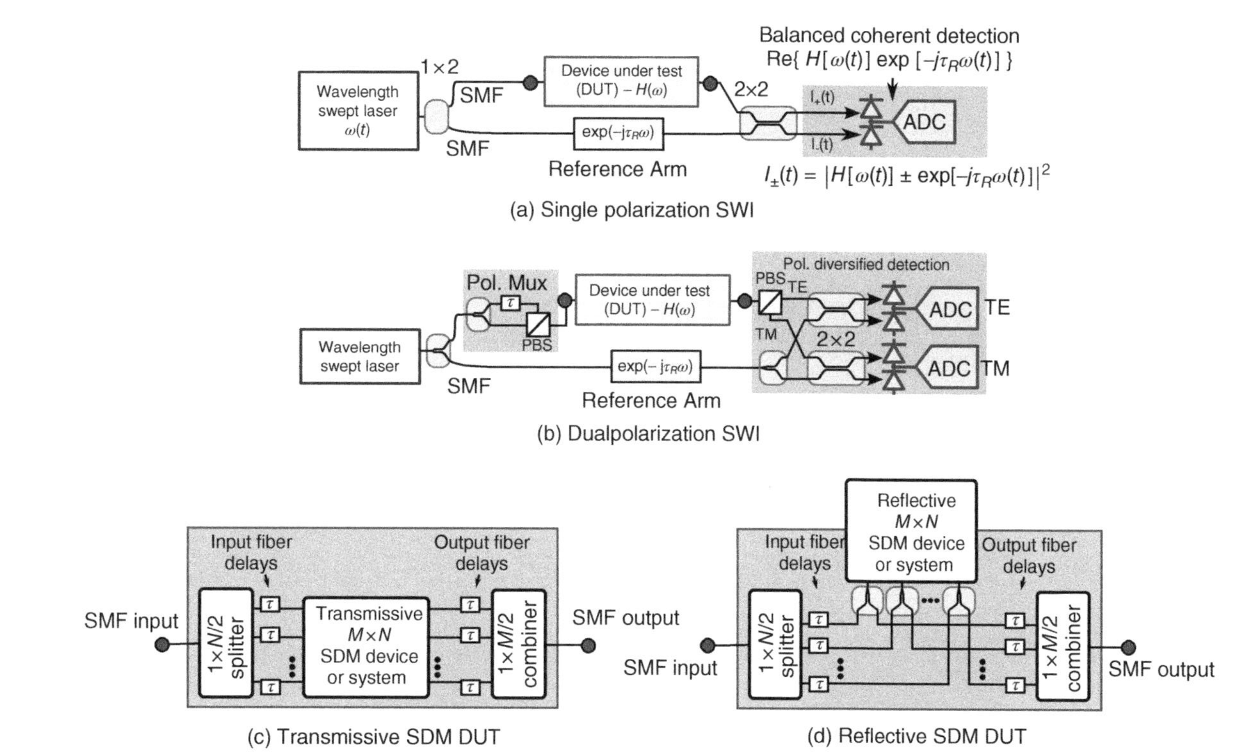

FIGURE 16.11 SWI configurations for (a) single-input mode and single-output mode, (b) both polarizations. Splitter box configurations for (c) SDM transmissive device and (d) reflective device.

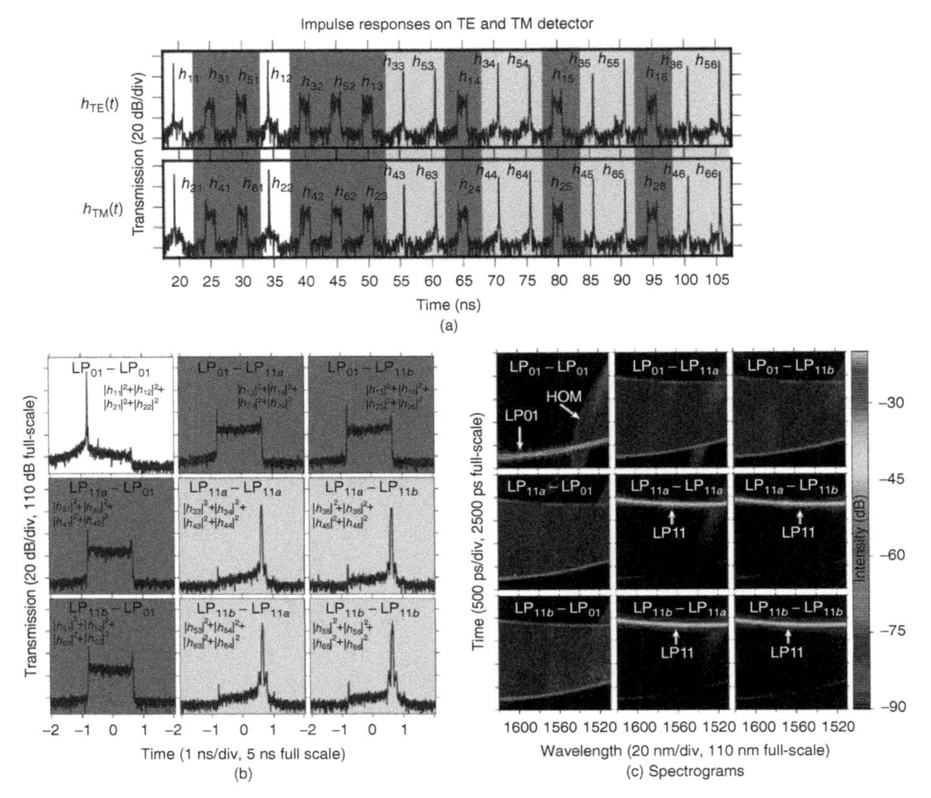

FIGURE 16.12 Exemplary SWI measurement of a 4.7 km graded-index few-mode fiber with three spatial modes interfaced with mode-selective phase-plate-based SMUXs. (a) Impulse responses measured on each received polarization. (b) Reorganization of the impulse responses into a matrix. (c) Spectrogram of the matrix.

on both polarizations). The impulses from Figure. 16.12(a) must be put into their proper spots in a matrix before computing MDL and IL through eigenanalysis. Figure 16.12(b) shows the rearrangement of the 36 matrix elements into a 3×3 intensity impulse response matrix. For simplicity, the intensity of the four polarization elements for each spatial mode are summed together to reduce the matrix from 6×6 to 3×3. The LP_{01} and LP_{11} groups are indicated by white and gray cells, respectively, and crosstalk cells are shaded dark gray.

The modes within each FMF mode group are nearly degenerate. Therefore, the four LP_{11} cells are nearly identical and contain a single peak. The mode-groups, (LP_{01} and LP_{11}), are weakly coupled because the propagation constant separation is large. The crosstalk cells have impulse responses shaped like a rectangle, which is bounded by the minimum and maximum group velocity.

Figure 16.12(d) shows the spectrogram of the transfer matrix. Spectrograms simultaneously display the time domain and frequency domain as a 2D plot and help

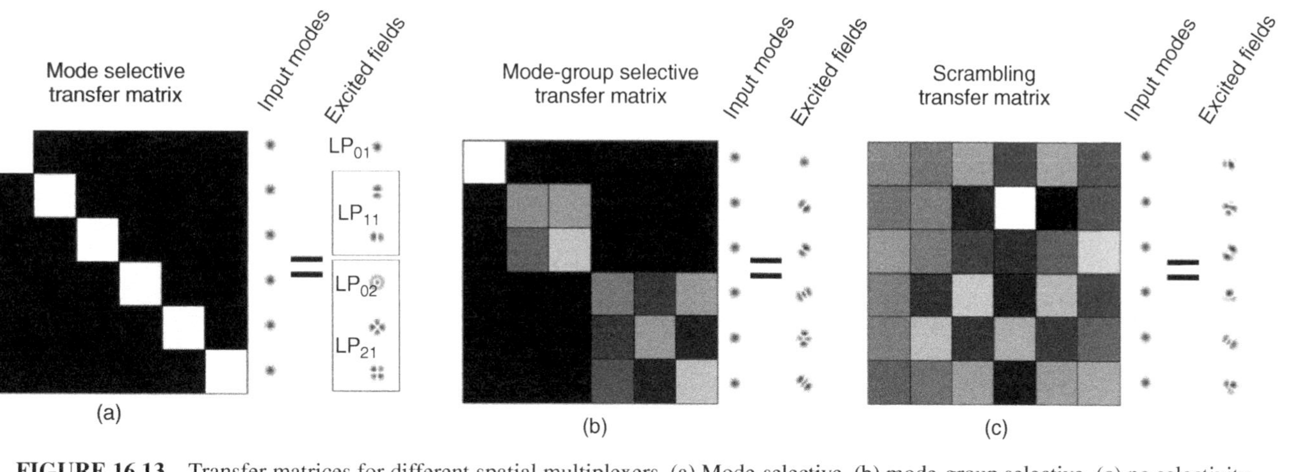

FIGURE 16.13 Transfer matrices for different spatial multiplexers. (a) Mode-selective, (b) mode-group selective, (c) no selectivity.

to identify modes, crosstalk, and dispersion. Each vertical stripe is calculated by filtering $H_{ij}(\omega)$ at the indicated frequency followed by an inverse Fourier transform. These spectrograms show that the two mode groups have different dispersions, and a higher-order mode (HOM) is guided at wavelengths below 1540 nm.

The entire transfer matrix was obtained in a single sweep lasting only 100 ms using two low-speed coherent receivers (50 MHz). In addition to the impulse response and spectrogram, MDL and IL can be calculated at each wavelength [107, 112]. SWI is a powerful technique for transfer matrix measurements of SDM components, systems, and fibers.

16.7.3 Spatial Multiplexers

Spatial multiplexers (SMUXs) are new and necessary components for SDM. Their function is to excite all modes uniformly, without introducing MDL. In addition to interfacing SDM fiber with transponders, SMUXs can also be used to build SDM devices and are essential to characterize other SDM components. In Section 16.7.5, we discuss how they can be used to build spatial channel equalizers or other devices based on spatial diversity.

Spatial multiplexers can be very simple, similar to the case of fiber ribbons or multielement fibers such as those shown in Figure 16.1(a)–(c). In general, SMUXs convert signals on one spatial mode basis set into modes on another basis set. Often the input basis set are N identical but spatially separated Gaussian like beams on N SMFs, which are connected to components such as transponders that are single-mode for high-speed operation. The output basis set are the modes of the SDM transmission fibers. Spatial channels in SDM fibers can be grouped into three types based on the level of coupling: (i) multiple modes/cores that are uncoupled, (ii) groups of modes/cores that couple strongly within a group but weakly between groups, and (iii) fully coupled modes/cores. SMUXs do not require mode-selectivity in MIMO systems, since MIMO performs the demultiplexing. The maximum amount of mode-selectivity in a SMUX, should match the number of coupled mode-groups of the fiber type. This eases fabrication and allows for scaling multiplexers to support more modes. Mode selectivity within mode groups is not helpful, as the fiber would mix the degenerate modes anyway.

In analogy to fibers, SMUX functionality can be grouped into three categories that match the coupling properties of the fibers: (i) mode-selective to match fibers with uncoupled spatial channels, (ii) mode-group selective to match fibers with groups of modes/cores that strongly mix, and (iii) scrambling multiplexers for fibers with completely scrambled modes. Figure 16.14 shows examples of these matrices for interfacing six SMFs to the six modes of a graded-index FMF. The boxes enclose modes within the same group (see also Figure 16.3). Modes within each group have unique spatial profiles, but strongly mix due to their similar propagation constants. The figures of merit for any multiplexers are insertion loss and MDL. Mode-selectivity can also be specified for the mode-group and mode-selective multiplexers.

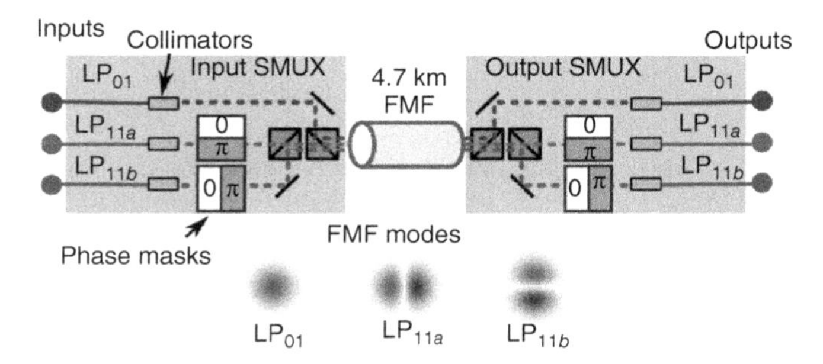

FIGURE 16.14 Free-space phase-mask-based spatial multiplexer/demultiplexer [64] for coupling to the spatial modes of an FMF.

The mode-selective SMUX transfer matrix is diagonal and each input excites exactly one fiber mode. It is the hardest to build due to the strict requirements on mode-selectivity. Often insertion loss is compromised for increased mode selectivity. An example is the phase-mask multiplexer, which uses phase masks to shape the phase front of N Gaussian beams to excite such different fiber modes [102]. Figure 16.13 shows two phase mask multiplexers interfacing with one 4.7 km FMF supporting three spatial modes. Beam splitters/combiners overlap the N-shaped beams on top of each other so that they can be simultaneously coupled into the FMF. These types of broadcast and select architectures have a theoretical minimum insertion loss of $1/N$ and are not suitable for a large number of modes. However, they are excellent for characterization because the phase-mask multiplexer is reconfigured by a simple replacement of the phase masks. Other broadcast and select spatial multiplexers are diffractive holograms implemented using SLMs [16, 20], or photonic integrated circuits [113, 114].

Using more advanced free-space techniques, the beam splitting losses can be avoided. Some recent examples thereof are multiplane light conversion, which implements a free-space unitary transform [115], free-space interferometers to capture the unused light from the beam splitters [116], and coordinate system remappers such as an orbital angular momentum demultiplexer [117]. Fiber devices can also achieve mode-selectivity without loss. Some examples include directional couplers [118], mode-selective photonic lanterns [119, 120], and velocity-tapered couplers [121].

Fibers with mode groups do not require a full mode-selective multiplexer since the modes in each mode-group are indistinguishable and mix immediately (i.e., degenerate). A mode-group-selective SMUX excites linear combinations of the modes within a particular mode group and Figure 16.14(b) shows a few simulated output fields (mode mixtures). In the fiber, each group can have different delays or attenuations; having access to the groups enables mode group equalization of attenuation or delay. Figure 16.18(b) shows an exemplary DGD equalizer built using a mode-group-selective SMUX, which is described in Section 16.7.5. This device was used to compensate for the DGD between the LP_{01} and LP_{11} mode of a 50-km MMF span and enabled 305-km transmission [79]. Mode-group-selective multiplexers are also easier to fabricate than fully mode-selective devices and we show an example of a mode-group-selective photonic lantern spatial multiplexer in Section 16.7.4.

Strongly coupled fibers scramble the modes completely so coupling to individual modes or mode groups do not add any functionality to the SDM system. Allowing the SMUX to excite scrambled orthogonal mixtures of modes greatly simplifies the multiplexer design and fabrication. Some examples are spot-based multiplexers [122] and standard photonic lantern multiplexers [104, 123]. Figure 16.14(c) shows that the output fields are totally scrambled—each output contains a little bit of all the modes. These multiplexers are ideally suited for MIMO systems, which already assume complete scrambling of the transmitted channels. SMUXs with scrambling can be beneficial also in conjunction with uncoupled or partially coupled fibers, as the transmitted signals are then distributed across multiple fiber modes and less susceptible to fiber-mode-dependent effects.

16.7.4 Photonic Lanterns

A "photonic lantern" (PL) is an adiabatic mode-converter, which merges N single-mode waveguides into a multimode waveguide that supports N modes [119, 120, 123–125]. Figures 16.15 and 16.16 show implementations of lanterns as

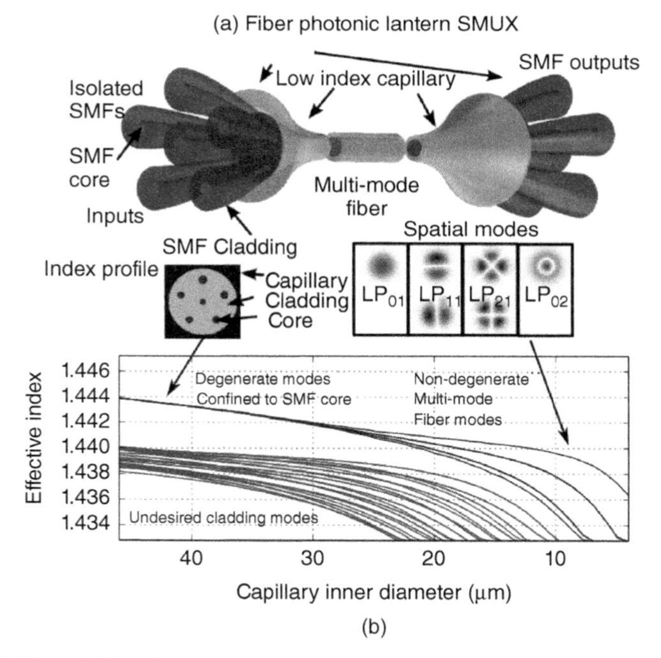

FIGURE 16.15 (a) Fiber-based photonic Lantern pair to couple into multimode fibers and (b) modal analysis of the taper region of the photonic lantern.

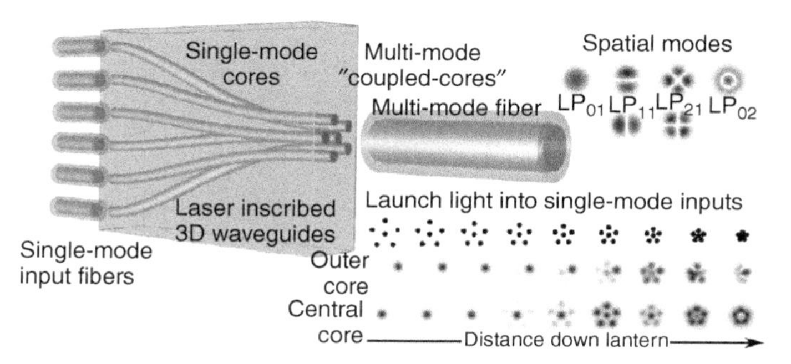

FIGURE 16.16 3D Waveguide base photonic lantern.

an all-fiber device and in three-dimensional waveguides. They offer significant advantages: Low IL and MDL, easy to package, scalable to large number of modes. Therefore, photonic lanterns are found in many MMF and FMF transmission experiments. The standard fiber lantern is built by placing identical fibers into a low refractive index capillary tube. The structure subsequently is reduced in size by a tapering process. At the early stages of the taper, the lantern acts like a pitch-reducing array and could be suitable for coupling to an MCF. The lantern taper is, however, more extreme, often reaching taper ratios larger than 1:20. At these ratios, the cores can no longer guide light, and the light becomes guided by the input fiber claddings and the low-index capillary.

They are adiabatic devices such that low MDL and IL are achieved when the taper length is longer than the adiabatic length, $L_{\min}$. In contrast, devices that rely on propagation constant matching such as directional couplers have stricter tolerances: the interaction length and the distance between the waveguides must be exact. The only fabrication constraint for lanterns is to ensure that the taper is longer than $L_{\min}$.

Historically, the photonic lantern was developed for astronomy, which required a way to collect light with a large aperture (i.e., MMF) and fan it out into high-resolution instrumentation, which required single-mode interfaces. In addition, these astro-photonic devices demonstrated spatial diversity (Section 16.7.5) well before the SDM field took off [126]. Lanterns have also been fabricated using laser-inscribed 3D waveguide technology where a femtosecond laser is used to change the refractive index properties inside a glass substrate. Laser-inscribed 3D waveguides allow for a flexible waveguide layout [127] and are preferable to couple SDM fibers to photonic lightwave circuits (PLCs) for spatial-diversity processing (see also Section 16.7.5). Devices for astronomical applications have been demonstrated with up to 61 fibers [128], or 121 3D waveguides [129]. 3D waveguides can be mass-produced by writing many lanterns on the same glass substrate.

While photonic lanterns designed for astro-photonics only require low insertion loss (IL), the lanterns for SDM must also have low MDL, which requires careful control of the waveguide arrangement in the taper [124]. Since mixing and scrambling occur in the MMF, amplifiers, and routing elements, it is not necessary for the photonic lantern SMUX to excite individual modes, which simplifies its design and enables scaling to a large number of modes. However, adding mode selectivity to lanterns only requires using dissimilar SMFs in the taper.

Devices supporting 12 spatial and polarization modes (i.e., six input SMFs) have enabled transmission over 177 km of FMF [24], 17 km transmission over a conventional 50 μm graded-index fiber [78], and 305 km transmission over the first three modes of a 50 μm graded-index fiber [79].

16.7.4.1 Operation Principle of Photonic Lanterns Photonic lanterns adiabatically convert a set of spatially separated Gaussian modes into a set of overlapping multimodes. Adiabatic devices are well understood through modal analysis of the taper region. To determine the minimum adiabatic taper length, L_{min} beam propagation techniques (BPM) are required to show how the launched fields evolve into the orthogonal combinations of the multiple modes.

Figure 16.17(a) shows the modal analysis of a photonic lantern with three identical fibers. Each lantern mode has an effective index and mode profile and the mode effective indexes are plotted against the inner diameter of the capillary. The inner diameter of the capillary is the new multimode core diameter. The undesired cladding modes curves are higher-order cladding modes confined between the capillary and SMF cladding material. Exciting these higher-order modes leads to loss. Launching into a single core would excite a superposition of all three lantern modes.

At each position along the taper, the lantern modes resemble the LP_{01}, or LP_{11} modes in an FMF. At large inner diameters the lantern modes are confined to the cores and are degenerate (i.e., the three overlapping blue curves). The lantern modes can be approximated by "spots" and launching light into one of these spots will evenly excite each lantern mode. As the inner diameter decreases, the cores begin to couple and the degeneracy between the lantern modes break (the LP01 and LP11 curves separate). At this point, the lantern modes move from the cores into the SMF claddings. During the final stages of the taper, the lantern modes are identical to the FMF modes. Adiabatic tapering ensures that light launched into a lantern mode remains in the lantern mode throughout the entire taper.

The identical fiber lantern has a scrambling transfer matrix since light launched into each core is a superposition of all three lantern modes. Since the propagation constants are degenerate, any perturbation will cause coupling between the three lantern modes, which only enhances the mode-scrambling. The boxes enclose the modes that strongly couple. At the input, the lantern modes strongly couple since the propagation constants are identical. At the middle and end of the taper, the degeneracy between the lantern LP_{01} and LP_{11} modes break (splitting mode index), and there is only strong coupling within the LP_{11} group.

Adding mode-selectivity to a photonic lantern is simple and only requires using fibers with dissimilar cores. Mode-selectivity requires (i) degenerate lantern modes through the entire taper and (ii) separation of the lantern modes' propagation constants in the entire taper. Figure 16.17(b) shows the modal analysis of the modes in the three core lantern with one slightly larger core. At the input, there are two groups of lantern modes: (i) one mode is confined entirely within the larger core, and (ii) two

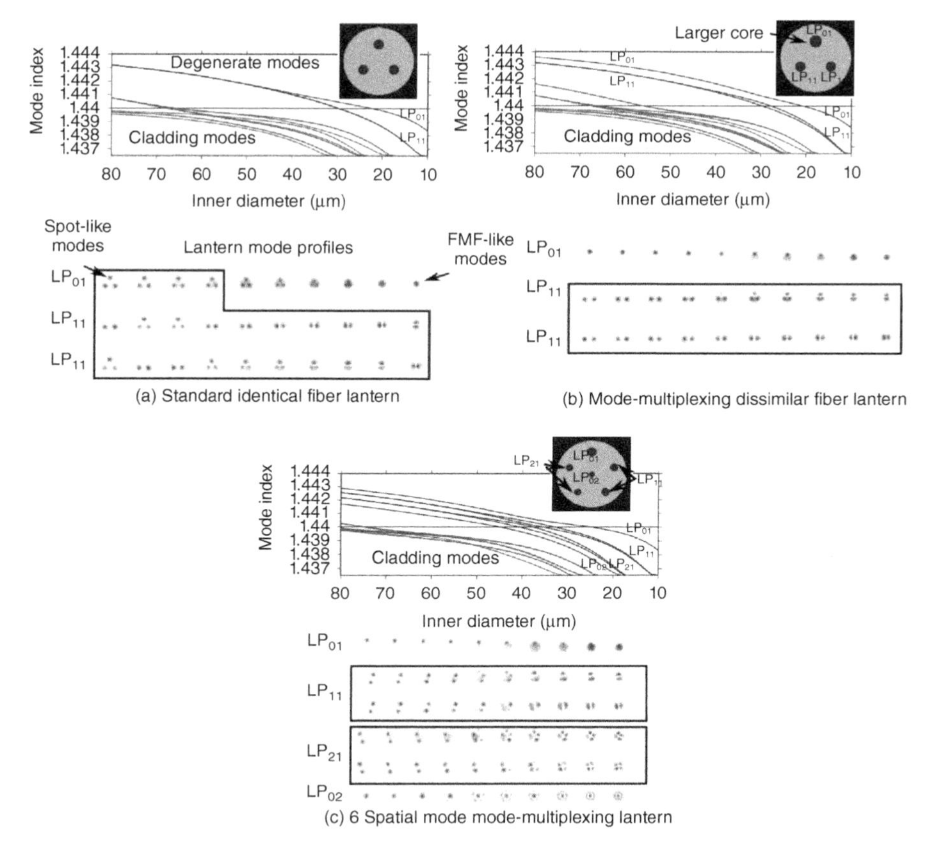

FIGURE 16.17 Modal analysis of three types of photonic lanterns. (a) Identical fiber lantern, (b) mode-group multiplexing dissimilar fiber lantern, and (c) mode-group multiplexing six-mode dissimilar fiber lantern.

degenerate modes are confined within the smaller cores. The two degenerate modes become the FMF LP_{11} modes, and the larger diameter core mode becomes the FMF LP_{01} mode. The difference in effective index inhibits coupling between the different groups. In addition, the mode profiles show clear evolution of the large diameter core mode into the FMF LP_{01} mode. The two smaller diameter core modes are strongly mixed, but also evolve only into orthogonal combinations of the FMF LP_{11} modes.

Lanterns are also easily scaled to support many modes, especially if the lantern does not require mode-selectivity. Figure 16.17(c) shows a six-mode lantern with mode-group selectivity. There are six cores with four different sizes whose propagation constants are selected to match the near degenerate mode-groups of a step index FMF. The largest core maps to the LP_{01} mode, and the smallest core maps to the LP_{02} mode. The two second largest cores map to the LP_{11} modes, and the two second smallest cores map to the LP_{21} mode. The propagation constants are separated during the taper, and the modal profiles corresponding to these propagation constants evolve from isolated core modes into the FMF modes. This device has been fabricated and has shown SMF to FMF coupling losses below 1.5 dB [130]. It is expected that future photonic lanterns will be scaled up to support the same number of modes in a 50 μm core graded-index fiber.

16.7.5 Spatial Diversity for SDM Components and Component sharing

SDM devices can offer significant cost savings over SMF components because a single SDM can process all modes simultaneously. In free space, a single bulk component such as a diffraction grating can process multiple modes at the same time. MCF amplifiers can amplify N modes in a single fiber and possibly with only one pump. To realize these savings, the components must not be degraded by MDL.

Mode-dependent losses and differential delays in SDM components result from the modes being non degenerate and, therefore, having a different spatial profile and a different propagation constant. For instance, a fiber Bragg grating inside an MMF will have a different center wavelength for each mode due to propagation constant differences between the modes. Free-space devices are affected by the spatial profile of the modes due to diffraction (each different mode also has a different diffraction pattern in angle space).

Spatial diversity is a technique to reformat the modes through a unitary transform such that each reformatted mode has an identical spatial-profile and propagation constant and effectively experiences the same transmission. A good spatial demultiplexer can implement such a unitary transform and convert N modes into separate Gaussian beams either in free space or in fiber. Figure 16.18(a) shows the black-box diagram of an MMF device with spatial diversity built using two spatial multiplexers and N identical SMFs. The spatial multiplexer demultiplexes the modes, then N identical single-mode components apply a function to all the demultiplexed modes evenly. These modes are then remultiplexed back onto the MMF. With mode-selective SMUXs, equalization could be applied as shown in the DGD equalizer of Figure 16.18(b).

All-fiber devices can be built by using an MCF, since each core has identical propagation constant and mode profile. An example is the astro-photonic device comprising two photonic lanterns built with MCFs, and a second MCF with fiber-Bragg gratings inscribed into each core as presented in Ref. [131].

Free-space-based devices can use fiber arrays with identical Gaussian beams and form the building blocks of some FMF-WSSs [98, 132]. Photonic lantern SMUXs have been used to add spatial diversity to routing elements, such as WSSs, to minimize modal dependencies by first demultiplexing the modes into separate free-space beams [98] or separate fibers [132] and then passing them through identical single-mode components. Figure 16.19 shows a spatial-diversity assembly fabricated in 3D glass waveguides. The spatial-diversity assembly contains a 3D waveguide with several photonic lantern SMUXs [98] to interface with an FMF array on one facet, and

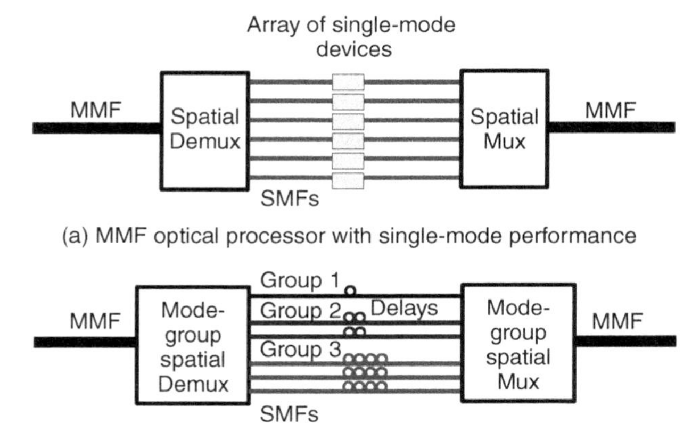

FIGURE 16.18 Building MMF components with spatial diversity using SMUXs. (a) Multimode component built from N single-mode devices. (b) Multimode component that applies different functions to different mode groups. From Ref. [130].

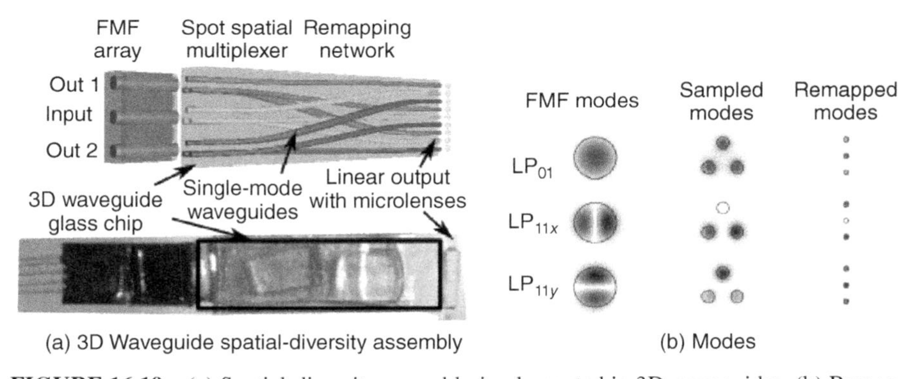

FIGURE 16.19 (a) Spatial-diversity assembly implemented in 3D waveguides. (b) Remapping of the FMF modes into identical Gaussian beams.

with free-space Gaussian beams on the other facet. The next section demonstrates an FMF-WSS using this spatial-diversity assembly to remove modal dependencies.

16.7.6 Wavelength-Selective Switches for SDM

Transparent wavelength routing and switching are the essential features of today's optical networks, and the functionality is enabled by WSSs. A WSS routes any wavelength on a single-input fiber to N-output fibers. WSS for SDM must also handle multiple modes or spatial paths and provide comparable performance at lower cost to current SMF-based devices. Obtaining a cost advantage over single-mode systems

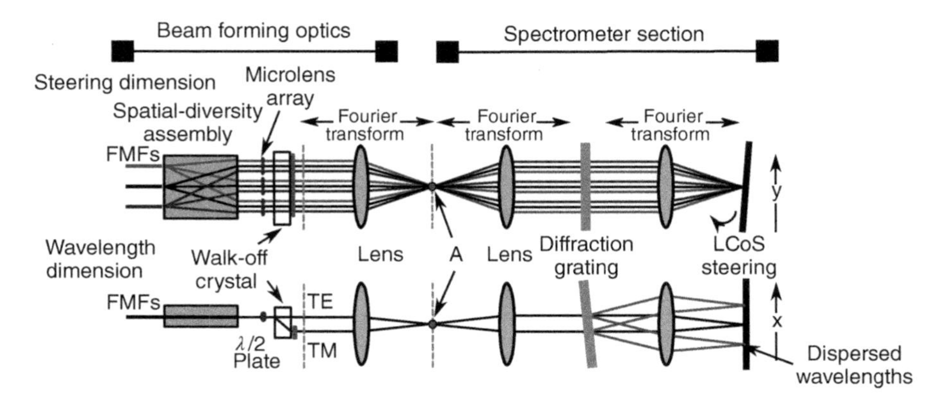

FIGURE 16.20 FMF wavelength-selective switch with wavelength dispersion in the x dimension, and fiber switching in the y direction.

necessitates that an SDM WSS must be able to handle the extra modes without requiring many additional/duplicated devices or components. Some SDM WSSs include a simple FMF ROADM built using collimators and thin-film filters [133], an MCF WSS built from a 21-fiber SMF WSS to act as a 1×2 WSS for 7-core fiber broken into 21 SMFs (i.e., 3 MCFs, 7 cores each) [44, 95], an FMF-WSS without spatial diversity [26], and a WSS with spatial diversity built by interfacing photonic lanterns to a commercial twin 1 × 20 [132].

WSSs exploit the two orthogonal dimensions perpendicular to the light-propagation direction to uncouple the beam steering axis from the wavelength separation axis. Figure 16.20 shows a typical SDM WSS construction with beam-forming optics and a spectrometer. The ports are arranged vertically in the steering dimension (y), and the wavelengths are dispersed in the wavelength/horizontal (x) dimension. The beam-forming optics reformat the FMF "modes" to optimize the beam sizes for spectral resolution and port count before entering the spectrometer section. In addition, they must distribute the modes across wavelength and steering axis because there are no additional dimensions to place the modes. The beam-forming optics can include (i) an optional spatial-diversity assembly to convert the FMF input modes into multiple Gaussian beams, (ii) a microlens array for collimation, (iii) polarization diversity optics to split the two polarizations, (iv) anamorphic optics using cylindrical lenses to optimize the spectral resolution and port count, and (v) a converging lens to transform the spatially separated beams into angularly separated beams converging to a single point (point A). The spectrometer section consists of a diffraction grating and a steering element placed in a double telecentric imaging system consisting of two lenses in a $4f$ configuration (distance between the lenses is $2f$, twice the focal length of the lenses and total distance between the imaging planes is then $4f$, hence the name "$4f$"). The $4f$ system images the angle multiplexed beams at point A onto the steering element, and disperses the wavelengths horizontally. Tilt angles applied at the steering element shift the beams vertically at the grating and microlens array planes. Note that there are different ways to build WSSs, but essentially they all contain equivalent beam-forming and a spectrometer section.

The most common steering element for WSSs consists of liquid crystal on silicon (LCoS) pixelated spatial light phase modulators [134]. LCoS can have up to 2 million pixels with programmable phase shifts. Mirrors are approximated by grouping pixels together, and their steering angle is defined through saw-tooth blazed gratings. This programmability enables flexible grid operation to support wavelength channels of varying width [135].

A WSS, or any switch, can be quantified by its throughput, and switching capacity. The switching capacity is the product of the individually controllable connections between the spatial channels and the wavelength channels (i.e., the number of switch states). The maximum switching capacity is bound by the steering element and is the product of the number of mirrors on the switching element and the number of positions the mirrors can direct a signal to with low crosstalk.

The throughput is defined as capacity transmitted through the switch at any given time. A 1×20 WSS with 100 wavelengths has 2000 unique switch states (switching capacity) and 100 wavelengths (throughput). In SDM, we add spatial channels that can be individually switchable or switched as a group (i.e., jointly switched). Switching between the N modes would require a steering element with N times larger switching capacity. The joint switching scheme can switch N modes identically using a single mirror and does not require any additional switching capacity [95, 136]. Without joint switching, there is no significant cost advantage of an SDM WSS over an equivalent-capacity SMF WSS with N times the fibers. This is because the larger steering element required for the N modes could be used to switch N times the fibers.

Joint switching enables an SDM switch to significantly increase the throughput by N with a negligible increases in costs. In addition, joint switching is ideally suited for systems that require MIMO DSP because all modes must travel together from the input to the output in order to undo the crosstalk. The joint switching principle is not new to SDM; conventional WSSs switch both polarizations jointly for two reasons: (i) polarizations strongly mix in SMF because of PMD and (ii) the steering element can be 2× smaller. In Section 16.7.6, we use spatial diversity to implement joint switching over modes to a WSS, whereas in the following section joint switching is implemented by steering a beam carrying multiple fiber modes.

16.7.6.1 Spatial Dependence of a Few-Mode-Fiber WSS

A FMF-WSS can be built by simply replacing the SMF collimator array with an FMF collimator array [26, 97, 137] and is a good example to demonstrate how easily an SMF component can be adapted to support SDM. Compared with an SMF WSS, this approach supports fewer input fibers since the higher-order modes require a larger steering angle, and also fewer wavelength channels due to the spatial mode-dependence of the passbands [136]. It still can obtain identical throughput; N times less fibers, but N times more modes. Nevertheless, the switching capacity of today's LCoS is underutilized, and this switch can still increase the throughput over an SMF version without increasing costs.

Each FMF mode has a different spatial profile and beam divergence. In an FMF-WSS, the different spatial profiles result in mode-dependent passband shapes and the non-uniform beam divergences require larger steering angles. As an example, Figure 16.21(a) compares the minimum steering angle required to move the LP_{01} past itself versus the LP_{11y} mode. The LP_{11y} has two lobes and can be thought of as being two smaller LP_{01} modes placed close together. Therefore, this mode diverges faster and requires a larger tilt angle from the steering element, which reduces the number of ports.

The different spatial profiles of the modes result in different passband shapes. Since modes mix, individual passband shapes do not matter—the frequency dependent MDL and IL will influence transmission performance. Frequency-dependent MDL can be simulated by building the transfer matrix of the switch at each wavelength. Mathematically, each cell is $H_{i,o}(f) = \int_x \int_y E_i(x - Df, y) E_o(x - Df, y) M(x, y) dx dy$, where $E_i(x, y)$ and $E_o(x, y)$ are the input and output mode complex amplitude profile of the scalar electrical fields, M is the beam-steering hologram, f is the optical frequency, and D is the spatial dispersion provided by the grating. Figure 16.22(a) shows the simulation of a few of these cells. The 01-01 transmission passband has the shape of a standard SMF WSS passband. The passband edges cause mixing between the different modes such as the 01-11x transfer function. In the center of the passband, each mode effectively sees a flat mirror and all light is reflected back without any mode mixing or MDL. From the transfer matrix, we can calculate the MDL and IL. Figure 16.22(b) shows the minimum and maximum transmission of a passband and its neighboring channel. Compared with the individual matrix elements, the minimum and maximum transmissions clearly indicate the extent of the MDL at the passband edges. For reference, the black lines are simulated passbands for an SMF-WSS where only that the LP_{01} mode is present. The modal dependence of the passband shapes require wider spectral guard-bands between channels, which reduces the number of wavelength channels in an FMF-WSS compared with an SMF WSS.

16.7.6.2 Few-Mode Fiber Wavelength-Selective Switch with Spatial Diversity and Reduced Steering Angle The mode dependencies of the simple FMF-WSS can be removed using the spatial diversity and the throughput can be increased using the spatial-diversity concepts introduced in Section 16.7.5. It can be implemented using a simple modification to a WSS by including the 3D waveguide spatial-diversity assembly from Figure 16.19. In a WSS, spatial diversity ensures that each modes effectively "sees" an identical passband and has an identical steering requirement [95, 98, 132].

Figure 16.19(b) shows how the assembly reformats the FMF modes into Gaussian beams arranged in a line. Identical Gaussian beams have the same spatial profile and beam divergence and, therefore, illuminate the same number of lines on the grating, and have the same shape on the LCoS. The order of the beams affects the maximum

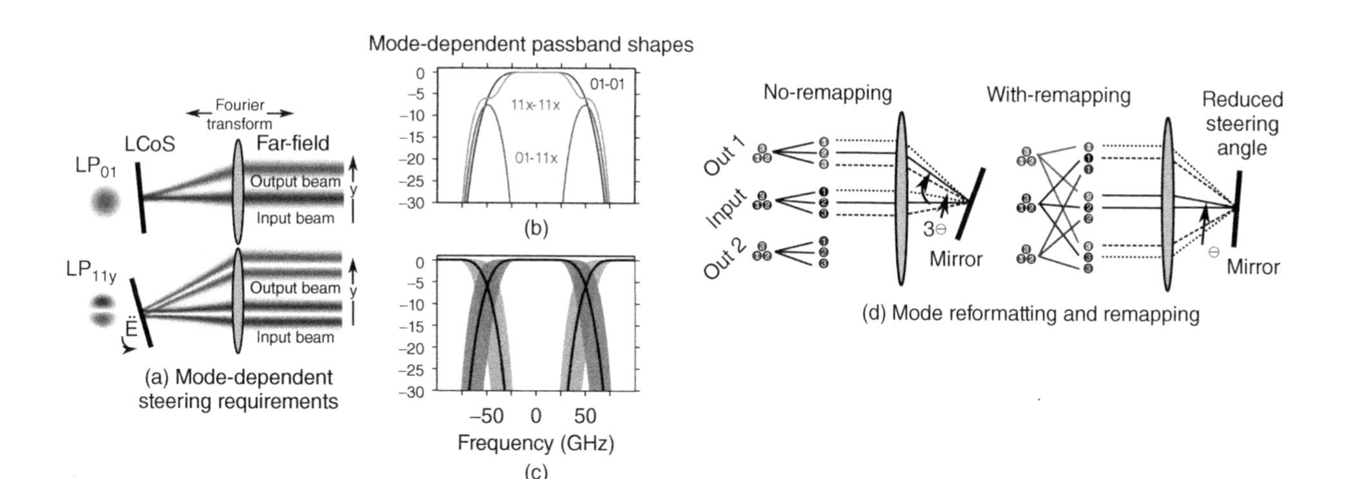

FIGURE 16.21 (a) Steering requirements for the LP_{01} mode and the LP_{11y} mode. (b) Steering requirements between the input fiber and output **1** for two possible arrangements of sampled modes on a linear array.

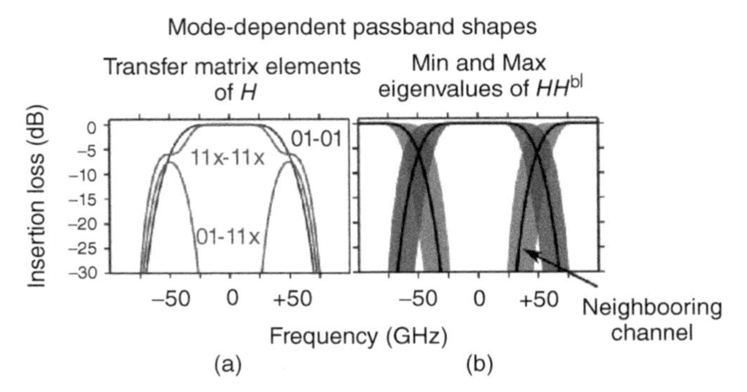

FIGURE 16.22 Passbands of FMF-WSS. (a) Cells of the transfer matrix corresponding to different modes. (b) Mode-dependent passbands displayed as minimum and maximum transmission. Black curve is the intensity transmission for an SMF-WSS.

required steering angle of the switching element. Minimizing this angle allows for increasing the throughput (i.e., more modes) without decreasing switching capacity. Figure 16.21(b) shows two methods for routing the demultiplexed modes to a linear array, and their respective steering angle requirements. The first scenario (left half of Figure 16.21(b)) groups the beams by fiber such that the modes originating from the same fiber are close together. The LCoS must steer a beam past three beams (i.e., the modes) to reach the next fiber. Alternatively, reshuffling the demultiplexed beam samples so the beams are grouped by modes (interleaved, shown in the right half of Figure 16.21(b)) reduces the steering angle range by a factor proportional to the number of modes. Grouping by modes places beams from different fibers close together. The LCoS only needs to steer past one beam to reach the next fiber, and all modes get routed together. The mapping of the beam samples between the inputs and outputs are 2-2, 1-3, and 3-1. This scrambling does not degrade transmission performance, as long as the spot multiplexers have near-unitary sampling. These spatial-diversity schemes require some more complex optics (lenses) to accommodate the increased field of view of the larger collimator array. However, lens costs are significantly less than the cost of the steering element. With the remapping network and spatial diversity, the fiber count can be equivalent to an SMF WSS with only a marginal increase in cost.

Figure 16.23 shows experimental results of an FMF-WSS built with and without spatial diversity [98]. Figure 16.23(a) shows how the switch is characterized from the three SMF inputs to three SMF outputs with the spot-multiplexers used to enter and exit the FMF. We use the SWI with spatial diversity to measure the 6×6 transfer matrix versus frequency between the six spatial and polarization inputs and outputs (Section 16.7.2). From the transfer matrix we obtain and plot the minimum and maximum transmission, similarly to the plots in Figure 16.21(b).

Figure 16.23(b) shows a measurement of the output of port 1 showing flat transmission with flexible bandwidth and mode-independent passbands across the entire C-Band. Figure 16.23(c) and (d) compares the minimum and maximum spectral

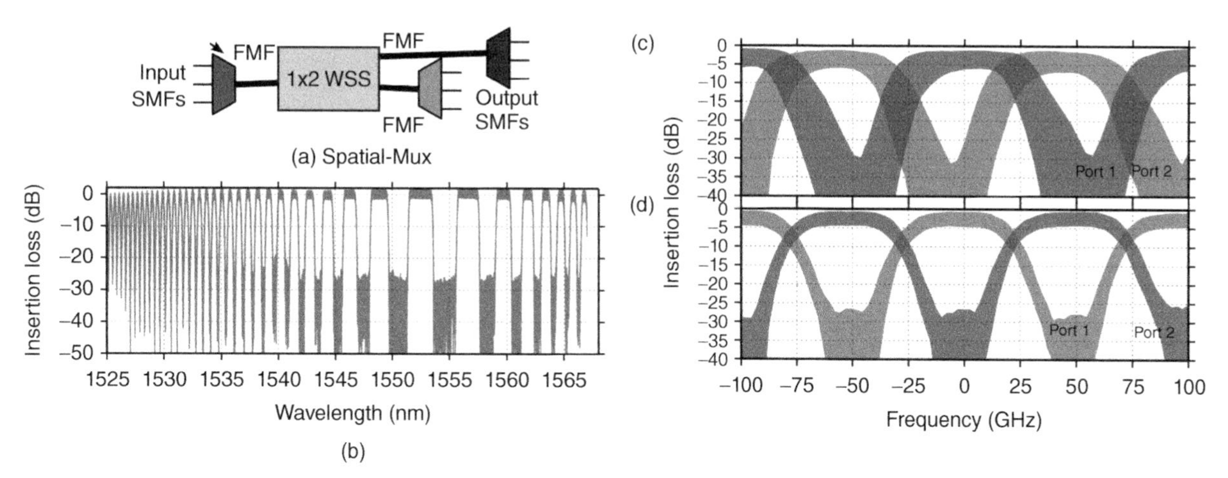

FIGURE 16.23 (a) Transfer matrix characterization experiment of WSS with spatial diversity. (b) Minimum and maximum transmission of SD-WSS across the C-Band with flexible grid passbands. 50-GHz interleaver sent between ports 1 and 2 for an FMF-WSS (c) without spatial diversity and (d) with spatial diversity.

transmission with and without spatial diversity for a 50-GHz interleaver between WSS outputs 1 and 2. At the passband peaks, the WSS without spatial diversity has larger MDL of 4.5 dB compared with spatial diversity 3.5 dB. Note, the MDL for the WSS with spatial diversity is consistent with four passes through the spot-multiplexers. Its insertion loss is around 14 dB, which is also consistent with two passes through the 3D waveguide (4–6 dB loss each pass), and 6-dB for losses within the switch optics (e.g., 3 dB from the LCoS, and 1 dB from the grating). The losses of the 3D waveguides are not representative for the technology. Typical loss in 3D waveguide can be as small as 1 dB when coupling from SMF to SMF. As expected, spatial diversity reduces the mode dependency, and can also be added with a simple design modification of the WSS.

16.7.7 SDM Fiber Amplifiers

SDM amplifiers [39, 97, 100, 101, 138] that directly support SDM fibers potentially offer cost and complexity savings through pump and component sharing. Component sharing and pump sharing is enabled by the closely spaced or even overlapping spatial channels. Some examples of erbium-doped fiber amplifiers (EDFA), in order of increasing integration (spatial density), are separate fibers bundled with a common coating material [139], multiple amplifying cores in one fiber pumped by separate pumps [9] and by a cladding pump [99], and MMF amplifiers [140]. Long transmission fibers (10 km or more) can themselves provide gain through Raman amplification [141, 142]. Providing equal gain to all modes is challenging and requires optimization of the pump spatial mode, doping spatial profile, and possible external gain flattening and/or equalization.

Figure 16.24 shows the differences between core pumping and cladding pumping for multicore and multimode fibers. In core-pumped amplifiers, the pump light is confined to the cores, often in the fundamental mode. In FMF core-pumped amplifiers, exciting higher-order pump modes can reduce mode-dependent gain. Some FMF amplifiers even use more than one pump in multiple modes to further reduce the mode-dependent gain [140, 143].

Cladding-pumped amplifiers [99, 101, 144] distribute the pump light in the fiber cladding, which is shared by all cores and modes of the MCFs. A low-index polymer coating surrounds the amplifying fiber to allow the cladding to act as a large multimode waveguide. Low-index polymers are used rather than a low-index glasses because of their much lower refractive index. Light is launched into hundreds to thousands of cladding modes, which uniformly illuminate the entire cladding. It is then possible to use multimode pump diodes, which are generally low cost, uncooled, and have >10 W output power, to illuminate the entire fiber. The multimode pumps can be coupled into the cladding without a wavelength combiner or disturbing the cores through side pumping [99, 145], v-groove coupling [146], or free space [147].

Figure 16.25 shows a complete amplifier that includes (i) pump combiner, (ii) amplifying fiber, (iii) isolator, and (iv) gain flattening filter. The pump combiner brings the pump and signal into the same fiber. Common pump wavelengths are 980 and 1480 nm. In core-pumped amplifiers, the pump light is in the same

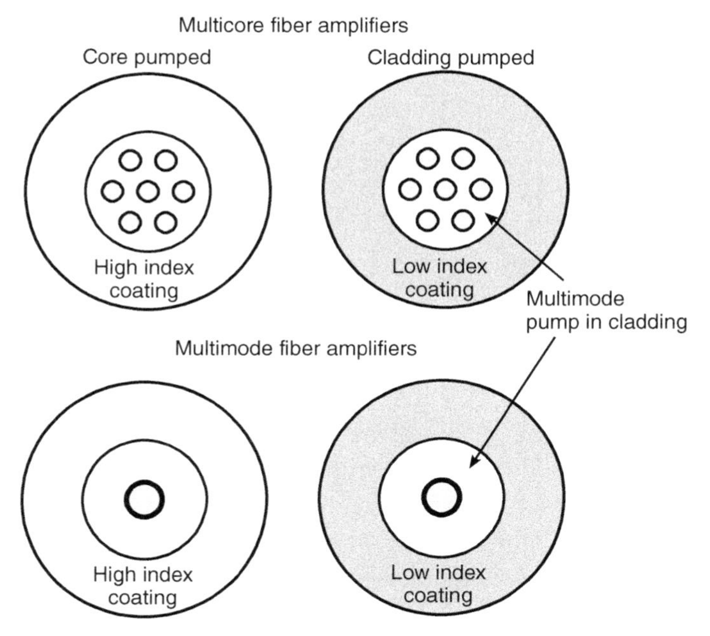

FIGURE 16.24 Doped fiber for SDM Amplification showing and spatial location of the pump for cladding pumping and core pumping.

spatial mode/modes as the signal and a wavelength combiner is necessary to avoid passive splitting/combining loss. Even though a wavelength combiner is required, component sharing can eliminate many components that would be duplicated in parallel single-mode amplifiers. For instance, the 19-core amplifier in Ref. [9] uses a single dichroic mirror to combine 19 pump lasers into the MCF and a single isolator to provide isolation to the 19 cores. Note, when building free-space components with more than one spatial mode/channel, it is important to use 4*f* imaging systems to ensure equal coupling of all modes. Cladding pump amplifiers use a different set of modes (i.e., cladding modes) to guide the pump [94, 99, 147]. This eliminates the wavelength combiner through techniques such as side pumping. In side pumping, an MMF that contains the multimode pump light is tapered and brought into contact with the amplifying fiber's cladding. The pump light can couple into the cladding with losses below 1 dB.

A challenge in cladding-pumped amplifiers is to strongly invert the gain [94, 148, 149]. An amplifier has a threshold pump intensity, I_{P0}, to reach transparency. In core-pumped amplifiers, the pump intensity can be increased by using smaller cores. The cladding area is about 100 times greater than the area of a single core, which means that the pump power of a cladding-pumped amplifier should be at least 100 times greater than that of a core-pumped amplifier to obtain equivalent performance. The signal intensity in the core can quickly become larger than the pump intensity, especially if the cores are very small (the case for core-pumped amplifiers). When

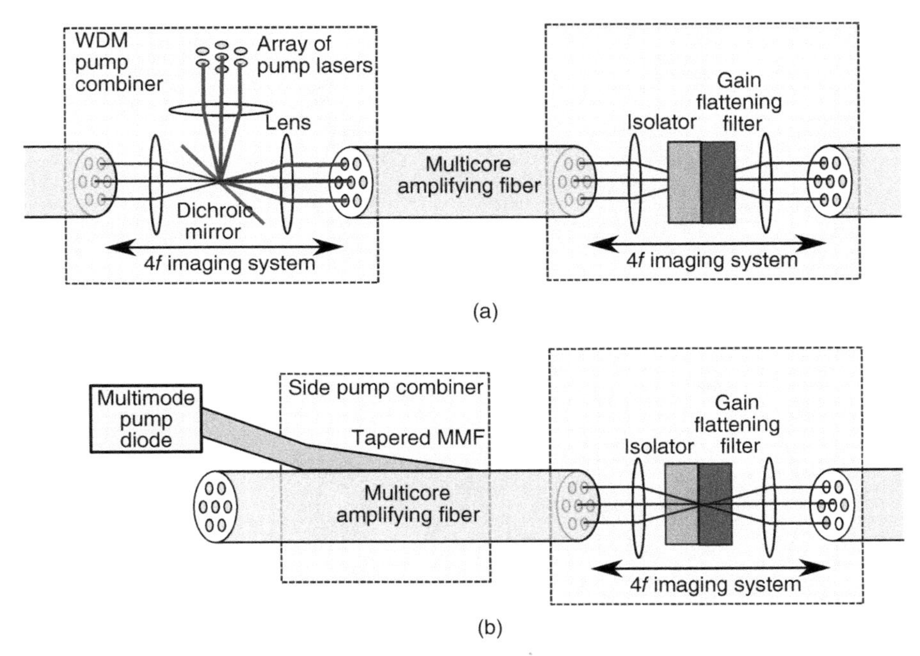

FIGURE 16.25 SDM amplifier and pump couplers for (a) core-pumped and (b) cladding-pumped MCF EDFA.

the signal intensity is nearly equal to the pump intensity, the inversion decreases. The lower inversion reduces the gain, requires longer amplifying fibers, which pushes the gain toward the L-band. However, these challenges should be solvable by decreasing the cladding area to increase the pump intensity, and increasing the size of the cores to reduce the signal intensity.

The simplicity of cladding-pumped amplifiers enables to scale the number of spatial channels without significantly increasing the amplifier cost or complexity. The main drawback of cladding-pumped amplifiers is the lack of control of the gain in individual cores. Reference [150] describes a per-core and spectral gain equalizer that is placed between two stages of a cladding-pumped amplifier. The gain equalizer is implemented in a free-space section and shares numerous optical components between all cores, such as the diffraction grating, the optical isolator, and a LCoS phased-array element used for attenuation. The scheme relies on the spatial modes being separated such that they can be demultiplexed onto different positions on the LCoS attenuating array.

Gain equalization between all modes/spatial channels must be considered for all types of SDM amplifiers. In amplifiers with well-separated modes, such as multicore amplifiers, the gain can be controlled through controlling the pump power in each core, or using an external equalizer [150] as described above. If the cores are very uniform, multicore amplifiers should produce similar gain for all cores without additional efforts. For amplifiers with spatial overlapping channels (i.e., FMF), external equalization is more difficult. Attempting to attenuate one particular mode without affecting the others modes is difficult without using a mode multiplexer [103]. Therefore, it is necessary to reduce the gain variations inside the amplifier. Notable gain equalization schemes include ring doping [97, 144, 151], pump modal control [151, 152], and microstructuring the gain region [153, 154]. The LP_{01} mode tends to get amplified the most, whereas the higher-order modes, whose fields extend further into the cladding, see less gain medium and are amplified the least. Ring-shaped doping profiles reduce the gain for the LP_{01} mode, while enhancing the gain of the LP_{11} modes. Higher-order modes tend to be ringlike, and launching into these modes can also enhance the gain for the higher-order modes while reducing the gain for the LP_{01} mode. The wavelength dependence of the gain is occasionally uniform for all modes, such that wavelength equalization can be applied to all modes independently of gain equalization [97].

16.8 CONCLUSION

SDM is the response to the limited optical transmission capacity of the SMF. Multiple technological options, including multiple SMFs, MCFs, and MMFs and the related components such as mode couplers or fan-in/fan-out devices are now available to solve the capacity problem. Cost reduction is expected to come mostly from integration, and in SDM systems integration will happen at various levels: the transponders, the fibers, the optical switches, and amplifiers.

Two possible paths for SDM are conceivable: (i) systems with low crosstalk between the spatial paths and (ii) systems with MIMO DSP for crosstalk mitigation. The second path requires more DSP, but can provide higher level of component integration and, therefore, lower the component cost.

Independently from the two paths, optical switches and optical amplifiers that support multiple spatial channels are going to play a key role in SDM systems. For optical switches, which are the key components defining optical meshed networks, the concept of "joint switching" will be essential to scale the switching capacity to the required levels.

ACKNOWLEDGMENTS

We acknowledge valuable discussions with P. J. Winzer, R. J. Essiambre, S. Randel, S. Bigo, S. Chandrasekhar, G. Charlet, A. R. Chraplyvy, C. R. Doerr, A. H. Gnauck, X. Liu, H. Kogelnik, S. K. Korotky, G. Raybon, M. Salsi, R. W. Tkach, C. Xie, Y. Sun, D. DiGiovanni, J. Fini, R. Lingle, D. Peckham, T. F. Taunay, L. Grüner-Nielsen, M. Hirano, T. Sasaki, and T. Hayashi, S. H. Chang, and H. S. Chung.

This work was (partly) supported by the IT R&D program of MSIP/IITP, Republic of Korea (10043383, Research of mode-division-multiplexing optical transmission technology over 10 km MMF).

REFERENCES

1. Essiambre R-J, Kramer G, Winzer PJ, Foschini GJ, Goebel B. Capacity limits of optical fiber networks. *J Lightwave Technol* 2010;28(4):662–701.
2. Winzer P. Spatial multiplexing in fiber optics: the 10x scaling of metro/core capacities. *Bell Labs Tech J* 2014;19:22–30.
3. Richardson DJ, Fini JM, Nelson LE. Space-division multiplexing in optical fibres. *Nat Photonics* 2013;7(5):354–362.
4. Li G, Bai N, Zhao N, Xia C. Space-division multiplexing: the next frontier in optical communication. *Adv Opt Photonics* 2014;6(4):413–487 [Online]. Available at http://aop.osa.org/abstract.cfm?URI=aop-6-4-413.
5. Jain S, Rancaño VJF, May-Smith TC, Petropoulos P, Sahu JK, Richardson DJ. Multi-element fiber technology for space-division multiplexing applications. *Opt Express* 2014;22(4):3787–3796 [Online]. Available at http://www.opticsexpress.org/abstract.cfm?URI=oe-22-4-3787.
6. Chandrasekhar S, Gnauck AH, Liu X, Winzer PJ, Pan Y, Burrows EC, Taunay T, Zhu B, Fishteyn M, Yan MF, Fini JM, Monberg E, Dimarcello F. WDM/SDM transmission of 10 x 128-Gb/s PDM-QPSK over 2688-km 7-core fiber with a per-fiber net aggregate spectral-efficiency distance product of 40,320 km.b/s/Hz. *Opt Express* 2012;20(2):706–711 [Online]. Available at http://www.opticsexpress.org/abstract.cfm?URI=oe-20-2-706.
7. Hayashi T, Taru T, Shimakawa O, Sasaki T, Sasaoka E. Design and fabrication of ultra-low crosstalk and low-loss multi-core fiber. *Opt Express* 2011;19(17):16576–16592 [Online]. Available at http://www.opticsexpress.org/abstract.cfm?URI=oe-19-17-16576.
8. Pepeljugoski P, Doany FE, Kuchta D, Lee B, Schow CL, Schares L. Connector performance analysis for d-shaped multi-core multi mode fiber. Optical Fiber Communication Conference; Optical Society of America; 2014. p. Th4J.4. [Online]. Available at http://www.opticsinfobase.org/abstract.cfm?URI=OFC-2014-Th4J.4. Accessed 2015 Sept 23.
9. Sakaguchi J, Puttnam BJ, Klaus W, Awaji Y, Wada N, Kanno A, Kawanishi T, Imamura K, Inaba H, Mukasa K, et al. 305 Tb/s space division multiplexed transmission using homogeneous 19-core fiber. *J Lightwave Technol* 2013;31(4):554–562.
10. Ryf R, Fontaine NK, Montoliu M, Randel S, Chang SH, Chen H, Chandrasekhar S, Gnauck AH, Essiambre R-J, Winzer PJ, Taru T, Hayashi T, Sasaki T. Space-division multiplexed transmission over 3×3 coupled-core multicore fiber. Optical Fiber Communications Conference and Exhibition (OFC); 2014. p 1–3. [Online]. Available at http://ieeexplore.ieee.org/stamp/stamp.jsp?arnumber=6886923 Accessed 2015 Sept 23.
11. Ryf R, Essiambre R-J, Randel S, Gnauck A, Winzer PJ, Hayashi T, Taru T, Sasaki T. MIMO-based crosstalk suppression in spatially multiplexed 3 x 56-Gb/s PDM-QPSK signals for strongly-coupled 3-core fiber. *Photonics Technol Lett* 2011;23(20):1469–1471.
12. Berdague S, Facq P. Mode division multiplexing in optical fibers. *Appl Opt* 1982;21(11):1950–1955.
13. Stuart HR. Dispersive multiplexing in multimode optical fiber. *Science* 2000;289(5477):281–283.
14. Franz B, Bulow H. Experimental evaluation of principal mode groups as high-speed transmission channels in spatial multiplex systems. *Photonics Technol Lett* 2012;

24(16):1363–1365. [Online]. Available at http://ieeexplore.ieee.org/stamp/stamp.jsp?arnumber=6210360.

15. Shi K, Gordon G, Paskov M, Carpenter J, Wilkinson T, Thomsen B. Degenerate mode-group division multiplexing using MIMO digital signal processing. Photonics Society Summer Topical Meeting Series, 2013 IEEE; July 2013. p 141–142.
16. Carpenter J, Wilkinson T. Characterization of multimode fiber by selective mode excitation. *J Lightwave Technol* 2012;30(10):1386–1392. [Online]. Available at http://ieeexplore.ieee.org/stamp/stamp.jsp?arnumber=6175092.
17. de Boer M, Tsekrekos CP, Martinez A, Kurniawan H, Bergmans JWM, Koonen AMJ, van den Boom HPA, Willems FMJ. A first demonstrator for a mode group diversity multiplexing communication system. Proceedings of IEE Seminar (Ref Optical Fibre Communications and Electronic Signal Processing No. 2005-11310); 2005.
18. Shah AR, Hsu RCJ, Tarighat A, Sayed AH, Jalali B. Coherent optical MIMO (COMIMO). *J Lightwave Technol* 2005;23(8):2410–2419.
19. Ryf R, Randel S, Gnauck AH, Bolle C, Essiambre R-J, Winzer PJ, Peckham DW, McCurdy A, Lingle R. Space-division multiplexing over 10 km of three-mode fiber using coherent 6 x 6 MIMO processing. Proceedings of the Optical Fiber Communication Conference (OFC); 2011. p PDPB10.
20. Salsi M, Koebele C, Sperti D, Tran P, Brindel P, Mardoyan H, Bigo S, Boutin A, Verluise F, Sillard P, Bigot-Astruc M, Provost L, Cerou F, Charlet G. Transmission at 2x100Gb/s, over two modes of 40km-long prototype few-mode fiber, using LCOS based mode multiplexer and demultiplexer. Optical Fiber Communication Conference/National Fiber Optic Engineers Conference 2011; Optical Society of America; Mar 2011. p PDPB9. [Online]. Available at http://www.opticsinfobase.org/abstract.cfm?URI=NFOEC-2011-PDPB9 Accessed 2015 Sept 23.
21. Li A, Al Amin A, Chen X, Shieh W. Reception of mode and polarization multiplexed 107Gb/s CO-OFDM signal over a two-mode fiber. Proceedings of the Optical Fiber Communication Conference (OFC); 2011. p. PDPB8.
22. Ip E, Bai N, Huang Y-K, Mateo E, Yaman F, Li M-J, Bickham S, Ten S, Linares J, Montero C, Moreno V, Prieto X, Tse V, Chung KM, Lau A, Tam H-Y, Lu C, Luo Y, Peng G-D, Li G. 88 x 3 x 112-Gb/s WDM transmission over 50 km of three-mode fiber with inline few-mode fiber amplifier. Proceedings, European Conference on Optical Communication (ECOC), Postdeadline; 2011. p Th.13.C.2.
23. Sleiffer VAJM, Jung Y, Veljanovski V, van Uden RGH, Kuschnerov M, Chen H, Inan B, Nielsen LG, Sun Y, Richardson DJ, Alam S, Poletti F, Sahu J, Dhar A, Koonen A, Corbett B, Winfield R, Ellis A, de Waardt H. 73.7 Tb/s (96 x 3 x 256-Gb/s) mode-division-multiplexed DP-16QAM transmission with inline MM-EDFA. *Opt Express* 2012;20(26):B428–B438. [Online]. Available at http://www.opticsexpress.org/abstract.cfm?URI=oe-20-26-B428.
24. Ryf R, Randel S, Fontaine NK, Montoliu M, Burrows E, Chandrasekhar S, Gnauck AH, Xie C, Essiambre R-J, Winzer P, Delbue R, Pupalaikis P, Sureka A, Sun Y, Gruner-Nielsen L, Jensen RV, Lingle R. 32-bit/s/Hz spectral efficiency WDM transmission over 177-km few-mode fiber. Optical Fiber Communication Conference/National Fiber Optic Engineers Conference 2013; Optical Society of America; 2013. p PDP5A.1. [Online]. Available at http://www.opticsinfobase.org/abstract.cfm?URI=OFC-2013-PDP5A.1. Accessed 2015 Sept 23.

25. Vuong J, Ramantanis P, Frignac Y, Salsi M, Genevaux P, Bendimerad DF, Charlet G. Mode coupling at connectors in mode-division multiplexed transmission over few-mode fiber. *Opt Express* 2015;23(2):1438–1455. [Online]. Available at http://www.opticsexpress.org/abstract.cfm?URI=oe-23-2-1438.

26. Ryf R, Fontaine NK, Dunayevsky J, Sinefeld D, Blau M, Montoliu M, Randel S, Liu C, Ercan B, Esmaeelpour M, Chandrasekhar S, Gnauck AH, Leon-Saval S, Bland-Hawthorn J, Salazar-Gil J, Sun Y, Gruner-Nielsen L, Lingle R, Marom D. Wavelength-selective switch for few-mode fiber transmission. Optical Communication (ECOC 2013), 39th European Conference and Exhibition on; 2013. p 1–3. [Online]. Available at http://ieeexplore.ieee.org/stamp/stamp.jsp?arnumber=6647874. Accessed 2015 Sept 23.

27. Ip E, Li M-J, Bennett K, Huang Y-K, Tanaka A, Korolev A, Koreshkov K, Wood W, Mateo E, Hu J, Yano Y. $146\lambda \times 6 \times 19$-Gbaud wavelength-and mode-division multiplexed transmission over 10×50-km spans of few-mode fiber with a gain-equalized few-mode EDFA. *J Lightwave Technol* 2014;32(4):790–797. [Online]. Available at http://jlt.osa.org/abstract.cfm?URI=jlt-32-4-790.

28. Xia C, Amezcua-Correa R, Bai N, Antonio-Lopez E, Arrioja DM, Schulzgen A, Richardson M, Liñares J, Montero C, Mateo E, et al. Hole-assisted few-mode multi-core fiber for high-density space-division multiplexing. *IEEE Photonics Technol Lett* 2012;24(21):1914–1917.

29. Mizuno T, Kobayashi T, Takara H, Sano A, Kawakami H, Nakagawa T, Miyamoto Y, Abe Y, Goh T, Oguma M, Sakamoto T, Sasaki Y, Ishida I, Takenaga K, Matsuo S, Saitoh K, Morioka T. 12-core x 3-mode dense space division multiplexed transmission over 40 km employing multi-carrier signals with parallel MIMO equalization. Optical Fiber Communication Conference: Postdeadline Papers; Optical Society of America; 2014. p Th5B.2. [Online]. Available at http://www.opticsinfobase.org/abstract.cfm?URI=OFC-2014-Th5B.2. Accessed 2015 Sept 23.

30. Zhu B, Taunay T, Yan M, Fishteyn M, Oulundsen G, Vaidya D. 70-Gb/s multicore multimode fiber transmissions for optical data links. *IEEE Photonics Technol Lett* 2010;22(22):1647–1649.

31. Qian D, Ip E, Huang M-F, Li MJ, Dogariu A, Zhang S, Shao Y, Huang Y-K, Zhang Y, Cheng X, Tian Y, Ji P, Collier A, Geng Y, Linares J, Montero C, Moreno V, Prieto X, Wang T. 1.05Pb/s transmission with 109b/s/Hz spectral efficiency using hybrid single- and few-mode cores. Frontiers in Optics 2012/Laser Science XXVIII; Optical Society of America; 2012. p FW6C.3. [Online]. Available at http://www.opticsinfobase.org/abstract.cfm?URI=FiO-2012-FW6C.3. Accessed 2015 Sept 23.

32. Winzer P, Gnauck A, Konczykowska A, Jorge F, Dupuy J-Y. Penalties from In-band crosstalk for advanced optical modulation formats. Proceedings of the Optical Fiber Communication Conference (OFC 2011). Optical Society of America; 2011. [Online]. Available at http://www.opticsinfobase.org/abstract.cfm?URI=ECOC-2011-Tu.5.B.7. Accessed 2015 Sept 23.

33. Randel S, Sierra A, Ryf R, Winzer PJ. Crosstalk tolerance of spatially multiplexed MIMO systems. European Conference and Exhibition on Optical Communication; Optical Society of America; 2012. p P4.08. [Online]. Available At http://www.opticsinfobase.org/abstract.cfm?URI=ECEOC-2012-P4.08. Accessed 2015 Sept 23.

34. Snyder AW. Coupled-mode theory for optical fibers. *J Opt Soc Am* 1972; 62(11):1267–1277. [Online]. Available at http://www.opticsinfobase.org/abstract.cfm?URI=josa-62-11-1267.

35. Huang WP, Li L. Coupled-mode theory for optical waveguides: an overview. *J Opt Soc Am A* 1994;11(3):963–983.
36. Fini JM, Zhu B, Taunay TF, Yan MF. Statistics of crosstalk in bent multicore fibers. *Opt Express* 2010;18(14):15 122–15 129.
37. Igarashi K, Tsuritani T, Morita I, Tsuchida Y, Maeda K, Tadakuma M, Saito T, Watanabe K, Imamura K, Sugizaki R, Suzuki M. Super-Nyquist-WDM transmission over 7,326-km seven-core fiber with capacity-distance product of 1.03 Exabit/s/km. *Opt Express* 2014;22(2):1220–1228. [Online]. Available at http://www.opticsexpress.org/abstract.cfm?URI=oe-22-2-1220.
38. Kobayashi T, Takara H, Sano A, Mizuno T, Kawakami H, Miyamoto Y, Hiraga K, Abe Y, Ono H, Wada M, Sasaki Y, Ishida I, Takenaga K, Matsuo S, Saitoh K, Yamada M, Masuda H, Morioka T. 2 x 344 Tb/s propagation-direction interleaved transmission over 1500-km MCF enhanced by multicarrier full electric-field digital back-propagation. Optical Communication (ECOC 2013), 39th European Conference and Exhibition on; Sept 2013.
39. Takahashi H, Tsuritani T, de Gabory ELT, Ito T, Peng WR, Igarashi K, Takeshima K, Kawaguchi Y, Morita I, Tsuchida Y, Mimura Y, Maeda K, Saito T, Watanabe K, Imamura K, Sugizaki R, Suzuki M. First demonstration of MC-EDFA-repeated SDM transmission of 40 x 128-Gbit/s PDM-QPSK signals per core over 6,160-km 7-core MCF. *Opt Express* 2013;21(1):789–795. [Online]. Available at http://www.opticsexpress.org/abstract.cfm?URI=oe-21-1-789.
40. Takara H, Sano A, Kobayashi T, Kubota H, Kawakami H, Matsuura A, Miyamoto Y, Abe Y, Ono H, Shikama K, Goto Y, Tsujikawa K, Sasaki Y, Ishida I, Takenaga K, Matsuo S, Saitoh K, Koshiba M, Morioka T. 1.01-Pb/s (12 SDM/222 WDM/456 Gb/s) crosstalk-managed transmission with 91.4-b/s/Hz aggregate spectral efficiency. European Conference and Exhibition on Optical Communication; Optical Society of America; 2012. p Th.3.C.1. [Online]. Available at http://www.opticsinfobase.org/abstract.cfm?URI=ECEOC-2012-Th.3.C.1. Accessed 2015 Sept 23.
41. Sakaguchi J, Awaji Y, Wada N, Kanno A, Kawanishi T, Hayashi T, Taru T, Kobayashi T, Watanabe M. Space division multiplexed transmission of 109-Tb/s data signals using homogeneous seven-core fiber. *J Lightwave Technol* 2012;30(4):658–665.
42. Nakazawa M, Yoshida M, Hirooka T. Nondestructive measurement of mode couplings along a multi-core fiber using a synchronous multi-channel OTDR. *Opt Express* 2012;20(11):12 530–12 540. [Online]. Available at http://www.opticsexpress.org/abstract.cfm?URI=oe-20-11-12530.
43. Matsushima Y. Prospective for deployment of 3M technologies. Photonics Society Summer Topical Meeting Series, 2014 IEEE; July 2014. p 144–145.
44. Nelson L, Feuer M, Abedin K, Zhou X, Taunay T, Fini J, Zhu B, Isaac R, Harel R, Cohen G, Marom D. Spatial superchannel routing in a two-span ROADM system for space division multiplexing. *IEEE/OSA J Lightwave Technol* 2014;32(4):783–789. [Online]. Available at http://ieeexplore.ieee.org/stamp/stamp.jsp?arnumber=6623116.
45. Feuer M, Nelson L, Zhou X, Woodward S, Isaac R, Zhu B, Taunay T, Fishteyn M, Fini J, Yan M. Joint digital signal processing receivers for spatial superchannels. *IEEE Photonics Technol Lett* 2012;24(21):1957–1960. [Online]. Available at http://ieeexplore.ieee.org/stamp/stamp.jsp?arnumber=6317137.
46. de Gabory ELT, Arikawa M, Hashimoto Y, Ito T, Fukuchi K. A shared carrier reception and processing scheme for compensating frequency offset and phase noise of

space-division multiplexed signals over multicore fibers. Optical Fiber Communication Conference/National Fiber Optic Engineers Conference 2013; Optical Society of America; 2013. p OM2C.2. [Online]. Available at http://www.opticsinfobase.org/abstract.cfm?URI=OFC-2013-OM2C.2. Accessed 2015 Sept 23.

47. Sakaguchi J, Awaji Y, Wada N. Development of carrier-phase synchronization swapper for space-division multiplexed self-homodyne optical networks. Optical Communication (ECOC), 2014 European Conference on; Sept 2014. p 1–3.
48. Puttnam BJ, Sakaguchi J, Mendinueta JMD, Klaus W, Awaji Y, Wada N, Kanno A, Kawanishi T. Investigating self-homodyne coherent detection in a 19 channel space-division-multiplexed transmission link. *Opt Express* 2013;21(2):1561–1566. [Online]. Available at http://www.opticsexpress.org/abstract.cfm?URI=oe-21-2-1561.
49. Puttnam BJ, Luis R, Delgado-Mendinueta J-M, Sakaguchi J, Klaus W, Awaji Y, Wada N, Kanno A, Kawanishi T. High-capacity self-homodyne PDM-WDM-SDM transmission in a 19-core fiber. *Opt Express* 2014;22(18):21 185–21 191. [Online]. Available at http://www.opticsexpress.org/abstract.cfm?URI=oe-22-18-21185.
50. Mendinueta JMD, Puttnam BJ, Sakaguchi J, Luís RS, Klaus W, Awaji Y, Wada N, Kanno A, Kawanishi T. Investigation of receiver DSP carrier phase estimation rate for self-homodyne space-division multiplexing communication systems. Optical Fiber Communication Conference/National Fiber Optic Engineers Conference 2013; Optical Society of America; 2013. p JTh2A.48. [Online]. Available at http://www.opticsinfobase.org/abstract.cfm?URI=NFOEC-2013-JTh2A.48. Accessed 2015 Sept 23.
51. Puttnam BJ, Delgado-Mendinueta J-M, Sakaguchi J, Luís RS, Klaus W, Awaji Y, Wada N, Kanno A, Kawanishi T. 105Tb/s transmission system using low-cost, MHz linewidth DFB lasers enabled by self-homodyne coherent detection and a 19-core fiber. Optical Fiber Communication Conference/National Fiber Optic Engineers Conference 2013. Optical Society of America; 2013. p OW1I.1. [Online]. Available at http://www.opticsinfobase.org/abstract.cfm?URI=OFC-2013-OW1I.1. Accessed 2015 Sept 23.
52. Liu X, Chandrasekhar S, Gnauck AH, Winzer PJ, Randel S, Corteselli S, Chraplyvy AR, Tkach RW, Zhu B, Taunay TF, Fishteyn M. Digital coherent superposition for performance improvement of spatially multiplexed coherent optical OFDM superchannels. *Opt Express* 2012;20(26):B595–B600. [Online]. Available at http://www.opticsexpress.org/abstract.cfm?URI=oe-20-26-B595. Accessed 2015 Sept 23.
53. Salz J. Digital transmission over cross-coupled linear channels. *AT&T Tech J* 1985;64:1147–1159.
54. Randel S, Sierra A, Mumtaz S, Tulino A, Ryf R, Winzer PJ, Schmidt C, Essiambre R. Adaptive MIMO signal processing for mode-division multiplexing. Proceedings and the National Fiber Optic Engineers Conference Optical Fiber Communication Conference and Exposition (OFC/NFOEC); 2012. p 1–3.
55. Chen X, He J, Li A, Ye J, Shieh W. Characterization of dynamic evolution of channel matrix in two-mode fibers. Optical Fiber Communication Conference/National Fiber Optic Engineers Conference 2013; Optical Society of America; 2013. p OM2C.3. [Online]. Available at http://www.opticsinfobase.org/abstract.cfm?URI=OFC-2013-OM2C.3. Accessed 2015 Sept 23.
56. Winzer PJ, Foschini GJ. MIMO capacities and outage probabilities in spatially multiplexed optical transport systems. *Opt Express* 2011;19(17):16 680–16 696.

57. Winzer PJ, Foschini GJ. Outage calculations for spatially multiplexed fiber links. Proceedings of the Optical Fiber Communication Conference (OFC). Los Angeles (CA): Optical Society of America: 2011. p OThO5.
58. Cvijetic N, Ip E, Prasad N, Li M-J. Experimental frequency-domain channel matrix characterization for SDM-MIMO-OFDM systems. Photonics Society Summer Topical Meeting Series; 2013 IEEE; July 2013. p 139–140.
59. Cvijetic N, Ip E, Prasad N, Li M-J, Wang T. Experimental time and frequency domain MIMO channel matrix characterization versus distance for 6×28Gbaud QPSK transmission over 40×25km few mode fiber. Optical Fiber Communication Conference; Optical Society of America; 2014. p Th1J.3. [Online]. Available at http://www.opticsinfobase.org/abstract.cfm?URI=OFC-2014-Th1J.3. Accessed 2015 Sept 24.
60. Marcuse D. In: Liao PF, Kelley PL, editors. Theory of Dielectric Optical Waveguides. Academic Press; 1991.
61. Juarez AA, Krune E, Warm S, Bunge CA, Petermann K. Modeling of mode coupling in multimode fibers with respect to bandwidth and loss. *J Lightwave Technol* 2014;32(8):1549–1558. [Online]. Available at http://jlt.osa.org/abstract.cfm?URI=jlt-32-8-1549.
62. Lobato A, Ferreira F, Kuschnerov M, van den Borne D, Jansen SL, Napoli A, Spinnler B, Lankl B. Impact of mode coupling on the mode-dependent loss tolerance in few-mode fiber transmission. *Opt Express* 2012;20(28):29 776–29 783. [Online]. Available at http://www.opticsexpress.org/abstract.cfm?URI=oe-20-28-29776.
63. Antonelli C, Mecozzi A, Shtaif M, Winzer PJ. Stokes-space analysis of modal dispersion in fibers with multiple mode transmission. *Opt Express* 2012;20(11):11 718–11 733. [Online]. Available at http://www.opticsexpress.org/abstract.cfm?URI=oe-20-11-11718.
64. Ryf R, Randel S, Gnauck A, Bolle C, Sierra A, Mumtaz S, Esmaeelpour M, Burrows E, Essiambre R-J, Winzer PJ, Peckham D, McCurdy A, Lingle R. Mode-division multiplexing over 96 km of few-mode fiber using coherent 6 x 6 MIMO processing. *J Lightwave Technol* 2012;30(4):521–531. [Online]. Available at http://ieeexplore.ieee.org/stamp/stamp.jsp?arnumber=6074912.
65. Kasahara M, Saitoh K, Sakamoto T, Hanzawa N, Matsui T, Tsujikawa K, Yamamoto F. Design of three-spatial-mode ring-core fiber. *J Lightwave Technol* 2014;32(7):1337–1343.
66. Fontaine N, Ryf R, Hirano M, Sasaki T. Experimental investigation of crosstalk accumulation in a ring-core fiber. Photonics Society Summer Topical Meeting Series, 2013 IEEE. Waikoloa, Hawaii: IEEE Photonic Society; 2013. p 111–112. [Online]. Available at http://ieeexplore.ieee.org/stamp/stamp.jsp?arnumber=6614511. Accessed 2015 Sept 24.
67. Grüner-Nielsen L, Sun Y, Nicholson JW, Jakobsen D, Jespersen KG, Robert Lingle J, Pálsdóttir B. Few mode transmission fiber with low DGD, low mode coupling, and low loss. *J Lightwave Technol* 2012;30(23):3693–3698. [Online]. Available at http://jlt.osa.org/abstract.cfm?URI=jlt-30-23-3693.
68. Peckham D, Sun Y, McCurdy A, Lingle R Jr. Few-mode fiber technology for spatial multiplexing. In: Kaminow I, Li T, Willner A, editors. Optical Fiber Telecommunication VIA. Elsevier; 2013.
69. Hanzawa N, Saitoh K, Sakamoto T, Matsui T, Tomita S, Koshiba M. Demonstration of mode-division multiplexing transmission over 10 km two-mode fiber with mode

coupler. Proceedings of the Optical Fiber Communication Conference (OFC); 2011. p OWA4.

70. Koebele C, Salsi M, Sperti D, Tran P, Brindel P, Mardoyan H, Bigo S, Boutin A, Verluise F, Sillard P, Astruc M, Provost L, Cerou F, Charlet G. Two mode transmission at 2x100Gb/s, over 40km-long prototype few-mode fiber, using LCOS-based programmable mode multiplexer and demultiplexer. *Opt Express* 2011;19(17):16 593–16 600. [Online]. Available at http://www.opticsexpress.org/abstract.cfm?URI=oe-19-17-16593.

71. Koebele C, Salsi M, Milord L, Ryf R, Bolle C, Sillard P, Bigo S, Charlet G, 40km transmission of five mode division multiplexed data streams at 100Gb/s with low MIMO-DSP complexity. Proceedings, European Conference on Optical Communication (ECOC), Postdeadline paper; 2011. p. Th.13.C.3.

72. Ryf R, Mestre MA, Gnauck AH, Randel S, Schmidt C, Essiambre R-J, Winzer PJ, Delbue R, Pupalaikis P, Sureka A, Hayashi T, Taru T, Sasaki T. Space-division multiplexed transmission over 4200-km 3-core microstructured fiber. Proceedings of the Optical Fiber Communication Conference (OFC); 2012. p PDP5C.2.

73. Ryf R, Fontaine NK, Montoliu M, Randel S, Ercan B, Chen H, Chandrasekhar S, Gnauck AH, Leon-Saval SG, Bland-Hawthorn J, Gil JRS, Sun Y, Lingle R. Photonic-lantern-based mode multiplexers for few-mode-fiber transmission. Optical Fiber Communication Conference; Optical Society of America; 2014. p W4J.2. [Online]. Available at http://www.opticsinfobase.org/abstract.cfm?URI=OFC-2014-W4J.2. Accessed 2015 Sept 24.

74. Ryf R, Randel S, Fontaine N, Palou X, Burrows E, Corteselli S, Chandrasekhar S, Gnauck A, Xie C, Essiambre R-J, Winzer PJ, Delbue R, Pupalaikis P, Sureka A, Sun Y, Gruner-Nielsen L, Jensen R, Lingle R. 708-km combined WDM/SDM transmission over few-mode fiber supporting 12 spatial and polarization modes. Optical Communication (ECOC 2013), 39th European Conference and Exhibition on; 2013. p 1–3. [Online]. Available at http://ieeexplore.ieee.org/stamp/stamp.jsp?arnumber=6647613. Accessed 2015 Sept 24.

75. Ho K-P, Kahn JM. Statistics of group delays in multimode fiber with strong mode coupling. *J Lightwave Technol* 2011;29(21):3119–3128.

76. Ho K-P, Kahn JM. Linear propagation effects in mode-division multiplexing systems. *J Lightwave Technol* 2014;32(4):614–628. [Online]. Available at http://jlt.osa.org/abstract.cfm?URI=jlt-32-4-614.

77. Arik SO, Askarov D, Kahn JM. Effect of mode coupling on signal processing complexity in mode-division multiplexing. *J Lightwave Technol* 2013;31(3):423–431.

78. Ryf R, Fontaine NK, Chen H, Guan B, Randel S, Sauer N, Yoo S, Koonen A, Delbue R, Pupalaikis P, Sureka A, Shubochkin R, Sun Y, Lingle R. 23 Tbit/s transmission over 17-km conventional 50 μm graded-index multimode fiber. Optical Fiber Communication Conference: Postdeadline Papers; Optical Society of America; 2014. p Th5B.1. [Online]. Available at http://www.opticsinfobase.org/abstract.cfm?URI=OFC-2014-Th5B.1. Accessed 2015 Sept 24.

79. Ryf R, Fontaine NK, Guan B, Huang B, Esmaeelpour M, Randel S, Gnauck AH, Chandrasekhar S, Adamiecki A, Raybon G, Tkach RW, Shubochkin R, Sun Y, Lingle RJ. 305-km combined wavelength and mode-multiplexed transmission over conventional graded-index multimode fibre. ECOC 2014, no. PD3.5; 2014.

80. Ryf R, Fontaine NK, Guan B, Essiambre R-J, Randel S, Gnauck AH, Chandrasekhar S, Adamiecki A, Raybon G, Ercan B, Scott R, Ben Yoo SJ, Hayashi T, Nagashima T, Sasaki T. 1705-km transmission over coupled-core fibre supporting 6 spatial modes. Optical Communication (ECOC), 2014 European Conference on; Sept 2014. p 1–3.
81. Ryf R, Fontaine NK, Chen H, Guan B, Huang B, Esmaeelpour M, Gnauck AH, Randel S, Yoo S, Koonen A, Shubochkin R, Sun Y, Lingle R. Mode-multiplexed transmission over conventional graded-index multimode fibers. *Opt Express* 2015;23(1):235–246. [Online]. Available at http://www.opticsexpress.org/abstract.cfm?URI=oe-23-1-235.
82. Chen Y, Lobato A, Jung Y, Chen H, Sleiffer VAJM, Kuschnerov M, Fontaine NK, Ryf R, Richardson DJ, Lankl B, Hanik N. 41.6 Tbit/s C-band SDM OFDM transmission through 12 spatial and polarization modes over 74.17 km few mode fiber. *J Lightwave Technol* 2015;PP(99):1–1.
83. Randel S, Ryf R, Sierra A, Winzer PJ, Gnauck AH, Bolle CA, Essiambre R-J, Peckham DW, McCurdy A, Lingle R. 6 x 56-Gb/s mode-division multiplexed transmission over 33-km few-mode fiber enabled by 6×6 MIMO equalization. *Opt Express* 2011;19(17):16 697–16 707.
84. Randel S, Winzer PJ, Montoliu M, Ryf R. Complexity analysis of adaptive frequency-domain equalization for MIMO-SDM transmission. Optical Communication (ECOC 2013), 39th European Conference and Exhibition on; Sept 2013. p 1–3.
85. Inan B, Spinnler B, Ferreira F, Lobato A, Adhikari S, Sleiffer V, van den Borne D, Hanik N, Jansen S. Equalizer complexity of mode division multiplexed coherent receivers. Optical Fiber Communication Conference and Exposition (OFC/NFOEC), 2012 and the National Fiber Optic Engineers Conference; 2012. p 1–3. [Online]. Available at http://ieeexplore.ieee.org/stamp/stamp.jsp?arnumber=6192152. Accessed 2015 Sept 24.
86. Faruk MS, Kikuchi K. Adaptive frequency-domain equalization in digital coherent optical receivers. *Opt Express* 2011;19(13):12 789–12 798. [Online]. Available at http://www.opticsexpress.org/abstract.cfm?URI=oe-19-13-12789.
87. Arik S, Askarov D, Kahn J. Adaptive frequency-domain equalization in mode-division multiplexing systems. *J Lightwave Technol* 2014;32(10):1841–1852.
88. Bai N, Li G. Adaptive frequency-domain equalization for mode-division multiplexed transmission. *IEEE Photonics Technol Lett* 2012;24(21):1918–1921.
89. Benvenuto N, Cherubini G. Algorithms for Communications Systems and their Applications. Chichester: John Wiley & Sons, Ltd; 2002.
90. van Uden RGH, Okonkwo CM, Sleiffer VAJM, de Waardt H, Koonen AMJ. MIMO equalization with adaptive step size for few-mode fiber transmission systems. *Opt Express* 2014;22(1):119–126. [Online]. Available at http://www.opticsexpress.org/abstract.cfm?URI=oe-22-1-119.
91. Sleiffer VAJM, Jung Y, Inan B, Chen H, van Uden RGH, Kuschnerov M, van den Borne D, Jansen SL, Veljanovski V, Koonen AMJ, Richardson DJ, Alam S, Poletti F, Sahu JK, Dhar A, Corbett B, Winfield R, Ellis AD, de Waardt H. Mode-division-multiplexed 3×112-Gb/s DP-QPSK transmission over 80-km few-mode fiber with inline MM-EDFA and blind DSP. Optical Communications (ECOC), 2012 38th European Conference and Exhibition on; 2012. p 1–3. [Online]. Available at http://ieeexplore.ieee.org/stamp/stamp.jsp?arnumber=6706137. Accessed 2015 Sept 24.
92. Kuschnerov M, Chouayakh M, Piyawanno K, Spinnler B, de Man E, Kainzmaier P, Alfiad M, Napoli A, Lankl B. Data-aided versus blind single-carrier coherent receivers. *IEEE Photonics J* 2010;2(3):387–403.

93. Li A, Amin AA, Chen X, Chen S, Gao G, Shieh W. Reception of dual-spatial-mode CO-OFDM signal over a two-mode fiber. *J Lightwave Technol* 2012;30(4):634–640.

94. Abedin KS, Taunay TF, Fishteyn M, DiGiovanni DJ, Supradeepa V, Fini JM, Yan MF, Zhu B, Monberg EM, Dimarcello F. Cladding-pumped erbium-doped multicore fiber amplifier. *Opt Express* 2012;20(18):20 191–20 200. [Online]. Available at http://www.opticsexpress.org/abstract.cfm?URI=oe-20-18-20191.

95. Feuer M, Nelson L, Abedin K, Zhou X, Taunay T, Fini J, Zhu B, Isaac R, Harel R, Cohen G, Marom D. ROADM system for space division multiplexing with spatial superchannels. Optical Fiber Communication Conference and Exposition and the National Fiber Optic Engineers Conference (OFC/NFOEC); 2013; March 2013. p 1–3.

96. Carpenter J, Leon-Saval SG, Salazar-Gil JR, Bland-Hawthorn J, Baxter G, Stewart L, Frisken S, Roelens MAF, Eggleton BJ, Schröder J. 1x11 few-mode fiber wavelength selective switch using photonic lanterns. *Opt Express* 2014;22(3):2216–2221. [Online]. Available at http://www.opticsexpress.org/abstract.cfm?URI=oe-22-3-2216.

97. Ip E, Gu RY, Li M-J, Huang Y-K, Kahn J. Experimental demonstration of a gain-flattening filter for few-mode fiber based on a spatial light modulator. Optical Fiber Communications Conference and Exhibition (OFC); 2014; March 2014. p 1–3.

98. Fontaine NK, Ryf R, Liu C, Ercan B, Salazar Gil JR, Leon-Saval SG, Bland-Hawthorn J, Neilson DT. Few-mode fiber wavelength selective switch with spatial-diversity and reduced-steering angle. Optical Fiber Communication Conference, Series OSA Technical Digest (online); Optical Society of America; Mar 2014. p Th4A.7. [Online]. Available at http://www.opticsinfobase.org/abstract.cfm?URI=OFC-2014-Th4A.7. Accessed 2015 Sept 24.

99. Abedin KS, Fini JM, Thierry TF, Zhu B, Yan MF, Bansal L, Dimarcello FV, Monberg EM, DiGiovanni DJ. Seven-core erbium-doped double-clad fiber amplifier pumped simultaneously by side-coupled multimode fiber. *Opt Lett* 2014;39(4):993–996. [Online]. Available at http://ol.osa.org/abstract.cfm?URI=ol-39-4-993.

100. Lim EL, Jung Y, Kang Q, May-Smith TC, Wong NHL, Standish R, Poletti F, Sahu JK, Alam S, Richardson DJ. First demonstration of cladding pumped few-moded EDFA for mode division multiplexed transmission. Optical Fiber Communication Conference, Series OSA Technical Digest (online); Optical Society of America; Mar 2014. p M2J.2. [Online]. Available at http://www.opticsinfobase.org/abstract.cfm?URI=OFC-2014-M2J.2. Accessed 2015 Sept 24.

101. Sakaguchi J, Klaus W, Puttnam BJ, Mendinueta JMD, Awaji Y, Wada N, Tsuchida Y, Maeda K, Tadakuma M, Imamura K, Sugizaki R, Kobayashi T, Tottori Y, Watanabe M, Jensen RV. 19-core MCF transmission system using EDFA with shared core pumping coupled via free-space optics. *Opt Express* 2013;22(1):90. [Online]. Available at http://www.opticsinfobase.org/abstract.cfm?URI=oe-22-1-90.

102. Ryf R, Fontaine NK, Mestre MA, Randel S, Palou X, Bolle C, Gnauck AH, Chandrasekhar S, Liu X, Guan B, Essiambre R-J, Winzer PJ, Leon-Saval S, Bland-Hawthorn J, Delbue R, Pupalaikis P, Sureka A, Sun Y, Grüner-Nielsen L, Jensen RV, Lingle R. 12 x 12 MIMO transmission over 130-km few-mode fiber. Frontiers in Optics 2012/Laser Science XXVIII; Optical Society of America; 2012. p FW6C.4. [Online]. Available at http://www.opticsinfobase.org/abstract.cfm?URI=FiO-2012-FW6C.4. Accessed 2015 Sept 24.

103. Blau M, Weiss I, Gerufi J, Sinefeld D, Bin-Nun M, Lingle R, Grüner-Nielsen L, Marom DM. Variable optical attenuator and dynamic mode group equalizer for few

mode fibers. *Opt Express* 2014;22(25):30 520–30 527. [Online]. Available at http://www.opticsexpress.org/abstract.cfm?URI=oe-22-25-30520.

104. Fontaine NK, Doerr CR, Mestre MA, Ryf R, Winzer PJ, Buhl L, Sun Y, Jiang X, Lingle R. Space-division multiplexing and all-optical MIMO demultiplexing using a photonic integrated circuit. Proceedings of the Optical Fiber Communication Conference (OFC); 2012. p PDP5B.1.
105. VanWiggeren G, Baney D. Swept-wavelength interferometric analysis of multiport components. *IEEE Photonics Technol Lett* 2003;15(9):1267–1269.
106. Gifford DK, Soller BJ, Wolfe MS, Froggatt ME. Optical vector network analyzer for single-scan measurements of loss, group delay, and polarization mode dispersion. *Appl Opt* 2005;44(34):7282–7286. [Online]. Available at http://ao.osa.org/abstract.cfm?URI=ao-44-34-7282.
107. Carpenter J, Eggleton BJ, Schröder J. Reconfigurable spatially-diverse optical vector network analyzer. *Opt Express* 2014;22(3):2706–2713. [Online]. Available at http://www.opticsexpress.org/abstract.cfm?URI=oe-22-3-2706.
108. Nicholson JW, Yablon AD, Ramachandran S, Ghalmi S. Spatially and spectrally resolved imaging of modal content in large-mode-area fibers. *Opt Express* 2008;16(10): 7233. [Online]. Available at http://www.opticsexpress.org/abstract.cfm?URI=oe-16-10-7233.
109. Demas J, Ramachandran S. Sub-second mode measurement of fibers using C^2 imaging. *Opt Express* 2014;22(19):23043. [Online]. Available at http://www.opticsinfobase.org/oe/fulltext.cfm?uri=oe-22-19-23043&id=301335.
110. Ma Y, Sych Y, Onishchukov G, Ramachandran S, Peschel U, Schmauss B, Leuchs G. Fiber-modes and fiber-anisotropy characterization using low-coherence interferometry. *Appl Phys B* 2009;96(2):345. [Online]. Available at http://www.springerlink.com/content/wp2550264g414433/abstract/.
111. Fatemi FK, Beadie G. Rapid complex mode decomposition of vector beams by common path interferometry. *Opt Express* 2013;21(26):32291. [Online]. Available at http://www.opticsinfobase.org/oe/fulltext.cfm?uri=oe-21-26-32291&id=276263.
112. Fontaine NK. Characterization of multi-mode fibers and devices for MIMO communications. SPIE, Volume 9009; 2013. p 90 090A–90–090A–8. [Online]. Available at http://dx.doi.org/10.1117/12.2044339. Accessed 2015 Sept 24.
113. Chen H, van Uden R, Okonkwo C, Koonen T. Compact spatial multiplexers for mode division multiplexing. *Opt Express* 2014;22(26):31582. [Online]. Available at http://www.opticsinfobase.org/oe/abstract.cfm?uri=oe-22-26-31582.
114. Ding Y, Ou H, Xu J, Peucheret C. Silicon photonic integrated circuit mode multiplexer. *IEEE Photonics Technol Lett* 2013;25(7):648–651.
115. Labroille G, Denolle B, Jian P, Genevaux P, Treps N, Morizur J-F. Efficient and mode selective spatial mode multiplexer based on multi-plane light conversion. *Opt Express* 2014;22(13):15 599–15 607. [Online]. Available at http://www.opticsexpress.org/abstract.cfm?URI=oe-22-13-15599.
116. Igarashi K, Souma D, Takeshima K, Tsuritani T. Selective mode multiplexer based on phase plates and Mach-Zehnder interferometer with image inversion function. *Opt Express* 2015;23(1):183–194. [Online]. Available at http://www.opticsexpress.org/abstract.cfm?URI=oe-23-1-183.

117. Berkhout GCG, Lavery MPJ, Courtial J, Beijersbergen MW, Padgett MJ. Efficient sorting of orbital angular momentum states of light. *Phys Rev Lett* 2010;105(15):153601. [Online]. Available at http://link.aps.org/doi/10.1103/PhysRevLett.105.153601.
118. Chang SH, Chung HS, Fontaine NK, Ryf R, Park KJ, Kim K, Lee JC, Lee JH, Kim BY, Kim YK. Mode division multiplexed optical transmission enabled by all-fiber mode multiplexer. *Opt Express* 2014;22(12):14 229–14 236. [Online]. Available at http://www.opticsexpress.org/abstract.cfm?URI=oe-22-12-14229.
119. Yerolatsitis S, Gris-Sánchez I, Birks TA. Adiabatically-tapered fiber mode multiplexers. *Opt Express* 2014;22(1):608–617. [Online]. Available at http://www.opticsexpress.org/abstract.cfm?URI=oe-22-1-608.
120. Leon-Saval SG, Fontaine NK, Salazar-Gil JR, Ercan B, Ryf R, Bland-Hawthorn J. Mode-selective photonic lanterns for space-division multiplexing. *Opt Express* 2014;22(1):1036–1044. [Online]. Available at http://www.opticsexpress.org/abstract.cfm?URI=oe-22-1-1036.
121. Riesen N, Gross S, Love JD, Withford MJ. Femtosecond direct-written integrated mode couplers. *Opt Express* 2014;22(24):29 855–29 861. [Online]. Available at http://www.opticsexpress.org/abstract.cfm?URI=oe-22-24-29855.
122. Ryf R, Fontaine NK, Essiambre R-J. Spot-based mode couplers for mode-multiplexed transmission in few-mode fiber. *Photonics Technol Lett* 2012;24(21):1973–1976.
123. Leon-Saval SG, Birks TA, Bland-Hawthorn J, Englund M. Multimode fiber devices with single-mode performance. *Opt Lett* 2005;30(19):2545–2547. [Online]. Available at http://ol.osa.org/abstract.cfm?URI=ol-30-19-2545.
124. Fontaine NK, Ryf R, Bland-Hawthorn J, Leon-Saval SG, et al. Geometric requirements for photonic lanterns in space division multiplexing. *Opt Express* 2012;20(24):27 123–27 132. [Online]. Available at http://www.opticsinfobase.org/abstract.cfm?uri=oe-20-24-27123.
125. Giles I, Chen R, Garcia-Munoz V. Fiber based multiplexing and demultiplexing devices for few mode fiber space division multiplexed communications. Optical Fiber Communication Conference, Series OSA Technical Digest (online); Optical Society of America; Mar 2014. p. Tu3D.1. [Online]. Available at http://www.opticsinfobase.org/abstract.cfm?URI=OFC-2014-Tu3D.1. Accessed 2015 Sept 24.
126. Bland-Hawthorn J, Lawrence J, Robertson G, Campbell S, Pope B, Betters C, Leon-Saval S, Birks T, Haynes R, Cvetojevic N, et al. PIMMS: photonic integrated multimode microspectrograph. SPIE Astronomical Telescopes + Instrumentation; International Society for Optics and Photonics; 2010. p 77 350N–77 350N. [Online]. Available at http://proceedings.spiedigitallibrary.org/proceeding.aspx?articleid=750728. Accessed 2015 Sept 24.
127. Mitchell P, Brown G, Thomson RR, Psaila N, Kar A. 57 channel (19x3) spatial multiplexer fabricated using direct laser inscription. Optical Fiber Communication Conference, Series OSA Technical Digest (online); Optical Society of America; Mar 2014. p M3K.5. [Online]. Available at http://www.opticsinfobase.org/abstract.cfm?URI=OFC-2014-M3K.5. Accessed 2015 Sept 24.
128. Noordegraaf D, Skovgaard PMW, Maack MD, Bland-Hawthorn J, Haynes R, Lgsgaard J. Multi-mode to single-mode conversion in a 61 port photonic lantern. *Opt Express* 2010;18(5):4673. [Online]. Available at http://www.opticsinfobase.org/oe/abstract.cfm?&uri=oe-18-5-4673. Accessed 2015 Sept 24.

129. Thomson RR, Harris RJ, Birks TA, Brown G, Allington-Smith J, Bland-Hawthorn J. Ultrafast laser inscription of a 121-waveguide fan-out for astrophotonics. *Opt Lett* 2012;37(12):2331–2333. [Online]. Available at http://ol.osa.org/abstract.cfm?URI=ol-37-12-2331.

130. Huang B, Fontaine NK, Ryf R, Guan B, Leon-Saval SG, Shubochkin R, Sun Y, Lingle R, Li G. All-fiber mode-group-selective photonic lantern using graded-index multimode fibers. *Opt Express* 2015;23(1):224–234. [Online]. Available at http://www.opticsexpress.org/abstract.cfm?URI=oe-23-1-224.

131. Birks TA, Mangan B, Díez A, Cruz J, Murphy D. "Photonic lantern" spectral filters in multi-core fibre. *Opt Express* 2012;20(13):13 996–14 008. [Online]. Available at http://www.opticsinfobase.org/abstract.cfm?uri=oe-20-13-13996.

132. Carpenter JA, Leon-Saval SG, Salazar Gil JR, Bland-Hawthorn J, Baxter G, Stewart L, Frisken S, Roelens MA, Eggleton BJ, Schröder J. 1x11 few-mode fiber wavelength selective switch using photonic lanterns. Optical Fiber Communication Conference; Optical Society of America; Mar 2014. p Th4A.2. [Online]. Available at http://www.opticsinfobase.org/abstract.cfm?URI=OFC-2014-Th4A.2. Accessed 2015 Sept 24.

133. Chen X, Li A, Ye J, Al Amin A, Shieh W. Reception of mode-division multiplexed superchannel via few-mode compatible optical add/drop multiplexer. *Opt Express* 2012;20(13):14 302–14 307. [Online]. Available at http://www.opticsexpress.org/abstract.cfm?URI=oe-20-13-14302.

134. Lazarev G, Hermerschmidt A, Krüger S, Osten S. LCOS spatial light modulators: trends and applications. In: Osten W, Reingand N, editors. Optical Imaging and Metrology; Wiley-VCH Verlag GmbH & Co. KGaA; 2012. p 1–29. [Online]. Available at http://onlinelibrary.wiley.com/doi/10.1002/9783527648443.ch1/summary.

135. Frisken S, Baxter G, Abakoumov D, Zhou H, Clarke I, Poole S. Flexible and grid-less wavelength selective switch using LCOS technology. Optical Fiber Communication Conference/National Fiber Optic Engineers Conference 2011, Series OSA Technical Digest (CD). Los Angeles (CA): Optical Society of America; Mar 2011. p OTuM3. [Online]. Available at http://www.opticsinfobase.org/abstract.cfm?URI=OFC-2011-OTuM3. Accessed 2015 Sept 24.

136. Fontaine NK, Ryf R, Neilson DT. Fiber-port-count in wavelength selective switches for space-division multiplexing. Optical Communication (ECOC 2013), 39th European Conference and Exhibition on; 2013. p 1–3. [Online]. Available at http://ieeexplore.ieee.org/stamp/stamp.jsp?arnumber=6647669. Accessed 2015 Sept 24.

137. Ho K-P, Kahn JM, Wilde JP. Wavelength-selective switches for mode-division multiplexing: scaling and performance analysis. *J Lightwave Technol* 2014;32(22):3724–3735. [Online]. Available at http://jlt.osa.org/abstract.cfm?URI=jlt-32-22-3724.

138. Salsi M, Ryf R, Le Cocq G, Bigot L, Peyrot D, Charlet G, Bigo S, Fontaine NK, Mestre MA, Randel S, Palou X, Bolle C, Guan B, Quiquempois Y. A six-mode erbium-doped fiber amplifier. Proceedings, European Conference on Optical Communication (ECOC), Volume. Postdeadline; 2012. p Th.3.A.6.

139. Watanabe K, Saito T, Tsuchida Y, Maeda K, Shiino M. Fiber bundle type fan-out for multicore Er doped fiber amplifier. 2013 18th OptoElectronics and Communications Conference held jointly with 2013 International Conference on Photonics in Switching; Optical Society of America; June 2013. p TuS1_5.

140. Bai N, Ip E, Huang Y-K, Mateo E, Yaman F, Li M-J, Bickham S, Ten S, Liñares JL, Montero C, Moreno V, Prieto X, Tse V, Chung KM, Lau APT, Tam H-Y, Lu C, Luo Y, Peng G-D, Li G, Wang T. Mode-division multiplexed transmission with inline few-mode fiber amplifier. *Opt Express* 2012;20(3):2668–2680.

141. Ryf R, Sierra A, Essiambre R-J, Randel S, Gnauck A, Bolle C, Esmaeelpour M, Winzer P, Delbue R, Pupalaikis P, Sureka A, Peckham DW, McCurdy A, Lingle R Jr. Mode-equalized distributed Raman amplification in 137-km few-mode fiber. Proceedings, European Conference on Optical Communication (ECOC), Postdeadline Paper; 2011. p. Th.13.K.5.

142. Antonelli C, Mecozzi A, Shtaif M. Modeling Raman amplification in multimode and multicore fibers. Optical Fiber Communication Conference; Optical Society of America; 2014. p W3E.1. [Online]. Available at http://www.opticsinfobase.org/abstract.cfm?URI=OFC-2014-W3E.1. Accessed 2015 Sept 24.

143. Nasiri Mahalati R, Askarov D, Kahn JM. Adaptive modal gain equalization techniques in multi-mode erbium-doped fiber amplifiers. *J Lightwave Technol* 2014;32(11):2133–2143. [Online]. Available at http://jlt.osa.org/abstract.cfm?URI=jlt-32-11-2133.

144. Jung Y, Lim EL, Kang Q, May-Smith TC, Wong NHL, Standish R, Poletti F, Sahu JK, Alam SU, Richardson DJ. Cladding pumped few-mode EDFA for mode division multiplexed transmission. *Opt Express* 2014;22(23):29 008–29 013. [Online]. Available at http://www.opticsexpress.org/abstract.cfm?URI=oe-22-23-29008.

145. Theeg T, Sayinc H, Neumann J, Overmeyer L, Kracht D. Pump and signal combiner for bi-directional pumping of all-fiber lasers and amplifiers. *Opt Express* 2012;20(27):28 125–28 141. [Online]. Available at http://www.opticsexpress.org/abstract.cfm?URI=oe-20-27-28125.

146. Goldberg L, Cole B, Snitzer E. V-groove side-pumped 1.5 μm fibre amplifier. *Electron Lett* 1997;33(25):2127–2129.

147. Ono H, Takenaga K, Ichii K, Matsuo S, Takahashi T, Masuda H, Yamada M. 12-core double-clad Er/Yb-doped fiber amplifier employing free-space coupling pump/signal combiner module. Optical Communication (ECOC 2013), 39th European Conference and Exhibition on; Sept 2013.

148. Krummrich PM, Akhtari S. Selection of energy optimized pump concepts for multi core and multi mode erbium doped fiber amplifiers. *Opt Express* 2014;22(24):30267. [Online]. Available at http://www.opticsinfobase.org/oe/fulltext.cfm?uri=oe-22-24-30267&id=305353.

149. Krummrich PM. Optical amplification and optical filter based signal processing for cost and energy efficient spatial multiplexing. *Opt Express* 2011;19(17):16 636–16 652.

150. Fontaine NK, Guan B, Ryf R, Chen H, Koonen A, Yoo S, Abedin KS, Fini JM, Taunay T, Neilson DT. Programmable gain equalizer for multicore fiber amplifiers. Optical Fiber Communication Conference: Postdeadline Papers; Optical Society of America; Mar 2014. p Th5C.5. [Online]. Available at http://www.opticsinfobase.org/abstract.cfm?URI=OFC-2014-Th5C.5. Accessed 2015 Sept 24.

151. Ip E. Gain equalization for few-mode fiber amplifiers with more than two propagating mode groups. Proceedings of IEEE Photonics Society Summer Topical Meeting Series; 2012. p 224–225.

152. Bai N, Ip E, Wang T, Li G. Multimode fiber amplifier with tunable modal gain using a reconfigurable multimode pump. *Opt Express* 2011;19(17):16 601–16 611.

153. Le Cocq G, Quiquempois Y, Bigot L. Gradient descent optimization for few-mode Er3+ doped fiber amplifiers with micro-structured core. 2014 IEEE Photonics Society Summer Topical Meeting Series; July 2014. p 150–151.

154. Le Cocq G, Bigot L, Le Rouge A, Bigot-Astruc M, Sillard P, Koebele C, Salsi M, Quiquempois Y. Modeling and characterization of a few-mode EDFA supporting four mode groups for mode division multiplexing. *Opt Express* 2012;20(24):27051. [Online]. Available at http://www.opticsinfobase.org/oe/fulltext.cfm?uri=oe-20-24-27051&id=246130.

INDEX

Note: Page numbers referring to figures are in *italics* and those referring to tables are in **bold**

Enabling Technologies for High Spectral-efficiency Coherent Optical Communication Networks, First Edition. Edited by Xiang Zhou and Chongjin Xie.
© 2016 John Wiley & Sons, Inc. Published 2016 by John Wiley & Sons, Inc.

WILEY SERIES IN MICROWAVE & OPTICAL ENGINEERING

Editor: Professor Kai Chang

Texas A&M University

Passive Macromodeling • *Stefano Grivet-Talocia, Bjorn Gustavsen*

Artificial Transmission Lines for RF and Microwave Applications • *Ferran Martín*

Radio-Frequency Integrated-Circuit Engineering • *Cam Nguyen*

Fundamentals of Microwave Photonics • *V. J. Urick, Keith J. Williams, Jason D. McKinney*

Radio Propagation and Adaptive Antennas for Wireless Communication Networks, 2nd Edition • *Nathan Blaunstein, Christos G. Christodoulou*

Microwave Noncontact Motion Sensing and Analysis • *Changzhi Li, Jenshan Lin*

Metamaterials with Negative Parameters: Theory, Design and Microwave Applications • *Ricardo Marqués, Ferran Martín, Mario Sorolla*

Photonic Sensing: Principles and Applications for Safety and Security Monitoring • *Gaozhi Xiao (Editor), Wojtek J. Bock (Editor)*

Fundamentals of Optical Fiber Sensors • *Zujie Fang, Ken Chin, Ronghui Qu, Haiwen Cai, Kai Chang (Series Editor)*

Compact Multifunctional Antennas for Wireless Systems • *Eng Hock Lim, Kwok Wa Leung*

Diode Lasers and Photonic Integrated Circuits, 2nd Edition • *Larry A. Coldren, Scott W. Corzine, Milan L. Mashanovitch*

Inverse Synthetic Aperture Radar Imaging With MATLAB Algorithms • *Caner Ozdemir*

Silica Optical Fiber Technology for Devices and Components: Design, Fabrication, and International Standards • *Kyunghwan Oh, Un-Chul Paek*

Microwave Bandpass Filters for Wideband Communications • *Lei Zhu, Sheng Sun, Rui Li*

Subsurface Sensing • *Ahmet S. Turk, Koksal A. Hocaoglu, Alexey A. Vertiy*

RF and Microwave Transmitter Design • *Andrei Grebennikov*

Microstrip Filters for RF / Microwave Applications, 2nd Edition • *Jia-Sheng Hong*

Fundamentals of Wavelets: Theory, Algorithms, and Applications, 2nd Edition • *Jaideva C. Goswami, Andrew K. Chan*

Radio Frequency Circuit Design, 2nd Edition • *W. Alan Davis*

Laser Diodes and Their Applications to Communications and Information Processing • *Takahiro Numai*

Time and Frequency Domain Solutions of EM Problems Using Integral Equations and a Hybrid Methodology • *B. H. Jung, T. K. Sarkar, Y. Zhang, Z. Ji, M. Yuan, M. Salazar-Palma, S. M. Rao, S. W. Ting, Z. Mei, A. De*

Fiber-Optic Communication Systems, 4th Edition • *Govind P. Agrawal*

Solar Cells and Their Applications, 2nd Edition • *Lewis M. Fraas, Larry D. Partain*

Microwave Imaging • *Matteo Pastorino*

EM Detection of Concealed Targets • *David J. Daniels*

Phased Array Antennas, 2nd Edition • *Robert C. Hansen*

Electromagnetic Simulation Techniques Based on the FDTD Method • *W. Yu*

Parallel Solution of Integral Equation-Based EM Problems in the Frequency Domain • *Y. Zhang, T. K. Sarkar*

Advanced Integrated Communication Microsystems • *Joy Laskar, Sudipto Chakraborty, Anh-Vu Pham, Manos M. Tantzeris*

Analysis and Design of Autonomous Microwave Circuits • *Almudena Suarez*

The Stripline Circulators: Theory and Practice • *J. Helszajn*

Physics of Multiantenna Systems and Broadband Processing • *T. K. Sarkar, M. Salazar-Palma, Eric L. Mokole*

Electromagnetic Shielding • *Salvatore Celozzi, Rodolfo Araneo, Giampiero Lovat*

Localized Waves • *Hugo E. Hernandez-Figueroa (Editor), Michel Zamboni-Rached (Editor), Erasmo Recami (Editor)*

High-Speed VLSI Interconnections, 2nd Edition • *Ashok K. Goel*

Optical Switching • *Georgios I. Papadimitriou, Chrisoula Papazoglou, Andreas S. Pomportsis*

Electron Beams and Microwave Vacuum Electronics • *Shulim E. Tsimring*

Asymmetric Passive Components in Microwave Integrated Circuits • *Hee-Ran Ahn*

Adaptive Optics for Vision Science: Principles, Practices, Design and Applications • *Jason Porter, Hope Queener, Julianna Lin, Karen Thorn, Abdul A. S. Awwal*

Phased Array Antennas: Floquet Analysis, Synthesis, BFNs and Active Array Systems • *Arun K. Bhattacharyya*

RF / Microwave Interaction with Biological Tissues • *Andre Vander Vorst, Arye Rosen, Youji Kotsuka*

History of Wireless • *T. K. Sarkar, Robert Mailloux, Arthur A. Oliner, M. Salazar-Palma, Dipak L. Sengupta*

Introduction to Electromagnetic Compatibility, 2nd Edition • *Clayton R. Paul*

Applied Electromagnetics and Electromagnetic Compatibility • *Dipak L. Sengupta, Valdis V. Liepa*

Multiresolution Time Domain Scheme for Electromagnetic Engineering • *Yinchao Chen, Qunsheng Cao, Raj Mittra*

Fundamentals of Global Positioning System Receivers: A Software Approach, 2nd Edition • *James Bao-Yen Tsui*

Microwave Ring Circuits and Related Structures, 2nd Edition
by Kai Chang, Lung-Hwa Hsieh

E-BOOK

Fundamentals of Global Positioning System Receivers: A Software Approach
by James Bao-Yen Tsui

Microstrip Filters for RF / Microwave Applications • *Jia-Shen G. Hong, M. J. Lancaster*

Arithmetic and Logic in Computer Systems • *Mi Lu*

Radio Frequency Circuit Design • *W. Alan Davis, Krishna Agarwal*

Smart Antennas • *T. K. Sarkar, Michael C. Wicks, M. Salazar-Palma, Robert J. Bonneau*

Wavelets in Electromagnetics and Device Modeling • *George W. Pan*

Planar Antennas for Wireless Communications • *Kin-Lu Wong*

RF and Microwave Circuit and Component Design for Wireless Systems • *Kai Chang, Inder Bahl, Vijay Nair*

Compact and Broadband Microstrip Antennas • *Kin-Lu Wong*

Spheroidal Wave Functions in Electromagnetic Theory • *Le-Wei Li, Xiao-Kang Kang, Mook-Seng Leong*

Theory and Practice of Infrared Technology for Nondestructive Testing • *Xavier P. Maldague*

Coplanar Waveguide Circuits, Components, and Systems • *Rainee N. Simons*

Electromagnetic Fields in Unconventional Materials and Structures • *Onkar N. Singh (Editor), Akhlesh Lakhtakia (Editor)*

Infrared Technology: Applications to Electro-Optics, Photonic Devices and Sensors • *Animesh R. Jha*

Analysis Methods for RF, Microwave, and Millimeter-Wave Planar Transmission Line Structures • *Cam Nguyen*

RF and Microwave Wireless Systems • *Kai Chang*

Analysis and Design of Integrated Circuit-Antenna Modules • *K. C. Gupta (Editor), Peter S. Hall (Editor)*

Electrodynamics of Solids and Microwave Superconductivity • *Shu-Ang Zhou*

Electromagnetic Optimization by Genetic Algorithms • *Yahya Rahmat-Samii (Editor), Eric Michielssen (Editor)*

Optical Filter Design and Analysis: A Signal Processing Approach • *Christi K. Madsen, Jian H. Zhao*

Optical Character Recognition • *Shunji Mori, Hirobumi Nishida, Hiromitsu Yamada*

SiGe, GaAs, and InP Heterojunction Bipolar Transistors • *Jiann S. Yuan*

Design of Nonplanar Microstrip Antennas and Transmission Lines • *Kin-Lu Wong*

Electromagnetic Propagation in Multi-Mode Random Media • *Harrison E. Rowe*

Optical Semiconductor Devices • *Mitsuo Fukuda*

Superconductor Technology: Applications to Microwave, Electro-Optics, Electrical Machines, and Propulsion Systems • *Animesh R. Jha*

Nonlinear Optical Communication Networks • *Eugenio Iannone, Francesco Matera, Antonio Mecozzi, Marina Settembre*

Introduction to Electromagnetic and Microwave Engineering • *Paul R. Karmel, Gabriel D. Colef, Raymond L. Camisa*

Advances in Microstrip and Printed Antennas • *Kai Fong Lee (Editor), Wei Chen (Editor)*

Optoelectronic Packaging • *Alan R. Mickelson (Editor), Nagesh R. Basavanhally (Editor), Yung-Cheng Lee (Editor)*

Active and Quasi-Optical Arrays for Solid-State Power Combining • *Robert A. York (Editor), Zoya B. Popovic (Editor)*

Phased Array-Based Systems and Applications • *Nicholas Fourikis*

Nonlinear Optics • *E. G. Sauter*

Integrated Active Antennas and Spatial Power Combining • *Julio A. Navarro, Kai Chang*

Finite Element Software for Microwave Engineering • *Tatsuo Itoh (Editor), Giuseppe Pelosi (Editor), Peter P. Silvester (Editor)*

Fundamentals of Microwave Transmission Lines • *Jon C. Freeman*

High-Frequency Electromagnetic Techniques: Recent Advances and Applications • *Asoke K. Bhattacharyya*

High-Frequency Analog Integrated Circuit Design • *Ravender Goyal*

Coherent Optical Communications Systems • *Silvello Betti, Giancarlo De Marchis, Eugenio Iannone*

Microwave Solid-State Circuits and Applications • *Kai Chang*

Microwave Devices, Circuits and Their Interaction • *Charles A. Lee, G. Conrad Dalman*

Microstrip Circuits • *Fred Gardiol*

Computational Methods for Electromagnetics and Microwaves • *Richard C. Booton*